Cytomegaloviruses

From Molecular Pathogenesis to Intervention

Volume I

Edited by

Matthias J. Reddehase

with the assistance of

Niels A.W. Lemmermann

Institute for Virology
University Medical Center of the Johannes Gutenberg-University
Mainz
Germany

Caister Academic Press

Copyright © 2013

Caister Academic Press
Norfolk, UK

www.caister.com

British Library Cataloguing-in-Publication Data
A catalogue record for this book is available from the British Library

ISBN for complete set of two volumes: 978-1-908230-18-8
ISBN of this volume: 978-1-908230-19-5

Cover image designed by Andrew Townsend, adapted from Figure I.6.1

Printed and bound in Malta by Gutenberg Press Ltd

Contents

Contents of Volume II

Current Books of Interest

www.caister.com

Contributors

Michael I. Abecassis
Comprehensive Transplant Center
Northwestern University
Feinberg School of Medicine
Chicago, IL
USA

mabecass@nmh.org

Allison Abendroth
Discipline of Infectious Diseases and Immunology
University of Sydney
Sydney
Australia

allison.abendroth@sydney.edu.au

Barbara Adler
Max von Pettenkofer Institute for Virology
Ludwig Maximilians University Munich
Munich
Germany

adlerb@lmb.uni-muenchen.de

Ana Angulo
Institut d'Investigacions Biomediques
August Pi i Sunyer (IDIBAPS)
Barcelona
Spain

aangulo@ub.edu

Selmir Avdic
Discipline of Infectious Diseases and Immunology
University of Sydney;
Centre for Virus Research
Westmead Millennium Institute
Sydney
Australia

selmir.avdic@sydney.edu.au

Jens-Bernhard Bosse
Department of Molecular Biology
Schultz Laboratory
Princeton University
Princeton, NJ
USA

jbosse@princeton.edu

Elke Bogner
Institute of Medical Virology
Charité Universitätsmedizin Berlin
Berlin
Germany

elke.bogner@charite.de

Eva Maria Borst
Department of Virology
Hannover Medical School
Hannover
Germany

borst.eva@mh-hannover.de

Wolfram Brune
Heinrich Pette Institute
Leibniz Institute for Experimental Virology
Hamburg
Germany

wolfram.brune@hpi.uni-hamburg.de

Julia K. Büttner
Institute for Virology
University Medical Center of the
 Johannes Gutenberg-University Mainz
Mainz
Germany

julia_buettner@uni-mainz.de

Patrizia Caposio
Vaccine and Gene Therapy Institute
Oregon Health and Science University
Beaverton, OR
USA

caposiop@ohsu.edu

Anamaris M. Colberg-Poley
Research Center for Genetic Medicine
Children's National Medical Center
Washington, DC
USA

acolberg-poley@childrensnational.org

Teresa Compton
Biogen Idec
Cambridge, MA
USA

teresa.compton@biogenidec.com

Charles H. Cook
Department of Surgery
The Ohio State University College of Medicine
Columbus, OH
USA

Charles.Cook@osumc.edu

Lisa P. Daley-Bauer
Department of Microbiology & Immunology
Emory Vaccine Center
Emory University School of Medicine
Atlanta, GA
USA

ldaley@emory.edu

Derrick J. Dargan
MRC – University of Glasgow Centre for Virus
  Research
Glasgow
UK

Andrew J. Davison
MRC – University of Glasgow Centre for Virus
  Research
Glasgow
UK

andrew.davison@glasgow.ac.uk

Aidan Dolan
MRC – University of Glasgow Centre for Virus
  Research
Glasgow
UK

Adam Feire
Immunocompromised Host Pathogens
Novartis Institutes for BioMedical Research
Emeryville, CA
USA

adam.feire@novartis.com

Kirsten Freitag
Institute for Virology
University Medical Center of the
  Johannes Gutenberg-University Mainz
Mainz
Germany

kfreitag@uni-mainz.de

Wade Gibson
Department of Pharmacology and Molecular Sciences
Johns Hopkins University School of Medicine
Baltimore, MD
USA

wgibson@jhmi.edu

Derek Gatherer
University of Glasgow Centre for Virus Research
Glasgow
UK

derek.gatherer@glasgow.ac.uk

Peter Ghazal
Division of Pathway Medicine
University of Edinburgh Medical School
Edinburgh
UK

pghazal@ed.ac.uk

Finn Grey
The Roslin Institute & Royal (Dick) School of
  Veterinary Studies
University of Edinburgh
Edinburgh
UK

finn.grey@roslin.ed.ac.uk

Marion Grießl
Institute for Virology
University Medical Center of the
  Johannes Gutenberg-University Mainz
Mainz
Germany

griessl@uni-mainz.de

Meaghan H. Hancock
Vaccine and Gene Therapy Institute
Oregon Health and Science University
Beaverton, OR
USA

hancocme@ohsu.edu

Gary S. Hayward
The Sidney Kimmel Comprehensive Cancer Center
The Johns Hopkins University School of Medicine
Baltimore, MD
USA

ghayward@jhmi.edu

Hartmut Hengel
Department of Virology
Institute for Medical Microbiology and Hygiene
Albert-Ludwigs-University
Freiburg
Germany

hartmut.hengel@uniklinik-freiburg.de

Marylouisa Holton
MRC – University of Glasgow Centre for Virus
  Research
Glasgow
UK

mary.holton@glasgow.ac.uk

Lauren M. Hook
Vaccine and Gene Therapy Institute
Oregon Health and Science University
Beaverton, OR
USA

hookl@ohsu.edu

Mary A. Hummel
Comprehensive Transplant Center
Northwestern University
Feinberg School of Medicine
Chicago, IL
USA

m-hummel@northwestern.edu

Robert F. Kalejta
Institute for Molecular Virology
McArdle Laboratory for Cancer Research
University of Wisconsin-Madison
Madison, WI
USA

rfkalejta@wisc.edu

Kai A. Kropp
Division of Pathway Medicine
University of Edinburgh Medical School
Edinburgh
UK

kai.kropp@ed.ac.uk

Igor Landais
Vaccine and Gene Therapy Institute
Oregon Health and Science University
Beaverton, OR
USA

landaisi@ohsu.edu

Niels A.W. Lemmermann
Institute for Virology
University Medical Center of the
  Johannes Gutenberg-University Mainz
Mainz
Germany

Lemmermann@uni-mainz.de

Xue-Feng Liu
Comprehensive Transplant Center
Northwestern University
Feinberg School of Medicine
Chicago, IL
USA

xue-liu@northwestern.edu

A. Lousie McCormick
MedImmune, LCC
Mountain View
USA

alm60co@yahoo.com

Alexander Mazein
Division of Pathway Medicine
University of Edinburgh Medical School
Edinburgh
UK

amazey@gmail.com

Jeffery Meier
Department of Internal Medicine
University of Iowa Hospitals and Clinics
Iowa City, Iowa
USA

jeffery-meier@uiowa.edu

Martin Messerle
Department of Virology
Hannover Medical School
Hannover
Germany

messerle.martin@mh-hannover.de

Edward S. Mocarski Jr
Department of Microbiology & Immunology
Emory Vaccine Center
Emory University School of Medicine
Atlanta, GA
USA

mocarski@emory.edu

Jay A. Nelson
Vaccine and Gene Therapy Institute
Oregon Health and Science University
Beaverton, OR
USA

nelsonj@ohsu.edu

Joshua D. Rabinowitz
Department of Chemistry
Lewis Sigler Institute
Princeton University
Princeton, NJ
USA

joshr@princeton.edu

Matthias J. Reddehase
Institute for Virology
University Medical Center of the
  Johannes Gutenberg-University Mainz
Mainz
Germany

matthias.reddehase@uni-mainz.de

Matthew Reeves
University of Cambridge
Department of Medicine
Cambridge
UK

mbr23@cam.ac.uk

Alec Redwood
UWA School of Pathology & Laboratory Medicine
Crawley, WA
Australia

alec.redwood@uwa.edu.au

Nina Reuter
Institute for Clinical and Molecular Virology
University Hospital Erlangen
Erlangen
Germany

nina.reuter@viro.med.uni-erlangen.de

Zsolt Ruzsics
Max von Pettenkofer Institute
Gene Center
Ludwig Maximilians University
Munich
Germany

ruzsics@imb.uni-muenchen.de

Christof K. Seckert
Institute for Virology
University Medical Center of the
  Johannes Gutenberg-University Mainz
Mainz
Germany

seckert@uni-mainz.de

Geoffrey Shellam
Microbiology and Immunology
University of Western Australia
Perth, WA
Australia

geoff.shellam@uwa.edu.au

Thomas Shenk
Department of Molecular Biology
Princeton University
Princeton, NJ
USA

tshenk@princeton.edu

Christian Sinzger
Institute of Virology
University Hospital Ulm
Ulm
Germany

Christian.sinzger@uniklinik-ulm.de

John Sinclair
University of Cambridge
Department of Medicine
Cambridge
UK

js@mole.bio.cam.ac.uk

Barry Slobedman
Discipline of Infectious Diseases and Immunology
University of Sydney
Sydney
Australia

barry.slobedman@sydney.edu.au

Lee Smith
Microbiology and Immunology
University of Western Australia
Perth, WA
Australia

lee.smith@ausdx.com

M. Shane Smith
Vaccine and Gene Therapy Institute
Oregon Health and Science University
Beaverton, OR
USA

ssmith@skgf.com

Deborah H. Spector
Department of Cellular and Molecular Medicine
University of California San Diego
La Jolla, CA
USA

dspector@ucsd.edu

Thomas Stamminger
Institute for Clinical and Molecular Virology
University Hospital Erlangen
Erlangen
Germany

thomas.stamminger@viro.med.uni-erlangen.de

Mark Stinski
Department of Microbiology
University of Iowa
Iowa City, Iowa
USA

mark-stinski@uiowa.edu

Daniel A. Streblow
Vaccine and Gene Therapy Institute
Oregon Health and Science University
Beaverton, OR
USA

streblow@ohsu.edu

Marco Thomas
Institute for Clinical and Molecular Virology
University Hospital Erlangen
Erlangen
Germany

marco.thomas@viro.med.uni-erlangen.de

Rebecca Tirabassi
Department of Medical Microbiology & Immunology
University of Wisconsin
Madison, WI
USA

rebecca.tirabassi@gmail.com

Mirko Trilling
Institute for Virology
University Hospital Essen
Essen
Germany

mirko.trilling@uk-essen.de

Steven Watterson
Division of Pathway Medicine
University of Edinburgh Medical School
Edinburgh
UK

s.watterson@ed.ac.uk

Chad Williamson
Laboratory of Cell Biology
National Heart, Lung, and Blood Institute
National Institutes of Health
Bethesda, MD
USA

chad.williamson@nih.gov

Preface to the First Edition (2006)

From Protozoan to Proteomics

On 23 September 2003, I happened to catch an infected e-mail that had passed through our university's firewall, spam filters and virus alert. It contained a particularly insidious and malicious virus that occupied a great deal of my time – and actually still does! As you may have already guessed, I'm talking about cytomegalovirus. On that fateful day when Annette Griffin introduced herself as an Acquisitions Editor with Horizon Scientific Press to ask whether I would be interested in editing a book on CMV, I was of course hesitant and colleagues warned me of the enormity of the task. What finally convinced me to undertake this project, however, was the proposed concept of the book. The aim was a publication that 'details the cutting-edge research and future potential in this increasingly important field', a publisher's view likely to meet with general approval in the scientific CMV community. I also agreed completely with the publisher's idea that the book 'should bring together recent research on human and animal cytomegaloviruses with chapters on infection models, pathogenic mechanisms, interactions with the host, genomics and molecular biology, latency and reactivation, and strategies for control'. At a time when parallel sessions are planned by conference organizers according to discipline rather than on the basis of pending problems that need to be solved by integrating different views and approaches, there is a risk that we will fail to understand CMV disease in all its complexity because molecular biologists and immunologists no longer listen to each other. Following my initial training in immunology, as a graduate student in the Koszinowski lab I profited a great deal from my mentor's strategy of combining the ideas and techniques of immunologists and molecular biologists. By editing this book, I have profited again from reading more than a thousand manuscript pages covering a broad spectrum of CMV research, and I certainly had plenty of new ideas for my own lab's to-do-list! There are many good monographs, articles in review series, and journal special issues on selected topics. Yet, to the best of my knowledge, this book is without precedent in that it provides an overview of current opinion and cutting-edge research on literally all aspects of CMV infections, although with a focus on basic science that was intended by the publisher. Rather than being asked to write a lecture book chapter or a comprehensive review of literature, the authors were encouraged to give opinions and to put forward hypotheses.

What I liked most was Horizon Press's plan to 'bring together recent research on human and animal cytomegaloviruses'. One cannot deny that there is some distance between these two groups. On the one hand, basic research in animal models sometimes ignores the pending clinical questions; on the other hand, information from animal models is sometimes ignored and rediscovered years later, often without fair referencing. This seems to have become something of a tradition. When I was a young postdoctoral scientist visiting Dr Monto Ho in the Cathedral of Learning at the University of Pittsburgh, he told me an anecdote that can also be found in his classic book *Cytomegalovirus Biology and Infection* (Ho, 1982). It actually dates back to a very honest report given by Dr Weller on the history of the isolation of human CMV (Weller, 1970): Dr Margaret Smith, well known in MCMV research for the isolation of the Smith strain, used the mouse salivary gland virus as a model to establish the methods for isolation of the agent that causes cytomegalic inclusion disease (CID) in humans. Her pioneering accomplishment in the mouse model was published in 1954 (Smith, 1954). It is less well known that she almost simultaneously succeeded in growing the corresponding agent from the salivary gland of an infant. She even discovered that the human agent does not produce cytopathic lesions in mouse tissue nor, indeed, the mouse one in human tissue. She was convinced – and rightly so, as became clear soon after – that each of the two agents produces cytopathic lesions only in homologous tissue, which reflects the species specificity of CMVs. Sadly, her submitted manuscript on the human agent was initially rejected by the journal's editors on the grounds that she might have propagated the mouse virus in human tissue. It therefore took another two years until human CMV became known to medical virologists as a result of back-to-back publications written by Rowe and colleagues and by Margaret Smith (Rowe et al., 1956; Smith, 1956), followed one year later by a publication from the Weller lab (Weller et al., 1957).

The search for the cradle of cytomegalovirus research takes us to Bonn, Germany, to a meeting of 14 members of the medical section of the Natural History Society of Prussian Rhineland and Westphalia that was chaired by Dr Leo on 27 June 1881 (Andrä, 1881). During this meeting, the pathologist Dr Ribbert gave a case report on a stillborn infant with lues (syphilis)-like symptoms and interstitial nephritis associated with the presence of tremendously enlarged 'cytomegalic' cells that were also characterized by an enlarged nucleus. It was not until 1904 (Jesionek and Kiolemenoglou) that these 'curious cells' were observed again in the kidneys, lungs and liver of a luetic fetus. Although the cytomegalic cells were then mistaken for protozoa, specifically for gregarines, the precision with which the cytopathic effect was described and illustrated deserves our highest respect (Fig. 1). Unfortunately, when Ribbert saw the report by Jesionek and Kiolemenoglou on the protozoan-like cells, he reinterpreted his original observation in support of the protozoan hypothesis (Ribbert, 1904). This led Monto Ho (1982) to conclude that 'he

[Ribbert] was unable to interpret his observation until he saw the report by Jesionek and Kiolemenoglou who had noted for the first time the presence of protozoan-like cells in lungs, kidneys and liver'. As a native German speaker, I have had the opportunity of reading the original minutes from the meeting of 27 June 1881, and for me there is no doubt at all that Ribbert was the first to correctly note the enlargement of cells and cytopathological alterations that are distinctively characteristic of CMV disease.

In 'homage' to the intranuclear inclusion body that is pathognomonic of CMV, we have compared the highly skilled sketch produced by Jesionek and Kiolemenoglou in 1904 (Fig. 1) with 'modern' in situ hybridization photographic images taken a century later (Fig. 2). Hybridization with virus-specific DNA probes identifies the intranuclear inclusion body as the site at which viral DNA is concentrated, and we know today that this is the site of viral DNA packaging into nucleocapsids. After an interval of a hundred years, it is instructive to compare the margination of cellular chromatin that is condensed in 'polar bodies', which, as we can show today, do not usually contain viral genomes.

There is another short scientific anecdote relating to Fig. 2. Here, the murine model was used to study in vivo coinfection with MCMV-WT and a mutant virus in which the gene of interest was deleted by replacement (Cicin-Sain et al., 2005; for a commentary, see Tremp, 2005). In the normal mouse liver, it is a relatively frequent event that lack of cytokinesis after mitosis leads to hepatocytes that possess two nuclei. The two viruses happened to coinfect such a binuclear hepatocyte; interestingly, MCMV-WT conquered one nucleus and the mutant virus the other. This finding implies that, in each of the two nuclei, a single viral genome molecule was successful in evading the host cell's epigenetic defence mechanisms.

This example brings us back to the original purpose of the book – to 'CMV today'. We leave behind the protozoan and turn instead to proteomics, genomics, and all the other topics of cutting-edge science.

Matthias J. Reddehase

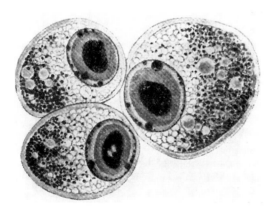

Figure 1 Intranuclear inclusion body as seen in the year 1904. Protozoan-like 'owl's eyes' cells from the left kidney of an alleged luetic fetus (Jesionek and Kiolemenoglou, 1904) portrayed from the Zeiss microscopic image after staining with haematoxylin and eosin. Reproduced by permission of the Münchner Medizinische Wochenschrift (MMW)-Fortschritte in der Medizin.

Acknowledgements

In the first place I have to thank Annette Griffin, Horizon Press's Acquisitions Editor, for her patience with me and all the other authors from the 'peculiar CMV community'. I have appreciated the cooperation of all the authors; they have invested their valuable time and were willing to share their scientific ideas with us. Special thanks go to Ulrich H. Koszinowski, whose lab I joined in 1980 as a student of immunology and whom I left in 1994 as a full professor and chair of virology. I

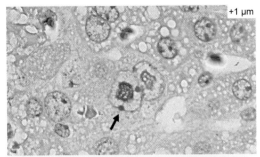

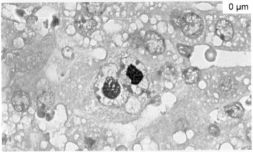

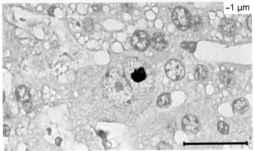

+1 μm

0 μm

−1 μm

Figure 2 100th anniversary of the intranuclear inclusion body in the year 2004. Binuclear hepatocyte in the liver of an immunocompromised mouse co-infected with MCMV-WT (red) and an M36 gene (Menard et al., 2003) deletion mutant (black). Shown are serial 1-μm sections analysed by two-colour *in situ* hybridization. (Top) DNA probe specific for MCMV-WT. The arrow points to displaced cellular chromatin that is condensed in a so-called 'polar body'. (Middle) DNA probes specific for MCMV-WT and the M36 gene deletion mutant. (Bottom) DNA probe specific for the mutant. Bar represents 25 μm. Courtesy of J. Podlech, Institute for Virology, Johannes Gutenberg-University, Mainz, Germany.

would not be editor of this book without him and the members of the 'early Koszinowski lab' in Tübingen: Günther Keil, Angelika Ebeling-Keil, our electron

microscopist Frank Weiland, Stipan Jonjic, Margarita Del Val, Martin Messerle, Brigitte Bühler, Konrad Münch, Mathias Fibi, Wolfgang Mutter, and so many others who accompanied me on the way. Finally, this book would not have been possible without the superb technical help of my Assistant Editor Niels Lemmermann. By defraying in part the costs for colour figures, printing of this book is supported by the Deutsche Forschungsgemeinschaft, Sonderforschungsbereich (Collaborative Research Grant) 490 'Invasion and Persistence in Infections'.

References

Andrä, C.J. (ed.). (1881). Verhandlungen des Naturhistorischen Vereines der preussischen Rheinlande und Westfalens. Achtunddreissigster Jahrgang. Vierte Folge: 8. Jahrgang. (Max Cohen & Sohn, Bonn), pp. 161–162.

Cicin-Sain, L., Podlech, J., Messerle, M., Reddehase, M.J., and Koszinowski, U.H. (2005). Frequent coinfection of cells explains functional in vivo complementation between cytomegalovirus variants in the multiply-infected host. J. Virol. 79, 9492–9502.

Ho, M. (1982). Cytomegalovirus: Biology and Infection. (Plenum Medical Book Company, New York).

Jesionek, A., and Kiolemenoglou, B. (1904). Ueber einen Befund von protozoënartigen Gebilden in den Organen eines hereditär-luetischen Fötus. Münchener Medizinische Wochenschrift 51, 1905–1907.

Menard, C., Wagner, M., Ruzsics, Z., Holak, K., Brune, W., Campbell, A.E., and Koszinowski, U.H. (2003). Role of murine cytomegalovirus US22 gene family members in replication in macrophages. J. Virol. 77, 5557–5570.

Ribbert, H. (1904). Über protozoenartige Zellen in der Niere eines syphilitischen Neugeborenen und in der Parotis von Kindern. Zbl. Allg. Pathol. 15, 945–948.

Rowe, W.P., Hartley, J.W., Waterman, S., Turner, H.C., and Huebner, R.J. (1956). Cytopathogenic agent resembling human salivary gland virus recovered from tissues cultured of human adenoids. Proc. Soc. Exp. Biol. Med. 92, 418–424.

Smith, M.G. (1954). Propagation of salivary gland virus of the mouse in tissue cultures. Proc. Soc. Exp. Biol. Med. 86, 435–440.

Smith, M.G. (1956). Propagation in tissue cultures of a cytopathogenic virus from human salivary gland virus (SGV) disease. Proc. Soc. Exp. Biol. Med. 92, 424–430.

Tremp, A. (2005). Fatal alliance. Nature Rev. Microbiol. 3, 669.

Weller, T.H. (1970). Cytomegaloviruses: the difficult years. J. Infect. Dis. 122, 532–539.

Weller, T.H., Macauley, J.C., Craig, J.M., and Wirth, P. (1957). Isolation of intranuclear inclusion producing agents from infants with illnesses resembling cytomegalic inclusion disease. Proc. Soc. Exp. Biol. Med. 94, 4–12.

Preface

From 'Omics' to Health Economy

When the first edition of our CMV book, then entitled *Cytomegaloviruses: Molecular Biology and Immunology*, was completed and when I kept a copy of it in my hands – fresh from the press and with Andrew Townsend's beautifully designed virion artwork on its cover – I liked it, but swore to myself that editing such a book will remain a 'once-in-a-lifetime experience'. I was curious to learn if the book would become a success and was wondering who, besides the authors, might spend time on reading such an opus, or even just parts of it. Sure, I got encouraging comments from colleagues and was told that graduate students working on a thesis in any facet of CMV research as well as postdoctoral fellows like the book, as did colleagues who wished to get a concise overview on the state of the art and current opinion in CMV topics neighbouring their own field – but this might have been just by courtesy.

There exist more reliable indicators to suggest the book was truly a success. First, the book was apparently no financial disaster, as otherwise the publisher, represented by Annette Griffin, would not have tried so hard to persuade me to break my own promise and accept editing of a second edition. This, however, was just an argument of economy, not necessarily indicating scientific quality. So, what counts more is the fact that all reinvited authors of the first book, despite having suffered from my pedantic editing, almost immediately agreed to update their chapters. What counts even more is the experience that newly invited authors enthusiastically expressed their willingness to join the team of authors for the second edition, which I take as an evidence that they were aware of the book and appreciated its quality. What counts most, in my view, is the fact that some authors helped to shape and further improve the book by suggesting new topics to be covered in new chapters and proposing authors who are most eligible to write them. The current edition has profited greatly from such an expert advice.

Admittedly, I was hesitant to accept the task of editing a second edition. At the time when I was asked, the wonderful book *Human Cytomegalovirus*, edited by our authors Thomas Shenk and Mark F. Stinski for *Current Topics in Microbiology and Immunology* (2008, volume 325) had been published recently, I myself had been guest editor for a special issue 'Cytomegalovirus' of *Medical Microbiology and Immunology* (2008, volume 197), and our authors John Sinclair and Mark Wills were planning as guest editors for a special issue 'Cytomegalovirus: Host–Pathogen Interactions' of *Virus Research* (2011, volume 157). So, was there really any need for updating Caister's *Cytomegaloviruses: Molecular Biology and Immunology*? Wouldn't both authors and readers be saturated with overviews on CMV? Would a new edition be able to compete with all these recent publications, and – what scared me most – would we be able to reach again the high quality standard set by the first edition?

The answer can only come from the critical readers. It was clear from the very beginning that the book must follow a concept that distinguishes it uniquely from related more recent publications. Whilst most of the 'special issues' on CMV focus on human CMV or on scientific disciplines such as the molecular biology or immunology of CMVs, this book aims at a multidisciplinary approach to understanding CMV disease, by taking up the tradition of Monto Ho's book *Cytomegaloviruses: Biology and Infection* in the Plenum Medical Book series *Current Topics in Infectious Disease* (1982) and the Excerpta Medica 'Proceedings of the Fourth International Cytomegalovirus Workshop', edited by Susan Michelson and by our author Stanley A. Plotkin (1993, International Congress Series 1032). It was an aim even of the first edition to detail the cutting-edge research on human CMV and animal CMV disease models, and to integrate views and approaches from different disciplines. Special emphasis in the first edition, however, was placed on basic science. Comments on the first edition, and in particular a book review written in *Clinical Infectious Diseases* by our new author Robin K. Avery, encouraged the publisher and me to expand

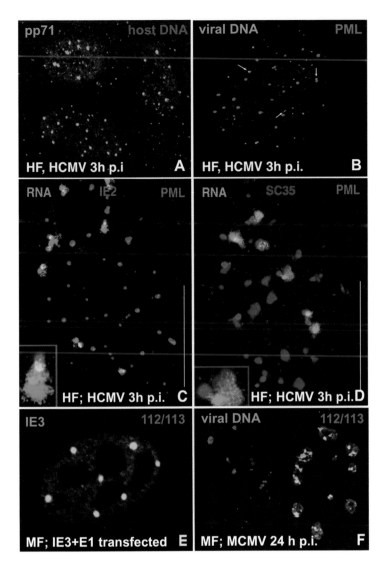

Figure 3 Distribution of viral and certain host proteins after infection. (A) An HCMV-infected fibroblast nucleus probed for the tegument protein pp71 (green) and DAPI (blue) to label nuclei. Green small dots represent single viral particles. Larger whitish dots in nuclei represent pp71 accumulation at ND10. (B) An HCMV-infected fibroblast nucleus at 3 hours post infection (p.i.) containing individual *in situ* hybridization signals (green) that correspond to viral genomes, only three of which are in the immediate vicinity of ND10 (red). (C) Same as B but stained for IE2 protein (blue) and *IE* transcripts (green). IE2 aggregates surround the *IE* transcripts (green) emanating from ND10 (red). Insert shows an immediate transcript environment at higher magnification (white line represents 10 μm). (D) An HCMV-infected fibroblast nucleus triple labelled for ND10 (red), *IE* transcripts (green), and the interchromosomal splicing component containing SC35 (blue). Transcripts emanate from ND10 and spread into the SC35 domain. Insert shows an immediate transcript environment at higher magnification (white line represents 10 μm). This image is represented in the centre of Fig. 4 in diagrammatic form. (E) Mouse fibroblast nucleus transfected with IE3 (green) and M112/113 (red) demonstrating that these two proteins colocalize. (F). Two nuclei infected at the same time with MCMV. Different stages of viral replication are present. Strong *in situ* hybridization (green) signals are present in a few replication compartments in one cell (right) but none in the other cell (left). Unit-strength signals at the surface of M112/113 domains in the nucleus (left) are interpreted as input genomes represented at the lower right of Fig. 4 in diagrammatic form. Reprinted from Tang and Maul (2006).

the book by integrating now also medical science to stimulate a lively dialogue between representatives of all disciplines involved in basic and clinical CMV research. In line with the idea of promoting interdisciplinarity, the grouping principle for the sequence of chapters is not by CMV species or by human disease

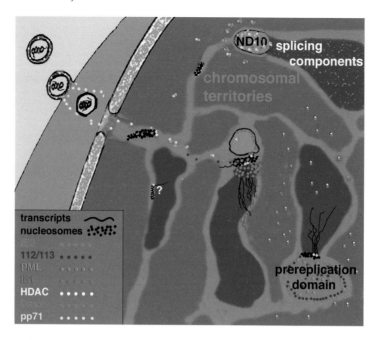

Figure 4 Diagrammatic representation of temporal and spatial aspects of early CMV infection. The interacting proteins are coded as double dots with the respective colours at the left lower corner. Reprinted from Tang and Maul (2006).

separated from animal models, but by the scientific and medical problems addressed.

The result is more than just an update of the first edition – it rather is an upgrade from previously 28 chapters in one volume to currently 46 chapters organized in two volumes. Accordingly, many topics that were missing or were only briefly touched in the first edition have now assumed the status of independent chapters. Though Volume 1, almost unavoidably, has a focus on basic science to lay the foundations for disease-oriented and clinical research emphasized in Volume 2, I am confident that neither of the two parts will be viewed as a 'stand-alone' book and that both volumes represent equally strong science presented by noted experts.

The new title, *Cytomegaloviruses: From Molecular Pathogenesis to Intervention*, was chosen to express this new commitment of the book, ranging now in topics from 'Omics' to Health Economy.

Writing this preface painfully reminded me of the inestimable loss for our scientific community when our previous author and noted Wistar scientist Gerd Maul left us on 23 August 2010, at the age of 70. He was not only an outstanding scientist who introduced cell biology into CMV research by discovering the 'nuclear domain 10' (Ascoli and Maul, 1991) as a key element of the host cell's innate defence against infection (reviewed by Tang and Maul, 2006; Maul, 2008; Maul

and Negorev, 2008; see also Chapter I.11) (Figs. 3 and 4), but also a philosopher of science with challenging ideas, and a sculptor of modern art with exhibitions at the 'Institute for Contemporary Art' at the University of Pennsylvania and the 'Highwire Gallery' in Philadelphia. Most of all, he was a wonderful colleague and friend. We miss him.

Finally, let me express my thanks to all authors who contributed to make this book a success and an invaluable source of information for all young scientists who are going to join us in our efforts to understand and manage CMV disease.

Matthias J. Reddehase

References

Ascoli, C.A., and Maul, G.G. (1991). Identification of a novel nuclear domain. J. Cell Biol. *112*, 785–795.

Maul, G.G. (2008). Initiation of cytomegalovirus infection at ND10. Curr. Top. Microbiol. Immunol. *325*, 117–132.

Maul, G.G., and Negorev, D. (2008). Differences between mouse and human cytomegalovirus interactions with their respective hosts at immediate-early times of the replication cycle. Med. Microbiol. Immunol. *197*, 241–249.

Tang, Q., and Maul, G.G. (2006). Immediate-early interactions and epigenetic defense mechanisms. In Cytomegaloviruses: Molecular Biology and Immunology, Reddehase, M.J., ed. (Caister Academic Press, Norfolk, UK), pp. 131–149.

Abbreviations

A

aa	amino acid(s)
Ab(s)	antibody(ies)
ACV	aciclovir, acyclovir
AIDS	acquired immune deficiency syndrome
APC	antigen-presenting cell
APC	anaphase-promoting complex
AT	adoptive (cell) transfer
ATCC	American Type Culture Collection
ATG	anti-thymocyte globulin
ATP	adenosine triphosphate (also ADP, AMP, CTP, UDP, etc.)

B

BAC	bacterial artificial chromosome(s)
BAL	bronchoalveolar lavage
BCG	bacillus Calmette–Guérin
BFU	burst-forming unit
BM	bone marrow
BMC	bone marrow cell(s)
BMT	bone marrow (cell) transplantation
bp	base pair(s)

C

CAV	cardiac allograft vasculopathy
CCL	CC chemokine ligand
CCMV	chimpanzee CMV
CCR	CC chemokine receptor
CD	cluster of differentiation (e.g. CD8)
cDNA	complementary DNA
CDV	cidofovir
CFU	colony-forming unit(s)
ChIP	chromatin immunoprecipitation
CI	confidence interval
CID	cytomegalic inclusion disease
CLP	common lymphoid progenitor
CLT	cytomegalovirus latency-associated transcript(s)
CMI	cell-mediated immunity
CMP	common myeloid progenitor

CMV	cytomegalovirus (in general; also used as synonym for human CMV)
CNS	central nervous system
CPE	cytopathic effect
CSF	colony-stimulating factor
CSF	cerebrospinal fluid
CTL	cytolytic (or cytotoxic) T lymphocyte(s)
CTLL	cytolytic T lymphocyte line(s)
CVD	cardiovascular disease

D

d	day(s)
D	donor (for transplantation)
DC	dendritic cell(s)
DISC	death-inducing signalling complex
DLI	donor lymphocyte infusion
DN	dominant negative
DNA	deoxyribonucleic acid
dpi (d p.i.)	day(s) post infection

E

E	early (phase of cytomegaloviral gene expression)
EC	endothelial cell(s)
ECM	extracellular matrix
EEC	early effector cell(s)
EGFP	enhanced green fluorescent protein
ELISA	enzyme-linked immunosorbent assay
ELISPOT	enzyme-linked immunospot
EM	electron microscope (microscopy)
ER	endoplasmic reticulum
ERGIC	endoplasmic reticulum Golgi intermediate compartment
E/T ratio	effector–target (cell) ratio
EYFP	enhanced yellow fluorescent protein

F

FACS	fluorescence-activated cell sorting (sorter)
FasL	Fas ligand
FcR	Fc receptor

FFU	focus-forming unit(s)
FL	fluorescence (intensity)
FLP	flippase (recombination enzyme)
FOS	foscarnet
FRT	FLP recognition target

G

gB	glycoprotein B (also gM, etc.)
GBM	glioblastoma multiforme
gc	glycoprotein complex(es)
G-CSF	granulocyte colony-stimulating factor
GCV	ganciclovir
GFP	green fluorescent protein
GLP	good laboratory practice
GM-CSF	granulocyte–macrophage colony-stimulating factor
GOI	gene of interest
gp	glycoprotein
GPCMV	guinea pig cytomegalovirus
GvH	graft-versus-host
GvHD	graft-versus-host-disease

H

h	hour(s)
H-2	major histocompatibility complex of the mouse
HAART	highly active anti-retroviral therapy
HC	haematopoietic cell(s)
HCMV	human cytomegalovirus
HCT	haematopoietic (stem) cell transplantation
HDAC	histone deacetylase(s)
HFF	human foreskin fibroblast(s)
HF	human fibroblast(s)
HHV	human herpesvirus
HLA	(human) histocompatibility leucocyte antigen(s)
HPLC	high-performance liquid chromatography
HPS	haematopoietic progenitor cell
HR	hazard ratio
HSC	(pluripotent, self-renewing) haematopoietic stem cell(s)
HSCT	haematopoietic stem cell transplantation
HSCTR	HSCT recipient
HUVEC	human umbilical vein endothelial cell(s)
HvG	host-versus-graft
HvGD	host-versus-graft disease
HvGR	host-versus-graft reaction

I

i.c.	intracerebral(ly), intracranial(ly)
i.fp.	intra-footpad, i.e. intraplantar(ily)
i.m.	intramuscular(ly)
i.n.	intranasal(ly)
i.p.	intraperitoneal(ly)

i.v.	intravenous(ly)
IDE	immunodominant epitope(s)
IE	immediate-early (phase of cytomegaloviral gene expression)
IFN	interferon(s) (e.g. IFN-γ)
Ig	immunoglobulin(s)
IHC	immunohistochemistry
IL	interleukin(s) (e.g. IL-2)
IM	inflammatory monocyte(s)
INM	inner nuclear membrane
IRF	IFN regulatory factor(s)
ISG	IFN stimulated gene(s)
ISH	*in situ* hybridization
ISRE	IFN-stimulated response element
ITAM	immunoreceptor tyrosine-based activation motif
ITIM	immunoreceptor tyrosine-based inhibitory motif
IVIG	intravenous immunoglobulin (plasma protein replacement therapy)

K

Kan	kanamycin
kb	kilobase(s)
kbp	kilobase pair(s)
kDa	kiloDalton
KIR	killer cell Ig-like receptor(s)
KLH	keyhole limpet haemocyanin
KO	knockout

L

L	late (phase of cytomegaloviral gene expression)
L	ligand (in abbreviations)
LAT	latency-associated transcript(s)
LC	Langerhans cell(s)
LD	lethal dose
LIR	leucocyte Ig-like receptor(s)
LN	lymph node(s)
LPS	lipopolysaccharide
LSEC	liver sinusoidal endothelial cell(s)
LTR	long terminal repeat

M

M	molar
mAb	monoclonal antibody(ies)
MAC	macrophage(s)
MBV	maribavir
MCMV	murine (or mouse) cytomegalovirus
MCP	major capsid protein
mCP	minor capsid protein
MDSC	monocyte-derived suppressor cell
MEF	mouse (murine) embryonic fibroblast(s)
MHC	major histocompatibility complex

MHC-I	MHC class I (genes or antigens)		PK	protein kinase (e.g. PKC)
MHC-II	MHC class II (genes or antigens)		PLN	popliteal lymph node
MIE	major immediate-early		PM	patrolling monocyte(s)
MIEP	MIE promoter (also promoter and enhancer)		pMHC-I	peptide-loaded major histocompatibility complex class I (protein)
min	minute(s)		PML	promyelocytic leukaemia protein (e.g. PML bodies)
miRNA	micro-RNA		PMNL	polymorphonuclear leucocyte(s)
mo	month(s)		POI	protein of interest
MOI	multiplicity of infection		poly(A)$^+$	polyadenylated (RNA)
mRNA	messenger RNA		pp	phosphoprotein
MS	mass spectrometry		PRR	pattern recognition receptor

N

n, N	number in study or group
n.d.	not determined
ND	nuclear domain (e.g. ND10)
NF	nuclear factor(s) (e.g. NF-κB)
NGS	next generation sequencing
NHP	non-human primate(s)
NK	natural killer (e.g. NK cells or NKT-cells)
NLS	nuclear localization signal
NOD	non-obese diabetic
NPC	nuclear pore complex
NPC	neuronal precursor (stem) cell(s); neuro-glial precursor stem cell(s)
NPLC	non-parenchymal liver cells
NSG	NOD SCID gamma
nt	nucleotide(s)
NT-Ab	(virus) neutralizing antibody

O

ONM	outer nuclear membrane
OR	odds ratio
ORF	open reading frame(s)
Ori	origin (e.g. oriLyt)
OVA	ovalbumin

P

p; P	probability
PAA	phosphonoacetic acid
PAGE	polyacrylamide gel electrophoresis
PAMP	pathogen associated molecular pattern
profAPC	professional antigen presenting cell(s)
PB	peripheral blood
PBL	peripheral blood lymphocyte(s)/leucocyte(s)
PBMC	peripheral blood mononuclear cell(s)
PCR	polymerase chain reaction
pDC	plasmacytoid DC(s)
PEC	peritoneal exudate cell(s)
PFA	phosphonoformic acid
PFU	plaque-forming unit(s)
p.i.	post infection
PISH	PCR in situ hybridization

Q

qPCR	quantitative polymerase chain reaction (equals real-time PCR)
qRT-PCR	quantitative reverse transcription PCR (equals RT-qPCR)

R

r	recombinant (e.g. rIFN-γ)
R	receptor (e.g. IL-2R)
R	recipient (of transplantation)
RAE-1	retinoic acid early inducible gene(s) 1
RCMV	rat cytomegalovirus
RFLP	restriction fragment length polymorphism
RFP	red fluorescent protein
RhCMV	rhesus (macaque) cytomegalovirus
RIP	radioimmunoprecipitation
RLN	regional lymph node
RNA	ribonucleic acid
RNAi	RNA interference
RT	reverse transcription (transcriptase), as in RT-PCR
RT-qPCR	reverse transcription quantitative PCR

S

s.c.	subcutaneous(ly)
SCF	stem cell factor
SCID	severe combined immunodeficiency
SCT	stem cell transplantation
SD	standard deviation
SE	standard error
SEM	standard error of the mean
SES	socioeconomic status
SG	salivary gland(s)
siRNA	short interfering RNA
SIV	simian immunodeficiency virus
SLEC	short-lived effector cell(s)
SLO	secondary lymphoid organ(s)
SMC	smooth muscle cell(s)
SOT	solid organ transplantation
SOTR	SOT recipient

STAT	signal transducer and activator of transcription

T

TAP	transporter associated with antigen processing
Tc	cytolytic (cytotoxic) T-cell(s)
TC	tissue culture
TCID	tissue culture infective dose
T_{CM}	central memory T-cell(s)
TCR	T-cell receptor for antigen
T_E	effector T-cell(s)
T_{EM}	effector-memory T-cell(s)
TEL	transcript(s) expressed in latency
TEM	transmission electron microscopy
TGF	transforming growth factor (e.g. TGF-β)
TGN	trans-Golgi network
Th	helper T-cell(s)
TLR	toll-like receptor(s)
T_M	memory T-cells
TMD	transmembrane domain
TNF	tumour necrosis factor (e.g. TNF-α)
TRAIL	TNF-related apoptosis-inducing ligand
Treg	regulatory T-cell(s)

U

U	unit(s)
UL	unique long (region)
US	unique short (region)
UTR	untranslated region
UV	ultraviolet (light)

V

V region	variable region of Ig or of TCR
valGCV	valganciclovir
VEGF	vascular endothelial cell growth factor
VGCV	valganciclovir
vol	volume
vRAP	viral regulator of (direct) antigen presentation

W

wk	week(s)
wt	weight
WT	wild-type (e.g. MCMV-WT)

Miscellaneous

β2m	$β_2$-microglobulin
$γ_C$	common γ-chain (of cytokine/interleukin receptors)
3D	three-dimensional

Comparative Genomics of Primate Cytomegaloviruses

Andrew J. Davison, Marylouisa Holton, Aidan Dolan, Derrick J. Dargan, Derek Gatherer and Gary S. Hayward

Abstract

The subfamily *Betaherpesvirinae* contains four genera and three species not yet assigned to genera. CMVs belong to this subfamily, and are usually reckoned to consist of primate viruses (such as HCMV) in the genus *Cytomegalovirus*, rodent viruses (such as MCMV) in the genus *Muromegalovirus*, and two of the unassigned viruses (GPCMV and tupaiid herpesvirus 1). In this chapter, we focus primarily on members of the genus *Cytomegalovirus*, which infect apes (including humans) and Old and New World monkeys. The genomes of nine of these viruses, representing six species, have been sequenced fully. They are distinguished by being among the largest and most complex in the family *Herpesviridae*, by exhibiting an extraordinary degree of interspecies and interstrain variation, and by all having suffered functional losses via mutation. These characteristics continue to contribute to the fascination and challenge of CMV genomics.

Introduction

Genomic studies are important because they provide the gene maps that undergird, implicitly or explicitly, nearly all molecular research on CMVs. In this chapter, we describe our current knowledge of the evolution, structures and genetic contents of CMV genomes, focusing primarily on primate CMVs. As well as providing factual detail, we emphasize that genomics is a nuanced art, and that, as a result, there are grey areas whose present implications are matters of debate. All maps, whether in this chapter or elsewhere, should be viewed circumspectly before expending research effort based on them; in short, let the buyer beware.

Further information on RhCMV is available in Chapter II.22. Detailed descriptions of non-primate CMVs may be found in Chapters I.2, II.5 and II.15.

Evolution of CMVs

The order *Herpesvirales* consists of three families: *Herpesviridae*, which contains viruses of mammals, birds and reptiles, *Alloherpesviridae*, which contains viruses of bony fish, and *Malacoherpesviridae*, which contains viruses of invertebrates (Pellett *et al.*, 2011). These families appear to have a very long history. On the basis of host range, their common ancestor is thought to have predated the divergence of vertebrates from invertebrates, some 900 million years ago (Mya). Moreover, evidence that the order *Herpesvirales* may be related to the order *Caudovirales*, which contains double-stranded DNA bacteriophages (such as T4 phage), implies the existence of an earlier ancestor over 2000 Mya (Davison, 1992; Baker *et al.*, 2005).

Fig. I.1.1A shows a phylogenetic tree, based on the DNA polymerase gene, for the family *Herpesviridae*, which comprises three subfamilies, the *Alphaherpesvirinae*, *Betaherpesvirinae* and *Gammaherpesvirinae*. Phylogenetic studies indicated that the common ancestor of these subfamilies existed approximately 200 Mya (McGeoch *et al.*, 1995), although this date was adjusted subsequently to 400 Mya (McGeoch and Gatherer, 2005). The corresponding age of the subfamily *Betaherpesvirinae*, to which CMVs belong, was estimated at approximately 180 (subsequently, 350) Mya. The subfamily *Betaherpesvirinae* is divided into four genera (*Cytomegalovirus*, *Muromegalovirus*, *Proboscivirus* and *Roseolovirus*) and three incompletely classified viruses. A list of the species in the subfamily, and the abbreviations for virus names used in this chapter, is provided in Table I.1.1. We have taken the liberty of including in the appropriate taxa a number of proposed species that are currently under consideration by the International Committee on Taxonomy of Viruses (ICTV).

Strictly, CMVs belong to the genus *Cytomegalovirus*, which contains viruses of apes (including humans), Old World monkeys (OWMs) and New World

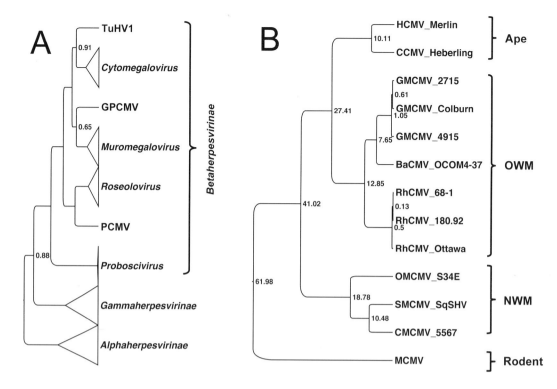

Figure I.1.1 (A) Bayesian phylogenetic tree for the family *Herpesviridae* based on aligned amino acid sequences for DNA polymerase with gapped sites removed. The names of the subfamilies *Alphaherpesvirinae* and *Gammaherpesvirinae* and the genera and viruses (abbreviations in Table I.1.1) in the subfamily *Betaherpesvirinae* are indicated. The program BEAST was run for 10 million iterations using the Jones-Taylor-Thornton model incorporating a site heterogeneity model allowing 5 gamma categories and invariant sites (JTT+G+I), with a relaxed, uncorrelated lognormal clock and a speciating Yule process as a tree prior. Trees were sampled every 5000 iterations, and the consensus tree was based on the final 25% of trees generated. The tree was rooted by using the subfamily *Alphaherpesvirinae* as the outgroup. Posterior probability values are shown at nodes only where < 1. (B) Bayesian phylogenetic tree for the genus *Cytomegalovirus* based on aligned DNA sequences for the DNA polymerase gene with gapped sites removed. Individual virus names (abbreviations are listed in Table I.1.1) followed by strain designations, and host type, are indicated. The program BEAST was run for 350 million iterations using the general time reversible (GTR) substitution model incorporating empirical base frequencies allowing 5 gamma categories, invariant sites, and substitution rate varying between the three codon positions, with a relaxed, uncorrelated, lognormal clock model and a speciating Yule process as a tree prior. Trees were sampled every 5,000 iterations, and the consensus tree was based on the final 25% of trees generated. The tree was rooted by using MCMV as the outgroup. The values for the mean time of most recent common ancestor (Mya) shown at nodes were calculated on the basis of setting the OWM-NWM bifurcation at 42.9 (±3.4) Mya, following Steiper and Young (2006). Posterior probability values were all 1, except for that at the node between GMCMV_2715 and GMCMV_Colburn, which was 0.97.

monkeys (NWMs). In practice, however, several other members of the subfamily *Betaherpesvirinae* are counted as CMVs. These include rodent viruses in the genus *Muromegalovirus*, and GPCMV and tupaiid herpesvirus 1 (TuHV1), which have not yet been classified into genera. A solution to the classification of these two viruses may lie in either placing them in two new genera or abolishing the genus *Muromegalovirus* and incorporating them and the dispossessed rodent viruses into the genus *Cytomegalovirus*. TuHV1 is of peripheral importance biologically, but was of interest initially from an evolutionary standpoint, because of

a suspected close relationship between tupaias (tree shrews) and primates. However, this relationship turned out not to be confirmed by molecular data (Schmitz *et al.*, 2000), and the position of TuHV1 in herpesvirus phylogeny also lends little support (Fig. I.1.1A). The two other genera (*Proboscivirus* and *Roseolovirus*) in the subfamily *Betaherpesvirinae* contain an elephant virus (EEHV) and human viruses (HHV6A, HHV6B and HHV7), respectively, which are not generally counted as CMVs. The former genus may end up containing several more elephant viruses (Latimer *et al.*, 2011), and the latter genus some additional ape

Table I.1.1 Members and proposed members of the subfamily *Betaherpesvirinae*

Species name[a]	Common name of virus[b]	Abbreviations of virus name[c]
Genus *Cytomegalovirus*		
Aotine herpesvirus 1[d]	Owl monkey CMV	OMCMV (AoHV1)
Cebine herpesvirus 1[d]	Capuchin monkey CMV	CMCMV (CbHV1)
Cercopithecine herpesvirus 5	Green monkey CMV	GMCMV (CeHV5)
Human herpesvirus 5	Human CMV	HCMV (HHV5)
Macacine herpesvirus 3	Rhesus CMV	RhCMV (McHV3)
Panine herpesvirus 2	Chimpanzee CMV	CCMV (PnHV2)
Papiine herpesvirus 3[d]	Baboon CMV	BaCMV (PaHV3)
Saimiriine herpesvirus 4[d]	Squirrel monkey CMV	SMCMV (SaHV3)
Genus *Muromegalovirus*		
Murid herpesvirus 1	Mouse CMV	MCMV (MuHV1)
Murid herpesvirus 2	Rat CMV strain Maastricht	RCMV (MuHV2)
Murid herpesvirus 8	Rat CMV strain England	RCMVE (MuHV8)
Genus *Proboscivirus*		
Elephantid herpesvirus 1	Elephant endotheliotropic herpesvirus	EEHV (ElHV1)
Genus *Roseolovirus*		
Human herpesvirus 6A	Human herpesvirus 6A	HHV6A
Human herpesvirus 6B	Human herpesvirus 6B	HHV6B
Human herpesvirus 7	Human herpesvirus 7	HHV7
Unassigned		
Caviid herpesvirus 2	Guinea pig CMV	GPCMV (CavHV2)
Suid herpesvirus 2	Pig CMV	PCMV (SuHV2)
Tupaiid herpesvirus 1	Tupaia herpesvirus 1	TuHV1

[a]A species is a formal, taxonomical grouping of viruses. Species names are always written in full, italicized and capitalized. Cercopithecine herpesvirus 3 strain SA6, which is listed in ICTV reports as an unassigned virus in the family *Herpesviridae*, belongs to the species *Cercopithecine herpesvirus 5* (Alcendor *et al.*, 2009). Aotine herpesvirus 3, which has been listed as an unassigned virus in the genus *Cytomegalovirus*, belongs to the proposed species *Aotine herpesvirus 1* (Ebeling *et al.*, 1983).
[b]Some viruses have more than one common (informal or vernacular) name. For example, GMCMV is frequently known as simian CMV. The single common name used in this chapter is listed.
[c]The abbreviation of the common name used in this chapter is given first, followed, where appropriate, in parentheses by the abbreviation (not used in this chapter) of the virus name derived from the species name.
[d]The creation and classification of this species is under consideration by the ICTV, and has not yet been accepted formally.

viruses (Lacoste *et al.*, 2005), as well as PCMV (Goltz *et al.*, 2000).

The most extensive phylogenetic analysis of primate CMVs is that of Leendertz *et al.* (2009). It incorporates well-characterized viruses, for which extensive sequence data are available, and viruses from chimpanzees, gorillas, orang-utans and OWMs, for which limited, novel data were derived for the study. The main conclusion was that primate CMVs have evolved cospeciationally with their hosts, but that horizontal transmission may have occurred between chimpanzees and gorillas

in both directions. Fig. I.1.1B shows a phylogenetic tree, based on the DNA polymerase gene, for primate CMVs relevant to this chapter. The data were obtained from the relevant genome sequence accessions in GenBank (Table I.1.2) and other accessions for partial sequences (GMCMV strain 4915, JQ264771; BaCMV strain OCOM4-37, AC090446; CMCMV strain 5567, JQ264772; MCMV strain Smith, U68299). Insufficient sequence information is available for the gorilla and orang-utan CMVs described in Leendertz *et al.* (2009) to include them in this tree. The primate viruses

Table I.1.2 Characteristics of genomes of members or proposed members of the genus *Cytomegalovirus*

Virus	Strain	GenBank accession no.	Sequenced genome (bp)	Wild-type genome (bp)[a]	No. of genes[b]	Mutated genes in sequenced genome[c]
HCMV	Merlin	AY446894	235646	235646	170	RL13[d] UL128
CCMV	Heberling	AF480884	241087	241087	170	UL128
GMCMV	2715	FJ483968	226205	226206	184	RL1 RL11C RL11D RL11O UL45 UL83A UL119
GMCMV	Colburn	FJ483969	219526			RL11D RL11G UL116 UL128 UL141 [US26 US27A US27B US27D US27E]
RhCMV	68–1	AY186194	221454[e]	224471[e,f]	185	RL11B RL11D RL11E O3 COX2 RL11G UL34 UL71 [UL128 UL130 UL146C UL146D UL146F UL146H] US18 US27D
RhCMV	180.92	DQ120516	215678[e]			RL11E O3 O4 COX2 RL11G RL11K UL116 UL130 [UL148 UL147A UL147 UL146B UL146C UL146D UL146F UL146H UL145 UL144 UL141 O11 O12 O13 O14] US28
RhCMV	Ottawa	JN227533	218041[e]			RL11B [RL11E] [COX2 RL11F RL11G] [RL11L] UL83B [OriLyt][g]
OMCMV	S34E	FJ483970	219474	219470	153	A6 A15 A16 A17 A19 A41 A42
SMCMV	SqSHV	FJ483967	196691	196692	141[h]	S10 UL116[i]

[a]Wild-type genome sequences were estimated on the basis of conceptual repairs made to the sequenced genomes. Deletions cannot be ruled out for viruses where only a single genome sequence is available (CCMV, OMCMV and SMCMV).
[b]Predicted protein-coding genes only are counted for the wild-type genomes (Figs. I.1.3–I.1.8).
[c]Genes are mutated by frameshifts, internal stop codons or (in square brackets) when affected in whole or part by deletions.
[d]Various mutations are present, and few or no genomes are wild-type.
[e]Genome ends may not have been identified accurately, and overall genome length is short owing to omission of the copy of TR at the right end.
[f]The wild-type genome sequence was onstructed by inverting 3902 bp (nt 163231–167132), inserting 1641 and 1425 bp sequences from strain CNPRC (Oxford *et al.*, 2008; EF990255) at the junctions, and repairing mutations elsewhere.
[g]The region containing the OriLyt-associated repeat is deleted by approximately 300 bp, probably as a result of sequence assembly problems.
[h]Counting only one of IRS1 and TRS1, which are identical to each other (Figure I.1.8).
[i]UL116 is mutated in approximately 75% of genomes.

fall into three main groups, infecting apes, OWMs and NWMs. In Fig. I.1.1B, the dates at which the various CMV lineages diverged were estimated on the assumption that the OWM and NWM branches cospeciated with their hosts. As in the analysis of Leendertz *et al.* (2009), the findings for subsequent lineages fit broadly with a cospeciational model (host divergence dates are available at www.timetree.org; Hedges *et al.*, 2006). However, divergence of the BaCMV and GMCMV lineages is somewhat later than might be anticipated, perhaps reflecting uncertainty about the relationships among baboon, green monkey and rhesus macaque, or an ancient interspecies transfer. Also, divergence of the HCMV and CCMV lineages is a little early, indicating that their ancestor may have speciated before their host.

Fig. I.1.1B supports the view that HCMV, CCMV, GMCMV, BaCMV, RhCMV, OMCMV, SMCMV and CMCMV represent different primate CMV species in the subfamily *Betaherpesvirinae*. The three strains of GMCMV are counted as being in the same species, divergence having commenced approximately 1 Mya. Strain Colburn was isolated ostensibly from a human, but is clearly a strain of GMCMV rather than HCMV (Huang *et al.*, 1978). Divergence of the three viruses denoted as RhCMV strains began approximately 0.5 Mya. However, unlike two of these strains (68-1 and 180.92), which were isolated from rhesus macaques, the third virus was isolated from a cynomolgus (crab-eating) macaque and originally named cynomolgus CMV (CyCMV) strain Ottawa (Marsh *et al.*, 2011). Although CyCMV lacks a few genes (RL11E, COX2, RL11F, RL11G and RL11L) that are present in

RhCMV strains 68-1 and 180.92, the estimated time of its divergence from the lineage leading to these strains is far less than that separating the two host macaque species (2.4 Mya), thus reflecting a probable interspecies transfer. We have denoted CyCMV as RhCMV strain Ottawa, as this nomenclature best reflects the genetic relationships. Insufficient data on the epidemiology and biology of macaque CMVs are available to permit the details of the proposed transfer to be discerned.

Structures of CMV genomes

CMV genomes are the largest among the members of the family *Herpesviridae*. The largest is that of CCMV, at a little over 240 kbp (Davison *et al.*, 2003b). The genomes of MCMV (Rawlinson *et al.*, 1996; Smith *et al.*, 2008; Cheng *et al.*, 2010), RCMV (Vink *et al.*, 2000) and GPCMV (Schleiss *et al.*, 2008; Kanai *et al.*, 2011) are approximately 230 kbp, that of TuHV1 a little smaller, at about 200 kbp (Bahr and Darai, 2001), and those of members of the genus *Roseolovirus* smaller still, at 150–160 kbp (Gompels *et al.*, 1995; Nicholas, 1996; Megaw *et al.*, 1998; Dominguez *et al.*, 1999; Isegawa *et al.*, 1999). The sizes and basic characteristics of the genomes of members of genus *Cytomegalovirus* are summarized in Table I.1.2. Complete, or essentially complete, sequences are available for HCMV strain Merlin (Dolan *et al.*, 2004; many other strains have also been sequenced), CCMV strain Heberling (Davison *et al.*, 2003b), GMCMV strains 2715 and Colburn, the latter supported by published data (Huang *et al.*, 1978; Jeang and Hayward, 1983), RhCMV strains 68-1 (Hansen *et al.*, 2003), 180.92 (Rivailler *et al.*, 2006) and Ottawa (Marsh *et al.*, 2011), OMCMV strain S34E, supported by published data (Ebeling *et al.*, 2003), and SMCMV strain SqSHV. Nucleotide composition ranges from 49.0 (RhCMV) to 61.7% G+C (CCMV), and may vary dramatically within a particular genome (Gatherer, 2008). We determined the sequences of OMCMV and SMCMV and the two GMCMV strains by standard Sanger sequencing of random plasmid libraries derived from purified virion DNA, and supplemented the SMCMV sequence from Illumina data. We deposited these sequences in GenBank in 2008, and parts have been used in publications (McGeoch *et al.*, 2006; Alcendor *et al.*, 2009; Leendertz *et al.*, 2009). We have recently updated the sequences and added the gene annotations summarized in this chapter.

Repeated sequences, in direct or inverted orientation, are characteristic of herpesvirus genomes, and various arrangements of these in relation to non-repeated (unique) sequences are found among members of the subfamily *Betaherpesvirinae* (Davison, 2007). The most frequently encountered arrangement

consists of a unique sequence (U) flanked by a direct terminal repeat (TR), giving an overall structure of TR-U-TR (Fig. I.1.2A). TR may vary greatly in size among the different viruses. It is 30 bp in MCMV (Marks and Spector, 1988), 504 bp in RCMV (Vink *et al.*, 1996), about 1 kbp in GPCMV (Gao and Isom, 1984) and several kbp in HHV6A, HHV6B and HHV7 (Lindquester and Pellett, 1991; Martin *et al.*, 1991; Dominguez *et al.*, 1996). This simple arrangement is also shared by some members of the genus *Cytomegalovirus*, namely RhCMV, in which TR is approximately 500 bp (Hansen *et al.*, 2003) and, from our data, GMCMV, in which it is 1130 bp (strain Colburn) or 1137 bp (strain 2715). The genome structures of HCMV and CCMV are more complex, consisting of long and short unique sequences (U_L and U_S) flanked by inverted repeats (TR_L and IR_L; TR_S and IR_S) (Weststrate *et al.*, 1980; Davison *et al.*, 2003b). These genomes also have a direct terminal repeat (the *a* sequence) of a few hundred bp (equivalent to TR), an inverted copy of which (the *a'* sequence) is present at the junction between IR_L and IR_S (Spaete and Mocarski, 1985; Tamashiro and Spector, 1986; Davison *et al.*, 2003b). Thus, the overall structure can be described as TR_L-U_L-IR_L-IR_S-U_S-TR_S or *ab*-U_L-*b'a'c'*-U_S-*ca* (Fig. I.1.2C). In addition, the HCMV and CCMV genomes exhibit a phenomenon known as segment inversion, which, via recombination between the inverted repeats in concatemeric DNA during DNA replication, results in virions containing equimolar amounts of four genome isomers differing in the relative orientations of U_L and U_S. The OMCMV and SMCMV genomes have a similar structure, with our data indicating the presence of the inverted repeats but not the *a/a'* sequence (for an exception, see below). For those genomes that have been studied in sufficient detail, the 3'-ends of the virion DNA strands consist of single, unpaired, complementary nucleotides that

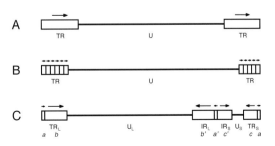

Figure I.1.2 Types of CMV genome structure (not to scale). Unique and repeated regions are shown as horizontal lines and rectangles, respectively, with the orientations of the latter shown by arrows. The nomenclature of unique and repeated regions in the three types (A, B and C) is indicated, and is explained in the text.

presumably anneal when the genome is circularized prior to DNA replication.

The descriptions of genome structures given above are simplified, and a number of additional features have been reported. Some of these may be due to the occurrence of alternative cleavage events during generation of unit-length genomes from replicated, concatemeric DNA, and others may reflect the fact that the termini of certain genomes have not been characterized adequately. Some GPCMV genomes lack the copy of TR at the right genome end, in effect not containing repeat sequences (McVoy *et al.*, 1997; Nixon and McVoy, 2002). Multiple *a/a'* sequences may be present in HCMV genomes at the left genome end and the junction between IR$_L$ and IR$_S$, and some genomes lack the *a* sequence at the right genome end (Tamashiro and Spector, 1986). Multiple copies of TR may be present at the ends of some RhCMV genomes (Hansen *et al.*, 2003; Fig. I.1.2B), and this may also be a feature of GMCMV, for which our sequence data imply the existence of a proportion of strain 2715 genomes with a slightly larger TR (1671 bp) due to alternative cleavage. For SMCMV, but not OMCMV, there is evidence in the sequence data for an *a/a'* sequence of 1340 bp in a proportion of genomes. It is not clear whether the genome ends of the RhCMV strains have been mapped properly, and the GenBank sequences are a little undersized because they lack the copy of TR at the right end (Table I.1.2). TuHV1 has been reported as lacking repeat sequences of the sort that characterize other CMV genomes (Albrecht *et al.*, 1985; Koch *et al.*, 1985), though this is not definitive, as the genome ends have not been studied directly.

Discovering the contents of CMV genomes

Sequencing herpesvirus genomes has been routine for many years, and enough is known about herpesviruses to make the process of annotating a new sequence largely straightforward. Nonetheless, some genes may be difficult to spot: those that are small or expressed by splicing, are poorly conserved or unique to a particular virus, have protein-coding regions that substantially overlap others, are compromised by sequencing errors or genuine mutations, utilize non-canonical initiation codons or esoteric translational mechanisms, or specify non-coding RNAs. As a result, competent analysis is bound to involve a balance of various criteria, and the picture of gene content obtained should not be viewed as fully accurate or complete. It is important to recognize that CMV genome maps are interpretations, and there are grey areas; thus the caveat emptor at the beginning of this chapter.

A basic criterion for defining genetic content involves identifying open reading frames (ORFs) above a certain size that are initiated by methionine codons (ATG) and terminated by stop codons (TAA, TGA or TAG). Having identified a set of candidate ORFs on this basis, comparative genomics, which recognizes the fact that genes are shaped by evolution, helps to distinguish ORFs that probably encode functional proteins from those that probably do not. The relevant computer tools either identify ORFs that are conserved in sequence (nucleotide or imputed amino acid) among related viruses or in other organisms, or assess how closely other sequence patterns (e.g. bias in codon usage or amino acid composition) fit those of known protein-coding regions. However, even with such assistance, some functional protein-coding ORFs might be missed, and other ORFs might be included that are non-functional. Moreover, the approaches described above cannot detect genes specifying RNAs that do not encode functional proteins (non-coding RNAs). Experimental studies on expression (RNA or protein) or function also have an important part to play in gene identification. However, these may vary in quality or importance and should not always be used to trump bioinformatic conclusions.

Development of the wild-type HCMV genome map to its present stage has been a protracted saga, not least because of the almost exclusive focus placed for many years on two laboratory strains, AD169 and Towne. These viruses have been passaged extensively in human fibroblast cells, and, as a result, are highly mutated (Cha *et al.*, 1996; Dargan *et al.*, 1997; Mocarski *et al.*, 1997; Prichard *et al.*, 2001; Bradley *et al.*, 2009). The determination of the genome sequence of one of these strains (AD169) resulted in the cataloguing of 189 protein-coding ORFs (Chee *et al.*, 1990). Later, the discounting of many of these ORFs as probably not encoding functional proteins, and the addition of other ORFs that had not been recognized originally, resulted in an overall reduction of this number by more than 40. This was achieved by taking into account data for HCMV published after Chee *et al.* (1990) and by analysing the genome sequence of CCMV, which is the closest relative of HCMV (Fig. I.1.1B), using comparative genomics as a major criterion while not excluding a number of non-conserved genes (Davison *et al.*, 2003a,b). Subsequently, analysis of the genome sequence of the low passage HCMV strain Merlin resulted in a gene map containing 165 functional protein-coding genes (Dolan *et al.*, 2004). Substantially higher numbers were put forward by others, mainly from the use of pattern-searching programmes that led to the incorporation of ORFs that are small, overlap larger ORFs, and are not conserved in CCMV

(Murphy et al., 2003a,b; Yu et al., 2003; Varnum et al., 2004; Murphy and Shenk, 2008). We have noted previously that pattern-searching programmes are at their least discriminating when analysing small ORFs (Dolan et al., 2004). Thus, our stance on the number of functional protein-coding genes remains conservative, with the latest estimate in wild-type HCMV standing at 170. However, we expect that more HCMV genes will be discovered. In this prospective context, it is worth highlighting two small ORFs that, despite bioinformatic shortcomings and lack of conservation in primate CMVs, might nonetheless encode functional proteins. These are the LUNA ORF, which is entirely antiparallel to part of the UL82 protein-coding region and whose encoded protein is recognized in appropriate serological assays by antibodies in HCMV-positive sera (Bego et al., 2005, 2011), and the RASCAL ORF, which overlaps the 5′-end of the US17 protein-coding region in another reading frame and may encode a protein that localizes to the nuclear lamina and cytoplasmic vesicles (Miller et al., 2010). It is important to point out that the absence of these ORFs from Fig. I.1.3 reflects our present position in regard to some of the grey areas of CMV genomics, rather than any final conclusion. Also, in addition to further functional protein-coding regions, it is likely that more non-coding RNAs will be mapped. This raises the challenge of finding out the functions of these transcripts.

Gene layout in CMV genomes

Our current map of the wild-type HCMV genome is shown in Fig. I.1.3. It is built on an interpretation of the strain Merlin sequence (Dolan et al., 2004), extended by subsequent discoveries with this and other strains (Mitchell et al., 2009; Scalzo et al., 2009; Davison, 2010; Gatherer et al., 2011). The map incorporates the 170 functional protein-coding genes mentioned above, and four genes specifying non-coding RNAs. It does not include genes encoding miRNAs, which are derived by cleavage of longer RNAs (see Chapter I.5), small RNAs from the region of the origin of DNA replication (OriLyt; Prichard et al., 1998), spliced RNAs of unknown significance (Gatherer et al., 2011; see Chapter I.11), and numerous ill-defined antisense RNAs (Zhang et al., 2007; Gatherer et al., 2011). Maps for CCMV, GMCMV, RhCMV, OMCMV and SMCMV are presented in Figs. I.1.4–I.1.8. The non-coding transcripts RNA2.7, RNA1.2, RNA4.9 and RNA5.0 have been mapped experimentally for HCMV (Fig. I.1.3) but not the other primate CMVs, and are consequently not included in Figs. I.1.4–I.1.8. The corresponding genome regions, where present, are highly diverged among the genomes, though the relevant

transcriptional signals are conserved in CCMV for all four RNAs and in the monkey CMVs for RNA4.9 and RNA5.0. It is not possible to go into great detail about these maps in this chapter or provide lists of gene functions, which have been published elsewhere for HCMV (Davison and Bhella, 2007). The relevant annotations may be found in the GenBank accessions (Table I.1.2), except for RhCMV, and a comprehensive collation of information, including RhCMV, is available from the authors.

It is remarkable that all fully sequenced HCMV strains that have been passaged in vitro are mutants, as well as some analysed directly from clinical material (Cha et al., 1996; Akter et al., 2003; Davison et al., 2003b; Dolan et al., 2004; Hahn et al., 2004; Cunningham et al., 2010; Dargan et al., 2010; Stanton et al., 2010). Particularly notable for mutating in vitro are the UL128 locus (UL128L), which consists of three genes (UL128, UL130 and UL131A) involved in promoting viral entry into non-fibroblast cells (see Chapter I.17), and RL13, whose function is unknown. The other fully sequenced primate CMVs are also mutants (Table I.1.2). Because of this, the maps in Figs. I.1.3–I.1.8 relate to the proposed wild-type genomes, as reconstructed conceptually from the sequenced genomes. Although some apparent mutations might reflect errors (e.g. those in UL34 and UL71 in RhCMV strain 68-1), most are likely to have arisen during growth in vivo or in vitro. The lesions listed in Table I.1.2 are manifested in various ways: by inversions, by large deletions that remove one or more genes in their entirety plus parts of flanking genes, by small deletions or insertions that cause frameshifts, and by single nucleotide substitutions that introduce in-frame stop codons. Subtler mutations, such as those resulting in deleterious amino acid substitutions, would not have been detected. Some lesions are substantial. For example, the RhCMV strain 68-1 genome has suffered inversion of a 3.9 kbp region, accompanied by deletion of flanking regions of 1.6 and 1.4 kbp, and RhCMV strain 180.92 has lost a region of 8.2 kbp (Oxford et al., 2008). Several genes are frameshifted in tracts of C:G residues: RL11D in GMCMV strain 2715, RL11B, RL11D, RL11E in RhCMV strain 68-1, RL11E in RhCMV strain 180.92, RL11B in RhCMV strain Ottawa, A6, A15, A16, A17, A19, A41 and A42 in OMCMV, S10 in SMCMV, and perhaps RL5A and RL6 in HCMV. Additional genes contain C:G tracts but are not frameshifted: RL11B and RL11D in RhCMV strain 180.92, RL11D in RhCMV strain Ottawa, and S1 in SMCMV. C:G tracts are inherently variable in length and therefore prone to generating frameshifts. It is not beyond the bounds of possibility that this might

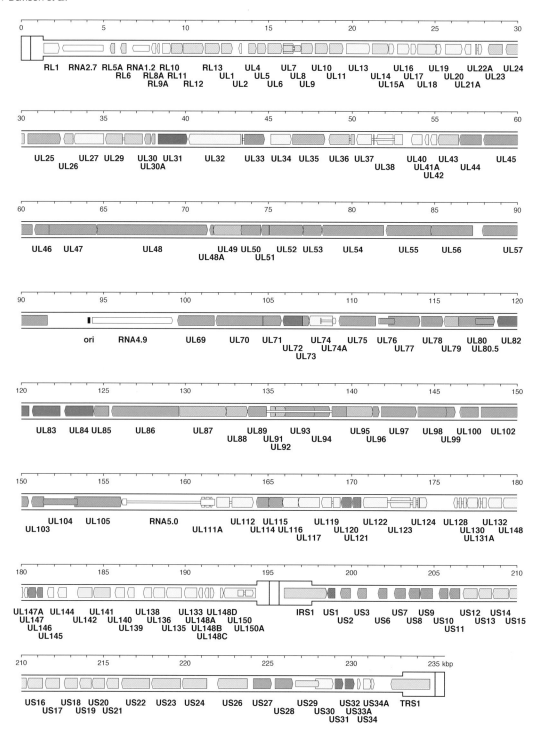

Figure I.1.3 Genome map of wild-type HCMV based on strain Merlin. The inverted repeats (TR$_L$, IR$_L$, IR$_S$ and TR$_S$) are shown in a thicker format than the unique regions (U$_L$ and U$_S$). See the legend of Fig. I.1.8 for further details. Modified from Dolan *et al.* (2004) with permission from the Society for General Microbiology.

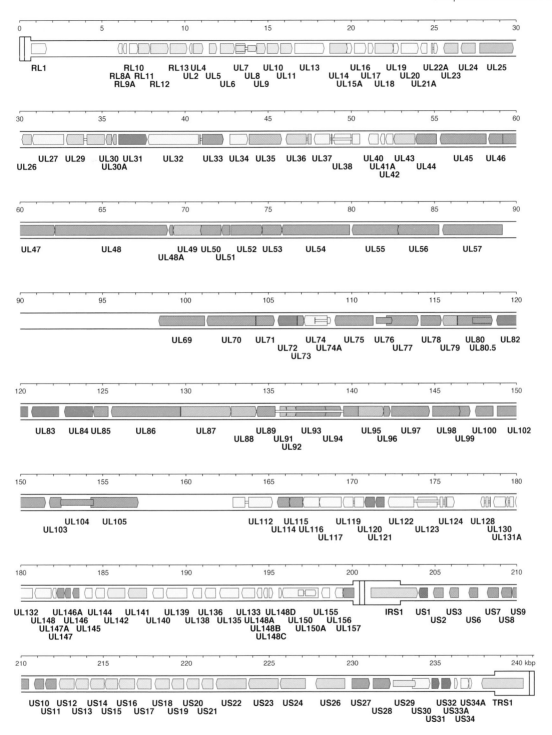

Figure I.1.4 Genome map of wild-type CCMV based on strain Heberling. The inverted repeats (TR_L, IR_L, IR_S and TR_S) are shown in a thicker format than the unique regions (U_L and U_S). See the legend of Fig. I.1.8 for further details. Modified from Davison *et al*. (2003b) with permission from the Society for General Microbiology.

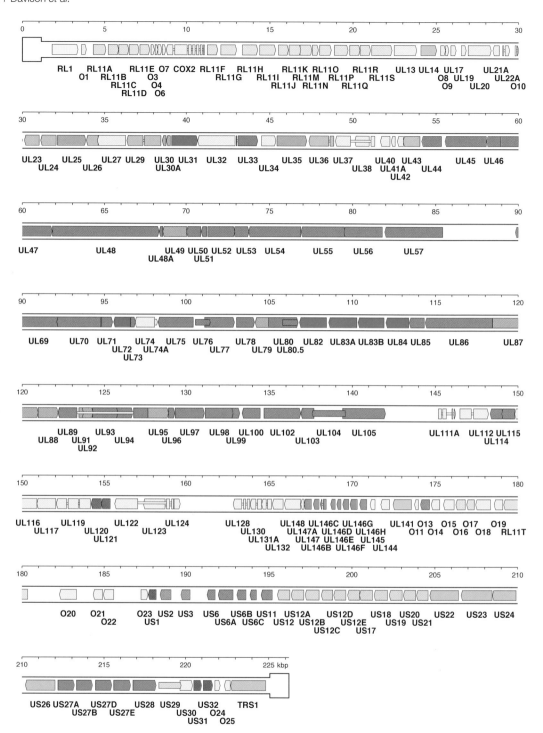

Figure I.1.5 Genome map of wild-type GMCMV based on strain 2715. The terminal direct repeat (TR) is shown in a thicker format than the unique region (U). See the legend of Fig. I.1.8 for further details.

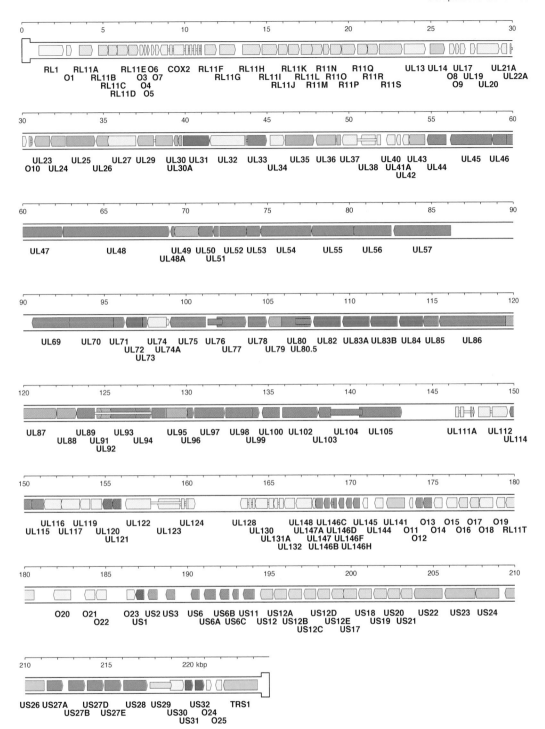

Figure I.1.6 Genome map of wild-type RhCMV based on strain 68-1, with absent regions inserted from strain CNPRC. The terminal direct repeat (TR) is shown in a thicker format than the unique region (U). See the legend of Fig. I.1.8 for further details.

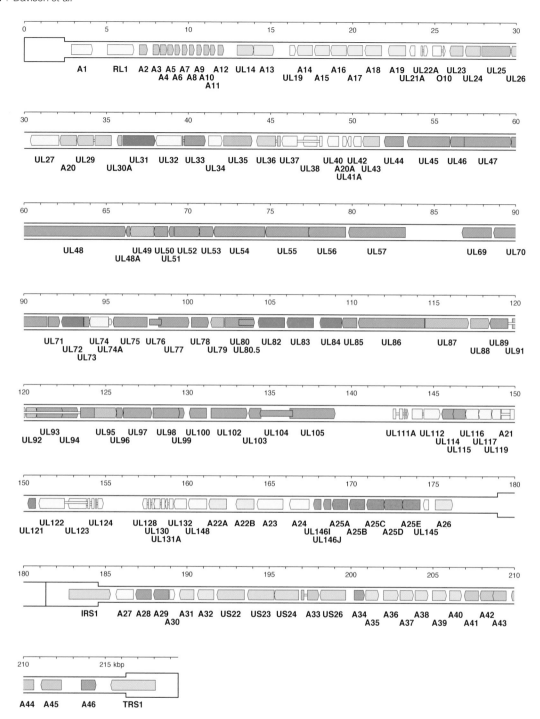

Figure I.1.7 Genome map of wild-type OMCMV based on strain S34E. The inverted repeats (TR$_L$, IR$_L$, IR$_S$ and TR$_S$) are shown in a thicker format than the unique regions (U$_L$ and U$_S$). See the legend of Fig. I.1.8 for further details.

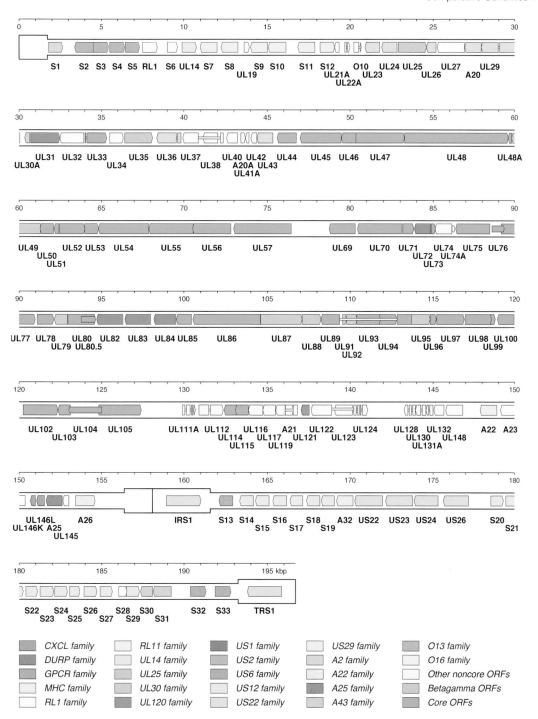

Figure I.1.8 Genome map of wild-type SMCMV based on strain SqSHV. The inverted repeats (TR$_L$, IR$_L$, IR$_S$ and TR$_S$) are shown in a thicker format than the unique regions (U$_L$ and U$_S$). Functional protein-coding regions are indicated in Figs. I.1.3–I.1.8 by coloured arrows grouped according to the key shown at the foot, and non-coding RNAs (Fig. I.1.3) as narrower, white-shaded arrows, with gene nomenclature below. Introns connecting protein-coding regions or exons in non-coding RNA5.0 (Fig. I.1.3) are shown as narrow white bars. The colours of protein-coding regions indicate conservation among members of the family *Herpesviridae* (core genes) or the subfamilies *Betaherpesvirinae* and *Gammaherpesvirinae* (betagamma genes), with subsets of the remaining non-core genes grouped into gene families. UL72 is both a core gene and a member of the deoxyuridine triphosphatase-related protein (DURP) gene family.

act *in vivo* as a means of turning certain functions on or off, via selection acting on heterogeneous genome populations.

In order to facilitate cross-referencing between the CMV genomes that form the focus of this chapter, we have applied a common gene nomenclature to Figs. I.1.3–I.1.8, whereby orthologous genes share the same name. The names fall into the following hierarchy: orthologues in HCMV (UL, US, RL, IRS and TRS prefixes), OWM (O prefix), OMCMV (A prefix) and SMCMV (S prefix), the prefix followed by a number. The principle that the purpose of a gene name is to provide a unique label, and not necessarily any other information, permits latitude in adding further descriptors (e.g. in gene families, see below) and newly discovered genes. However, no system is perfect, and particular difficulties may arise in circumstances where orthology is difficult to determine. These usually involve low sequence conservation, and may be compounded by the presence in different viruses of different numbers of genes in a gene family. Gene families are very prominent in CMVs (see the key to Fig. I.1.8), being manifested as sets of related (and often distantly related) genes that were presumably generated by gene duplication. Gene families may demonstrate complex evolution involving multiple duplications, deletions and translocations (Sahagun-Ruiz *et al.*, 2004; Arav-Boger *et al.*, 2005; Davison and Stow, 2005; Alcendor *et al.*, 2009). An example where nomenclature is problematic in these circumstances is the US6 gene family, which has six members in HCMV and CCMV (Figs. I.1.3 and I.1.4) and five in GMCMV and RhCMV (Figs. I.1.5 and I.1.6; Pande *et al.*, 2005). Two of the OWM genes (US6 and US11, which encode immune evasion functions) have clear orthologues in HCMV and CCMV, and hence are named after them. However, the relationships of the remaining three genes to US7-US10 in HCMV and CCMV are ambiguous, and consequently these are named after the gene family by use of a further descriptor (US6A, US6B and US6C), so that precise orthology is not implied. US6 family members are also present in the U_S region in OMCMV and SMCMV (Figs. I.1.7 and I.1.8), with some additional members located near the left end of U_L in the latter, but, since their orthologous relationships are not at all discernible, they take virus-specific names.

The central region of CMV genomes (in the section UL44–UL115) contains genes that are inferred as having been present in the ancestor of the family *Herpesviridae* (marked as core genes in Figs. I.1.3–I.1.8) or the ancestor of the subfamilies *Betaherpesvirinae* and *Gammaherpesvirinae* (marked as betagamma genes). A broader subset of genes characterized the ancestor of the subfamily *Betaherpesvirinae* (in the section UL23-UL124, plus possibly US22, US26 and TRS1), and an even broader one the ancestor of the genus *Cytomegalovirus* (in the section UL21A-UL145, plus RL1, UL14 and UL19, and members of the US6, US12, US22 and US29 gene families). CMV genomes are impressively divergent near the genome ends. For example, the ape and OWM CMVs have large arrays of RL11 gene family members near the left end (as many as 19 in RhCMV strains 68-1 and 180.92), plus another member elsewhere, whereas the NWM CMVs lack this family and instead have a prominent family encoding MHC-related proteins. To the right of conserved gene UL145, the ape CMVs have an array of leftward-oriented genes (UL138-UL148B) that encode proteins containing one or two potential transmembrane domains near the N terminus. The OWM CMVs lack obvious orthologues of these genes, but they do have an array of leftward-oriented genes (O11-O15, O18-O19 and O21-O22) that encode proteins with similarly located domains. Thus, these genes might represent highly diverged orthologues of genes in the UL138-UL148B region. However, the NWM CMV genomes lack orthologues, whether detectable or highly diverged, to any of these genes. Many of the HCMV genes that lack orthologues in other CMVs, or have them only in CCMV, are known or thought to be involved in specific interactions with the host, particularly modulating the immune response (see Chapter II.7 and Chapter II.8) or having potential roles in latency (see Chapter I.20). Although it would be unexceptional to find that immunomodulatory genes are not conserved among primate CMVs, it would be more problematic to find this for genes that have pivotal roles in latency. Thus, in our view, non-conserved HCMV genes proposed as functioning in latency, such as forms of UL111A (encoding an interleukin 10 homologue and absent from CCMV), the LUNA ORF (conserved only, and atypically, in CCMV) and UL138 (not conserved in NWM CMVs and possibly not in OWM CMVs), are unlikely to have central roles, any effects on latency probably being exerted indirectly (Bego *et al.*, 2005; Goodrum *et al.*, 2007; Avdic *et al.*, 2011). Rather, we speculate that insights into the key determinants of latency will eventually come from other directions entirely (see Chapter I.19).

Variation among CMV strains

Different strains of a particular CMV are very similar to each other in the great majority of their genome sequences, but may exhibit substantial, localized variation (often termed hypervariation). This may involve the entire length or, more commonly, a portion of a protein-coding region, and may also be reflected by gross variations in the numbers of genes in a specific location (apparent deletions), including gene families (Sahagun-Ruiz et al., 2004; Arav-Boger et al., 2005; Alcendor et al., 2009). By far the greatest effort on CMV genetic variation has focused on HCMV, and an extensive literature has been summarized at intervals by excellent reviews (Pignatelli et al., 2004; Puchhammer-Stöckl and Görzer, 2011). Fig. I.1.9 illustrates the extent of sequence variation across the HCMV genome as determined for 23 strains, calculated in terms of nucleotide substitutions. Most genes are highly conserved, but some exhibit substantial variation, which is thought to have occurred early in human evolution (Bradley et al., 2008). The finding that variable genes are stable in individuals accords with the view that HCMV evolves relatively slowly in comparison with RNA viruses (Hassan-Walker et al., 2004; Stanton et al., 2005), although there is evidence for a degree of genome-wide variability from deep sequencing studies (Renzette et al., 2011). Variable genes encode proteins that are secreted or associated with membranes and therefore are (or were) presumably under strong immune selection. Many, but not all, are dispensable for growth in cell culture, and are thought to encode proteins that are important for interactions with the host, with involvement in processes such as cell tropism, immune evasion or viral spread. Recombination has also been an important generator of diversity in HCMV, and, as a result, a great many strains are in circulation (Haberland et al., 1999; Rasmussen et al., 2003).

In the terms by which Fig. I.1.9 was derived, the most variable HCMV genes are UL146, RL12 and RL13. UL146 is a member of the CXCL gene family (Fig. I.1.3; encoding known or predicted CXC- or β-chemokines) and exists as 14 distinct variants (Dolan et al., 2004; Arav-Boger et al., 2005, 2006b; Bradley et al., 2008). RL12 and RL13 are members of the RL11 gene family (Fig. I.1.3), and encode predicted membrane glycoproteins of unknown function. Additional variable members of the RL11 gene family are RL6, UL1, UL9 and UL11, and (less variable) RL5A, UL4, UL6, UL7, UL8 and UL10 (Hitomi et al., 1997; Alderete et al., 1999; Bar et al., 2001; Davison et al., 2003a; Sekulin et al., 2007; Engel et al., 2011). Other highly variable genes include UL40 (encoding glycoprotein gp40), UL73 (virion glycoprotein N), UL74 (virion glycoprotein O), UL120 (related to cellular OX-2), UL147 (a member of the CXCL gene family) and UL139 (ostensibly related to CD24) (Pignatelli et al., 2001, 2003; Paterson et al., 2002; Rasmussen et al., 2003; Mattick et al., 2004; Arav-Boger et al., 2005, 2006b; He et al., 2006; Lurain et al., 2006; Qi et al., 2006). Several other genes are also variable, but less so. These include UL20 (containing an Ig-related ectodomain),

UL33 (a G protein-coupled receptor (GPCR)), UL37 (an immediate-early glycoprotein), UL55 (virion glycoprotein B), UL116, UL119 (possibly related to cellular OX-2) and UL144 (similar to cellular tumour necrosis-related factor) (Chou and Dennison, 1991; Meyer-Konig et al., 1998; Shepp et al., 1998; Lurain et al., 1999; Trincado et al., 2000; Bale et al., 2001; Hayajneh et al., 2001a,b; Arav-Boger et al., 2002; Coaquette et al., 2004; Picone et al., 2005; Deckers et al., 2009; Jelcic et al., 2011). Levels of variation that are lower, but still greater than those in the majority of HCMV genes, characterize additional genes that are not listed here. Variation also occurs in some regions that apparently do not encode proteins, such as the a sequence (Bale et al., 2001) and the region between US34A and TRS1 (Fig. I.1.9). Length polymorphisms in short tandem repeats, whether protein-coding or not, have also been noted (Davis et al., 1999).

Many studies of variation of HCMV genes have also examined potential links with the outcomes of HCMV infection. In our view, some of this research were conducted on marginal premises because it dealt with genes that are not particularly variable, or, in some cases, that no longer feature in genome annotations. However, some genes with low variability are of interest, in that they mutate under specific circumstances. These include two core genes, UL54 (DNA polymerase) and UL97 (protein kinase), in which mutations are selected in response to antiviral drugs (Hakki and Chou, 2011; see Chapter II.19), and UL128L, in which mutations are selected when virus is passaged in fibroblasts. However, these events are limited to situations that do not occur generally *in vivo*. For hypervariable genes, studies that support links between particular variants and clinical outcome are in the minority, and, in retrospect, a significant number lack the requisite power (Pignatelli et al., 2004; Bradley et al., 2008). Among this minority, UL73 variants have been differentiated from each other in their effects during congenital infection (Pignatelli et al., 2010), whereas links between UL144 variants and clinical sequelae have proved controversial (Picone et al., 2005, 2007; Arav-Boger et al., 2006a; Heo et al., 2008).

One factor that impinges on studies of pathogenesis is the occurrence of infections with multiple HCMV strains, which are possibly advantageous for viral growth (Čičin-Šain et al., 2005). Standard approaches based on PCR and sequencing are not optimal for detecting all variants of a gene, and can, in principle, result in a pathogenic variant going undetected. Mixed UL55 variants can increase disease progression (Coaquette et al., 2004) and confer delayed clearance of virus during antiviral treatment (Manuel et al., 2009), and mixed variants of UL75 (not a particularly variable

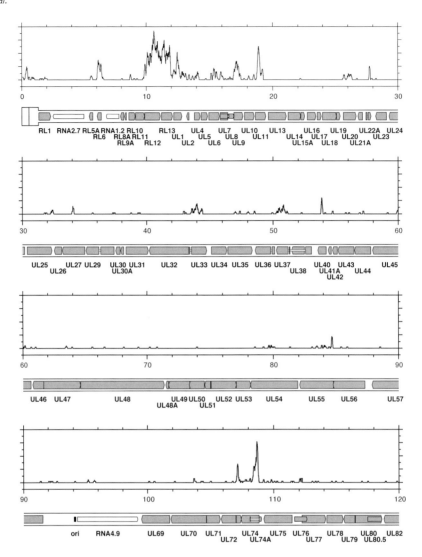

Figure I.1.9 Sequence variation across the HCMV genome. An alignment of the genome sequences of 23 HCMV strains was generated, and all gaps were removed (when in strain Merlin) or replaced with the consensus sequence (when in other strains). Thus, the alignment registered nucleotide substitutions, but not insertions or deletions, in relation to a consensus sequence that was the length of the strain Merlin genome.

gene, encoding virion glycoprotein H) have also been linked to more severe disease in renal transplant patients (Ishibashi *et al.*, 2007).

Concluding remarks

Many factors make CMV genomics interesting: the size and complexity of the genomes involved, the impressive degree of variation, which is presumably due to interactions with the host that are much more extensive than those of other herpesviruses, the instability of wild-type virus *in vitro*, the continuing discovery of new genes and gene functions, and the convoluted way in which the research history has developed. CMV genomics is important because it provides a necessary foundation for functional studies aimed at improving clinical practice, whether these studies concern the most important virus in the group, HCMV, or the various CMVs that are employed as experimental models.

Acknowledgements

This work was supported by the UK Medical Research Council.

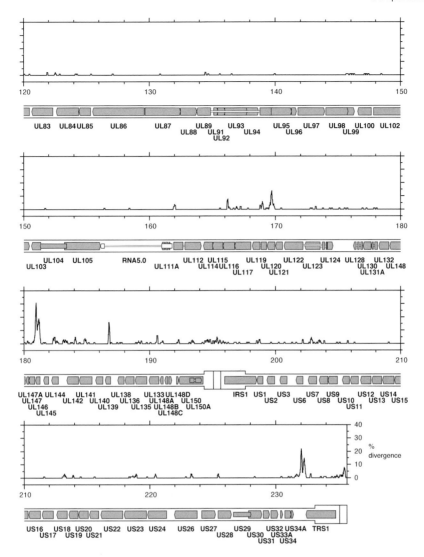

Figure I.1.9 (continued) A nucleotide position was counted as variant when >2 strains differed from the consensus. The plot shows per cent divergence calculated by counting the number of variant nucleotides in a 100 nucleotide window shifted by increments of 10 nucleotides. Protein-coding regions (see Fig. I.1.3) are aligned below the plot.

References

Akter, P., Cunningham, C., McSharry, B.P., Dolan, A., Addison, C., Dargan, D.J., Hassan-Walker, A.F., Emery, V.C., Griffiths, P.D., Wilkinson, G.W.G., *et al.* (2003). Two novel spliced genes in human cytomegalovirus. J. Gen. Virol. *84*, 1117–1122.

Albrecht, M., Darai, G., and Flügel, R.M. (1985). Analysis of the genomic termini of tupaia herpesvirus DNA by restriction mapping and nucleotide sequencing. J. Virol. *56*, 466–474.

Alcendor, D.J., Zong, J., Dolan, A., Gatherer, D., Davison, A.J., and Hayward, G.S. (2009). Patterns of divergence in the vCXCL and vGPCR gene clusters in primate cytomegalovirus genomes. Virology *395*, 21–32.

Alderete, J.P., Jarrahian, S., and Geballe, A.P. (1999). Translational effects of mutations and polymorphisms in a repressive upstream open reading frame of the human cytomegalovirus UL4 gene. J. Virol. *73*, 8330–8337.

Arav-Boger, R., Willoughby, R.E., Pass, R.F., Zong, J.-C., Jang, W.-J., Alcendor, D., and Hayward, G.S. (2002). Polymorphisms of the cytomegalovirus (CMV)-encoded tumor necrosis factor-α and β-chemokine receptors in congenital CMV disease. J. Infect. Dis. *186*, 1057–1064.

Arav-Boger, R., Zong, J.-C., and Foster, C.B. (2005). Loss of linkage disequilibrium and accelerated protein divergence in duplicated cytomegalovirus chemokine genes. Virus Genes *31*, 65–72.

Arav-Boger, R., Battaglia, C.A., Lazzarotto, T., Gabrielli, L., Zong, J.C., Hayward, G.S., Diener-West, M., and Landini,

M.P. (2006a). Cytomegalovirus (CMV)-encoded UL144 (truncated tumor necrosis factor receptor) and outcome of congenital CMV infection. J. Infect. Dis. *194*, 464–473.

Arav-Boger, R., Foster, C.B., Zong, J.-C., and Pass, R.F. (2006b). Human cytomegalovirus-encoded α-chemokines exhibit high sequence variability in congenitally infected newborns. J. Infect. Dis. *193*, 788–791.

Avdic, S., Cao, J.Z., Cheung, A.K.L., Abendroth, A., and Slobedman, B. (2011). Viral interleukin-10 expressed by human cytomegalovirus during the latent phase of infection modulates latently infected myeloid cell differentiation. J. Virol. *85*, 7465–7471.

Bahr, U., and Darai, G. (2001). Analysis and characterization of the complete genome of tupaia (tree shrew) herpesvirus. J. Virol. *75*, 4854–4870.

Baker, M.L., Jiang, W., Rixon, F.J., and Chiu, W. (2005). Common ancestry of herpesviruses and tailed DNA bacteriophages. J. Virol. *79*, 14967–14970.

Bale, J.F., Jr., Petheram, S.J., Robertson, M., Murph, J.R., and Demmler, G. (2001). Human cytomegalovirus *a* sequence and UL144 variability in strains from infected children. J. Med. Virol. *65*, 90–96.

Bar, M., Shannon-Lowe, C., and Geballe, A.P. (2001). Differentiation of human cytomegalovirus genotypes in immunocompromised patients on the basis of UL4 gene polymorphisms. J. Infect. Dis. *183*, 218–225.

Bego, M., Maciejewski, J., Khaiboullina, S., Pari, G., and St Jeor, S. (2005). Characterization of an antisense transcript spanning the UL81–82 locus of human cytomegalovirus. J. Virol. *79*, 11022–11034.

Bego, M.G., Keyes, L.R., Maciejewski, J., and St Jeor, S.C. (2011). Human cytomegalovirus latency-associated protein LUNA is expressed during HCMV infections in vivo. Arch. Virol. *156*, 1847–1851.

Bradley, A.J., Kovács, I.J., Gatherer, D., Dargan, D.J., Alkharsah, K.R., Chan, P.K.S., Carman, W.F., Dedicoat, M., Emery, V.C., Geddes, C.C., *et al.* (2008). Genotypic analysis of two hypervariable human cytomegalovirus genes. J. Med. Virol. *80*, 1615–1623.

Bradley, A.J., Lurain, N.S., Ghazal, P., Trivedi, U., Cunningham, C., Baluchova, K., Gatherer, D., Wilkinson, G.W.G., Dargan, D.J., and Davison, A.J. (2009). High-throughput sequence analysis of variants of human cytomegalovirus strains Towne and AD169. J. Gen. Virol. *90*, 2375–2380.

Cha, T.-A., Tom, E., Kemble, G.W., Duke, G.M., Mocarski, E.S., and Spaete, R.R. (1996). Human cytomegalovirus clinical isolates carry at least 19 genes not found in laboratory strains. J. Virol. *70*, 78–83.

Chee, M.S., Bankier, A.T., Beck, S., Bohni, R., Brown, C.M., Cerny, R., Horsnell, T., Hutchison, C.A. III, Kourazides, T., Martignetti, J.A., *et al.* (1990). Analysis of the protein-coding content of the sequence of human cytomegalovirus strain AD169. Curr. Top. Microbiol. Immunol. *154*, 125–169.

Cheng, T.P., Valentine, M.C., Gao, J., Pingel, J.T., and Yokoyama, W.M. (2010). Stability of murine cytomegalovirus genome after *in vitro* and *in vivo* passage. J. Virol. *84*, 2623–2628.

Coaquette, A., Bourgeois, A., Dirand, C., Varin, A., Chen, W., and Herbein, G. (2004). Mixed cytomegalovirus glycoprotein B genotypes in immunocompromised patients. Clin. Infect. Dis. *39*, 155–161.

Chou, S.W., and Dennison, K.M. (1991). Analysis of interstrain variation in cytomegalovirus glycoprotein B sequences encoding neutralization-related epitopes. J. Infect. Dis. *163*, 1229–1234.

Čičin-Šain, L., Podlech, J., Messerle, M., Reddehase, M.J., and Koszinowski, U.H. (2005). Frequent coinfection of cells explains functional in vivo complementation between cytomegalovirus variants in the multiply infected host. J. Virol. *79*, 9492–9502.

Cunningham, C., Gatherer, D., Hilfrich, B., Baluchova, K., Dargan, D.J., Thomson, M., Griffiths, P.D., Wilkinson, G.W.G., Schulz, T.F., and Davison, A.J. (2010). Sequences of complete human cytomegalovirus genomes from infected cell cultures and clinical specimens. J. Gen. Virol. *91*, 605–615.

Dargan, D.J., Douglas, E., Cunningham, C., Jamieson, F., Stanton, R.J., Baluchova, K., McSharry, B.P., Tomasec, P., Emery, V.C., Percivalle, E., *et al.* (2010). Sequential mutations associated with adaptation of human cytomegalovirus to growth in cell culture. J. Gen. Virol. *91*, 1535–1546.

Dargan, D.J., Jamieson, F.E., Maclean, J., Dolan, A., Addison, C., and McGeoch, D.J. (1997). The published DNA sequence of the human cytomegalovirus strain AD169 lacks 929 base pairs affecting genes UL42 and UL43. J. Virol. *71*, 9833–9836.

Davis, C.L., Field, D., Metzgar, D., Saiz, R., Morin, P.A., Smith, I.L., Spector, S.A., and Wills, C. (1999). Numerous length polymorphisms at short tandem repeats in human cytomegalovirus. J. Virol. *73*, 6265–6270.

Davison, A.J. (1992). Channel catfish virus: a new type of herpesvirus. Virology *186*, 9–14.

Davison, A.J. (2007). Introduction: definition and classification of the human herpesviruses: comparative analysis of the genomes. In Human Herpesviruses: Biology, Therapy and Immunoprophylaxis, Arvin, A., Campadelli-Fiume, G., Mocarski, E., Moore, P.S., Roizman, B., Whitley, R., and Yamanishi, K., eds. (Cambridge University Press, Cambridge, UK), pp. 10–26.

Davison, A.J. (2010). Herpesvirus systematics. Vet. Microbiol. *143*, 52–69.

Davison, A.J., and Bhella, D.J. (2007). Basic virology and viral gene effects on host cell functions: betaherpesviruses: comparative genome and virion structure. In Human Herpesviruses: Biology, Therapy and Immunoprophylaxis, Arvin, A., Campadelli-Fiume, G., Mocarski, E., Moore, P.S., Roizman, B., Whitley, R., and Yamanishi, K., eds. (Cambridge University Press, Cambridge, UK), pp. 177–203.

Davison, A.J., and Stow, N.D. (2005). New genes from old: redeployment of dUTPase by herpesviruses. J. Virol. *79*, 12880–12892.

Davison, A.J., Akter, P., Cunningham, C., Dolan, A., Addison, C., Dargan, D.J., Hassan-Walker, A.F., Emery, V.C., Griffiths, P.D., and Wilkinson, G.W.G. (2003a). Homology between the human cytomegalovirus RL11 gene family and human adenovirus E3 genes. J. Gen. Virol. *84*, 657–663.

Davison, A.J., Dolan, A., Akter, P., Addison, C., Dargan, D.J., Alcendor, D.J., McGeoch, D.J., and Hayward, G.S. (2003b). The human cytomegalovirus genome revisited: comparison with the chimpanzee cytomegalovirus genome. J. Gen. Virol. *84*, 17–28.

Deckers, M., Hofmann, J., Kreuzer, K.-A., Reinhard, H., Edubio, A., Hengel, H., Voigt, S., and Ehlers, B. (2009). High genotypic diversity and a novel variant of human cytomegalovirus revealed by combined UL33/UL55 genotyping with broad-range PCR. Virol. J. 6, 210.

Dolan, A., Cunningham, C., Hector, R.D., Hassan-Walker, A.F., Lee, L., Addison, C., Dargan, D.J., McGeoch, D.J., Gatherer, D., Emery, V.C., et al. (2004). Genetic content of wild-type human cytomegalovirus. J. Gen. Virol. 85, 1301–1312.

Dominguez, G., Black, J.B., Stamey, F.R., Inoue, N., and Pellett, P.E. (1996). Physical and genetic maps of the human herpesvirus 7 strain SB genome. Arch. Virol. 141, 2387–2408.

Dominguez, G., Dambaugh, T.R., Stamey, F.R., Dewhurst, S., Inoue, N., and Pellett, P.E. (1999). Human herpesvirus 6B genome sequence: coding content and comparison with human herpesvirus 6A. J. Virol. 73, 8040–8052.

Ebeling, A., Keil, G., Nowak., B., Fleckenstein, B., Berthelot, N., and Sheldrick, P. (1983). Genome structure and virion polypeptides of the primate herpesviruses *Herpesvirus aotus* types 1 and 3: comparison with human cytomegalovirus. J. Virol. 45, 715–726.

Engel, P., Pérez-Carmona, N., Albà, M.M., Robertson, K., Ghazal, P., and Angulo, A. (2011). Human cytomegalovirus UL7, a homologue of the SLAM-family receptor CD229, impairs cytokine production. Immunol. Cell Biol. 89, 753–766.

Gao, M., and Isom, H.C. (1984). Characterization of the guinea pig cytomegalovirus genome by molecular cloning and physical mapping. J. Virol. 52, 436–447.

Gatherer, D. (2008). Evolution of the G+C frontier in the rat cytomegalovirus genome. Virology: Research and Treatment 1, 75–86.

Gatherer, D., Seirafian, S., Cunningham, C., Holton, M., Dargan, D.J., Baluchova, K., Hector, R.D., Galbraith, J., Herzyk, P., Wilkinson, G.W.G., et al. (2011). High-resolution human cytomegalovirus transcriptome. Proc. Natl. Acad. Sci. U.S.A. 108, 19755–19760.

Goltz, M., Widen, F., Banks, M., Belak, S., and Ehlers, B. (2000). Characterization of the DNA polymerase loci of porcine cytomegaloviruses from diverse geographic origins. Virus Genes 21, 249–255.

Gompels, U.A., Nicholas, J., Lawrence, G., Jones, M., Thomson, B.J., Martin, M.E.D., Efstathiou, S., Craxton, M., and Macaulay, H.A. (1995). The DNA sequence of human herpesvirus-6: structure, coding content, and genome evolution. Virology 209, 29–51.

Goodrum, F., Reeves, M., Sinclair, J., High, K., and Shenk, T. (2007). Human cytomegalovirus sequences expressed in latently infected individuals promote a latent infection in vitro. Blood 110, 937–945.

Haberland, M., Meyer-König, U., and Hufert, F.T. (1999). Variation within the glycoprotein B gene of human cytomegalovirus is due to homologous recombination. J. Gen. Virol. 80, 1495–1500.

Hahn, G., Revello, M.G., Patrone, M., Percivalle, E., Campanini, G., Sarasini, A., Wagner, M., Gallina, A., Milanesi, G., Koszinowski, U., et al. (2004). Human cytomegalovirus UL131–128 genes are indispensable for virus growth in endothelial cells and virus transfer to leukocytes. J. Virol. 78, 10023–10033.

Hakki, M., and Chou, S. (2011). The biology of cytomegalovirus drug resistance. Curr. Opin. Infect. Dis. 24, 605–611.

Hansen, S.G., Strelow, L.I., Franchi, D.C., Anders, D.G., and Wong, S.W. (2003). Complete sequence and genomic analysis of rhesus cytomegalovirus. J. Virol. 77, 6620–6636.

Hassan-Walker, A.F., Okwuadi, S., Lee, L., Griffiths, P.D., and Emery, V.C. (2004). Sequence variability of the α-chemokine UL146 from clinical strains of human cytomegalovirus. J. Med. Virol. 74, 573–579.

Hayajneh, W.A., Colberg-Poley, A.M., Skaletskaya, A., Bartle, L.M., Lesperance, M.M., Contopoulos-Ioannidis, D.G., Kedersha, N.L., and Goldmacher, V.S. (2001a). The sequence and antiapoptotic functional domains of the human cytomegalovirus UL37 exon 1 immediate-early protein are conserved in multiple primary strains. Virology 279, 233–240.

Hayajneh, W.A., Contopoulos-Ioannidis, D.G., Lesperance, M.M., Venegas, A.M., and Colberg-Poley, A.M. (2001b). The carboxyl terminus of the human cytomegalovirus UL37 immediate-early glycoprotein is conserved in primary strains and is important for transactivation. J. Gen. Virol. 82, 1569–1579.

He, R., Ruan, Q., Qi, Y., Ma, Y.-P., Huang, Y.-J., Sun, Z.-R., and Ji, Y.-H. (2006). Sequence variability of human cytomegalovirus UL146 and UL147 genes in low-passage clinical isolates. Intervirology 49, 215–223.

Hedges, S.B., Dudley, J., and Kumar, S. (2006). TimeTree: a public knowledge-base of divergence times among organisms. Bioinformatics 22, 2971–2972.

Heo, J., Petheram, S., Demmler, G., Murph, J.R., Adler, S.P., Bale, J., and Sparer, T.E. (2008). Polymorphisms within human cytomegalovirus chemokine (*UL146/UL147*) and cytokine receptor genes (*UL144*) are not predictive of sequelae in congenitally infected children. Virology 378, 86–96.

Hitomi, S., Kozuka-Hata, H., Chen, Z., Sugano, S., Yamaguchi, N., and Watanabe, S. (1997). Human cytomegalovirus open reading frame UL11 encodes a highly polymorphic protein expressed on the infected cell surface. Arch. Virol. 142, 1407–1427.

Huang, E.-S., Kilpatrick, B., Lakeman, A., and Alford, C.A. (1978). Genetic analysis of a cytomegalovirus-like agent isolated from human brain. J. Virol. 26, 718–723.

Isegawa, Y., Mukai, T., Nakano, K., Kagawa, M., Chen, J., Mori, Y., Sunagawa, T., Kawanishi, K., Sashihara, J., Hata, A., et al. (1999). Comparison of the complete DNA sequences of human herpesvirus 6 variants A and B. J. Virol. 73, 8053–8063.

Ishibashi, K., Tokumoto, T., Tanabe, K., Shirakawa, H., Hashimoto, K., Kushida, N., Yanagida, T., Inoue, N., Yamaguchi, O., Toma, H., et al. (2007). Association of the outcome of renal transplantation with antibody response to cytomegalovirus strain-specific glycoprotein H epitopes. Clin. Infect. Dis. 45, 60–67.

Jeang, K.-T., and Hayward, G.S. (1983). A cytomegalovirus DNA sequence containing tracts of tandemly repeated CA dinucleotides hybridizes to highly repetitive dispersed elements in mammalian cell genomes. Mol. Cell. Biol. 3, 1389–1402.

Jelcic, I., Reichel, J., Schlude, C., Treutler, E., Sinzger, C., and Steinle, A. (2011). The polymorphic HCMV glycoprotein UL20 is targeted for lysosomal degradation by multiple cytoplasmic dileucine motifs. Traffic 12, 1444–1456.

Kanai, K., Yamada, S., Yamamoto, Y., Fukui, Y., Kurane, I., and Inoue, N. (2011). Re-evaluation of the genome sequence of guinea pig cytomegalovirus. J. Gen. Virol. 92, 1005–1020.

Koch, H.-G., Delius, H., Matz, B., Flügel, R.M., Clarke, J., and Darai, G. (1985). Molecular cloning and physical mapping of the tupaia herpesvirus genome. J. Virol. 55, 86–95.

Lacoste, V., Verschoor, E.J., Nerrienet, E., and Gessain, A. (2005). A novel homologue of *Human herpesvirus 6* in chimpanzees. J. Gen. Virol. 86, 2135–2140.

Latimer, E., Zong, J.-C., Heaggans, S.Y., Richman, L.K., and Hayward, G.S. (2011). Detection and evaluation of novel herpesviruses in routine and pathological samples from Asian and African elephants: identification of two new proboscivirus (EEHV5 and EEHV6) and two new gammaherpesviruses (EGHV3B and EGHV5). Vet. Microbiol. 147, 28–41.

Leendertz, F.H., Deckers, M., Schempp, W., Lankester, F., Boesch, C., Mugisha, L., Dolan, A., Gatherer, D., McGeoch, D.J., and Ehlers, B. (2009). Novel cytomegaloviruses in free-ranging and captive great apes: phylogenetic evidence for bidirectional horizontal transmission. J. Gen. Virol. 90, 2386–2394.

Lindquester, G.J., and Pellett, P.E. (1991). Properties of the human herpesvirus 6 strain Z29 genome: G + C content, length, and presence of variable-length directly repeated terminal sequence elements. Virology 182, 102–110.

Lurain, N.S., Fox, A.M., Lichy, H.M., Bhorade, S.M., Ware, C.F., Huang, D.D., Kwan, S.-P., Garrity, E.R., and Chou, S. (2006). Analysis of the human cytomegalovirus genomic region from UL146 through UL147A reveals sequence hypervariability, genotypic stability, and overlapping transcripts. Virol. J. 3, 4.

Lurain, N.S., Kapell, K.S., Huang, D.D., Short, J.A., Paintsil, J., Winkfield, E., Benedict, C.A., Ware, C.F., and Bremer, J.W. (1999). Human cytomegalovirus UL144 open reading frame: sequence hypervariability in low-passage clinical isolates. J. Virol. 73, 10040–10050.

McGeoch, D.J., and Gatherer, D. (2005). Integrating reptilian herpesviruses into the family *Herpesviridae*. J. Virol. 79, 725–731.

McGeoch, D.J., Cook, S., Dolan, A., Jamieson, F.E., and Telford, E.A.R. (1995). Molecular phylogeny and evolutionary timescale for the family of mammalian herpesviruses. J. Mol. Biol. 247, 443–458.

McGeoch, D.J., Rixon, F.J., and Davison, A.J. (2006). Topics in herpesvirus genomics and evolution. Virus Res. 117, 90–104.

McVoy, M.A., Nixon, D.E., and Adler, S.P. (1997). Circularization and cleavage of guinea pig cytomegalovirus genomes. J. Virol. 71, 4209–4217.

Manuel, O., Åsberg, A., Pang, X., Rollag, H., Emery, V.C., Preiksaitis, J.K., Kumar, D., Pescovitz, M.D., Bignamini, A.A., Hartmann, A., *et al.* (2009). Impact of genetic polymorphisms in cytomegalovirus glycoprotein B on outcomes in solid-organ transplant recipients with cytomegalovirus disease. Clin. Infect. Dis. 49, 1160–1166.

Marks, J.R., and Spector, D.H. (1988). Replication of the murine cytomegalovirus genome: structure and role of the termini in generation and cleavage of concatenates. J. Virol. 162, 98–107.

Marsh, A.K., Willer, D.O., Ambagala, A.P.N., Dzamba, M., Chan, J.K., Pilon, R., Fournier, J., Sandstrom, P., Brudno, M., and MacDonald, K.S. (2011). Genomic sequencing and characterization of cynomolgus macaque cytomegalovirus. J. Virol. 85, 12995–13009.

Martin, M.E.D., Thomson, B.J., Honess, R.W., Craxton, M.A., Gompels, U.A., Liu, M.-Y., Littler, E., Arrand, J.R.,

Teo, I., and Jones, M.D. (1991). The genome of human herpesvirus 6: maps of unit-length and concatemeric genomes for nine restriction endonucleases. J. Gen. Virol. 72, 157–168.

Mattick, C., Dewin, D., Polley, S., Sevilla-Reyes, E., Pignatelli, S., Rawlinson, W., Wilkinson, G., Dal Monte, P., and Gompels, U.A. (2004). Linkage of human cytomegalovirus glycoprotein gO variant groups identified from worldwide clinical isolates with gN genotypes, implications for disease associations and evidence for N-terminal sites of positive selection. Virology 318, 582–597.

Megaw, A.G., Rapaport, D., Avidor, B., Frenkel, N., and Davison, A.J. (1998). The DNA sequence of the RK strain of human herpesvirus 7. Virology 244, 119–132.

Meyer-König, U., Schrage, B., Huzly, D., Bongarts, A., and Hufert, F.T. (1998). High variability of cytomegalovirus glycoprotein B gene and frequent multiple infections in HIV-infected patients with low CD4 T-cell count. AIDS 12, 2228–2230.

Miller, M.S., Furlong, W.E., Pennell, L., Geadah, M., and Hertel, L. (2010). RASCAL is a new human cytomegalovirus-encoded protein that localizes to the nuclear lamina and in cytoplasmic vesicles at late times postinfection. J. Virol. 84, 6483–6496.

Mitchell, D.P., Savaryn, J.P., Moorman, N.J., Shenk, T., and Terhune, S.S. (2009). Human cytomegalovirus UL28 and UL29 open reading frames encode a spliced mRNA and stimulate accumulation of immediate-early RNAs. J. Virol. 83, 10187–10197.

Mocarski, E.S., Prichard, M.N., Tan, C.S., and Brown, J.M. (1997). Reassessing the organization of the UL42-UL43 region of the human cytomegalovirus strain AD169 genome. Virology 239, 169–175.

Murphy, E., and Shenk, T. (2008). Human cytomegalovirus genome. Curr. Top. Microbiol. Immunol. 325, 1–19.

Murphy, E., Rigoutsos, I., Shibuya, T., and Shenk, T.E. (2003a). Reevaluation of human cytomegalovirus coding potential. Proc. Natl. Acad. Sci. U.S.A. 100, 13585–13590.

Murphy, E., Yu, D., Grimwood, J., Schmutz, J., Dickson, M., Jarvis, M.A., Hahn, G., Nelson, J.A., Myers, R.M., and Shenk, T.E. (2003b). Coding potential of laboratory and clinical strains of human cytomegalovirus. Proc. Natl. Acad. Sci. U.S.A. 100, 14976–14981.

Nicholas, J. (1996). Determination and analysis of the complete nucleotide sequence of human herpesvirus 7. J. Virol. 70, 5975–5989.

Nixon, D.E., and McVoy, M.A. (2002). Terminally repeated sequences on a herpesvirus genome are deleted following circularization but are reconstituted by duplication during cleavage and packaging of concatemeric DNA. J. Virol. 76, 2009–2013.

Oxford, K.L., Eberhardt, M.K., Yang, K.-W., Strelow, L., Kelly, S., Zhou, S.-S., and Barry, P.A. (2008). Protein coding content of the ULb' region of wild-type rhesus cytomegalovirus. Virology 373, 181–188.

Pande, N.T., Powers, C., Ahn, K., and Früh, K. (2005). Rhesus cytomegalovirus contains functional homologues of US2, US3, US6, and US11. J. Virol. 79, 5786–5798.

Paterson, D.A., Dyer, A.P., Milne, R.S.B., Sevilla-Reyes, E., and Gompels, U.A. (2002). A role for human cytomegalovirus glycoprotein O (gO) in cell fusion and a new hypervariable locus. Virology 293, 281–294.

Pellett, P.E., Davison, A.J., Eberle, R., Ehlers, B., Hayward, G.S., Lacoste, V., Minson, A.C., Nicholas, J., Roizman, B., Studdert, M.J., *et al.* (2011). Herpesviridae. In Virus

Taxonomy, Ninth Report of the International Committee on Taxonomy of Viruses, King, A.M.Q., Adams, M.J., Carstens, E.B., and Lefkowitz, E.J., eds. (Elsevier Academic Press, London, UK), pp. 111–122.

Picone, O., Costa, J.-M., Chaix, M.-L., Ville, Y., Rouzioux, C., and Leruez-Ville, M. (2005). Human cytomegalovirus UL144 gene polymorphisms in congenital infections. J. Clin. Microbiol. *43*, 25–29.

Picone, O., Costa, J.-M., Chaix, M.-L., Ville, Y., Rouzioux, C., and Leruez-Ville, M. (2007). Comments on 'cytomegalovirus (CMV)-encoded UL144 (truncated tumor necrosis factor receptor) and outcome of congenital CMV infection'. J. Infect. Dis. *196*, 1719–1720.

Pignatelli, S., Dal Monte, P., and Landini, M.P. (2001). gpUL73 (gN) genomic variants of human cytomegalovirus isolates are clustered into four distinct genotypes. J. Gen. Virol. *82*, 2777–2784.

Pignatelli, S., Dal Monte, P., Rossini, G., Chou, S., Gojobori, T., Hanada, K., Guo, J.J., Rawlinson, W., Britt, W., Mach, M., *et al.* (2003). Human cytomegalovirus glycoprotein N (gpUL73-gN) genomic variants: identification of a novel subgroup, geographical distribution and evidence of positive selective pressure. J. Gen. Virol. *84*, 647–655.

Pignatelli, S., Dal Monte, P., Rossini, G., and Landini, M.P. (2004). Genetic polymorphisms among human cytomegalovirus (HCMV) wild-type strains. Rev. Med. Virol. *14*, 383–410.

Pignatelli, S., Lazzarotto, T., Gatto, M.R., Dal Monte, P., Landini, M.P., Faldella, G., and Lanari, M. (2010). Cytomegalovirus gN genotypes distribution among congenitally infected newborns and their relationship with symptoms at birth and sequelae. Clin. Infect. Dis. *51*, 33–41.

Prichard, M.N., Jairath, S., Penfold, M.E.T., St Jeor, S., Bohlman, M.C., and Pari, G.S. (1998). Identification of persistent RNA–DNA hybrid structures within the origin of replication of human cytomegalovirus. J. Virol. *72*, 6997–7004.

Prichard, M.N., Penfold, M.E.T., Duke, G.M., Spaete, R.R., and Kemble, G.W. (2001). A review of genetic differences between limited and extensively passaged human cytomegalovirus strains. Rev. Med. Virol. *11*, 191–200.

Puchhammer-Stöckl, E., and Görzer, I. (2011). Human cytomegalovirus: an enormous variety of strains and their possible clinical significance in the human host. Future Virology *6*, 259–271.

Qi, Y., Mao, Z.-Q., Ruan, Q., He, R., Ma, Y.-P., Sun, Z.-R., Ji, Y.-H., and Huang, Y. (2006). Human cytomegalovirus (HCMV) UL139 open reading frame: sequence variants are clustered into three major genotypes. J. Med. Virol. *78*, 517–522.

Rasmussen, L., Geissler, A., and Winters, M. (2003). Inter- and intragenic variations complicate the molecular epidemiology of human cytomegalovirus. J. Infect. Dis. *187*, 809–819.

Rawlinson, W.D., Farrell, H.E., and Barrell, B.G. (1996). Analysis of the complete DNA sequence of murine cytomegalovirus. J. Virol. *70*, 8833–8849.

Renzette, N., Bhattacharjee, B., Jensen, J.D., Gibson, L., and Kowalik, T.F. (2011). Extensive genome-wide variability of human cytomegalovirus in congenitally infected infants. PLoS Pathog. *7*, e1001344.

Rivailler, P., Kaur, A., Johnson, R.P., and Wang, F. (2006). Genomic sequence of rhesus cytomegalovirus 180.92: insights into the coding potential of rhesus cytomegalovirus. J. Virol. *80*, 4179–4182.

Sahagun-Ruiz, A., Sierra-Honigmann, A.M., Krause, P., and Murphy, P.M. (2004). Simian cytomegalovirus encodes five rapidly evolving chemokine receptor homologues. Virus Genes *28*, 71–83.

Scalzo, A.A., Forbes, C.A., Smith, L.M., and Loh, L.C. (2009). Transcriptional analysis of human cytomegalovirus and rat cytomegalovirus homologues of the M73/M73.5 spliced gene family. Arch. Virol. *154*, 65–75.

Schleiss, M.R., McGregor, A., Choi, K.Y., Date, S.V., Cui, X., and McVoy, M.A. (2008). Analysis of the nucleotide sequence of the guinea pig cytomegalovirus (GPCMV) genome. Virol. J. *5*, 139.

Schmitz, J., Ohme, M., and Zischler, H. (2000). The complete mitochondrial genome of *Tupaia belangeri* and the phylogenetic affiliation of scandentia to other eutherian orders. Mol. Biol. Evol. *17*, 1334–1343.

Sekulin, K., Görzer, I., Heiss-Czedik, D., and Puchhammer-Stöckl, E. (2007). Analysis of the variability of CMV strains in the RL11D domain of the RL11 multigene family. Virus Genes *35*, 577–583.

Shepp, D.H., Match, M.E., Lipson, S.M., and Pergolizzi, R.G. (1998). A fifth human cytomegalovirus glycoprotein B genotype. Res. Virol. *149*, 109–114.

Smith, L.M., McWhorter, A.R., Masters, L.L., Shellam, G.R., and Redwood, A.J. (2008). Laboratory strains of murine cytomegalovirus are genetically similar to but phenotypically distinct from wild strains of virus. J. Virol. *82*, 6689–6696.

Spaete, R.R., and Mocarski, E.S. (1985). The *a* sequence of the cytomegalovirus genome functions as a cleavage/packaging signal for herpes simplex virus defective genomes. J. Virol. *54*, 817–824.

Stanton, R.J., Baluchova, K., Dargan, D.J., Cunningham, C., Sheehy, O., Seirafian, S., McSharry, B.P., Neale, M.L., Davies, J.A., Tomasec, P., *et al.* (2010). Reconstruction of the complete human cytomegalovirus genome in a BAC reveals RL13 to be a potent inhibitor of replication. J. Clin. Invest. *120*, 3191–3208.

Stanton, R., Westmoreland, D., Fox, J.D., Davison, A.J., and Wilkinson, G.W.G. (2005). Stability of human cytomegalovirus genotypes in persistently infected renal transplant recipients. J. Med. Virol. *75*, 42–46.

Steiper, M.E., and Young, N.M. (2006). Primate molecular divergence dates. Mol. Phylogenet. Evol. *41*, 384–394.

Tamashiro, J.C., and Spector, D.H. (1986). Terminal structure and heterogeneity in human cytomegalovirus strain AD169. J. Virol. *59*, 591–604.

Trincado, D.E., Scott, G.M., White, P.A., Hunt, C., Rasmussen, L., and Rawlinson, W.D. (2000). Human cytomegalovirus strains associated with congenital and perinatal infections. J. Med. Virol. *61*, 481–487.

Varnum, S.M., Streblow, D.N., Monroe, M.E., Smith, P., Auberry, K.J., Paša-Tolić, L., Wang, D., Camp, D.G. II, Rodland, K., Wiley, S., *et al.* (2004). Identification of proteins in human cytomegalovirus (HCMV) particles: the HCMV proteome. J. Virol. *78*, 10960–10966.

Vink, C., Beuken, E., and Bruggeman, C.A. (1996). Structure of the rat cytomegalovirus genome termini. J. Virol. *70*, 5221–5229.

Vink, C., Beuken, E., and Bruggeman, C.A. (2000). Complete DNA sequence of the rat cytomegalovirus genome. J. Virol. *74*, 7656–7665.

Weststrate, M.W., Geelen, J.L., and van der Noordaa, J. (1980). Human cytomegalovirus DNA: physical maps for restriction endonucleases *Bgl*II, *Hin*dIII and *Xba*I. J. Gen. Virol. *49*, 1–21.

Yu, D., Silva, M.C., and Shenk, T. (2003). Functional map of human cytomegalovirus AD169 defined by global mutational analysis. Proc. Natl. Acad. Sci. U.S.A. *100*, 12396–12401.

Zhang, G., Raghavan, B., Kotur, M., Cheatham, J., Sedmak, D., Cook, C., Waldman, J., and Trgovcich, J. (2007). Antisense transcription in the human cytomegalovirus transcriptome. J. Virol. *81*, 11267–11281.

Molecular Evolution of Murine Cytomegalovirus Genomes

Alec J. Redwood, Geoffrey R. Shellam and Lee M. Smith

Abstract

Cytomegaloviruses have co-evolved with their hosts since the mammalian radiation. The MCMV genome appears to be highly conserved and unlike the HCMV genome contains no large-scale deletions and rearrangements following serial *in vitro* passage. The genome of MCMV is both highly conserved and highly variable. The central regions of the genome, containing the betaherpesvirus and herpesvirus conserved genes, are highly conserved. However, significant variation occurs in species-specific genes at genomic termini, where the known and putative immune evasion genes reside. Variation in the MCMV genome consists of presence/absence polymorphisms in individual genes, grouping of genes into specific genotypes and random nucleotide diversity across the genome. However, much of the genotypic variation is under strong purifying selection indicating that these genotypes are fixed and conserved at the population level. The individual genotypes are either known to, or are likely to, target variant host gene products. Consequently, *in vivo* replication of MCMV is likely to be viral strain dependent and reflect the particular repertoire of genes encoded by the infecting strain. A total of 22 MCMV genes are genotypic, indicating considerable potential for variation in the MCMV population. This variation likely reflects genetic heterogeneity in the target population and suggests exquisite adaptation of the virus to its host.

Introduction

Herpesviruses are classified in three separate families within the order *Herpesvirales*. The family *Herpesviridae* contains the mammalian, bird and reptile viruses (Davison *et al.*, 2009). Murine cytomegalovirus (MCMV – Murid herpesvirus 1), a member of the *Betaherpesvirinae* subfamily, is the type species of the genus *Muromegalovirus*. The natural hosts for MCMV are members of the house mouse complex (*Mus musculus*).

Cytomegaloviruses (CMV) exhibit strict specificity for their host species. Consequently, because human CMV (human herpesvirus-5, HCMV) cannot be studied in experimental models, MCMV infection of mice has become the most widely used model for studying the principles of CMV infection and pathogenesis. The genetics of HCMV is described in Chapter I.1.

The MCMV genome

MCMV was first isolated in 1954 (Smith, 1954) and the genome was fully sequenced more than 40 years later in 1996 (Rawlinson *et al.*, 1996). Comparisons with the sequence of HCMV indicated that the genomes of the two viruses were essentially co-linear over the central region, a feature later shown to be common for all CMV genomes. MCMV (Smith strain) was originally described as having a genome of 230,278 bp, with a GC content of 58.7% and encoding 170 open reading frames (ORFs). For identification of coding sequences, ORFs were required to be at least 300 bp in length (100 amino acids), and overlap other coding sequences by less than 60%. Genes were numbered from left to right on the genome, with the 78 genes identified as containing significant amino acid sequence homology to genes found in HCMV named with the prefix M, whilst ORFs with no homology to those in HCMV were named with the prefix m.

Since the initial mapping of the MCMV genome (Ebeling *et al.*, 1983; Mercer *et al.*, 1983) and the subsequent publication of the full genome sequence (Rawlinson *et al.*, 1996) our knowledge of the coding potential of this virus has increased greatly, changing with new sequencing technologies, gene prediction algorithms and the discovery of micro RNAs. Fig. I.2.1 shows our current knowledge of the coding potential of the MCMV genome based on the original study by Rawlinson and colleagues (1996) with modifications based on subsequent sequence and functional studies. The annotation includes an additional 13 genes that

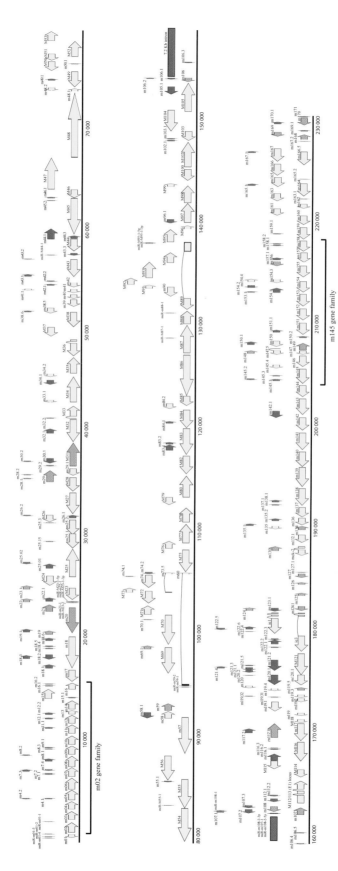

Figure I.2.1 Annotation of the MCMV genome. The annotation of the MCMV genome is based on the original Smith strain sequence. Shown in yellow are ORFs which have homologues in other herpesviruses or for which functional data exists. Shown in blue are ORFs predicted by bioinformatic analyses, and in grey those predicted in the original annotation of the Smith strain genome which have either no known function or no known homologues in other herpesvirus genes. The region encoding the large 7.2 kb stable intron is depicted in green and the areas encoding the pre microRNAs are shown in red.

have been identified experimentally. These are: m33.1 (Baluchova *et al.*, 2008), m38.5 (McCormick *et al.*, 2003, 2005) and m41.1 (Brocchieri *et al.*, 2005; Cam *et al.*, 2009), m60 and m73.5 (Scalzo *et al.*, 2004), m34.2, m84.2, m132.1, m145.4, m154.3 and m154.4 (Tang *et al.*, 2006), m147.5 (Loewendorf *et al.*, 2004) and m166.5 (Kattenhorn *et al.*, 2004). Subsequent analysis has also led to the removal of two previously annotated ORFs. The m30 ORF has been removed as several studies have noted single bp insertions in m30 (Ahasan and Sweet, 2007), M31 (Kattenhorn *et al.*, 2004) or both (Schumacher *et al.*, 2010; Jordan *et al.*, 2011). The single bp insertion in m30 leads to readthrough from m30 to M31 as previously described (Ahasan and Sweet, 2007; Smith *et al.*, 2008) and re-establishes the 3′ and 5′ length homology of M31 with RCMV (Maastricht strain) R31, rather than simply the 5′ homology as described for the single bp insertion in M31 (Kattenhorn *et al.*, 2004). Moreover, all 12 fully sequenced strains of MCMV (Table I.2.1) encode only the longer M31 ORF (Smith *et al.*, 2008, 2012; Cheng *et al.*, 2010). Similarly, M45.1 has been removed as it is part of the M45 reading frame (Brune *et al.*, 2001). Finally the ORF m59, whilst included in the annotation, is found only in the Smith strain of MCMV (Smith *et al.*, 2008, 2012; Cheng *et al.*, 2010).

Computational approaches have also been used to assess the coding potential of the MCMV genome (Brocchieri *et al.*, 2005). This study analysed the sequence homology between Smith strain MCMV (Rawlinson *et al.*, 1996) and Maastricht strain RCMV (Vink *et al.*, 2000), and predicted coding sequences based on the G+C distribution in different codon positions in high GC genomes. By shortening the minimum gene size to 20 codons, removing the restrictions on overlap between genes, and not requiring an ATG start codon (therefore allowing for the prediction of spliced genes), at least 34 and possibly up to 227 previously un-annotated coding regions were identified in the MCMV genome. These computationally predicted genes, unless shown by other studies to be functional, are shown in blue in Fig. I.2.1. ORFs which are shown in grey were originally predicted by Rawlinson and colleagues (1996) but have either no homology with other CMV ORFs or still have no known function.

Sequences for non-protein-coding RNAs have also been shown to be important components of the MCMV genome. MCMV has been shown to encode sequences for multiple pre-microRNAs which regulate both host and viral RNA targets as is the case for other herpesviruses (reviewed in detail in Chapter I.5). These are dispersed throughout the genome, but there are three clusters (m01, m21/22/23, and m106/107) in which 11 out of 18 pre-microRNAs are encoded.

Sequences encoding pre-microRNAs are shown in red in Fig. I.2.1. Additionally, the region encoding the large 7.2 kb 'stable intron' (Kulesza and Shenk, 2006) between the m106 and m107 gene sequences has also been included in the annotation (shown in green on Fig. I.2.1).

Fig. I.2.1 is likely to represent a minimal estimate of the coding potential of MCMV. This is because the MCMV genome is transcriptionally complex with multiple spliced transcripts, many of which share 5′ or 3′ termini, located throughout the genome including the M112/113 (E1) (Bühler *et al.*, 1990; Ciocco-Schmitt *et al.*, 2002), ie1 (Keil *et al.*, 1987a,b) and the m70 (Loh, 1989; Scalzo *et al.*, 2004) regions. It is likely that this is also the case for other regions of the genome suggesting that the coding potential for this virus far exceeds the present state of knowledge.

Origins of MCMV strains

In order to understand the genetics and the evolution of MCMV it is useful to have an understanding of the origins of the MCMV strains in common usage and the genetic backgrounds of the mice from which these strains were isolated. The natural host of MCMV is the house mouse, however the house mouse complex (*Mus musculus*) is polytypic and comprises *M. musculus musculus*, *M. m. domesticus*, *M. m. castaneus*, *M. m. molossinus* (a hybrid of *M. m. musculus* and *M. m. castaneus*) and *M. m. gentilulus*. These lineages appear to be undergoing speciation and are considered either subspecies of *M. musculus* or separate species (reviewed in Tucker, 2007).

Laboratory strains of MCMV

Two laboratory strains of MCMV, Smith and K181, are in widespread use. The Smith strain of MCMV was isolated in 1954 by Margaret Smith from the salivary glands of an infected laboratory mouse (Smith, 1954). Smith strain MCMV was deposited into the ATCC (VR-1399, reaccessioned from VR-194) in October 1967 by Wallace Rowe from the National Institutes of Health in the USA, who obtained the strain directly from Margaret Smith (Brodsky and Rowe, 1958). The K181 strain of MCMV was isolated in the 1970s by June Osborne (cited in Hudson *et al.*, 1976). Whilst the strains of mice used to isolate Smith and K181 were not recorded, laboratory mice – including the outbred Swiss mice in widespread use at the time – are predominantly of the *M. m. domesticus* genetic background (Frazer *et al.*, 2007; Yang *et al.*, 2007, 2011).

The K181 strain of MCMV was originally described as a more virulent variant of Smith, and was isolated

Table I.2.1 Fully sequenced MCMV strains

Strain	Location of isolation	Genome size – bp	Accession number	BAC	References
Smith	St. Louis, USA	230,278	NC_004065	Yes	Smith (1954), Messerle et al. (1997), Wagner et al. (1999), Jordan et al. (2011)
K181	Wisconsin, USA	230,301	AM886412	Yes	Hudson et al. (1976), Misra and Hudson (1980), Redwood et al. (2005), Timoshenko et al. (2009)
C4A	Canberra, Australia	230,111	EU579861	No	Gorman et al. (2006), Smith et al. (2008)
C4B	Canberra, Australia	230,154	HE610452	No	Gorman et al. (2006)
C4C	Canberra, Australia	229,924	HE610453	No	Gorman et al. (2006)
C4D	Canberra, Australia	229,935	HE610456	No	Gorman et al. (2006)
G4	Geraldton, Australia	230,227	EU579859	Yes	Booth et al. (1993), Smith et al. (2008)
WP15B	Walpeup, Australia	230,118	EU579860	No	Smith et al. (2008)
N1	Nannup, Australia	229,884	HE610454	No	Booth et al. (1993)
AA18D	Macquarie Island, Australia	229,543	HE610451	No	Smith et al., 2012
N07	Beacon Island, Australia	229,452	HE610455	No	Smith et al., 2012
WT1	USA[a]	230,408	GU305914	No	Cheng et al. (2010)

[a]WT1 is a variant of Smith strain MCMV – possibly a recombinant of Smith and K181 strains.

from the salivary glands of mice after serial *in vivo* passage of the Smith strain (Hudson *et al.*, 1976). In cell culture K181 was reported to produce smaller plaques and to replicate to lower titres than Smith (Hudson *et al.*, 1988). *In vivo* however, K181 was reported to reach higher titres in the salivary glands than Smith (Hudson *et al.*, 1976; Misra and Hudson, 1980), and to result in enhanced mortality in young mice in comparison to the Smith strain (Hudson *et al.*, 1988). There have been no recent reports comparing the virulence of K181 to Smith strain MCMV. Sequence analysis of the K181[Perth] bacterial artificial chromosome (BAC) (Redwood *et al.*, 2005) and Smith suggest that K181, whilst closely related to Smith, is indeed a different strain of MCMV (Rawlinson *et al.*, 1996; Smith *et al.*, 2008). K181[Perth] was originally obtained by our laboratory from Dr D. Lang of Duke University (Durham, NC, U.S.A) with the belief that it was the Smith strain of MCMV (Chalmer *et al.*, 1977). However, later RFLP analysis identified the strain as K181 and it was given the designation K181[Perth] (Xu *et al.*, 1992). Comparisons of the Smith derived clone of MCMV, WT1 (Cheng *et al.*, 2010) with the sequence of Smith (Rawlinson *et al.*, 1996) and K181 (Smith *et al.*, 2008) suggest that WT1 is likely to be a recombinant between Smith and K181 (Smith *et al.*, 2012). Taken together these data suggest that early stocks of Smith were a mix of both Smith and K181 strains.

Both the Smith (Messerle *et al.*, 1997; Wagner *et al.*, 1999) and the K181 (Redwood *et al.*, 2005) strains have been cloned as BACs. The Smith strain MCMV BAC is actually chimeric, with the sequences 207529 to 218461 (accession number NC_004065) derived from K181. The K181 strain BAC was produced from the K181[Perth] strain of MCMV. It is reported that the Smith BAC derived virus is somewhat less virulent than the K181 BAC derived virus, which may reflect inherent differences in the virulence of the parental viruses used for BAC construction. However, a single bp deletion in the MCK-2 gene of the original Smith BAC (pSM3fr) has recently been repaired, restoring the salivary gland replication of this virus (Jordan *et al.*, 2011). With the advent of BAC cloning of MCMV strains and access to the complete sequences of these clones (Redwood *et al.*, 2005; Jordan *et al.*, 2011) it could be argued that the BAC clones represent an ideal opportunity for investigators to standardize the viral strains in use.

Low-passage wild-derived strains of MCMV

Recently, a series of low-passage strains of MCMV have been isolated from free-living wild mice and are proving useful in expanding our understanding of the genetic and phenotypic behaviour of this virus (Booth *et al.*, 1993; Lyons *et al.*, 1996; Rawlinson *et al.*, 1997; Fairweather *et al.*, 1998; Farroway *et al.*, 2002; Voigt *et al.*, 2003; Gorman *et al.*, 2006, 2008; Smith *et al.*, 2006, 2008 Corbett *et al.*, 2007, 2011; Nikolovski *et al.*, 2009). The wild derived strains of MCMV were isolated from mice caught on mainland Australia, Australian offshore islands or the sub-Antarctic islands Macquarie Island (Australian Territory) and Kerguelen Island (French Territory). Fig. I.2.2 shows the distribution pattern of the house mouse complex with asterisks denoting locations of mice used to isolate these strains. To date the majority of low passage strains of MCMV were derived from *M. m. domesticus*. The house mouse was probably introduced to Australia at the time of first settlement in 1788. Free-living mice on Macquarie and Kerguelen Island are believed to be distinct from those on mainland Australia, having been introduced by sealers or from shipwrecks early in the nineteenth century (Kidder, 1876; Cumpston, 1968). Therefore strains of MCMV isolated from Macquarie or Kerguelen Island are temporally and geographically separated from those isolated from mainland Australia. Finally, Wirral strain MCMV, reported as sequence information in GenBank (m144 – accession number AY337609, m152 – accession number AY337610 and m157 – accession number AY337611), was isolated from wild mice caught near Liverpool in the UK, where mice are also *M. m. domesticus*.

Strains of MCMV from subspecies other than *M. m. domesticus* have not been described until recently. Ehlers and colleagues (2007) reported the existence of novel rodent herpesviruses from *M. musculus* and 15 other rodent species including *Rattus norvegicus* from Germany, the UK and Thailand. These included an MCMV strain from Germany, 17 other novel betaherpesviruses and 21 novel gammaherpesviruses (Ehlers *et al.*, 2007). Ehlers and colleagues (2007) stated that the MCMV strain was detected in *M. musculus*. However, both *M. m. musculus* and *M. m. domesticus* subspecies are found in Germany, with southern Germany being a hybrid transition zone between the two subspecies (Sage *et al.*, 1986). The exact subspecies of mouse that harboured this virus is therefore unclear.

The vast majority of MCMV strains so far described have come exclusively from *M. m. domesticus* or from chimeric laboratory mice, which are predominantly of *M. m. domesticus* background. However, since the Smith strain replicates well in *M. m. musculus* and *M. m. castaneus* background mice (Adam *et al.*, 2006), it is likely that the natural host of MCMV is broadly the house mouse complex, rather than any of the single subspecies. However, it is not known whether natural MCMV infection can be transmitted between the individual

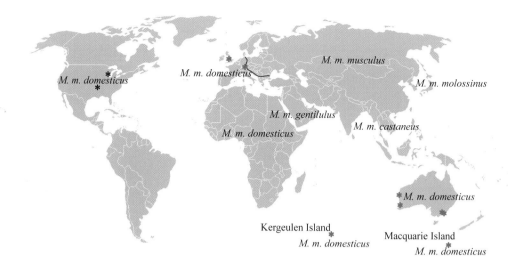

Figure I.2.2 Origins of MCMV strains. Shown is the global distribution of members of the house mouse complex. Asterisks denote the location of mice used to isolate laboratory strains (black) and wild derived strains (red) of MCMV. The black line in Western Europe denotes the hybrid transition zone between *M. m. musculus* and *M. m. domesticus*. To data all of the MCMV strains isolated from free-living mice were circulating in *M. m. domesticus* populations.

musculus subspecies of the genus *Mus*. The lineages of *M. musculus* are considered by most to be subspecies, however there is some contention about the exact taxonomy of the house mouse complex. For instance, there is some reproductive incompatibility between *M. m. musculus* and *M. m. domesticus* (reviewed in Tucker, 2007). This is not entirely an esoteric point as it raises questions about the definition of species specificity with regard to these viruses. Also, if the house mouse complex is undergoing speciation, is this also true of MCMV? It is worth noting that we know nothing about the genetics of MCMV circulating in the other subspecies of the house mouse complex. Indeed our knowledge of MCMV genetics, with the exception of K181 and Smith, is derived almost entirely from strains of MCMV isolated from Australia or Macquarie Island.

Molecular evolution of the MCMV genome

Effects of long-term passage on MCMV genomes

Long-term, and even short-term passage of HCMV in fibroblasts results in cell culture-related adaptations which typically result in the loss of epithelial and endothelial cell tropism (reviewed in Chapter I.17). The laboratory strains of MCMV, Smith and K181, have been passaged extensively *in vitro* and *in vivo* since their first isolation. Whilst the exact number of *in vivo*

passages are unknown for these viruses, records at the ATCC state that prior to submission of Smith strain MCMV, the virus was passaged five times in MEFs after many passages in adult mice made by the subcutaneous inoculation of salivary gland suspension (ATCC, personal communication). At this time, *in vivo* passage was extensively used for the production of viral stocks. One report published only seven years after isolation, stated that Smith strain MCMV was obtained as a 60th mouse passage stock and then passaged another 20 times in mouse embryonic tissue culture before use after another 1–18 passages *in vivo* (Mannini and Medearis, 1961). The passage history of K181 since its isolation in the 1970s is even less clear as it has not been deposited into ATCC and is held as serially *in vitro* and *in vivo* passaged stocks in several laboratories around the world.

Serial *in vitro* passage is known to affect the stability of MCMV. Vancouver strain, an attenuated mutant of Smith strain MCMV, was produced following repeated *in vitro* passage, and has a 9.4 kbp deletion in the *Hin*dIII E region of the genome and a 900 bp insertion in the *Eco*RI K fragment (Mercer *et al.*, 1983; Boname and Chantler, 1992). Serial *in vivo* passage can also affect genome stability with as few as seven passages of K181 through congenic Ly49H⁺ mice leading to the accumulation of loss of function mutations in the MCMV gene m157 (Voigt *et al.*, 2003). Mutations in m157 occur even more rapidly in T-cell-deficient Ly49H⁺ C57BL/6 mice (French *et al.*, 2004). Notwithstanding these studies, the MCMV genome appears to be highly stable following serial *in vitro* and *in vivo* passage, with no evidence of the major genomic rearrangements or deletions (Smith *et al.*, 2008) seen during *in vitro* HCMV passage (reviewed in Chapter I.17). MCMV also exhibits low rates of mutation accumulation due to genetic drift (Cheng *et al.*, 2010). Genome-wide, pairwise nucleotide identities range from 96.71% to 99.06%, whilst overall nucleotide diversity (Π) is 0.01470. This is approximately half the level of variability seen in alignments of the UL region from HCMV sequences ($\Pi = 0.02818$) (Smith *et al.*, 2012). The stability of the MCMV genome is further demonstrated by the remarkable conservation of genome size. Even given some level of uncertainty relating to sequence assembly at the genomic termini, genome sizes range from 229,451 bp (NO7 strain) to 230,281 bp (Smith strain), a variation of only 830 bp or 0.36% (Table I.2.1). Whilst indicating that the genomes are highly stable, these data also suggest that the genome is close to the maximum size that can be packaged within the mature capsid. These data also indicate that highly passaged laboratory strains of MCMV do not differ significantly from less passaged wild derived strains, and remain a useful tool for CMV research. A caveat of this statement is that even low passage strains of MCMV have been passaged on fibroblasts, albeit minimally (less than ten *in vitro* passages prior to sequencing), allowing for the possibility of very rapid and complete selection in cell culture. It is also worth noting that no single MCMV genome can reflect the entire pan-genome of MCMV, and that different strains of MCMV are likely to be useful in addressing the subtleties of host/pathogen interaction.

Molecular evolution revealed by strain-to-strain genomic comparisons

To date, the genomes of 12 individual strains of MCMV have been completely sequenced (Table I.2.1). Comparing these genomes, as well as individual gene sequences from other strains submitted to DNA databases, provides a powerful means of determining the structure and the evolutionary pressure on this large genome. Most immediately notable is that the central region of the MCMV genome, a region which encodes the herpesvirus and betaherpesvirus conserved genes is, as expected, highly conserved (Fig. I.2.3). Five genes (M46, M85, M92, M96 and m135) have 100% amino acid identity across all 12 viral strains, and for a further six genes, a single viral strain contained a single amino acid change. These single amino acid changes are mostly confined to the Smith sequence, possibly indicating errors in the original sequencing. Variability between strains is mostly confined to the genomic termini in

locations containing the species-specific CMV genes. Many of these genes are known or putative immune evasion genes, suggesting inherent differences in the capacity of MCMV strains to modulate host immunity, possibly reflecting differences in the host population structure, discussed below. Variability within the MCMV genome appears to exist at three levels – presence/absence polymorphisms in individual genes, genes existing as distinct genotypes, and nucleotide diversity.

Presence/absence polymorphisms

Whilst the coding potential of the MCMV genome is broadly conserved across all of the sequenced strains, presence/absence polymorphisms due to indels (insertions or deletions) causing frameshifts or point mutations resulting in premature termination codons, are seen within the MCMV genome. These types of polymorphisms are of great interest, particularly where these genes have definitively been shown to exist, as they not only provide information about the evolution of the MCMV genome, but are also obvious candidates for the observed phenotypic differences between viral strains. The gene m145.4 was discovered by sequencing proteins incorporated within the virion (Kattenhorn *et al.*, 2004), and was subsequently shown to be transcribed at 24 hours post infection (Tang *et al.*, 2006). However, it only appears to be present within the laboratory Smith and K181 strains, as no other sequenced strain has an appropriate ATG start codon. Similarly, ORFs m154.4, m156 and m21 are only found in a subset of the viral strains. Two strains of virus for which there is limited sequence data, W8 and W8211, contain apparent duplications of the m03/m03.5 gene (Corbett *et al.*, 2007), the only such gene duplication thus far seen in the MCMV genome. Gene families such as the m02 and m145 gene family in MCMV, and the US22 gene family in HCMV, are thought to have arisen by processes of gene duplication and therefore the duplication of m03/m03.5 may be a further expansion of the m02 gene family. Interestingly, W8 and W8211 were both isolated from mice trapped in Walpeup, Victoria (Australia), suggesting geographically localized viral genome evolution, possibly reflecting the genetic nature of the host in this location.

Genotypic variation

A number of genes within the MCMV genome exist as multiple, distinct genotypes, and it has been postulated that the presence of these genotypes provides the virus greater flexibility, at the population level, to interact with polymorphic components of the host immune

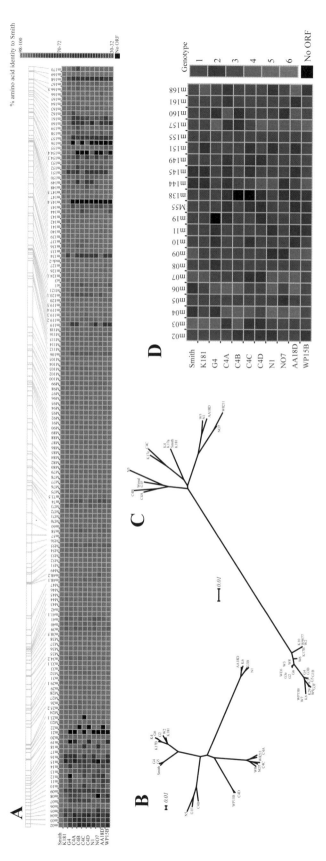

Figure I.2.3 Variation of the MCMV genome. Amino acid conservation in the MCMV genome compiled by analysis of the known genes of the fully sequenced MCMV genomes (A). The data are represented as a heat map with all strains compared to Smith. Highest amino acid homologies are shown in red and lowest in blue. It is clearly evident that the MCMV genome is highly conserved and variation is largely restricted to the genomic termini. Much of the variation in the MCMV genome is present as distinct genotypes such that shown for the Neighbour-Joining trees of the codon-aligned amino acid sequences of m04 (B) and m157 (C). In total, 22 MCMV genes exist as distinct genotypes as shown in the genotype plot (D). All 12 strains of MCMV have distinct patterns of viral genotypes in these 22 genes. The potential for variation in the MCMV population is significant. Data from WT1 are not shown here, as it is a Smith strain variant.

system (Voigt *et al.*, 2003; Smith *et al.*, 2006; Corbett *et al.*, 2011). By examining all of the available MCMV sequences, either complete genomes or single gene sequences, 22 genes within the MCMV population can be identified as forming distinct genotypes (Fig. I.2.3D). Although the determination of a genotype without evidence of phenotype variability is subjective, these were assigned predominantly based on Maximum Likelihood phylogenies of gene sequences, allowing us to identify up to six genotypes for individual genes (see Fig. I.2.3B and C for examples). One gene (m06) was assigned genotypes based on the pattern of three conserved indels within the gene; seven of the eight putative m06 genotypes have been detected in laboratory and low passage wild strains of MCMV. Given that each of the fully sequenced strains of MCMV has a unique pattern of genotypes for these 22 genes (Fig. I.2.3D), the potential for variability in the MCMV population is obvious.

Nucleotide diversity

Not all variability in the MCMV genome leads to the creation of distinct genotypes of individual genes. However all variability provides a useful tool to assess the drivers of MCMV evolution. Ratios of non-synonymous to synonymous amino acid substitutions (dN/dS ratios) are widely used as a measure of the selective pressures operating on an organism's genome. dN/dS ratios of over 1 are indicative of positive selection whilst dN/dS ratios under 1 are evidence of negative or purifying selection. When applied to MCMV, over 90% of the coding sequences analysed had dN/dS ratios of less than 0.6, suggesting almost all MCMV genes are under purifying selection. Only one gene, m120.1, which co-localizes to the mitochondria and whose function is unclear (Brocchieri *et al.*, 2005), has been shown to have a global dN/dS of greater than 1, indicative of positive selection. Codon-based analyses of selection on MCMV coding sequences indicate that only 0.98% of codons analysed (from a total of 116 genes) demonstrate some evidence of positive selection. Conversely, 11.87% of the same codons demonstrated some evidence of purifying selection (Smith *et al.*, 2012).

Selective pressures may be more evident in genes that either target the host immune system or are themselves targets of host immunity. The down regulation of the cell surface expression of MHC class I molecules is a common feature of herpesviruses, and is consistent with the importance of CD8 T-cells in controlling viral replication during both primary infection and reactivation from latency. Of all the experimentally described nonameric CTL epitopes of MCMV, either H-2[b] (Gold *et al.*, 2002; Munks *et al.*, 2006) or H-2[d]

restricted (Reddehase *et al.*, 1989; Holtappels *et al.*, 2000a,b, 2001, 2002a,b, 2006, 2008; see also Chapter II.17), 25% are variable in at least one virus (first shown for the example of the IE1 peptide in Lyons *et al.*, 1996), compared to the background level of nonamer variation within these genes (15.6%). This possibly indicates some selective pressure to avoid host CD8 T-cell control. However, comparing the variability seen in the 15mer I-A[b] restricted CD4 T-cell epitopes (Arens *et al.*, 2008; Walton *et al.*, 2008), 25% (5 out of 20) are variable, compared with the innate 15mer variation within these epitope-containing genes of 29.5% (Smith *et al.*, 2012). Whilst limited in scope, this analysis suggests limited pressure or capacity to evade CD4 T-cell control. It is also worth noting that this analysis was performed with a single MHC class II allele and thus could simply reflect the lack of I-A[b] alleles in the mice from which these viruses were collected.

MCMV encodes multiple structural glycoproteins including gB, gM, gN, gH, gL and gO that, as surface antigens, are either known or putative targets of antibodies. Sequence analysis has demonstrated significant heterogeneity in the major structural glycoprotein gB, resulting in the identification of three distinct genotypes. Two of the suggested neutralizing antibody binding sites (Xu *et al.*, 1996) in the gB protein are reasonably conserved, but the third antibody binding region (amino acids 17–79 of the Smith strain gene) contains significant sequence variation, including several non-synonymous amino acid substitutions. The level of sequence variation within this region correlates with the ability of antisera against the Smith strain protein to neutralize individual MCMV strains (Xu *et al.*, 1996).

The gM gene has almost complete identity between all 12 fully sequenced MCMV strains, with the exception of a single amino acid change (G218R) in gM of WP15B. The HCMV protein gN forms a complex with gM and exists as at least four distinct genotypes in viral populations, with two of these genotypes (gN3 and gN4) containing evidence of positive selection as they contain regions where dN/dS > 1 (Pignatelli *et al.*, 2003). gN is also a target for both gN subtype-variable and gN subtype-independent neutralizing antibodies within HCMV-infected individuals (Burkhardt *et al.*, 2009; see also Chapter II.10). Interestingly, these gN subtype-variable neutralizing antibodies were found to neutralize the gN4 subtype more effectively than the other gN subtypes. It is possible therefore that the positive selection seen in gN4 sequences may be an attempt at antibody escape by the virus. In contrast to HCMV however, MCMV gN sequences are highly conserved between viral strains, with no evidence of positive selection and an overall dN/dS ratio of 0.43.

Genes of the gcIII complex (gL, gO and gH) are highly variable between HCMV strains, with two genotypes of gH, four genotypes of gL and seven genotypes of gO. Whilst admittedly many fewer sequences are available for these genes for MCMV, the gcIII complex appears to be much more conserved in the murine viruses, with only two genotypes of gH and two or three genotypes of gO. gL is highly conserved, with a minimum amino acid similarity between strains of 98%, and no stratification into genotypes.

For other herpesviruses, recombination is a major driver of viral evolution, and this also appears to be the case for MCMV. Early reports of recombination between strains of HCMV identified not only intra-patient recombinants but also demonstrated recombination occurring in cell culture (Chou, 1989; Haberland *et al.*, 1999). Since then there have been multiple reports of HCMV recombination, with evidence from conflicting phylogenetic signals within variable viral genes (Steininger *et al.*, 2005, 2007; Garrigue *et al.*, 2007; Yan *et al.*, 2008; Faure-Della Corte *et al.*, 2010). However, little is known about recombination between MCMV strains. We have previously identified evidence of recombination within the m144 gene (Smith *et al.*, 2006), and in more recent studies we have performed statistical tests of recombination across the aligned genomes of 11 strains of MCMV. Bioinformatic analysis (SplitsTree) clearly identifies a reticulate phylogenetic network for the aligned MCMV genomes, and frequent recombination events were identified across the MCMV genome by RDP and Bootscan (Martin *et al.*, 2010), with 86 recombination sites identified by more than five statistical analyses (Smith *et al.*, 2012).

In common with many other herpesviruses (Grey and Nelson, 2008), the MCMV genome encodes multiple microRNAs (Dölken *et al.*, 2007). Approximately 60% of these are localized within three miRNA clusters, the m01-, m21/m22/M23- and m107/m108 miRNA clusters. The remaining miRNAs are scattered throughout the genome. All 21 miRNAs thus far identified are identical in the available sequenced genomes – presumably reflecting the invariant nature of the miRNA target (either host or viral). The single nucleotide change (U→C) in the Smith strain miR-m108-1 is a sequencing error (Dölken *et al.*, 2007; Schumacher *et al.*, 2010).

Viral genetics and host resistance

Several genetically controlled innate host resistance mechanisms to MCMV have been identified in inbred mice, including *Cmv1* (Scalzo *et al.*, 1990; Brown *et al.*, 2001; Daniels *et al.*, 2001; Lee *et al.*, 2001), *Cmv2* (Rodriguez *et al.*, 2004), *Cmv3* (Desrosiers *et al.*, 2005) and *Cmv4* (Adam *et al.*, 2006). These heritable resistance

mechanisms to MCMV are all mediated by enhanced NK cell responses (see also Chapter II.9). Two MCMV encoded genes, m157 (*Cmv1*) and m04 (*Cmv3*), are known to directly contribute to heritable resistance to MCMV. *Cmv1* mediated resistance in C57BL/6 mice is due to direct interaction between m157 and host encoded Ly49H, an activation receptor expressed exclusively on NK cells. m157 is an MHC class I homologue and remains the only known ligand for Ly49H. m157 expression at the cell surface is independent of peptide and is sufficient for direct interaction with Ly49H (Arase *et al.*, 2002; Smith *et al.*, 2002).

The interaction of m04 with Ly49 molecules is more complex. In MA/My mice, resistance to MCMV is linked to the expression of Ly49P and H-2D^k (Desrosiers *et al.*, 2005). Ly49P specifically recognizes MCMV H-2D^k-infected cells via a mechanism that is dependent on the expression of the MCMV gene m04 and another as yet unidentified MCMV gene (Kielczewska *et al.*, 2009). Since this initial description of Ly49P-mediated resistance it has become clear that other strains of mice also regulate MCMV replication by targeting m04. The NK cell activating receptors, Ly49P^{NOD}, Ly49L^{BALB} and Ly49D2^{PWK} all recognize MCMV infected cells via m04 and H-2 dependent and allele specific mechanisms (Pyzik *et al.*, 2011). Resistance of BALB.H2^k mice to MCMV is mediated by Ly49L in conjunction with m04 and H-2^k (Pyzik *et al.*, 2011). Inhibitory Ly49 receptors are also engaged by MHC class I molecules that have been restored to the cell surface by interactions with m04 (Babic *et al.*, 2010).

The utility of the mouse model is that the effects of viral genetics on phenotype can be readily assessed *in vivo*. Despite this, little is known about the effects of genetic variation on the behaviour of the virus in its natural host. Both m04 and m157 exist in the MCMV population as distinct genotypes – phylogenetic trees are shown in Fig. I.2.3B and C, respectively (Voigt *et al.*, 2003; Smith *et al.*, 2006; Corbett, 2007, 2011) – and both are under strong purifying selection (Smith *et al.*, 2012). These are therefore evolutionarily stable genotypes that are likely to target either different host proteins or allelic variants. Recent studies have supported this hypothesis, as it is clear that many m157 gene products are able to ligate alternate Ly49 molecules. The m157 genes of Smith and K181 are identical and their products target Ly49H as well as the inhibitory NK cell receptors Ly49I and Ly49C (Arase *et al.*, 2002; Corbett *et al.*, 2011). The Ly49 family is not only polygenic but also polymorphic, and there is allelic specificity to the interaction of m157 with Ly49 molecules. For instance m157^{Smith} binds Ly49I^{129} but not Ly49I^{B6} (Arase *et al.*, 2002). Alternative genotypes of m157 also have different specificities for Ly49 molecules than the prototypical m157[Smith]. Variant m157 gene products interact with the inhibitory receptors Ly49C[BALB/c], Ly49C[B6] and Ly49C[NZB] as well as the activating receptor Ly49H[NZB] (Corbett *et al.*, 2011). Failure to engage Ly49H[B6] by variant m157 gene products allows for enhanced MCMV replication in Ly49H+ mice (Voigt *et al.*, 2003). Finally, a large number of the m157 gene products have no known target (Corbett *et al.*, 2011).

The effects of m04 variation on local host–pathogen interactions are also unknown. However, given that the three proteins (and possibly another as yet unidentified MCMV protein) involved in this axis, gp34 (m04), MHC class I and Ly49 family proteins are all highly variable, the complexity of this system is self evident. Interestingly, the DNA sequence of m157 from the Wirral strain (accession number AY337611), which was isolated in the UK, is identical to that present in G1F (accession number AY228653), a strain of MCMV isolated near Geraldton, Australia. In contrast, identity between m04 sequences has only been seen in MCMV strains isolated from similar locations. This suggests a tighter linkage between host and pathogen genetics for m04 than for m157, possibly reflecting less flexibility for the virus in the genotype of m04 encoded compared to m157. In support of this, the deletion of m04 has profound effects on replication *in vivo* (Babic *et al.*, 2010), whereas deletion of m157 has only a modest effect on *in vivo* replication when the target of the protein is an inhibitory Ly49 molecule (Bubic *et al.*, 2004). We would predict that, like m157, different genotypes of m04 target different Ly49 molecules and perhaps allelic variants. Therefore, variation in these and other MCMV genes are likely to have profound implications for the replication of individual strains of MCMV in infected hosts. Furthermore, such viral gene variation provides opportunity for adaptation of the virus to host variability at the population level.

Conclusion

In conclusion, the MCMV population is highly variable, built on a combination of presence/absence polymorphisms, genotypic variation and individual nucleotide variation. Despite this level of variability the genome is also remarkably stable. The genome size is highly conserved between strains and there is little variation between genes located in the central regions of the genome. The bulk of the variation between MCMV strains is located at the left and right genomic termini where known and putative immune evasion genes are located. Despite this level of variation, individual genotypes present within the MCMV population are highly

stable and under strong purifying selection. Thus MCMV is well adapted to its host at both the individual and population level.

Acknowledgements
We would like to thank members of the laboratory past and present whose work has contributed to this chapter. We would like to apologize to colleagues in the field whose work could not be cited due to space limitations. The authors are supported by the National Health and Medical Research Council, Commonwealth Government of Australia.

References
Adam, S.G., Caraux, A., Fodil-Cornu, N., Loredo-Osti, J.C., Lesjean-Pottier, S., Jaubert, J., Bubic, I., Jonjic, S., Guenet, J.L., Vidal, S.M., *et al.* (2006). Cmv4, a new locus linked to the NK cell gene complex, controls innate resistance to cytomegalovirus in wild-derived mice. J. Immunol. 176, 5478–5485.

Ahasan, M.M., and Sweet, C. (2007). Murine cytomegalovirus open reading frame m29.1 augments virus replication both in vitro and in vivo. J. Gen. Virol. 88, 2941–2951.

Arase, H., Mocarski, E.S., Campbell, A.E., Hill, A.B., and Lanier, L.L. (2002). Direct recognition of cytomegalovirus by activating and inhibitory NK cell receptors. Science 296, 1323–1326.

Arens, R., Wang, P., Sidney, J., Loewendorf, A., Sette, A., Schoenberger, S.P., Peters, B., and Benedict, C.A. (2008). Murine cytomegalovirus induces a polyfunctional CD4 T-cell response. J. Immunol. 180, 6472–6476.

Babic, M., Pyzik, M., Zafirova, B., Mitrovic, M., Butorac, V., Lanier, L.L., Krmpotic, A., Vidal, S.M., and Jonjic, S. (2010). Cytomegalovirus immunoevasin reveals the physiological role of 'missing self' recognition in natural killer cell dependent virus control in vivo. J. Exp. Med. 207, 2663–2673.

Baluchova, K., Kirby, M., Ahasan, M.M., and Sweet, C. (2008). Preliminary characterization of murine cytomegaloviruses with insertional and deletional mutations in the M34 open reading frame. J. Med. Virol. 80, 1233–1242.

Boname, J.M., and Chantler, J.K. (1992). Characterization of a strain of murine cytomegalovirus which fails to grow in the salivary glands of mice. J. Gen. Virol. 73, 2021–2029.

Booth, T.W., Scalzo, A.A., Carrello, C., Lyons, P.A., Farrell, H.E., Singleton, G.R., and Shellam, G.R. (1993). Molecular and biological characterization of new strains of murine cytomegalovirus isolated from wild mice. Arch. Virol. 132, 209–220.

Brocchieri, L., Kledal, T.N., Karlin, S., and Mocarski, E. (2005). Predicting coding potential from genome sequence: application to betaherpesviruses infecting rats and mice. J. Virol. 79, 7570–7596.

Brodsky, I., and Rowe, W.P. (1958). Chronic subclinical infection with mouse salivary gland virus. Proc. Soc. Exp. Biol. Med. 99, 654–655.

Brown, M.G., Dökun, A.O., Heusel, J.W., Smith, H.R., Beckman, D.L., Blattenberger, E.A., Dubbelde, C.E., Stone, L.R., Scalzo, A.A., and Yokoyama, W.M. (2001).

Vital involvement of a natural killer cell activation receptor in resistance to viral infection. Science 292, 934–937.

Brune, W., Menard, C., Heesemann, J., and Koszinowski, U.H. (2001). A ribonucleotide reductase homolog of cytomegalovirus and endothelial cell tropism. Science 291, 303–305.

Bubic, I., Wagner, M., Krmpotic, A., Saulig, T., Kim, S., Yokoyama, W.M., Jonjic, S., and Koszinowski, U.H. (2004). Gain of virulence caused by loss of a gene in murine cytomegalovirus. J. Virol. 78, 7536–7544.

Bühler, B., Keil, G.M., Weiland, F., and Koszinowski, U.H. (1990). Characterization of the murine cytomegalovirus early transcription unit e1 that is induced by immediate-early proteins. J. Virol. 64, 1907–1919.

Burkhardt, C., Himmelein, S., Britt, W., Winkler, T., and Mach, M. (2009). Glycoprotein N subtypes of human cytomegalovirus induce a strain-specific antibody response during natural infection. J. Gen. Virol. 90, 1951–1961.

Cam, M., Handke, W., Picard-Maureau, M., and Brune, W. (2009). Cytomegaloviruses inhibit Bak- and Bax-mediated apoptosis with two separate viral proteins. Cell Death Differ. 17, 655–665.

Chalmer, J.E., Mackenzie, J.S., and Stanley, N.F. (1977). Resistance to murine cytomegalovirus linked to the major histocompatibility complex of the mouse. J. Gen. Virol. 37, 107–114.

Cheng, T.P., Valentine, M.C., Gao, J., Pingel, J.T., and Yokoyama, W.M. (2010). Stability of murine cytomegalovirus genome after in vitro and in vivo passage. J. Virol. 84, 2623–2628.

Chou, S.W. (1989). Reactivation and recombination of multiple cytomegalovirus strains from individual organ donors. J. Infect. Dis. 160, 11–15.

Ciocco-Schmitt, G.M., Karabekian, Z., Godfrey, E.W., Stenberg, R.M., Campbell, A.E., and Kerry, J.A. (2002). Identification and characterization of novel murine cytomegalovirus M112–113 (e1) gene products. Virology 294, 199–208.

Corbett, A.J., Forbes, C.A., Moro, D., and Scalzo, A.A. (2007). Extensive sequence variation exists among isolates of murine cytomegalovirus within members of the m02 family of genes. J. Gen. Virol. 88, 758–769.

Corbett, A.J., Coudert, J.D., Forbes, C.A., and Scalzo, A.A. (2011). Functional consequences of natural sequence variation of murine cytomegalovirus m157 for Ly49 receptor specificity and NK cell activation. J. Immunol. 186, 1713–1722.

Cumpston, J.S. (1968). Macquarie Island. Publication No. 93. ANARE Science Report Series A(1) Australia Department of External Affairs, Antarctic Division, Canberra.

Daniels, K.A., Devora, G., Lai, W.C., O'Donnell, C.L., Bennett, M., and Welsh, R.M. (2001). Murine cytomegalovirus is regulated by a discrete subset of natural killer cells reactive with monoclonal antibody to Ly49H. J. Exp. Med. 194, 29–44.

Davison, A.J., Eberle, R., Ehlers, B., Hayward, G.S., McGeoch, D.J., Minson, A.C., Pellett, P.E., Roizman, B., Studdert, M.J., and Thiry, E. (2009). The order Herpesvirales. Arch. Virol. 154, 171–177.

Desrosiers, M.P., Kielczewska, A., Loredo-Osti, J.C., Adam, S.G., Makrigiannis, A.P., Lemieux, S., Pham, T., Lodoen, M.B., Morgan, K., Lanier, L.L., *et al.* (2005). Epistasis between mouse Klra and major histocompatibility complex class I loci is associated with a new mechanism of natural killer cell-mediated innate resistance to cytomegalovirus infection. Nat. Genet. 37, 593–599.

Dölken, L., Perot, J., Cognat, V., Alioua, A., John, M., Soutschek, J., Ruzsics, Z., Koszinowski, U., Voinnet, O., and Pfeffer, S. (2007). Mouse cytomegalovirus microRNAs dominate the cellular small RNA profile during lytic infection and show features of posttranscriptional regulation. J. Virol. 81, 13771–13782.

Ebeling, A., Keil, G.M., Knust, E., and Koszinowski, U.H. (1983). Molecular cloning and physical mapping of murine cytomegalovirus DNA. J. Virol. 47, 421–433.

Ehlers, B., Kuchler, J., Yasmum, N., Dural, G., Voigt, S., Schmidt-Chanasit, J., Jakel, T., Matuschka, F.R., Richter, D., Essbauer, S., *et al.* (2007). Identification of novel rodent herpesviruses, including the first gammaherpesvirus of Mus musculus. J. Virol. 81, 8091–8100.

Fairweather, D., Lawson, C.M., Chapman, A.J., Brown, C.M., Booth, T.W., Papadimitriou, J.M., and Shellam, G.R. (1998). Wild isolates of murine cytomegalovirus induce myocarditis and antibodies that cross-react with virus and cardiac myosin. Immunology 94, 263–270.

Farroway, L.N., Singleton, G.R., Lawson, M.A., and Jones, D.A. (2002). The impact of murine cytomegalovirus (MCMV) on enclosure populations of house mice (Mus domesticus). Wildlife Res. 29, 11–17.

Faure-Della Corte, M., Samot, J., Garrigue, I., Magnin, N., Reigadas, S., Couzi, L., Dromer, C., Velly, J.-F., Déchanet-Merville, J., Fleury, H.J.A., *et al.* (2010). Variability and recombination of clinical human cytomegalovirus strains from transplantation recipients. J. Clin. Virol. 47, 161–169.

Frazer, K.A., Eskin, E., Kang, H.M., Bogue, M.A., Hinds, D.A., Beilharz, E.J., Gupta, R.V., Montgomery, J., Morenzoni, M.M., Nilsen, G.B., *et al.* (2007). A sequence-based variation map of 8.27 million SNPs in inbred mouse strains. Nature 448, 1050–1053.

French, A.R., Pingel, J.T., Wagner, M., Bubic, I., Yang, L., Kim, S., Koszinowski, U., Jonjic, S., and Yokoyama, W.M. (2004). Escape of mutant double-stranded DNA virus from innate immune control. Immunity 20, 747–756.

Garrigue, I., Corte, M.F.-D., Magnin, N., Couzi, L., Capdepont, S., Rio, C., Merville, P., Dechanet-Merville, J., Fleury, H., and Lafon, M.-E. (2007). Variability of UL18, UL40, UL111a and US3 immunomodulatory genes among human cytomegalovirus clinical isolates from renal transplant recipients. J. Clin. Virol. 40, 120–128.

Gold, M.C., Munks, M.W., Wagner, M., Koszinowski, U.H., Hill, A.B., and Fling, S.P. (2002). The murine cytomegalovirus immunomodulatory gene m152 prevents recognition of infected cells by M45-specific CTL but does not alter the immunodominance of the M45-specific CD8 T-cell response in vivo. J. Immunol. 169, 359–365.

Gorman, S., Harvey, N.L., Moro, D., Lloyd, M.L., Voigt, V., Smith, L.M., Lawson, M.A., and Shellam, G.R. (2006). Mixed infection with multiple strains of murine cytomegalovirus occurs following simultaneous or sequential infection of immunocompetent mice. J. Gen. Virol. 87, 1123–1132.

Gorman, S., Lloyd, M.L., Smith, L.M., McWhorter, A.R., Lawson, M.A., Redwood, A.J., and Shellam, G.R. (2008). Prior infection with murine cytomegalovirus (MCMV) limits the immunocontraceptive effects of an MCMV vector expressing the mouse zona-pellucida-3 protein. Vaccine 26, 3860–3869.

Grey, F., and Nelson, J. (2008). Identification and function of human cytomegalovirus microRNAs. J. Clin. Virol. 41, 186–191.

Haberland, M., Meyer-Konig, U., and Hufert, F.T. (1999). Variation within the glycoprotein B gene of human cytomegalovirus is due to homologous recombination. J. Gen. Virol. 80, 1495–1500.

Holtappels, R., Thomas, D., Podlech, J., Geginat, G., Steffens, H.P., and Reddehase, M.J. (2000a). The putative natural killer decoy early gene m04 (gp34) of murine cytomegalovirus encodes an antigenic peptide recognized by protective antiviral CD8 T-cells. J. Virol. 74, 1871–1884.

Holtappels, R., Thomas, D., and Reddehase, M.J. (2000b). Identification of a K(d)-restricted antigenic peptide encoded by murine cytomegalovirus early gene M84. J. Gen. Virol. 81, 3037–3042.

Holtappels, R., Podlech, J., Grzimek, N.K., Thomas, D., Pahl-Seibert, M.F., and Reddehase, M.J. (2001). Experimental preemptive immunotherapy of murine cytomegalovirus disease with CD8 T-cell lines specific for ppM83 and pM84, the two homologs of human cytomegalovirus tegument protein ppUL83 (pp65). J. Virol. 75, 6584–6600.

Holtappels, R., Grzimek, N.K., Thomas, D., and Reddehase, M.J. (2002a). Early gene m18, a novel player in the immune response to murine cytomegalovirus. J. Gen. Virol. 83, 311–316.

Holtappels, R., Thomas, D., Podlech, J., and Reddehase, M.J. (2002b). Two antigenic peptides from genes m123 and m164 of murine cytomegalovirus quantitatively dominate CD8 T-cell memory in the H-2d haplotype. J. Virol. 76, 151–164.

Holtappels, R., Gillert-Marien, D., Thomas, D., Podlech, J., Deegen, P., Herter, S., Oehrlein-Karpi, S.A., Strand, D., Wagner, M., and Reddehase, M.J. (2006). Cytomegalovirus encodes a positive regulator of antigen presentation. J. Virol. 80, 7613–7624.

Holtappels, R., Simon, C.O., Munks, M.W., Thomas, D., Deegen, P., Kühnapfel, B., Däubner, T., Emde, S.F., Podlech, J.r., Grzimek, N.K.A., et al. (2008). Subdominant CD8 T-Cell epitopes account for protection against cytomegalovirus independent of immunodomination. J. Virol. 82, 5781–5796.

Hudson, J.B., Misra, V., and Mosmann, T.R. (1976). Cytomegalovirus infectivity: analysis of the phenomenon of centrifugal enhancement of infectivity. Virology 72, 235–243.

Hudson, J.B., Walker, D.G., and Altamirano, M. (1988). Analysis in vitro of two biologically distinct strains of murine cytomegalovirus. Arch. Virol. 102, 289–295.

Jordan, S., Krause, J., Prager, A., Mitrovic, M., Jonjic, S., Koszinowski, U.H., and Adler, B. (2011). Virus progeny of MCMV bacterial artificial chromosome pSM3fr shows reduced growth in salivary glands due to a fixed mutation of MCK-2. J. Virol. 85, 10346–10353.

Kattenhorn, L.M., Mills, R., Wagner, M., Lomsadze, A., Makeev, V., Borodovsky, M., Ploegh, H.L., and Kessler, B.M. (2004). Identification of proteins associated with murine cytomegalovirus virions. J. Virol. 78, 11187–11197.

Keil, G.M., Ebeling-Keil, A., and Koszinowski, U.H. (1987a). Immediate-early genes of murine cytomegalovirus:

location, transcripts, and translation products. J. Virol. 61, 526–533.

Keil, G.M., Ebeling-Keil, A., and Koszinowski, U.H. (1987b). Sequence and structural organization of murine cytomegalovirus immediate-early gene 1. J. Virol. 61, 1901–1908.

Kidder, J.H. (1876). The Natural History of Kerguelen Island. Amer. Nat. 10, 481–484.

Kielczewska, A., Pyzik, M., Sun, T., Krmpotic, A., Lodoen, M.B., Munks, M.W., Babic, M., Hill, A.B., Koszinowski, U.H., Jonjic, S., et al. (2009). Ly49P recognition of cytomegalovirus-infected cells expressing H2-Dk and CMV-encoded m04 correlates with the NK cell antiviral response. J. Exp. Med. 206, 515–523.

Kulesza, C.A., and Shenk, T. (2006). Murine cytomegalovirus encodes a stable intron that facilitates persistent replication in the mouse. Proc. Natl. Acad. Sci. U.S.A. 103, 18302–18307.

Lee, S.H., Girard, S., Macina, D., Busa, M., Zafer, A., Belouchi, A., Gros, P., and Vidal, S.M. (2001). Susceptibility to mouse cytomegalovirus is associated with deletion of an activating natural killer cell receptor of the C-type lectin superfamily. Nat. Genet. 28, 42–45.

Loewendorf, A., Kruger, C., Borst, E.M., Wagner, M., Just, U., and Messerle, M. (2004). Identification of a mouse cytomegalovirus gene selectively targeting CD86 expression on antigen-presenting cells. J. Virol. 78, 13062–13071.

Loh, L.C. (1989). Synthesis and processing of a 22–26K murine cytomegalovirus glycoprotein recognized by a neutralizing monoclonal antibody. Virology 169, 474–478.

Lyons, P.A., Allan, J.E., Carrello, C., Shellam, G.R., and Scalzo, A.A. (1996). Effect of natural sequence variation at the H-2Ld-restricted CD8+ T-cell epitope of the murine cytomegalovirus ie1-encoded pp89 on T-cell recognition. J. Gen. Virol. 77, 2615–2623.

McCormick, A.L., Skaletskaya, A., Barry, P.A., Mocarski, E.S., and Goldmacher, V.S. (2003). Differential function and expression of the viral inhibitor of caspase 8-induced apoptosis (vICA) and the viral mitochondria-localized inhibitor of apoptosis (vMIA) cell death suppressors conserved in primate and rodent cytomegaloviruses. Virology 316, 221–233.

McCormick, A.L., Meiering, C.D., Smith, G.B., and Mocarski, E.S. (2005). Mitochondrial cell death suppressors carried by human and murine cytomegalovirus confer resistance to proteasome inhibitor-induced apoptosis. J. Virol. 79, 12205–12217.

Mannini, A., and Medearis, D.N. (1961). Mouse salivary gland virus infections. Am. J. Hyg. 73, 329–343.

Martin, D.P., Lemey, P., Lott, M., Moulton, V., Posada, D., and Lefeuvre, P. (2010). RDP3: a flexible and fast computer program for analyzing recombination. Bioinformatics 26, 2462–2463.

Mercer, J.A., Marks, J.R., and Spector, D.H. (1983). Molecular cloning and restriction endonuclease mapping of the murine cytomegalovirus genome (Smith Strain). Virology 129, 94–106.

Messerle, M., Crnkovic, I., Hammerschmidt, W., Ziegler, H., and Koszinowski, U.H. (1997). Cloning and mutagenesis of a herpesvirus genome as an infectious bacterial artificial chromosome. Proc. Natl. Acad. Sci. U.S.A. 94, 14759–14763.

Misra, V., and Hudson, J.B. (1980). Minor base sequence differences between the genomes of two strains of murine cytomegalovirus differing in virulence. Arch. Virol. *64*, 1–8.

Munks, M.W., Gold, M.C., Zajac, A.L., Doom, C.M., Morello, C.S., Spector, D.H., and Hill, A.B. (2006). Genome-wide analysis reveals a highly diverse CD8 T-cell response to murine cytomegalovirus. J. Immunol. *176*, 3760–3766.

Nikolovski, S., Lloyd, M.L., Harvey, N., Hardy, C.M., Shellam, G.R., and Redwood, A.J. (2009). Overcoming innate host resistance to vaccination: employing a genetically distinct strain of murine cytomegalovirus avoids vector-mediated resistance to virally vectored immunocontraception. Vaccine *27*, 5226–5232.

Pignatelli, S., Dal Monte, P., Rossini, G., Chou, S., Gojobori, T., Hanada, K., Guo, J.J., Rawlinson, W., Britt, W., Mach, M., *et al.* (2003). Human cytomegalovirus glycoprotein N (gpUL73-gN) genomic variants: identification of a novel subgroup, geographical distribution and evidence of positive selective pressure. J. Gen. Virol. *84*, 647–655.

Pyzik, M., Charbonneau, B., Gendron-Pontbriand, E.-M., Babic, M., Krmpotic, A., Jonjic, S., and Vidal, S.M. (2011). Distinct MHC class I-dependent NK cell-activating receptors control cytomegalovirus infection in different mouse strains. J. Exp. Med. *208*, 1105–1117.

Rawlinson, W.D., Farrell, H.E., and Barrell, B.G. (1996). Analysis of the complete DNA sequence of murine cytomegalovirus. J. Virol. *70*, 8833–8849.

Rawlinson, W.D., Zeng, F., Farrell, H.E., Cunningham, A.L., Scalzo, A.A., Booth, T.W., and Scott, G.M. (1997). The murine cytomegalovirus (MCMV) homolog of the HCMV phosphotransferase (UL97(pk)) gene. Virology *233*, 358–363.

Reddehase, M.J., Rothbard, J.B., and Koszinowski, U.H. (1989). A pentapeptide as minimal antigenic determinant for MHC class I-restricted T lymphocytes. Nature *337*, 651–653.

Redwood, A.J., Messerle, M., Harvey, N.L., Hardy, C.M., Koszinowski, U.H., Lawson, M.A., and Shellam, G.R. (2005). Use of a murine cytomegalovirus, K181-derived bacterial artificial chromosome as a vaccine vector for immunocontraception. J. Virol. *79*, 2998–3008.

Rodriguez, M., Sabastian, P., Clark, P., and Brown, M.G. (2004). Cmv1-independent antiviral role of NK cells revealed in murine cytomegalovirus-infected New Zealand white mice. J. Immunol. *173*, 6312–6318.

Sage, R.D., Whitney, J.B., 3rd, and Wilson, A.C. (1986). Genetic analysis of a hybrid zone between domesticus and musculus mice (Mus musculus complex): hemoglobin polymorphisms. Curr. Top. Microbiol. Immunol. *127*, 75–85.

Scalzo, A.A., Fitzgerald, N.A., Simmons, A., La Vista, A.B., and Shellam, G.R. (1990). Cmv-1, a genetic locus that controls murine cytomegalovirus replication in the spleen. J. Exp. Med. *171*, 1469–1483.

Scalzo, A.A., Dallas, P.B., Forbes, C.A., Mikosza, A.S., Fleming, P., Lathbury, L.J., Lyons, P.A., Laferte, S., Craggs, M.M., and Loh, L.C. (2004). The murine cytomegalovirus M73.5 gene, a member of a 3′ co-terminal alternatively spliced gene family, encodes the gp24 virion glycoprotein. Virology *329*, 234–250.

Schumacher, U., Handke, W., Jurak, I., and Brune, W. (2010). Mutations in the M112/M113-coding region facilitate murine cytomegalovirus replication in human cells. J. Virol. *84*, 7994–8006.

Smith, H.R., Heusel, J.W., Mehta, I.K., Kim, S., Dorner, B.G., Naidenko, O.V., Iizuka, K., Furukawa, H., Beckman, D.L., Pingel, J.T., *et al.* (2002). Recognition of a virus-encoded ligand by a natural killer cell activation receptor. Proc. Natl. Acad. Sci. U.S.A. *99*, 8826–8831.

Smith, L.M., Shellam, G.R., and Redwood, A.J. (2006). Genes of murine cytomegalovirus exist as a number of distinct genotypes. Virology *352*, 450–465.

Smith, L.M., McWhorter, A.R., Masters, L., Shellam, G.R., and Redwood, A.J. (2008). Laboratory strains of murine cytomegalovirus are genetically similar to but phenotypically distinct from wild strains of virus. J. Virol. *82*, 6689–6696.

Smith, L.M., McWhorter, A.R., Shellam, G.R., and Redwood, A.J. (2012). The genome of murine cytomegalovirus is shaped by purifying selection and extensive recombination. Virology pii, S0042–6822(12)00422–9.

Smith, M.G. (1954). Propagation of salivary gland virus of the mouse in tissue cultures. Proc. Soc. Exp. Biol. Med. *86*, 435–440.

Steininger, C. (2007). Clinical relevance of cytomegalovirus infection in patients with disorders of the immune system. Clin. Microbiol. Infect. *13*, 953–963.

Steininger, C., Schmied, B., Sarcletti, M., Geit, M., and Puchhammer-Stockl, E. (2005). Cytomegalovirus genotypes present in cerebrospinal fluid of HIV-infected patients. AIDS *19*, 273–278.

Tang, Q., Murphy, E.A., and Maul, G.G. (2006). Experimental confirmation of global murine cytomegalovirus open reading frames by transcriptional detection and partial characterization of newly described gene products. J. Virol. *80*, 6873–6882.

Timoshenko, O., Al-Ali, A., Martin, B.A.B., and Sweet, C. (2009). Identification of mutations in a temperature-sensitive mutant (tsm5) of murine cytomegalovirus using complementary genome sequencing. J. Med. Virol. *81*, 511–518.

Tucker, P.K. (2007). Systematics of the genus Mus. In The Mouse in Biomedical Research, Fox, J.G., Barthold, S.W., Davisson, M.T., Newcomer, C.E., Quimby, F.W., and Smith, A.L., eds. (Academic Press, Amsterdam), pp. 13–24.

Vink, C., Beuken, E., and Bruggeman, C.A. (2000). Complete DNA sequence of the rat cytomegalovirus genome. J. Virol. *74*, 7656–7665.

Voigt, V., Forbes, C.A., Tonkin, J.N., Degli-Esposti, M.A., Smith, H.R., Yokoyama, W.M., and Scalzo, A.A. (2003). Murine cytomegalovirus m157 mutation and variation leads to immune evasion of natural killer cells. Proc. Natl. Acad. Sci. U.S.A. *100*, 13483–13488.

Wagner, M., Jonjic, S., Koszinowski, U.H., and Messerle, M. (1999). Systematic excision of vector sequences from the BAC-cloned herpesvirus genome during virus reconstitution. J. Virol. *73*, 7056–7060.

Walton, S.M., Wyrsch, P., Munks, M.W., Zimmermann, A., Hengel, H., Hill, A.B., and Oxenius, A. (2008). The dynamics of mouse cytomegalovirus-specific CD4 T-cell responses during acute and latent infection. J. Immunol. *181*, 1128–1134.

Xu, J., Dallas, P.B., Lyons, P.A., Shellam, G.R., and Scalzo, A.A. (1992). Identification of the glycoprotein H gene of murine cytomegalovirus. J. Gen. Virol. *73*, 1849–1854.

Xu, J., Lyons, P.A., Carter, M.D., Booth, T.W., Davis-Poynter, N.J., Shellam, G.R., and Scalzo, A.A. (1996). Assessment of antigenicity and genetic variation of glycoprotein B of murine cytomegalovirus. J. Gen. Virol. 77, 49–59.

Yan, H., Koyano, S., Inami, Y., Yamamoto, Y., Suzutani, T., Mizuguchi, M., Ushijima, H., Kurane, I., and Inoue, N. (2008). Genetic linkage among human cytomegalovirus glycoprotein N (gN) and gO genes, with evidence for recombination from congenitally and post-natally infected Japanese infants. J. Gen. Virol. 89, 2275–2279.

Yang, H., Bell, T.A., Churchill, G.A., and Pardo-Manuel de Villena, F. (2007). On the subspecific origin of the laboratory mouse. Nat. Genet. 39, 1100–1107.

Yang, H., Wang, J.R., Didion, J.P., Buus, R.J., Bell, T.A., Welsh, C.E., Bonhomme, F., Yu, A.H.-T., Nachman, M.W., Pialek, J., et al. (2011). Subspecific origin and haplotype diversity in the laboratory mouse. Nat. Genet. 43, 648–655.

Manipulating CMV Genomes by BAC Mutagenesis: Strategies and Applications

Zsolt Ruzsics, Eva M. Borst, Jens B. Bosse, Wolfram Brune and Martin Messerle

Abstract

Cloning of a CMV genome as a BAC was reported for the first time more than a decade ago, and since then this approach has virtually opened the avenue for unrestricted mutagenesis of CMVs. This chapter gives an overview of recent developments in BAC-based mutagenesis techniques and their application to specific questions of CMV biology. One focus is on mutagenesis of essential CMV genes and the design of complementation strategies, as well as on conditional CMVs that allow the analysis of this class of viral genes *in vitro* and *in vivo*.

Introduction

Several properties of CMVs such as the slow replication kinetics, their tight cell-association and the large DNA genome have hampered the generation and isolation of CMV mutants in tissue culture. Many of these limitations have now been overcome by cloning of CMV genomes as BAC and the application of techniques of bacterial genetics for mutagenesis. This approach is now widely used in many laboratories and has been so successful that we hardly read of CMV mutants any more that were generated by traditionally used mutagenesis techniques.

Cloning of the CMV genomes (Fig. I.3.1A) is the most tedious step of the whole BAC-based mutagenesis process. Fortunately, this step has to be done only once for each CMV strain to be cloned. Recombination plasmids were constructed that contained the BAC replicon flanked by DNA sequences homologous to the designated insertion site in the viral genome (Messerle *et al.*, 1997; Borst *et al.*, 1999). The linearized recombination plasmid was introduced into infected cells to enable recombination with the CMV genome. Since such recombination events occur rarely and cannot be controlled, recombinant viruses carrying the BAC vector have to be enriched by utilizing suitable selection markers or by applying plaque purification or limiting dilution techniques. In an alternative approach, *loxP* recognition sites for Cre recombinase were anchored in the CMV genome in a first recombination step, followed by Cre-mediated insertion of the BAC vector (Chang and Barry, 2003). The second step is highly specific and probably quite efficient, but the approach still relies on homologous recombination in the first step and thus is hardly less laborious. It provides, however, a means to excise the vector sequences upon reconstitution of virus progeny in cell culture (see below). Following amplification of the recombinant CMV, circular replication intermediates of the viral genomes were isolated from infected cells and transferred into *E. coli*. Bacterial clones harbouring CMV BAC were selected by means of the antibiotic resistance encoded on the vector backbone. Owing to the *repE*, *oriS*, and *parA*, *B* and *C* elements of the BAC vector (Shizuya *et al.*, 1992), CMV BAC can be propagated with remarkable stability in suitable *E. coli* hosts (Messerle *et al.*, 1997; Borst *et al.*, 1999). The restriction profile of the CMV BAC remained unchanged, and upon transfection into permissive cells, they gave rise to plaques and infectious particles (Fig. I.3.1B). Recent sequencing of HCMV and MCMV BAC that were passaged in *E. coli* for a substantial time period confirmed the high stability of the cloned CMV genomes in bacteria (Stanton *et al.*, 2010; Jordan *et al.*, 2011). This property laid the basis for targeted manipulation of the CMV BAC in *E. coli*.

Full-length CMV genomes containing the BAC replicon are oversized, necessitating the development of strategies to eliminate the vector sequences upon reconstitution of infectious virus in order to prevent spontaneous deletions. One approach was to flank the BAC replicon with *loxP* sites to allow recombinase Cre-mediated excision (Messerle *et al.*, 1997; Hobom *et al.*, 2000). By inserting an intron-containing *cre* gene into the BAC vector, Yu *et al.* (2002) adopted an elegant strategy for the self-excision of the BAC vector in eukaryotic cells, and the same approach was subsequently used for an RhCMV BAC (Chang and

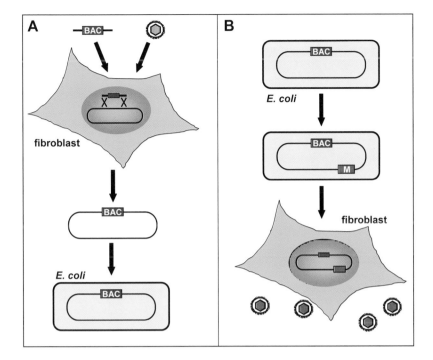

Figure I.3.1 Cloning and mutagenesis of a CMV genome as BAC. (A) The BAC replicon (blue), flanked by viral sequences, is transfected into fibroblasts, which are subsequently infected with CMV. Viral genomes that have incorporated the BAC cassette by homologous recombination are isolated and transferred to *E. coli*. (B) The CMV genome maintained as BAC in *E. coli* is used as a substrate to introduce a mutation (M, red). The modified BAC is then transfected into fibroblasts to reconstitute mutant virus.

Barry, 2003) and the HCMV Merlin BAC (Stanton *et al.*, 2010). The second strategy for excision relies on the duplication of viral DNA sequences flanking the BAC replicon (Wagner *et al.*, 1999; Hobom *et al.*, 2000). It was observed that homologous recombination in BAC-transfected cells leads to spontaneous excision of the BAC replicon from the viral genome, and in combination with the selection-pressure against packaging of oversized genomes this principle allowed to generate CMV genomes indistinguishable from the WT CMV genome, i.e. no remnants of the BAC replicon remained in the genomes. This approach appears especially suitable for the generation of mutants of animal CMVs that are used for *in vivo* studies and should differ from the WT virus only by the intended specific mutation.

Since the first BAC cloning of the MCMV Smith strain (Messerle *et al.*, 1997) and of the HCMV AD169 laboratory strain (Borst *et al.*, 1999), the genomes of several animal CMV and of additional HCMV strains including those of clinical isolates have been cloned (McGregor and Schleiss, 2001; Marchini *et al.*, 2001; Hahn *et al.*, 2002,2003b; Yu *et al.*, 2002; Chang and Barry, 2003; Murphy *et al.*, 2003b; Redwood *et al.*, 2005; Sinzger *et al.*, 2008; Stanton *et al.*, 2010). BAC cloning was also successful for the β-herpesvirus

HHV-6A (Borenstein and Frenkel, 2009; Tang *et al.*, 2010). Many of the CMV BAC constructed early on did not contain the full-length CMV genomes and had some non-essential genes replaced by the vector sequences. This was done on purpose, because it seemed that enlargement of the CMV genomes due to the inserted BAC replicon interfered with the packaging capacity of the CMV capsid. In other cases, some viral genomic sequences got lost, probably during insertion of the BAC vector by non-homologous, unexpected recombination events (Murphy *et al.*, 2003b; Sinzger *et al.*, 2008). However, once the genomes were successfully transferred to *E. coli*, the missing viral genes could be re-inserted, restoring the full genetic complement (Wagner *et al.*, 1999; Hobom *et al.*, 2000; Stanton *et al.*, 2010). When Yu *et al.* (2002) later managed to clone the full-length HCMV AD169 genome, it turned out that the virus retaining the BAC vector in its genome displayed delayed growth kinetics and produced reduced yields, supporting the view that there is an upper limit for the CMV genome to be packaged. A more thorough study performed with GPCMV showed that the genome size can be increased only modestly (Cui *et al.*, 2009).

In some cases it turned out that the BAC-cloned

CMV genomes carried defective genes (Murphy *et al.*, 2003b; Lilja and Shenk, 2008; Sinzger *et al.*, 2008; Stanton *et al.*, 2010; Jordan *et al.*, 2011). Most likely, this was not due to propagation of the BAC in *E. coli*, but rather it seems that variants were cloned that were already present in the tissue-grown virus population. In retrospect, this may be not surprising, since it is now well documented that mutations in the CMV genome rapidly accumulate due to adaptation of CMVs to growth in cell culture (Barry and Chang, 2007; Dargan *et al.*, 2010; Jordan *et al.*, 2011), and many strains have been passaged for decades. Stanton *et al.* (2010) cloned the HCMV Merlin strain at a stage when it had undergone only a few passages in tissue culture, and this strain was therefore expected to come close to a WT isolate of HCMV. Interestingly, this early isolate already carried a couple of mutations and the colleagues invested substantial effort to repair all of these mutations to generate a BAC that corresponds to a kind of prototypic HCMV WT genome. This BAC may now represent the gold standard for further research. As expected, the virus derived from this BAC has many properties in common with clinical HCMV isolates, including slow growth and cell association (Stanton *et al.*, 2010). However, we have to be aware that these features may hamper speedy studies and moreover, the strain may rapidly re-acquire adaptive mutations as soon as it is propagated in cell culture. Paradoxically and contrary to early concerns, BAC cloning in *E. coli* seems to be a safe haven for CMV genomes, and it is now rather the tissue culture systems that limit our studies of CMV. Although some of the BAC-cloned CMV strains (e.g. the laboratory strains AD169 and Towne) lack a couple of genes and may carry a few defective ones, they nevertheless represent excellent tools to address many important biological questions. We just have to keep in mind that they – as any experimental system – have limitations and hence may not be suitable for analysing all aspects of CMV biology. Actually, for certain studies these strains may even be superior to slowly growing clinical isolates.

Mutagenesis of BAC-cloned CMV genomes

The establishment of methods to manipulate the CMV BAC in *E. coli* was the next logical step after successful cloning. Owing to the large size of the BAC, traditional modification techniques (e.g. treatment with restriction enzymes followed by re-ligation) cannot be applied, simply because almost every enzyme cuts several times in the BAC molecules. However, sophisticated methods, established before to manipulate large DNA molecules by homologous recombination in *E. coli*, could be adopted to mutate the CMV BAC.

Scientists working on mouse and human genetics have further refined these techniques during recent years, and virologists could profit from their efforts. On the other hand, at least one of the novel techniques, *en passant* mutagenesis, was developed by herpes virologists (Tischer *et al.*, 2006, 2010) and is now utilized by other geneticists, too. Meanwhile, there is a wealth of techniques that facilitate the rapid and efficient introduction of any kind of mutation into cloned CMV genomes. Below we describe the advantages as well as potential limitations of these techniques.

Reverse genetics

Reverse genetics of CMV is the most common application of the BAC techniques. Manipulation of CMV genomes usually aims at the generation of different alleles of a given gene, which typically results in a distinct phenotype of the respective virus. The possible alterations range from the complete removal or inactivation of a gene (null allele) to the exchange of a single nucleotide only (point mutation). In this way, we can study the functional consequences of the loss of a viral protein (or of a protein domain) or of non-coding RNAs such as viral miRNAs. We can also examine the functional relevance of *cis*-acting DNA sequences, e.g. transcription factor binding sites in viral promoters, elements of origins of replication, or of sequences involved in packaging of the viral genomes. Another important aim is the visualization of viral proteins and their biochemical characterization, e.g. by tagging such proteins with fluorescent proteins or with epitopes that can be recognized by antibodies. This allows the monitoring of the temporal expression characteristics and subcellular localization of viral proteins, their interaction with other viral or cellular proteins, assembly and maturation of capsids, and also trafficking of capsids and virions, either upon entry into or egress from cells.

RecA-mediated allelic exchange

RecA-mediated allelic exchange was one of the first techniques used for modification of CMV BAC. This technique makes use of the general recombination machinery of *E. coli*, with the RecA protein and exonuclease V of *E. coli* being important components of this pathway. RecA is able to perform strand invasion and can anneal single-stranded ends of a recombination substrate, usually generated by exonuclease V, with the homologous sequences in the target molecule. In many *E. coli* strains the *recA* gene is inactivated, preventing recombination and thereby providing enhanced stability of cloned DNA fragments. To render such strains recombination-proficient, *recA* expression must be turned on, at least transiently. This increases the risk of instability of the BAC-cloned CMV genomes; fortunately, however, the CMV BAC seem to be remarkably stable under such conditions (Messerle *et al.*, 1997; Borst *et al.*, 1999; Yu *et al.*, 2002). Nevertheless, we recommend to perform mutagenesis in an *E. coli* strain that can be transiently rendered *recA+* (Borst *et al.*, 2007), to minimize the risk of acquiring adventitious mutations.

The mutant allele is provided by a circular shuttle plasmid and is flanked by homology arms of substantial size (1.5 to 2 kbp on each side) to increase the chance of homologous recombination. The product of the recombination event is a co-integrate between the two molecules. A second recombination event leads to the resolution of the co-integrate, resulting in either a BAC carrying the mutation or one with the WT configuration. The introduction of a mutation has to be verified by restriction analysis, PCR, or sequencing. The use of positive and negative selection markers and appropriate selection schemes made the construction of mutants with this technique convenient and highly reliable. We and others still use it very successfully (Borst and Messerle, 2005; Lemmermann *et al.*, 2010, 2011). Admittedly, the construction of the shuttle plasmid usually involves several cloning steps and can be time-consuming. However, when several different mutations have to be introduced into the same region of the viral genome, or when one specific mutation has to be transferred to the genomes of different CMV strains and mutants, this technique is still of value.

Given the differences between RecA and Redβ [which is the key player of the now more popular Red recombineering technique (see below)], in particular the ability for strand invasion by RecA, the potential of RecA-mediated recombination may be underestimated. Red and RecA-driven recombination proceeds at least in part by different mechanisms. Notably, there seems to be a size limitation in the insertion of foreign sequences by Red recombineering (Kuhlman and Cox, 2010; Maresca *et al.*, 2010). Such limitations might be overcome by advancement of RecA-based techniques or by combined use of the Red and RecA functions (Wang *et al.*, 2006).

Red recombineering

The most widely utilized technique for reverse genetics of CMV BAC is now Red recombineering that depends on the recombination genes *red-α, -β, -γ* of bacteriophage λ. This method owes its popularity both to the ease of application and its efficiency. Typically, a linear DNA fragment carrying the genetic alteration plus a selectable marker (e.g. a Kan resistance gene, *knR*)

flanked by DNA sequences homologous to the target site is generated and introduced into an *E. coli* strain expressing the recombination functions and containing the BAC with the viral genome to be mutated (Fig. I.3.2A). *red-a* encodes an exonuclease that is acting on the DNA strands of the linear fragment in a 5′–3′-direction, leaving single-stranded 3′-overhanging ends. Redβ binds to the single-stranded DNA, protects it from further degradation and mediates the annealing with the homologous sequences in the BAC. Redγ must be co-expressed with Redα and -β to prevent the degradation of the linear DNA fragment by other nucleases present in *E. coli*. It was proposed that the linear DNA is joined to the BAC via two independent recombination events, i.e. by double cross-over (Fig. I.3.2A). However, even more than 10 years after development of the mutagenesis technique (Zhang *et al.*, 1998; Datsenko and Wanner, 2000; Yu *et al.*, 2000), the exact mechanism by which the Red proteins mediate recombination is still a matter of debate, although there is now growing evidence that a single-stranded intermediate and active replication of the target molecule play a central role (Maresca *et al.*, 2010; Mosberg *et al.*, 2010). Since the Red functions require only 30–50 nucleotides of homologous sequences to initiate recombination, the linear recombination substrates can easily be generated by PCR with the homology regions being incorporated into the primers used for the amplification of a selectable marker (Fig. I.3.2A). Alternatively, preassembled linear fragments isolated from plasmids work equally well for recombination. Since the homology arms and also the site of mutation can be freely chosen, the method is not reliant on the availability of appropriate restriction sites and mutations can be introduced at virtually any position of the viral genome. In principle, Red recombineering can get along without a selection marker, and sometimes, especially when small alterations are introduced through single-stranded oligonucleotides, recombination can occur with such a high frequency that the identification of mutated BAC becomes feasible by PCR screening (Holtappels *et al.*, 2004). In practice, however, a selection marker strongly simplifies and speeds up the identification of successfully modified clones. Since selection is straightforward and recombination is highly precise, modified genomes can be generated within a few days. Highly useful tools such as low copy vectors encoding the Red functions (Datsenko and Wanner, 2000) and *E. coli* strains carrying a defective λ prophage (Yu *et al.*, 2000) are now available, and help to express adequate levels of the recombination proteins.

Initially the method was widely used to delete single genes from the CMV BAC by replacing viral sequences with a selectable marker (Borst *et al.*, 2001;

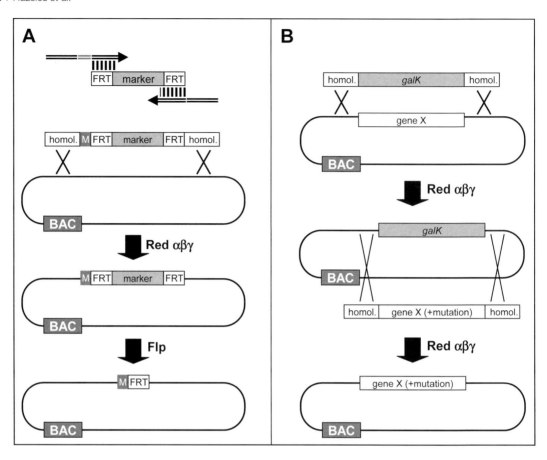

Figure I.3.2 BAC mutagenesis using linear DNA fragments. (A) A selectable marker flanked by FRT sites and sequences carrying the mutation (M) are amplified by PCR. About 50 bp of homologous sequences at either end are introduced through the PCR primers. The linear PCR product is electroporated into an *E. coli* host that expresses the *red*α, -β, and -γ genes of bacteriophage λ. The Red proteins perform the homologous recombination. In a second step, the selectable marker can be removed using Flp recombinase. The mutation and a single FRT site remain in the mutant genome. (B) Two-step replacement strategy. A viral gene is first replaced by a selectable and a counter-selectable marker cassette (e.g. *galK*). In a second step, the marker is replaced by a mutated gene sequence.

Wagner *et al.*, 2002; Dunn *et al.*, 2003; Ménard *et al.*, 2003). In addition, mutants with large deletions covering dozens of genes were attained (Loewendorf *et al.*, 2004; Čičin-Šain *et al.*, 2007). The technique was also successfully used to insert additional sequences into the BAC, e.g. to tag CMV proteins or to express reporter genes (Marquardt *et al.*, 2011). If the additional sequences are included in the PCR primer, the size of the insertion is defined by the technical limitations that apply to the synthesis of long oligonucleotides. The insertion of larger DNA segments such as reporter genes requires the assembly of the corresponding DNA sequences in a plasmid and their cloning next to a selection marker (Borst *et al.*, 2001; Marquardt *et al.*, 2011). The linear fragment required for recombination can then be obtained either by

release from the plasmid with restriction enzymes or by PCR amplification.

The selection marker represents a foreign sequence that may potentially interfere with the expression of other viral genes, and also increases the genome size. Already early on measures were therefore taken to remove selection markers after successful mutagenesis. One possibility is to flank the marker with *loxP* and FRT recognition sites for the Cre and Flp recombinases, respectively. Marker excision with recombinases leaves only 34 bp of these recognition sequences behind (and some nucleotides originating from the priming regions) (Fig. I.3.2A). These few additional nucleotides may be tolerable in cases in which hundreds of nt of foreign sequences are inserted, e.g. reporter genes, or if complete ORF are deleted. Nevertheless, one should keep in mind that the *loxP* and FRT sites can form a stem–loop structure and may therefore potentially destabilize transcripts or impair their translation. Likewise, such recognition sites are not tolerable within an ORF, e.g. when the codon(s) for an aa of a protein have to be changed.

In general there is now a trend towards seamless mutagenesis methods that do not leave any extra changes except the designated ones (see below). Also, with the increasing knowledge of the transcriptome of CMVs, we become aware that transcription of the CMV genome is highly complex. It is not just that ORF are overlapping and that many transcripts are identical at their 3′-ends due to the sharing of polyadenylation signal sequences. Recent studies indicate that large parts of the CMV genomes are transcribed from both strands and that there are much more splicing events, some spanning large areas, than previously anticipated (Zhang *et al.*, 2007; Gatherer *et al.*, 2011; Lacaze *et al.*, 2011). Accordingly, mutagenesis of one specific viral gene, e.g. the deletion of a complete ORF, may have unforeseen impact on the expression of neighbouring genes or even long ranging effects on distantly located genes and thus, we cannot be sure that the phenotype of such a mutant reflects the disruption of the specific gene only. Now that much more sophisticated mutagenesis techniques have become available, we can take such considerations into account and design the genetic alterations more carefully. For instance, mutagenesis of ATG start codons or the insertion of a short 'stop cassette', interrupting an ORF, are more appropriate strategies to address the consequences of the loss of a specific gene.

*gal*K-based mutagenesis

To construct a BAC that carries only the desired genetic change without residual operational sequences an approach was developed utilizing two consecutive steps of Red recombination (Warming *et al.*, 2005). A metabolic marker gene, *galK,* was used that allows both positive and negative selection in the *E. coli* strain SW102, which is deficient in its genomic *galK* locus. The *galK* gene codes for galactokinase, which is an essential enzyme of the galactose metabolism. If the *E. coli* SW102 strain is grown on galactose as the exclusive carbon source, the galactokinase is essential for bacterial growth. Thus, successful delivery of the *galK* marker by Red recombination can be selected by plating the recombinants on minimal media containing galactose as the only carbon source. This first recombination step should be designed in a way that the *galK* marker replaces the region of the viral genome that is to be modified (Fig. I.3.2B). The galactokinase

can also phosphorylate galactose antimetabolites such as 2-deoxygalactose (DOG). The resulting molecules are toxic to *E. coli* and this way *galK* can be used as a negative selection marker. In a second round of Red recombineering, the *galK* marker is replaced by the desired mutated sequences. To select for incorporation of the mutation, the recombinants are plated on medium containing an alternative carbon source and DOG. Only bacteria carrying BAC in which the *galK* marker is replaced with the mutated sequences can grow, because the unchanged BAC carry a toxic gene. As a result, a BAC mutant is constructed in which only the desired mutation but no operational sequences are left. However, both selection steps require special media and are time-consuming, because the recombinants must be propagated in minimal media. To simplify the experimental burden, we fused the *galK* marker to a *knR* cassette, which allowed us to carry out the first positive selection step on normal media. Only in the second step did we utilize *galK* as a negative selection marker (Dölken *et al.*, 2010). This approach allowed us to introduce single nucleotide changes and to add epitope tags to the MCMV M50 and M53 genes at their native genomic location (Dölken *et al.*, 2010; Jordan *et al.*, 2011; Pogoda *et al.*, 2012; Lemnitzer *et al.*, in preparation). Unfortunately, the counter-selection step is not working with 100% efficiency and therefore, this method is work-intensive and time-consuming. Nevertheless, *galK*-based recombineering is highly useful if seamless replacement or introduction of large inserts are required.

en passant mutagenesis

The so-called *en passant* mutagenesis technique is a similarly appealing method that allows the introduction of all kinds of mutations (deletions, insertions, point mutations) into a BAC without leaving behind any additional foreign sequences in the viral genomes. Usually, no elaborate cloning steps are required, and the procedure is as fast as Red recombineering, on which the technique is based. En passant mutagenesis was first established with an equine herpesvirus-1 BAC (Tischer *et al.*, 2006), and, in the meantime, it has been used for the generation of numerous CMV mutants underlining its utility and convenience (Jiang *et al.*, 2008; Schuessler *et al.*, 2008, 2010, 2011; Jordan *et al.*, 2011; E.M. Borst, unpublished). The concept relies on two consecutive Red recombination steps, ideally performed in the specifically engineered *E. coli* strain GS1783 (Tischer *et al.*, 2010). This strain carries the Red genes and the I-SceI meganuclease gene anchored in the bacterial chromosome, and these functions can be expressed in an inducible manner. Alternatively, the

genes encoding the mentioned enzymes can be introduced into a recombination-deficient *E. coli* strain using
a plasmid, but in our hands the use of the GS1783 strain
is more convenient and efficient than the approach with
the plasmid-borne functions.

The procedure basically follows the Red recombineering scheme, except that by virtue of the primers
a part of the flanking homology region (boxes B and C
in Fig. I.3.3A) and the sequences carrying the desired
mutation are duplicated. Also, a recognition site for
the meganuclease I-SceI is included next to the positive selection marker. In the first step, recombination
between the linear PCR fragment and the BAC occurs
via the terminal homology sequences (boxes A/B and
C/D), resulting in the duplication of the mentioned
sequences in the intermediate BAC (Fig. I.3.3A).

Transient expression of I-SceI leads to the linearization
of the BAC, followed by Red-mediated intramolecular
recombination within the BAC via the duplicated
sequences. As a result, the selection marker is excised,
leaving the seamless insertion of the mutation. Seamless deletions of an ORF can be easily accomplished by
providing duplications of homology regions flanking
the genomic locus of interest (Tischer *et al.*, 2006). For
the scarless insertion of larger sequences, for instance
to generate fusions between a fluorescent reporter
protein and a viral protein of interest (POI), the positive selection marker and the I-SceI site are cloned into
the sequences to be inserted (in the example shown
in Fig. I.3.3B into the EGFP ORF). Simultaneously,
a ~ 50 nt duplication of the sequence adjacent to the
insertion site has to be introduced together with the

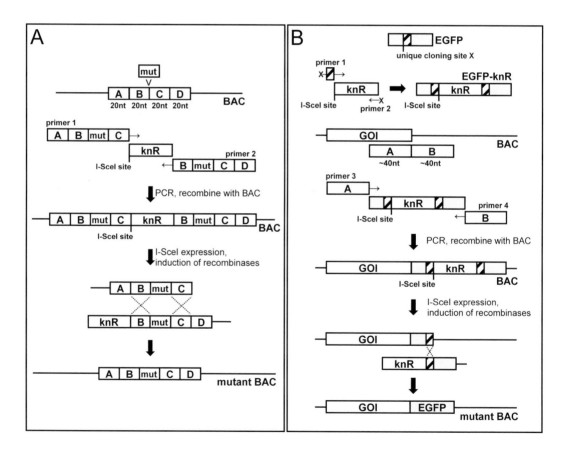

Figure I.3.3 Introduction of point mutations or epitope tags using *en passant* mutagenesis. (A) The *knR* marker is
PCR-amplified using primers 1 and 2 that carry the mutation plus sequences A/B/C or B/C/D homologous to the
insertion site in the BAC. After insertion of the linear DNA fragment, expression of I-SceI and of the recombination
functions Redα, -β, -γ results in linearization of the BAC, followed by recombination via the duplicated sequences.
(B) Insertion of larger DNA fragments, e.g. tagging of a GOI with EGFP. *knR* is cloned into the EGFP ORF with
a simultaneous sequence duplication (~50 bp; hatched boxes). The EGFP–*knR* construct then serves as a PCR
template with primers 3 and 4, harbouring homology regions A or B. Recombination, followed by subsequent
induction of I-SceI and the Red functions, results in a BAC carrying the GOI fused to the EGFP encoding sequence.

selection marker in order to provide the substrate for the second recombination step. Following insertion of the sequences for EGFP and the selection marker, I-SceI induced Red recombination via the duplicated sequences yields a repaired BAC harbouring the EGFP ORF fused to the viral ORF of interest.

Since its invention, the *en passant* mutagenesis technique has greatly contributed to the analysis of CMV gene functions. For instance, it was used to examine the role of the UL128, UL130 and UL131A gene products in HCMV EC tropism (Schuessler *et al.*, 2008, 2010, 2011), as well as to gain insight into the function of gpO (Jiang *et al.*, 2008). We have successfully used the method to introduce various point mutations as well as sequences for epitope and fluorescent tags into the HCMV AD169 genome (E.M. Borst, unpublished). A crucial point to consider is the quality of the oligonucleotides used for this mutagenesis strategy. Since the required oligonucleotides are usually rather long and often include coding sequences, they have to be subjected to the highest purification procedure to eliminate oligonucleotides with synthesis errors. In our hands problems with this technique were almost exclusively connected to flaws in primer synthesis, whereas Red recombination always worked with perfect accuracy (E.M. Borst, unpublished). Sequencing of the mutated region is therefore mandatory. Improvement in gene synthesis services will minimize such issues in the future, further increasing the power of the *en passant* technique for many applications.

Forward genetics

Forward genetics is a classical genetic procedure that starts out with a physical or measurable characteristic of an organism (the phenotype) and aims at the identification of the gene (or genes) that are responsible for it. Prior to BAC cloning of CMV genomes, this was difficult to do as a substantial number of mutant viruses needed to be made and analysed. Considering the large but still manageable size of CMV genomes, there are two reasonable ways to identify genes by their function using BAC mutagenesis. One involves the construction of mutants with large deletions, the other requires a library of transposon insertion mutants.

Transposon mutagenesis

Transposons are mobile genetic elements that can relocate and insert at any genomic location in a virtually random manner. In bacteria they have been utilized widely as random insertion mutagens both at the genomic level and in the analysis of individual genes (reviewed by Hayes, 2003). With their cloning as BAC, CMV genomes also became accessible to transposon mutagenesis. The power of the technique lies in its easy and rapid application and in particular in its capacity to generate a large number of mutants within a short time. Several different transposon systems have been tested for their suitability to manipulate CMV genomes (Table I.3.1). They can be divided into two categories: *in vitro* systems, in which transposition takes place in a test tube, and *in vivo* systems, in which transposition occurs in the bacterial host. Because of the potential vulnerability of the large BAC molecules to shearing *in vitro*, transposon mutagenesis in *E. coli* (*in vivo*) is the preferred method for BAC mutagenesis. The first transposon system for the mutagenesis of an entire herpesvirus genome was based on Tn*1721* (a member of the Tn3 family), which displays a preference for

Table I.3.1 Transposon systems used for BAC mutagenesis

Transposon	Properties	References
Tn*1721* (Tn*Max*)	*In vivo* system, preferential insertion into BAC, high efficiency	Brune *et al.* (1999, 2001, 2002, 2003), Hobom *et al.* (2000), Ménard *et al.* (2003), Yu *et al.* (2003), Zimmermann *et al.* (2005)
Tn5	*In vivo* system, preferential insertion into bacterial genome, requires re-transformation	Smith and Enquist (1999), Yu *et al.* (2003)
Tn5[a]	*In vitro* system, results in high percentage of incomplete BAC	McGregor *et al.* (2004)
Tn5[a] (transposome)	*In vivo* system, requires transposon–transposase complex (transposome)	McGregor *et al.* (2004)
Tn7 [a]	*In vitro* system, results in high percentage of incomplete BAC	McGregor *et al.* (2004)
Tn7	*In vivo* system, insertion into defined attachment site (non-random)	Hahn *et al.* (2003a)

[a]Commercially available through Epicentre and NEB, respectively.

insertion into negatively supercoiled plasmids (Brune *et al.*, 1999). The transposon is provided on a donor plasmid with a temperature-sensitive replication mode. The transposition control unit with the transposase gene is also located on the donor plasmid but separately from the transposable element and is therefore not translocated. Accordingly, secondary transposition events can be prevented by eliminating the donor plasmid after mutagenesis (Brune, 2002). Transposition into the bacterial chromosome occurred in only 5–10% of the clones, obviating a need for enrichment of clones with mutant BAC (Brune *et al.*, 1999; Hobom *et al.*, 2000). Almost all of the BAC contained a single insertion, which remained stable in the genome after reconstitution of viruses. Primer binding sites at the respective termini of the transposon allowed the determination of the insertion site by direct sequencing. This method was first used to identify essential and non-essential genes of MCMV (Brune *et al.*, 1999). Later on, mutants of whole families of CMV genes, namely of the HCMV envelope gp genes and of the MCMV genes homologous to the HCMV US22 family members, were isolated from BAC libraries in order to study their function (Hobom *et al.*, 2000; Ménard *et al.*, 2003). Although the distribution of the transposon insertions did not seem to be completely random (Gutermann *et al.*, 2002), the libraries were of sufficient complexity to find mutants of each member of the gene families.

Smith and Enquist (1999) adopted a transposon mutagenesis procedure based on the Tn5 transposon, using a donor plasmid with an R6K origin of replication. Such plasmids need the phage λ protein π for propagation and cannot replicate in the recipient bacteria that contain the CMV BAC. Upon transfer of the Tn5 donor plasmid by conjugation or electroporation the transposon is inserted either into the BAC or the bacterial chromosome. Since the latter occurs in a substantial proportion of the transposition events, one needs to isolate the BAC and retransform fresh bacteria in order to enrich for the mutant BAC. Yu *et al.* (2003) used both previously described transposon systems (i.e. the Tn5- and the Tn1721-derived transposons) to create a comprehensive library of HCMV AD169 mutants that enabled them to define non-essential and essential ORF of HCMV. In addition to the selection marker (*knR*), both transposons contained reporter genes (*gfp*, *lacZ*) for eukaryotic as well as for prokaryotic cells. The *lacZ* marker facilitated the generation of revertant HCMV BAC by allelic exchange and blue-white screening. Another interesting property of one of the transposons is the ability to tag the truncated proteins in four of the six possible frames in order to identify proteins that are expressed by an ORF carrying the insertion.

Two commercially available *in vitro* mutagenesis systems that are based on Tn5 and Tn7, respectively, were tested by McGregor *et al.* (2004). In principle both systems were considered to be suitable for the manipulation of a GPCMV BAC. However, since handling the BAC molecules cannot be avoided during the *in vitro* reaction and the subsequent electroporation into *E. coli*, the majority of the resulting BAC either displayed deletions or rearrangements. This problem was elegantly overcome by the use of so-called Tn5 'transposomes'. Transposomes are complexes between the transposon molecule and the transposase, which are formed *in vitro* in the absence of Mg^{2+} ions. Once transferred into bacteria carrying the BAC, the complex is activated by the Mg^{2+} ions inside the cells, resulting in the insertion of the transposon into a target molecule. Since the complex is self-limiting there is no risk for secondary transposition events. Unfortunately, Tn5 is not able to discriminate between the BAC episome and the bacterial chromosome and thus there remains a need to isolate the mutated BAC and retransform them into new *E. coli* bacteria in order to separate them from clones that received an insertion into the bacterial genome (McGregor *et al.*, 2004).

Considering the versatility of the transposon-based techniques and their relatively easy, rapid and inexpensive application it becomes obvious that these techniques have a considerable potential for the analysis of CMV gene functions. A prerequisite of forward genetics is the generation of a comprehensive library of mutants, ideally by saturating random mutagenesis. The Tn1721-based transposon system is probably the most suitable one for a forward-genetic screen because literally thousands of mutants can be obtained overnight (Brune *et al.*, 1999, 2001). Yet, there was the problem of effectively converting the mutated genomes into mutant viruses. This obstacle was overcome by endowing the *E. coli* bacteria with the invasion of *Yersinia pseudotuberculosis* and the haemolysin O protein of *Listeria monocytogenes*. These proteins enabled the bacteria to invade fibroblasts, which was followed by the release of the CMV BAC into the cells and the consequent initiation of the viral replication cycle. Thus, the bacterial clones had simply to be inoculated onto fibroblasts in order to produce a collection of viral mutants. The expression of GFP, which was encoded by the transposable element, facilitated the identification of the mutant viruses. Using this procedure a library of approximately 600 mutant MCMV genomes was generated, yielding about 200 viable mutants (Brune *et al.*, 2001). There was no need to characterize the individual mutants prior to screening. The first application was to screen the library for mutants that displayed a growth deficit in EC. To this end, fibroblasts and EC were infected in parallel with individual mutants and monitored for the spread of the infection. Whereas all viruses grew on fibroblasts, six of them failed to grow on EC. By means of the transposon insertions, the mutations could be easily mapped in the viral genomes and were found to reside in one single MCMV gene, M45 (Brune *et al.*, 2001). The protein encoded by M45 is a constituent of the viral tegument and shows homology to the large subunit of the cellular ribonucleotide reductase, but does not represent a functional enzyme subunit (Lembo *et al.*, 2004). Infection experiments with the M45 mutants revealed that M45 is required for protecting EC and macrophages from viral infection-induced cell death (Brune *et al.*, 2001). Mechanistically, the M45 protein inhibits the induction of programmed necrosis (a caspase-independent form of programmed cell death) by interacting with receptor-interacting protein 1 (RIP1) and RIP3 (Mack *et al.*, 2008; Upton *et al.*, 2008, 2010). The same library was used to identify another viral gene with cell death-inhibiting function (Brune *et al.*, 2003) and to find a viral inhibitor of IFN-γ receptor signalling (Zimmermann *et al.*, 2005). These three examples highlight the power of the forward genetic approach. Screening a library for a particular phenotype led directly to the identification of the responsible gene.

Mutants with large deletions

Mutants with large deletions may generally speed up the identification of genes responsible for specific viral properties and reduce the effort needed to screen a complete library. Assuming that immunomodulatory genes, which are not essential for replication in cell culture, are mainly located towards the termini of CMV genomes, Loewendorf *et al.* (2004) constructed a set of MCMV mutants with deletions of about 10–15 kbp. Flow-cytometric analysis of macrophages infected with the deletion mutants rapidly led to the definition of a genomic region encoding a protein that interferes with the expression of the costimulatory molecule CD86 on professional antigen presenting cells (APC). By constructing mutants with successively smaller deletions the responsible gene was finally identified. Screening of a series of these deletion mutants also led to the identification of three MCMV genes that inhibit the expression of cellular ligands for the NKG2D receptor of NK cells (Hasan *et al.*, 2005; Krmpotic *et al.*, 2005; Lenac *et al.*, 2006). Following this strategy there is, however, a risk of missing a gene if it is located between essential genes within the central region of the CMV genomes.

Thus, in the future, it will probably make sense to perform a more systematic screening by combining mutants that carry large deletions and mutants with a disruption of defined genes. The comprehensive libraries of HCMV mutants constructed by Yu *et al.* (2003) and Dunn *et al.* (2003) may represent a starting point.

Analysis of essential CMV genes: complementation strategies and conditional mutants

Of the 170–200 potential ORF of a CMV genome (Murphy *et al.*, 2003a; Dolan *et al.*, 2004) (see Chapter I.1 and Chapter I.2), about one-quarter are essential for growth of the viruses in cell culture (Dunn *et al.*, 2003; Yu *et al.*, 2003). Propagation of a null mutant virus on a *trans*-complementing cell line and subsequent infection of normal cells was for a long time considered the gold standard for how the function of the often poorly characterized essential CMV proteins can be studied. By exploring at which stage the viral life cycle is blocked, one can come to conclusions about the role of the respective essential protein during CMV infection. However, the generation of complementing cell lines for CMV mutants remains a challenging task and consequently only a few have been reported so far (Mocarski *et al.*, 1996; Angulo *et al.*, 2000; Borst *et al.*, 2008; Sanders *et al.*, 2008; Isaacson and Compton, 2009; Mohr *et al.*, 2010). The paucity of mutants in essential genes applies particularly to HCMV, because it replicates efficiently only in primary human fibroblasts, which are difficult to manipulate. These difficulties in CMV genetics forced the development of alternative approaches such as the conditional expression of POI to avoid the cumbersome construction of such cell lines.

Ectopic *cis*-complementation

Detailed genetic analysis of essential genes can not only be done by studying the null phenotype, but also by investigating mutant alleles in the viral context as a valuable alternative. This approach involves deletion of the WT gene followed by its replacement at the original site with mutated alleles or by ectopic re-insertion of mutated and reverted alleles at a neutral other site of the viral genome. A series of mutated CMV genomes was constructed by ectopic re-insertion using the Flp/FRT system.

To achieve this, an FRT site is inserted into the CMV WT genome at a position supposed to not affect the functions under study (Bubeck *et al.*, 2004; Borst

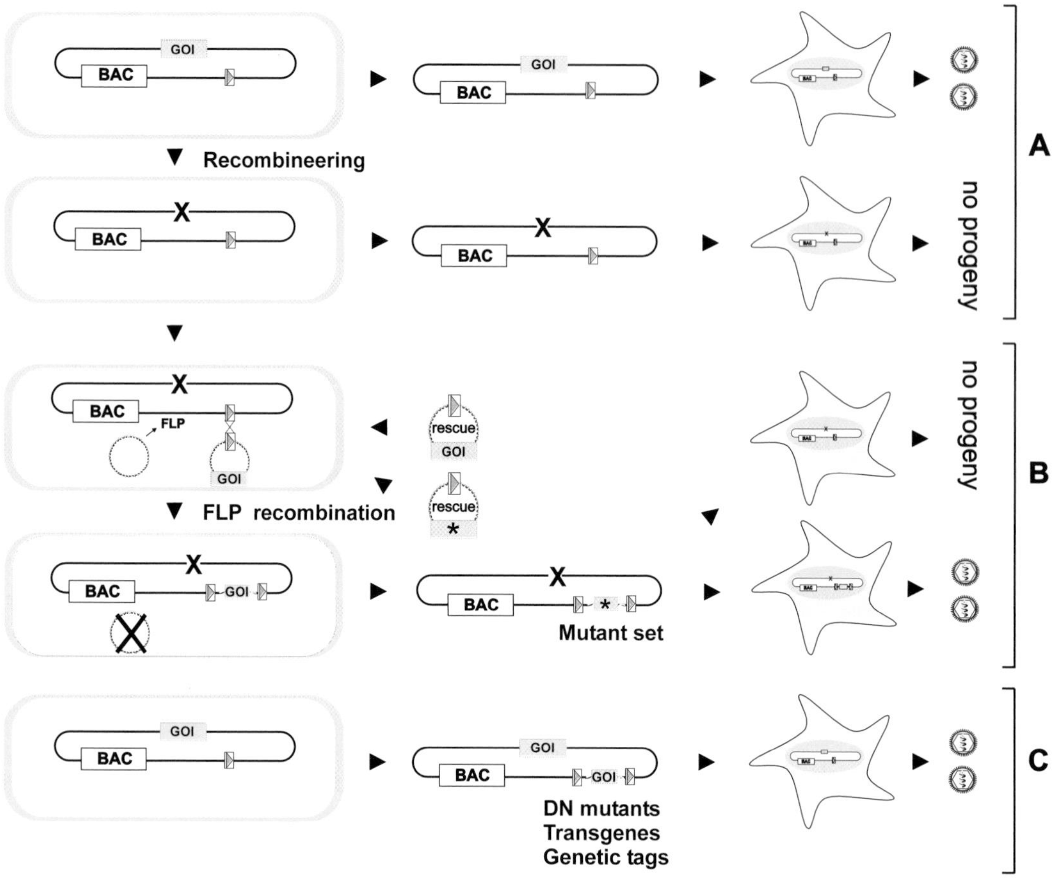

Figure I.3.4 Ectopic *cis*-complementation by the Flp/FRT system. (A) The GOI (grey box) can be subjected to rescue experiments by ectopic *cis*-complementation if its specific deletion results in a growth defect. (B) The GOI is subcloned into a rescue plasmid containing one FRT site (grey triangle). This plasmid or plasmids carrying mutant alleles can be reintroduced into the deletion BAC by Flp recombination. Subsequently, BAC DNA is transfected into eukaryotic cells for virus reconstitution, and cell cultures are screened for plaque formation. The WT gene should rescue the null phenotype (*cis*-complementation). If mutant alleles carry a functionally important alteration, the growth defect is maintained. (C) The Flp/FRT system can also be used for insertion of mutant alleles, transgenes, genetic tags and markers into WT BAC.

et al., 2005; Maninger *et al.*, 2011). At first, these WT-like viral BAC are then used to generate deletion mutants (Fig. I.3.4A). The null phenotype of this deletion construct must be fully rescuable by reinsertion of the WT gene. To this end, a rescue plasmid needs to be constructed by cloning of the WT gene into a transfer plasmid that carries an FRT site and a conditional origin of replication. The conditional nature of its maintenance in *E. coli* allows the insertion of this rescue plasmid into the CMV BAC without disturbing the copy control of the BAC. Transient expression of the Flp recombinase can induce intermolecular recombination between the two FRT sites, one on the rescue plasmid and the other on the target BAC, resulting in the unification of the two circular DNA molecules in the transformed bacteria (Bubeck *et al.*,

2004; Borst *et al.*, 2007). The recombinants can be selected by the combined antibiotic resistance of the target BAC and the rescue plasmid. If the re-insertion of the WT allele at the ectopic position results in the reversion of the null phenotype (which means that the deletion of the native locus is *cis*-complemented), the system can be used to study the effect of mutations in the target gene (Fig. I.3.4B). The easiness to construct such BAC recombinants allowed the analysis of mutant alleles of a viral gene that were generated by random mutagenesis, i.e. the WT coding sequence on the rescue plasmid was subjected to *in vitro* transposon mutagenesis (Bubeck *et al.*, 2004; Lötzerich *et al.*, 2006; Maninger *et al.*, 2011). A selected set of this transposon insertion mutants were inserted into the respective deletion BAC and provided sets of viral genomes that were *cis*-complemented by the mutant alleles. These sets of genomes carrying different mutations in the target gene were then tested for viability in virus reconstitution. Mutations in the target gene that affected an essential function did not support rescue, whereas mutations which hit an irrelevant site permitted virus reconstitution. In combination with biochemical and cell biological assays, this procedure allowed mapping of functionally important sites of essential CMV genes at high resolution (Bubeck *et al.*, 2004; Lötzerich *et al.*, 2006; Maninger *et al.*, 2011). Most of these studies were done in MCMV, but the applicability of the ectopic rescue based system has also been confirmed for HCMV (Adler *et al.*, 2006). Not least, the ectopic insertion of functional foreign genes is also an easy and straightforward approach to construct recombinant CMVs (Čičin-Šain *et al.*, 2008).

The ectopic insertion of mutant alleles or gene variants, however, is not restricted to rescue studies only. A number of studies were published based on the ectopic insertion of a second copy of the gene of interest (GOI) into the WT CMV genome. In these constructs, the WT gene under study was left intact at its native position and the ectopically inserted second copy was modified. First, this approach was used to show that some N-terminal GFP fusions of the small capsid protein (SCP) of MCMV are dominant-negative (DN) (Borst *et al.*, 2001). Testing the inhibitory potential of a second mutant copy of the GOI is still the major application of ectopic insertion into WT-like genomes (Fig. I.3.4C). Large mutant sets of the M50, M53, and M94 MCMV genes were tested for their inhibitory potential after insertion into viable genomes, and a number of DN alleles of them were identified by this type of screens (Rupp *et al.*, 2007; Popa *et al.*, 2010; Maninger *et al.*, 2011).

Non-viable alleles of essential genes can be very useful not only if they are inhibitors but also if they carry beneficial features. A special sort of mutants of this kind are constructs carrying genetic tags. Fusion of any protein to another polypeptide may affect its function. Still, the advantageous feature of this somewhat deficient fusion protein may be utilized in the presence of the WT copy if the latter is capable of complementing the functional defect of the fusion protein. For example, special fusion proteins between MCMV SCP and fluorescent proteins could recently be generated that are by themselves non-functional but also non-inhibitory. The ectopic insertion into WT BAC coding for WT SCP resulted in viruses producing fluorescent capsids. This allowed the study of dynamic events in CMV entry and capsid morphogenesis for the first time (Bosse *et al.*, 2012).

Conditional expression of essential genes and mutants regulated by the tetracycline-dependent repressor system

A further alternative to deletion mutagenesis, one that avoids *trans*-complementation, is the conditional expression of a GOI. The aim is here the construction of a recombinant CMV, in which the GOI can be switched on or off by regulating its expression. The most extensively utilized conditional gene expression system is based on the various elements of the bacterial *tet* operon, which were tailored to operate in different genetic systems including mammalian cells (Gossen *et al.*, 1995). To regulate gene expression in MCMV, the Tet repressor (TetR)-based mammalian conditional expression system was adapted to the viral context (Rupp *et al.*, 2005). The default state of this regulatory cassette is the off state (Fig. I.3.5A): The TetR proteins are bound to the *tet* operator (tetO) sequences positioned within the promoter element that regulates the GOI, thereby interfering with the assembly of the transcription initiation complex. TetR binding is controlled by tetracycline or its analogue, doxycycline, which causes a drastic reduction in the affinity of TetR to tetO. Loss of TetR binding to tetO releases the transcriptional block on the promoter (Fig. I.3.5A). The regulation cassettes designed for use in MCMV carried all necessary elements of the *tet*-system, including the tetO-containing conditional transcription unit and a constitutive expression cassette for TetR (Fig. I.3.5B). The entire regulation cassette was introduced into the MCMV BAC by Flp/FRT-mediated ectopic insertion. This design allowed the construction of conditional viruses, which operated independently of special cell lines and required only the administration of doxycycline in order to induce gene expression. The regulation cassette was applied in MCMV for regulating the expression of reporter genes and essential viral genes (Rupp *et al.*, 2005).

Most *tet*-regulated conditional CMVs were, however, constructed to study the effect of DN mutants of viral genes. DN mutants are loss-of-function mutants that inhibit the development of the WT phenotype even in the presence of the WT allele. This is why DN mutants provide a shortcut to the analysis of the viral phenotype in the absence of a specific gene function. The DN alleles, which were identified by the above discussed inhibitory screens, were inserted into the MCMV BAC via the *tet*-based regulation cassette, and recombinant viruses were reconstituted in the absence of doxycycline, i.e. when the expression of the ORF in the cassette was off. The expression of the DN alleles could be induced by addition of doxycycline. Many different DN variants derived from MCMV genes were

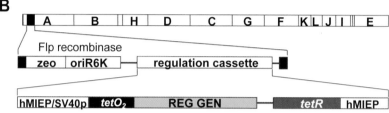

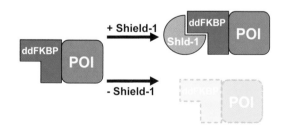

Figure I.3.5 Principle of the conditional *tet*-regulated viral transcription unit. (A) The cassette contains both the regulated transcription unit (REG GEN, blue) and the regulator gene *tet*R (red). Two *tet* operator sequences (*tetO₂*) inserted into a promoter (pro) downstream of the TATA box are bound by Tet repressor (TetR) dimers, leading to a failure in the binding of the transcription initiation complex (green) and thus in transcription initiation (Off). After binding of doxycycline (dox), TetR undergoes a conformational change and loses the affinity to *tetO*. The GOI is transcribed depending on the characteristics of the individual promoter. (B) Displayed is the MCMV genome (*Hind*III map) containing an FRT site (black box). The insertion vector contains the regulation cassette flanked by the origin of replication (oriR6K), zeocin resistance gene (zeo), and FRT sites. Modifications of the regulation cassette can be generated in the context of the insertion vector and can then be introduced into the BAC by Flp-based recombination.

analysed by this system, which apparently was tight enough to prevent the unwanted expression of a DN allele when the recombinant viruses were reconstituted from the respective BAC and propagated for further studies. By the induction of DN alleles of SCP, M50, M53 and M94, MCMV growth could be regulated in a range of 1000- to million-fold (Rupp *et al.*, 2007; Popa *et al.*, 2010; Maninger *et al.*, 2011). The same *tet*-system was also shown to be operational in HCMV for analysing the effect of a DN UL50 allele (Rupp *et al.*, 2007).

Conditional CMV mutants depending on the destabilization/stabilization of essential viral proteins

A further alternative to study essential viral gene functions is by induction of instability followed by degradation of a POI. A novel strategy to control protein amounts in living cells was described by Banaszynski *et al.* (2006). It is based on mutant forms of the cellular FKBP12 protein that are unstable and confer this instability to other proteins upon fusion to these destabilizing FKBP domains (ddFKBP; Fig. I.3.6),

with destabilization being dependent on proteasome function. The addition of a small synthetic ligand called Shield-1 stabilizes the fusion proteins, and protein levels can be tuned by using different concentrations of the ligand. Importantly, the system is reversible, i.e. protein function can be restored by re-adding Shield-1 to cells kept in the absence of the ligand. Since Shield-1

Figure I.3.6 The ddFKBP/Shield-1 principle. A destabilizing version of the FKBP12 protein (ddFKBP) is fused to the POI. In the presence of the small synthetic ligand Shield-1, the fusion protein is stabilized and functional, whereas the lack of Shield-1 results in degradation of the fusion protein.

binds basically to FKBP12 only and has a more than 1000-fold higher affinity for ddFKBP than for the endogenous cellular WT FKBP12 protein, side effects on cellular metabolism are minimal (Maynard-Smith et al., 2007). The ddFKBP/Shield-1 system has been successfully applied to regulate various cellular proteins (Banaszynski et al., 2006).

We analysed the applicability of this system to generate conditionally replicating CMV mutants, whose growth is dependent on the Shield-1 ligand (Glass et al., 2009). The ddFKBP moiety was fused to the major IE proteins of MCMV and HCMV, respectively, as well as the two hitherto uncharacterized essential HCMV proteins pUL51 and pUL77. ddFKBP was added to the N-termini of the respective viral proteins, because levels of N-terminal fusions were described to become better adjustable than fusions to the C-terminus (Banaszynski et al., 2006). For the UL51 and UL77 mutants the respective ORF at the original genomic positions had to be disrupted, and the genes encoding the ddFKBP variants were inserted at an ectopic genomic position, because the authentic ORF overlap with neighbouring essential genes, thereby preventing their specific modification at the native position. Upon transfection of permissive fibroblasts, efficient plaque formation was observed in the presence of Shield-1 with all of the genomes except of the ddFKBP-UL77 mutant, thus suggesting that manipulating the N-terminus of pUL77 is not compatible with viral growth. Yet, upon fusing ddFKBP to the C-terminus of pUL77, a conditional mutant was obtained. The successful reconstitution of virus from the CMV BAC expressing ddFKBP fusion proteins proved paradigmatically that the conditional approach of protein destabilization can be extended to viruses.

Further evaluation of the conditional system revealed that the HCMV mutants grew to titres that were comparable to those of the parental virus when the recommended concentration of Shield-1 (1 μM) was added, although with slightly delayed kinetics in case of the UL51 and UL77 mutants (Glass et al., 2009). For efficient complementation of the MCMV IE mutant higher Shield-1 concentrations were required (A. Busche, unpublished), indicating that the optimal Shield-1 concentration has to be determined empirically for each mutant. Most mutants did not produce virus progeny in the absence of the ligand, underlining the excellent tightness of the system. The essential IE fusion proteins as well as the UL51 and UL77 fusion proteins were destabilized efficiently when Shield-1 was lacking, and their levels were adjustable by varying the concentration of the ligand. This allowed controlling the size of the plaques formed by the mutants using different Shield-1 amounts. Finally, we could prove the

reversibility of the system through re-adding Shield-1 to cells that were initially kept without the ligand and demonstrating restoration of protein expression over time. In summary, the ddFKBP/Shield-1 system proved to be a promising new strategy to investigate essential CMV gene functions, circumventing the need to construct complementing cell lines.

In contrast to other conditional approaches, this one is based on post-translational regulation. Potentially, it can be combined with other principles such as the tet-system to further improve its utility (Almogy and Nolan, 2009). A major advantage is the specificity of the method, because solely the ddFKBP-tagged protein will be degraded, so that off-target effects can largely be excluded. In the meanwhile, the ddFKBP/Shield-1 system has successfully been applied to regulate other CMV proteins (Qian et al., 2010; Perng et al., 2011). However, the approach may not be applicable to all viral proteins; specifically, viral proteins with per se high or low stability are a concern, and there is at least one report of a study in which the technique failed (Fehr and Yu, 2011). Our results suggest that a drastic decline in protein levels can be sufficient to block the CMV infection cycle, and that a complete KO may not be required. Notably, in the CMV mutants analysed by us we did not observe any reversion to the WT sequence of the P106 residue in ddFKBP that is crucial for protein destabilization. Although the destabilizing domain is rather small (107 aa), caution is needed in terms of possible impairment of protein function upon fusion to ddFKBP. For instance, it was not compatible with virus growth when fused to the N-terminus of UL77.

An intriguing aspect is the applicability of the approach to controlling virus replication in vivo. The pharmacokinetic properties of the ligand have been optimized, at least for application in cell cultures (Banaszynski et al., 2006), and it was shown that the method can be exploited to regulate protein function in living animals, including proteins delivered by viral vectors (Banaszynski et al., 2008). Pilot experiments in our laboratory confirmed that the approach can be used to control growth of MCMV mutants in mice (A. Busche and M. Messerle, unpublished results), although there remains a need to study the bioavailability of the ligand in vivo, and to optimize the mode of delivery.

It is noteworthy that the ddFKBP/Shield-1 stabilizing system can be reversed. Bonger et al. (2011) developed a cryptic degradation domain (degron) that can be fused to POI and induces degradation of the fusion protein only upon addition of the Shield-1 ligand. In addition, DN variants of viral POI (such as those described above) can be fused to ddFKBP. In the absence of Shield-1 the corresponding DN protein will be degraded, thus allowing propagation of the conditional CMV mutant, whereas upon addition of the ligand the DN protein will be stabilized and will disrupt the CMV replication cycle. Combining different conditional approaches will increase our future understanding of CMV gene functions and the interactions between essential CMV proteins; knowledge that is for instance pivotal for the rational design of novel antiviral substances.

Conditional CMV mutants operating *in vivo*

The Cre/*loxP*-system was the first established, site-specific recombination system for reverse genetic studies of gene functions (Lewandoski, 2001). The Cre enzyme, a product of the *cre* gene of bacteriophage P1, recognizes a 34-bp sequence called *loxP* and catalyses DNA recombination between pairs of *loxP* sites, resulting in the excision of the intervening DNA sequences. The Cre/*loxP* system is extensively used for construction of conditional alleles in mammalian genetics. To silence the GOI, an essential region of the gene is flanked by *loxP* sites (*floxed*). Mating mice that carry such an allele with mice expressing Cre under the control of cell type specific promoters will result in progeny in which the *floxed* allele is deleted in selected tissues. In contrast, activation of the GOI can be achieved if *floxed* transcriptional stop sequences are placed between the promoter and the coding sequence for preventing transcription of the coding region. Cell type specific Cre expression then leads to the excision of the inhibitory sequences, and this will activate the transcription of the target gene.

Using BAC technology, we generated an MCMV reporter virus conditionally expressing EGFP under the control of the HCMV MIE promoter (Sacher *et al.*, 2008). A *floxed* stop cassette was inserted between the promoter and the EGFP ORF for preventing the expression of the reporter (Fig. I.3.7A). When this reporter virus infected normal cells, the reporter gene remained silenced. However, if a cell type that expressed Cre was infected, the stop cassette was excised by the recombinase and the reporter gene was activated. Since the recombined locus is stably maintained in the viral progeny, it was possible to monitor virus transmission to cells that did not express Cre. Thus, virus dissemination between different cell types and the colonization of distant organs can be studied by keeping track of recombined EGFP-expressing virus.

The amount of infectious progeny produced by a specific cell type can be quantified in a standard plaque assay (Fig. I.3.7B). While the total number of plaques allows the calculation of the virus load of an organ, the amount of EGFP$^+$ plaques gives an estimate of the contribution of the Cre-expressing cells to the viral load. Depending on the aim of the study, it is important to choose the proper time point after infection for this assay. The proportion of EGFP$^+$ plaques reveals the productivity of the Cre-expressing cell type only if the assay is carried out immediately after the first viral replication cycle. At any later time points, viruses with a recombined reporter locus allow the tracking of spread from the Cre-expressing cell type to other cell types or organs (Fig. I.3.7C). This method allowed us to figure out that MCMV does not spread from parenchymal organs to other sites of the body via secondary viraemia, and pointed to the importance of organ specific aspects of the virus produced in EC (Sacher *et al.*, 2008). We could also shed light on tissue and cell type specific aspects of immune control in the MCMV model (Sacher *et al.*, 2012). Furthermore, we set up new models based on the Cre/*loxP* system to study MCMV dissemination in transplant settings (Sacher *et al.*, 2011). A promising new application of this conditional gene expression system is to regulate expression of an essential gene with Cre recombination. This allowed us to turn on the essential M94 gene selectively in Cre-expressing cells of transgenic mice, thereby studying MCMV dissemination from hepatocytes *in vivo* (Sacher *et al.*, 2012; T. Sacher, unpublished).

Perspectives of BAC mutagenesis

Looking back 10 years, when the first CMV BAC were generated, it was amazing to witness a flourishing in CMV genetics and research. Generation of CMV mutants is no longer a limiting factor, and we can now introduce any kind of mutation into CMV, ranging from targeted point mutations to random insertions, and even essential viral genes are now accessible to genetic analysis as pointed out in this chapter. Thus, we can now focus our studies on important questions of CMV pathogenesis. There are, however, still possibilities or even needs for improvement of the mutagenesis techniques, and we would not be surprised if we see further refinements of these methods in the near future.

Generation of a CMV BAC is still arduous and not at all a routine technique. This may have hampered the BAC cloning of additional CMV strains with unique properties. Furthermore, the need for insertion of the BAC vector by homologous recombination in tissue culture bears the risk that adaptive mutations accumulate in the CMV genome during the required propagation time. Ideally, we would like to isolate CMV genomes directly *ex vivo* from tissues and recombine them with the vector sequences either *in vitro* or in recombination-proficient *E. coli*. Cloning by recombination in *E. coli* has been shown for the somewhat

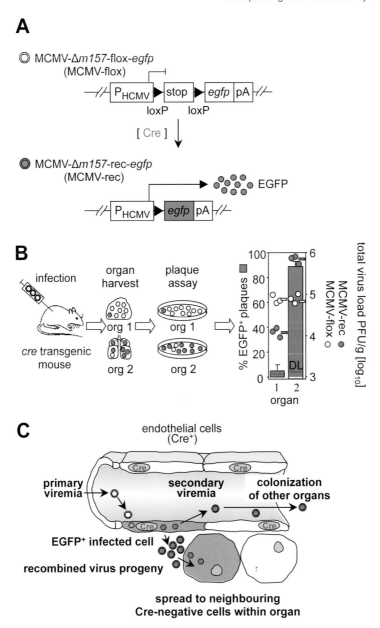

Figure I.3.7 Tracking of cell type specific progeny by conditional virus recombination. (A) A Cre-inducible expression cassette was inserted into the MCMV *m157* gene, resulting in MCMV-Δ*m157*-flox-*egfp* (MCMV-flox). Cre-mediated removal of a transcriptional stop sequence flanked by *loxP* sites (*floxed*) led to expression of EGFP under control of the HCMV MIE promoter (MCMV-rec). (B) Schematic setup of experiments and analysis: Transgenic mice expressing Cre recombinase under control of a cell type specific promoter are infected with the MCMV-flox reporter virus. At different time points p.i., the amounts (PFU) of MCMV-rec and MCMV-flox in various organs (org1, org2) were determined by plaque assay. The graph depicts virus loads as absolute amounts of MCMV-rec (green circles) and MCMV-flox (open circles) per gram organ for individual mice (logarithmic scale on the right-hand side). Horizontal bars mark mean values and dotted lines indicate detection limit. Green columns refer to the linear scale on the left and show the mean percentage of EGFP⁺ plaques (±SD). (C) Cell type specific Cre-mediated recombination of viral genomes in *cre* transgenic mice allows identification and quantification of labelled infected cells and recombined virus progeny produced by defined cell types (e.g. EC). Spread of labelled virus to Cre-negative neighbouring cells and dissemination to other organs can be analysed. Reprinted from *Cell Host & Microbe*, Vol. 3, Sacher, T., Podlech, J., Mohr, C.A., Jordan, S., Ruzsics, Z., Reddehase, M.J., and Koszinowski, U.H., The major virus-producing cell type during murine cytomegalovirus infection, the hepatocyte, is not the source of virus dissemination in the host, pp. 263–272, Copyright (2008), with permission from Elsevier.

smaller genomes of adenoviruses (Ruzsics *et al.*, 2006), and should in principle also be possible for CMVs. However, it might be that the structure of the CMV genomes (terminal and internal repeats) and their large size are not compatible with such a strategy. Transposon-mediated insertion of the BAC vector into CMV genomes, as described for γ-herpesviruses (Zhou *et al.*, 2009), circumvents the recombination step usually needed for joining the vector and the CMV genome. Yet, it still requires propagation of the recombinant virus in cell culture and potentially also the repair of the BAC-cloned genome in *E. coli*.

As expected, the mutagenesis techniques underwent refinement during recent years and we saw a shift from RecA-mediated recombination towards the use of the bacteriophage λ encoded Red recombination functions. Furthermore, suitable *E. coli* strains were constructed that facilitate the mutagenesis process (Yu *et al.*, 2000; Warming *et al.*, 2005; Tischer *et al.*, 2010) and, due to inducible and transient expression of the recombination functions, guarantee improved stability of the BAC-cloned CMV genomes. During propagation of the BAC in *E. coli* the recombination functions are shut off, hence minimizing the risk of acquiring unwanted mutations. However, currently most of the BAC are kept in the *E. coli* strain DH10B (or derivatives thereof), which was reported to display an approximately 10-fold higher mutation rate than an *E. coli* K-12 wild-type strain, mainly due to transposition of insertion sequence elements (Durfee *et al.*, 2008). Thus, there might be a need to develop further optimized bacterial strains for BAC mutagenesis. Instability of the cloned genomes was and remains a concern, although the risk seems to be low. We recommend to commence mutagenesis with the original CMV BAC clones whenever possible and to avoid unnecessarily extended propagation of the BAC in *E. coli*. Good laboratory practice is mandatory, including the thorough characterization of mutated genomes by restriction analysis. Complete sequencing of the genomes of newly generated mutants is now possible within days and costs are decreasing. Consequently, characterization of mutants by sequencing may become a routine procedure soon.

Reverse genetics will remain important in constructing selected mutants to answer specific questions about individual genes. It will also allow us to define the biological relevance of a CMV gene by performing *in vivo* experiments with CMV mutants in a relevant animal model. Elegant studies by the Koszinowski group, using conditional mutants as described herein, taught us a lot about dissemination of CMV in its host as well as the organ specific effects of the immune control and of viral immune evasion. The combination of mouse and virus genetics appears to be a promising approach to shed more light on the sophisticated interaction of viral and host functions (see also Chapter II.12).

Forward genetics is pivotal in linking individual viral genes with specific characteristics and functions of CMV. This requires however, the set-up of manageable read-out systems in order to perform efficient screening of large virus libraries. The limited availability of suitable cell culture systems for HCMV as well as its rapid adaption to growth in fibroblast cultures remain a concern, but also a challenge in defining the various functions of this virus. BAC-based mutagenesis techniques will also be instrumental in developing a CMV vaccine. From the technical point of view the door is now open, and recent process in this direction (Hansen *et al.*, 2010; Schleiss, 2010) represent the first steps to tackle this goal.

Acknowledgements
We apologize to all colleagues whose publications have not been included owing to space constraints. Our work has been supported by grants from the Deutsche Forschungsgemeinschaft.

References
Adler, B., Scrivano, L., Ruzcics, Z., Rupp, B., Sinzger, C., and Koszinowski, U. (2006). Role of human cytomegalovirus UL131A in cell type-specific virus entry and release. J. Gen. Virol. *87*, 2451–2460.

Almogy, G., and Nolan, G.P. (2009). Conditional protein stabilization via the small molecules Shld-1 and rapamycin increases the signal-to-noise ratio with tet-inducible gene expression. Biotechniques *46*, 44–50.

Angulo, A., Ghazal, P., and Messerle, M. (2000). The major immediate-early gene ie3 of mouse cytomegalovirus is essential for viral growth. J. Virol. *74*, 11129–11136.

Banaszynski, L.A., Chen, L.C., Maynard-Smith, L.A., Ooi, A.G., and Wandless, T.J. (2006). A rapid, reversible, and tunable method to regulate protein function in living cells using synthetic small molecules. Cell *126*, 995–1004.

Banaszynski, L.A., Sellmyer, M.A., Contag, C.H., Wandless, T.J., and Thorne, S.H. (2008). Chemical control of protein stability and function in living mice. Nat. Med. *14*, 1123–1127.

Barry, P.A., and Chang, W.L. (2007). Primate betaherpesviruses. In Human Herpesviruses: Biology, Therapy, and Immunoprophylaxis, Arvin, A.M., Campadelli-Fiume, G., Mocarski, E.S., Moore, P.S., Roizman, B., Whitley, R., and Yamanishi, K., eds. (Cambridge University Press, Cambridge, UK), pp. 1051–1075.

Bonger, K.M., Chen, L.C., Liu, C.W., and Wandless, T.J. (2011). Small-molecule displacement of a cryptic degron causes conditional protein degradation. Nat. Chem. Biol. *7*, 531–537.

Borenstein, R., and Frenkel, N. (2009). Cloning human herpes virus 6A genome into bacterial artificial chromosomes and study of DNA replication intermediates. Proc. Natl. Acad. Sci. U.S.A. *106*, 19138–19143.

Borst, E.M., and Messerle, M. (2005). Analysis of human cytomegalovirus oriLyt sequence requirements in the context of the viral genome. J. Virol. *79*, 3615–3626.

Borst, E.M., Hahn, G., Koszinowski, U.H., and Messerle, M. (1999). Cloning of the human cytomegalovirus (HCMV) genome as an infectious bacterial artificial chromosome in *Escherichia coli*: a new approach for construction of HCMV mutants. J. Virol. *73*, 8320–8329.

Borst, E.M., Mathys, S., Wagner, M., Muranyi, W., and Messerle, M. (2001). Genetic evidence of an essential role for cytomegalovirus small capsid protein in viral growth. J. Virol. *75*, 1450–1458.

Borst, E.M., Benkartek, C., and Messerle, M. (2007). Use of bacterial artificial chromosomes in generating targeted mutations in human and mouse cytomegaloviruses. In Current Protocols in Immunology, Coligan, J.E., Bierer, B., Margulies, D.H., Shevach, E.M., Strober, W., and Coico, R., eds. (John Wiley & Sons, New York, NY), pp. 10.32.1–10.32.30.

Borst, E.M., Wagner, K., Binz, A., Sodeik, B., and Messerle, M. (2008). The essential human cytomegalovirus gene UL52 is required for cleavage-packaging of the viral genome. J. Virol. *82*, 2065–2078.

Bosse, J.B., Bauerfeind, R., Popilka, L., Marcinowski, L., Taeglich, M., Jung, C., Striebinger, H., von Einem, J., Gaul, U., Walther, P., *et al.* (2012). A beta-herpesvirus with fluorescent capsids to study transport in living cells. PLoS One *7*, e40585.

Brune, W. (2002). Random transposon mutagenesis of large DNA molecules in *Escherichia coli*. Methods Mol. Biol. *182*, 165–171.

Brune, W., Ménard, C., Hobom, U., Odenbreit, S., Messerle, M., and Koszinowski, U.H. (1999). Rapid identification of essential and nonessential herpesvirus genes by direct transposon mutagenesis. Nat. Biotechnol. *17*, 360–364.

Brune, W., Ménard, C., Heesemann, J., and Koszinowski, U.H. (2001). A ribonucleotide reductase homolog of cytomegalovirus and endothelial cell tropism. Science *291*, 303–305.

Brune, W., Nevels, M., and Shenk, T. (2003). Murine cytomegalovirus m41 open reading frame encodes a Golgi-localized antiapoptotic protein. J. Virol. *77*, 11633–11643.

Bubeck, A., Wagner, M., Ruzsics, Z., Lötzerich, M., Iglesias, M., Singh, I.R., and Koszinowski, U.H. (2004). Comprehensive mutational analysis of a herpesvirus gene in the viral genome context reveals a region essential for virus replication. J. Virol. *78*, 8026–8035.

Chang, W.L., and Barry, P.A. (2003). Cloning of the full-length rhesus cytomegalovirus genome as an infectious and self-excisable bacterial artificial chromosome for analysis of viral pathogenesis. J. Virol. *77*, 5073–5083.

Čičin-Šain, L., Bubić, I., Schnee, M., Ruzsics, Z., Mohr, C., Jonjić, S., and Koszinowski, U.H. (2007). Targeted deletion of regions rich in immune-evasive genes from the cytomegalovirus genome as a novel vaccine strategy. J. Virol. *81*, 13825–13834.

Čičin-Šain, L., Ruzsics, Z., Podlech, J., Bubić, I., Ménard, C., Jonjić, S., Reddehase, M.J., and Koszinowski, U.H. (2008). Dominant-negative FADD rescues the in vivo fitness of a cytomegalovirus lacking an antiapoptotic viral gene. J. Virol. *82*, 2056–2064.

Cui, X., McGregor, A., Schleiss, M.R., and McVoy, M.A. (2009). The impact of genome length on replication and genome stability of the herpesvirus guinea pig cytomegalovirus. Virology *386*, 132–138.

Dargan, D.J., Douglas, E., Cunningham, C., Jamieson, F., Stanton, R.J., Baluchova, K., McSharry, B.P., Tomasec, P., Emery, V.C., Percivalle, E., *et al.* (2010). Sequential mutations associated with adaptation of human cytomegalovirus to growth in cell culture. J. Gen. Virol. *91*, 1535–1546.

Datsenko, K.A., and Wanner, B.L. (2000). One-step inactivation of chromosomal genes in *Escherichia coli* K-12 using PCR products. Proc. Natl. Acad. Sci. U.S.A. *97*, 6640–6645.

Dolan, A., Cunningham, C., Hector, R.D., Hassan-Walker, A.F., Lee, L., Addison, C., Dargan, D.J., McGeoch, D.J., Gatherer, D., Emery, V.C., *et al.* (2004). Genetic content of wild-type human cytomegalovirus. J. Gen. Virol. *85*, 1301–1312.

Dölken, L., Krmpotić, A., Kothe, S., Tuddenham, L., Tanguy, M., Marcinowski, L., Ruzsics, Z., Elefant, N., Altuvia, Y., Margalit, H., *et al.* (2010). Cytomegalovirus microRNAs facilitate persistent virus infection in salivary glands. PLoS Pathog. *6*, e1001150.

Dunn, W., Chou, C., Li, H., Hai, R., Patterson, D., Stolc, V., Zhu, H., and Liu, F. (2003). Functional profiling of a human cytomegalovirus genome. Proc. Natl. Acad. Sci. U.S.A. *100*, 14223–14228.

Durfee, T., Nelson, R., Baldwin, S., Plunkett, G., III, Burland, V., Mau, B., Petrosino, J.F., Qin, X., Muzny, D.M., Ayele, M., *et al.* (2008). The complete genome sequence of *Escherichia coli* DH10B: insights into the biology of a laboratory workhorse. J. Bacteriol. *190*, 2597–2606.

Fehr, A.R., and Yu, D. (2011). Human cytomegalovirus early protein pUL21a promotes efficient viral DNA synthesis and the late accumulation of immediate-early transcripts. J. Virol. *85*, 663–674.

Gatherer, D., Seirafian, S., Cunningham, C., Holton, M., Dargan, D.J., Baluchova, K., Hector, R.D., Galbraith, J., Herzyk, P., Wilkinson, G.W., *et al.* (2011). High-resolution human cytomegalovirus transcriptome. Proc. Natl. Acad. Sci. U.S.A. *108*, 19755–19760.

Glass, M., Busche, A., Wagner, K., Messerle, M., and Borst, E.M. (2009). Conditional and reversible disruption of essential herpesvirus proteins. Nat. Methods *6*, 577–579.

Gossen, M., Freundlieb, S., Bender, G., Müller, G., Hillen, W., and Bujard, H. (1995). Transcriptional activation by tetracyclines in mammalian cells. Science *268*, 1766–1769.

Gutermann, A., Bubeck, A., Wagner, M., Reusch, U., Ménard, C., and Koszinowski, U.H. (2002). Strategies for the identification and analysis of viral immune-evasive genes – cytomegalovirus as an example. Curr. Top. Microbiol. Immunol. *269*, 1–22.

Hahn, G., Khan, H., Baldanti, F., Koszinowski, U.H., Revello, M.G., and Gerna, G. (2002). The human cytomegalovirus ribonucleotide reductase homolog UL45 is dispensable for growth in endothelial cells, as determined by a BAC-cloned clinical isolate of human cytomegalovirus with preserved wild-type characteristics. J. Virol. *76*, 9551–9555.

Hahn, G., Jarosch, M., Wang, J.B., Berbes, C., and McVoy, M.A. (2003a). Tn7-mediated introduction of DNA sequences into bacmid-cloned cytomegalovirus genomes for rapid recombinant virus construction. J. Virol. Methods *107*, 185–194.

Hahn, G., Rose, D., Wagner, M., Rhiel, S., and McVoy, M.A. (2003b). Cloning of the genomes of human cytomegalovirus strains Toledo, TownevarRIT3, and Towne long as BACs and site-directed mutagenesis using a PCR-based technique. Virology *307*, 164–177.

Hansen, S.G., Powers, C.J., Richards, R., Ventura, A.B., Ford, J.C., Siess, D., Axthelm, M.K., Nelson, J.A., Jarvis, M.A., Picker, L.J., *et al.* (2010). Evasion of CD8+ T-cells is critical for superinfection by cytomegalovirus. Science *328*, 102–106.

Hasan, M., Krmpotić, A., Ruzsics, Z., Bubić, I., Lenac, T., Halenius, A., Loewendorf, A., Messerle, M., Hengel, H., Jonjić, S., *et al.* (2005). Selective down-regulation of the NKG2D ligand H60 by mouse cytomegalovirus m155 glycoprotein. J. Virol. *79*, 2920–2930.

Hayes, F. (2003). Transposon-based strategies for microbial functional genomics and proteomics. Annu. Rev. Genet. *37*, 3–29.

Hobom, U., Brune, W., Messerle, M., Hahn, G., and Koszinowski, U.H. (2000). Fast screening procedures for random transposon libraries of cloned herpesvirus genomes: mutational analysis of human cytomegalovirus envelope glycoprotein genes. J. Virol. *74*, 7720–7729.

Holtappels, R., Podlech, J., Pahl-Seibert, M.F., Jülch, M., Thomas, D., Simon, C.O., Wagner, M., and Reddehase, M.J. (2004). Cytomegalovirus misleads its host by priming of CD8 T-cells specific for an epitope not presented in infected tissues. J. Exp. Med. *199*, 131–136.

Isaacson, M.K., and Compton, T. (2009). Human cytomegalovirus glycoprotein B is required for virus entry and cell-to-cell spread but not for virion attachment, assembly, or egress. J. Virol. *83*, 3891–3903.

Jiang, X.J., Adler, B., Sampaio, K.L., Digel, M., Jahn, G., Ettischer, N., Stierhof, Y.D., Scrivano, L., Koszinowski, U., Mach, M., *et al.* (2008). UL74 of human cytomegalovirus contributes to virus release by promoting secondary envelopment of virions. J. Virol. *82*, 2802–2812.

Jordan, S., Krause, J., Prager, A., Mitrović, M., Jonjić, S., Koszinowski, U.H., and Adler, B. (2011). Virus progeny of murine cytomegalovirus bacterial artificial chromosome pSM3fr show reduced growth in salivary glands due to a fixed mutation of MCK-2. J. Virol. *85*, 10346–10353.

Krmpotić, A., Hasan, M., Loewendorf, A., Saulig, T., Halenius, A., Lenac, T., Polić, B., Bubić, I., Kriegeskorte, A., Pernjak-Pugel, E., *et al.* (2005). NK cell activation through the NKG2D ligand MULT-1 is selectively prevented by the glycoprotein encoded by mouse cytomegalovirus gene m145. J. Exp. Med. *201*, 211–220.

Kuhlman, T.E., and Cox, E.C. (2010). Site-specific chromosomal integration of large synthetic constructs. Nucleic Acids Res. *38*, e92.

Lacaze, P., Forster, T., Ross, A., Kerr, L.E., Salvo-Chirnside, E., Lisnic, V.J., López-Campos, G.H., García-Ramirez, J.J., Messerle, M., Trgovcich, J., *et al.* (2011). Temporal profiling of the coding and noncoding murine cytomegalovirus transcriptomes. J. Virol. *85*, 6065–6076.

Lembo, D., Donalisio, M., Hofer, A., Cornaglia, M., Brune, W., Koszinowski, U., Thelander, L., and Landolfo, S. (2004). The ribonucleotide reductase R1 homolog of murine cytomegalovirus is not a functional enzyme subunit but is required for pathogenesis. J. Virol. *78*, 4278–4288.

Lemmermann, N.A., Gergely, K., Böhm, V., Deegen, P., Däubner, T., and Reddehase, M.J. (2010). Immune evasion proteins of murine cytomegalovirus preferentially affect cell surface display of recently generated peptide presentation complexes. J. Virol. *84*, 1221–1236.

Lemmermann, N.A., Kropp, K.A., Seckert, C.K., Grzimek, N.K., and Reddehase, M.J. (2011). Reverse genetics modification of cytomegalovirus antigenicity and immunogenicity by CD8 T-cell epitope deletion and insertion. J. Biomed. Biotechnol. *2011*, e812742.

Lenac, T., Budt, M., Arapović, J., Hasan, M., Zimmermann, A., Simić, H., Krmpotić, A., Messerle, M., Ruzsics, Z., Koszinowski, U.H., *et al.* (2006). The herpesviral Fc receptor fcr-1 down-regulates the NKG2D ligands MULT-1 and H60. J. Exp. Med. *203*, 1843–1850.

Lewandoski, M. (2001). Conditional control of gene expression in the mouse. Nat. Rev. Genet. *2*, 743–755.

Lilja, A.E., and Shenk, T. (2008). Efficient replication of rhesus cytomegalovirus variants in multiple rhesus and human cell types. Proc. Natl. Acad. Sci. U.S.A. *105*, 19950–19955.

Loewendorf, A., Krüger, C., Borst, E.M., Wagner, M., Just, U., and Messerle, M. (2004). Identification of a mouse cytomegalovirus gene selectively targeting CD86 expression on antigen-presenting cells. J. Virol. *78*, 13062–13071.

Lötzerich, M., Ruzsics, Z., and Koszinowski, U.H. (2006). Functional domains of murine cytomegalovirus nuclear egress protein M53/p38. J. Virol. *80*, 73–84.

McGregor, A., Liu, F., and Schleiss, M.R. (2004). Identification of essential and non-essential genes of the guinea pig cytomegalovirus (GPCMV) genome via transposome mutagenesis of an infectious BAC clone. Virus Res. *101*, 101–108.

McGregor, A., and Schleiss, M.R. (2001). Molecular cloning of the guinea pig cytomegalovirus (GPCMV) genome as an infectious bacterial artificial chromosome (BAC) in *Escherichia coli*. Mol. Genet. Metab. *72*, 15–26.

Mack, C., Sickmann, A., Lembo, D., and Brune, W. (2008). Inhibition of proinflammatory and innate immune signaling pathways by a cytomegalovirus RIP1-interacting protein. Proc. Natl. Acad. Sci. U.S.A. *105*, 3094–3099.

Maninger, S., Bosse, J.B., Lemnitzer, F., Pogoda, M., Mohr, C.A., von Einem, J., Walther, P., Koszinowski, U.H., and Ruzsics, Z. (2011). M94 is essential for the secondary envelopment of murine cytomegalovirus. J. Virol. *85*, 9254–9267.

Marchini, A., Liu, H., and Zhu, H. (2001). Human cytomegalovirus with IE-2 (UL122) deleted fails to express early lytic genes. J. Virol. *75*, 1870–1878.

Maresca, M., Erler, A., Fu, J., Friedrich, A., Zhang, Y., and Stewart, A.F. (2010). Single-stranded heteroduplex intermediates in lambda Red homologous recombination. BMC Mol. Biol. *11*, 54.

Marquardt, A., Halle, S., Seckert, C.K., Lemmermann, N.A., Veres, T.Z., Braun, A., Maus, U.A., Förster, R., Reddehase, M.J., Messerle, M., *et al.* (2011). Single cell detection of latent cytomegalovirus reactivation in host tissue. J. Gen. Virol. *92*, 1279–1291.

Maynard-Smith, L.A., Chen, L.C., Banaszynski, L.A., Ooi, A.G., and Wandless, T.J. (2007). A directed approach for engineering conditional protein stability using biologically silent small molecules. J. Biol. Chem. *282*, 24866–24872.

Ménard, C., Wagner, M., Ruzsics, Z., Holak, K., Brune, W., Campbell, A.E., and Koszinowski, U.H. (2003). Role of murine cytomegalovirus US22 gene family members in replication in macrophages. J. Virol. 77, 5557–5570.

Messerle, M., Crnković, I., Hammerschmidt, W., Ziegler, H., and Koszinowski, U.H. (1997). Cloning and mutagenesis of a herpesvirus genome as an infectious bacterial artificial chromosome. Proc. Natl. Acad. Sci. U.S.A. 94, 14759–14763.

Mocarski, E.S., Kemble, G.W., Lyle, J.M., and Greaves, R.F. (1996). A deletion mutant in the human cytomegalovirus gene encoding IE1(491aa) is replication defective due to a failure in autoregulation. Proc. Natl. Acad. Sci. U.S.A. 93, 11321–11326.

Mohr, C.A., Arapović, J., Muhlbach, H., Panzer, M., Weyn, A., Dölken, L., Krmpotić, A., Voehringer, D., Ruzsics, Z., Koszinowski, U., et al. (2010). A spread-deficient cytomegalovirus for assessment of first-target cells in vaccination. J. Virol. 84, 7730–7742.

Mosberg, J.A., Lajoie, M.J., and Church, G.M. (2010). Lambda red recombineering in Escherichia coli occurs through a fully single-stranded intermediate. Genetics 186, 791–799.

Murphy, E., Rigoutsos, I., Shibuya, T., and Shenk, T.E. (2003a). Reevaluation of human cytomegalovirus coding potential. Proc. Natl. Acad. Sci. U.S.A. 100, 13585–13590.

Murphy, E., Yu, D., Grimwood, J., Schmutz, J., Dickson, M., Jarvis, M.A., Hahn, G., Nelson, J.A., Myers, R.M., and Shenk, T.E. (2003b). Coding potential of laboratory and clinical strains of human cytomegalovirus. Proc. Natl. Acad. Sci. U.S.A. 100, 14976–14981.

Perng, Y.C., Qian, Z., Fehr, A.R., Xuan, B., and Yu, D. (2011). The human cytomegalovirus gene UL79 is required for the accumulation of late viral transcripts. J. Virol. 85, 4841–4852.

Pogoda, M., Bosse, J.B., Wagner, F.M., Schauflinger, M., Walther, P., Koszinowski, U.H., and Ruzsics, Z. (2012). Characterization of conserved region two (CR2)-deficient mutants of the cytomegalovirus egress protein pM53. J. Virol. 86, 12512–12524.

Popa, M., Ruzsics, Z., Lötzerich, M., Dölken, L., Buser, C., Walther, P., and Koszinowski, U.H. (2010). Dominant negative mutants of the murine cytomegalovirus M53 gene block nuclear egress and inhibit capsid maturation. J. Virol. 84, 9035–9046.

Qian, Z., Leung-Pineda, V., Xuan, B., Piwnica-Worms, H., and Yu, D. (2010). Human cytomegalovirus protein pUL117 targets the mini-chromosome maintenance complex and suppresses cellular DNA synthesis. PLoS Pathog. 6, e1000814.

Redwood, A.J., Messerle, M., Harvey, N.L., Hardy, C.M., Koszinowski, U.H., Lawson, M.A., and Shellam, G.R. (2005). Use of a murine cytomegalovirus K181-derived bacterial artificial chromosome as a vaccine vector for immunocontraception. J. Virol. 79, 2998–3008.

Rupp, B., Ruzsics, Z., Sacher, T., and Koszinowski, U.H. (2005). Conditional cytomegalovirus replication in vitro and in vivo. J. Virol. 79, 486–494.

Rupp, B., Ruzsics, Z., Buser, C., Adler, B., Walther, P., and Koszinowski, U.H. (2007). Random screening for dominant-negative mutants of the cytomegalovirus nuclear egress protein M50. J. Virol. 81, 5508–5517.

Ruzsics, Z., Wagner, M., Osterlehner, A., Cook, J., Koszinowski, U., and Burgert, H.G. (2006). Transposon-assisted cloning and traceless mutagenesis of adenoviruses: development of a novel vector based on species D. J. Virol. 80, 8100–8113.

Sacher, T., Podlech, J., Mohr, C.A., Jordan, S., Ruzsics, Z., Reddehase, M.J., and Koszinowski, U.H. (2008). The major virus-producing cell type during murine cytomegalovirus infection, the hepatocyte, is not the source of virus dissemination in the host. Cell Host Microbe 3, 263–272.

Sacher, T., Andrassy, J., Kalnins, A., Dölken, L., Jordan, S., Podlech, J., Ruzsics, Z., Jauch, K.W., Reddehase, M.J., and Koszinowski, U.H. (2011). Shedding light on the elusive role of endothelial cells in cytomegalovirus dissemination. PLoS Pathog. 7, e1002366.

Sacher, T., Mohr, C.A., Weyn, A., Schlichting, C., Koszinowski, U.H., and Ruzsics, Z. (2012). The role of cell types in cytomegalovirus infection in vivo. Eur. J. Cell. Biol. 91, 70–77.

Sanders, R.L., Clark, C.L., Morello, C.S., and Spector, D.H. (2008). Development of cell lines that provide tightly controlled temporal translation of the human cytomegalovirus IE2 proteins for complementation and functional analyses of growth-impaired and nonviable IE2 mutant viruses. J. Virol. 82, 7059–7077.

Schleiss, M.R. (2010). Can we build it better? Using BAC genetics to engineer more effective cytomegalovirus vaccines. J. Clin. Invest. 120, 4192–4197.

Schuessler, A., Sampaio, K.L., and Sinzger, C. (2008). Charge cluster-to-alanine scanning of UL128 for fine tuning of the endothelial cell tropism of human cytomegalovirus. J. Virol. 82, 11239–11246.

Schuessler, A., Sampaio, K.L., Scrivano, L., and Sinzger, C. (2010). Mutational mapping of UL130 of human cytomegalovirus defines peptide motifs within the C-terminal third as essential for endothelial cell infection. J. Virol. 84, 9019–9026.

Schuessler, A., Sampaio, K.L., Straschewski, S., and Sinzger, C. (2012). Mutational mapping of pUL131A of human cytomegalovirus emphasizes its central role for endothelial cell tropism. J. Virol. 86, 504–512.

Shizuya, H., Birren, B., Kim, U.J., Mancino, V., Slepak, T., Tachiiri, Y., and Simon, M. (1992). Cloning and stable maintenance of 300-kilobase-pair fragments of human DNA in Escherichia coli using an F-factor-based vector. Proc. Natl. Acad. Sci. U.S.A. 89, 8794–8797.

Sinzger, C., Hahn, G., Digel, M., Katona, R., Sampaio, K.L., Messerle, M., Hengel, H., Koszinowski, U., Brune, W., and Adler, B. (2008). Cloning and sequencing of a highly productive, endotheliotropic virus strain derived from human cytomegalovirus TB40/E. J. Gen. Virol. 89, 359–368.

Smith, G.A., and Enquist, L.W. (1999). Construction and transposon mutagenesis in Escherichia coli of a full-length infectious clone of pseudorabies virus, an alphaherpesvirus. J. Virol. 73, 6405–6414.

Stanton, R.J., Baluchova, K., Dargan, D.J., Cunningham, C., Sheehy, O., Seirafian, S., McSharry, B.P., Neale, M.L., Davies, J.A., Tomasec, P., et al. (2010). Reconstruction of the complete human cytomegalovirus genome in a BAC reveals RL13 to be a potent inhibitor of replication. J. Clin. Invest. 120, 3191–3208.

Tang, H., Kawabata, A., Yoshida, M., Oyaizu, H., Maeki, T., Yamanishi, K., and Mori, Y. (2010). Human herpesvirus 6 encoded glycoprotein Q1 gene is essential for virus growth. Virology 407, 360–367.

Tischer, B.K., von Einem, J., Kaufer, B., and Osterrieder, N. (2006). Two-step red-mediated recombination for versatile high-efficiency markerless DNA manipulation in *Escherichia coli*. Biotechniques *40*, 191–197.

Tischer, B.K., Smith, G.A., and Osterrieder, N. (2010). En passant mutagenesis: a two step markerless red recombination system. Methods Mol. Biol. *634*, 421–430.

Upton, J.W., Kaiser, W.J., and Mocarski, E.S. (2008). Cytomegalovirus M45 cell death suppression requires receptor-interacting protein (RIP) homotypic interaction motif (RHIM)-dependent interaction with RIP1. J. Biol. Chem. *283*, 16966–16970.

Upton, J.W., Kaiser, W.J., and Mocarski, E.S. (2010). Virus inhibition of RIP3-dependent necrosis. Cell Host Microbe 7, 302–313.

Wagner, M., Jonjić, S., Koszinowski, U.H., and Messerle, M. (1999). Systematic excision of vector sequences from the BAC-cloned herpesvirus genome during virus reconstitution. J. Virol. 73, 7056–7060.

Wagner, M., Gutermann, A., Podlech, J., Reddehase, M.J., and Koszinowski, U.H. (2002). Major histocompatibility complex class I allele-specific cooperative and competitive interactions between immune evasion proteins of cytomegalovirus. J. Exp. Med. *196*, 805–816.

Wang, J., Sarov, M., Rientjes, J., Fu, J., Hollak, H., Kranz, H., Xie, W., Stewart, A.F., and Zhang, Y. (2006). An improved recombineering approach by adding RecA to lambda Red recombination. Mol. Biotechnol. *32*, 43–53.

Warming, S., Costantino, N., Court, D.L., Jenkins, N.A., and Copeland, N.G. (2005). Simple and highly efficient BAC recombineering using galK selection. Nucleic Acids Res. *33*, e36.

Yu, D., Ellis, H.M., Lee, E.C., Jenkins, N.A., Copeland, N.G., and Court, D.L. (2000). An efficient recombination system for chromosome engineering in *Escherichia coli*. Proc. Natl. Acad. Sci. U.S.A. *97*, 5978–5983.

Yu, D., Smith, G.A., Enquist, L.W., and Shenk, T. (2002). Construction of a self-excisable bacterial artificial chromosome containing the human cytomegalovirus genome and mutagenesis of the diploid TRL/IRL13 gene. J. Virol. *76*, 2316–2328.

Yu, D., Silva, M.C., and Shenk, T. (2003). Functional map of human cytomegalovirus AD169 defined by global mutational analysis. Proc. Natl. Acad. Sci. U.S.A. *100*, 12396–12401.

Zhang, G., Raghavan, B., Kotur, M., Cheatham, J., Sedmak, D., Cook, C., Waldman, J., and Trgovcich, J. (2007). Antisense transcription in the human cytomegalovirus transcriptome. J. Virol. *81*, 11267–11281.

Zhang, Y., Buchholz, F., Muyrers, J.P., and Stewart, A.F. (1998). A new logic for DNA engineering using recombination in *Escherichia coli*. Nat. Genet. *20*, 123–128.

Zhou, F., Li, Q., and Gao, S.J. (2009). A sequence-independent in vitro transposon-based strategy for efficient cloning of genomes of large DNA viruses as bacterial artificial chromosomes. Nucleic Acids Res. *37*, e2.

Zimmermann, A., Trilling, M., Wagner, M., Wilborn, M., Bubić, I., Jonjić, S., Koszinowski, U., and Hengel, H. (2005). A cytomegaloviral protein reveals a dual role for STAT2 in IFN-γ signaling and antiviral responses. J. Exp. Med. *201*, 1543–1553.

Human Cytomegalovirus Metabolomics

I.4

Joshua D. Rabinowitz and Thomas Shenk

Abstract

Metabolic changes at the whole organism and cellular levels have been described for many diseases, and the alterations often underlie disease progression. Known metabolites can be quantified by using liquid chromatography to fractionate a complex mixture of compounds with analysis of the output by mass spectrometry. This approach has been applied to quantify steady state levels of metabolites as well as to monitor the flux of isotopically labelled metabolites through pathways in human cytomegalovirus-infected fibroblasts. Cytomegalovirus hijacks cellular metabolism, markedly inducing flux through much of central carbon metabolism, including glycolysis, nucleotide metabolism, the tricarboxylic acid cycle and the fatty acid metabolic enzyme acetyl-CoA carboxylase. This chapter details the metabolic changes that accompany infection, discusses the current understanding of mechanisms underlying the changes, and considers the physiological roles of the changes in human cytomegalovirus replication and spread.

Introduction

The network of chemical reactions involving the derivation of energy currency and biosynthetic building blocks from nutrient inputs consists of about 1000 compounds and about 2000 interconverting reactions (Ouzounis and Karp, 2000), and all viruses require the products of this metabolic network. It has long been known that HCMV induces global RNA (Tanaka *et al.*, 1975) and protein (Furukawa *et al.*, 1973) synthesis, as well as glucose uptake (Landini, 1984). Recently, improved technologies for measuring enzyme activities and their products have enabled in-depth analysis of the HCMV-host cell metabolic interaction. A large number of intracellular metabolites can now be identified and quantified by using liquid chromatography coupled to mass spectrometry (LC-MS) (Want *et al.*, 2005; Rabinowitz *et al.*, 2011; Reaves and Rabinowitz,

2011). However, interpretation of the biology underlying changes in the levels of metabolites is confounded by the fact that the steady state level of a metabolite can be elevated for two completely different reasons: production might be increased or consumption might be reduced. These ambiguities can be overcome by kinetic flux profiling in which cells are switched from unlabelled medium to medium containing isotope-labelled nutrient. This results in labelling of downstream metabolites, with the labelling rates directly proportional to flux through the metabolic pathway (Yuan *et al.*, 2008).

Both steady state and fluxomic analysis have now been applied to HCMV-infected cells (Munger *et al.*, 2006, 2008; Vastag *et al.*, 2011). HCMV disrupts cellular metabolic homeostasis and institutes its own metabolic programme with markedly elevated flux of carbon from glucose and glutamine through central metabolic pathways. In this chapter we review the discovery and consequences of these changes to the HCMV replication cycle. There are several other excellent reviews that consider key aspects of this topic (Yu *et al.*, 2011a,b).

An HCMV-specific metabolic programme

The first hint that HCMV alters central metabolism came years ago, when it was found that infection significantly elevates glucose uptake in fibroblasts (Landini, 1984). An increase was first evident at 24 hpi, and uptake continued to accelerate through the late phase of infection. The mechanism supporting the increase in glucose uptake was described much more recently. Infection induces and activates the glucose transporter type 4 (GLUT4) (Yu *et al.*, 2011c). The HCMV IE1 protein actively eliminates GLUT1, the normal fibroblast transporter; and infection induces GLUT4 by an unknown mechanism that circumvents the known Akt-mediated translocation pathway. Indinavir, a drug that inhibits GLUT4, but not GLUT1, substantially reduces the production of HCMV progeny, demonstrating the dependence of the virus on GLUT4 activity.

The effect of HCMV infection and elevated glucose uptake on metabolism was quantified by LC-MS. HCMV markedly increased the levels of numerous metabolites, including intermediates involved in glycolysis, the tricarboxylic acid (TCA; also known as citric acid) cycle and pyrimidine biosynthesis (Munger *et al.*, 2006). Although the kinetics of accumulation varied, the levels of most metabolites increased throughout the late phase of infection, consistent with the timing of glucose uptake. The magnitudes of the changes were impressive, with many metabolites increasing by more

than 5-fold, and the peak levels of many metabolites exceeded those observed for either quiescent or rapidly growing cells, demonstrating that the virus disrupts cellular metabolic homeostasis. This new programme was conserved when a laboratory strain (AD169) or clinical isolate (TB40/E) was analysed (Fig. I.4.1), and it occurred in two cell types (fibroblasts and epithelial cells) where HCMV undergoes active replication (Vastag *et al.*, 2011). Another herpesvirus, herpes simplex virus type 1, also markedly altered the metabolism of fibroblasts. However, the perturbations were largely distinct from those instituted by HCMV. Thus, the changes following infection are not generic

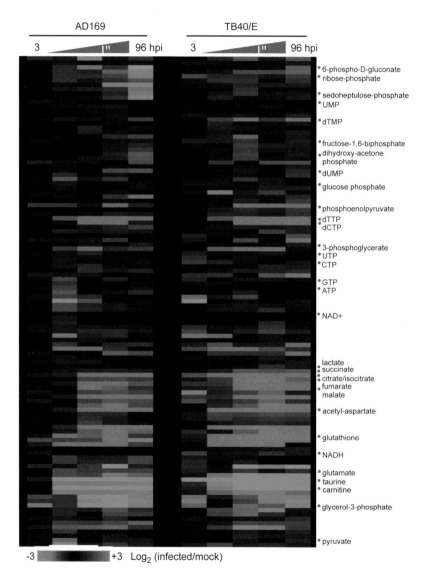

Figure I.4.1 Altered metabolite levels in fibroblasts infected with the AD169 laboratory strain or the TB40/E clinical isolate of HCMV. Cells were infected and sampled at 3, 24, 48, 72 and 96 hpi. Values are the averages of duplicate independent biological experiments. Modified from Vastag *et al.* (2011).

cellular responses to stress. Rather, HCMV hijacks cellular metabolism to execute its own, specific programme.

To better understand the dynamics leading to changes in metabolite levels, fluxomic analysis was used to quantify carbon flux in mock-infected and HCMV-infected fibroblasts (Munger *et al.*, 2008). This analytical method employs LC-MS to monitor the flow of ^{13}C-labelled forms of glucose and glutamine through metabolic pathways (Yuan *et al.*, 2008). The study confirmed the increased uptake of glucose after infection and demonstrated that glutamine uptake was also elevated with concomitant increases in the excretion of lactate and glutamate. HCMV infection caused a nearly global increase in metabolic flux (Fig. I.4.2). Glycolytic flux was elevated by a factor of about two, nucleotide biosynthetic flux was increased about three-fold, core TCA cycle fluxes increased to an even greater extent, and the efflux of two carbon units from the cycle via citrate to malonyl CoA increased by a

factor of about 20. Thus, HCMV markedly up-regulates flux through much of central carbon metabolism with efflux of carbon from both glucose and glutamine to the fatty acid biosynthetic pathway. Malonyl CoA is the committed intermediate in fatty acid synthesis and its concentration increased from undetectable in uninfected cells to an amount 10-fold above the detection limit following infection. Importantly, pharmacological inhibition of two enzymes in the fatty acid biosynthetic pathway, acetyl-CoA carboxylase or fatty acid synthase, inhibited the production of viral progeny, revealing the dependence of the virus on *de novo* lipid synthesis. HCMV not only induces synthesis of fatty acids, but it also blocks their catabolic consumption at the expense of ATP production. A viral immediate-early protein, pUL37x1, has been shown to relocalize the interferon-inducible viperin protein to mitochondria, where viperin binds to the β-subunit of the mitochondrial trifunctional protein, blocking β-oxidation of fatty acids (Seo *et al.*, 2011).

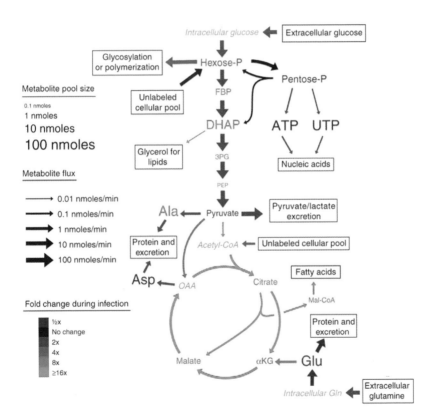

Figure I.4.2 HCMV reprogrammes host cell metabolism, utilizing carbon for fatty acid synthesis. Fibroblasts were infected with HCMV or mock-infected, and their metabolism analysed based on (i) nutrient uptake and waste excretion, (ii) metabolite pool sizes by LC-MS, (iii) dynamics of ^{13}C-glucose and ^{13}C-glutamine labelling by LC-MS, and (iv) steady-state labelling after feeding 1,2-^{13}C-glucose. The map shows the changes in the concentrations of metabolites and flux induced by HCMV infection. Font sizes correspond to metabolite pool sizes in uninfected cells, arrow widths show net fluxes in uninfected cells, and the colour indicates the change in concentrations and fluxes induced by HCMV in fibroblasts. Metabolites in gray were not measured. Modified from Munger *et al.* (2008).

Consistent with its elevated uptake (Munger *et al.*, 2008), glutamine is required for ATP production in infected but not uninfected fibroblasts (Chambers *et al.*, 2010). Glutaminase and glutamate dehydrogenase, enzymes required for the conversion of glutamine to α-ketoglutarate, increased in activity after infection. Infected cells starved for glutamine were depleted for ATP and failed to produce infectious progeny. Both defects were rescued by supplementation of glutamine-free medium with α-ketoglutarate, conclusively demonstrating the anaplerotic use of glutamine within HCMV-infected cells.

Multiple regulatory mechanisms underlie the metabolic changes

The AMP-activated protein kinase (AMPK) is altered by infection and contributes importantly to HCMV-induced changes in core metabolism. AMPK, a sensor of cellular energy homeostasis, is comprised of a catalytic subunit, AMPKα, and two regulatory subunits, AMPKβ and AMPKγ. Activation of the kinase requires cooperative AMP binding to AMPγ, which occurs stochastically with shifts in the AMP:ATP ratio, and phosphorylation of AMPKα at Thr172 (Hardie, 2007; Canto and Auwerx, 2010). AMPK controls GLUT4 relocalization to the plasma membrane (Hardie, 2007), and it phosphorylates and activates 6-phosphofructo-1-kinase (PFK-1) (Hue and Rider, 1987), a key upstream regulator of glycolytic flow. Both of these activities link the kinase to altered metabolism in HCMV-infected cells, where GLUT4 (Yu *et al.*, 2011c) and PFK-1 (Munger *et al.*, 2006) are induced. Inhibition of AMPK by treating infected fibroblasts with siRNA or the drug compound C reduces glucose uptake, the induction of central metabolism and the yield of HCMV (McArdle *et al.*, 2012; Terry *et al.*, 2012).

Three kinases are known to phosphorylate Thr172 of AMPKα and activate the kinase: Ca^{2+}/calmodulin-dependent kinase kinase (CaMKK), TGF-β-activated kinase 1 (TAK1), and liver kinase B1 (LKB1). The level of CaMKK protein is elevated after infection, and pharmacological inhibition of CaMKK or treatment with a dominant-negative CaMKK variant reduces the production of progeny (McArdle *et al.*, 2011). Drug treatment also reduces the accumulation of labelled fructose bisphosphate when cells are fed ^{13}C-labelled glucose, indicating that glycolytic activity is reduced. Drug-mediated inhibition of CaMKK after infection blocks AMPK activation (McArdle *et al.*, 2012), consistent with the view that CaMKK exerts its effect on metabolism via activation of AMPK. In a global siRNA screen of cellular kinases, CaMKK, but not TAK1 or LKB1, was found to be needed for viral growth (Terry *et al.*, 2012), so it is possible that it is the only kinase serving to elevate AMPK activity in infected cells. Induced CaMKK could be activated during HCMV replication by the pUL37x1 protein. It induces the release of calcium stores into the cytosol (Sharon-Friling *et al.*, 2006), which would be expected to activate the calcium-dependent enzyme.

It is noteworthy that AMPK modulates targets that would be expected to interfere with HCMV replication. For example, AMPK phosphorylates the TSC2 protein, a subunit of the tuberous sclerosis protein complex (TSC1/2). When phosphorylated, TSC1/2 is a negative regulator of mammalian target of rapamycin complex 1 (mTORC1) and thereby inhibits translation and cellular growth. However, the HCMV pUL38 protein binds to TSC2 and blocks its ability to control mTORC1, preserving translational activity in infected cells (Moorman *et al.*, 2008). AMPK also phosphorylates and inhibits acetyl-CoA carboxylase (ACC) (Munday, 2002), which catalyses the carboxylation of acetyl-CoA to form malonyl-CoA, the first committed step in fatty acid biosynthesis. There are two isoforms of this enzyme, ACC1 and ACC2, and the enzyme not only remains active in HCMV-infected cells, but its inhibition substantially reduces the yield of viral progeny (Munger *et al.*, 2008). One or more immediate-early or early viral proteins increase the expression and specific activity of ACC1 (Spencer *et al.*, 2011). The ACC1 promoter contains a sterol regulatory element (SRE), and the SRE binding protein 2 (SREBP2) transcription factor is activated by infection. This presumably leads to enhanced mRNA and, subsequently, protein accumulation. It is intriguing that the specific activity of ACC1 increases in spite of the fact that the AMPK-mediated inhibitory phosphorylation of ACC1 at Ser79 increases after infection (Spencer *et al.*, 2011). Perhaps a viral protein interacts with ACC1 to prevent its regulation by AMPK, as occurs for TSC2.

SREBP1 and 2 are a members of a family of transcription factors that control cellular lipid homeostasis (Goldstein *et al.*, 2006). When inactive, they are anchored in the endoplasmic reticulum membrane. They are activated by cleavage of the SREBP precursor after which the mature form of the transcription factor moves to the nucleus and binds to SREs, activating transcription (Espenshade and Hughes, 2007). HCMV antagonizes the normal sterol-mediated feedback that down regulates SREBP activation (Yu *et al.*, 2012). Cleavage becomes constitutive with elevated expression of a wide range of SREBP-regulated lipogenic enzymes including: ACC1, as discussed above; acetyl CoA lyase, which produces cytosolic acetyl-CoA from citrate; sterol CoA desaturase, which controls the synthesis of monounsaturated fatty acids, and hydroxymethylglutaryl (HMG) CoA reductase, a rate limiting enzyme for cholesterol synthesis. The constitutive activation requires viral gene expression, but the critical gene product and its mode of action remain to be identified.

In addition to up-regulating enzymes involved in key aspects of metabolism, HCMV appears to induce mitochondrial biogenesis. Infection has long been known to induce mitochondrial DNA synthesis (Furukawa *et al.*, 1976). Mitochondrial DNA is increased by a factor of about 200; mitochondrial respiration is induced; and the peroxisomal-activated receptor gamma coactivator 1α (PGC-1α), a transcription factor that plays a central role in mitochondrial biogenesis, is activated following infection (Kaarbo *et al.*, 2011). Of interest, activation of PGC-1α is mediated by AMPK (Jager *et al.*, 2007), so it is likely that changes in mitochondrial function are yet another consequence of AMPK activation.

In sum, several mechanisms have been identified for the activation of metabolism by HCMV: the transcriptional activation of genes encoding lipogenic enzymes and the post-translational activation of downstream AMPK targets.

Consequences of metabolic alterations during HCMV infection

The physiological consequences of most alterations in HCMV-infected cells can be predicted from the known functions of metabolites. For example, nucleotides are elevated to support intense viral DNA and RNA synthesis, and amino acids increase to sustain translation. However, there are some changes that remain enigmatic. N-acetyl-aspartate is a case in point. This modified amino acid is the most powerfully up-regulated metabolite of those monitored in infected fibroblasts. Although it is known to be an abundant constituent of the brain, where it is synthesized in mitochondria of neurons, its function remains uncertain. It is, as yet, unclear how or why it is up-regulated in HCMV-infected fibroblasts.

Perhaps the most intriguing change to metabolism is the enhanced synthesis of fatty acids, even though the infected cells are not actively dividing. Newly synthesized fatty acids can be catabolized to generate energy, serve signalling functions, used as substrates for modification of proteins, or incorporated into lipids. As discussed above, the ability of pUL37x1 to relocalize viperin to mitochondria where it blocks β-oxidation of fatty acids (Seo *et al.*, 2011),

argues that fatty acids are probably not degraded to a significant extent after infection. Protein modification, signalling and lipid synthesis are all possible uses for fatty acids within infected cells.

The relative amounts of polar and neutral lipids were shown to change with HCMV infection some years ago (Abrahamsen *et al.*, 1996). More recently, MS analysis (Ivanova *et al.*, 2007) was used to identify and quantify numerous glycerophospholipid (GP) species differing in head group, acyl chain length and degree of saturation within HCMV-infected fibroblasts and virions (Liu *et al.*, 2011). GPs serve signalling functions and they are, of course, major constituents of membranes. Total cell GPs assayed changed to only a modest extent following infection, with phosphatidic acid (PA) species increasing to the greatest extent. PAs serve a membrane structural role, but also signal to regulate critical cell functions, including cell survival, proliferation and vesicular trafficking (Wang *et al.*, 2006). PAs can induce mitogen-activated protein kinase (MAPK) activity, which is induced in two waves during HCMV infection (Johnson *et al.*, 2001), one of which coincides with the elevation of PAs. PA also interacts with GLUT4 (Heyward *et al.*, 2008) and is thought to facilitate its translocation to the cell surface, and, thus, it could contribute to the dramatic increase in GLUT4 activity following HCMV infection. Thus, changes to PA species following infection could influence signalling through multiple pathways following HCMV infection. In contrast to whole cells, which showed subtle changes with infection, the HCMV virion envelope exhibited a unique signature (Liu *et al.*, 2011). Its differentiation from the whole cell argues that the virion envelope is derived from a distinct cellular compartment. There is strong evidence that the virion acquires its envelope at a membrane-rich assembly compartment (Sanchez *et al.*, 2000), which forms adjacent to the nucleus during the late phase of infection. It is likely that *de novo* synthesized lipids are used to construct or modify membranes at this site, thereby supplying virion envelopes with a distinct lipid composition. One of the GPs enriched in virions is phosphatidylethanolamine (PE). PEs can impose curvature when concentrated on the inner leaflet of a lipid bilayer (Yang *et al.*, 1997), which might support formation of the compact spherical shape of the virion envelope.

Thus, newly synthesized fatty acids support *de novo* lipid biosynthesis. These lipids have potential to modulate signalling within infected cells, and they are likely incorporated into membranes within the viral assembly compartment and ultimately contribute to the lipid composition of virions.

One more argument for a possible role of HCMV in tumorigenesis

Many years ago, HCMV was reported to transform human cells (Geder et al., 1976) to a tumorigenic phenotype in nude mice (Geder et al., 1977). However, viral DNA was not detected in transformed cells, leading to speculation that HCMV transforms by a 'hit and run' mechanism (McDougall et al., 1984), which is still unproven. For some time it was assumed that HCMV would not transform cells because it blocks cell cycle progression in fibroblasts. We now know that it stimulates progression in other cell types (Maussang et al., 2006; Bentz and Yurochko, 2008). Further, HCMV encodes gene products that block apoptosis (Goldmacher, 2005), disrupt the function of three tumour suppressors (Alwine, 2008), induce DNA damage and disrupt DNA repair systems (Shen et al., 1997; Fortunato et al., 2000; Gaspar and Shenk, 2006; Luo et al., 2007), avoid immune surveillance (Powers et al., 2008) and reactivate telomerase (Straat et al., 2009). Finally, although initially controversial, it is becoming accepted that HCMV DNA and proteins are often present in human tumours (Cobbs et al., 2002; Harkins et al., 2002; Samanta et al., 2003; Prins et al., 2008), and there are indications that inhibition of viral replication can interfere with tumour growth in mice carrying HCMV-infected tumour xenografts (Baryawno et al., 2011).

Metabolic perturbations mark an additional cancer-like change to HCMV-infected cells. Almost 90 years ago, Warburg and his colleagues reported that tumour cells metabolize glucose differently than cells in normal tissues (Warburg et al., 1924). Although cancer cells and normal cells both convert glucose to pyruvate via glycolysis, they use the pyruvate differently (Vander Heiden et al., 2009). Normal cells generally use pyruvate to feed the TCA cycle in mitochondria, where it is catabolized through oxidative phosphorylation. In contrast, cancer cells convert a substantial portion of pyruvate to lactate, irrespective of the availability of oxygen. This observation was surprising and remained an enigma for many years, because it appeared that cancer cells used glucose inefficiently, at least as regards ATP production. Aerobic glycolysis generates only two molecules of ATP per molecule of glucose. If the pyruvate from glucose was completely oxidized in the in the TCA cycle, it would generate another 36 molecules of ATP.

It is now known that a portion of the pyruvate produced from glucose in tumour cells does indeed enter mitochondria, where it is converted to acetyl-CoA that combines with oxaloacetetic acid to form citrate, which can be shuttled back to the cytoplasm (Farfari et al., 2000). In the cytoplasm, citrate is converted back to oxaloacetic acid and acetyl CoA, and the acetyl CoA is then available for use as a substrate for fatty acid synthesis. Thus, a portion of the carbon from glucose is used biosynthetically to support cell growth through the generation of biomass. In addition to the role for acetyl-CoA in fatty acid production, glycolytic intermediates are used to generate non-essential amino acids and ribose is used to produce nucleotides. How do cancer cells feed the TCA cycle? They utilize exogenous glutamine, which is converted to α-ketoglutarate through a process termed glutaminolysis, feeding carbon to the TCA cycle as well as supplying additional citrate that can move to the cytoplasm (DeBerardinis et al., 2007).

Strikingly, the two signature metabolic alterations to tumour cells also occur in HCMV infected cells. Just as cancer cells depend on carbon from glucose for biosynthetic use, HCMV-infected cells require glucose for the synthesis of fatty acids, presumably to build lipids for the virion envelope. Also like cancer cells, HCMV-infected cells employ glutaminolysis to feed the TCA cycle. It has been suggested that HCMV may function as an oncomodulator, increasing malignancy of tumour cells (Cinatl et al., 1996; Cobbs et al., 2007), enhancing resistance to anti-cancer drugs (Cinatl et al., 1998), and inducing secretion of proangiogenic factors (Dumortier et al., 2008). Perhaps the metabolic changes driven by HCMV also serve an oncomodulatory function, instituting a metabolic state in infected tumour cells that promotes their growth by modulating the use of glucose and glutamine.

Of course, none of these associations prove that HCMV has a role in tumorigenesis. But, when its presence in tumours is viewed in the context of the malignancy-promoting activities of HCMV proteins and the cancer-like metabolic changes that HCMV induces, it becomes clear that a role for HCMV in human cancer is a possibility that must be considered.

Questions and perspectives

The realization that HCMV hijacks core metabolism and depends on de novo fatty acid synthesis raises many questions, which have only begun to be explored. We now appreciate the central role that AMPK and SREBP activation play in the metabolic changes, but only a few of the many potential downstream targets of these effectors have been evaluated at this point. It will be important to explore the full range and roles of these downstream targets in the metabolic hijacking, and, of course, it is likely that additional regulatory pathways are modulated that await discovery. We also know very little about which viral gene products execute the hijacking. It's likely that pUL37x1 contributes to the CaMKK-mediated activation of AMPK and it also probably helps to prevent the β-oxidation of fatty acids through its interaction with viperin. However, it is not yet clear how the virus preserves ACC1 activity in spite of its modification with inhibitory phosphorylations, and it's not known how the virus induces constitutive SREBP cleavage and activation. We also do not understand why the virus needs to induce *de novo* fatty acid synthesis rather than rely on pre-existing lipid stores, and we know very little about how the newly synthesized fatty acids and lipids support the production of infectious progeny. Finally, we know that profound metabolic changes occur during active HCMV replication, but possible metabolic changes have not yet been explored in cells that do not support active viral growth. Does the virus transiently or permanently alter metabolism during latency, and if so, how do the perturbations contribute to the process? Does the virus help to drive metabolic changes in tumour cells, facilitating the production of biomass needed for rapid cell growth? The new understanding of the productively infected cell metabolome provides a framework for exploring these questions and many others.

One final point deserves mention. So far, the infected-cell metabolome has been described for three viruses: HCMV (Munger *et al.*, 2006, 2008; Vastag *et al.*, 2011), herpes simplex virus type 1 (Vastag *et al.*, 2011) and influenza A virus (Rabinowitz *et al.*, 2011). Each of these viruses establishes a different programme following infection of human fibroblasts. In influenza virus-infected cells, the largest increase was evident for acetylneuraminic acid, which is cleaved from glycoproteins by the virus-coded neuraminidase. Herpes simplex virus markedly increased levels of deoxynucleotides. Intriguingly, the metabolic response to infection with these viruses mirrors the targets of their preferred therapeutics: oseltamivir inhibits the influenza virus neuraminidase and acyclovir is a nucleotide analogue that inhibits the herpes simplex virus polymerase. HCMV infection elevated numerous metabolites, but its changes were not directed primarily at deoxynucleotides. The relatively modest induction of deoxynucleotides correlates with a poor therapeutic index of ganciclovir, an analogue of deoxyguanosine that is the preferred treatment for HCMV infections. The levels of citrate and acetyl-CoA, both of which support fatty acid biosynthesis, are elevated to a much greater extent than deoxynucleotides. Thus, the metabolomic response to HCMV infection identifies enzymes of fatty acid and lipid biosynthesis as promising targets for improved therapeutic interventions.

References

Abrahamsen, L.H., Clay, M.J., Lyle, J.M., Zink, J.M., Fredrikson, L.J., DeSiervo, A.J., and Jerkofsky, M.A. (1996). The effects of cytomegalovirus infection on polar lipids and neutral lipids in cultured human cells. Intervirology 39, 223–229.

Alwine, J.C. (2008). Modulation of host cell stress responses by human cytomegalovirus. Curr. Top. Microbiol. Immunol. 325, 263–279.

Baryawno, N., Rahbar, A., Wolmer-Solberg, N., Taher, C., Odeberg, J., Darabi, A., Khan, Z., Sveinbjornsson, B., FuskevAg, O.M., Segerstrom, L., *et al.* (2011). Detection of human cytomegalovirus in medulloblastomas reveals a potential therapeutic target. J. Clin. Invest. 121, 4043–4055.

Bentz, G.L., and Yurochko, A.D. (2008). Human CMV infection of endothelial cells induces an angiogenic response through viral binding to EGF receptor and beta1 and beta3 integrins. Proc. Natl. Acad. Sci. U.S.A. 105, 5531–5536.

Canto, C., and Auwerx, J. (2010). AMP-activated protein kinase and its downstream transcriptional pathways. Cell. Mol. Life. Sci. 67, 3407–3423.

Chambers, J.W., Maguire, T.G., and Alwine, J.C. (2010). Glutamine metabolism is essential for human cytomegalovirus infection. J. Virol. 84, 1867–1873.

Cinatl, J., Jr., Cinatl, J., Vogel, J.U., Rabenau, H., Kornhuber, B., and Doerr, H.W. (1996). Modulatory effects of human cytomegalovirus infection on malignant properties of cancer cells. Intervirology 39, 259–269.

Cinatl, J., Jr., Cinatl, J., Vogel, J.U., Kotchetkov, R., Driever, P.H., Kabickova, H., Kornhuber, B., Schwabe, D., and Doerr, H.W. (1998). Persistent human cytomegalovirus infection induces drug resistance and alteration of programmed cell death in human neuroblastoma cells. Cancer Res. 58, 367–372.

Cobbs, C.S., Harkins, L., Samanta, M., Gillespie, G.Y., Bharara, S., King, P.H., Nabors, L.B., Cobbs, C.G., and Britt, W.J. (2002). Human cytomegalovirus infection and expression in human malignant glioma. Cancer Res. 62, 3347–3350.

Cobbs, C.S., Soroceanu, L., Denham, S., Zhang, W., Britt, W.J., Pieper, R., and Kraus, M.H. (2007). Human cytomegalovirus induces cellular tyrosine kinase signaling and promotes glioma cell invasiveness. J. Neurooncol. 85, 271–280.

DeBerardinis, R.J., Mancuso, A., Daikhin, E., Nissim, I., Yudkoff, M., Wehrli, S., and Thompson, C.B. (2007). Beyond aerobic glycolysis: transformed cells can engage in glutamine metabolism that exceeds the requirement for protein and nucleotide synthesis. Proc. Natl. Acad. Sci. U.S.A. 104, 19345–19350.

Dumortier, J., Streblow, D.N., Moses, A.V., Jacobs, J.M., Kreklywich, C.N., Camp, D., Smith, R.D., Orloff, S.L., and Nelson, J.A. (2008). Human cytomegalovirus secretome contains factors that induce angiogenesis and wound healing. J. Virol. 82, 6524–6535.

Espenshade, P.J., and Hughes, A.L. (2007). Regulation of sterol synthesis in eukaryotes. Annu. Rev. Genet. 41, 401–427.

Farfari, S., Schulz, V., Corkey, B., and Prentki, M. (2000). Glucose-regulated anaplerosis and cataplerosis in pancreatic beta-cells: possible implication of a pyruvate/citrate shuttle in insulin secretion. Diabetes 49, 718–726.

Fortunato, E.A., Dell'Aquila, M.L., and Spector, D.H. (2000). Specific chromosome 1 breaks induced by human cytomegalovirus. Proc. Natl. Acad. Sci. U.S.A. 97, 853–858.

Furukawa, T., Fioretti, A., and Plotkin, S. (1973). Growth characteristics of cytomegalovirus in human fibroblasts with demonstration of protein synthesis early in viral replication. J. Virol. 11, 991–997.

Furukawa, T., Sakuma, S., and Plotkin, S.A. (1976). Human cytomegalovirus infection of WI-38 cells stimulates mitochondrial DNA synthesis. Nature 262, 414–416.

Gaspar, M., and Shenk, T. (2006). Human cytomegalovirus inhibits a DNA damage response by mislocalizing checkpoint proteins. Proc. Natl. Acad. Sci. U.S.A. 103, 2821–2826.

Geder, K.M., Lausch, R., O'Neill, F., and Rapp, F. (1976). Oncogenic transformation of human embryo lung cells by human cytomegalovirus. Science 192, 1134–1137.

Geder, L., Kreider, J., and Rapp, F. (1977). Human cells transformed in vitro by human cytomegalovirus: tumorigenicity in athymic nude mice. J. Natl. Cancer Inst. 58, 1003–1009.

Goldmacher, V.S. (2005). Cell death suppression by cytomegaloviruses. Apoptosis 10, 251–265.

Goldstein, J.L., DeBose-Boyd, R.A., and Brown, M.S. (2006). Protein sensors for membrane sterols. Cell 124, 35–46.

Hardie, D.G. (2007). AMP-activated/SNF1 protein kinases: conserved guardians of cellular energy. Nat. Rev. Mol. Cell. Biol. 8, 774–785.

Harkins, L., Volk, A.L., Samanta, M., Mikolaenko, I., Britt, W.J., Bland, K.I., and Cobbs, C.S. (2002). Specific localisation of human cytomegalovirus nucleic acids and proteins in human colorectal cancer. Lancet 360, 1557–1563.

Heyward, C.A., Pettitt, T.R., Leney, S.E., Welsh, G.I., Tavare, J.M., and Wakelam, M.J. (2008). An intracellular motif of GLUT4 regulates fusion of GLUT4-containing vesicles. BMC Cell Biol. 9, 25.

Hue, L., and Rider, M.H. (1987). Role of fructose 2,6-bisphosphate in the control of glycolysis in mammalian tissues. Biochem. J. 245, 313–324.

Ivanova, P.T., Milne, S.B., Byrne, M.O., Xiang, Y., and Brown, H.A. (2007). Glycerophospholipid identification and quantitation by electrospray ionization mass spectrometry. Methods Enzymol. 432, 21–57.

Jager, S., Handschin, C., St-Pierre, J., and Spiegelman, B.M. (2007). AMP-activated protein kinase (AMPK) action in skeletal muscle via direct phosphorylation of PGC-1alpha. Proc. Natl. Acad. Sci. U.S.A. 104, 12017–12022.

Johnson, R.A., Ma, X.L., Yurochko, A.D., and Huang, E.S. (2001). The role of MKK1/2 kinase activity in human cytomegalovirus infection. J. Gen. Virol. 82, 493–497.

Kaarbo, M., Ager-Wick, E., Osenbroch, P.O., Kilander, A., Skinnes, R., Muller, F., and Eide, L. (2011). Human cytomegalovirus infection increases mitochondrial biogenesis. Mitochondrion 11, 935–945.

Landini, M.P. (1984). Early enhanced glucose uptake in human cytomegalovirus-infected cells. J. Gen. Virol. 65, 1229–1232.

Liu, S.T., Sharon-Friling, R., Ivanova, P., Milne, S.B., Myers, D.S., Rabinowitz, J.D., Brown, H.A., and Shenk, T. (2011). Synaptic vesicle-like lipidome of human cytomegalovirus virions reveals a role for SNARE machinery in virion egress. Proc. Natl. Acad. Sci. U.S.A. 108, 12869–12874.

Luo, M.H., Rosenke, K., Czornak, K., and Fortunato, E.A. (2007). Human cytomegalovirus disrupts both ataxia telangiectasia mutated protein (ATM)- and ATM-Rad3-related kinase-mediated DNA damage responses during lytic infection. J. Virol. 81, 1934–1950.

McArdle, J., Schafer, X.L., and Munger, J. (2011). Inhibition of calmodulin-dependent kinase kinase blocks human cytomegalovirus-induced glycolytic activation and severely attenuates production of viral progeny. J. Virol. 85, 705–714.

McArdle, J., Moorman, N.J., and Munger, J. (2012). HCMV Targets the Metabolic Stress Response through Activation of AMPK Whose Activity Is Important for Viral Replication. PLoS Pathog. 8, e1002502.

McDougall, J.K., Nelson, J.A., Myerson, D., Beckmann, A.M., and Galloway, D.A. (1984). HSV, CMV, and HPV in human neoplasia. J. Invest. Dermatol. 83, 72s-76s.

Maussang, D., Verzijl, D., van Walsum, M., Leurs, R., Holl, J., Pleskoff, O., Michel, D., van Dongen, G.A., and Smit, M.J. (2006). Human cytomegalovirus-encoded chemokine receptor US28 promotes tumorigenesis. Proc. Natl. Acad. Sci. U.S.A. 103, 13068–13073.

Moorman, N.J., Cristea, I.M., Terhune, S.S., Rout, M.P., Chait, B.T., and Shenk, T. (2008). Human cytomegalovirus protein UL38 inhibits host cell stress responses by antagonizing the tuberous sclerosis protein complex. Cell Host Microbe 3, 253–262.

Munday, M.R. (2002). Regulation of mammalian acetyl-CoA carboxylase. Biochem. Soc. Trans. 30, 1059–1064.

Munger, J., Bajad, S.U., Coller, H.A., Shenk, T., and Rabinowitz, J.D. (2006). Dynamics of the cellular metabolome during human cytomegalovirus infection. PLoS Pathog. 2, e132.

Munger, J., Bennett, B.D., Parikh, A., Feng, X.J., McArdle, J., Rabitz, H.A., Shenk, T., and Rabinowitz, J.D. (2008). Systems-level metabolic flux profiling identifies fatty acid synthesis as a target for antiviral therapy. Nat. Biotechnol. 26, 1179–1186.

Ouzounis, C.A., and Karp, P.D. (2000). Global properties of the metabolic map of Escherichia coli. Genome Res. 10, 568–576.

Powers, C., DeFilippis, V., Malouli, D., and Fruh, K. (2008). Cytomegalovirus immune evasion. Curr. Top. Microbiol. Immunol. 325, 333–359.

Prins, R.M., Cloughesy, T.F., and Liau, L.M. (2008). Cytomegalovirus immunity after vaccination with autologous glioblastoma lysate. N. Engl. J. Med. 359, 539–541.

Rabinowitz, J.D., Purdy, J.G., Vastag, L., Shenk, T., and Koyuncu, E. (2011). Metabolomics in drug target discovery. Cold Spring Harb Symp. Quant. Biol. 76, 235–246.

Reaves, M.L., and Rabinowitz, J.D. (2011). Metabolomics in systems microbiology. Curr. Opin. Biotechnol. 22, 17–25.

Samanta, M., Harkins, L., Klemm, K., Britt, W.J., and Cobbs, C.S. (2003). High prevalence of human cytomegalovirus in prostatic intraepithelial neoplasia and prostatic carcinoma. J. Urol. 170, 998–1002.

Sanchez, V., Greis, K.D., Sztul, E., and Britt, W.J. (2000). Accumulation of virion tegument and envelope proteins in a stable cytoplasmic compartment during human cytomegalovirus replication: characterization of a potential site of virus assembly. J. Virol. 74, 975–986.

Seo, J.Y., Yaneva, R., Hinson, E.R., and Cresswell, P. (2011). Human cytomegalovirus directly induces the antiviral protein viperin to enhance infectivity. Science 332, 1093–1097.

Sharon-Friling, R., Goodhouse, J., Colberg-Poley, A.M., and Shenk, T. (2006). Human cytomegalovirus pUL37x1 induces the release of endoplasmic reticulum calcium stores. Proc. Natl. Acad. Sci. U.S.A. 103, 19117–19122.

Shen, Y., Zhu, H., and Shenk, T. (1997). Human cytomagalovirus IE1 and IE2 proteins are mutagenic and mediate 'hit-and-run' oncogenic transformation in cooperation with the adenovirus E1A proteins. Proc. Natl. Acad. Sci. U.S.A. *94*, 3341–3345.

Spencer, C.M., Schafer, X.L., Moorman, N.J., and Munger, J. (2011). Human cytomegalovirus induces the activity and expression of acetyl-coenzyme A carboxylase, a fatty acid biosynthetic enzyme whose inhibition attenuates viral replication. J. Virol. *85*, 5814–5824.

Straat, K., Liu, C., Rahbar, A., Zhu, Q., Liu, L., Wolmer-Solberg, N., Lou, F., Liu, Z., Shen, J., Jia, J., *et al.* (2009). Activation of telomerase by human cytomegalovirus. J. Natl. Cancer Inst. *101*, 488–497.

Tanaka, S., Furukawa, T., and Plotkin, S.A. (1975). Human cytomegalovirus stimulates host cell RNA synthesis. J. Virol. *15*, 297–304.

Terry, L.J., Vastag, L., Rabinowitz, J.D., and Shenk, T. (2012). Human kinome profiling identifies a requirement for AMP-activated protein kinase during human cytomegalovirus infection. Proc. Natl. Acad. Sci. U.S.A. *109*, 3071–3076.

Vander Heiden, M.G., Cantley, L.C., and Thompson, C.B. (2009). Understanding the Warburg effect: the metabolic requirements of cell proliferation. Science *324*, 1029–1033.

Vastag, L., Koyuncu, E., Grady, S.L., Shenk, T.E., and Rabinowitz, J.D. (2011). Divergent effects of human cytomegalovirus and herpes simplex virus-1 on cellular metabolism. PLoS Pathog. *7*, e1002124.

Wang, X., Devaiah, S.P., Zhang, W., and Welti, R. (2006). Signaling functions of phosphatidic acid. Prog. Lipid Res. *45*, 250–278.

Want, E.J., Cravatt, B.F., and Siuzdak, G. (2005). The expanding role of mass spectrometry in metabolite profiling and characterization. Chembiochem. *6*, 1941–1951.

Warburg, O., Posener, K., and Negelein, E. (1924). Über den Stoffwechsel der Tumoren. Biochem. Zeitschrift *152*, 319–344.

Yang, Q., Guo, Y., Li, L., and Hui, S.W. (1997). Effects of lipid headgroup and packing stress on poly(ethylene glycol)-induced phospholipid vesicle aggregation and fusion. Biophys. J. *73*, 277–282.

Yu, Y., Clippinger, A.J., and Alwine, J.C. (2011a). Viral effects on metabolism: changes in glucose and glutamine utilization during human cytomegalovirus infection. Trends Microbiol. *19*, 360–367.

Yu, Y., Clippinger, A.J., Pierciey, F.J., Jr., and Alwine, J.C. (2011b). Viruses and metabolism: alterations of glucose and glutamine metabolism mediated by human cytomegalovirus. Adv. Virus Res. *80*, 49–67.

Yu, Y., Maguire, T.G., and Alwine, J.C. (2011c). Human cytomegalovirus activates glucose transporter 4 expression to increase glucose uptake during infection. J. Virol. *85*, 1573–1580.

Yu, Y., Maguire, T.G., and Alwine, J.C. (2012). Human cytomegalovirus infection induces adipocyte-like lipogenesis through activation of Sterol Regulatory Element Binding Protein 1. J. Virol. *86*, 2942–2949.

Yuan, J., Bennett, B.D., and Rabinowitz, J.D. (2008). Kinetic flux profiling for quantitation of cellular metabolic fluxes. Nat. Protoc. *3*, 1328–1340.

Cytomegalovirus-encoded miRNAs

Meaghan H. Hancock, Igor Landais, Lauren M. Hook,
Finn Grey, Rebecca Tirabassi and Jay A. Nelson

Abstract

MicroRNAs (miRNAs) represent an important class of small regulatory RNAs that regulate cellular processes including development and malignancies. Since the discovery that viruses encode miRNAs over 240 viral miRNAs have been identified primarily in the herpesvirus family. The cytomegalovirus (CMV) family encodes multiple miRNAs encoded throughout the viral genome. Recent work has shown that the CMV miRNAs regulate expression of both viral and cellular genes including the CMV immediate-early gene 1 (IE1) and cellular genes involved in immune recognition and cell cycle regulation. In this chapter we will review our current knowledge of the targets and function of CMV encoded miRNAs.

Introduction

The discovery that animal cells encode small regulatory RNA molecules known as microRNAs (miRNAs) is one of the more important findings in biological research. Initially identified in forward genetic screens for developmental genes in *Caenorhabditis elegans*, the miRNA lin-4 was observed to regulate lin-14 (Lee *et al.*, 1993). Since this discovery, the miRNA family has expanded exponentially, aided by the development of more sensitive high-throughput identification techniques. Many of these miRNAs are conserved across species and regulate important biological processes including cellular differentiation, DNA repair, metabolism, apoptosis, immunity, ageing, and cancer (Bartel, 2009).

The canonical pathway for cellular biogenesis of miRNAs is depicted in Fig. I.5.1 (Bartel, 2009; Yang and Lai, 2011). Briefly, miRNA is initially produced from a single-stranded precursor, termed primary miRNA (pri-miRNA) that is derived from RNA polymerase II and is 5′-capped and 3′-polyadenylated and folded into one or several hairpin structures. These hairpins are processed by the microprocessor complex that is composed of an RNase III enzyme (Drosha) and non-catalytic subunit (DGCR8), and binds at the double-stranded (ds) RNA- single-stranded (ss) RNA junction of the hairpin. Cleavage of the pri-miRNA generates a 60–100 nucleotide (nt) precursor miRNA (pre-miRNA) that is exported to the cytoplasm by Exportin 5. In the cytoplasm the pre-miRNA is cleaved by a second RNase III enzyme (Dicer) that generates a 21–22 (nt) double-stranded miRNA duplex. To yield the single stranded miRNA the thermodynamically less stable strand (called the guide strand) dissociates from the thermodynamically more stable strand (the passenger strand) and is incorporated for stabilization into the RNA-induced silencing complex (RISC) that contains Argonaute (Ago) proteins (predominantly Ago2). The passenger strand is usually degraded but in some instances may also be incorporated into RISC generating two RISC populations from a single miRNA duplex (Bartel, 2009). Other pathways for miRNA biogenesis exist that do not require processing by Drosha but have not been described for HCMV miRNAs (reviewed in Cullen, 2011).

In general miRNAs post-transcriptionally regulate cellular gene expression through the binding of RISC to the mRNA target. The fate of the target mRNA depends on whether the miRNA incorporated into RISC is fully or partially complementary with the target. Extensive complementarity generally leads to Ago-mediated mRNA cleavage, an siRNA-like mechanism predominant in plants but rare in animals (Huntzinger and Izaurralde, 2011). In animals, miRNAs more commonly target mRNAs via partial complementarity. Of particular importance is the seed sequence, a 6–8 residue stretch located between nucleotides 2 and 8 of the miRNA (Bartel, 2009). Mismatches in the seed sequence (often a G:U wobble base pair) can in some case be compensated by additional 3′ compensatory pairing at nt 13–16 (Fig. I.5.1). Alternatively, centred pairing (11–12 contiguous Watson–Crick nt pairs at the centre of the miRNA) can also substitute for the

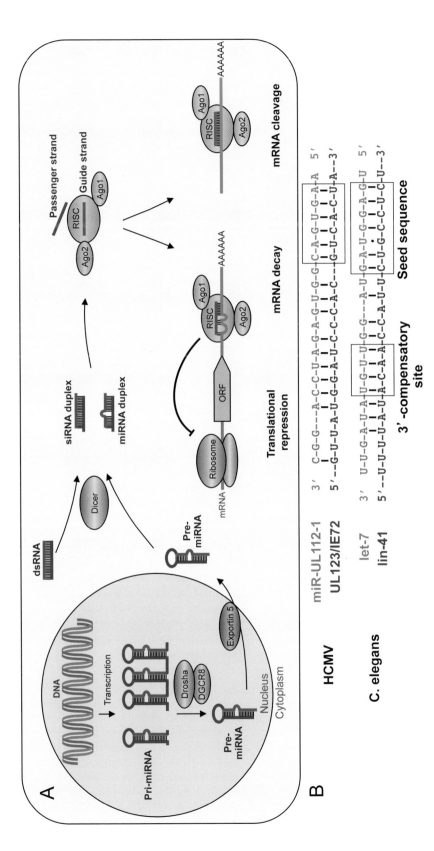

Figure I.5.1 Major pathway of microRNA biogenesis and microRNA/siRNA-mediated down-regulation of target mRNAs. (A) Expression, maturation, incorporation into RISC complex and mode of action of miRNAs and siRNAs (see text for details). (B) Examples of base complementarity between a miRNA (red) and its mRNA target (blue). Seed sequences and additional 3′-complementarity sites are boxed.

absence of seed sequence complementarity (Shin *et al.*, 2010). Partial pairing between a miRNA and the mRNA target leads to translational repression, mRNA decay or both. The mechanisms responsible for each of these outcomes have not been fully elucidated, but the finding that RISC-bound mRNAs locate in processing bodies (P-bodies) which are discrete regions in the cytoplasm involved in mRNA turnover, suggests the involvement of decapping and deadenylation processes in degradation (Huntzinger and Izaurralde, 2011). The miRNA target sites in mRNAs are most commonly located in the 3′UTR, although sites in the 5′UTR and the coding sequence have also been observed (Bartel, 2009; Grey *et al.*, 2010). The presence of several closely spaced (13–35 nt) miRNA target sites or a combination of different miRNA target sites in the mRNA may increase the efficiency of miRNA translational repression in a synergistic manner (Saetrom *et al.*, 2007; Tirabassi *et al.*, 2011).

β-Herpesvirus-encoded miRNAs

SV40 was the first mammalian virus shown to encode a single miRNA regulating the expression of T-antigen (Sullivan *et al.*, 2005). Since this observation miRNAs have been identified in a number of double-stranded, nuclear-replicating DNA viruses including several papovaviruses other than SV40, adenovirus, and multiple members of the herpesvirus family (Table I.5.1). Of the reported 240+ virally encoded miRNAs, the vast majority are encoded by the herpesviruses. Epstein–Barr Virus (EBV), a γ-herpesvirus, was the first to be shown to encode multiple miRNAs that are clustered in the region of the genome that expresses genes involved in viral latency (Pfeffer *et al.*, 2004). Since this discovery miRNAs have been characterized in other herpesvirus subfamilies including the α-herpesviruses (Herpes Simplex Viruses 1 and 2, and Marek's Disease Virus), β-herpesviruses (HCMV, Mouse CMV, and Rat CMV), and other γ-herpesviruses [Kaposi's Sarcoma associated Herpesvirus (KSHV), Rhesus rhadinovirus, Rhesus lymphocryptovirus, and MHV-68] (Cai *et al.*, 2005, 2006; Grey *et al.*, 2005; Pfeffer *et al.*, 2005; Samols *et al.*, 2005; Buck *et al.*, 2007; Dölken *et al.*, 2007; Morgan *et al.*, 2008; Umbach *et al.*, 2008, 2010; Tang *et al.*, 2009; Yao *et al.*, 2009; Jurak *et al.*, 2010; Reese *et al.*, 2010; Riley *et al.*, 2010; Meyer *et al.*, 2011). The α-herpesviruses also express miRNAs that are clustered in the region associated with viral latency (Umbach *et al.*, 2008; Tang *et al.*, 2009; Jurak *et al.*, 2010). In contrast to the α- and γ-herpesviruses, the β-herpesvirus miRNAs are located throughout the viral genome as single miRNAs and small clusters of two to nine miRNAs (Fig. I.5.2). This difference

in the genomic location of miRNA expression for the herpesviruses may reflect the stage at which miRNA expression was detected during the viral life cycle. For the α- and γ- herpesviruses latently infected cells were used to analyse miRNA expression while acutely infected cells were used for CMV. Comparison across the herpesviruses family reveals low miRNA sequence conservation despite the fact that members of the α- and γ-herpesviruses harbour miRNAs in conserved genomic locations (synteny).

HCMV miRNAs

A variety of approaches have been used over the years to identify miRNAs expressed from the HCMV genome during acute infection of cells. In early work Pfeffer and colleagues (2005) used both a prediction algorithm and small RNA cloning to identify nine HCMV pre-miRNAs (Table I.5.2). In parallel studies performed by our group (Grey *et al.*, 2005), a comparative bioinformatics method was used to predict the presence of miRNAs in the HCMV genome. A computer algorithm called stem–loop finder (SLF) (Combimatrix) was used to predict potential RNA transcripts from the HCMV genome that could form stem–loop secondary structures. To further refine this method the predicted stem–loop sequences were compared to the closely related chimpanzee cytomegalovirus genome as we predicted that genuine miRNA sequences would demonstrate evolutionary conservation. Additional analysis of the conserved stem loops was performed using an online algorithm, called miRscan, that compares two conserved sequences and determines the likelihood that they encode a miRNA based on a number of structural aspects such as the ability to form a stem loop, symmetry of bulge loops and conservation of the predicted miRNA sequence (Lim *et al.*, 2003). This analysis led to the identification of 13 candidate miRNA sequences encoded throughout the viral genome. Expression of 5 of the 13 miRNAs, miR-UL36-1, miR-UL70-1, miR-US4-1, miR-US5-1 and miR-US5-2, was confirmed through Northern blot analysis (Grey *et al.*, 2005). Following this study, and those of two other groups, a total of 11 miRNAs were identified, including miR-UL22A-1, miR-UL112-1, miR-UL148D-1, miR-US25-1, miR-US25-2, miR-US33-1 and the five previously described (Table I.5.2) (Dunn *et al.*, 2005; Pfeffer *et al.*, 2005). Consistent with mammalian species many of the HCMV miRNAs were encoded within intergenic regions, and in at least one case, miR-UL36-1, encoded within the spliced intron of a coding gene. A number of the miRNAs are encoded within open reading frames (Fig. I.5.2). miR UL112-1 is encoded directly antisense to the viral uracil DNA glycosylase,

Table I.5.1 Virus miRNAs

Virus family	Virus species	Host	# of pre-miRNAs	Literature
α-herpesvirus	HSV-1	H	16	Umbach *et al.* (2008), Jurak *et al.* (2010)
	HSV-2	H	18	Jurak *et al.* (2010)
	Herpes B virus	S	3	Besecker *et al.* (2009)
	BHV1	B	10	Glazov *et al.* (2010
	Marek's disease virus type 1	A	14	Burnside *et al.* (2008), Yao *et al.* (2008)
	Marek's disease virus type 2	A	18	Yao *et al.* (2007), Waidner *et al.* (2009)
	Infectious laryngotracheitis virus	A	7	Rachamadugu *et al.* (2009), Waidner *et al.* (2009)
β-herpesvirus	HCMV	H	13	Table I.5.2
	RhCMV	S	17	Hancock *et al.* (2012)
	MCMV	M	18	Buck *et al.* (2007), Dölken *et al.* (2007)
	RCMV	M	24	Meyer *et al.* (2011)
γ-herpesvirus	EBV	H	25	Pfeffer *et al.* (2004), Cai *et al.* (2006)
	KSHV	H	13	Cai *et al.* (2005), Pfeffer *et al.* (2005), Samols *et al.* (2005)
	Rhesus monkey rhadinovirus	S	7	Umbach *et al.* (2010)
	Rhesus lymphocryptovirus	S	36	Cai *et al.* (2006), Riley *et al.* (2010)
	Mouse γ-herpesvirus 68	M	15	Pfeffer *et al.* (2005), Reese *et al.* (2010)
Polyomavirus	SV40	S	1	Sullivan *et al.* (2005)
	SA12	S	1	Cantalupo *et al.* (2005)
	JCV, BKV	H	1	Seo *et al.* (2008)
	Merkel cell polyomavirus	H	1	Seo *et al.* (2009)
	Mouse polyomavirus	M	1	Sullivan *et al.* (2009)
Bandicoot papillomatosis carcinomatosis virus	Types 1 and 2	Mar	1	Chen *et al.* (2011)
Adenovirus	Types 2 and 5	H	1	Andersson *et al.* (2005)
Ascovirus	HvAV	I	1	Hussain *et al.* (2008)
Baculovirus	BMNPV	I	4	Singh *et al.* (2010)

H, human; S, simian; B, bovine; A, avian; M, murine; Mar, marsupial; I, insect.

a situation that could theoretically lead to the cleavage of the transcript and negative regulation of the gene. Although one group reported that miR-UL112-1 was able to cleave the UL114 transcript (Stern-Ginossar *et al.*, 2009), studies by our group found that the mRNA was resistant to cleavage (unpublished observation). In contrast, miR-UL112-1 was able to target fully complementary sequences placed in luciferase reporter constructs, suggesting that flanking sequences within UL114 protect the sequence from miRNA directed cleavage.

Recently, using next-generation sequencing (NGS)

of small RNAs from human fibroblasts infected with an HCMV clinical strain (TR), our group identified three additional species: miR-UL59-1, miR-UL148D 5p (miRNA passenger strand) and miR-US29-1 (unpublished observations). These observations have been extended by another report that used NGS of HCMV Towne strain small RNAs generated during infection of human fibroblasts. Using this approach, these investigators expanded the number of HCMV-encoded mature miRNAs to 24 including one novel precursor pre-miR-US22 (Stark *et al.*, 2011) (Table I.5.2; Fig. I.5.2). Some of these new HCMV miRNAs are expressed

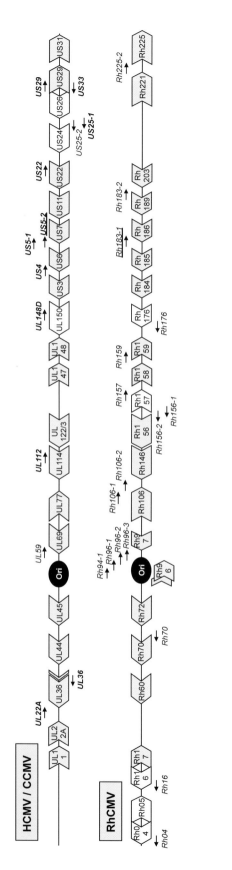
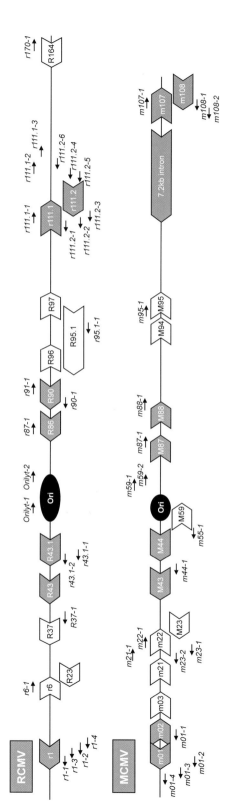

Figure I.5.2 Comparative genomic organization of miRNAs between CMV species. Position and transcriptional direction of CMV miRNAs are represented by black arrows. Block arrows and black boxes indicate open reading frames and the origin of replication, respectively. Grey backgrounds indicate ORF conservation between CMV species. In the HCMV/CCMV schematic, miRNAs conserved between the two species are indicated by bold lettering. Underlines indicate conservation between HCMV and RhCMV miRNAs. For clarity, only ORFs in the vicinity of miRNA sites are represented (adapted from Tuddenham and Pfeffer, 2011).

Table I.5.2 HCMV miRNAs

miRNA	Detected in:					Our data	Cells[a]
	Pfeffer et al. (2005)	Grey et al. (2005)	Dunn et al. (2005)	Stern-Ginossar et al. (2009)	Stark et al. (2011)		
miR-UL22A-5p	C	NB	C, NB	RT-PCR	DS	DS, RT-PCR	U373MG, RPE, F, EC, THP-1
miR-UL22A-3p			C, NB		DS		U373MG, RPE, F, EC, THP-1
miR-UL36–5p	C	NB		RT-PCR	DS	DS, RT-PCR	F, EC, THP-1
miR-UL36–3p					DS		F
miR-UL59–5p						DS, RT-PCR	F, EC, THP-1
miR-UL112–3p	C	NB		RT-PCR	DS	DS, RT-PCR	F, EC, THP-1, serum
miR-UL112–5p					DS		F
miR-UL148D-3p	C			RT-PCR	DS	DS, RT-PCR	F
miR-UL148D-5p						DS, RT-PCR	F, EC, THP-1
miR-US4–5p		NB		RT-PCR	DS	RT-PCR	F
miR-US4–3p					DS		F
miR-US5–1–3p	C	NB		RT-PCR	DS	DS, RT-PCR	F, EC, THP-1
miR-US5–2–3p	C	NB		RT-PCR	DS	DS, RT-PCR	F, EC, THP-1
miR-US5–2–5p					DS		F
miR-US22–5p					DS, NB	RT-PCR	F
miR-US22–3p					DS		F
miR-US25–1–5p	C	NB	C, NB	RT-PCR	DS	DS, RT-PCR	U373MG, RPE, F, EC, THP-1
miR-US25–1–3p					DS		F
miR-US25–2–5p	C	NB		RT-PCR	DS	DS, RT-PCR	F, EC, THP-1
miR-US25–2–3p					DS	DS, RT-PCR	F, EC, THP-1
miR-US29–5p					DS, NB	DS, RT-PCR	F, EC, THP-1
miR-US29–3p					DS		F
miR-US33–5p	C			RT-PCR	DS	RT-PCR	F, EC, THP-1
miR-US33–3p					DS		F

C, cloning; NB, northern blot; DS, deep sequencing.
[a]U373MG, astrocytoma cells; RPE, retinal pigment epithelial cells; EC, human endothelial cells; F, fibroblasts (HFF and/or NHDF); THP-1, human acute monocytic leukaemia cells.

at extremely low or negligible levels that may explain why these species were not detected in earlier studies. Whether these low level miRNAs are expressed at higher levels in different tissues is unknown but recent work with Rhesus CMV (RhCMV) miRNAs suggests that this might be the case (Hancock *et al.*, 2012). Additionally these studies were performed using the Towne HCMV strain, a laboratory isolate that lacks a 19-kbp region of the viral genome. Subsequent analysis of a clinical isolate with this 19-kbp region using these techniques may reveal yet more undiscovered HCMV miRNAs. One of the more interesting results of the Stark *et al.* (2011) study was the identification of a novel miR-US4 isoform called US4-5P. The sequence for this miRNA was shifted 5 bp from the 5′ end compared with

the previously identified miR-US4-1. Whether this miR-US4 variant is HCMV strain specific is unknown.

Expression of HCMV transcripts can be grouped into three kinetic classes, immediate-early (IE), early (E), or late (L), based on their requirement for expression of viral protein and viral DNA replication (DeMarchi *et al.*, 1980). Analysis of HCMV miRNA expression indicated that the majority of the viral miRNAs are expressed with E replication kinetics accumulating gradually over the course of infection (Grey *et al.*, 2005), with the exception of miR-UL70-1 3p and miR-US22A-1 3p that are expressed with IE kinetics. However, miR-UL70-1 3p has not been verified as an authentic miRNA, and the initial characterization by Northern blot analysis showed probe hybridization with a small RNA moiety in uninfected cells suggesting that this potential viral miRNA was an artefact (Grey *et al.*, 2005; Stark *et al.*, 2011). Interestingly, miR-UL36-1 expression was blocked by cycloheximide but not fos-carnet, suggesting that this miRNA exhibits E kinetics of expression (Grey *et al.*, 2005). This observation was surprising, as the UL36 transcript that contains the intron encoding the UL36-1 miRNA exhibits IE kinetics (Tenney and Colberg-Poley, 1991). Furthermore, the ~80-base species detected in the initial Northern blot assay was detected at elevated levels in the cycloheximide sample. One potential explanation for the block in the production of mature miRNAs would be that the extended cycloheximide treatment depleted proteins, such as Dicer, required for the processing of mature miRNAs.

MCMV miRNAs

Two independent groups published the discovery of Mouse CMV (MCMV) miRNAs in back-to back publications in 2007 (Buck *et al.*, 2007; Dölken *et al.*, 2007). Cloning of small RNA libraries from MCMV-infected mouse fibroblasts and bone marrow-derived macrophages identified 18 miRNA loci (Fig. I.5.2) expressing 27 miRNA species. Validation by Northern blot confirmed the expression of 21 miRNAs. Similar to HCMV, most MCMV miRNAs are expressed with IE or E kinetics with the exception of miRNAs of the m108 cluster, which require viral DNA replication for expression (Buck *et al.*, 2007). Dölken *et al.* (2007) assessed miRNA expression in the liver, spleen and lungs of mice infected with wild-type (WT) MCMV. Using RNase protection assays, the authors determined that miR-m01-4 and miR-M44-1 levels decreased in the liver and spleen from 3 to 5 days post infection (p.i.), while miR-m01-4 increased in expression in the lung over that time period. These trends roughly correlated with viral titres in these tissues (Dölken *et al.*, 2007).

The mouse model of MCMV infection was the first to assess the importance of viral miRNAs *in vivo* (Dölken *et al.*, 2010). CMV infects many cell types *in vivo* and must evade the host immune response, thus the importance of viral miRNAs in this context is critical to address. C57BL/6 and BALB/c mice were infected with wild type virus, a miR-M23-2/m21-1 knockout virus or its revertant. At 3 days p.i., no differences in viral replication in the liver, lung or kidneys were observed in any animals. However, at 14 days p.i. an ~100-fold attenuation of viral loads was observed in the salivary glands of C57BL/6 mice infected with the miRNA knockout virus. Under the same experimental conditions the attenuation of miR-M23-2/m21-2 knockout virus growth was only 2-fold in the salivary

glands of BALB/c mice. Lowering the viral dose 20-fold increased the attenuation of viral replication in the salivary glands to 5-fold in these mice. C57BL6 mice express the NK cell receptor Ly49H, which recognizes the viral protein m157 and elicits an effective NK cell response (Arase *et al.*, 2002) (see also Chapter II.9). BALB/c mice do not express this receptor, and thus are more susceptible to MCMV infection, which correlates with the increased replication of the miRNA mutants in this mouse strain. These data indicate that viral load and genetic background are significant contributors to the attenuation of viral miRNA mutant replication in the salivary glands. The miRNA mutants were also attenuated for viral growth in several other mouse strains, including 129SvJ.IFNγR–/–, indicating that the observed effect is independent of IFNγ. Elimination of virus from the salivary gland depends on the actions of both CD4[+] T-cells and NK cells. Depletion of both, but not each individually was required to enhance miR-M23-2/m21-2 knockout virus replication. Interestingly, a significant difference was observed between titres of WT and knockout virus in salivary glands after combined NK and CD4[+] T-cell depletion, suggesting that other factors are also involved in limiting viral replication in this tissue in the absence of miR-M23-2/m21-1.

RCMV miRNAs

In order to identify miRNAs encoded by Rat CMV (RCMV), small RNA was isolated from virus infected fibroblasts as well as salivary gland tissue from persistently infected animals and analysed by NGS (Meyer *et al.*, 2011). Results from these analyses indicated that RCMV expressed 37 mature miRNAs (Fig I.5.2) that were located at 24 loci (24 guide strands and 13 passenger strands). Up to five isoforms were detected for each miRNA with variations occurring at the 3′, and less frequently at the 5′ end of the sequence, similar to what was reported for MCMV (Buck *et al.*, 2007; Dölken *et al.*, 2007). Interestingly, two short miRNAs (17 bases) were only detected in salivary gland tissue, suggesting that the virus may process pre-miRNAs differently in different tissues. In support of this idea, some isoforms were differentially represented in the infected fibroblasts and salivary gland samples. The significance of such differential expression of miRNA species and isoforms between tissues and/or cell types is unknown. The miRNAs located in the r111.1 gene cluster of RCMV are differentially expressed in fibroblasts, with some miRNAs species expressed with E kinetics, while others are expressed with L kinetics. Interestingly, the r111.1 cluster in RCMV is syntenic with the m108 cluster in MCMV (Fig. I.5.2), which are expressed with L kinetics. These observations suggest that this conserved miRNA cluster may play a role in persistence of these rodent CMVs. Meyer et al. (2011) hypothesized that similar to expression of RCMV genes, miRNAs may be differentially expressed in tissues during the acute and persistent stages of infection. RT-PCR was used to assess the expression of miR-R87-1 and miR-R111.1-2 in various tissues at 7 and 28 days post infection. Both miR-R87-1 and miR-R111.1-2 were expressed in all tissues examined at 7 days, corresponding with acute infection. At 28 days p.i., miR-R111.1-2 was highly up-regulated in the salivary gland, but undetected in all other tissues. In contrast, miR-R87-1 was down-regulated in the salivary gland at this time point. These data suggest that expression of RCMV miRNAs is highly regulated during different stages of infection *in vivo* (see also Chapter II.15).

RhCMV miRNAs

Recently, our lab performed an analysis of small RNA species obtained from Rhesus CMV (RhCMV) infected cells by NGS that revealed the presence of 17 viral miRNAs (Hancock et al, 2012). Similar to RCMV, RhCMV miRNA isoforms were detected with sequence variability mostly at the 3′ end and are dispersed throughout the viral genome (Fig. I.5.2). Analysis of RhCMV infected fibroblasts by RT-PCR revealed the presence of thirteen of the seventeen RhCMV miRNAs. Unlike rat, mouse and human CMV miRNAs, which accumulate throughout the course of infection, some RhCMV miRNAs were expressed at their highest levels early in infection. Three patterns of expression were observed in these cells: (1) peak expression at 24–48 h p.i. followed by a decline; (2) peak expression at the earliest time-points tested followed by a sharp decline by 48 hours; and (3) gradual accumulation over the course of infection similar to the other species of CMVs. In contrast with the differential expression observed in fibroblasts, all miRNAs gradually increased in abundance during infection of rhesus endothelial cells. However, viral replication was delayed and DNA levels were significantly lower in this cell type, suggesting that the observed miRNA expression kinetics could reflect the delayed expression patterns of transcripts from the viral genome. RhCMV miRNA expression was also examined in salivary gland tissue obtained from RhCMV persistently infected rhesus macaques. Of the 17 RhCMV miRNAs, only six were detected in the salivary glands of infected animals, demonstrating differential expression of the RhCMV miRNAs *in vivo* and suggesting that only some viral transcripts are being transcribed and/or processed in the salivary glands to form mature miRNAs. Similar to

the observations with MCMV (Dölken et al., 2010), these miRNAs could play an important role in viral persistence in this tissue. Comparison between RhCMV and HCMV miRNA sequences revealed that RhCMV miR-Rh183-1 is highly similar to HCMV miR-5-2. These miRNAs share similar seed sequences and are located in the 3′UTR of orthologous genes (Rh186 and US7, respectively; Fig. I.5.2). Moreover, luciferase assays revealed that both miRNAs down-regulate expression of the corresponding gene. The homology between these two viral miRNAs suggests significant evolutionary pressure to maintain this seed sequence and thus the ability to target common cellular and viral genes. The importance of these homologous miRNAs during CMV replication and persistence will be tested in the rhesus macaque (*Macaca mulatta*) model of RhCMV infection (see also Chapter II.22).

Targets of HCMV miRNAs

The main thrust of the herpesvirus miRNA field since the discovery that viruses encode miRNAs has been to identify cellular and viral mRNA targets. Initial approaches used bioinformatic algorithms utilizing miRNA seed sequences to predict mRNA targets. This method combined with luciferase reporter assays to validate these targets was tedious and had a high rate of false positives. Although *in silico* prediction of miRNA targets using new algorithms has improved, newer biochemical methods have provided greater accuracy in target identification. The basis of the biochemical methods involves the ribonucleoprotein immunoprecipitation (RIP) of the RISC complexes using either antibodies to Argonaute 2 or a myc-tagged version of the protein followed by isolation of the mRNA and analysis on a microarray chip containing probes for all the known genes in the genome of the host (RIP-Chip) (Keene et al., 2006; Easow et al., 2007; Karginov et al., 2007; Tan et al., 2009). While this technique was generated to identify mRNAs enriched in cells expressing the miRNA of interest, a more direct approach uses streptavidin-coated microbeads to precipitate a biotinylated miRNA and its associated mRNAs (Grey et al., 2010). Comparison of results from the two methods has provided a powerful approach to reliably identify mRNA targets (Grey et al., 2010) (see Tables I.5.3 and I.5.4). Newer biochemical methods have been developed to identify miRNA targets that are based on either high-throughput sequencing of RNAs isolated by chemical cross-linking immunoprecipitation (HITS-CLIP) (Licatalosi et al., 2008; Chi et al., 2009) or photoactivatable-ribonucleoside-enhanced CLIP (Hafner et al., 2010) that should significantly expand the number of potential CMV mRNA targets.

Table I.5.3 Validated viral targets of CMV miRNAs

Virus	miRNA	Target	Description	Validation method	Literature
HCMV	miR-UL112	IE72	MIE IE72	Luc, WB (infection)	Grey *et al.* (2007a)
				Luc, WB, KO	Murphy *et al.* (2008)
		UL120/1	MIE region exons	Luc	Grey *et al.* (2007a)
		UL112/3	Viral DNA synthesis	Luc	Grey *et al.* (2007a)
		UL114	Viral DNA glycosylase	Luc, WB	Stern-Ginossar *et al.* (2009)
		UL17/18	MHC class I homologue	Luc	Huang *et al.* (2010)
	miR-US5–1	US7	Unknown function	Luc, WB (KO)	Tirabassi *et al.* (2011)
	miR-US5–2	US7	Unknown function	Luc, WB (KO)	Tirabassi *et al.* (2011)
RhCMV	miR-Rh183-1	Rh186	Unknown function	Luc	Hancock *et al.* (2012)

Luc, luciferase assay; WB, Western blot; KO, infection with knockout virus.

Table I.5.4 Validated cellular targets of CMV miRNAs

Virus	miRNA	Target	Description	Validation method	Literature
HCMV	miR-US25–1	CCNE2	Cyclin E2	Luc, WB (KO)	Grey *et al.* (2010)
		H3F3B	Histone H3 variant	Luc	Grey *et al.* (2010)
		TRIM28	Transcriptional corepressor	Luc, WB (KO)	Grey *et al.* (2010)
	miR-UL112	MICB	NK cell activating receptor	Luc, KO	Stern-Ginossar *et al.* (2007)
		ZFP36L1	Zinc-finger protein	Luc	Huang *et al.* (2010)
		Transportin 1	Subunit of karyopherin complex	Luc	Huang *et al.* (2010)
		L7a	Ribosomal protein	Luc	Huang *et al.* (2010)
		IL32	Cytokine	Luc	Huang *et al.* (2010)
		NFκB activating protein	Unknown	Luc	Huang *et al.* (2010)
	miR-US4–1	ERAP1	Aminopeptidase	Luc, WB, KO	Kim *et al.* (2011)
MCMV	miR-M23–2	CXCL16	Chemokine	Luc, KO	Dölken *et al.* (2010)

Luc, luciferase assay; WB, Western blot; KO, infection with knockout virus.

HCMV miRNA targets in the viral genome

miR-UL112-1 targets HCMV IE1

Some of the first studies to identify viral miRNA targets focused on the highly expressed HCMV miRNA miR-UL112-1. Our group used a comparative bioinformatics-based approach to identify potential transcripts regulated by HCMV-encoded miRNAs (Grey *et al.*, 2007). Interestingly, 14 viral gene transcripts were found to contain potential target sequences for miR-UL112-1 including a cluster of three potential targets within the major immediate-early (MIE) region of the virus: UL120, UL121, and UL123 (IE72), also known as IE1. The MIE of HCMV encodes a number of regulatory

proteins that coordinate viral gene expression during infection, including the major trans-activators IE72 and IE86 (see also Chapter I.10). Disruption of IE72 results in a significant attenuation of viral replication following low multiplicity infections (Mocarski *et al.*, 1996). Three target sites were identified within the 3′UTR sequences of UL120, UL121, and UL123. When cloned down-stream of the reporter gene all three target sites effectively down-regulated luciferase expression in the presence of miR-UL112-1. In addition, expression of miR-UL112-1 in combination with a vector containing the MIE region significantly reduced IE72 expression. Since IE72 is an important trans-activator that is required for efficient viral replication, Grey and colleagues (2007) hypothesized that aberrant expression of miR-UL112-1 may result in attenuation of viral replication. HCMV DNA replication was inhibited up to 5-fold in cells that had been transfected with synthetic miR-UL112-1 RNA duplex prior to infection, indicating that expression of miR-UL112-1 has the potential to attenuate acute replication of HCMV. This observation may have important implications for latency control of HCMV. Subsequently, Murphy *et al.* (2008) developed a miRNA-target prediction algorithm to identify viral targets of four herpesviruses including HCMV. High probability targets included IE transactivators for each of the four viruses: ICP0 of HSV-1; IE1 of HCMV; BZLF1 and BRLF1 of EBV; and Rta and Zta of KSHV, offering confirmation of the data published by Grey *et al.* (2007). In addition, Murphy and colleagues found that mutant viruses unable to express miR-UL112-1 or that encoded IE1 lacking the miR-UL112-1 target sites expressed higher levels of IE1 protein during infection. The determination that herpesviruses express miRNAs that target their transactivator genes is highly significant, and suggests that the viruses may utilize these miRNAs to establish and/or maintain viral latency.

miR-UL112-1 targets the HCMV uracil DNA glycosylase

Stern-Ginnosar and colleagues identified the viral uracil DNA glycosylase (UL114) that is encoded antisense to miR-UL112-1 as a second viral gene target of miR-UL112-1 (Stern-Ginossar *et al.*, 2009). The uracil DNA glycosylase associates with the DNA polymerase processing factor ppUL44 within the DNA replication complex and increases the efficacy of both E and L phase viral DNA synthesis (Prichard *et al.*, 1996). The authors demonstrated that miR-UL112-1 reduces UL114 protein levels and that this had moderate effects on the ability of the virus to properly excise uracil residues from viral DNA. The authors hypothesized that the miR-UL112-1-mediated suppression of UL114 may be used to inhibit DNA replication during the establishment of latency. In addition to miR-UL112-1, four additional miRNAs miR-US33-1, miR-UL63-1, miR-UL70-1 and miR-UL148D-1 are encoded complementary to ORFs of known viral genes (US29, UL62, UL70 and UL150, respectively) and would be predicted to regulate their expression, as would miRNAs encoded antisense to the 3′UTRs of viral genes.

miRs US5-1 and US5-2 target the US7 gene of HCMV

A recent study observed that two HCMV miRNAs (miR-US5-1 and miR-US5-2) encoded antisense to the 3′UTR of US7 were able to down-regulate the viral gene (Tirabassi *et al.*, 2011). In this study miR-US5-1 and miR-US5-2 were observed to regulate US7 in a highly synergistic manner even at very low miRNA concentrations. The regulation of US7 by the HCMV miRNAs was mediated through three functional miRNA binding sites in the 3′ UTR of which two were completely complementary and one was an imperfect match. Interestingly, inhibition of US7 expression mediated by the two complementary sites involved degradation of the mRNA while the site with the imperfect match appeared to inhibit expression through a translational mechanism. This result suggests that translational repression due to incomplete miRNA:mRNA pairing can exert an effect equal to that of repression resulting from complete pairing. This data provided the first demonstration of a gene that is regulated through both mechanisms. In addition, these observations were the first evidence that HCMV miRNAs can act cooperatively to enhance the down-regulation of targets. This result has significant implications in determining the phenotypic effects mediated by HCMV miRNAs and the necessity to determine all of the potential miRNAs that target an individual mRNA. Viruses with mutations that inactivate miR-US5-1 and/or miR-US5-2 displayed an increase in US7 protein expression. Although the exact function of US7 is currently unknown, the ORF is located in a region that encodes proteins involved in MHC down-regulation (Hansen and Bouvier, 2009). Since RhCMV also encodes a miR-US-5-2 homologue that also targets the RhCMV US7 homologue (see above), the conservation suggests that regulation of this gene is important in the life-cycle of the virus.

Cellular targets of CMV miRNAs

HCMV miR-UL112-1 targets MICB

One of the first HCMV miRNA cellular targets identified was the MHC I polypeptide related sequence B (MICB) (Stern-Ginossar *et al.*, 2007), a stress-induced ligand for the NK cell activating receptor NKG2D (Strong, 2002) (see also Chapter II.8). MICB was discovered as a miR-UL112-1 target using the bioinformatic algorithm RepTar, which identifies repetitive elements within 3′ UTRs and evaluates their potential for miRNA binding sites (Elefant *et al.*, 2010). Historically, host specificity has hampered identification of cellular targets of CMV miRNAs, but RepTar is unique in that this algorithm does not require evolutionary conservation of miRNA target sites. The identification of miR-UL112-1 binding sites within the MICB 3′ UTR suggests that HCMV uses multiple mechanisms to evade the NKG2D-mediated NK cell response, as the UL16 protein also down-regulates MICB surface expression during infection (Welte *et al.*, 2003). Expression of miR-UL112-1 in various tumour cell lines resulted in down-regulation of MICB surface expression (Stern-Ginossar *et al.*, 2007). Interestingly, the related NK cell activating receptor MICA differs from MICB in the miR-UL112-1 binding site by only one nucleotide; however MICA was not affected by miR-UL112-1 expression. To confirm that miR-UL112-1 acts to down-regulate MICB through binding the predicted target sites within the 3′ UTR, the MICB WT or mutant seed sequence 3′ UTRs were cloned into a luciferase reporter. In these studies HCMV miR-UL112-1 was capable of decreasing luciferase expression from the WT but not mutant MICB 3′ UTR. MICB mRNA levels were not changed in the presence of miR-UL112-1, suggesting that MICB is targeted for translational repression rather than mRNA degradation. This group also demonstrated that WT but not mutant miR-UL112-1 viruses more efficiently down-regulated MICB surface expression that resulted in decreased NK cell killing.

MICB protein levels are also modulated by at least nine cellular miRNAs (Stern-Ginossar *et al.*, 2008; Nachmani *et al.*, 2009, 2010). The miR-UL112-1 binding site overlaps that of the cellular miRNA miR-373, which may be crucial to prevent this site from being mutated by the host (Nachmani *et al.*, 2010). Surprisingly, these authors determined that over–expressing combinations of cellular miRNAs often diminishes the down-regulation of MICB. In contrast, combining the cellular miRNA miR-376a with the viral miR-UL112-1 resulted in synergistic down-regulation of MICB. Nachmani and colleagues also demonstrated that the proximity of the cellular and viral miRNA target sites (24 nucleotides from the 5′ ends of each miRNA) was essential to the synergistic response, as moving the sites further from one another (> 850 nucleotides) resulted in only additive down-regulation of MICB. MICB mRNA levels are not altered during the synergistic response, suggesting that translational repression can be enhanced under these conditions. Infection of cells expressing miRNA sponges (constructs that contain many complementary binding sites for a miRNA of interest and can titrate miRNAs from the cellular environment; Ebert and Sharp, 2010) for both miR-UL112-1 and miR-376a resulted in synergistic increases in MICB protein levels at 72 h p.i. compared to cells expressing each sponge individually. The increased MICB levels also resulted in enhanced NK cell killing. These data suggest that HCMV may have evolved to cooperate with cellular miRNAs to synergistically down-regulate target genes.

HCMV miR-US4-1 targets ERAP1 to inhibit CD8+ T-cell responses

Recently, the HCMV miR-US4-1 was shown to target the ERAP1 aminopeptidase involved in trimming peptides in the ER for loading onto MHC class I molecules for antigen presentation (Kim *et al.*, 2011). In this report miR-US4-1 was shown to target the 3′ UTR of ERAP1 mRNA in a dose-dependent manner as well as significantly decrease ERAP1 mRNA and protein levels. Using a well-characterized ovalbumin peptide T-cell antigen, the authors determined that expression of miR-US4-1 significantly decreased CD8+ T-cell recognition of the antigenic ovalbumin peptide. These studies suggest another mechanism used by HCMV to evade recognition by the human immune system. Interestingly, although miR-US4-1 was identified as one of the original HCMV encoded miRNAs that is highly expressed during infection, as mentioned above a recent study using high-throughput sequencing of small RNAs from HCMV infected cells identified a miRNA that is shifted five bases from the 5′ end (miR-US4–5P). The Kim *et al.* (2011) studies overexpressed miR-US-4-1 using an expression construct with the annotated sequence of miR-US4-1 as well as a miRNA mutant in which three point mutations were made within the first seven nucleotides. One of these three mutations was within the miR-US4–5P seed sequence making it unclear which miRNA species caused the down-regulation of ERAP1.

HCMV miR-US25-1 targets genes involved in the cell cycle

The Ago-2 RIP-Chip method was used in combination with RISC immunoprecipitation of biotinylated miRNA to identify cellular mRNA targets of one of the most highly expressed HCMV miRNAs, miR-US25-1 (Grey *et al.*, 2010). A number of the mRNAs enriched in these assays encoded proteins involved in cell cycle control including cyclin E2, BRCC3, EID1, MAPRE2 and CD147 as well as histone proteins suggesting that miR-US25-1 is targeting genes within a pathway (see also Chapter I.14). Surprisingly, the majority of the cellular transcripts enriched in these methods encoded miR-US-25-1 seed sequences in the 5′UTR rather than the 3′ UTR of targeted mRNAs. Using luciferase assays the 5′ UTR sequences were found to be functional in down-regulating expression from these constructs. In addition mutation of miR-US25-1 within a replication competent virus resulted in a significant increase in cyclin E2. These observations provided the first example of a viral miRNA that was able to repress protein expression through 5′ UTR sequences. Infection with HCMV has long been known to manipulate the cell cycle by altering the expression of cyclin-dependent kinases (CDKs) and their associated cyclin subunits (Payton and Coats, 2002). Cyclin E proteins are expressed early in G1 phase when they bind to and activate CDK2, resulting in progression into S phase. Previous studies have demonstrated that HCMV induces resting G0 cells to enter the cell cycle whereupon the virus blocks further progression at the G1/S boundary (Kalejta and Shenk, 2002). By blocking the cell cycle at the G1/S phase the virus creates a cellular environment conducive for DNA replication. HCMV induced expression of cyclin E1 is thought to play an important role in driving cells into the G1/S phase (Jault *et al.*, 1995). Although deletion of US25-1 did not result in a phenotypic effect on viral replication following infection of primary human fibroblast cells, regulation of the target genes identified may be important in other cell types, such as endothelial cells or macrophage cells, or during the latent or persistent phase of the virus life cycle. These observations provide the first evidence that HCMV miRNAs are targeting cellular pathways.

HCMV miRNAs target several members of the endocytic pathway

Further evidence that HCMV miRNAs are targeting cellular gene pathways is derived from our unpublished observations that multiple HCMV miRNAs cooperatively target multiple cellular mRNA transcripts that encode proteins involved in cytokine release through the endocytic compartment (Hook *et al.*, manuscript in preparation). Using RISC-IP with miR-UL112-1, we observed that two of the top ten enriched cellular transcripts included vesicle-associated membrane protein 3 (VAMP3) and a member of the RAS oncogene family (RAB5C). VAMP3 is a vesicular R-SNARE (soluble NSF attachment protein receptor) on recycling endosomes that plays an important role in secretion of TNF-α, while RAB5C is small guanosine triphosphatase (GTPase) that regulates membrane trafficking between the recycling endosome and the cellular membrane. *In silico* analyses indicated that the 3′UTRs of both VAMP3 and RAB5C each contain seed sequences for both miR-UL112-1 and miR-US5-1. We observed that both miR-UL112-1 and miR-US5-1 were able to down-regulate luciferase reporters containing the 3′ UTRs of VAMP3 and RAB5C as well as expression of protein, with both miRNAs required for maximal effect. *In silico* analyses of the secretion pathway indicated that several other members may also be targeted by the HCMV miRNAs including RAB11A, SNAP23, syntaxin 6 (STX6), and syntaxin 7 (STX7).

We have verified that RAB11A and SNAP23 are targets of miR-US5-1 and miR-UL112-1 among others and are currently examining STX6 and STX7. Lastly, we also observed that miR-UL112-1 and miR-US5-1 transfection significantly reduced the release of inflammatory cytokines in cells stimulated by LPS demonstrating that these HCMV miRNAs can functionally regulate the inflammatory cytokine. Importantly, this reduction was greater than when the individual cellular proteins VAMP3 and RAB5C were knocked down using siRNA suggesting that miR-UL112-1 and miR-US5-1 were each targeting multiple cellular genes in the secretion pathway and possibly others for maximal effect. HCMV miRNA regulation of release of the inflammatory cytokines may represent a novel mechanism for viral immune evasion. Taken together all the above data suggest that HCMV miRNAs cooperatively target multiple genes and cellular pathways to create a proviral environment.

MCMV targets the CXCL16 chemokine

Currently the only cellular target identified for an MCMV miRNA is CXCL16 (Dölken *et al.*, 2010). In these studies the RepTar algorithm was used to identify CXCL16 as a possible target of miR-M23-2 and miR-m21-1. CXCL16 is a chemokine with soluble and transmembrane forms, and binds CXCR6 to attract Th1, Tc1 and NK cells to infection (Deng *et al.*, 2010). In luciferase assays, miR-M23-2 was shown to down-regulate expression of luciferase from constructs containing the CXCL16 3′ UTR. In addition, revertant but not mutant miR-M23-2/miR-m21-1 viruses were able to reduce expression of luciferase from a reporter containing the 3′ UTR of CXCL16. Importantly the mutant but not revertant miR-M23-2/miR-m21-1 viruses exhibited reduced growth in the salivary glands of infected mice. This study was the first report of a virus growth phenotype associated with a mutant viral miRNA.

Effects of CMV infection on cellular miRNAs

In addition to encoding their own miRNAs, viruses can alter the expression of cellular miRNAs in a manner beneficial to the virus. In one of the earliest studies evaluating the regulation of cellular miRNAs by HCMV, authors Wang *et al.* (2008) identified several cellular miRNAs that were significantly up or down-regulated following infection. Those most profoundly altered included miR-17, -20, -106, and -219 (up-regulated, $n=4$) and miR-21, -99, -100, -101, -155, -181, -213, -222, -223, and -320 (down-regulated, $n=10$). Of those down-regulated during HCMV infection two miRNAs, miR-100 and miR-101, were predicted by the miRNA target prediction algorithms MIRANDA and TargetScan to target members belonging to the mammalian target of rapamycin (mTOR) pathway. The mTOR pathway plays an important role in a variety of cellular processes including regulation of the cell cycle, cell survival, protection from apoptosis, as well as being important for HCMV replication (Mamane *et al.*, 2006). miR-100 was found to target the mTOR Complex 1 (mTORC1) subunit raptor, while the mTOR Complex 2 (mTORC2) subunit rictor contains two target sites for miR-101. Wang *et al.* (2008) hypothesized that if a cellular miRNA is down-regulated upon HCMV infection, overexpressing the miRNA exogenously would interfere with viral replication. Overexpression of either the miR-100 or miR-101 in HCMV-infected cells reduced the yield of infectious HCMV (~70% reduction at 50 nM of either mimic). When combined, the two mimics had a greater effect than either alone (33-fold reduction with 25 nM of each mimic). Importantly, the data suggest that HCMV alters the expression of cellular miRNAs to directly benefit its own replication.

Poole and colleagues assessed cellular miRNA expression patterns in CD34+ progenitor cells (a potential reservoir of latent virus) experimentally infected with HCMV (Poole *et al.*, 2011). They discovered that miR-let-7a, miR-let7b, miR-297 and miR-92a were decreased 2- to 3-fold in two independent experiments. Target prediction software suggested GATA-2, a transcription factor involved in survival and proliferation of haematopoietic progenitor cells (Pan *et al.*, 2000), was a potential target of miR-92a. Luciferase assays confirmed the 3′ UTR of GATA-2 is targeted for down-regulation by miR-92a, and Poole *et al.* (2011) demonstrate that GATA-2 protein levels are increased in HCMV-infected CD34+ progenitor cells. Cellular IL-10 is an important antiapoptotic protein that is regulated by GATA-2 (Shin *et al.*, 2003). ELISA assays indicate that myelomonocytic cell lines transfected with antagomirs to miR-92a secreted increased amounts of IL-10. Knocking down expression of GATA-2 with siRNAs abrogated the effect of the antagomir to miR-92a, suggesting a direct link between IL-10 levels and miR-92a regulation of GATA-2. Interestingly, infection of CD34+ progenitor cells with HCMV in the presence of neutralizing antibodies to IL-10 resulted in loss of viral genomes over 10 days of infection, potentially due to increased apoptosis of this cell population. Thus, Poole and colleagues suggest that HCMV down-regulates the expression of miR-92a for increasing GATA-2 levels and IL-10 production

in order to prevent apoptosis and maintain the viral genome during latent infection of CD34+ progenitor cells.

The cellular miRNA miR-132 was observed to be highly induced following infection of cells by a number of herpesviruses including HCMV, HSV-1, and KSHV. Induction was the result of virus binding and/or viral entry and not viral gene expression, as miR-132 was induced after exposure of lymphatic ECs (LECs) to ultraviolet irradiated, but not protease-treated virus. Inhibition of miR-132 increased IFN-β mRNA levels during infection of THP-1 monocytes with HCMV, HSV-1, or KSHV, and suppressed HCMV and KSHV replication as measured by the cell-associated virus load in THP-1 cells. Using a combination of the PITA and TargetScan algorithms, the authors identified the transcriptional co-activator, p300 as a potential target of miR-132. p300 and CREB-binding protein (CBP) are involved in the initiation of antiviral innate immunity (Merika *et al.*, 1998). Luciferase assays were used to confirm that p300 is a *bona fide* target of miR-132. Transfection with mimics caused only small changes in p300 mRNA levels, suggesting miR-132 primarily blocks p300 translation. The model proposed suggests that upon herpesvirus infection, kinases MAPK, SAPK, and possibly PKA are activated, leading to the phosphorylation and activation of CREB inducing expression of miR-132. The authors propose that miR-132 targets the 3′UTR of p300, resulting in reduced transcription of IFN- and NFκB-inducible genes. Whether HCMV utilizes protein- or miRNA-based mechanisms to potentiate miR-132 expression is unknown.

In order to identify cellular miRNAs that were involved in herpesvirus replication, 286 unique cellular miRNAs were systematically overexpressed or inhibited during infection with HSV-1, MCMV, HCMV, or MHV-68 (Santhakumar *et al.*, 2010). In these studies the miRNAs miR-30b, miR-30d, and miR-93 were observed to enhance infection for each virus while miR-24, miR-103, miR-214, and miR-199-3p exhibited antiviral effects. Interestingly, miR-214 and miR-199-3p also proved effective in limiting the replication of Semliki forest RNA virus. Worthy of note, both MCMV and HCMV down-regulated miR-199-3p and miR-214 expression during infection. This observation suggests that CMVs can modulate the expression of cellular miRNAs in order to interfere with the cellular antiviral response.

Using HITS-CLIP methods to determine why individual miRNAs are rapidly down-regulated during MCMV infection, specific viral RNA segments were identified within the RISC complex that were associated with miRNA degradation (Libri *et al.*, 2012). In this work the 3′ UTR of a previously uncharacterized

MCMV ORF, m169, was associated with miR-27 that is rapidly degraded during infection. The 3'UTR of m169 was observed to inhibit miR-27 functional activity in luciferase assays and inhibition of m169 also correlated with an increase in miR-27 during infection. The authors speculate that viral regulation of cellular miRNA expression may be an important mechanism for adapting the host cell for viral growth and latency.

Conclusions

These above studies have clearly demonstrated that CMV miRNAs are able to regulate both cellular and viral genes. The question remains as to why CMV has evolved miRNA regulation of genes especially given that a number of functions associated with viral miRNAs are redundant with virally expressed proteins. For example both miR-UL112-1 and UL16 target MICB to avoid NK cell killing (Dunn *et al.*, 2003; Stern-Ginossar *et al.*, 2007). One explanation for the redundancy may be that the expression of UL16 protein may be a more robust mechanism of targeting MICB, required during acute replication of the virus, when the host immune response is likely to be more vigorous. However, targeting of MICB by miR-UL112-1 may be more suitable for long term regulation during persistent or latent infection where MICB is at lower levels and miRNA regulation would be adequate, while avoiding the production of potentially immunogenic viral proteins. Another example of redundancy is regulation of the cell cycle by miR-US25-1 and multiple HCMV encoded proteins (Wiebusch and Hagemeier, 1999; McElroy *et al.*, 2000; Murphy *et al.*, 2000; Kalejta and Shenk, 2002, 2003; Kim *et al.*, 2003; Grey *et al.*, 2010). Clearly, regulation of the cell cycle by CMV proteins is critical for optimal replication of the virus (Castillo and Kowalik, 2004; Sanchez and Spector, 2008). In this case miR-US25-1 may function as a rheostat regulator, modulating expression of cyclin E2 to generate the correct balance in protein induction. This situation may contribute to the virus's ability to block cell cycle progression at the G1/S phase, or to protect the infected cell against toxicity. Overexpression of cyclin E2 has been linked to sensitivity to apoptosis and unchecked induction of cyclin E2 may be detrimental to the virus (Mazumder *et al.*, 2000; Ugland *et al.*, 2008). Alternatively, miR-US25-1 function may be unrelated to cell cycle control. Recent studies have suggested that herpesvirus miRNAs may be important during persistent or latent infection (Grey *et al.*, 2007; Murphy *et al.*, 2008; Umbach *et al.*, 2008; Ziegelbauer *et al.*, 2009; Lei *et al.*, 2010). A myeloid progenitor cell is considered to be the latent site of CMV (Sinclair and Sissons, 1996; Soderberg-Naucler *et al.*, 1997) (see also Chapters

I.19 and I.20). By targeting myeloid progenitor genes involved in cell cycle progression and differentiation, the virus could manipulate the production of cells generated by latently infected progenitors to favour certain cell types such as monocytes and macrophages. Although deletion of miR-US25-1 did not result in a phenotypic effect on the replication following infection of primary human fibroblast cells, regulation of the target genes identified may be important in other cell types, such as endothelial cells or macrophage cells, or during the latent or persistent phase of the virus life cycle.

Another potential function for CMV miRNAs is exemplified by miR-UL112-1 targeting of the HCMV transcriptional activator protein IE72 (IE1) gene (Grey *et al.*, 2007). In this case regulation of this important HCMV gene may be specifically required for the establishment or maintenance of latent or persistent infection rather than during acute replication of the virus. CMVs establish lifelong infections of their hosts, during which time the virus maintains a restricted replication profile either as a persistent infection, with production of low levels of virus, or a true latent infection, in which virus is only produced following a reactivation stimulus. To achieve this, the virus must employ strict regulation of viral gene expression and mechanisms to evade the host immune system. The non-immunogenic nature of miRNAs and their ability to target multiple genes, both host and viral, would make them ideal agents of gene regulation during latent or persistent infection. Expression of viral miRNAs may allow the virus to restrict the expression of immunogenic viral genes involved in acute replication, while at the same time modulating the host immune response. The MIE genes, including IE72 (IE1), have been suggested to play pivotal roles in controlling latency and reactivation as both genes are important in driving the expression of E and L genes required for acute replication of the virus. Studies have also suggested that expression of the MCMV functional homologues of IE72 and IE86, MCMV IE1 and IE3, respectively, may be important triggers of viral reactivation in a lung model of latent infection (Simon *et al.*, 2006). miR-UL112-1 may therefore play a significant role during HCMV latency by restricting reactivation of the virus through negative regulation of IE72 expression.

Clinically, the discovery of HCMV miRNAs may provide a tool to develop new therapies. Currently small molecule inhibitors of miRNAs are available and are highly effective and specific. Targeting of viral miRNAs using such inhibitors may drive the virus towards acute replication, while at the same time disrupting viral immune modulation resulting in possible immune clearance of the virus. Although this approach

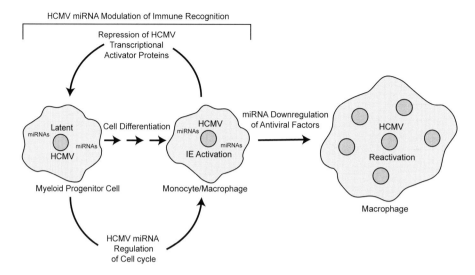

Figure I.5.3 The role of HCMV miRNAs in latency and reactivation.

is attractive, the treatment may have potential risks of pathological effects caused by acutely replicating HCMV, especially in immunocompromised patients—which are also the patients most in need of antiviral therapy. A second approach based on the idea that viral miRNAs restrict viral replication would be to mimic the expression of miRNA using synthetic miRNAs. Unlike many antivirals that artificially block viral processes, delivery of endogenous viral miRNAs could exploit the virus's own mechanisms to subdue replication. Not only might this approach be effective, but also the treatment would be less prone to the problems of viral escape and resistance as the virus has evolved to maintain these mechanisms. Potential drawbacks include off-target effects of delivering high levels of synthetic miRNAs and also the risk of immune suppression due to regulation of immune regulatory molecules. Given the potential for the regulation of multiple viral and cellular genes, additional studies will be necessary to demonstrate important regulatory roles for each of the HCMV miRNAs. An important direction in the future will be to investigate the potential role of HCMV miRNAs in the establishment and maintenance of persistent and latent infections and the possible development of therapeutic agents based on the function of viral miRNAs. Proposed roles for miRNAs in HCMV latency and reactivation are illustrated in Fig. I.5.3. In the current model HCMV latently infected myeloid progenitor cells express miRNAs such as miR-US25-1 that regulate cell cycle proteins that determine either cell division or differentiation into monocytes and macrophages. At this stage miRNAs that repress replication such as miR-UL112-1, either suppress reactivation of

virus to maintain latency or target antiviral factors to allow viral replication in macrophages. Future work in the CMV field will elucidate the role of miRNAs in the viral life cycle.

Acknowledgement

This work is supported by grants from the National Institutes of Health (AI 21640 to J.A.N.).

References

Andersson, M.G., Haasnoot, P.C., Xu, N., Berenjian, S., Berkhout, B., and Akusjarvi, G. (2005). Suppression of RNA interference by adenovirus virus-associated RNA. J. Virol. *79*, 9556–9565.

Arase, H., Mocarski, E.S., Campbell, A.E., Hill, A.B., and Lanier, L.L. (2002). Direct recognition of cytomegalovirus by activating and inhibitory NK cell receptors. Science *296*, 1323–1326.

Bartel, D.P. (2009). MicroRNAs: target recognition and regulatory functions. Cell *136*, 215–233.

Besecker, M.I., Harden, M.E., Li, G., Wang, X.J., and Griffiths, A. (2009). Discovery of herpes B virus-encoded microRNAs. J. Virol. *83*, 3413–3416.

Buck, A.H., Santoyo-Lopez, J., Robertson, K.A., Kumar, D.S., Reczko, M., and Ghazal, P. (2007). Discrete clusters of virus-encoded micrornas are associated with complementary strands of the genome and the 7.2-kilobase stable intron in murine cytomegalovirus. J. Virol. *81*, 13761–13770.

Burnside, J., Ouyang, M., Anderson, A., Bernberg, E., Lu, C., Meyers, B.C., Green, P.J., Markis, M., Isaacs, G., Huang, E., *et al.* (2008). Deep sequencing of chicken microRNAs. BMC Genomics *9*, 185.

Cai, X., Lu, S., Zhang, Z., Gonzalez, C.M., Damania, B., and Cullen, B.R. (2005). Kaposi's sarcoma-associated herpesvirus expresses an array of viral microRNAs in latently infected cells. Proc. Natl. Acad. Sci. U.S.A. *102*, 5570–5575.

Cai, X., Schafer, A., Lu, S., Bilello, J.P., Desrosiers, R.C., Edwards, R., Raab-Traub, N., and Cullen, B.R. (2006). Epstein–Barr virus microRNAs are evolutionarily conserved and differentially expressed. PLoS Pathog. *2*, e23.

Cantalupo, P., Doering, A., Sullivan, C.S., Pal, A., Peden, K.W., Lewis, A.M., and Pipas, J.M. (2005). Complete nucleotide sequence of polyomavirus SA12. J. Virol. *79*, 13094–13104.

Castillo, J.P., and Kowalik, T.F. (2004). HCMV infection: modulating the cell cycle and cell death. Int. Rev. Immunol. *23*, 113–139.

Chen, C.J., Kincaid, R.P., Seo, G.J., Bennett, M.D., and Sullivan, C.S. (2011). Insights into Polyomaviridae microRNA function derived from study of the bandicoot papillomatosis carcinomatosis viruses. J. Virol. *85*, 4487–4500.

Chi, S.W., Zang, J.B., Mele, A., and Darnell, R.B. (2009). Argonaute HITS-CLIP decodes microRNA–mRNA interaction maps. Nature *460*, 479–486.

Cullen, B.R. (2011). Viruses and microRNAs: RISCy interactions with serious consequences. Genes Dev. *25*, 1881–1894.

DeMarchi, J.M., Schmidt, C.A., and Kaplan, A.S. (1980). Patterns of transcription of human cytomegalovirus in permissively infected cells. J. Virol. *35*, 277–286.

Deng, L., Chen, N., Li, Y., Zheng, H., and Lei, Q. (2010). CXCR6/CXCL16 functions as a regulator in metastasis and progression of cancer. Biochim. Biophys. Acta *1806*, 42–49.

Dölken, L., Perot, J., Cognat, V., Alioua, A., John, M., Soutschek, J., Ruzsics, Z., Koszinowski, U., Voinnet, O., and Pfeffer, S. (2007). Mouse cytomegalovirus microRNAs dominate the cellular small RNA profile during lytic infection and show features of posttranscriptional regulation. J. Virol. *81*, 13771–13782.

Dölken, L., Krmpotic, A., Kothe, S., Tuddenham, L., Tanguy, M., Marcinowski, L., Ruzsics, Z., Elefant, N., Altuvia, Y., Margalit, H., et al. (2010). Cytomegalovirus microRNAs facilitate persistent virus infection in salivary glands. PLoS Pathog. *6*, e1001150.

Dunn, C., Chalupny, N.J., Sutherland, C.L., Dosch, S., Sivakumar, P.V., Johnson, D.C., and Cosman, D. (2003). Human cytomegalovirus glycoprotein UL16 causes intracellular sequestration of NKG2D ligands, protecting against natural killer cell cytotoxicity. J. Exp. Med. *197*, 1427–1439.

Dunn, W., Trang, P., Zhong, Q., Yang, E., van Belle, C., and Liu, F. (2005). Human cytomegalovirus expresses novel microRNAs during productive viral infection. Cell. Microbiol. *7*, 1684–1695.

Easow, G., Teleman, A.A., and Cohen, S.M. (2007). Isolation of microRNA targets by miRNP immunopurification. RNA *13*, 1198–1204.

Ebert, M.S., and Sharp, P.A. (2010). MicroRNA sponges: progress and possibilities. RNA *16*, 2043–2050.

Elefant, N., Berger, A., Shein, H., Hofree, M., Margalit, H., and Altuvia, Y. (2010). RepTar: a database of predicted cellular targets of host and viral miRNAs. Nucleic Acids Res. *39*, D188–D194.

Glazov, E.A., Horwood, P.F., Assavalapsakul, W., Kongsuwan, K., Mitchell, R.W., Mitter, N., and Mahony, T.J. (2010).

Characterization of microRNAs encoded by the bovine herpesvirus 1 genome. J. Gen. Virol. *91*, 32–41.

Grey, F., Antoniewicz, A., Allen, E., Saugstad, J., McShea, A., Carrington, J.C., and Nelson, J. (2005). Identification and characterization of human cytomegalovirus-encoded microRNAs. J. Virol. *79*, 12095–12099.

Grey, F., Meyers, H., White, E.A., Spector, D.H., and Nelson, J. (2007). A human cytomegalovirus-encoded microRNA regulates expression of multiple viral genes involved in replication. PLoS Pathog. *3*, e163.

Grey, F., Tirabassi, R., Meyers, H., Wu, G., McWeeney, S., Hook, L., and Nelson, J.A. (2010). A viral microRNA down-regulates multiple cell cycle genes through mRNA 5′UTRs. PLoS Pathog. *6*, e1000967.

Hafner, M., Landthaler, M., Burger, L., Khorshid, M., Hausser, J., Berninger, P., Rothballer, A., Ascano, M., Jr., Jungkamp, A.C., Munschauer, M., et al. (2010). Transcriptome-wide identification of RNA-binding protein and microRNA target sites by PAR-CLIP. Cell *141*, 129–141.

Hancock, M.H., Tirabassi, R.S., and Nelson, J.A. (2012). Rhesus cytomegalovirus encodes seventeen microRNAs that are differentially expressed in vitro and in vivo. Virology *425*, 133–142.

Hansen, T.H., and Bouvier, M. (2009). MHC class I antigen presentation: learning from viral evasion strategies. Nat. Rev. Immunol. *9*, 503–513.

Huang, Y., Qi, Y., Ruan, Q., Ma, Y., He, R., Ji, Y., and Sun, Z. (2010). A rapid method to screen putative mRNA targets of any known microRNA. Virol. J. *8*, 8.

Huntzinger, E., and Izaurralde, E. (2011). Gene silencing by microRNAs: contributions of translational repression and mRNA decay. Nat. Rev. Genet. *12*, 99–110.

Hussain, M., Taft, R.J., and Asgari, S. (2008). An insect virus-encoded microRNA regulates viral replication. J. Virol. *82*, 9164–9170.

Jault, F.M., Jault, J.M., Ruchti, F., Fortunato, E.A., Clark, C., Corbeil, J., Richman, D.D., and Spector, D.H. (1995). Cytomegalovirus infection induces high levels of cyclins, phosphorylated Rb, and p53, leading to cell cycle arrest. J. Virol. *69*, 6697–6704.

Jurak, I., Kramer, M.F., Mellor, J.C., van Lint, A.L., Roth, F.P., Knipe, D.M., and Coen, D.M. (2010). Numerous conserved and divergent microRNAs expressed by herpes simplex viruses 1 and 2. J. Virol. *84*, 4659–4672.

Kalejta, R.F., and Shenk, T. (2002). Manipulation of the cell cycle by human cytomegalovirus. Front. Biosci. *7*, d295–d306.

Kalejta, R.F., and Shenk, T. (2003). The human cytomegalovirus UL82 gene product (pp71) accelerates progression through the G1 phase of the cell cycle. J. Virol. *77*, 3451–3459.

Karginov, F.V., Conaco, C., Xuan, Z., Schmidt, B.H., Parker, J.S., Mandel, G., and Hannon, G.J. (2007). A biochemical approach to identifying microRNA targets. Proc. Natl. Acad. Sci. U.S.A. *104*, 19291–19296.

Keene, J.D., Komisarow, J.M., and Friedersdorf, M.B. (2006). RIP-Chip: the isolation and identification of mRNAs, microRNAs and protein components of ribonucleoprotein complexes from cell extracts. Nat. Protoc. *1*, 302–307.

Kim, J., Kwon, Y.J., Park, E.S., Sung, B., Kim, J.H., Park, C.G., Hwang, E.S., and Cha, C.Y. (2003). Human cytomegalovirus (HCMV) IE1 plays role in resistance to apoptosis with etoposide in cancer cell line by Cdk2 accumulation. Microbiol. Immunol. *47*, 959–967.

Kim, S., Lee, S., Shin, J., Kim, Y., Evnouchidou, I., Kim, D., Kim, Y.K., Kim, Y.E., Ahn, J.H., Riddell, S.R., *et al.* (2011). Human cytomegalovirus microRNA miR-US4-1 inhibits CD8(+) T-cell responses by targeting the aminopeptidase ERAP1. Nat. Immunol. *12*, 984–991.

Lee, R.C., Feinbaum, R.L., and Ambros, V. (1993). The C. elegans heterochronic gene lin-4 encodes small RNAs with antisense complementarity to lin-14. Cell *75*, 843–854.

Lei, X., Bai, Z., Ye, F., Xie, J., Kim, C.G., Huang, Y., and Gao, S.J. (2010). Regulation of NF-kappaB inhibitor IkappaBalpha and viral replication by a KSHV microRNA. Nat. Cell. Biol. *12*, 193–199.

Libri, V., Helwak, A., Miesen, P., Santhakumar, D., Borger, J.G., Kudla, G., Grey, F., Tollervey, D., and Buck, A.H. (2012). Murine cytomegalovirus encodes a miR-27 inhibitor disguised as a target. Proc. Natl. Acad. Sci. U.S.A. *109*, 279–284.

Licatalosi, D.D., Mele, A., Fak, J.J., Ule, J., Kayikci, M., Chi, S.W., Clark, T.A., Schweitzer, A.C., Blume, J.E., Wang, X., *et al.* (2008). HITS-CLIP yields genome-wide insights into brain alternative RNA processing. Nature *456*, 464–469.

Lim, L.P., Glasner, M.E., Yekta, S., Burge, C.B., and Bartel, D.P. (2003). Vertebrate microRNA genes. Science *299*, 1540.

McElroy, A.K., Dwarakanath, R.S., and Spector, D.H. (2000). Dysregulation of cyclin E gene expression in human cytomegalovirus-infected cells requires viral early gene expression and is associated with changes in the Rb-related protein p130. J. Virol. *74*, 4192–4206.

Mamane, Y., Petroulakis, E., LeBacquer, O., and Sonenberg, N. (2006). mTOR, translation initiation and cancer. Oncogene *25*, 6416–6422.

Mazumder, S., Gong, B., and Almasan, A. (2000). Cyclin E induction by genotoxic stress leads to apoptosis of hematopoietic cells. Oncogene *19*, 2828–2835.

Merika, M., Williams, A.J., Chen, G., Collins, T., and Thanos, D. (1998). Recruitment of CBP/p300 by the IFN beta enhanceosome is required for synergistic activation of transcription. Mol. Cell. *1*, 277–287.

Meyer, C., Grey, F., Kreklywich, C.N., Andoh, T.F., Tirabassi, R.S., Orloff, S.L., and Streblow, D.N. (2011). Cytomegalovirus microRNA expression is tissue specific and is associated with persistence. J. Virol. *85*, 378–389.

Mocarski, E.S., Kemble, G.W., Lyle, J.M., and Greaves, R.F. (1996). A deletion mutant in the human cytomegalovirus gene encoding IE1(491aa) is replication defective due to a failure in autoregulation. Proc. Natl. Acad. Sci. U.S.A. *93*, 11321–11326.

Morgan, R., Anderson, A., Bernberg, E., Kamboj, S., Huang, E., Lagasse, G., Isaacs, G., Parcells, M., Meyers, B.C., Green, P.J., *et al.* (2008). Sequence conservation and differential expression of Marek's disease virus microRNAs. J. Virol. *82*, 12213–12220.

Murphy, E., Vanicek, J., Robins, H., Shenk, T., and Levine, A.J. (2008). Suppression of immediate-early viral gene expression by herpesvirus-coded microRNAs: implications for latency. Proc. Natl. Acad. Sci. U.S.A. *105*, 5453–5458.

Murphy, E.A., Streblow, D.N., Nelson, J.A., and Stinski, M.F. (2000). The human cytomegalovirus IE86 protein can block cell cycle progression after inducing transition into the S phase of permissive cells. J. Virol. *74*, 7108–7118.

Nachmani, D., Stern-Ginossar, N., Sarid, R., and Mandelboim, O. (2009). Diverse herpesvirus microRNAs target the stress-induced immune ligand MICB to escape recognition by natural killer cells. Cell Host Microbe *5*, 376–385.

Nachmani, D., Lankry, D., Wolf, D.G., and Mandelboim, O. (2010). The human cytomegalovirus microRNA miR-UL112 acts synergistically with a cellular microRNA to escape immune elimination. Nat. Immunol. *11*, 806–813.

Pan, X., Minegishi, N., Harigae, H., Yamagiwa, H., Minegishi, M., Akine, Y., and Yamamoto, M. (2000). Identification of human GATA-2 gene distal IS exon and its expression in hematopoietic stem cell fractions. J. Biochem. *127*, 105–112.

Payton, M., and Coats, S. (2002). Cyclin E2, the cycle continues. Int. J. Biochem. Cell Biol. *34*, 315–320.

Pfeffer, S., Zavolan, M., Grasser, F.A., Chien, M., Russo, J.J., Ju, J., John, B., Enright, A.J., Marks, D., Sander, C., *et al.* (2004). Identification of virus-encoded microRNAs. Science *304*, 734–736.

Pfeffer, S., Sewer, A., Lagos-Quintana, M., Sheridan, R., Sander, C., Grasser, F.A., van Dyk, L.F., Ho, C.K., Shuman, S., Chien, M., *et al.* (2005). Identification of microRNAs of the herpesvirus family. Nat. Meth. *2*, 269–276.

Poole, E., McGregor Dallas, S.R., Colston, J., Joseph, R.S., and Sinclair, J. (2011). Virally induced changes in cellular microRNAs maintain latency of human cytomegalovirus in CD34 progenitors. J. Gen. Virol. *92*, 1539–1549.

Prichard, M.N., Duke, G.M., and Mocarski, E.S. (1996). Human cytomegalovirus uracil DNA glycosylase is required for the normal temporal regulation of both DNA synthesis and viral replication. J. Virol. *70*, 3018–3025.

Rachamadugu, R., Lee, J.Y., Wooming, A., and Kong, B.W. (2009). Identification and expression analysis of infectious laryngotracheitis virus encoding microRNAs. Virus Genes *39*, 301–308.

Reese, T.A., Xia, J., Johnson, L.S., Zhou, X., Zhang, W., and Virgin, H.W. (2010). Identification of novel microRNA-like molecules generated from herpesvirus and host tRNA transcripts. J. Virol. *84*, 10344–10353.

Riley, K.J., Rabinowitz, G.S., and Steitz, J.A. (2010). Comprehensive analysis of Rhesus lymphocryptovirus microRNA expression. J. Virol. *84*, 5148–5157.

Saetrom, P., Heale, B.S., Snove, O., Jr., Aagaard, L., Alluin, J., and Rossi, J.J. (2007). Distance constraints between microRNA target sites dictate efficacy and cooperativity. Nucleic Acids Res. *35*, 2333–2342.

Samols, M.A., Hu, J., Skalsky, R.L., and Renne, R. (2005). Cloning and identification of a microRNA cluster within the latency-associated region of Kaposi's sarcoma-associated herpesvirus. J. Virol. *79*, 9301–9305.

Sanchez, V., and Spector, D.H. (2008). Subversion of cell cycle regulatory pathways. Curr. Top. Microbiol. Immunol. *325*, 243–262.

Santhakumar, D., Forster, T., Laqtom, N.N., Fragkoudis, R., Dickinson, P., Abreu-Goodger, C., Manakov, S.A., Choudhury, N.R., Griffiths, S.J., Vermeulen, A., *et al.* (2010). Combined agonist-antagonist genome-wide functional screening identifies broadly active antiviral microRNAs. Proc. Natl. Acad. Sci. U.S.A. *107*, 13830–13835.

Seo, G.J., Fink, L.H., O'Hara, B., Atwood, W.J., and Sullivan, C.S. (2008). Evolutionarily conserved function of a viral microRNA. J. Virol. *82*, 9823–9828.

Seo, G.J., Chen, C.J., and Sullivan, C.S. (2009). Merkel cell polyomavirus encodes a microRNA with the ability to autoregulate viral gene expression. Virology *383*, 183–187.

Shin, C., Nam, J.W., Farh, K.K., Chiang, H.R., Shkumatava, A., and Bartel, D.P. (2010). Expanding the microRNA targeting code: functional sites with centered pairing. Mol. Cell *38*, 789–802.

Shin, H.D., Park, B.L., Kim, L.H., Jung, J.H., Kim, J.Y., Yoon, J.H., Kim, Y.J., and Lee, H.S. (2003). Interleukin 10 haplotype associated with increased risk of hepatocellular carcinoma. Hum. Mol. Genet. *12*, 901–906.

Simon, C.O., Holtappels, R., Tervo, H.M., Böhm, V., Daubner, T., Oehrlein-Karpi, S.A., Kühnapfel, B., Renzaho, A., Strand, D., Podlech, J., *et al.* (2006). CD8 T-cells control cytomegalovirus latency by epitope-specific sensing of transcriptional reactivation. J. Virol. *80*, 10436–10456.

Sinclair, J., and Sissons, P. (1996). Latent and persistent infections of monocytes and macrophages. Intervirology *39*, 293–301.

Singh, J., Singh, C.P., Bhavani, A., and Nagaraju, J. (2010). Discovering microRNAs from Bombyx mori nucleopolyhedrosis virus. Virology *407*, 120–128.

Soderberg-Naucler, C., Fish, K.N., and Nelson, J.A. (1997). Reactivation of latent human cytomegalovirus by allogeneic stimulation of blood cells from healthy donors. Cell *91*, 119–126.

Stark, T.J., Arnold, J.D., Spector, D.H., and Yeo, G.W. (2011). High-resolution profiling and analysis of viral and host small RNAs during human cytomegalovirus infection. J. Virol. *86*, 226–235.

Stern-Ginossar, N., Elefant, N., Zimmermann, A., Wolf, D.G., Saleh, N., Biton, M., Horwitz, E., Prokocimer, Z., Prichard, M., Hahn, G., *et al.* (2007). Host immune system gene targeting by a viral miRNA. Science *317*, 376–381.

Stern-Ginossar, N., Gur, C., Biton, M., Horwitz, E., Elboim, M., Stanietsky, N., Mandelboim, M., and Mandelboim, O. (2008). Human microRNAs regulate stress-induced immune responses mediated by the receptor NKG2D. Nat. Immunol. *9*, 1065–1073.

Stern-Ginossar, N., Saleh, N., Goldberg, M.D., Prichard, M., Wolf, D.G., and Mandelboim, O. (2009). Analysis of human cytomegalovirus-encoded microRNA activity during infection. J. Virol. *83*, 10684–10693.

Strong, R.K. (2002). Asymmetric ligand recognition by the activating natural killer cell receptor NKG2D, a symmetric homodimer. Mol. Immunol. *38*, 1029–1037.

Sullivan, C.S., Grundhoff, A.T., Tevethia, S., Pipas, J.M., and Ganem, D. (2005). SV40-encoded microRNAs regulate viral gene expression and reduce susceptibility to cytotoxic T-cells. Nature *435*, 682–686.

Sullivan, C.S., Sung, C.K., Pack, C.D., Grundhoff, A., Lukacher, A.E., Benjamin, T.L., and Ganem, D. (2009). Murine Polyomavirus encodes a microRNA that cleaves early RNA transcripts but is not essential for experimental infection. Virology *387*, 157–167.

Tan, L.P., Seinen, E., Duns, G., de Jong, D., Sibon, O.C., Poppema, S., Kroesen, B.J., Kok, K., and van den Berg, A. (2009). A high throughput experimental approach to identify miRNA targets in human cells. Nucleic Acids Res. *37*, e137.

Tang, S., Patel, A., and Krause, P.R. (2009). Novel less-abundant viral microRNAs encoded by herpes simplex virus 2 latency-associated transcript and their roles in regulating ICP34.5 and ICP0 mRNAs. J. Virol. *83*, 1433–1442.

Tenney, D.J., and Colberg-Poley, A.M. (1991). Expression of the human cytomegalovirus UL36–38 immediate-early region during permissive infection. Virology *182*, 199–210.

Tirabassi, R., Hook, L., Landais, I., Grey, F., Meyers, H., Hewitt, H., and Nelson, J. (2011). Human cytomegalovirus US7 is regulated synergistically by two virally-encoded miRNAs and by two distinct mechanisms. J. Virol. *85*, 11938–11944.

Ugland, H., Boquest, A.C., Naderi, S., Collas, P., and Blomhoff, H.K. (2008). cAMP-mediated induction of cyclin E sensitizes growth-arrested adipose stem cells to DNA damage-induced apoptosis. Mol. Biol. Cell *19*, 5082–5092.

Umbach, J.L., Kramer, M.F., Jurak, I., Karnowski, H.W., Coen, D.M., and Cullen, B.R. (2008). MicroRNAs expressed by herpes simplex virus 1 during latent infection regulate viral mRNAs. Nature *454*, 780–783.

Umbach, J.L., Strelow, L.I., Wong, S.W., and Cullen, B.R. (2010). Analysis of rhesus rhadinovirus microRNAs expressed in virus-induced tumors from infected rhesus macaques. Virology *405*, 592–599.

Waidner, L.A., Morgan, R.W., Anderson, A.S., Bernberg, E.L., Kamboj, S., Garcia, M., Riblet, S.M., Ouyang, M., Isaacs, G.K., Markis, M., *et al.* (2009). MicroRNAs of Gallid and Meleagrid herpesviruses show generally conserved genomic locations and are virus-specific. Virology *388*, 128–136.

Wang, F.Z., Weber, F., Croce, C., Liu, C.G., Liao, X., and Pellett, P.E. (2008). Human cytomegalovirus infection alters the expression of cellular microRNA species that affect its replication. J. Virol. *82*, 9065–9074.

Welte, S.A., Sinzger, C., Lutz, S.Z., Singh-Jasuja, H., Sampaio, K.L., Eknigk, U., Rammensee, H.G., and Steinle, A. (2003). Selective intracellular retention of virally induced NKG2D ligands by the human cytomegalovirus UL16 glycoprotein. Eur. J. Immunol. *33*, 194–203.

Wiebusch, L., and Hagemeier, C. (1999). Human cytomegalovirus 86-kilodalton IE2 protein blocks cell cycle progression in G(1). J. Virol. *73*, 9274–9283.

Yang, J.S., and Lai, E.C. (2011). Alternative miRNA biogenesis pathways and the interpretation of core miRNA pathway mutants. Mol. Cell. *43*, 892–903.

Yao, Y., Zhao, Y., Xu, H., Smith, L.P., Lawrie, C.H., Sewer, A., Zavolan, M., and Nair, V. (2007). Marek's disease virus type 2 (MDV-2)-encoded microRNAs show no sequence conservation with those encoded by MDV-1. J. Virol. *81*, 7164–7170.

Yao, Y., Zhao, Y., Xu, H., Smith, L.P., Lawrie, C.H., Watson, M., and Nair, V. (2008). MicroRNA profile of Marek's disease virus-transformed T-cell line MSB-1: predominance of virus-encoded microRNAs. J. Virol. *82*, 4007–4015.

Yao, Y., Zhao, Y., Smith, L.P., Lawrie, C.H., Saunders, N.J., Watson, M., and Nair, V. (2009). Differential expression of microRNAs in Marek's disease virus-transformed T-lymphoma cell lines. J. Gen. Virol. *90*, 1551–1559.

Ziegelbauer, J.M., Sullivan, C.S., and Ganem, D. (2009). Tandem array-based expression screens identify host mRNA targets of virus-encoded microRNAs. Nat. Genet. *41*, 130–134.

Cytomegalovirus Proteomics

Patrizia Caposio, Daniel N. Streblow and Jay A. Nelson

Abstract

Proteomics is the large-scale study of proteins, particularly their structure and interaction. In this chapter we will discuss the results obtained using a proteomic approach to analyse what is secreted from cytomegalovirus infected cells: the composition of the viral and subviral particles as well as the cellular factors that are involved in the viral pathogenesis. In the previous edition we described the viral and cellular proteins that compose the infectious HCMV virion, the entry competent, non-replicating viral particles such as dense bodies (DBs) and non-infectious enveloped particles (NIEPs). Using a gel-free 2-D capillary liquid chromatography (LC)-MS/MS and a Fourier transform ion cyclotron resonance (FTICR) mass spectrometry we were able to identify the relative abundance of viral and cellular proteins in purified HCMV particles. The first part of this chapter will be an update of the literature that has been published in these last few years on the structure and composition of the viral particles. The second part of the chapter will be dedicated to the analysis of the cellular factors secreted from infected cells that act in a paracrine fashion to enhance wound healing (WH) and angiogenesis (AG) associated with the development of long term diseases like atherosclerosis, transplant vascular sclerosis (TVS), chronic allograft rejection (CR) and glioblastoma.

Introduction

Viruses are obligate intracellular parasites that rely on the host cell for many of the processes essential for their replication. Viruses manipulate cellular pathways detrimental to virus growth and divert complex cellular machinery necessary for efficient viral replication. During the co-evolution with its host, cytomegalovirus (CMV) has developed mechanisms to subvert the host cell's regulatory system to create an environment conducive to productive infection. Understanding the molecular events at the interface of the virus–host relationship is crucial for the identification of targets for future therapeutics.

One of the most complex processes in the viral life cycle is the assembly of infectious viral particles. The herpesvirus family encompasses a group of large complex enveloped viruses that are 150 nm to 300 nm in size and are ubiquitous in almost every species of vertebrates in nature. The human herpesviruses constitute some of the most important known human viral pathogens. Herpesviruses are grouped together based on virion structure that includes a capsid containing a large double-stranded DNA, a tegument composed for the majority of phosphoproteins that surrounds the capsid and an envelope with multiple glycoprotein complexes that mediate the attachment to the cells. Subfamilies of the herpesviruses are based on biological properties and structure of the viral genome and include α-, β- and γ-viruses.

Human cytomegalovirus (HCMV) is a prototypic β-herpesvirus that encodes over 200 predicted open reading frames (ORFs) (Chee et al., 1990; Davison et al., 2003; Dunn et al., 2003; Murphy et al., 2003a,b; Yu et al., 2003). The mature HCMV virion is 150–200 nm in diameter. It is composed of a 100-nm icosahedral capsid that contains a linear 230-kbp double-stranded DNA genome with attached proteins, and a large tegument component, surrounded by the envelope that contains a cellular lipid bilayer with viral glycoproteins (Mocarski and Courcelle, 2001) (Fig. I.6.1). HCMV-infected cells generate three different types of particles including infectious mature virions, non-infectious enveloped particles (NIEPs), and dense bodies (DBs). NIEPs are composed of the same viral proteins as infectious virions and possess a capsid but lack viral DNA. So, by electron microscopy NIEPs can be distinguished from mature virions by their lack of an electron dense DNA core (Mocarski and Courcelle, 2001). DBs are uniquely characteristic of HCMV and simian CMV infections and are non-replicating, fusion-competent enveloped particles composed primarily of

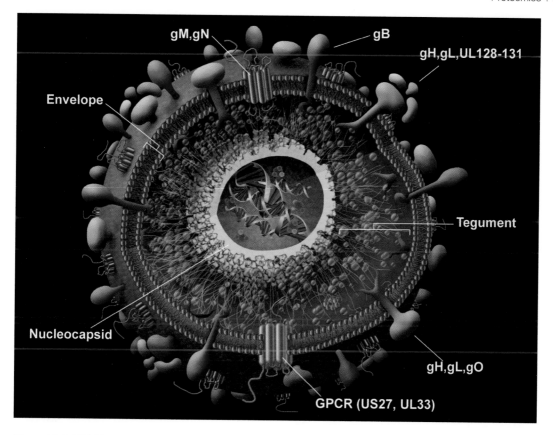

Figure I.6.1 HCMV structure. HCMV virions are comprised of three major layers. The first layer is the nucleocapsid containing the linear double-stranded viral DNA genome, which is surrounded by a proteinaceous tegument layer. The tegumented capsids are enveloped by a host-derived lipid bi-layer that is studded with viral glycoproteins.

the tegument protein pp65 (UL83). The quantities of these different HCMV particles are dependent on the viral strain and the multiplicity of infection. MCMV infected cells do not produce DBs, instead, this virus forms multicapsid virions, which are not observed in HCMV-infected cells.

The HCMV capsid is assembled in the nucleus from five viral proteins that are encoded by UL86, UL85, UL80, UL48.5, and UL46. The capsid is composed of 162 capsomers made up of hexons and pentons arranged in a T = 16 icosahedral lattice structure (Fig. I.6.2). The capsomers are studded with the smallest capsid protein (UL48.5). The capsid is surrounded by the tegument, which is acquired in both the nucleus and cytoplasm of the infected cells. There are 20–25 virion-associated tegument proteins, many of these are phosphorylated and have unknown function (Mocarski and Courcelle, 2001; Phillips and Bresnahan, 2011; To et al., 2011). Some of the more prominent tegument proteins include UL83 (pp65), UL82 (pp71), UL99 (pp28), UL32 (pp150), UL48, UL69, UL82, TRS1, and IRS1. Cytoplasmic viral capsids containing tegument acquire

the envelope by budding into the Trans Golgi Network (TGN) or a closely apposed cellular compartment. The envelope contains virally encoded glycoproteins, including gB (UL55), gM (UL100), gH (UL75), gL (UL115), gO (UL74), gN (UL73), gp48 (UL4), UL33 and in the clinical strains also UL128, UL130 and UL131 (Ryckman et al., 2008).

Using a gel-free two-dimensional capillary liquid chromatography-tandem mass spectrometry (MS/MS) and a Fourier transform ion cyclotron resonance MS we identified 71 HCMV AD169-encoded proteins that included 12 proteins encoded by known viral open reading frames (ORFs) previously not associated with virions and 12 proteins from novel viral ORFs. Analysis of the relative abundance of HCMV proteins indicated that the predominant virion protein was the pp65 tegument protein and that gM rather than gB was the most abundant glycoprotein. We have also identified over 70 host cellular proteins in HCMV virions, which include cellular structural proteins, enzymes, and chaperones. In addition, analysis of HCMV dense bodies indicated that these viral particles are composed of 29 viral

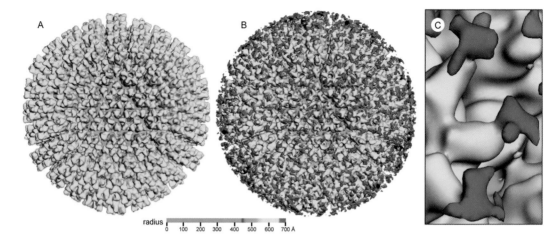

radius 0 100 200 300 400 500 600 700 A

Figure I.6.2 3D reconstructions of the HCMV capsid. (A) Capsid decorated by antibodies against the smallest capsid protein pORF48.5. (B) The structure was determined by electron cryomicroscopy to 22-Å resolution and coloured according to radius (see colour bar) so that densities are shown in purple and are attached only to major capsid protein (MCP, pORF86) subunits of hexons, not pentons. (C) Close-up of a region clearly showing attachment of antibody to the tips of hexon subunits. (Modified by Z. H. Zhou from Yu *et al.*, 2003, with permission from the publisher.)

proteins with a reduced quantity of cellular proteins in comparison to HCMV virions.

One of the hallmarks of CMV infection is the ability of the virus to hijack part of the host cellular machinery towards the secretion of factors that can promote long-term diseases such as vascular disorders and cancer. Our group and others have demonstrated that HCMV infection alters the types and quantities of bioactive proteins released from infected cells, which we designate as the HCMV secretome. Many of these factors have important roles in vascular disease, and we hypothesize that a major role of CMV infection in the acceleration of atherosclerosis, restenosis and chronic graft rejection after transplantation is through the increased production of wound healing and angiogenic factors. However, neither the complete proteome of the HCMV secretome nor its effects on WH and AG were known. Therefore, in order to determine the effects of HCMV infection on the extracellular milieu, we generated secretomes from HCMV-infected and mock-infected fibroblasts and determined their protein contents (proteomes) by gel-free LC-MS/MS and by specific protein arrays. In our LC-MS/MS analysis of the HCMV-infected and mock-infected secretomes, we identified more than 1200 proteins, with 800 having two or more peptide hits. Of the proteins found by MS/MS, more than 1000 were specific or highly enriched in the HCMV secretome. Pathway analysis indicated that many WH/AG proteins were present in the secretome. The results from the MS analysis have been confirmed and expanded using an RayBio® Human Cytokine Array

G Series 2000 antibody array. This approach allowed us to compare secretomes from different CMV-infected cells and it has been useful for the identification of critical factors involved in angiogenesis such as interleukin 6 (IL-6).

HCMV proteins in virion particles

We determined the protein composition of the HCMV particles purified from the supernatant culture fluids of human fibroblasts infected with HCMV strain AD169 utilizing high-throughput mass spectroscopy (Varnum *et al.*, 2004). This HCMV strain was selected at that time because of the availability of the complete sequence of the viral genome (Chee *et al.*, 1990) that is essential in the analysis of the peptides encoded by the viral genome. HCMV particles used for MS analysis were obtained from AD169 infected supernatants containing virions and DBs purified to greater than 96% by sequential sedimentation and density ultracentrifugation gradients as determined by electron microscopy. Electron micrographs of the purified HCMV virion and DB particles are shown in Fig. I.6.3A, as well as SDS-PAGE analysis of the same preparations in Fig. I.6.3B.

To determine the peptide content of the HCMV particles, purified viral preparations were digested to yield a complex mixture of polypeptides that was analysed by a two-stage mass spectroscopy approach. The first stage employed both one-dimensional and two-dimensional LC (mudPIT) coupled to MS/MS

A

Virion

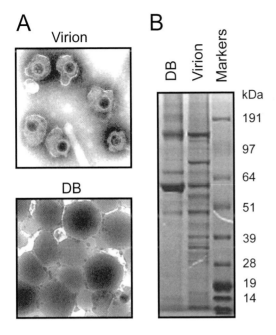

DB

Figure I.6.3 Characterization of HCMV virion and dense body preparation. (A) Transmission electron microscopy of HCMV virion and dense body (DB) preparations (magnification = ×8400). (B) Analysis of the proteins that constitute the purified HCMV particle preparations. Proteins were separated using NuPAGE MOPS gradient gels and visualized by Comassie blue staining.

TRL14. Nine of these HCMV-encoded polypeptides (UL38, UL50, UL71, UL79, UL93, UL96, UL102, US23, US24) are required for efficient virus growth in cultured fibroblasts (Dunn *et al.*, 2003; Yu *et al.*, 2003).

Our results were further expanded by Phillips and Bresnahan (2011) and To *et al.* (2011) who identified protein–protein interactions in the virion performing a yeast two-hybrid screening. Their study revealed 13 novel interactions, including UL25–UL25, UL25–UL26, UL32–UL35, UL43–UL83, UL45–UL25, UL45–UL45, UL45–UL69, UL48–UL45, UL48–UL88, UL69–UL88, UL82–UL94, UL94–US22 and UL46–UL86. Several of these interactions were subsequently confirmed in coimmunoprecipitation experiments following coexpression in transient overexpression studies (UL25–UL25, UL25–UL26, UL32–UL45, UL45–UL25, UL45–UL69, UL69–UL88, UL82–UL94 and UL94–US22). Three of the interactions identified in the yeast two-hybrid screen were also confirmed in the context of HCMV infection (UL99–UL94, UL88–UL69 and UL48–UL88). These coimmunoprecipitation studies validate the yeast two-hybrid results and suggest the importance of these interactions during HCMV infection.

The MS approach has been also extremely helpful for us to analyse the content of large nuclear and cytoplasmic aggregates, called aggresomes, that sequester considerable quantities of viral proteins in cells infected with HCMV in the absence of UL97 kinase activity (Prichard *et al.*, 2008). The aggresomes were denatured and digested with trypsin, and the complex mixture of peptides was analysed by two-dimensional liquid chromatography coupled to MS/MS. Identified peptides were compared to those in an HCMV-FASTA database. This analysis revealed that the aggresomes contain large quantities of viral structural proteins (Table I.6.2). These included the capsid proteins UL46 (minor capsid binding protein), UL48A (smallest capsid protein), UL80 (assembly protein), UL85 (minor capsid protein), and UL86 (major capsid protein) as well as a number of tegument proteins, including UL25, UL26, UL32, UL35, UL47, UL48, UL82, UL83, UL94, and US22. In addition, a number of proteins involved in transcription and DNA replication were also present, including IRS1, UL31, UL34, UL44, UL57, UL69, UL84, UL98, UL104, and UL122. Overall, the ratios of viral proteins present in the aggresomes resembled dense bodies rather than virions (Varnum *et al.*, 2004) and might suggest that defective capsids were sequestered in aggresomes prior to genome packaging. This is consistent with potential packaging defects reported previously (Wolf *et al.*, 2001) as well as the apparent association of defective capsids with aggregates in electron micrographs (Prichard *et al.*, 2005).

(Link *et al.*, 1999; Adkins *et al.*, 2002). The results from the LC-MS/MS analysis were verified and extended by employing high-accuracy mass measurements using LC-FTICR combined with chromatographic elution time information. Using this approach we identified 59 proteins, including 12 proteins encoded by known HCMV ORFs that were not previously shown to reside in virions (Table I.6.1). The known virion proteins identified included five capsid proteins (UL46, UL48–49, UL80, UL85, and UL86), nineteen tegument proteins (UL24, UL25, UL26, UL32, UL35, UL43, UL47, UL48, UL71, UL82, UL83, UL88, UL94, UL96, UL99, UL103, US22, US23, and US24), 19 glycoproteins (RL10, TRL14, UL5, UL22A, UL33, UL38, UL41A, UL50, UL55, UL73, UL74, UL75, UL77, UL93, UL100, UL115, UL119, UL132, and US27), 16 proteins involved in DNA replication and transcription (IRS1, TRS1, UL44, UL45, UL51, UL54, UL57, UL69, UL72, UL79, UL84, UL89, UL97, UL104, UL112, and UL122) and two G-protein coupled proteins (UL33 and US27). This analysis also identified twelve HCMV-encoded polypeptides not previously associated with the virion including UL5, UL38, UL50, UL71, UL79, UL93, UL96, UL103, UL132, US23, US24, and

Another interesting method to study binding partners for viral proteins is the target genetic approach that involves immunoaffinity purification of virus–host protein complexes followed by MS analysis of co-isolated proteins. The use of GFP-tagged pUL99 and pUL32 co-expressed during viral infection allowed Moorman (Moorman *et al.*, 2010) to integrate information regarding their localization and interactions during the progression of virion assembly. Using cryogenic cell lysis, affinity purifications on magnetic beads conjugated with anti GFP-antibodies, and mass spectrometry they identified numerous novel interactions. In particular, the presence of two host proteins, ubiquitin and clathrin in isolates of pUL99 and pUL32, respectively, led the authors to propose distinct parallel processes during HCMV virion assembly: (a) ubiquitin-mediated trafficking of pUL99, (b) pUL32 trafficking through clathrin-associated vesicles from a non-Golgi source, and (c) gB trafficking in yet a third compartment. These distinct compartments proceed to merge as the infection progresses.

HCMV proteins in dense body (DB) particles

Mass spectroscopy analysis of DB preparations indicated that these particles were composed of 24 viral proteins including five capsid proteins, 13 tegument proteins, three glycoproteins, and three proteins involved in virus transcription and replication (Table I.6.3). As previously described, UL83 (pp65) was the predominant protein present in DBs (Table I.6.4). However, representing 60% of the total protein it was less abundant than previously predicted in other studies (Irmiere and Gibson, 1983; Baldick and Shenk, 1996). Interestingly, UL25 was significantly more abundant in DBs than in virions (13% in BDs vs. 2.2% in virions) (Baldick and Shenk, 1996). Another abundant protein was UL26 that represented 2.4% of the protein in BDs, compared with 0.1% of protein in virion. The tegument phosphoprotein UL32 (pp150) was detected in HCMV DBs as well. In contrast to virions in which pp150 was 9.1% of the particle protein mass, UL32 was 2% of the DB particles. While the amount of UL32 in the virion was lower than previously estimated, our analysis indicates that pp150 is preferentially incorporated into virions than DBs (Benko *et al.*, 1988). Additionally, the presence of five nucleocapsid proteins, UL46, UL48A, UL80, UL85 and UL86, in the DB preparations is perhaps surprising, given that DBs are reported to lack a nucleocapsid (Sarov and Abady, 1975; Gibson, 1983). However, at least two of these proteins, UL85 and UL86, have previously been detected in DBs (Baldick and Shenk, 1996). The five

capsid proteins combined represent a relatively small portion of the total DB protein (~7.8%), whereas these same five proteins represent ~ 24% of the total protein isolated from the virion particles.

In regards to HCMV glycoproteins detected in DBs, UL55, UL100, and UL132 were decreased in relative abundance by 4-fold, 2-fold, and 2-fold respectively, while UL75 and UL115 were equivalent in comparison to the abundance of these glycoproteins in virions. The total percentage of viral glycoproteins in the DB preparations was 5.2% compared with 13% in virion preparations. The discrepancy may reflect the variation in size as well as high tegument composition (85.5%) of the DBs. Interestingly, TRL10, UL22A, UL33, UL77, UL119, and US27 that were present in virion particles were not detected in the DB preparations. These observations suggest that these glycoproteins are present in low concentrations in DBs and may not be essential for the functional properties of DBs.

Host proteins associated with HCMV particles

Another area of controversy in the herpesvirus field centres on the host proteins that are incorporated into HCMV particles. In our studies of HCMV virions a total of 71 host cell proteins were detected with high confidence significantly increasing the number of previously identified virion proteins (Table I.6.5) (Grundy *et al.*, 1987; Stannard, 1989; Wright *et al.*, 1995; Baldick and Shenk, 1996; Giugni *et al.*, 1996; Michelson *et al.*, 1996; Mocarski and Courcelle, 2001). These HCMV cellular virion proteins included cytoskeletal proteins, such as α- and β-actin, tubulin, several annexins, α-actinin, and vimentin, as well as cellular proteins involved in translational control, including initiation and elongation factors. Other cellular proteins identified in HCMV virion preparations include clathrin and ADP-ribosylation factor 4. These proteins are involved in vesicular trafficking in the endoplasmic reticulum and Golgi, suggesting a role for these proteins in viral envelopment and/or egress. In addition to the above mentioned cellular virion proteins, four isoforms of the signal transduction protein 14-3-3 were also identified in HCMV preparations, together with other signalling proteins such as RasGAP, casein kinase 2, and β_2-GTO-binding regulatory protein. The cellular protein β_2-microglobulin was previously reported to be a component of the HCMV virion (Grundy *et al.*, 1987; Stannard, 1989). However, this cellular protein was present in low amounts in virion preparations and was eliminated from the cellular protein list, based on the conservative criteria employed in this study. The

Table I.6.1 HCMV proteins found in virions. HCMV proteins isolated from virions that were identified by LC-MS/MS and FTICR

Group	HCMV ORF	Comments	LCG MS/MS		FTICR	
			No. of different peptides	Max Xcorr	No. of different peptides	Per cent coverage
Capsid	UL46	Minor capsid binding protein	20	5.3	14	44.8
	UL48–49	Smallest capsid protein	8	6.52	5	54.7
	UL80	Assembly precursor	37	6.36	30	35.6
	UL85	Minor capsid protein	21	6.73	22	63.1
	UL86	Major capsid protein	149	3.97	123	71
Tegument	UL24	Tegument protein	8	5.06	9	38.3
	UL25	Tegument protein; UL25 family	60	7.04	59	59.2
	UL26	Tegument protein, US22 family	9	4.77	10	53.7
	UL32	Tegument protein, pp150	135	3.01	100	70.5
	UL35	Tegument phosphoprotein, UL25 family	42	6.27	40	56.1
	UL43	Tegument protein, US22 family	7	5.5	10	28.1
	UL47	Tegument protein	53	6.1	64	57.5
	UL48	Large tegument protein	111	4.29	109	56.8
	UL71[a]	Tegument protein	12	6.32	11	40.4
	UL82	Upper matrix phosphoprotein, pp71	70	6.39	47	69.3
	UL83	Lower matrix phosphoprotein, pp65	123	5.44	86	92
	UL88	Tegument protein	14	6.8	17	33.6
	UL94	Tegument protein	10	5.08	12	26.4
	UL96	Tegument protein	1	4.46	1	19.7
	UL99	Tegument protein, pp28	8	5.87	9	64.7
	UL103[a]	Tegument protein	8	5.18	8	37
	US22	Tegument protein; US22 family	2	3.16	2	5.4
	US23[a]	Tegument protein, US22 family	1	2.61	1	4.6
	US24[a]	Tegument protein, US22 family	1	4.83	2	7
Glycoproteins	RL10		5	2.36	4	22.8
	TRL14[a]		a		1	7.5
	UL5[a]	RL11 family	a		1	5.4
	UL22A	Secreted glycoprotein	1	5.04	1	19.4
	UL33	G-protein coupled receptor	4	6.11	4	14.1
	UL38[a]		a		1	5.7
	UL41A		2	5.72	2	25.6
	UL50	Membrane protein involved in nuclear capsid egress	1	2.82	4	10.6
	UL55	gB	21	6.16	23	24.8
	UL73	gN	2	3.47	2	6.5
	UL74	gO	4	5.07	4	13.5
	UL75	gH	21	6.15	22	35.7
	UL77	Pyruvoyl decarboxylase; DNA packaging protein	14	5.65	12	31.2
	UL93		15	5.35	14	31.7

Table I.6.1 Continued

Group	HCMV ORF	Comments	LCG MS/MS		FTICR	
			No. of different peptides	Max Xcorr	No. of different peptides	Per cent coverage
	UL100	gM	13	5.24	7	15.9
	UL115	gL	11	4.73	9	47.1
	UL119	IgG Fc-binding glycoprotein	2	2.23	1	4.6
	UL132[a]		8	5.89	8	47
	US27	G-protein coupled receptor	4	4.25	2	7.7
Transcription/ replication machinery	IRS1	Transcriptional transactivator; US22 family	15	6.01	17	25.8
	TRS1	Viral gene transactivator	10	6.92	23	34.7
	UL44	DNA processivity factor	1	4.32	9	31
	UL45	Ribonucleotide reductase homologue (enzymatically inactive)	43	5.85	52	52.2
	UL51	Terminase component	[a]		1	3.2
	UL54	DNA polymerase	[a]		1	1.6
	UL57	ssDNA-binding protein	[a]		1	0.4
	UL69	Posttranscriptional regulator of gene expression	6	4.17	7	19
	UL72	Deoxyuridine triphosphatase homologue (enzymatically inactive)	[a]		1	4.6
	UL79*	DNA replication	[a]		1	10.9
	UL84	Transdominant inhibitor of IE2-mediated transactivation	1	2.5	3	12.8
	UL89	Terminase component	[a]		1	3.1
	UL97	Phosphotransferase	13	5.95	9	32.1
	UL104	DNA packaging protein; capsid portal protein	9	4.68	9	23
	UL112	DNA replication	1	3.3	4	4.7
	UL122	Viral and cellular gene transactivator, IE2	2	4.26	4	11.7

[a]Denotes proteins newly identified associated with HCMV virions.

presence of these cellular proteins in virion preparations may be attributed to co-purification of cellular components in virion preparations, purified virions sticking to cellular proteins, or that the cellular proteins are constituents of the viral particles. However, we believe that the majority of the cellular proteins in virion preparations are integral parts of the particle, since some of these cellular proteins have already been identified in virion preparations by electron microscopy in the apparent absence of detectable amounts of cellular organelles and debris.

In regards to the analysis of cellular proteins in DBs, only a small number of host cell proteins were detected in these HCMV particles including glyceraldehyde-3-phosphate dehydrogenase, annexin A2, β-actin and heat shock 70-kDa proteins. Interestingly, the cellular protein composition of the DBs differed markedly from the cellular proteins in virion preparations. A possible explanation for this observation is that DB formation and egress is fundamentally different from envelopment and egress of virion particles.

Comparison of the HCMV and MCMV proteomes

Concurrent with our analysis of the HCMV virion proteome a group led by Drs. Kessler and Ploegh at Harvard University identified the proteins that compose the Smith strain of murine CMV (MCMV) virion (Kattenhorn *et al.*, 2004). Since the MCMV genome was not annotated, a putative MCMV ORF database was generated using the gene prediction algorithm,

Table I.6.2 Viral proteins identified by liquid chromatography MS/MS in cytoplasmic and nuclear aggresomes

HCMV ORF	Cytoplasmic aggresomes		Nuclear aggresomes		Description
	Max Xcorr	No. unique peptides	Max Xcorr	No. unique peptides	
IRS1	5.12	2	4.73	3	Transcriptional transactivator; US22 family
UL25	4.97	14	5.54	25	Tegument protein; UL25 family
UL26	3.46	2	4.91	4	Tegument protein; US22 family
UL31	ND	ND	3.97	3	Hypothetical protein
UL32	4.71	4	4.38	4	Tegument protein, pp150
UL34	ND	ND	4.34	2	Transcriptional repressor
UL35	ND	ND	6.58	3	Tegument phosphoprotein, UL25 family
UL44	5.57	11	5.16	7	DNA processivity factor
UL46	ND	ND	4.18	3	Minor capsid binding protein
UL47	ND	ND	3.5	1	Tegument protein
UL48	4.3	1	5.68	5	Large tegument protein
UL48A	7.17	5	7.03	2	Capsid protein located at tips of hexons
UL50	6.23	1	ND	ND	Membrane protein involved in nuclear capsid egress
UL57	1.91	1	ND	ND	ssDNA-binding protein
UL69	ND	ND	3.28	2	Post-transcriptional regulator of gene expression
UL71	5.09	2	ND	ND	Tegument protein
UL77	ND	ND	4.58	1	Pyruvoyl decarboxylase; DNA packaging protein
UL80	5.6	9	4.1	2	Assembling precursor
UL82	ND	ND	4.82	5	Upper matrix phosphoprotein, pp71
UL83	6	60	6.56	82	Lower matrix phosphoprotein, pp65
UL84	5.05	4	5.65	3	Transdominant inhibitor of IE2-mediated transactivation
UL85	3	2	2.99	3	Minor capsid protein
UL86	5.59	23	6.41	21	Major capsid protein
UL94	ND	ND	3.83	1	Tegument protein
UL98	ND	ND	3.32	1	DNase
UL104	ND	ND	3.84	2	DNA packaging protein; capsid portal protein
UL112	1.97	1	ND	ND	DNA replication
UL115	4.16	1	ND	ND	Envelope glycoprotein associated with gH and gO
UL122	4.24	3	3.23	1	Immediate-early transcriptional regulator IE2
US22	2.48	1	3.53	1	Tegument protein; US22 family

ND, not determined.

GeneMark. Similar to HCMV, the MCMV genome is 230 kbp and contains 170 predicted ORFs of > 100 aa. However, at the genome level MCMV and HCMV only share about 42.5% sequence identity, and of the 170 ORFs only 78 have significant amino acid identity with proteins from HCMV. One distinguishing feature of MCMV is the fact that unlike HCMV, MCMV does not produce DBs. Since HCMV mutants lacking pp65 do not form DBs, the inability of MCMV to generate

DBs suggests there exist functional differences between HCMV pp65 and its MCMV homologues M83 and M84, which is highlighted by their low sequence similarity.

In the Harvard study, purified MCMV preparations were denatured and analysed using two different mass spectrometry methods including the traditional SDS-PAGE separation followed by in-gel digestion and the tryptic digestion in-solution. The

Table I.6.3 HCMV proteins found in dense bodies. HCMV proteins isolated from dense bodies that were identified by LC-MS/MS and FTICR

Viral protein group	HCMV ORF	#Unique Peptides	Max Xcorr	FTICR different peptides identified
Capsid	UL46	1	3.6	6
	UL48A	1	5.8	1
	UL80	1	6.1	2
	UL85	4	5	4
	UL86	22	5	19
Tegument	UL24	1	1.9	2
	UL25	17	6.3	13
	UL26	3	3.6	3
	UL32	11	5.4	15
	UL35	5	5.6	9
	UL43	1	3.9	1
	UL47	2	4.3	6
	UL48	7	5.4	12
	UL71	2	5.2	ND
	UL82	9	5.1	6
	UL83	40	6.3	14
	UL88	a	a	1
	UL94	1	4.3	1
	UL99	1	2.4	ND
	UL103	a	a	1
Glycoproteins	UL55	a	a	1
	UL74	1	4.2	ND
	UL75	4	5.6	2
	UL100	a	a	3
	UL132	a	a	1
	US27	a	a	1
	UL69	1	3.4	ND
Transcription/ replication machinery	UL45	2	4.3	6
	UL97	a	a	4
	TRS1	1	4.7	5
	IRS1	3	5.6	2

[a]Identified peptides do not meet minimal criteria for positive identification.
ND: not detected by FTICR.

MCMV tryptic peptides were separated by nano-flow liquid chromatography and analysed by tandem mass spectrometry (LC-MS/MS). The two methods of peptide preparation yielded highly different levels of detection with the in-solution digestion method identifying 58 viral proteins in contrast to only 19 MCMV proteins detected by in-gel digestions. This study confirms our findings that the sensitivity of detecting proteins associated with viral preparations is far better when using the in-solution digestion approach. A total of 38 MCMV proteins of known function were identified in this approach including capsid, tegument, envelope, replication and immunomodulatory family members. Peptides encoded by 20 other MCMV ORFs without a known function were also identified in this study.

Table I.6.4 Average abundance for HCMV proteins associated with virions and dense bodies

HCMV ORF	Virion average abundance	Dense bodies average abundance
UL83	15.4	60.2
UL48–49	12.6	4.4
UL100	9.2	4.4
UL32	9.1	2
UL82	8.9	1.7
UL48	8.8	0.8
UL80	7.7	1.1
UL86	6	1.5
UL45	4.7	0.9
UL85	2.8	1
UL25	2.2	12.7
UL46	1.5	0.3
UL47	1.5	0.3
UL55	1.4	0.3
UL94	1.2	1
IRS1	0.8	0.3
UL88	0.7	0.1
UL77	0.6	<0.1
TRS1	0.6	0.3
UL75	0.6	0.6
UL35	0.5	1
UL115	0.5	0.3
UL119	0.5	<0.1
UL132	0.4	0.2
UL104	0.4	<0.1
US27	0.2	<0.1
RL10	0.2	0.1
UL72	0.2	<0.1
UL22A	0.2	<0.1
UL97	0.1	0.5
UL71	0.1	<0.1
UL26	0.1	2.4
US22	0.1	<0.1
UL103	0.1	0.1
UL24	0.1	0.5
UL44	0.1	0.4
UL33	0.1	<0.1
UL43	<0.1	0.2
UL69	<0.1	0.1
UL93	<0.1	0.1
UL99	<0.1	0.5

Abundances are based upon a percentage of the total proteins present derived from integrated peptide MS peak intensities. Detection limit was ≤0.1.

Table I.6.5 Selected host proteins identified by category

ATP binding	DEAD/H (Asp-Glu-Ala-Asp/His) box polypeptide 1, sodium/potassium-transporting ATPase
Ca2+ binding	Annexin I, annexin V, annexin VI, annexin A2, calreticulin, alpha-actinin 1
Chaperone	Cyclophilin A, glucose-regulated protein, heat shock 70 kDa protein, heat-shock protein 90 kDa, tumour rejection antigen
Cytoskeleton	α-Actin,_β-actin, cofilin, filamin, keratin, moesin,_α-tubulin,_β-tubulin, vimentin
enzymes	Aminopeptidase N, transketolase, vinculin
Glycolysis	Enolase, glyceraldehyde-3-phosphate dehydrogenase, lactate dehydrogenase, phosphoglycerate kinase, serine/threonine protein phosphatase PP1, triosephosphate isomerase
Protein transport	ADP-ribosylation factor 4, Clathrin, polyubiquitin 3, β-RAB GDP dissociation inhibitor
Signal transduction	α1-Casein kinase 2, 14–3–3 protein (four isoforms)
Transcription/ translation	Eukaryotic translation elongation factor 2, eukaryotic translation elongation factor 1, enhancer protein, eukaryotic translation initiation factor 4A

According to this analysis, MCMV contains four capsid proteins (m48.2, the smallest capsid protein; M85, the minor capsid protein; M86, the major capsid protein; and M46, the minor capsid binding protein) in contrast to five HCMV capsid proteins. The MCMV virion proteins detected in this study that have homologues of HCMV virion proteins detected in our studies are listed in Table I.6.6. For convenience and consistency with previous publications we have designated the MCMV ORFs that have homologues in HCMV with an-upper case 'M', whereas those without HCMV counterparts are designed with a lower case 'm'. Similar to our UL80 in analysis of HCMV, the viral assembly protein/protease M80 was detected in MCMV virions. Likewise, 10 tegument proteins were also detected in virion preparations including M25 (homologue of UL25), M32 (homologue of pp150), M35, M47 (high molecular weight binding protein), M48 (high molecular weight tegument protein), M51, M82 (homologue of pp71), M83 (homologue of pp65), M94, and M99 (homologue of pp28). In MCMV virions four glycoproteins were detected including M55 (gB), M74 (gO), M75 (gH), and M100 (gM). Similar to HCMV a number of MCMV replication proteins were also detected in virion preparations including M54 (DNA polymerase), M44 (polymerase accessory protein), M57 (major DNA-binding protein), M69 (IE transactivator), M70, M102, and M105 (helicase/primase complex). In addition, other MCMV genes involved in DNA replication and packaging were also detected including the DNA packaging protein M56, the viral dUTPase M72, the viral protein kinase M97, and the viral exonuclease M98. The independent detection of replication enzymes in both HCMV and MCMV virions suggests that CMV may have the potential

to replicate the viral DNA template prior to the early phase of replication, when these genes are expressed by the virus. Interestingly, three different MCMV proteins with immunomodulatory activity were observed in MCMV preparations including m138 (Fc receptor), M43, and M45 (antiapoptotic protein). The HCMV homologues of M43 and M45 were also observed in our HCMV virion preparations. In total, 30 virion proteins were detected in both MCMV and HCMV preparations of the 40 MCMV proteins detected in virions with homologues in HCMV in the Harvard study (Table I.6.6).

A number of annotated MCMV proteins with unknown function were also found associated with virion particles including: m18, m25.2, M28, M31, M35, m39, M71, M87, m90, M95, m107, M121, m150, m151, m163, and m165. All of these proteins were detected by the in-solution tryptic method and not by the in-gel analysis suggesting that they are either low copy number or that they are highly insoluble. The Harvard group also detected peptides corresponding to small ORFs previously not described including m166.5 and ORF 105,932–106,072 (44 aa). Expression of m166.5 protein was confirmed in infected cells by Western blot. In addition to finding novel ORFs, the 3′ proximal ends of m20 and M31 were remapped based on the GeneMark sequence data analysis that was confirmed by the proteomics approach. Previous sequencing errors mapped the 3′ end of m20 out of frame that in this study was correctly extended an extra 226bp. The combined use of new sequencing methods with proteomics demonstrates the utility of these techniques to correctly define ORFs.

A number of MCMV virion proteins that have homologues to HCMV ORFs were not found in

Table I.6.6 Homologous viral proteins detected by MS analysis in HCMV and MCMV

Virus detected	MCMV ORF	HCMV ORF	Comments
MCMV/HCMV	M25	UL25	Tegument protein
	m25.2	US22	Tegument protein
	M28	UL28	
	M31	UL31	
	M32	UL32	Tegument protein, pp150
	M35	UL35	Tegument protein, UL25 family
	M43	UL43	Tegument protein, US22 family
	M44	UL44	DNA processivity factor
	M45	UL45	Ribonucletide reductase
	M46	UL46	Minor capsid binding protein
	M47	UL47	Tegument protein
	M48	UL48	Large tegument protein
	m48.2	UL48-49	Smallest capsid protein
	M51	UL51	Terminase component
	M54	UL54	DNA polymerase
	M55	UL55	gB
	M57	UL57	ssDNA binding protein
	M69	UL69	Viral transactivator
	M71	UL71	Tegument protein
	M72	UL72	dUTPase homolog
	M74	UL74	gO
	M75	UL75	gH
	M77	UL77	Pyruvoyl decarboxylase
	M80	UL80	Assembly/Protease
	M82	UL82	Upper matrix phosphoproteins, pp71
	M83	UL83	Lower matrix phosphoprotein, pp83
	M85	UL85	Minor capsid protein
	M86	UL86	Major capsid protein
	M88	UL88	Tegument protein
	M94	UL94	Tegument protein
	M95	UL95	
	M97	UL97	Phosphotransferase
	M98	UL98	Alkaline nuclease
	M99	UL99	Tegument protein, pp28
	M100	UL100	gM
	M102	UL102	Helicase-primase subunit
	M104	UL104	Structural protein
	M105	UL105	Helicase-primase subunit
	M116	UL116	
	M121	UL121	

HCMV virions. These unique MCMV virion proteins include M28, M31, m39, M70, M87, M95, M98, M102, M105, M116, and M121. Only a few of these unique MCMV virion proteins have putative functions including M56 (terminase) M70 (helicase/primase subunit), M98 (alkaline nuclease), and M105 (helicase/primase subunit). Comparison of the HCMV and MCMV proteomes also indicated that HCMV contained a number of unique virion proteins with MCMV homologues. These unique HCMV proteins include UL24 (tegument; an endothelial cell tropism determinant), UL26, UL33 (vGPCR), UL38, UL50, UL73 (gN), UL79, UL84, UL89, UL93, UL96, UL103, UL112, UL115 (gL), US23, and US24. Interestingly, the HCMV homologue (UL115) of M115 (gL) forms a complex with gH and gO that was detected in HCMV virions at levels similar to that of gH. Similarly, HCMV gN (UL73), a chaperone for the most abundant glycoprotein gM, was detected in HCMV but not MCMV preparations. Lastly, although both MCMV and HCMV contain virally encoded GPCRs, only HCMV virions contained these viral proteins that included UL33 and US27. The virally encoded GPCR UL78 or the MCMV homologue M78 was not found in either virion preparation, although M78 has been reported in MCMV virions in previous studies (Oliveira and Shenk, 2001).

A number of HCMV-associated proteins that do not have homologues in MCMV were found in HCMV virions in our studies including UL5, UL22A, UL41A, UL119, UL122, UL132, US27 (vGPCR), IRS1, TRS1, RL10, and TRL14. This observation may reflect the sequence divergence between the two genomes and thus we would predict that there would be a number of MCMV-associated proteins without proper positional homologues in HCMV including: m02, m18, m20, m39, m107, m117.1, m147, m150, m151, m163, m165, and m165.5. Whether these proteins are functional homologues to the HCMV-associated proteins that lack MCMV positional homologues is yet to be determined but studies in this area may prove interesting.

The Harvard group also identified a number of cellular proteins in MCMV virion preparations including: actin-γ, annexin I/IV, cofilin, histone H2A, the translation factor EF1α, glyceraldehyde 3-phosphate dehydrogenase, cadherin, and the RhoGDP dissociation factor. The majority of these cellular proteins were also found in HCMV AD169 virion preparations confirming our studies in an independent study. Whether these proteins have a role in virus replication is yet to be determined but a subject of intense study.

Analysis of HCMV secretome in different cells types

In vivo results from our rat heterotopic solid organ transplantation model clearly demonstrate that CMV infection plays a role in the vascular disease process, leading to the acceleration of graft failure and ultimately to organ rejection (Streblow *et al.*, 2008). The virus-mediated acceleration of TVS occurs through altered regulation of inflammation and wound healing processes (Streblow *et al.*, 2007). Many clinical studies on human transplants have linked HCMV infection to the development of arterial restenosis following angioplasty, atherosclerosis, and solid organ TVS (Melnik *et al.*, 1983; Speir *et al.*, 1994; Melnick *et al.*, 1998). To provide solid evidences that HCMV infection modified many of the host's cellular functions that promote tissue repair/angiogenesis we decided to use the technology of the gel-free liquid chromatography (LCQ)–MS–MS to study the protein composition of supernatant culture fluids of human fibroblast cells (secretome) infected with HCMV strains AD169 and TR (Dumortier *et al.*, 2008). We identified more then 1200 proteins with 800 havingtwo2 or more peptide hits (Table I.6.7). Of the proteins identified by MS/MS, more than 1000 were specific or highly enriched in the HCMV secretome, more than 260 proteins were common to both the HCMV- and mock-infected secretomes, and more than 225 were specific for the mock-infected secretome. We detected ten viral proteins in the HCMV secretome of which only four were identified with more than one peptide, including UL32 (pp150), UL44, UL122, and UL123. Interestingly, pp150, the sole structural protein detected in the viral secretome, was detected by Western blotting as a 65-kDa species, suggesting that the secreted protein may be a cleavage product of the larger 150-kDa species. Pathway analysis indicated that many (~ 100 proteins) of the proteins present in the HCMV secretome play an import role in WH and AG. It was noted that a cluster of proteins were involved in integrin signalling and that this cluster was enriched for laminins. Laminins are widely distributed extracellular matrix (ECM) proteins involved in cell adhesion signalling and have been implicated in playing a fundamental role in angiogenesis by directly affecting gene and protein expression profiles (Folkman, 2003). TGF-β signalling and AG pathways were also identified in this initial screening.

To validate and expand the MS studies we assayed for changes in 174 cytokines/growth factors present in the HCMV secretome using RayBio® Human Cytokine Array G Series 2000 antibody arrays. We analysed and compared secretomes from HCMV- and mock-infected fibroblasts and human umbilical vein endothelial cells (HUVECs).

Table I.6.7 Numbers of cellular proteins identified by LC-Q-MS-MS in the HCMV AD169 and TR secretome[a]

Name(s)	Reference sequence	Mock spectra	Mock peptide	AD169 Spectra	AD169 peptide	TR spectra	TR peptide
Fibronectin		66	16	261	69	534	84
Thrombospondin-1	NP_003237	34	15	302	44	380	53
Prepro-alpha 2(I) collagen	NP_000080	3	2	77	22	213	40
Complement C3	NP_000055	26	12	139	50	185	57
Plasma protease C1 inhibitor	NP_000053	4	2	90	14	185	20
Glia-derived nexin	NP_006207	5	3	95	23	153	22
Collagen alpha1(VI) chain	NP_001839	3	3	126	24	152	27
Stromelysin-1, MMP3	NP_002413	21	7	42	18	147	21
Alpha-1-antichymotrypsin	NP_001076	14	2	62	13	103	14
Plasminogen activator inhibitor-1	NP_000593	10	4	78	23	94	21
Laminin alpha-4 chain	NP_002281	5	4	52	24	91	38
72 kDa type IV collagenase, MMP2	NP_004521	21	6	51	15	90	23
Galectin-1	NP_002296	3	1	119	9	90	11
Galectin-3 binding protein	NP_005558	9	4	64	17	82	19
Tenascin	NP_002151	9	5	39	28	77	33
Interstitial collagenase, MMP1	NP_002412	13	5	65	20	67	20
Laminin gamma-1 chain	NP_002284	5	4	39	22	66	28
Sparc	NP_003109	7	2	27	7	65	12
TNF receptor superfamily member 11b	NP_002537	1	1	17	6	65	16
Insulin-like growth factor BP 7	NP_001544	2	1	22	6	64	12
TGF-β-induced protein IG-H3	NP_000349	3	3	26	14	64	24
Thrombospondin-2		1	1	48	19	62	22
Insulin-like growth factor BP 5	NP_000590	1	1	28	5	61	10
Laminin beta-1 chain	NP_002282	1	1	52	30	60	24
Cathepsin B	NP_680093			23	5	57	9
Insulin-like growth factor BP 6	NP_002169	2	1	12	3	49	5
Metalloproteinase inhibitor 1	NP_003245	1	1	18	4	40	7
Urokinase-type plasminogen activator	NP_002649	1	1	11	7	36	11
Macrophage colony-stimulating factor-1	NP_757351	1	1	16	6	31	9
Extracellular matrix protein 1	NP_004416	1	1	7	6	19	9
Stromelysin-2, MMP10	NP_002416			10	3	19	3
Clusterin	NP_976084	3	1	9	8	19	9
Metalloproteinase inhibitor 2, TIMP2	NP_003246	7	3	12	3	17	4
Cathepsin K	NP_000387	1	1	11	5	17	7
Midkine	NP_002382					16	4
Stem cell growth factor; lymphocyte secreted C-type lectin	NP_002966			7	6	16	8
Latent TGF beta BP, isoform 1l				5	3	16	4
Insulin-like growth factor BP 4	NP_001543					15	3
Cathepsin D	NP_001900			12	8	14	11
Cathepsin L	NP_666023			10	6	14	7
Growth-regulated protein alpha, CXCL1	NP_001502			1	1	14	6

Table I.6.7 Continued

Name(s)	Reference sequence	Mock spectra	Mock peptide	AD169 Spectra	AD169 peptide	TR spectra	TR peptide
Dickkopf related protein-1	NP_036374	2	1			14	4
Matrix metalloproteinase-19	NP_002420			12	7	13	11
Laminin, alpha 2	NP_000417			8	6	13	11
Aminopeptidase N	NP_001141			17	12	13	12
Dickkopf related protein-3	NP_056965	5	2	7	3	12	6
Tumour endothelial marker 1, CD248	NP_065137			4	3	12	6
Angiotensinogen	NP_000020			6	5	12	7
Adamts-1	NP_008919	1	1	7	5	11	7
Isoform beta of poliovirus receptor				7	4	11	4
Growth/differentiation factor 15	NP_004855	1	1	10	6	11	5
Serpin I2	NP_006208			10	1	10	1
Insulin-like growth factor BP 3	NP_000589	1	1	4	3	10	5
Cathepsin Z	NP_001327			4	3	9	3
Laminin beta-2	NP_002283			7	6	9	7
Metalloproteinase inhibitor 3, TIMP3	NP_000353			2	2	5	2
Xaa -Pro dipeptidase	NP_000276			4	4	5	3
Laminin gamma-3 chain						4	1
Vascular endothelial growth factor C	NP_005420			7	4	4	3
Monocyte differentiation antigen, CD14	NP_000582			5	4	4	3
Macrophage metalloelastase, MMP12	NP_002417			5	4	4	3
Latent TGF beta BP 1	NP_000618					4	3
Thrombospondin-3	NP_009043			4	2	4	2
Intercellular adhesion molecule-1	NP_000192			3	3	3	3
Bone morphogenic protein 1	NP_006120			3	3	3	3
Endothelial protein C receptor	NP_006395					3	3
Vascular endothelial growth factor receptor 1, Flt1	NP_002010			4	4	3	3
Low-affinity Ig gamma Fc region receptor III-B	NP_000561					3	1
HLA class I, A-1 alpha chain	NP_002107	1	1	3	2	3	2
Neural cell adhesion molecule L1	NP_000416			2	2	3	2
Alpha 3 type VI collagen isoform 3	NP_476506			1	1	3	1
Latent TGF beta BP 3				1	1	3	2
Latent TGF beta BP 2				4	3	3	2
TGF beta receptor III	NP_003234			1	1	3	3
Cathepsin F	NP_003784			1	1	2	2
Beta platelet-derived growth factor receptor	NP_002600					2	2
Low-density lipoprotein receptor-related protein1	NP_002323			4	4	2	2
Vascular non-inflammatory molecule 3	NP_060869			1	1	2	1
Apolipoprotein E	NP_000032			2	2	2	2
EGF-containing fibulin-like ECM protein1	NP_004096			2	1	2	2

Name(s)	Reference sequence	Mock spectra	Mock peptide	AD169 Spectra	AD169 peptide	TR spectra	TR peptide
Interleukin-4-induced protein 1	NP_690863			4	3	2	2
Macrophage migration inhibitory factor	NP_002406			3	1	2	1
Cathepsin L2	NP_001324			1	1	1	1
TGF beta1	NP_000651			1	1	1	1
E-cadherin				2	1	1	1
Vasoactive intestinal peptide	NP_003372	2	1	4	1	1	1
Urokinase receptor-associated protein	NP_006030			5	4	1	1
Plasma kallikreins	NP_000883			1	1	1	1
Vascular endothelial growth factor A						1	1
Angiopoietin-1	NP_001137					1	1
Small inducible cytokine B10, CXCL10	NP_001556					1	1
Platelet-derived growth factor C	NP_057289			2	2	1	1
Interleukin-8						1	1
Insulin-like growth factor BP 2	NP_000588	1	1			1	1
Cathepsin S	NP_004070			1	1	1	1
Plasminogen activator inhibitor-2	NP_002566			1	1		
Plasma serine protease inhibitor	NP_000615			1	1		
Interleukin-7 receptor alpha chain				2	2		
TGF-beta 2				1	1		

[a]Values in columns labelled 'spectra' are the total number of peptides detected by MS/MS for each protein. Values in columns labelled 'peptides' indicate the number of different peptides detected by MS-MS for each protein.

In the fibroblasts we detected 144 of the 174 factors in the HCMV secretome when our cut-off value was set at an average intensity of 500. Of these 144 proteins detected, 41 factors were significantly induced over the mock-infected secretome. The most highly abundant WH/AG-associated cellular factors identified in the HCMV secretomes included cytokines/chemokines (IL-6, osteoprotegerin, MIP-1α/CCL3, RANTES/CCL5, MCP-3/CCL7, MIP-3α/CCL20, GROα/CXCL1, ENA-78/CXCL5, and CXCL16), receptors (TNF-RI and II and ICAM-1), growth factors (TGF-β1, HGF and GM-CSF), ECM modifiers (MMP-1, TIMP-1, TIMP-2, TIMP-4), and the angiogenic RNase angiogenin. Interestingly, many of the genes identified as up-regulated by microarray analysis in the rat allograft hearts were also found in the HCMV secretome, suggesting that the factors involved in HCMV-induced angiogenesis and wound healing are similar to those expressed in the RCMV-infected rat organ allografts (Streblow et al., 2008). Analysis of the secretome from HCMV-infected HUVEC identified similar cytokines (IL-6, GCP-2/CXCL6, MPIF-1/CCL23, GM-CSF, osteoprotegerin, TNFα) and chemokines (RANTES/CCL5, MCP-2/CCL8, MIP-3α/CCL20, IP-10/CXCL11, MIP-1α/CCL3 and MIP-1β/CCL4, MCP-3/CCL7, I-TAC/CXCL11, GRO-α/CXCL1, IL-8/CXCL8) that can indirectly contribute to switch the growth/differentiation process from a natural healing to a pathological process (Botto et al., 2011).

Our approach was also used by Fiorentini et al. (2011) to study the effect of HCMV infection on lymphatic endothelial cells (LEC). In their recent publication the authors showed the complexity of the HCMV-induced LEC-derived secretome and the presence of several factors involved in the promotion of WH/AG. A comparison between the array results from HCMV infected fibroblasts (NHDF) and endothelial cells (HUVEC and LEC) identified IL-6, GM-CSF, ICAM-1, MIP-1β/CCL4, MCP-3/CCL7, MIP-3α/CCL20, and I-TAC as the most highly abundant WH/AG-associated factors present in the HCMV secretome (Table I.6.8).

Table I.6.8 Most abundant factors[a] present in the HCMV secretome from mock-infected and infected NHDFs, HUVECs and LECs

Protein	Mock NHDF average intensity ±SD	Infected NHDF average intensity ±SD	Fold Induction	Mock HUVEC average intensity ±SD	Infected HUVEC average intensity ±SD	Fold Induction	Mock LEC average intensity ±SD	Infected LEC average intensity ±SD	Fold Induction
IL-6	1037±447	24617±2641	27.7	6769±395	43424±4339	6.4	1449±79	53183±1337	36.7
GM-CSF	28±5	5661±831	20.2	1422±102	4384±507	3.1	654±37	9144±928	13.9
ICAM-1	325±185	725±267	2.2	544±28	1636±209	3	240±39	1144±172	4.7
MIP-1_	435±89	7058±238	16.2	982±42	3580±947	3.6	35±2	694±107	19.8
MCP-3	26±9	102±22	3.9	8184±963	28152±3962	3.4	305±33	1370±71	4.5
MIP-3_	17±12	896±92	5.2	5815±369	47202±6571	8.1	8053±1229	24162±4666	3
I-TAC	251±189	1115±227	4.4	2436±219	7626±996	3.1	934±19	1437±100	1.5

[a]Factor values (in arbitrary units) were determined by RayBiotech protein array analysis. Averages ± standard deviation (SD) are shown from mock and infected intensity values.

Role of IL-6 in HCMV induced angiogenesis and survival of endothelial cells

Angiogenesis is a physiological process involving growth of new blood vessels from pre-existing ones and consists of a growth phase followed by a stabilization phase (Auerbach *et al.*, 2003; Guidolin *et al.*, 2004). Growth phase events include proteolytic digestion of the basement membrane (BM) and extracellular matrix (ECM) of the existing vessel, migration and proliferation of ECs, lumen formation within the EC sprout, and anastomosis of sprouts to form neovessels. Stabilization involves arrest of EC proliferation, EC differentiation, intercellular adhesion and remodelling of the BM/ECM network to create an immature capillary. Key steps in both phases of angiogenesis can be modelled using an *in vitro* assay: the matrigel assay for capillary-like tubular formation (Wegener *et al.*, 2000; Xiao *et al.*, 2002). The extent of tubule formation and stabilization depends on factors produced by the EC themselves in coordination with exogenous angiogenic agonists or antagonists in the culture media. We utilized the matrigel assay to test the angiogenic activity of the HCMV secretome. In our first publication we have shown that HCMV secretome from infected-fibroblasts contains factors that promote AG and allow stabilization of neovessels and that generation of this active secretome requires HCMV replication (Dumortier *et al.*, 2008). In a recent publication (Botto *et al.*, 2011) we extended our studies to the identification of the cellular factors secreted from infected endothelial cells involved in the angiogenic process. Using the RayBiotech human cytokines/growth factors array we identified IL-6, GM-CSF and IL-8/CXCL8 as the most abundant pro-angiogenic factors present in the secretome of infected fibroblasts

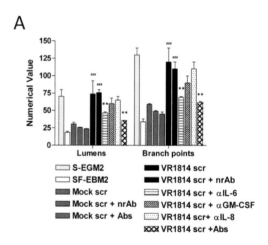

Figure I.6.4 Neutralization of IL-6 decreases angiogenesis. Quantification of matrigel tube formation assay after 24 hours in the presence of mock secretome (Mock scr), mock secretome with a non-related antibody (Mock scr+nrAb), mock secretome with anti-IL-6, anti-GM-CSF, and anti-IL-8/CXCL8 Abs (Mock scr +Abs), HCMV VR1814 secretome (VR1814 scr), HCMV VR1814 secretome with a non-related antibody (VR1814 scr+nrAb), HCMV VR1814 secretome with anti-IL-6 antibody (VR1814 scr+αIL-6), HCMV VR1814 secretome with anti-GM-CSF antibody (VR1814 scr+αGM-CSF), HCMV VR1814 secretome with anti-IL-8/CXCL8 antibody (VR1814 scr+αIL-8), and HCMV VR1814 secretome with anti-IL-6, anti-GM-CSF, and anti-IL-8/CXCL8 antibodies (VR1814 scr+Abs). The results are expressed as mean ± SD, [###] $P<0.001$ versus mock, [**]$P<0.01$ versus HCMV VR1814 secretome.

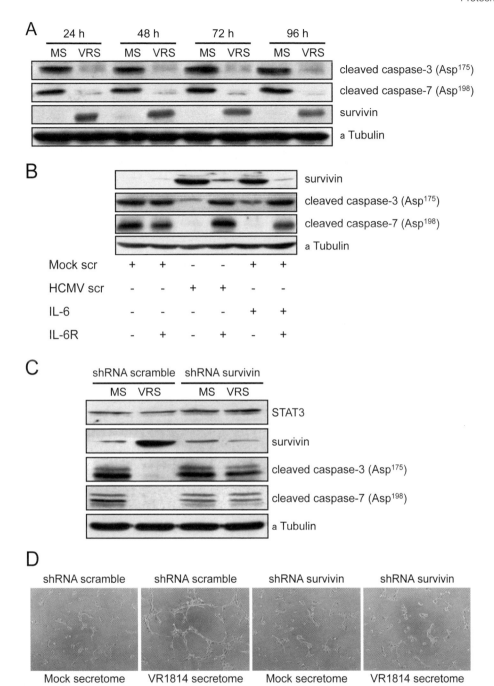

Figure I.6.5 HCMV secretome stimulates survivin expression through IL-6 receptor blocking apoptosis and promoting angiogenesis. (A) HCMV secretome inhibits caspase-3 and -7 activity and increases survivin expression. Cell lysates from HUVEC incubated with mock-secretome (MS) and HCMV VR1814 (VRS) were subjected to sodium dodecyl sulfate-polyacrylamide gel electrophoresis (SDS-PAGE) followed by Western blotting to evaluate the expression of activated/cleaved caspase-3 and -7 and survivin. αTubulin served as an internal control. (B) HUVEC cells were treated with anti-IL6R antibody and stimulated with the secretome for 48 hours. Expression of survivin, cleaved caspase-3 and -7 was determined by Western blotting. (C) HUVEC were infected with recombinant retroviruses expressing either a survivin-directed shRNA or a non-specific control (scramble) shRNA followed by stimulation with the HCMV secretome Expression of STAT3, survivin, cleaved caspase-3 and -7 was determined by Western blotting. (D) Representative examples of matrigel tube formation of each culture condition as above (magnification = ×10).

and HUVECs. To investigate the role of IL-6, GM-CSF, and IL-8/CXCL8 in AG, secretomes from mock-infected and HCMV infected-HUVECs were pre-incubated with anti–IL-6, anti–GM-CSF, and anti-IL-8/CXCL8 antibodies alone or in combination and then tested for their ability to induce tubule formation on matrigel. Controls included HUVECs grown in serum-free endothelial basal media (SF-EBM2) (negative control), in complete endothelial growth media (S-EGM2) (positive control), or in secretome-treated with a non-related antibody IL-1R. Neutralization of IL-6 activity resulted in 50% reduction of AG either with secretome from infected-HUVECs (Fig. I.6.4) or from infected-NHDF (data not shown), while GM-CSF and IL-8/CXCL8 neutralization did not significantly affect the angiogenic response. Moreover, the angiogenic effect observed 24 hours after secretome stimulation persisted for up to 1 week after treatment. The tubules formed in the presence of HCMV secretome remained intact, while those induced by mock secretome or HCMV secretome treated with IL-6 neutralizing antibodies had degenerated severely.

To determine the mechanism through which the HCMV secretome induces vessel stabilization, we analysed the effect of virus-free supernatants on the apoptosis pathway. Immunoblot analysis has clearly shown that HCMV secretome is able to block the activation of caspase-3 and -7, exerting an antiapoptotic effect on the endothelial cells (Fig. I.6.5A). Moreover, neutralization of IL-6 but not of GM-CSF or IL8/CXCL-8 increased caspase -3 and -7 activity.

The binding of IL-6 to its receptor has been reported to result in STAT3 phosphorylation, which subsequently regulates the expression of nuclear genes involved in apoptosis, including survivin (Yu *et al.*, 2004; Hodge *et al.*, 2005). It has been shown that inhibition of STAT3 signalling induces apoptosis and diminishes survivin expression in primary effusion lymphoma (Aoki *et al.*, 2003). Therefore, to determine whether survivin was involved in secretome-mediated apoptosis, we examined levels of this protein in ECs treated with the HCMV secretome. As shown in Fig. I.6.5A, HCMV-secretome significantly increased survivin expression at each time point analysed. Neutralization of IL-6 receptor on ECs abolished the ability of HCMV secretome to increase survivin and caused the activation of caspase-3 and -7 (Fig. I.6.5B). These results suggest that the HCMV secretome promotes survival of ECs through the induction of survivin via IL-6 receptor–mediated signalling. Finally, using an shRNA against survivin we were able to demonstrate that suppression of survivin induces apoptosis and rapid regression of tubule

capillary networks in ECs stimulated with HCMV secretome (Fig. I.6.5C,D).

In agreement with our findings, Fiorentini *et al.* (2011) identified IL-6 in the secretome of HCMV-infected LECs as one of main mediators of vessel neoformation. However, they observed a significant GM-CSF-dependent haemangiogenic and lymphangiogenic effect in the same secretome. This different result can be explained by the lower concentration of GM-CSF in the HUVEC-derived secretome than that measured in the HCMV-infected LEC secretome.

Viral genes involved in the secretion of pro-angiogenic factors

CMV encodes chemokines and chemokine receptor homologues that recruit and stimulate a multitude of host cellular infiltrates involved in the inflammatory response.

Among these homologues there is a G protein-coupled receptor (GPCR), which is structurally similar to the human chemokine receptor CCR1, the viral chemokine receptor US28 (Gao and Murphy, 1994). US28 binds various chemokines, including CCL2, CCL5, and CX3CL1 (Bodaghi *et al.*, 1998), and may thereby suppress the host immune response (Randolph-Habecker *et al.*, 2002). In addition, US28 also signals constitutively and shows G protein promiscuity, traits that enable it to hijack the host cell's signalling machinery (Casarosa *et al.*, 2001).

We have observed that the HCMV-encoded chemokine receptor US28 can stimulate the migration of vascular smooth muscle cells (SMC) and macrophages in a ligand-dependent manner. However, US28-mediated migration does not occur through the same pathways in both cell types (Vomaske *et al.*, 2009). SMC migration is mediated by binding to CC-chemokines and is blocked by the CX3C-chemokine Fractalkine (Streblow *et al.*, 1999; Vomaske *et al.*, 2009). In contrast, macrophage migration induced by US28 occurs after ligation with Fractalkine but not CC-chemokines. This suggests a dual role for US28 in migration that is both cell type and ligand specific. In addition to its role in directly promoting cellular migration, US28 induces cyclooxygenase-2 (COX-2) expression via activation of nuclear factor-κB, driving the production of vascular endothelial growth factor (VEGF), which promotes angiogenesis (Maussang *et al.*, 2009). In their most recent publication, Singler *et al.* (2011) showed in a glioblastoma tumour model that US28 induces cell proliferation through NF-κB and IL-6–STAT3 signalling pathways. Analysis of a set of secreted growth factors, chemokines, and cytokines enabled them to confirm increased VEGF secretion by US28-expressing cells. Notably, the concentration of IL-6 was increased in the medium of US28-expressing cells. The rise in IL-6 production and secretion in US28 expressing cells was associated with increased activation of STAT3 through the upstream activation of JAK1. The presence of US28 and STAT3 phosphorylation in cells lining the blood vessels in primary glioblastoma tumours suggests that US28 may be involved in the formation and maintenance of these tumours within the vascular niche. The IL-6 receptor is present in glioblastomas, and IL-6 triggers proliferation and migration in cerebral endothelial cells (Goswami *et al.*, 1998; Yao *et al.*, 2007). Cells expressing US28 may also influence neighbouring cells in a paracrine manner, effectively reprogramming these cells to display a more malignant phenotype.

The CMV family also encodes chemokine homologues. HCMV encodes one CC-chemokine (UL128) and two CXC-chemokines (UL146 and UL147) (Cha *et al.*, 1996). UL146 or vCXCL1 was shown to induce the migration of neutrophils by binding to the cellular CXCR2 receptor (Penfold *et al.*, 1999; Sparer *et al.*, 2004). The most well characterized CMV-encoded CC-chemokine is the MCMV m129/131 chemokine or MCK-2, which recruits immature myelomoncytic leucocytes from the bone marrow to sites of MCMV infection and promotes viral dissemination (Noda *et al.*, 2006).

Preliminary results from our group identify a region near the left terminus of the HCMV genome involved in the secretion of proangiogenic factors. These data are supported by the requirement of viral replication to produce an effective HCMV secretome, since UV-inactivated virus failed to induce wound healing and angiogenesis (Dumortier *et al.*, 2008; Fiorentini *et al.*, 2011). Moreover, HCMV secretome produced in the presence of PFA (foscarnet), a viral polymerase inhibitor, significantly reduced the angiogenic effect suggesting that late or early-late viral genes can be involved in this process (Caposio, unpublished results).

Conclusions

In this chapter we described the detailed analysis of the viral and cellular proteins that compose HCMV particles as well as the cellular proteins secreted from infected cells and their effect on the host environment.

Although a significant amount of information has been accumulated over the years concerning the protein content of HCMV particles, the high-throughput mass spectrometry data analysis presented in this review indicates that HCMV virions and DBs are much more complex than previously predicted from other studies. Previous work using conventional approaches has suggested that 30–50 ORFs encode proteins that are necessary for the formation of an infectious HCMV virion (Baldick and Shenk, 1996). However, mass spectroscopy analysis of purified virion preparations indicates that at least 71 ORFs encode proteins composing the virion. In addition, a significantly higher number of cellular proteins were also present in virion preparations with at least 75 proteins meeting our criteria for significance in mass spectroscopy analysis. The major concern in this observation is whether the proteins that we detected are actually part of the virion or a co-purification contaminant with the particles. Two interesting findings argue against the contaminant theory: first is the fact that different cellular proteins were observed in the DB preparations that were prepared by similar purification procedures, second, the mass spectrometry analysis of the aggresomes revealed the presence of cellular proteins like heat shock proteins, aurora-related kinase 1, nucleophosmin and many others (Prichard *et al.*, 2008). Whether the cellular proteins are an integral constituent of the purified particles or not, the virion with these proteins forms an infectious unit and these cellular proteins must be considered part of this complex. Another unexpected observation was the presence of enzymes associated with replication of the viral DNA template in the virion. This observation was also noted in the analysis of the MCMV proteome. These findings are intriguing with respect to the hypothesis that HCMV may undergo an initial round of DNA template replication prior to the synthesis of the early genes that encode the replication enzymes. The role of these enzymes in the virion is unknown and they may represent remnants of the encapsidation of viral templates undergoing DNA synthesis. However, the enzymes may be important to amplify the viral genome upon initial infection.

The analysis of the secretome from different HCMV-infected cells has provided new insights into the pathophysiology of HCMV-accelerated vascular diseases and tumours. As we discussed in the second part of this review the virus is able to manipulate the host cell response inducing the secretion of factors that promote angiogenesis and wound healing, which are important processes that drive HCMV-associated vasculopathy, chronic allograft rejection and glioblastoma. The proteomic approach has been helpful to identify IL-6 as an important pro-angiogenic and antiapoptotic factor present in the HCMV secretome. The increase of IL-6 in the secretome following HCMV infection of EC parallels the observation of increased IL-6 in CMV-infected tissues (Humbert *et al.*, 1993). Studies by our group have shown that IL-6 mRNA is significantly up-regulated in RCMV-infected heart tissues both at 21 and 28 days post transplantation in a rat allograft model during the development of TVS (Orloff *et al.*, 2002; Streblow *et al.*, 2003). Thus, the findings presented in this review together with the observations for the RCMV allograft model provide a possible mechanism of CMV infection accelerated TVS through secretion of IL-6 from infected cells, which involves induction of a pathological tissue-remodelling process that leads to endothelial cell proliferation and survival.

One question that arises from these studies is how would HCMV induction of cellular cytokines that block apoptosis be beneficial for the virus? HCMV is known to encode at least two antiapoptotic proteins, vICA and vMIA, with the latter gene being indispensable for viral replication (Goldmacher, 2005). In addition, HCMV has been reported to induce the antiapoptotic cellular proteins AKT, Bcl-2, and Np73 (Michaelis *et al.*, 2004). Clearly, these observations indicate that blocking cellular apoptosis is an essential characteristic of the virus. Therefore, one explanation for the HCMV induction of IL-6 mediated anti-apoptosis described in this report is that this pathway may represent a redundant mechanism for the virus to maintain cellular survival of HCMV persistently infected endothelial cells. However, a by-product of this process is the stimulation of angiogenesis that can lead to pathological consequences such as atherosclerosis or TVS.

Finally, US28 studies suggest a possible role of viral genes involved in the secretion of proliferative cellular factors suggesting future research meant to identify the specific HCMV genes that would make possible prevention or abrogation of the HCMV-associated vasculopathy, chronic allograft rejection and glioblastoma.

Acknowledgements

Portions of this research were supported in part by the U.S. Department of Energy (DOE), Office of Biological and Environmental Research, LDRD under the Biomolecular Systems Initiative at the Pacific Northwest National Laboratory, NIH National Center for Research Resources (RR18522), and the Environmental Molecular Science Laboratory (a US Department of Energy user facility located at the Pacific Northwest National Laboratory). Battelle Memorial Institute for the US Department of Energy under contract DE-AC06-76RLO-1830 operates Pacific Northwest National Laboratory. The work at OHSU was supported in part by a Public Service Grant from the National Institutes of Health (AI 21640) (J.A.N.). D.N.S. is supported by an AHA Scientist Development Grant. The proteomics study of the secretome was support by NIH grants: J.A. Nelson (HL 088603) and D.N. Streblow (HL 083194).

References

Adkins, J.N., Varnum, S.M., Auberry, K.J., Moore, R.J., Angell, N.H., Smith, R.D., Springer, D.L., and Pounds, J.G. (2002). Toward a human blood serum proteome: analysis by multidimensional separation coupled with mass spectrometry. Mol. Cell. Proteomics *1*, 947–955.

Aoki, Y., Feldman, G.M., and Tosato, G. (2003). Inhibition of STAT3 signaling induces apoptosis and decreases survivin expression in primary effusion lymphoma. Blood *101*, 1535–1542.

Auerbach, R., Lewis, R., Shinners, B., Kubai, L., and Akhtar, N. (2003). Angiogenesis assays: a critical overview. Clin. Chem. *49*, 32–40.

Baldick, C.J., Jr., and Shenk, T. (1996). Proteins associated with purified human cytomegalovirus particles. J. Virol. *70*, 6097–6105.

Benko, D.M., Haltiwanger, R.S., Hart, G.W., and Gibson, W. (1988). Virion basic phosphoprotein from human cytomegalovirus contains O-linked N-acetylglucosamine. Proc. Natl. Acad. Sci. U.S.A. *85*, 2573–2577.

Bodaghi, B., Jones, T.R., Zipeto, D., Vita, C., Sun, L., Laurent, L., Arenzana-Seisdedos, F., Virelizier, J.L., and Michelson, S. (1998). Chemokine sequestration by viral chemoreceptors as a novel viral escape strategy: Withdrawal of chemokines from the environment of cytomegalovirus-infected cells. J. Exp. Med. *188*, 855–866.

Botto, S., Streblow, D.N., DeFilippis, V., White, L., Kreklywich, C.N., Smith, P.P., and Caposio, P. (2011). IL-6 in Human Cytomegalovirus secretome promotes angiogenesis and survival of endothelial cells through the stimulation of survivin. Blood *117*, 352–361.

Casarosa, P., Bakker, R.A., Verzijl, D., Navis, M., Timmerman, H., Leurs, R., and Smit, M.J. (2001). Constitutive signaling of the human cytomegalovirus-encoded chemokine receptor US28. J. Biol. Chem. *276*, 1133–1137.

Cha, T.A., Tom, E., Kemble, G.W., Duke, G.M., Mocarski, E.S., and Spaete, R.R. (1996). Human cytomegalovirus clinical isolates carry at least 19 genes not found in laboratory strains. J. Virol. *70*, 78–83.

Chee, M.S., Bankier, A.T., Beck, S., Bohni, R., Brown, C.M., Cerny, R., Horsnell, T., Hutchison, C.A., 3rd, Kouzarides, T., Martignetti, J.A., *et al.* (1990). Analysis of the protein-coding content of the sequence of human cytomegalovirus strain AD169. Curr. Top. Microbiol. Immunol. *154*, 125–169.

Davison, A.J., Dolan, A., Akter, P., Addison, C., Dargan, D.J., Alcendor, D.J., McGeoch, D.J., and Hayward, G.S. (2003). The human cytomegalovirus genome revisited: comparison with the chimpanzee cytomegalovirus genome. J. Gen. Virol. *84*, 17–28.

Dumortier, J., Streblow, D.N., Moses, A.V., Jacobs, J.M., Kreklywich, C.N., Camp, D., Smith, R.D., Orloff, S.L., and Nelson, J.A. (2008). Human cytomegalovirus secretome contains factors that induce angiogenesis and wound healing. J. Virol. *82*, 6524–6535.

Dunn, W., Chou, C., Li, H., Hai, R., Patterson, D., Stolc, V., Zhu, H., and Liu, F. (2003). Functional profiling of a human cytomegalovirus genome. Proc. Natl. Acad. Sci. U.S.A. *100*, 14223–14228.

Fiorentini, S., Luganini, A., Dell'Oste, V., Lorusso, B., Cervi, E., Caccuri, F., Bonardelli, S., Landolfo, S., Caruso, A., and Gribaudo, G. (2011). Human cytomegalovirus productively infects lymphatic endothelial cells and induces a secretome that promotes angiogenesis and lymphangiogenesis through interleukin-6 and granulocyte–macrophage colony-stimulating factor. J. Gen. Virol. *92*, 650–660.

Folkman, J. (2003). Fundamental concepts of the angiogenic process. Curr. Mol. Med. *3*, 643–651.

Gao, J.L., and Murphy, P.M. (1994). Human cytomegalovirus open reading frame US28 encodes a functional β chemokine receptor. J. Biol. Chem. *269*, 28539–28542.

Gibson, W. (1983). Protein counterparts of human and simian cytomegaloviruses. Virology *128*, 391–406.

Giugni, T.D., Soderberg, C., Ham, D.J., Bautista, R.M., Hedlund, K.O., Moller, E., and Zaia, J.A. (1996). Neutralization of human cytomegalovirus by human CD13-specific antibodies. J. Infect. Dis. *173*, 1062–1071.

Goldmacher, V.S. (2005). Cell death suppression by cytomegalovirus. Apoptosis *10*, 251–265.

Goswami, S., Gupta, A., and Sharma, S.K. (1998). Interleukin-6-mediated autocrine growth promotion in human glioblastoma multiforme cell line U87MG. J. Neurochem. *71*, 1837–1845.

Grundy, J.E., McKeating, J.A., and Griffiths, P.D. (1987). Cytomegalovirus strain AD169 binds beta 2 microglobulin in vitro after release from cells. J. Gen. Virol. *68*, 777–784.

Guidolin, D., Vacca, A., Nussdorfer, G.G., and Ribatti, D. (2004). A new image analysis method based on topological and fractal parameters to evaluate the angiostatic activity of docetaxel by using the Matrigel assay *in vitro*. Microvasc. Res. *67*, 117–124.

Hodge, D.R., Hurt, E.M., and Farrar, W.L. (2005). The role of IL-6 and STAT3 in inflammation and cancer. Eur. J. Cancer *41*, 2502–2512.

Humbert, M., Delattre, R.M., Fattal, S., Rain, B., Cerrina, J., Dartevelle, P., Simonneau, G., Duroux, P., Galanaud, P., and Emilie, D. (1993). *In situ* production of interleukin-6 within human lung allografts displaying rejection or cytomegalovirus pneumonia. Transplantation. *56*, 623–627.

Irmiere, A., and Gibson, W. (1983). Isolation and characterization of a noninfectious virion-like particle released from cells infected with human strains of cytomegalovirus. Virology *130*, 118–133.

Kattenhorn, L.M., Mills, R., Wagner, M., Lomsadze, A., Makeev, V., Borodovsky, M., Ploegh, H.L., and Kessler, B.M. (2004). Identification of proteins associated with murine cytomegalovirus virions. J. Virol. *78*, 11187–11197.

Link, A.J., Eng, J., Schieltz, D.M., Carmack, E., Mize, G.J., Morris, D.R., Garvik, B.M., and Yates, J.R., 3rd (1999). Direct analysis of protein complexes using mass spectrometry. Nature Biotechnol. *17*, 676–682.

Maussang, D., Langemeijer, E., Fitzsimons, C.P., Stigter-van Walsum, M., Dijkman, R., Borg, M.K., Slinger, E., Schreiber, A., Michel, D., Tensen, C.P., et al. (2009). The human cytomegalovirus-encoded chemokine receptor US28 promotes angiogenesis and tumor formation via cyclooxygenase-2. Cancer Res. *69*, 2861–2869.

Melnick, J.L., Petrie, B.L., Dreesman, G.R., Burek, J., McCollum, C.H., and DeBakey, M.E. (1983). Cytomegalovirus antigen within human arterial smooth muscle cells. Lancet *322*, 644–647.

Melnick, J.L., Adam, E., and DeBakey, M.E. (1998). The link between CMV and atherosclerosis. Infect. Med. *15*, 479–486.

Michaelis, M., Kotchetkov, R., Vogel, J.U., Doerr, H.W., and Cinatl, J., Jr. (2004). Cytomegalovirus infection block apoptosis in cancer cells. Cell. Mol. Life Sci. *61*, 1307–1316.

Michelson, S., Turowski, P., Picard, L., Goris, J., Landini, M.P., Topilko, A., Hemmings, B., Bessia, C., Garcia, A., and Virelizier, J.L. (1996). Human cytomegalovirus carries serine/threonine protein phosphatases PP1 and a host cell derived PP2A. J. Virol. *70*, 1415–1423.

Mocarski, E.S., and Courcelle, C.T. (2001). Cytomegalovirus and their replication. In Fields Virology, Straus, S.E., ed. (Lippincott-Raven, Philadelphia, PA), pp. 2629–2673.

Moorman, N.J., Sharon-Friling, R., Shenk, T., and Cristea, I.M. (2010). A targeted spatial-temporal proteomics approach implicates multiple cellular trafficking pathways in human cytomegalovirus virion maturation. Mol. Cell. Proteomics *9*, 851–860.

Murphy E., Yu, D., Grimwood, J., Schmutz, J., Dickson, M., Jarvis, M.A., Hahn, G., Nelson, J.A., Myers, R.M., and Shenk, T.E. (2003a). Coding potential of laboratory and clinical strains of human cytomegalovirus. Proc. Natl. Acad. Sci. U.S.A. *100*, 14976–14981.

Murphy, E., Rigoutsos, I., Shibuya, T., and Shenk, T.E. (2003b). Reevaluation of human cytomegalovirus coding potential. Proc. Natl. Acad. Sci. U.S.A. *100*, 13585–13590.

Noda, S., Aguirre, S.A., Bitmansour, A., Brown, J.M., Sparer, T.E., Huang, J., and Mocarski, E.S. (2006). Cytomegalovirus MCK-2 controls mobilization and recruitment of myeloid progenitor cells to facilitate dissemination. Blood *107*, 30–38.

Oliveira, S.A., and Shenk, T.E. (2001). Murine cytomegalovirus M78 protein, a G protein-coupled receptor homologue, is a constituent of the virion and facilitates accumulation of immediate-early viral mRNA. Proc. Natl. Acad. Sci. U.S.A. *98*, 3237–3242.

Orloff, S.L., Streblow, D.N., Soderberg-Naucler, C., Yin, Q., Kreklywich, C., Corless, C.L., Smith, P.A., Loomis, C.B., Mills, L.K., Cook, J.W., et al. (2002). Elimination of donor-specific alloreactivity prevents cytomegalovirus-accelerated chronic rejection in rat small bowel and heart transplants. Transplantation *73*, 679–688.

Penfold, M.E., Dairaghi, D.J., Duke, G.M., Saederup, N., Mocarski, E.S., Kemble, G.W., and Schall, T.J. (1999). Cytomegalovirus encodes a potent alpha chemokine. Proc. Natl. Acad. Sci. U.S.A. *96*, 9839–9844.

Phillips, S.L., and Bresnahan, W.A. (2011). Identification of binary interactions between human cytomegalovirus virion proteins. J. Virol. *85*, 440–447.

Prichard, M.N., Britt, W.J., Daily, S.L., Harline, C.B., and Kern, E.R. (2005). Human cytomegalovirus UL97 kinase is required for the normal intranuclear distribution of pp65 and virion morphogenesis. J. Virol. *79*, 15494–15502.

Prichard, M.N., Sztul, E., Daily, S.L., Perry, A.L., Frederick, S.L., Gill, R.B., Hartline, C.B., Streblow, D.N., Varnum, S.M., Smith, R.D., and Kern, E.R. (2008). Human cytomegalovirus UL97 kinase activity is required for the hyperphosphorylation of retinoblastoma protein and inhibits the formation of nuclear aggresomes. J. Virol. *82*, 5054–5067.

Randolph-Habecker, J.R., Rahill, B., Torok-Storb, B., Vieira, J., Kolattukudy, P.E., Rovin, B.H., and Sedmak, D.D. (2002). The expression of the cytomegalovirus chemokine receptor homolog US28 sequesters biologically active CC chemokines and alters IL-8 production. Cytokine *19*, 37–46.

Ryckman, B.J., Chase, M.C., and Johnson, D.C. (2008). HCMV gH/gL/UL128–131 interferes with virus entry into epithelial cells: evidence for cell type-specific receptors. Proc. Natl. Acad. Sci. U.S.A. *105*, 14118–14123.

Sarov, I., and Abady, I. (1975). The morphogenesis of human cytomegalovirus. Isolation and polypeptide characterization of cytomegalovirions and dense bodies. Virology *66*, 464–473.

Slinger, E., Maussang, D., Schreiber, A., Siderius, M., Rahbar, A., Fraile-Ramos, A., Lira, S.A., Söderberg-Nauclér, C., and Smit, M.J. (2010). HCMV-encoded chemokine receptor US28 mediates proliferative signaling through the IL–6-STAT3 axis. Sci. Signal. *3*, ra58.

Sparer, T.E., Gosling, J., Schall, T.J., and Mocarski, E.S. (2004). Expression of human CXCR2 in murine neutrophils as a model for assessing cytomegalovirus chemokine vCXCL-1 function in vivo. J. Interferon Cytokine Res. *24*, 611–620.

Speir, E., Modali, R., Huang, E.S., Leon, M.B., Shawl, F., Finkel, T., and Epstein, S.E. (1994). Potential role of human cytomegalovirus and p53 interaction in coronary restenosis. Science *265*, 391–394.

Stannard, L.M. (1989). Beta 2 microglobulin binds to the tegument of cytomegalovirus: an immunogold study. J. Gen. Virol. *70*, 2179–2184.

Streblow, D.N., Söderberg-Nauclér, C., Vieira, J., Smith, P., Wakabayashi, E., Rutchi, F., Mattison, K., Altschuler, Y., and Nelson, J.A. (1999). The human cytomegalovirus chemokine receptor US28 mediates vascular smooth muscle cell migration. Cell *99*, 511–520.

Streblow, D.N., Kreklywich, C., Yin, Q., De La Melena, V.T., Corless, C.L., Smith, P.A., Brakebill, C., Cook, J.W., Vink, C., Bruggeman, C.A., *et al.* (2003). Cytomegalovirus-mediated up-regulation of chemokine expression correlates with the acceleration of chronic rejection in rat heart transplants. J. Virol. *77*, 2182–2194.

Streblow, D.N., Orloff, S.L., and Nelson, J.A. (2007). Acceleration of allograft failure by cytomegalovirus. Curr. Opin. Immunol. *19*, 577–582.

Streblow, D.N., Kreklywich, C.N., Andoh, T., Moses, A.V., Dumortier, J., Smith, P.P., Defilippis, V., Fruh, K., Nelson, J.A., and Orloff, S.L. (2008). The role of angiogenic and wound repair factors during CMV-accelerated transplant vascular sclerosis in rat cardiac transplants. Am. J. Transplant. *8*, 277–287.

To, A., Bai, Y., Shen, A., Gong, H., Umamoto, S., Lu, S., and Liu, F. (2011). Yeast two hybrid analyses reveal novel binary interaction between human cytomegalovirus-encoded virion proteins. PLoS One *6*, e177966.

Varnum, S.M., Streblow, D.N., Monroe, M.E., Smith, P., Auberry, K.J., Pasa-Tolic, L., Wang, D., Camp, D.G., 2nd, Rodland, K., Wiley, S., *et al.* (2004). Identification of proteins in human cytomegalovirus (HCMV) particles: the HCMV proteome. J. Virol. *78*, 10960–10966.

Vomaske, J., Melnychuk, R.M., Smith, P.P., Powell, J., Hall, L., DeFilippis, V., Fruh, K., Smit, M., Schlaepfer, D.D., Nelson, J.A., *et al.* (2009). Differential ligand binding to a human cytomegalovirus chemokine receptor determines cell type-specific motility. PLoS Pathog *5*, e1000304.

Wegener, J., Keese, C.R., and Giaever, I. (2000). Electric cell-substrate impedance sensing (ECIS) as a noninvasive means to monitor the kinetics of cell spreading to artificial surfaces. Exp. Cell. Res. *259*, 158–166.

Wolf, D.G., Courcelle, C.T., Prichard, M.N., and Mocarski, E.S. (2001). Distinct and separate roles for herpesvirus-conserved UL97 kinase in cytomegalovirus DNA synthesis and encapsidation. Proc. Natl. Acad. Sci. U.S.A. *98*, 1895–1900.

Wright, J.F., Kurosky, A., Pryzdial, E.L., and Wasi, S. (1995). Host cellular annexin II is associated with cytomegalovirus particles isolated from cultured human fibroblasts. J. Virol. *69*, 4784–4791.

Xiao, C., Lachance, B., Sunahara, G., and Luong, J.H. (2002). An in-depth analysis of electric cell-substrate impedance sensing to study the attachment and spreading of mammalian cells. Anal. Chem. *74*, 1333–1339.

Yao, J.S., Zhai, W., Fan, Y., Lawton, M.T., Barbaro, N.M., Young, W.L., and Yang, G.Y. (2007). Interleukin-6 up-regulates expression of KDR and stimulates proliferation of human cerebrovascular smooth muscle cells. J. Cereb. Blood Flow Metab. *27*, 510–520.

Yu, D., Silva, M.C., and Shenk, T. (2003). Functional map of human cytomegalovirus AD169 defined by global mutational analysis. Proc. Natl. Acad. Sci. U.S.A. *100*, 12396–12401.

Yu, H., and Jove, R. (2004). The STATs of cancer-new molecular targets come of age. Nat. Rev. Cancer *4*, 97–105.

A Systems Pathway View of Cytomegalovirus Infection

Peter Ghazal, Alexander Mazein, Steven Watterson, Ana Angulo and Kai A. Kropp

Abstract

This chapter discusses the use of systems biology towards understanding the combinatorially complex set of molecular interactions that underpin the infection process by CMV. A hallmark of systems biology is the elucidation of pathways rather than single gene or protein activities. This generally involves the use of bioinformatics and computational modelling to analyse unbiased high-throughput data such as those derived from whole genome sequencing, genome-wide transcriptomics, proteomics and metabolomics. The emerging studies in the area of CMV systems biology have to date underscored the requirement for host dependencies on transcription factor networks, cell signalling, metabolism and cellular trafficking. Here we consider at the systems pathway level the importance of host dependency and host protection pathways in regulating the CMV transcription–replication cycle.

Introduction

Systems biology is emerging as a powerful approach for tackling high-throughput 'omic' data and unravelling the complexity of biological systems. High-throughput technologies have enabled sequencing of complete viral and host genomes, and the description of the metabolome and proteome upon infection. Such technological advancements have delivered data on the genome-wide temporal gene expression patterns, metabolic alterations, the complete proteome of the virus particle and host-viral protein interaction networks. Herpesviruses and in particular cytomegaloviruses have been at the forefront of these investigations including the development of the first whole genome shotgun sequencing approach (Chee et al., 1990), the first viral microarray for expression profiling (Chambers et al., 1999), the first complete DNA–virus proteome (Varnum et al., 2004), and the first description of intraviral and host–virus interactomes (protein–protein interactions) (Uetz et al., 2006). More recently, these systems level studies have been especially insightful towards uncovering new relationships between the virus and host in metabolic, transcriptional and signal transduction pathways.

Historically CMV biology has been dominated by applying a reductionist approach focusing on specific viral or cellular components at a single gene or single protein level. Such low-throughput studies are exclusively hypothesis driven and inherently biased in experimental design but have nevertheless generated a wealth of information, nearly always in a confined biological context. With the development of powerful 'omic' technologies, we now have an opportunity of combining hypothesis driven research with data driven discovery that is inherently unbiased in its experimental testing. New insights in metabolic and gene regulatory pathways playing crucial roles in processes of the viral transcription-replication cycle, response to external stimuli and host cell metabolism, are beginning to emerge. The ultimate grand challenge is to be able to predict the outcome of infection based on a formal (mathematically correct) assembly of the processes and the components together to create an integrated view of whole virus–host interactions.

As for all herpesviruses, the viral gene expression of CMV is strictly regulated in a cascaded manner and is divided in the immediate-early (IE), early (E) and late (L) phases. The infection cycle is absolutely dependent on both viral and host components. Importantly, this includes the sufficient activation and expression of the viral major IE genes, encoding regulatory proteins of viral or host gene expression and key biosynthetic and metabolic pathways of the cell (Spector, 1996; Sanchez and Spector, 2006; Yu et al., 2011; see also Chapter I.14). Thus as key checkpoints of control, infection only progresses to the early and late phases if the IE gene products are present and there is sufficient metabolic and biosynthetic capacity. Therefore, IE-gene expression and replication are coupled and both are entangled in the host networks controlling the activity of cellular factors.

In this regard the first section of this chapter highlights the significance of host dependency and reaffirms the *cellular clockwork model* of infection. In the next section we consider the pathway biology and logic of the regulation of the major immediate-early promoter. The last section discusses recent systems level studies that provide evidence disclosing how CMV and the cellular response to infection dramatically changes host cell metabolism.

Host dependency and the cellular clockwork model of infection

As obligatory intracellular parasites viruses are strictly dependent on host cell pathways and accordingly have evolved highly effective mechanisms to harness cellular processes to ensure production of progeny virus. Cytomegaloviruses have a comparably long replication cycle and are dependent to a varying degree on the host transcriptional machinery to express viral genes as well as on cellular metabolic and biosynthetic pathways for their own synthesis (Mocarski *et al.*, 2007). Therefore, despite their ability to effectively control the cell, these viruses are themselves hostages of their host system, and as a consequence the viral infection cycle is constrained by homeostatic control pathways (Ghazal *et al.*, 2000a,b).

The nature and consequences of the conditional dependencies a virus has with its host has led to a view of the viral transcription–replication cycle giving equal importance to both the virus and the host cell. This view has been previously referred to by us as a *cellular clockwork model* of infection (Fig. I.7.1) (Ghazal *et al.*, 2000a). The model involves a temporal checkpoint control that inter-depends on the cellular milieu (through two-way interaction pathways) at key kinetic stages (IE, E and L times of infection) of the viral cycle (Ghazal *et al.*, 2000b). This scenario, shown schematically in Fig. I.7.1, starts with the entry of the viral cycle to initiate IE gene expression and as it progresses through its cascade there exists the potential for multiple host checkpoints (indicated by the host-virus interacting pathways in Fig. I.7.1) mediated by transcription factor and cell signalling networks, energy metabolism and biosynthetic pathways, and in late morphogenesis stages of the viral cycle, cellular trafficking pathways. In this connection both the virus and the host innate and ultimately the adaptive immune responses kinetically alter the intracellular milieu of the infected cell.

Investigating the complexity of the biological responses suggested by the cellular clockwork model requires in addition to the traditional gene or protein analytic approach a systems investigation at the pathway biology level. Pathway biology approaches involve

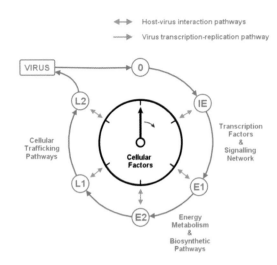

Figure I.7.1 Cellular clockwork model of infection. The clock handle indicates clockwise the changes in the viral expression co-ordinately linked with host cell changes. IE indicates immediate-early, E1 early, E2 delayed early, L1 leaky late and L2 true late viral gene expression, respectively. These phases depict the viral transcription-replication pathway (unidirectional arrows) and its interdependent interaction with host-regulatory pathways by bidirectional arrows. The corresponding host-pathways playing a pivotal role as checkpoints during the different phases of replication are indicated.

elucidating the network of cause–effect relationships of multiple genes, proteins and metabolites by combining experimental and computational methodology (Kellam, 2001; Jenner and Young, 2005; Forst, 2006; König *et al.*, 2008; Watterson *et al.*, 2008).

The cellular clockwork model considers that an infected cell can be pre-set or reset to 'zero', in effect resulting in a latent infection. In latently infected cells reactivation or initiating the viral transcription-replication cycle requires the appropriate cell signalling and transcription factor activation for the clock handle to move towards the IE gene activation stage. In the case of a primary infection, the transition of the clockwork from '0' to 'IE', is facilitated in HCMV-infected cells by the pp71 protein (product of UL82), which is incorporated into the virion and therefore is present at 'pre'-immediate-early times. CMV genomes are shuttled into the vicinity of PML bodies in the cell nucleus, a highly suppressive environment for gene expression. PML bodies are enriched with inhibitory proteins, such as Daxx and HDACs, which are known to suppress MIEP activity. The pp71 protein interacts with Daxx and induces its proteasomal degradation relieving the Daxx-HDAC mediated repression of the MIEP (Kalejta, 2008; see also Chapter I.9).

It is worth noting that the Pattern Recognition Receptors (PRRs; for CMV, TLR 3, 7, 9 and 2, are discussed in detail in Chapters I.8 and II.9) trigger a strong antiviral response that co-ordinately follows the clock handle to the E viral kinetic class. At this time point, viruses need to effectively counteract the innate immune response (including intrinsic factors) of the host to successfully move towards completing their transcription-replication cycle. In most cases viruses accomplish this by encoding viral effector molecules that work as, either structural or functional, mimics of host proteins, e.g. viral IL-10 (Jenkins et al., 2004, 2008) or antagonize host protein function, e.g. M27 (Abenes et al., 2001; Lee et al., 2002; Khan et al., 2004; Zimmermann et al., 2005) and M45 (Mack et al., 2008). From early times of infection the virus expresses a battery of viral 'immune evasion genes' and for CMV this accounts for a substantial part of the encoded ORFs (Del Val et al., 1989; Kavanagh et al., 2001; Mocarski, 2002a; Reddehase, 2002; Krmpotic et al., 2003; Voigt et al., 2003; Park et al., 2004; Besold et al., 2007; Hansen and Bouvier, 2009; Lemmermann et al., 2011). These genes mostly belong to the early kinetic classes (E_1 or E_2, as shown in Fig. I.7.1) and provide some of the critical viral countermeasures to the cellular checkpoints. In the final (L) phases of virion morphogenesis and maturation appropriate activity of cellular biosynthetic pathways and trafficking systems are essential.

Notably, at immediate-early times, viral inhibitors to prevent activation of signalling pathways associated with anti-viral factors are not yet available. At this immediate-early time of infection the intracellular environment encountered by the virus is characterized by activated 'pro-inflammatory' signalling pathways and transcription factors (TFs). This therefore provides a critical point of control marked by the absolute dependency of the viral enhancer elements of the major immediate-early promoter (MIEP) on the host transcription factors (see also Chapter I.10). In the following sections we discuss the pathway biology mapping of the MIEP and how CMV opportunistically copes with the pro-inflammatory status of the cell and its underlying systems logic. We then consider the E (early) to L (late) transitions of the clock handle and discuss how CMV alters the energy metabolism and biosynthetic pathways of the cell to its advantage while the innate immune response modulates lipid metabolic pathways as a counter checkpoint.

Pathway biology and logic of the major immediate-early promoter

The initial step of productive infection is the onset of viral IE-gene expression, controlled by cellular transcription factors (TFs) (Ghazal et al., 1987, 1988a,b; see also Chapter I.10), whose activity is regulated by a complex network of signalling cascades. To interact with cellular transcription factors the viral genome contains enhancer elements that compete with cellular enhancers for the pool of RNA polymerase II in the cell. In this context the HCMV MIEP-enhancer binds a complex array of more than 15 different TFs (Stinski and Isomura, 2008) and is one of the strongest transcription enhancers known (Boshart et al., 1985). The MIEP of HCMV controls the expression of the crucial viral trans-activators IE1 and IE2 and is, therefore, one of the rate limiting factors for replication after viral entry. The viral enhancer represents a kind of regulatory 'synapse' where the input of different cellular regulatory networks is integrated and determines the transcription level output of the IE-genes. Owing to the coupling of IE-gene expression and viral replication this is a major factor determining the permissiveness of the cell. The importance of the enhancer for viral replication has been demonstrated in different CMVs, e.g. the murine (Angulo et al., 1998; Grzimek et al., 1999; Ghazal et al., 2003; Podlech et al., 2010), the human (Meier and Pruessner, 2000; Meier et al., 2002; Isomura et al., 2004; DeFilippis, 2007), simian (Chang et al., 1990) or the rat CMV (Sandford et al., 2001), were deletion of the enhancer massively impairs replication or abolishes it almost completely.

The cellular signalling landscape is shaped during the viral entry process

The process of infection triggers toll-like receptor (TLR) signalling and activates the cytoplasmic DNA sensor molecules ZBP-1 (DAI) and AIM2 (Boyle et al., 1999; Compton et al., 2003; Boehme and Compton, 2006; Juckem et al., 2008; Stinski and Isomura, 2008; DeFilippis et al., 2010; Rathinam et al., 2010; see also Chapter I.8). The activity of these signalling pathways leads to marked production of type I interferons (see Chapter I.16) and other pro-inflammatory cytokines that confer an antiviral state to neighbouring cells and also eventually activate the innate and adaptive immune response (Tabeta et al., 2004; Beutler et al., 2005a,b, 2006; DeFilippis et al., 2006). Significantly, however, CMV is able to replicate despite the strong activation of these antiviral factors. To overcome antiviral host factors CMV has to counteract the activity of the innate immune system allowing it to eventually escape to latency and persisting lifelong in its host (DeFilippis, 2007; Mocarski et al., 2007). Comparison of the cellular immune enhancers and the CMV enhancer show that binding motifs of TFs controlling expression of innate immune genes are also present in the CMV enhancer. Below we consider how the MIEP might opportunistically take advantage of a pro-inflammatory state of the cell.

Opportunistic role of the MIEP in harnessing TF networks associated with immune gene expression

In studies of HCMV it has been shown that signalling events triggered by IFN-α or IFN-γ may be involved in activation of *ie*1 expression in human fibroblasts. Netterwald *et al.* (2005) demonstrated that ISGs and HCMV share common signalling factors for application of inhibitors targeting the tyrosine kinases or PKC led to reduced mRNA levels of *ie*1 upon infection of fibroblasts. They also showed that short term pre-treatment of fibroblasts with IFNs resulted in increased *ie*1 mRNA levels. This result is somewhat contradictory to studies that have shown that CMV disrupts the IFN triggered JAK/STAT signalling (Baron and Davignon, 2008; Le *et al.*, 2008) or that long term IFN-γ pre-treatment is inhibiting *ie*1 expression in macrophages and mouse embryonic fibroblasts (Gribaudo *et al.*, 1993, 1995; Presti *et al.*, 2001; Kropp *et al.*, 2011). It is worth noting that the inhibition of the JAK/STAT signalling by the IE1 protein seems to target the last element of the signalling pathway, the ISGF3 complex by interacting with STAT2 and STAT1. The exact mechanism of this interference is unknown. However, IE1 does not trigger the degradation of STAT1, while blocking its ability to activate innate immunity genes (Poma *et al.*, 1996; Paulus *et al.*, 2006). It is possible that IE1 abrogates the ability of ISGF3 to activate cellular targets and changes its binding specificities to favour the viral IE enhancer element.

TLRs also seem to play contradictory roles in CMV infections. In the mouse system it has been shown that genetic knock-out mutants lacking TLR3 and TLR9 are highly susceptible to MCMV infection (Tabeta *et al.*, 2004). Therefore at the level of the whole organism the activity of the TLR signalling pathways is necessary for survival of the host. At a single cell level, however, evidence suggests that an activity of the TLR signalling might be beneficial for the virus. In RAW 264.7 cells (monocytic cell line) it was shown that treatment with ligands for TLR4 and TLR9, LPS or CpG oligomers, respectively, increased the transcription from HCMV MIEP driven reporter-gene plasmids and this activation was dependent on the presence of NF-κB and AP-1 sites in the enhancer (Lee *et al.*, 2004). Recently it has been demonstrated that TLR9 activation by CpGs enhanced HCMV replication in different types of fibroblasts, depending on the time point of stimulation (Iversen *et al.*, 2009). In contradiction to the work of

Lee and colleagues the stimulation of fibroblasts with LPS did inhibit viral replication in plaque assays. The involvement of IFN and TLR signalling in activating the viral enhancer is further supported by the fact that the viral enhancer shares many transcription factor binding sites with regulatory elements of cellular innate immune genes, like *Ifnb*1 or *Tnfa* (Stinski and Isomura, 2008). The HCMV enhancer and the *Tnfa* enhancer (Drouet *et al.*, 1991; Kraemer *et al.*, 1995) share binding sites for ETS, NF-κB, SP1 and AP-1, while NF-κB and AP-1 are known to bind the *Ifnb*1 (Honda *et al.*, 2006) and the HCMV enhancer.

However, the role of NF-κB binding in viral gene expression is still unclear. It has been demonstrated with expression plasmids containing the viral MIEP that deletion of NF-κB binding sites in the MIEP reduces reporter gene expression (Lee *et al.*, 2004). However, an HCMV enhancer in an MCMV backbone, with deleted NF-κB binding is insensitive to this mutation in the context of viral infection, suggesting a level of redundancy for IE-gene activation (Benedict *et al.*, 2004; Gustems *et al.*, 2006). In contrast to this it has been demonstrated that NF-κB seems to be essential for HCMV replication and MIEP activity in quiescent HUVEC cells, indicating that in the absence of signalling events triggered by serum, NF-κB activation upon infection becomes crucial (Caposio *et al.*, 2007). In this regard it has been very recently shown that a combinatorial knock-out of NF-κB and AP-1 sites in the MIEP lead to a reduced MCMV replication in organs of newborn mice (Isern *et al.*, 2011). AP-1 is another important factor for expression of immune genes and is activated by TLR signalling via MAPK-activation or prostaglandin E_2 cAMP-related signalling by CMV infection (Mocarski, 2002b; Zhu *et al.*, 2002). AP-1 is also one of the factors probably responsible for the serum dependent activation of the MIEP, for serum starved cells show low expression levels of AP-1 which are strongly induced upon serum treatment (Kovary and Bravo, 1991). Taken together with the fact that combinatorial knock-out of AP-1 and NF-κB motifs reduces viral replication this indicates that AP-1 and NF-κB represent two host factors central to CMV gene expression.

Another example for the complexity of the CMV gene expression is demonstrated by the importance of CREB/ATF and SP1 for the MIEP activity. SP1 is a common transcription factor in development but also controls, in concert with other TFs, expression of many immune genes (Ueda *et al.*, 1994; Look *et al.*, 1995; Wright *et al.*, 1995; Tone *et al.*, 2000, 2002). In the context of the HCMV MIEP it has been demonstrated that two SP1/SP3 binding sites act synergistically on IE-gene expression and that at least one of these sites needs to be present to maintain significant IE-gene expression levels (Isomura *et al.*, 2005). Furthermore, deletion of the upstream CREB binding motif indicated that CREB binding plays an important role in assembling the different TFs interacting with the proximal enhancer of HCMV (Lashmit *et al.*, 2009). These studies show that the enhancer has a high level of complexity and redundancy which is controlled by a robust network of TFs that shows a great overlap with the network controlling expression of immune factors. It is therefore difficult to establish the level of importance of individual TFs in the context of infection and their logic relation and interaction without consideration of combinatorial approaches.

Logic modelling of CMV MIEP regulation

Owing to the complexity of the MIEP–TF interactions and large number of signalling components involved, it is almost impossible to intuitively understand the behaviour of TF networks that control the MIEP. Mathematical modelling and computer simulation techniques can be useful for understanding the complexity and dynamics of such transcription factor networks. But before we can begin to model, we must first assemble a description of the pathway biology of the enhancer from previous research findings. This in itself is non-trivial. We must compile, integrate and visualize the components and interactions along a pathway using a standardized synthesis methodology (Kohn and Aladjem, 2006; Watterson and Ghazal, 2010). This generally follows a multistage process and can be represented using a variety of graphical tools that subsequently can be translated into a formal model. This process has been applied in a non-exhaustive way to the MIEP and is graphically shown in Fig. I.7.2 using a logic representation (Watterson *et al.*, 2008) that can be readily transformed to algebraic expression. In this connection the interactions in the MIEP–TF complex form a robust signalling network that can be represented by a series of Boolean logic gates, in particular the robust input for the CMV MIEP can be interpreted as Boolean logic OR gates (Fig. I.7.3). However, as discussed above, coordinate control of MIEP activity, as demonstrated for AP-1, NF-κB and CREB, does indicate an underlying AND logic for these TFs in this system.

Furthermore a standard notation that is being developed for pathway simulation work is the Systems Biology Graphical Notation (SBGN, see Fig. I.7.4 for notation key) (Novère *et al.*, 2009) and an exemplary representation of the elements and transitions compiling the upstream signalling network activating NF-κB,

AP-1, ATF and other TFs interacting with the MIEP is shown in Fig. I.7.2. The purpose of notation systems is to avoid lengthy text and they are intended to be unambiguous with regard to the transition states of the various components so that the graphical outputs are both human and machine readable using SBML (Hucka *et al.*, 2003; Czauderna *et al.*, 2010). There is, however, a degree of caution to pathway information assembled from the established literature. Such graphical models assume that the current state of the literature reflects an adequate level of understanding for the purpose in hand. Accordingly, it is important to consider how best to independently validate the pathway constructed for the system under investigation, for instance by combining pathway modelling and hypothesis generation with experimental tools such as the use of RNAi to target specific TFs using both directed and unbiased screening approaches. In this case the results of such experiments can be correlated between experimental and computational results as a means to validate the obtained pathway information and suggest new models. This work is under current intense investigation in the lab and is revealing a high level of robustness of the enhancer to alterations in single TF knock down and would be consistent with an 'OR' logic gate for many of the TFs (Fig. I.7.3).

The pathway models can be used to generate experimentally verifiable hypotheses and provide valuable insights into the working and general principles of organization of the enhancer. An important aspect of modelling TF-promoter activity is the occurrence of stochastic or random events that operates at a basal level of gene expression. Many recent studies have confirmed the inherent stochasticity of gene expression (Elowitz *et al.*, 2002; Swain *et al.*, 2002; Raj *et al.*, 2006; Maheshri and O'Shea, 2007) including the CMV-MIEP (Kurz *et al.*, 1999; Grzimek *et al.*, 2001; Simon *et al.*, 2007; Reddehase *et al.*, 2008; Podlech *et al.*, 2010). Future modelling approaches, mathematical formalisms and simulation algorithms will be required to account for this.

Importance of host cell metabolic pathways

As the transcription-replication cycle progresses towards the later stages of infection there will inevitably be an increased demand on host cell energy metabolism and biosynthetic pathways. The area of host cell metabolism has been particularly under-investigated in the past but has recently come to the fore through the application of systems approaches (see also Chapter I.4). These have included the elegant studies involving metabolic flux analysis (Munger *et al.*, 2006, 2008)

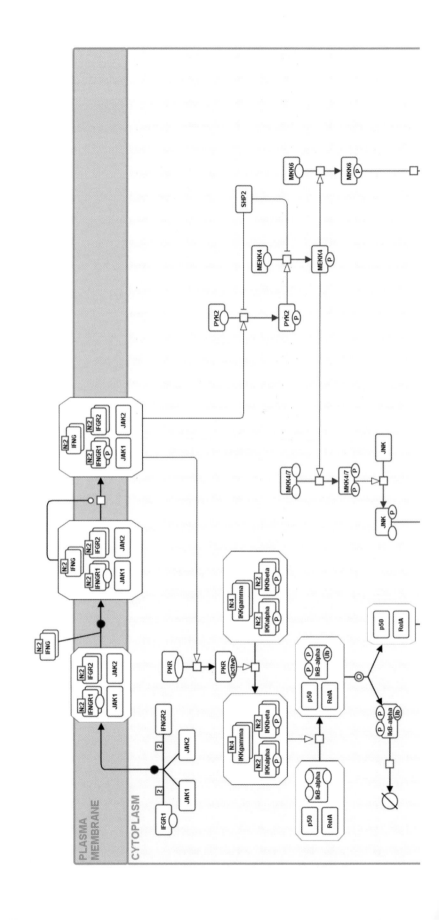

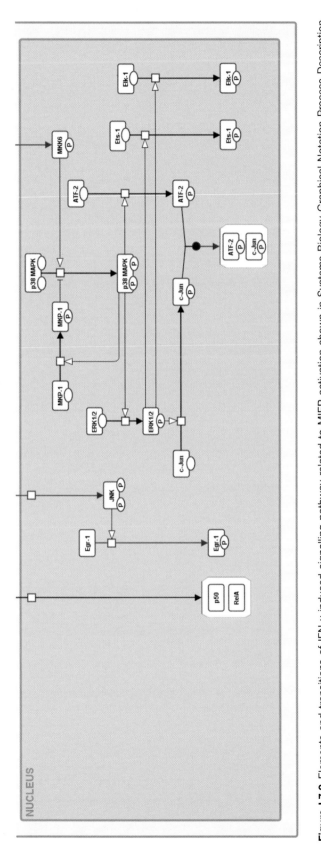

Figure I.7.2 Elements and transitions of IFN-γ-induced signalling pathway related to MIEP activation shown in Systems Biology Graphical Notation Process Description language (Novère et al., 2009; SBGN, www.sbgn.org). The diagram includes IFN-γ receptor (IFNGR) complex formation and activation by IFN-γ (IFNG), NF-κB pathway activation via Protein kinase RNA-activated (PKR), c-Jun N-terminal kinase (JNK) and mitogen-activated protein kinase p38 (p38 MAPK) pathways activation via protein–tyrosine kinase 2 (PYK2).

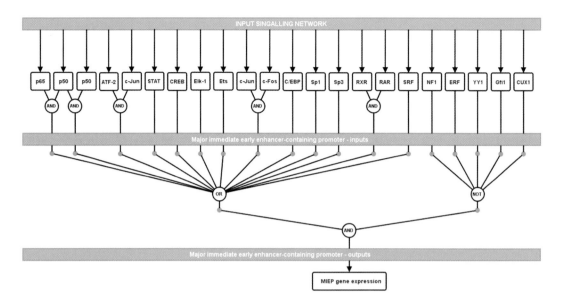

Figure I.7.3 Logic representation of the interaction between major immediate-early enhancer/promoter (MIEP) and the transcription factors involved in MIEP activated gene expression. Each line between the inputs and outputs represents separate interactions using the logic notation described in detail by Watterson *et al.* (2008).

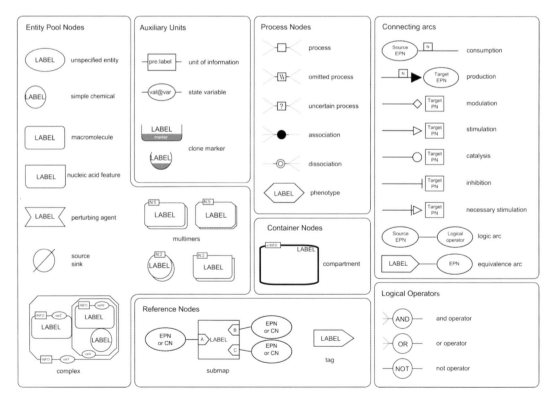

Figure I.7.4 Systems Biology Graphical Notation process description reference card level 1 version 1.1 (SBGN, www.sbgn.org).

using liquid chromatography–tandem mass spectrometry analysis, microarray analysis and ^{13}C-labelled glucose. These systems level studies showed that glucose in HCMV-infected cells, unlike normal cells, is not completely broken down in the tricarboxylic acid (TCA) cycle to promote oxidative phosphorylation but is diverted instead so that it serves as a carbon source that is mainly used for increasing fatty acid biosynthesis. The diversion of glucose carbon for fatty acid synthesis is essential for HCMV as blocking fatty acid biosynthesis with drugs significantly inhibits viral growth (Munger *et al.*, 2008). CMVs are enveloped viruses and the virion egress and assembly process requires fatty acid for its membrane formation and possibly for the viral assembly complex (vAC). The vAC is composed of viral proteins as well as components from the endoplasmic reticulum, Golgi, trans-Golgi network and early endosomes. Our current understanding of HCMV egress and assembly suggests in short that nucleocapsids exit the nucleus and move through the vAC where the virion tegument layer and envelope are acquired (Sanchez *et al.*, 2000a,b; Homman-Loudiyi *et*

al., 2003; Gibson, 2006; Das *et al.*, 2007; Seo and Britt, 2007; see also Chapter I.13).

At the metabolic pathway level illustrated in Fig. I.7.5, an increase of glucose uptake takes place through the induction of the glucose transporter GLUT4 and an up-regulation of glycolytic enzymes. However, glucose is not completely broken down by the TCA cycle where it proceeds through glycolysis to pyruvate, and is converted to lactate in the cytoplasm and to AcetylCoA in the mitochondrion. The AcetylCoA combines with oxaloacetate (OAA) to form citrate, which shuttles out of the mitochondrion and is converted back to OAA and AcetylCoA in the cytoplasm, thereby supplying increased cytoplasmic AcetylCoA for both fatty acid and sterol synthesis. To increase the flux of glucose carbon from the TCA cycle to the AcetylCoA pool, HCMV affects glutamine metabolism by increasing exogenous glutamine uptake and inducing glutaminolysis enzymes to convert the imported glutamine to alpha-ketoglutarate to anaplerotically maintain the TCA cycle (Chambers *et al.*, 2010; Vastag *et al.*, 2011; Yu *et al.*, 2011).

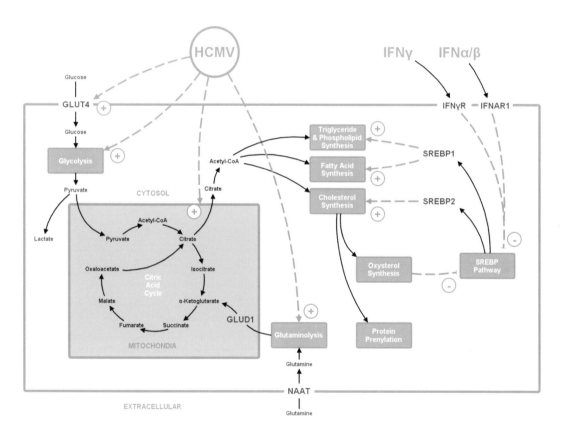

Figure I.7.5 Interactions between CMV and host cell metabolism upon infection. A schematic representation of pro-viral metabolic changes induced by CMV and the antiviral counteraction by IFNs. CMV increases Acetyl-CoA flux towards fatty acid synthesis and the sterol-prenylation arm. IFNs inhibit the master-regulator SREBP2 activity, negatively regulating cholesterol metabolism and protein prenylation.

While HCMV infection dramatically alters cellular metabolism of glucose and glutamine and fatty acid synthesis (Munger *et al.*, 2006; Spencer *et al.*, 2011), recent systems level studies with MCMV have shown that infection also results in a marked alteration of sterol biosynthesis (Blanc *et al.*, 2011), schematically shown in Fig. I.7.5. These studies involved using a pathway biology strategy integrating genomic, lipidomic, and biochemical approaches with computational bioinformatics. Genetic studies showed a coupling of the type I interferon response upon viral infection to the sterol pathway and specifically identified the mevalonate-isoprenoid arm as playing a pivotal role in antiviral functions (Blanc *et al.*, 2011). The isoprenoid arm of the sterol synthesis pathway provides the substrate for protein-prenylation (Omer and Gibbs, 1994; Greenwood *et al.*, 2006). A pathway biology compliant model of the prenylation biosynthesis is shown in Fig. I.7.6. Protein prenylation is essential for membrane localization and proper function of otherwise cytosolic proteins (Vancura *et al.*, 1994; Zeng *et al.*, 2000; Calero *et al.*, 2003). The precise number of prenylated proteins, the 'prenyl-ome' of the cell, is to date not

known but estimates derived from genomic predictions suggest hundreds of potentially prenylated target proteins (Greenwood *et al.*, 2006). Interestingly, the analyses of HCMV and MCMV genomic sequences do not predict CMV to encode prenylated proteins (unpublished observation). Taken together with the fact that down-regulation of the prenylation arm results in inhibition of viral replication, this suggests direct or indirect dependency of CMV replication on prenylated host-proteins.

As described earlier in the chapter upon detecting an infection, cells produce high levels of type I and type II interferon as part of the innate-immune response, which in turn signals, autocrine as well as paracrine, through the interferon receptor to regulate a TF-network resulting in lowering enzyme levels on the cholesterol pathway (Fig. I.7.5). Genetic ablation of *Ifnar1* completely eliminates the ability of the host to reduce cholesterol metabolism in response to the infection. This effect is seen with a range of viruses, including MCMV and HCMV (Blanc *et al.*, 2011; and unpublished work). The control mechanism involves regulation by interferon of an essential

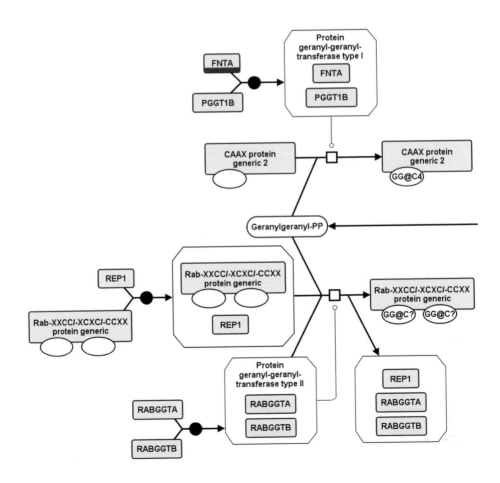

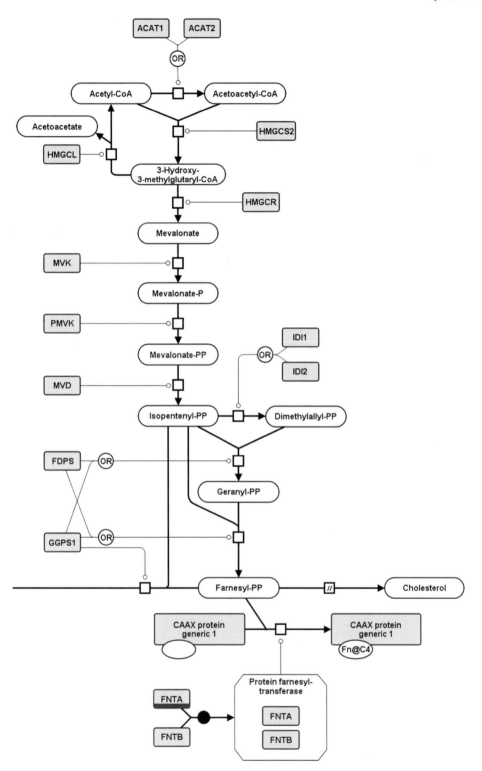

Figure I.7.6 SBGN representations of the farnesyl diphosphate and geranylgeranyl diphosphate synthesis and protein prenylation pathways (for notation key refer to Fig. I.7.4). Acetyl-CoA enters the pathway and is converted into farnesyl-PP by several enzymatic reactions. Farnesyl-PP represents the branching point were metabolites in form of geranylgeranyl-PP are diverted towards the protein prenylation arm. Prenylation reactions of different geranylgeranyl-transferase classes and the respective active enzyme–substrate complexes are shown.

master transcription factor, named SREBP-2, which coordinates the gene activity of multiple genes associated with the cholesterol pathway (Blanc *et al.*, 2011). Initial investigations into how lowered cholesterol might protect against viral infection reveals that the protection is not due to a requirement of the virus for cholesterol itself but instead involves the prenylation side-branch of the pathway that chemically links lipids to protein. Drugs such as statins and small interfering RNAs that block this part of the pathway were also shown to protect against CMV infection of cells in culture and in mice. A definitive link to sterol metabolism that is independent of cholesterol is established by the observation that the antiviral effect of down-regulating the sterol pathway upon infection is completely blocked if cells are provided with an excess of mevalonate but not cholesterol. Furthermore, the antiviral potency of type I interferon is severely diminished in the presence of excess geranylgeraniol, a prenylation metabolite, highlighting a requirement of the mevalonate–isoprenoid branch as part of an interferon mechanism for protecting against infection (Fig. I.7.5 and I.7.6).

Altogether these studies have uncovered a key checkpoint role for cellular metabolism in either promoting or blocking CMV growth and which has led to previously unappreciated physiological roles for diverting energy and fatty acid metabolism in infection to a new role for the cholesterol metabolic pathway in protecting against infection.

Conclusion and future perspective

Systems biology methodologies applied to CMV infection can help reconcile the combinatorial difficulties of dealing with high dimensional 'omics' data and the complex biological process of infection. These methods have enormous potential to improve our understanding of host–virus interaction pathway biology. In other scientific disciplines this blend of theoretical abstraction and prediction and experiment is well established. In the near future we are likely to see an increasing number of systems level studies of CMV biology and this will provide a new level of understanding. We have discussed here the early efforts in this area for CMV, although not exhaustive, they illustrate the issues of dealing with high levels of complexity, raising the question of whether a common logic code can be identified, thereby highlighting the strong interconnections between energetic and metabolic pathways of the host cell and a productive viral infection.

Acknowledgements

This work was supported by the Welcome Trust (WT0066784/Z/02/Z) and by the BBSRC (R36269) and the BBSRC/EPSRC (BB/D019621/1) to P.G. The Centre for Systems Biology at Edinburgh is a Centre for Integrated Systems Biology (CISB) supported by the BBSRC and EPSRC. The RNAi Global Initiative provided reagent support for RNAi studies. A.A. holds a grant from the Spanish Ministry of Science and Innovation (SAF 2008–00382). K.A.K. is supported by the German Research Foundation (DFG; KR 3890/1-1). We would like to thank members of the Ghazal Lab, and all of our colleagues, for many helpful discussions.

References

Abenes, G., Lee, M., Haghjoo, E., Tong, T., Zhan, X., and Liu, F. (2001). Murine cytomegalovirus open reading frame M27 plays an important role in growth and virulence in mice. J. Virol. 75, 1697–1707.

Angulo, A., Messerle, M., Koszinowski, U.H., and Ghazal, P. (1998). Enhancer requirement for murine cytomegalovirus growth and genetic complementation by the human cytomegalovirus enhancer. J. Virol. 72, 8502–8509.

Baron, M., and Davignon, J.L. (2008). Inhibition of IFN-γ-Induced STAT1 tyrosine phosphorylation by human CMV is mediated by SHP2. J. Immunol. 181, 5530–5536.

Benedict, C.A., Angulo, A., Patterson, G., Ha, S., Huang, H., Messerle, M., Ware, C.F., and Ghazal, P. (2004). Neutrality of the canonical NFκB-dependent pathway for human and murine cytomegalovirus transcription and replication in vitro. J. Virol. 78, 741–750.

Besold, K., Frankenberg, N., Pepperl-Klindworth, S., Kuball, J., Theobald, M., Hahn, G., and Plachter, B. (2007). Processing and MHC class I presentation of human cytomegalovirus pp65-derived peptides persist despite gpUS2-11-mediated immune evasion. J. Gen. Virol. 88, 1429–1439.

Beutler, B., Crozat, K., Koziol, J.A., and Georgel, P. (2005a). Genetic dissection of innate immunity to infection: the mouse cytomegalovirus model. Curr. Opin. Immunol. 17, 36–43.

Beutler, B., Georgel, P., Rutschmann, S., Jiang, Z., Croker, B., and Crozat, K. (2005b). Genetic analysis of innate resistance to mouse cytomegalovirus (MCMV). Brief. Funct. Genomic. Proteomic. 4, 203–213.

Beutler, B., Jiang, Z., Georgel, P., Crozat, K., Croker, B., Rutschmann, S., Du, X., and Hoebe, K. (2006). Genetic analysis of host resistance: toll-like receptor signaling and immunity at large. Annu. Rev. Immunol. 24, 353–389.

Blanc, M., Yuan, W., Roberstson, K., Watterson, S., Shui, G., Lacaze, P.A., Khondoker, M., Dickinson, P., Sing, G., Rodriguez-Martin, S., *et al.* (2011). Host defense against viral infection involves interferon mediated down-regulation of sterol biosynthesis. PLoS Biol. 9, e1000598.

Boehme, K.W., and Compton, T. (2006). Virus entry and activation of innate immunity. In Cytomegaloviruses: Molecular Biology and Immunity, Reddehase, M.J., and Lemmermann, N., eds. (Caister Academic Press, Norfolk, UK), pp. 111–130.

Boshart, M., Weber, F., Jahn, G., Dorsch-Häsler, K., Fleckenstein, B., and Schaffner, W. (1985). A very strong enhancer is located upstream of an immediate-early gene of human cytomegalovirus. Cell *41*, 521–530.

Boyle, K.A., Pietropaolo, R.L., and Compton, T. (1999). Engagement of the cellular receptor for glycoprotein B of human cytomegalovirus activates the interferon-responsive pathway. Mol. Cell. Biol. *19*, 3607–3613.

Calero, M., Chen, C.Z., Zhu, W., Winand, N., Havas, K.A., Gilbert, P.M., Burd, C.G., and Collins, R.N. (2003). Dual prenylation is required for Rab protein localization and function. Mol. Biol. Cell *14*, 1852–1867.

Caposio, P., Luganini, A., Hahn, G., Landolfo, S., and Gribaudo, G. (2007). Activation of the virus-induced IKK/NF-κB signalling axis is critical for the replication of human cytomegalovirus in quiescent cells. Cell. Microbiol. *9*, 2040–2054.

Chambers, J., Angulo, A., Amaratunga, D., Guo, H., Jiang, Y., Wan, J.S., Bittner, A., Frueh, K., Jackson, M.R., Peterson, P.A., et al. (1999). DNA microarrays of the complex human cytomegalovirus genome: Profiling kinetic class with drug sensitivity of viral gene expression. J. Virol. *73*, 5757–5766.

Chambers, J.W., Maguire, T.G., and Alwine, J.C. (2010). Glutamine metabolism is essential for human cytomegalovirus infection. J. Virol. *84*, 1867–1873.

Chang, Y.N., Crawford, S., Stall, J., Rawlins, D.R., Jeang, K.T., and Hayward, G.S. (1990). The palindromic series I repeats in the simian cytomegalovirus major immediate-early promoter behave as both strong basal enhancers and cyclic AMP response elements. J. Virol. *64*, 264–277.

Chee, M.S., Bankier, A.T., Beck, S., Bohni, R., Brown, C.M., Cerny, R., Horsnell, T., Hutchison, C.A., Kouzarides, T., Martignetti, J.A., et al. (1990). Analysis of the protein-coding content of the sequence of human cytomegalovirus strain AD169. Curr. Top. Microbiol. Immunol. *154*, 125–169.

Compton, T., Kurt-Jones, E.A., Boehme, K.W., Belko, J., Latz, E., Golenbock, D.T., and Finberg, R.W. (2003). Human cytomegalovirus activates inflammatory cytokine responses via CD14 and Toll-like receptor 2. J. Virol. *77*, 4588–4596.

Czauderna, T., Klukas, C., and Schreiber, F. (2010). Editing, validating and translating of SBGN maps. Bioinformatics *26*, 2340–2341.

Das, S., Vasanji, A., and Pellett, P.E. (2007). Three-dimensional structure of the human cytomegalovirus cytoplasmic virion assembly complex includes a reoriented secretory apparatus. J. Virol. *81*, 11861–11869.

DeFilippis, V.R. (2007). Induction and evasion of the type I interferon response by cytomegaloviruses. Adv. Exp. Med. Biol. *598*, 309–324.

DeFilippis, V.R., Robinson, B., Keck, T.M., Hansen, S.G., Nelson, J.A., and Früh, K.J. (2006). Interferon regulatory factor 3 is necessary for induction of antiviral genes during human cytomegalovirus infection. J. Virol. *80*, 1032–1037.

DeFilippis, V.R., Alvarado, D., Sali, T., Rothenburg, S., and Früh, K. (2010). Human cytomegalovirus induces the interferon response via the DNA sensor ZBP1. J. Virol. *84*, 585–598.

Del Val, M., Muench, K., Reddehase, M.J., and Koszinowski, U.H. (1989). Presentation of CMV immediate-early antigen to cytolytic T lymphocytes is selectively prevented by viral genes expressed in the early phase. Cell *58*, 305–315.

Drouet, C., Shakhov, A.N., and Jongeneel, C.V. (1991). Enhancers and transcription factors controlling the inducibility of the tumor necrosis factor-alpha promoter in primary macrophages. J. Immunol. *147*, 1694–1700.

Elowitz, M.B., Levine, A.J., Siggia, E.D., and Swain, P.S. (2002). Stochastic gene expression in a single cell. Science *297*, 1183–1186.

Forst, C.V. (2006). Host-pathogen systems biology. Drug Discov. Today *11*, 220–227.

Ghazal, P., Lubon, H., Fleckenstein, B., and Hennighausen, L. (1987). Binding of transcription factors and creation of a large nucleoprotein complex on the human cytomegalovirus enhancer. Proc. Natl. Acad. Sci. U.S.A. *84*, 3658–3662.

Ghazal, P., Lubon, H., and Hennighausen, L. (1988a). Multiple sequence-specific transcription factors modulate cytomegalovirus enhancer activity in vitro. Mol. Cell. Biol. *8*, 1809–1811.

Ghazal, P., Lubon, H., and Hennighausen, L. (1988b). Specific interactions between transcription factors and the promoter-regulatory region of the human cytomegalovirus major immediate-early gene. J. Virol. *62*, 1076–1079.

Ghazal, P., Gonzalez, A.J.C., Garcia-Ramirez, J.J., Kurz, S., and Angulo, A. (2000a). Viruses: hostages to the cell. Virology *275*, 233–237.

Ghazal, P., Garcia-Ramirez, J.J., Gonzalez, A.J.C., Kurz, S., and Angulo, A. (2000b). Principles of homeostasis in governing virus activation and latency. Immunol. Res. *21*, 219–223.

Ghazal, P., Messerle, M., Osborn, K., and Angulo, A. (2003). An essential role of the enhancer for murine cytomegalovirus in vivo growth and pathogenesis. J. Virol. *77*, 3217–3228.

Gibson, W. (2006). Assembly and maturation of the capsid. In Cytomegaloviruses: Molecular Biology and Immunity, Reddehase, M.J., and Lemmermann, N., eds. (Caister Academic Press, Norfolk, UK), pp. 231–244.

Greenwood, J., Steinman, L., and Zamvil, S.S. (2006). Statin therapy and autoimmune disease: from protein prenylation to immunomodulation. Nat. Rev. Immunol. *6*, 358–370.

Gribaudo, G., Ravaglia, S., Caliendo, A., Cavallo, R., Gariglio, M., Martinotti, M.G., and Landolfo, S. (1993). Interferons inhibit onset of murine cytomegalovirus immediate-early gene transcription. Virology *197*, 303–311.

Gribaudo, G., Ravaglia, S., Gaboli, M., Gariglio, M., Cavallo, R., and Landolfo, S. (1995). Interferon-α inhibits the murine cytomegalovirus immediate-early gene expression by down-regulating NF-κB activity. Virology *211*, 251–260.

Grzimek, N.K., Podlech, J., Steffens, H.P., Holtappels, R., Schmalz, S., and Reddehase, M.J. (1999). In vivo replication of recombinant murine cytomegalovirus driven by the paralogous major immediate-early promoter-enhancer of human cytomegalovirus. J. Virol. *73*, 5043–5055.

Grzimek, N.K., Dreis, D., Schmalz, S., and Reddehase, M.J. (2001). Random, Asynchronous, and asymmetric transcriptional activity of enhancer-flanking major immediate-early genes ie1/3 and ie2 during murine cytomegalovirus latency in the lungs. J. Virol. *75*, 2692–2705.

Gustems, M., Borst, E., Benedict, C.A., Perez, C., Messerle, M., Ghazal, P., and Angulo, A. (2006). Regulation of the transcription and replication cycle of human cytomegalovirus is insensitive to genetic elimination of the cognate NF-κB binding sites in the enhancer. J. Virol. *80*, 9899–9904.

Hansen, T.H., and Bouvier, M. (2009). MHC class I antigen presentation: learning from viral evasion strategies. Nat. Rev. Immunol. *9*, 503–513.

Homman-Loudiyi, M., Hultenby, K., Britt, W., and Soderberg-Naucler, C. (2003). Envelopment of human cytomegalovirus occurs by budding into golgi-derived vacuole compartments positive for gB, Rab 3, trans-golgi network 46, and mannosidase II. J. Virol. *77*, 3191–3203.

Honda, K., Takaoka, A., and Taniguchi, T. (2006). Type I Interferon gene induction by the interferon regulatory factor family of transcription factors. Immunity *25*, 349–360.

Hucka, M., Finney, A., Sauro, H.M., Bolouri, H., Doyle, J.C., Kitano, H., and the rest of the SBML Forum. (2003). The systems biology markup language (SBML): a medium for representation and exchange of biochemical network models. Bioinformatics *19*, 524–531.

Isern, E., Gustems, M., Messerle, M., Borst, E., Ghazal, P., and Angulo, A. (2011). The activator protein 1 binding motifs within the human cytomegalovirus major immediate-early enhancer are functionally redundant and act in a cooperative manner with the NF-κB sites during acute infection. J. Virol. *85*, 1732–1746.

Isomura, H., Tsurumi, T., and Stinski, M.F. (2004). Role of the proximal enhancer of the major immediate-early promoter in human cytomegalovirus replication. J. Virol. *78*, 12788–12799.

Isomura, H., Stinski, M.F., Kudoh, A., Daikoku, T., Shirata, N., and Tsurumi, T. (2005). Two Sp1/Sp3 binding sites in the major immediate-early proximal enhancer of human cytomegalovirus have a significant role in viral replication. J. Virol. *79*, 9597–9607.

Iversen, A.C., Steinkjer, B., Nilsen, N., Bohnhorst, J., Moen, S.H., Vik, R., Stephens, P., Thomas, D.W., Benedict, C.A., and Espevik, T. (2009). A proviral role for CpG in cytomegalovirus infection. J. Immunol. *182*, 5672–5681.

Jenkins, C., Abendroth, A., and Slobedman, B. (2004). A novel viral transcript with homology to human interleukin-10 is expressed during latent human cytomegalovirus infection. J. Virol. *78*, 1440–1447.

Jenkins, C., Garcia, W., Abendroth, A., and Slobedman, B. (2008). Expression of a human cytomegalovirus latency-associated homolog of interleukin-10 during the productive phase of infection. Virology *370*, 285–294.

Jenner, R.G., and Young, R.A. (2005). Insights into host responses against pathogens from transcriptional profiling. Nat. Rev. Microbiol. *3*, 281–294.

Juckem, L.K., Boehme, K.W., Feire, A.L., and Compton, T. (2008). Differential initiation of innate immune responses induced by human cytomegalovirus entry into fibroblast cells. J. Immunol. *180*, 4965–4977.

Kalejta, R.F. (2008). Tegument proteins of human cytomegalovirus. Microbiol. Mol. Biol. Rev. *72*, 249–265.

Kavanagh, D.G., Gold, M.C., Wagner, M., Koszinowski, U.H., and Hill, A.B. (2001). The multiple immune-evasion genes of murine cytomegalovirus are not redundant. J. Exp. Med. *194*, 967–978.

Kellam, P. (2001). Post-genomic virology: the impact of bioinformatics, microarrays and proteomics on investigating host and pathogen interactions. Rev. Med. Virol. *11*, 313–329.

Khan, S., Zimmermann, A., Basler, M., Groettrup, M., and Hengel, H. (2004). A cytomegalovirus inhibitor of gamma interferon signaling controls immunoproteasome induction. J. Virol. *78*, 1831–1842.

Kohn, K.W., and Aladjem, M.I. (2006). Circuit diagrams for biological networks. Mol. Syst. Biol. 2.2006.0002.

König, R., Zhou, Y., Elleder, D., Diamond, T.L., Bonamy, G.M.C., Irelan, J.T., Chiang, C., Tu, B.P., De Jesus, P.D., and Lilley, C.E. (2008). Global analysis of host–pathogen interactions that regulate early-stage HIV-1 replication. Cell *135*, 49–60.

Kovary, K., and Bravo, R. (1991). Expression of different Jun and Fos proteins during the G0-to-G1 transition in mouse fibroblasts: in vitro and in vivo associations. Mol. Cell. Biol. *11*, 2451–2459.

Kraemer, B., Wiegmann, K., and Kroenke, M. (1995). Regulation of the human TNF promoter by the transcription factor Ets. J. Biol. Chem. *270*, 6577–6583.

Krmpotic, A., Bubic, I., Polic, B., Lucin, P., and Jonjic, S. (2003). Pathogenesis of murine cytomegalovirus infection. Microbes Infect. *5*, 1263–1277.

Kropp, K.A., Robertson, K.A., Sing, G., Rodriguez-Martin, S., Blanc, M., Lacaze, P., Hassim, M.F.B.N., Khondoker, M.R., Busche, A., Dickinson, P., et al. (2011). Reversible inhibition of murine cytomegalovirus replication by gamma interferon (IFN-γ) in primary macrophages involves a primed type I IFN-signaling subnetwork for full establishment of an immediate-early antiviral state. J. Virol. *85*, 10286–10299.

Kurz, S.K., Rapp, M., Steffens, H.P., Grzimek, N.K., Schmalz, S., and Reddehase, M.J. (1999). Focal transcriptional activity of murine cytomegalovirus during latency in the lungs. J. Virol. *73*, 482–494.

Lashmit, P., Wang, S., Li, H., Isomura, H., and Stinski, M.F. (2009). The CREB site in the proximal enhancer is critical for cooperative interaction with the other transcription factor binding sites to enhance transcription of the major intermediate-early genes in human cytomegalovirus-infected cells. J. Virol. *83*, 8893–8904.

Le, V.T.K., Trilling, M., Wilborn, M., Hengel, H., and Zimmermann, A. (2008). Human cytomegalovirus interferes with signal transducer and activator of transcription (STAT) 2 protein stability and tyrosine phosphorylation. J. Gen. Virol. *89*, 2416–2426.

Lee, M., Abenes, G., Zhan, X., Dunn, W., Haghjoo, E., Tong, T., Tam, A., Chan, K., and Liu, F. (2002). Genetic analyses of gene function and pathogenesis of murine cytomegalovirus by transposon-mediated mutagenesis. J. Clin. Virol. *25(Suppl. 2)*, 111–122.

Lee, Y., Sohn, W.J., Kim, D.S., and Kwon, H.J. (2004). NF-KB and c-Jun-dependent regulation of human cytomegalovirus immediate-early gene enhancer/promoter in response to lipopolysaccharide and bacterial CpG-oligodeoxynucleotides in macrophage cell line RAW 264.7. Eur. J. Biochem. *271*, 1094–1105.

Lemmermann, N.A.W., Böhm, V., Holtappels, R., and Reddehase, M.J. (2011). In vivo impact of cytomegalovirus evasion of CD8 T-cell immunity: facts and thoughts based on murine models. Virus Res. *157*, 161–174.

Look, D.C., Pelletier, M.R., Tidwell, R.M., Roswit, W.T., and Holtzman, M.J. (1995). Stat1 depends on transcriptional synergy with Sp1. J. Biol. Chem. *270*, 30264–30267.

Mack, C., Sickmann, A., Lembo, D., and Brune, W. (2008). Inhibition of proinflammatory and innate immune signaling pathways by a cytomegalovirus RIP1-interacting protein. Proc. Natl. Acad. Sci. U.S.A. 105, 3094–3099.

Maheshri, N., and O'Shea, E.K. (2007). Living with noisy genes: How cells function reliably with inherent variability in gene expression. Annu. Rev. Biophys. Biomol. Struct. 36, 413–434.

Meier, J.L., and Pruessner, J.A. (2000). The human cytomegalovirus major immediate-early distal enhancer region is required for efficient viral replication and immediate-early gene expression. J. Virol. 74, 1602–1613.

Meier, J.L., Keller, M.J., and McCoy, J.J. (2002). Requirement of multiple cis-acting elements in the human cytomegalovirus major immediate-early distal enhancer for viral gene expression and replication. J. Virol. 76, 313–326.

Mocarski, E.S.J. (2002a). Immunomodulation by cytomegaloviruses: manipulative strategies beyond evasion. Trends Microbiol. 10, 332–339.

Mocarski, E.S.J., Shenk, T., and Pass, R.F. (2007). Cytomegaloviruses. In Fields Virology, Knipe, D.M., and Howley, P.M., eds. (Lippincott Williams & Wilkins, Philadelphia, PA), pp. 2701–2772.

Mocarski, E.S. (2002b). Virus self-improvement through inflammation: no pain, no gain. Proc. Natl. Acad. Sci. U.S.A. 99, 3362–3364.

Munger, J., Bajad, S.U., Coller, H.A., Shenk, T., and Rabinowitz, J.D. (2006). Dynamics of the cellular metabolome during human cytomegalovirus infection. PLoS Pathog. 2, e132.

Munger, J., Bennett, B.D., Parikh, A., Feng, X.J., McArdle, J., Rabitz, H.A., Shenk, T., and Rabinowitz, J.D. (2008). Systems-level metabolic flux profiling identifies fatty acid synthesis as a target for antiviral therapy. Nat. Biotech. 26, 1179–1186.

Netterwald, J., Yang, S., Wang, W., Ghanny, S., Cody, M., Soteropoulos, P., Tian, B., Dunn, W., Liu, F., and Zhu, H. (2005). Two gamma interferon-activated site-like elements in the human cytomegalovirus major immediate-early promoter/enhancer are important for viral replication. J. Virol. 79, 5035–5046.

Novère, N.L., Hucka, M., Mi, H., Moodie, S., Schreiber, F., Sorokin, A., Demir, E., Wegner, K., Aladjem, M.I., Wimalaratne, S.M., et al. (2009). The systems biology graphical notation. Nat. Biotech. 27, 735–741.

Omer, C.A., and Gibbs, J.B. (1994). Protein prenylation in eukaryotic microorganisms: genetics, biology and biochemistry. Mol. Microbiol. 11, 219–225.

Park, B., Kim, Y., Shin, J., Lee, S., Cho, K., Früh, K., Lee, S., and Ahn, K. (2004). Human cytomegalovirus inhibits tapasin-dependent peptide loading and optimization of the MHC class I peptide cargo for immune evasion. Immunity 20, 71–85.

Paulus, C., Krauss, S., and Nevels, M. (2006). A human cytomegalovirus antagonist of type I IFN-dependent signal transducer and activator of transcription signaling. Proc. Natl. Acad. Sci. U.S.A. 103, 3840–3845.

Podlech, J., Pintea, R., Kropp, K.A., Fink, A., Lemmermann, N.A.W., Erlach, K.C., Isern, E., Angulo, A., Ghazal, P., and Reddehase, M.J. (2010). Enhancerless cytomegalovirus is capable of establishing a low-level maintenance infection in severely immunodeficient host tissues but fails in exponential growth. J. Virol. 84, 6254–6261.

Poma, E.E., Kowalik, T.F., Zhu, L., Sinclair, J.H., and Huang, E.S. (1996). The human cytomegalovirus IE1-72 protein interacts with the cellular p107 protein and relieves p107-mediated transcriptional repression of an E2F-responsive promoter. J. Virol. 70, 7867–7877.

Presti, R.M., Popkin, D.L., Connick, M., Paetzold, S., and Virgin, H.W. (2001). Novel cell type-specific antiviral mechanism of interferon-gamma action in macrophages. J. Exp. Med. 193, 483–496.

Raj, A., Peskin, C.S., Tranchina, D., Vargas, D.Y., and Tyagi, S. (2006). Stochastic mRNA synthesis in mammalian cells. PLoS Biol. 4, e309.

Rathinam, V.A.K., Jiang, Z., Waggoner, S.N., Sharma, S., Cole, L.E., Waggoner, L., Vanaja, S.K., Monks, B.G., Ganesan, S., Latz, E., et al. (2010). The AIM2 inflammasome is essential for host defense against cytosolic bacteria and DNA viruses. Nat. Immunol. 11, 395–402.

Reddehase, M.J., Simon, C.O., Seckert, C.K., Lemmermann, N., and Grzimek, N.K.A. (2008). Murine model of cytomegalovirus latency and reactivation. In: Human Cytomegalovirus, Shenk, T.E., and Stinski, M.F., eds. (Springer, Berlin), pp. 315–331.

Reddehase, M.J. (2002). Antigens and immunoevasins: opponents in cytomegalovirus immune surveillance. Nat. Rev. Immunol. 2, 831–844.

Sanchez, V., and Spector, D.H. (2006). Exploitation of host cell cycle regulatory pathways by HCMV. In Cytomegaloviruses: Molecular Biology and Immunity, Reddehase, M.J., and Lemmermann, N., eds. (Caister Academic Press, Norfolk, UK), pp. 205–230.

Sanchez, V., Sztul, E., and Britt, W.J. (2000a). Human cytomegalovirus pp28 (UL99) localizes to a cytoplasmic compartment which overlaps the endoplasmic reticulum-golgi-intermediate compartment. J. Virol. 74, 3842–3851.

Sanchez, V., Greis, K.D., Sztul, E., and Britt, W.J. (2000b). Accumulation of virion tegument and envelope proteins in a stable cytoplasmic compartment during human cytomegalovirus replication: characterization of a potential site of virus assembly. J. Virol. 74, 975–986.

Sandford, G.R., Brock, L.E., Voigt, S., Forester, C.M., and Burns, W.H. (2001). Rat cytomegalovirus major immediate-early enhancer switching results in altered growth characteristics. J. Virol. 75, 5076–5083.

Seo, J.Y., and Britt, W.J. (2007). Cytoplasmic envelopment of human cytomegalovirus requires the postlocalization function of tegument protein pp28 within the assembly compartment. J. Virol. 81, 6536–6547.

Simon, C.O., Kühnapfel, B., Reddehase, M.J., and Grzimek, N.K.A. (2007). Murine cytomegalovirus major immediate-early enhancer region operating as a genetic switch in bidirectional gene pair transcription. J. Virol. 81, 7805–7810.

Spector, D.H. (1996). Activation and regulation of human cytomegalovirus early genes. Intervirol. 39, 361–377.

Spencer, C.M., Schafer, X.L., Moorman, N.J., and Munger, J. (2011). Human cytomegalovirus induces the activity and expression of acetyl-coenzyme A carboxylase, a fatty acid biosynthetic enzyme whose inhibition attenuates viral replication. J. Virol. 85, 5814–5824.

Stinski, M.F., and Isomura, H. (2008). Role of the cytomegalovirus major immediate-early enhancer in acute infection and reactivation from latency. Med. Microbiol. Immunol. 197, 223–231.

Swain, P.S., Elowitz, M.B., and Siggia, E.D. (2002). Intrinsic and extrinsic contributions to stochasticity in gene expression. Proc. Natl. Acad. Sci. U.S.A. 99, 12795–12800.

Tabeta, K., Georgel, P., Janssen, E., Du, X., Hoebe, K., Crozat, K., Mudd, S., Shamel, L., Sovath, S., Goode, J., *et al.* (2004). Toll-like receptors 9 and 3 as essential components of innate immune defense against mouse cytomegalovirus infection. Proc. Natl. Acad. Sci. U.S.A. *101*, 3516–3521.

Tone, M., Powell, M.J., Tone, Y., Thompson, S.A.J., and Waldmann, H. (2000). IL-10 gene expression is controlled by the transcription factors Sp1 and Sp3. J. Immunol. *165*, 286–291.

Tone, M., Tone, Y., Babik, J.M., Lin, C.Y., and Waldmann, H. (2002). The role of Sp1 and NF-κB in regulating CD40 gene expression. J. Biol. Chem. *277*, 8890–8897.

Ueda, A., Okuda, K., Ohno, S., Shirai, A., Igarashi, T., Matsunaga, K., Fukushima, J., Kawamoto, S., Ishigatsubo, Y., and Okubo, T. (1994). NF-κB and Sp1 regulate transcription of the human monocyte chemoattractant protein-1 gene. J. Immunol. *153*, 2052–2063.

Uetz, P., Dong, Y.A., Zeretzke, C., Atzler, C., Baiker, A., Berger, B., Rajagopala, S.V., Roupelieva, M., Rose, D., Fossum, E., *et al.* (2006). Herpesviral protein networks and their interaction with the human proteome. Science *311*, 239–242.

Vancura, A., Sessler, A., Leichus, B., and Kuret, J. (1994). A prenylation motif is required for plasma membrane localization and biochemical function of casein kinase I in budding yeast. J. Biol. Chem. *269*, 19271–19278.

Varnum, S.M., Streblow, D.N., Monroe, M.E., Smith, P., Auberry, A.J., Pasa-Tolic, L., Wang, D., Camp II, D.G., Rodland, K., Wiley, S., *et al.* (2004). Identification of proteins in human cytomegalovirus (HCMV) particles: the HCMV proteome. J. Virol. *78, 20*, 10960–10966.

Vastag, L., Koyuncu, E., Grady, S.L., Shenk, T.E., and Rabinowitz, J.D. (2011). Divergent effects of human cytomegalovirus and Herpes Simplex Virus-1 on cellular metabolism. PLoS Pathog. *7*, e1002124.

Voigt, V., Forbes, C.A., Tonkin, J.N., gli-Esposti, M.A., Smith, H.R.C., Yokoyama, W.M., and Scalzo, A.A. (2003). Murine cytomegalovirus m157 mutation and variation leads to immune evasion of natural killer cells. Proc. Natl. Acad. Sci. U.S.A. *100*, 13483.

Watterson, S., and Ghazal, P. (2010). Use of logic theory in understanding regulatory pathway signaling in response to infection. Fut. Microbiol. *5*, 163–176.

Watterson, S., Marshall, S., and Ghazal, P. (2008). Logic models of pathway biology. Drug Discov. Today *13*, 447–456.

Wright, K.L., Moore, T.L., Vilen, B.J., Brown, A.M., and Jenny, P.-Y. (1995). Major histocompatibility complex class II-associated invariant chain gene expression is up-regulated by cooperative interactions of Sp1 and NF-Y. J. Biol. Chem. *270*, 20978–20986.

Yu, Y., Clippinger, A.J., and Alwine, J.C. (2011). Viral effects on metabolism: changes in glucose and glutamine utilization during human cytomegalovirus infection. Trends Microbiol. *19*, 360–367.

Zeng, Q., Si, X., Horstmann, H., Xu, Y., Hong, W., and Pallen, C.J. (2000). Prenylation-dependent association of protein-tyrosine phosphatases PRL-1, -2, and -3 with the plasma membrane and the early endosome. J. Biol. Chem. *275*, 21444–21452.

Zhu, H., Cong, J.P., Yu, D., Bresnahan, W.A., and Shenk, T.E. (2002). Inhibition of cyclooxygenase 2 blocks human cytomegalovirus replication. Proc. Natl. Acad. Sci. U.S.A. *99*, 3932–3937.

Zimmermann, A., Trilling, M., Wagner, M., Wilborn, M., Bubic, I., Jonjic, S., Koszinowski, U., and Hengel, H. (2005). A cytomegaloviral protein reveals a dual role for STAT2 in IFN-γ signaling and antiviral responses. J. Exp. Med. *201*, 1543–1553.

Virus Entry and Activation of Innate Defence

I.8

Adam L. Feire and Teresa Compton

Abstract

All viruses must deliver their genomes to host cells to initiate infection. The cell plasma membrane serves as an initial barrier that must be crossed if an infection is going to take place. This chapter will summarize what is known about the entry pathway of human cytomegalovirus (HCMV) with certain parallels and commonalities noted between HCMV and other betaherpesviruses. The roles of HCMV envelope glycoproteins in mediating critical virus entry events such as attachment and fusion as well as the current knowledge of the identification of cellular receptors that serve as entry mediators will also be described. This chapter will also discuss the entry-associated innate antiviral response and the emerging role of signalling pathways in the early events in infection. Lastly, we will examine how virus entry and innate antiviral response may be coordinated.

Introduction

In the simplest context, entry requires that all enveloped viruses, including HCMV, adhere to the cell surface and trigger fusion between the virus envelope and a cellular membrane that results in the deposition of virion components into the cytoplasm (Fig. I.8.1). Following content delivery to the cytoplasm, certain tegument proteins and genome-containing capsids translocate to the nucleus in a process known as uncoating. For these structurally complex viruses whose envelopes contain as many as 20 proteins and glycoproteins, attachment is a multistep process typically involving more than one envelope glycoprotein interacting with a series of cell surface molecules that serve as receptors and co-receptors. A likely consequence of these virus–cell interactions is receptor-activated conformational changes in the envelope glycoproteins that play roles in membrane fusion. Here too, multiple envelope glycoproteins are required to fuse membranes. Another consequence of these initial virus–host interactions

may be the formation and/or delivery of bound virions to specialized membrane domains or compartments that are optimal for fusion and activation of signal transduction cascades. The pathway leading to HCMV entry is also accompanied by innate antiviral defence. This considerably heightens the complexity of the molecular events occurring during the early steps in HCMV infection.

HCMV entry

To begin a discussion of virus entry at the cellular level, one must first consider the basis of cellular tropism since receptors involved in entry are expressed on permissive cells. In the human host, HCMV causes systemic infection and exhibits a tropism for fibroblasts, endothelial cells, epithelial cells, monocytes/macrophages, smooth muscle cells, stromal cells, neuronal cells, neutrophils, and hepatocytes (Myerson et al., 1984; Ibanez et al., 1991; Sinzger et al., 2000). This exceptionally broad cellular tropism is the root of HCMV disease manifestation of most organ systems and tissue types in the immunocompromised host. Although HCMV is considered to have a very restricted cell tropism *in vitro*, entry into target cells is very promiscuous, as HCMV is able to bind, fuse and initiate replication in all tested vertebrate cell types. Productive *in vitro* replication is only supported by primary fibroblasts, endothelial and differentiated myeloid cells as well as certain astrocyte cell lines (Ibanez et al., 1991; Nowlin et al., 1991). The ability of HCMV to enter such a wide range of cells is highly indicative of multiple cell specific receptors, broadly expressed receptors or a complex entry pathway in which a combination of both cell specific and broadly expressed cellular receptors are utilized.

History of cellular receptors for HCMV

It has been known for some time that HCMV initiates infection via a tethering interaction of virions and

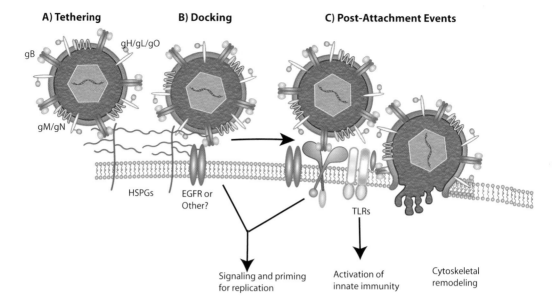

A) Tethering **B) Docking** **C) Post-Attachment Events**

gB

gH/gL/gO

gM/gN

HSPGs

EGFR or
Other?

TLRs

Signaling and priming
for replication

Activation of
innate immunity

Cytoskeletal
remodeling

Figure I.8.1 Working model for HCMV entry into cells. HCMV initially attaches in a tethering step to heparan sulfate proteoglycans (HSPGs) via gM/gN and/or gB. The gB protein interacts in a stable docking step to epidermal growth factor receptor (EGFR) on many HCMV permissive cell types or to as yet unidentified receptors in haematopoietic cell types. Other interactions between HCMV envelope glycoproteins and cellular integrins promote receptor clustering. At least one of these interactions triggers fusion that leads to internalization of virion components. Signal transduction events are initiated via EGFR and/or integrins and these events are hypothesized to prime and facilitate downstream steps in the virus life cycle such as nuclear translocation of the capsid and efficient viral gene expression. Toll-like receptors (TLRs) detect an HCMV-displayed pathogen-associated molecular pattern during virus entry leading to distinct signalling events and activation of innate defence.

cell surface heparan sulfate proteoglycans (HSPGs) (Compton *et al.*, 1993) (Fig. I.8.1). HSPG engagement is a relatively conserved feature of herpesvirus entry pathways and is thought to also play a role in HHV-6 and HHV-7 interactions with lymphoid cells (Conti *et al.*, 2000) At least in cell culture systems, HCMV engagement of HSPGs is thought to play a crucial role in recruiting virions to the cell surface and enhancing the engagement of other receptors (Compton *et al.*, 1993). This hypothesis is further supported by biochemical analysis of HCMV binding, which indicates biphasic binding properties with multiple distinct affinities (Boyle and Compton, 1998).

The broad ability of HCMV to enter cells in culture has hampered efforts to identify cellular receptors using modern molecular approaches such as expression cloning. Over the past 20 years, numerous receptor candidates have been put forward. These candidate receptor molecules were selected on the basis of sound logic and solid initial criteria but none of these molecules have held up to commonly accepted parameters of virus receptors. HCMV virions were initially shown to bind β_2-microglobulin (β_2m) in urine samples (McKeating *et al.*, 1986, 1987; Grundy *et al.*, 1988). This observation led to numerous

binding studies concluding that the HCMV envelope binds β_2m as it is released from cells (Grundy *et al.*, 1987b; McKeating *et al.*, 1987; Stannard, 1989). This β_2m–HCMV complex was then thought to associate with the alpha chain of HLA class I antigens (Grundy *et al.*, 1987a,beersma *et al.*, 1990, 1991; Browne *et al.*, 1990). These data led to a model in which β_2m-coated HCMV bound MHC class I molecules and displaced β_2m from the MHC class I α chain. However, it was later determined that β_2m expression had no correlation with *in vitro* entry or *in vivo* spread of infectivity (Beersma *et al.*, 1991; Wu *et al.*, 1994; Polic *et al.*, 1996).

Virus-cell overlay blots were used to globally analyse cellular proteins that could bind to HCMV virions. These studies identified a cell surface protein of approximately 30 kDa, whose expression correlated with cells permissive for entry (Taylor and Cooper, 1990; Nowlin *et al.*, 1991). This protein was later identified as annexin II, a phospholipid and calcium binding protein that has membrane bridging function (Wright *et al.*, 1993, 1995). Annexin II was also shown to bind HCMV virions (Wright *et al.*, 1994, 1995). Upon further study it was found that gB was able to directly interact with annexin II and that this protein was able to enhance HCMV binding and fusion to phospholipid membranes (Pietropaolo and Compton, 1997; Raynor *et al.*, 1999). However, cells devoid of annexin II are fully permissive for entry and initiation of infection (Pietropaolo and Compton, 1999). What role, if any, annexin II plays in the life cycle of HCMV is unknown but given its membrane bridging activity, it remains formally possible that this enhances entry, cell–cell spread and/or maturation and egress.

CD13, human aminopeptidase N has also been implicated as an HCMV receptor. This hypothesis was based on the fact that only human peripheral blood mononuclear cells (PBMCs) that were CD13 positive supported productive infection (Soderberg *et al.*, 1993a,b; Larsson *et al.*, 1998). This led to a more thorough study of this possibility in which CD13-specific antibodies and chemical inhibitors of CD13 activity were both shown to inhibit HCMV binding and entry (Soderberg *et al.*, 1993a). Excitement from this report was dampened by later reports that CD13 antibodies could bind and neutralize virus before contact with cells, coupled with normal entry of HCMV into CD13-depleted cells (Giugni *et al.*, 1996). More recently, an interaction between HCMV and CD13 was shown to be important in inhibition of differentiation of monocytes into macrophages suggesting this may be a strategy for interference with cellular differentiation pathways (Gredmark *et al.*, 2004).

A consideration of HCMV-induced signalling cascades led Wang *et al.* (2003) to hypothesize a role for epidermal growth factor receptor (EGFR) as an HCMV receptor. It was found that EGFR became phosphorylated in response to HCMV and this phosphorylation event correlated with the activation of phosphatidylinositol 3-kinase (PI-3 kinase) and Akt, as well as the mobilization of intracellular Ca^{2+}. Signalling events were blocked by EGFR antibodies. In addition, chemically cross-linked virus provided evidence for a gB–EGFR interaction. A limitation of the study, however, was that there was no experimental evidence that EGFR functioned in an entry event *per se*. It was also not directly shown that EGFR was required for the delivery of virion components across the plasma membrane. These data are confounded by conflicting results in the literature. Fairley and colleagues demonstrated that HCMV promoted inactivation of EGFR phosphorylation and signalling (Fairley *et al.*, 2002). In an ironic twist, studies investigating CD13 as an HCMV receptor used antibodies to EGFR as a 'negative control'. In these experiments, EGFR polyclonal antibodies had no effect on HCMV entry (Soderberg *et al.*, 1993a). It is important to note that EGFR is not typically thought to be expressed on all HCMV permissive cells, including those of

haematopoietic lineage. However, one recent study suggests that human peripheral blood monocytes do express EGFR, unlike other leucocytes, and HCMV infection in these cells coincides with EGFR-activated motility (Chan *et al.*, 2009). To clarify these discrepancies, a study was performed that directly demonstrated no discernible role for EGFR in virus entry or signalling in a range of cell types including fibroblasts, epithelial cells, and endothelial cells (Isaacson *et al.*, 2007).

With increasing scepticism for a role for EGFR in HCMV entry and virus signalling, a study was published suggesting that a different growth factor receptor was in fact the receptor in question (Soroceanu *et al.*, 2008). This report suggested that platelet-derived growth factor receptor alpha (PDGFR-a) allowed for virus entry and activated signalling consequent with HCMV entry of host cells; likely through a direct interaction with glycoprotein B. While no report has been published to directly refute these claims as in the case for EGFR, no additional reports have been published supporting or expounding on these findings for several years. Combined, these observations and the history of HCMV entry receptors engender caution. There is a need for careful duplication of the data and verification of these molecules as receptors.

Cellular integrins serve as co-receptors for betaherpesviruses

The first and foremost observation about HCMV biology was its namesake characteristic, cytomegaly, or cell enlargement. *In vitro* studies initially demonstrated a unique cytopathogenic effect (CPE) of infected cells, with HCMV infected cells appearing round and enlarged with intracellular viral inclusion bodies (Albrecht and Weller, 1980). Infection proceeded with two waves of cell rounding, the first beginning as early as 1 hour post infection, corresponding to the entry event, and another at approximately 24 hours post infection. The cause of this phenomenon was widely speculated upon; however theories for HCMV-induced cell rounding included cation influx, suppression of fibronectin synthesis and integrin down-regulation (Albrecht and Weller, 1980; Ihara *et al.*, 1982; Albrecht *et al.*, 1983; Warren *et al.*, 1994). Cellular integrins are ubiquitously expressed cell surface receptors that, when activated, lead to major reorganization of the cytoskeleton. Integrins exist on the plasma membrane as a non-covalently linked heterodimer consisting of an α and β subunit, which conveys specificity in cell–cell and cell–ECM (extracellular matrix) attachment, immune cell recruitment, extravasation, and signalling (Cary *et al.*, 1999; Berman and Kozlova, 2000; Berman *et al.*, 2003). In addition, integrins have emerged as receptors for a broad range of pathogens including pathogenic plant spores, bacteria and viruses. Feire *et al.* (2004) tested the hypothesis that integrins were involved in the HCMV entry pathway. Analysis of the effects of various neutralizing antibodies showed α2 β1, α6 β1, and αᵥ β3 were involved in HCMV entry (Feire *et al.*, 2004). Furthermore, cells devoid of beta 1 integrins exhibited little to no infection by HCMV or mouse cytomegalovirus while entry and spread were restored when the expression of β1 integrin was re-introduced into the cells. The integrin antibodies had no effect on virus attachment but specifically blocked the delivery of a virion tegument protein, pp65 (pUL83), suggesting that integrins function at a post-attachment stage of infection, possibly at the level of membrane fusion. The fact that multiple integrin heterodimers are utilized is consistent with integrin biology in that many natural integrin ligands, such as extracellular matrix proteins, bind to multiple heterodimers. Indeed, most integrin-binding viruses frequently interact with one to three individual integrin molecules. Integrins are capable of engaging ligands through a number of identified ECM protein motifs, the most common of which contain the amino acid sequence RGD. However, there are a number of RGD-independent integrin binding motifs, including the disintegrin domain proteins of the ADAM (A Disintegrin and A Metalloprotease) family of proteins. After inspection of all HCMV structural glycoproteins, the only homology to an integrin-binding domain was a disintegrin-like consensus sequence (RX₅₋₇DLXXF/L) (Wolfsberg *et al.*, 1995; Stone *et al.*, 1999; Eto *et al.*, 2002) on the amino-terminus of gB. Sequence alignments confirmed that the gB disintegrin loop was more than 98% identical among 44 clinical isolates analysed. The role of this sequence in entry was confirmed since synthetic peptides to this sequence inhibited both HCMV and MCMV entry, but not the disintegrin-loop-lacking HSV. In later studies, a larger 90-amino acid fragment of the gB disintegrin-like loop was generated, which more naturally mirrors the full disintegrin-like domain found in the ADAM family of proteins (Feire *et al.*, 2010). This prokaryotic produced gB disintegrin-like domain (gB-DLD) fragment bound permissive cells in a specific and dose-dependent manner and was able to inhibit HCMV entry into a broad range of cell types at concentrations much lower than the previously characterized synthetic peptides. Additionally, a direct interaction between gB-DLD and beta 1 integrin was demonstrated. The HCMV gB disintegrin-loop was conserved throughout most members of the gamma and all of the betaherpesvirus subfamilies, but not in the alphaherpesvirus subfamily where there are previously identified RGD sequences. The presence of integrin binding sequences among conserved *Herpesviridae* glycoproteins strongly suggests that integrins may be important for entry and signalling throughout this virus family. Interestingly, EGFR has been shown to become phosphorylated and signal indirectly, as a result of integrin activation through src family kinases or focal adhesion kinase (FAK) (Miyamoto *et al.*, 1996; Jones *et al.*, 1997; Moro *et al.*, 1998). Future work will no doubt be aimed at an analysis of the integrin-triggered signalling events and defining their roles in entry and infection.

Entry-activated cell signalling

It has been apparent for many years that cells respond to HCMV virions by activation of numerous cell signalling pathways including changes in Ca^{2+} homeostasis, activation of phospholipases C and A2, as well as increased release of arachidonic acid and its metabolites (for review, see Fortunato *et al.*, 2000). All of these changes can be triggered by UV-inactivated virions, suggesting that structural components of the virus are responsible for activation during virus-cell contact and/or virus entry. Virus-cell contact also results in the activation of transcription factors such as cfos/jun, myc, NF-κB, SP-1, as well as mitogen-activated protein (MAP) kinase ERK1/2 and p38 (Kowalik *et al.*, 1993; Yurochko *et al.*, 1995; Boyle *et al.*, 1999). These virally induced cellular physiological changes are associated with a profound effect on host cell gene expression. Thousands of transcripts are altered in their expression and most transcriptional changes do not require viral gene expression (Zhu *et al.*, 1998; Browne *et al.*, 2001; Simmen *et al.*, 2001). These data are consistent with the interpretation that HCMV engages a cellular receptor(s) that activate(s) signal transduction pathways culminating in reprogramming of cellular transcription.

Activation of innate defence by cytomegaloviruses

Cytomegaloviruses trigger strong anti-pathogen responses upon infection. Effective control of HCMV *in vivo* requires cooperation between many facets of both the innate and adaptive arms of the host immune system. The discussion here is limited to activation of innate responses triggered during entry of the virus into host cells, focusing on recent advances in the induction of interferon-α/β responses by HCMV (see also Chapter I.16) and the emerging role of Toll-like receptors in detection of HCMV virions.

Activation of the interferon-α/β response by HCMV

Like most viruses, HCMV elicits an extremely potent interferon-α/β response from their host. Interferon-α/β confers an antiviral state on cells, causing them to become refractory to viral infection and limiting virus replication at the site of infection (Stark et al., 1998). Interferon-α/β is also a critical component of the cytokine cocktail that activates T-cell responses. The cardinal role of interferon-α/β in the immune response to CMVs is evidenced by the observation that mice lacking the interferon-α receptor (both interferon- α and interferon-β utilize the same receptor complex) are highly susceptible to MCMV infection as compared to wild-type mice (Salazar-Mather et al., 2002). CMVs have long been known to elicit interferon-α/β; however the mechanism(s) by which these responses are induced have only recently begun to be defined.

A growing body of evidence indicates that HCMV activates innate defence during binding and entry of virions to the host cell. Transcriptional profiling studies demonstrated that HCMV is a potent inducer of many indicators of innate defence, including inflammatory cytokines and interferon-α/β (Zhu et al., 1997, 1998; Browne et al., 2001; Simmen et al., 2001). The induction is rapid with transcriptional up-regulation of innate markers observed within 2 hours of infection (Browne et al., 2001). In addition, a variety of cell signalling pathways are activated upon infection. NF-κB, a central player in a variety of host cell defences, is activated as early as 15 minute post infection, and IRF3, the key transcriptional regulator of interferon-α/β responses, is also activated upon HCMV infection (Yurochko et al., 1995, 1997a,b; Navarro et al., 1998; Yurochko and Huang, 1999; Preston et al., 2001; Guerrero et al., 2004). Furthermore, activation of innate effectors by HCMV is unaffected by the presence of the protein synthesis inhibitor cycloheximide, indicating that the production of additional secreted factors such as interferon-α/β or inflammatory cytokines is not required to elicit these responses (Zhu et al., 1997, 1998; Navarro et al., 1998; Browne et al., 2001; Preston et al., 2001; Boehme et al., 2004). In conclusion, the rapid and direct induction of innate responses by HCMV suggests that they are triggered during virus binding and entry into its target cell.

Overview of the interferon pathway

The interferon-α/β response to viral infection can be divided into two phases: an activation phase and an amplification phase (see also Chapter I.16; Taniguchi et al., 2001; Taniguchi and Takaoka, 2002). The activation phase begins with the initial detection of the virus and propagation of intracellular signals that culminate in the secretion of interferon-α/β (Stark et al., 1998). The mechanisms by which cells detect viruses and initiate the interferon response have not been completely defined, however one means by which cells can activate interferon-α/β responses is through specific members of the Toll-like receptor (TLR) family (TLRs 3, 4, 7, 8, and 9). An increasing amount of evidence indicates that these TLRs play a critical role in the interferon response to herpes viral infection (Alexopoulou et al., 2001; Lund et al., 2003, 2004; Diebold et al., 2004; Heil et al., 2004; Krug et al., 2004a,b) and the specific roles of TLRs in the interferon response to CMVs will be discussed below. Other mechanisms for induction of interferon-α/β responses to viruses exist, but these processes remain poorly understood.

Viral detection leads to activation of the interferon regulatory factor 3 (IRF3) transcription factor via phosphorylation of specific serine and threonine residues in its carboxy-terminal domain by the cellular kinases IKKε and TBK1 (Servant et al., 2001; Fitzgerald et al., 2003; Sharma et al., 2003). IRF3 normally resides in the cytoplasm, however upon phosphorylation it translocates to the nucleus where it complexes with p300/CBP to drive expression of a subset of interferon stimulated genes (ISGs), as well as interferon-β (Yoneyama et al., 1998; Suhara et al., 2002). Secreted interferon–β acts in an autocrine/paracrine manner through the IFN-α/β receptor and the well defined Janus kinases and signal transducers and activators of transcription (JAK-STAT) pathway to promote expression of the full complement of ISGs (Stark et al., 1998). Among these ISGs is IRF7 which, like IRF3, is activated by phosphorylation (presumably by IKKε and TBK1) and complexes with IRF3 to drive the expression the IFN-α genes (Sato et al., 1998, 2000). IFN-α, like IFN-β, acts in an autocrine/paracrine manner to induce ISG expression. This amplification loop provides a rapid and effective mechanism by which host cells can respond to viral infection.

Interestingly, the specialized cell type primarily responsible for secreting type I interferon in response to a wide variety of viruses, plasmacytoid dendritic cells (pDCs), selectively produce interferon-α as opposed to interferon-β (Perussia et al., 1985; Feldman et al., 1994; Poltorak et al., 1998b; Cella et al., 1999; Siegal et al., 1999; Asselin-Paturel et al., 2001; Dalod et al., 2002, 2003; Fonteneau et al., 2003; see also Chapter II.11). New evidence indicates that IRF7 is initially activated in pDCs, as opposed to IRF3 (Dai et al., 2004; Honda et al., 2004; Kawai et *al.*, 2004). The specificity of IRF7 for the interferon-α genes provides the mechanism by which pDCs induce secretion of interferon-α, but not interferon-β which is dependent upon IRF3. Functional differences exist between the different interferon-α subtypes and interferon– β and it is likely that these differences are the reason for the preference of interferon-α exhibited by pDCs (Evinger *et al.*, 1981; Weck *et al.*, 1981; Foster *et al.*, 1996).

Role of Toll-like receptors in innate immune activation

As described above, TLRs are one way in which cells can detect viruses and initiate interferon- α/β responses. TLRs are not limited to induction of interferon responses, but rather serve a larger function as general pathogen recognition receptors (PRRs) that detect and initiate immune responses to a myriad of bacteria, fungi, and viruses (Takeda and Akira, 2003). The primary consequences of TLR activation are inflammatory cytokine secretion, expression of immune co-stimulatory molecules, dendritic cell maturation, and for certain TLRs, the production of interferon-α/β (Takeuchi and Akira, 2001). Together these factors limit viral replication at the site of infection, elicit infiltration of immune cells to the site of infection, and initiate and modulate adaptive immune responses by B and T-cells (see also Chapter II.11).

TLRs are type I transmembrane proteins comprising leucine-rich repeat (LRR) ectodomains and cytoplasmic regions with Toll-interleukin-1 receptor (TIR) domains (Rock *et al.*, 1998). To date eleven TLRs have been identified in humans and, although all cells express at least a subset of TLRs, they are expressed predominantly by immune sentinel cells such as macrophages and DCs (Medzhitov *et al.*, 1997; Rock *et al.*, 1998; Takeuchi *et al.*, 1999; Chuang and Ulevitch, 2000, 2001; Du *et al.*, 2000; Hornung *et al.*, 2002; Zarember and Godowski, 2002; Zhang *et al.*, 2004). TLRs recognize a wide variety of microbial pathogens on the basis of pathogen associated molecular patterns (PAMPs). PAMPs are structural elements found uniquely on microbes, but not normally present in the host organism. Some examples of TLR ligands are lipopolysaccharide (LPS) and peptidoglycan from bacteria, dsRNA from viruses, and unmethylated CpG DNA motifs from bacteria and viruses (Poltorak *et al.*, 1998a; Qureshi *et al.*, 1999; Hemmi *et al.*, 2000; Alexopoulou *et al.*, 2001; Bauer *et al.*, 2001). An expanded list of TLR PAMPs can be found in Table I.8.1. Ligand-induced TLR clustering brings the cytoplasmic tails of the receptors into close proximity, thereby creating a docking site for intracellular signalling intermediates. TLRs activate NF-κB and mitogen-activated protein kinases, however a subset of TLRs (TLRs 3, 7, 8, and 9) concomitantly activate interferon-α/β secretion by activating members of the interferon regulator factor family (Wagner, 2004). A comprehensive discussion of TLR signalling is beyond the scope of this chapter, however a number of reviews are available detailing various aspects of TLR biology (Akira, 2003; Akira and Hemmi, 2003; O'Neill *et al.*, 2003; Takeda and Akira, 2003; Boehme and Compton, 2004).

Table I.8.1 Viral ligands of Toll-like receptors

Toll-like receptor	Viral ligand
TLR2	Peptidoglycan Zymosan Measles virus (haemagglutinin) Human cytomegalovirus (gB) HSV-1
TLR3	dsRNA MCMV
TLR4	LPS RSV (F protein) MMTV (envelope protein)
TLR7 and TLR8	Influenza A (genomic RNA) HIV-1 (synthetic RNA oligonucleotide from U5 region of genome) VSV
TLR9	HSV-1 HSV-2 (genomic DNA) MCMV

Roles of HCMV envelope glycoproteins in virus entry

Composition of the HCMV envelope

The HCMV envelope is exceedingly complex and currently incompletely defined. The HCMV genome encodes open reading frames (ORFs) to at least 57 putative glycoproteins; far more than other herpesviruses, however, the extent of transcription, translation and function of the majority of these glycoproteins remains unknown. Biochemical studies of HCMV virions have revealed that 14 glycoproteins are structural; eight of these experimentally shown to reside in the envelope (Britt et al., 2004). Among structural glycoproteins, a group homologous to other herpesviruses and thus thought to serve conserved functions in entry exists. These include glycoproteins B (gB), H (gH), L (gL), O (gO), M (gM), and N (gN), while others (gpTRL10, gpTRL11, gpTRL12 and gpUL132) are HCMV-specific (Table I.8.2).

For many years, the large genome and complicated reverse genetics system have made the creation of HCMV knockout and mutant viruses difficult. Recently,

a system capable of such mutations was developed whereby HCMV is maintained as an infectious bacterial artificial chromosome (BAC) within *Escherichia coli* (Borst et al., 1999). This development has greatly hastened the process of mutating individual ORFs (see also Chapter I.3) and will generate much information regarding both the structure and function of many envelope glycoproteins. In fact, the BAC system has demonstrated the requirement for several glycoprotein genes in the production of replication-competent virus (Hobom, 2000; Yu et al., 2002; Dunn et al., 2003). The HCMV glycoprotein homologues gB, gM, gN, gH, gL, have been shown to be essential for growth, while gO knockout virus remained viable with a small plaque phenotype (Hobom, 2000). Genes for all the currently identified HCMV-specific envelope glycoproteins, including UL4 (gp48), TRL10 (gpTRL10), TRL11 (gpTRL11), TRL12 (gpTRL12), US27, UL33, and UL132 have been shown to be non-essential and play no known roles in the entry pathway (Dunn et al., 2003). The HCMV-encoded chemokine receptor US28 is present in the virion envelope and has been shown to promote cell–cell fusion mediated by HIV and VSV viral proteins, however the gene has been

Table I.8.2 Envelope proteins of HCMV

ORF	Protein name	Essential	Complex partner	Role in entry
UL4	gpUL4; gp48	No	None known	None known
UL33	UL33	No	None known	None known
UL55	gB	Yes	Forms homodimers	Receptor binding, fusion, signal transduction, innate immune activation
UL73	gN	Not determined	UL100; gM	None known
UL74	gO	No; defect in cell-to-cell spread	UL75; gH UL115; gL	Enhancer of cell-to-cell spread
UL75	gH	Yes	UL74; gO UL115; gL	Fusion, receptor binding (?), innate immune activation
UL100	gM	Yes	UL73; gN	HSPG binding
UL115	gL	Yes	UL75; gH UL74; gO	Required for gH activity
UL128	UL128	In many cell types	gH, gL, UL130, UL131a	Required for epithelial, endothelial and myeloid cells
UL130	UL130	In many cell types	gH, gL, UL1128, UL131a	Required for epithelial, endothelial and myeloid cells
UL131	UL131	In many cell types	gH, gL, UL1128, UL130	Required for epithelial, endothelial and myeloid cells
TRL10	gpTRL10	Not determined	None known	None known
TLR12	gpTRL12	Not determined	None known	None known
US27	US27	No	None known	None known
US28	US28	No	None known	None known

shown to be non-essential and there is no evidence for a role for gpUS28 in either HCMV–cell or cell–cell fusion (Pleskoff *et al.*, 1997, 1998; Dunn *et al.*, 2003).

The essential and abundant HCMV envelope glycoproteins conserved throughout the *Herpesviridae* (gB, gM, gN, gH, and gL) were originally classified as three distinct disulfide-linked high molecular weight complexes (gCI-gCIII) (Gretch *et al.*, 1988). The gCI complex is composed of homodimers of glycoprotein B (gB) (Britt, 1984; Britt and Auger, 1986). The gCII heterodimeric complex is composed of glycoprotein M (gM) and glycoprotein N (gN) (Mach *et al.*, 2000). Finally, the gCIII complex is a heterotrimeric complex composed of glycoprotein H (gH), glycoprotein L (gL), and glycoprotein O (gO) (Huber and Compton, 1997, 1998; Li *et al.*, 1997).

More recently, it was observed that a specific region of the genome containing UL131, UL130, and UL128 was required for growth in endothelial cells and for transfer of genomes from endothelial cells to leucocytes (Hahn *et al.*, 2004). It was later shown that these ORFs encode for a previously undiscovered glycoprotein complex comprised of gH/gL/UL128/UL130/UL131a found in the envelope of the virus that was essential for virus entry into epithelial cells, endothelial cells, and leucocytes (Wang and Shenk, 2005; Ryckman *et al.*, 2006, 2008). The implications of the discovery of this complex for an alternate mode of entry and for cell-type tropism will be further described and discussed in Chapter I.17.

Envelope glycoproteins that bind to receptors

At least two glycoprotein complexes have heparin binding ability. Both gB and the gM component of gCII can bind to soluble heparin suggesting a critical role in the initial tethering event (Kari and Gehrz, 1992, 1993; Carlson *et al.*, 1997). Functional redundancy in particular with respect to HSPG binding is also a shared property with other herpesviruses. The gB protein appears to be the primary receptor binding protein. Soluble forms of gB exhibit biphasic cell binding properties, and cells treated with gB are refractory to infection suggesting that gB is tying up critical receptor sites used by the virus (Boyle and Compton, 1998). One of the binding sites for gB is HSPGs in that cells lacking HSPGs had a single component Scatchard plot as compared to a biphasic plot for HSPG-bearing cells. As noted above, it now seems clear that a second binding partner is an integrin (Feire *et al.*, 2004) but much work remains to formally prove the disintegrin hypothesis and confirm the role of this domain in receptor engagement. The gB protein may also engage

EGFR at least in certain cell types; however, it is not yet known if this interaction is a consequence of initial integrin binding. The gH complex may also have a distinct receptor. Syngeneic monoclonal anti-idiotypic antibodies were created that bear the 'image' of this glycoprotein complex (Keay *et al.*, 1988). This reagent led to a putative gH receptor; a phosphorylated 92.5-kDa cell surface glycoprotein that appears to mediate virus–cell fusion, but not attachment and Ca^{2+} influxes at the plasma membrane (Keay *et al.*, 1989, 1995; Keay and Baldwin, 1991, 1992, 1996). The data set relies heavily on a single reagent, (anti-idiotypic antibodies), and has led to only a partially sequenced receptor clone with no identified homology to any known human protein (Keay and Baldwin, 1996; Baldwin, 2000). Since HCMV gH is essential, infectivity can be neutralized with anti-gH antibodies. Anti-idiotypic antibodies can also neutralize infectivity, presumably by occupying the receptor for gH. The closest relative of HCMV, HHV-6, contains a homologous complex gH/gL/gQ with CD46 identified as a putative receptor (Santoro *et al.*, 1999), however the identity of an HCMV gH/gL/gO receptor remains unknown, yet quite possible.

Roles of envelope glycoproteins in membrane fusion

Membrane fusion remains a big black box in herpes virology. Unlike orthomyxoviruses, paramyxoviruses, filoviruses and retroviruses that use a single envelope glycoprotein for membrane fusion, herpesviruses typically employ multi-component fusion machines frequently consisting of gB, gH, and gL (Spear and Longnecker, 2003). Both the HCMV gB and gH-containing complex trigger neutralizing antibodies that block infection at a post-attachment stage of entry, presumably at the level of fusion (Britt, 1984; Utz *et al.*, 1989; Keay and Baldwin, 1991; Tugizov *et al.*, 1994; Bold *et al.*, 1996). One limitation of these conclusions, however, is the lack of a direct fusion assay and thus a role for these glycoproteins in fusion is only inferred. Despite the complexity of multi-component fusion machines, it is very likely that there are strong parallels to single component fusion proteins. Alpha-helical coiled coils are considered critical structural domains involved in fusion that function to drive the energetic folding of membranes together. Conformational changes in fusogenic proteins bearing these coiled coils are also a defining paradigm. Using an algorithm to detect potential coiled coils, Lopper and Compton (2004) identified heptad repeat regions in gB and gH that were predicted to form coiled coils. Synthetic peptides to these motifs substantially inhibited HCMV entry including virion content delivery suggesting that these motifs play a fundamental role in membrane fusion. Genetic analysis of these motifs in the context of HCMV virions will be required to further analyse the importance of alpha-helical coiled coils in HCMV entry. Another fundamental question will be to determine if the gB integrin interaction is a trigger of conformational change that leads to exposure of membrane fusion domains. Intriguingly, disintegrin-bearing cellular proteins in the ADAM family are known to trigger fusion via integrin interaction in a variety of processes including sperm–egg fusion and myoblast fusion (White, 2003). Development of a reliable fusion assay is also greatly needed to begin a dissection of the biophysical properties of HCMV fusion glycoproteins.

Role of the envelope glycoproteins in activation of innate defence

Transcriptionally incompetent virions are equally efficient, if not more robust, in their ability to induce expression of genes involved in innate antiviral responses (Zhu et al., 1997,1998; Navarro et al., 1998; Browne et al., 2001; Preston et al., 2001; Simmen et al., 2001). These observations suggest that one or more virion structural components are responsible for initiating innate responses. The rapidity of the response coupled with the ability of a virion structural component to elicit the responses pointed to the possibility that glycoproteins displayed on the surface of the virion were activating innate antiviral defence by interacting with host cell receptors.

Glycoprotein B is a component capable of activating the interferon pathway

The ability of the envelope glycoprotein B (gB), the primary viral ligand, to directly bind cells make it a candidate for direct activation of innate receptors (Boyle and Compton, 1998). A soluble form of gB comprised of the ectodomain of gB and retaining cell binding properties consistent with virion-associated gB was utilized to test the ability of gB to activate innate antiviral responses independent of other virion components (Boyle et al., 1999). Microarray analysis revealed remarkable similarities with respect to ISG induction between HCMV infected cells and cells treated with soluble gB (Simmen et al., 2001). In addition to direct induction of ISG transcription, soluble gB activates IRF3, NF-κB, and elicits the production of IFNβ (Yurochko et al., 1997a; Boyle and Compton, 1998; Chin and Cresswell, 2001; Simmen et al., 2001; Boehme et al., 2006). Together these data suggest that virus entry is not strictly required for the induction of interferon-α/β responses to CMV, but rather that

interaction of gB with a cell surface receptor during virus binding is sufficient to elicit antiviral responses. Recently we tested the ability of HCMV to activate interferon responses in cells expressing a dominant negative form of TLR2 (DN-TLR2), a receptor known to trigger inflammatory cytokine responses to HCMV. Our results showed that the interferon pathway was intact in cells expressing DN-TLR2 suggesting that it is not the receptor involved in activation of the interferon response (Boehme et al., 2006).

Although the specific receptor responsible for activation of the interferon pathway by HCMV has not been identified, other members of the herpesvirus family may provide clues into mechanisms by which HCMV can activate the interferon-α/β pathway. MCMV, HSV-1, and HSV-2 elicit interferon-α secretion via TLR9 (Lund et al., 2003; Krug et al., 2004a,b; Tabeta et al., 2004). Each of these viruses has a CpG motif-rich genome that constitutes the molecular basis for recognition of these viruses by TLR9 (Lund et al., 2003; Krug et al., 2004a). HCMV also contains a CpG-rich genome and is therefore also likely subject to innate sensing by TLR9, however this has not been formally established. Recognition of viral genomic DNA by TLR9, which resides in intracellular compartments, would likely require degradation of the virion within an appropriate compartment in order to release the viral genome from its capsid housing and make it available for detection by TLR9 (Takeshita et al., 2001; Ahmad-Nejad et al., 2002). In addition, TLR3 is critical for MCMV-induced interferon-α production (Alexopoulou et al., 2001; Tabeta et al., 2004). dsRNA species resulting from bidirectional transcription of the viral genome during the late stages of MCMV infection are hypothesized to activate TLR3 (Tabeta et al., 2004). Such a mechanism might also be important for the interferon-α/β response to HCMV. Finally, RNAs in the tegument of the HCMV virion may also render it subject to sensing by TLR7 and/or TLR8, which detect ssRNA from several different viruses (Diebold et al., 2004; Heil et al., 2004; Lund et al., 2004). A growing possibility is that HCMV activates innate responses by several different means. Dissection of the multiple mechanisms by which HCMV may induce interferon responses will undoubtedly be the focus of intense future study.

Envelope glycoproteins activate TLR-2 mediated inflammatory cytokine induction

Envelope glycoproteins including gB are also likely molecular triggers for the TLR2-mediated inflammatory cytokine induction. Similar to the interferon-α/β response, UV-inactivated virions elicit TLR2-dependent responses indicating that gene expression is not required for TLR activation, but rather a structural component(s) of the virion is responsible for triggering TLR2 (Compton *et al.*, 2003). Envelope glycoprotein B appears to be the trigger for TLR2 as the two physically associate with one another and can be co-immunoprecipitated from cells. Furthermore, recombinant gB has intrinsic ability to induce IL-6 and IL-8 in a TLR2-dependent manner (Boehme *et al.*, 2006). The gH-containing complex may also be a trigger. Neutralizing antibodies to gH impair the ability of HCMV to activate cytokine production (Boehme *et al.*, 2006). The ability of glycoproteins to activate TLRs is intriguing because TLRs are theorized to recognize structures not normally found in the host, such as LPS or CpG DNA. However, viral glycoproteins are produced by the host cell's own protein synthesis and glycosylation machinery and therefore do not intuitively harbour modifications or moieties that differ significantly from the host. It is possible that these proteins may possess unique structural conformations that are not assumed by host cell proteins. Similar to HCMV, respiratory syncytial virus activates TLR4 through its envelope protein, Env (Rassa, 2002). A common property of Env and gB is that they play critical roles in the binding and entry of their respective viruses. This observation suggests that viruses are detected at the earliest stages of infection, simply via contact between glycoproteins on the viral envelope and TLRs on the cell surface. The ability of TLRs to recognize viral proteins that mediate cell entry is highly advantageous to the host because it allows the cell to begin mounting the appropriate response before the virus has actually invaded.

Additionally, innate activation may occur as a result of the fusion process itself. Virus–cell fusion is typically driven by helical bundle formation by viral glycoproteins. These structures could be subject to recognition by TLRs, or other PRRs. Both HCMV gB and gH are important for HCMV fusion (E. Kinzler and T. Compton, unpublished results) and contain putative coiled coil domains which are the basis of helical bundle formation (Lopper and Compton, 2004). A soluble gB comprised of the amino-terminal segment of the molecule (residues 1–460), but lacking the carboxy-terminal domain where the predicted coiled coil domain resides, is unable to activate IRF3 or induce ISG transcription (Boehme *et al.*, 2006). These data support a role for fusion-generated structures as important determinants for innate defence. Intriguingly, a small molecule inhibitor of HCMV entry that allows cell binding but prevents entry inhibits the induction of interferon-α/β responses (Netterwald *et al.*, 2004). However, the inhibitor targets gB and may disrupt critical interactions between gB and receptor(s) that initiate the interferon-α/β response (Jones *et al.*, 2004). Further study will be required to fully elucidate the variety of mechanisms by which HCMV activates innate defence.

Concluding thoughts: coordination of entry and innate antiviral defence

We are left with an apparent dichotomy. As HCMV enters cells to establish infection, the host recognizes virions and activates innate defence responses, including the innate cellular immune response. How are the two processes coordinated or are they at all? At this time, there is no apparent role for TLRs in driving an actual entry event. Rather at some point in the entry process, host immune sensors detect a pathogen-associated molecular pattern displayed by HCMV envelope glycoproteins and activate host innate defences (Fig. I.8.1). One possibility is that entry receptors (EGFR, integrins and signalling accessory molecules) and innate defence machinery (TLR2, its membrane-associated partners, cytoplasmic adaptors and signalling machinery) coalesce into specialized membrane microdomains with integrins playing central ligating role. Concentration of all of these cell surface receptors into a defined platform likely facilitates cell signalling events, some of which are optimal for replication and others of which are clearly hostile to the virus. Intriguingly integrins have also been shown to associate with TLR2 and to partition into cholesterol-rich lipid rafts (Ogawa *et al.*, 2002; Triantafilou *et al.*, 2002). The complexity of events at the cell surface during the initial encounter of HCMV and cells represents an exciting opportunity to better understand the molecular underpinnings of the early virus–host interactions. The recent identification of cell surface molecules involved in the early steps of infection has greatly enhanced our knowledge of entry events in infection. Yet much remains to be done to elucidate aspects of mechanism of entry events and the corresponding innate immune activation.

References
Ahmad-Nejad, P., Hacker, H., Rutz, M., Bauer, S., Vabulas, R.M., and Wagner, H. (2002). Bacterial CpG-DNA and lipopolysaccharides activate Toll-like receptors at distinct cellular compartments. Eur. J. Immunol. 32, 1958–1968.
Akira, S. (2003). Toll-like receptor signaling. J. Biol. Chem. 278, 38105–38108.

Akira, S., and Hemmi, H. (2003). Recognition of pathogen-associated molecular patterns by TLR family. Immunol. Lett. 85, 85–95.

Albrecht, T., Speelman, D.J., and Steinsland, O.S. (1983). Similarities between cytomegalovirus-induced cell rounding and contraction of smooth muscle cells. Life Sci. 32, 2273–2278.

Albrecht, T., and Weller, T.H. (1980). Heterogeneous morphologic features of plaques induced by five strains of human cytomegalovirus. Am. J. Clin. Pathol. 73, 648–654.

Alexopoulou, L., Holt, A.C., Medzhitov, R., and Flavell, R.A. (2001). Recognition of double-stranded RNA and activation of NF-kappaB by Toll-like receptor 3. Nature 413, 732–738.

Asselin-Paturel, C., Boonstra, A., Dalod, M., Durand, I., Yessaad, N., Dezutter-Dambuyant, C., Vicari, A., O'Garra, A., Biron, C., Briere, F., et al. (2001). Mouse type I IFN-producing cells are immature APCs with plasmacytoid morphology. Nat. Immunol. 2, 1144–1150.

Baldwin, B.R., Zhang, C., and Keay, S. (2000). Cloning and epitope mapping of a functional partial fusion receptor for human cytomegalovirus gH. J. Gen. Virol. 81, 27–35.

Bauer, S., Kirschning, C.J., Hacker, H., Redecke, V., Hausmann, S., Akira, S., Wagner, H., and Lipford, G.B. (2001). Human TLR9 confers responsiveness to bacterial DNA via species-specific CpG motif recognition. Proc. Natl. Acad. Sci. U.S.A. 98, 9237–9242.

Beersma, M.F., Wertheim, van, D.P., and Feltkamp, T.E. (1990). The influence of HLA-B27 on the infectivity of cytomegalovirus for mouse fibroblasts. Scand. J. Rheumatol. Suppl. 87, 102–103.

Beersma, M.F., Wertheim, van, D.P., Geelen, J.L., and Feltkamp, T.E. (1991). Expression of HLA class I heavy chains and beta 2-microglobulin does not affect human cytomegalovirus infectivity. J. Gen. Virol. 72, 2757–2764.

Berman, A.E., and Kozlova, N.I. (2000). Integrins: structure and functions. Membr. Cell Biol. 13, 207–244.

Berman, A.E., Kozlova, N.I., and Morozevich, G.E. (2003). Integrins: structure and signaling. Biochemistry 68, 1284–1299.

Boehme, K.W., and Compton, T. (2004). Innate sensing of viruses by toll-like receptors. J. Virol. 78, 7867–7873.

Boehme, K.W., Singh, J., Perry, S.T., and Compton, T. (2004). Human cytomegalovirus elicits a coordinated cellular antiviral response via envelope glycoprotein B. J. Virol. 78, 1202–1211.

Boehme, K.W., Guerrero, M., and Compton, T. (2006). Human cytomegalovirus glycoproteins B and H are necessary for TLR2 activation in permissive cells. J. Immunol. 177, 7094–7102.

Bold, S., Ohlin, M., Garten, W., and Radsak, K. (1996). Structural domains involved in human cytomegalovirus glycoprotein B- mediated cell–cell fusion. J. Gen. Virol. 77, 2297–2302.

Borst, E.M., Hahn, G., Koszinowski, U.H., and Messerle, M. (1999). Cloning of the human cytomegalovirus (HCMV) genome as an infectious bacterial artificial chromosome in Escherichia coli: a new approach for construction of HCMV mutants. J. Virol. 73, 8320–8329.

Boyle, K.A., and Compton, T. (1998). Receptor-binding properties of a soluble form of human cytomegalovirus glycoprotein B. J. Virol. 72, 1826–1833.

Boyle, K.A., Pietropaolo, R.L., and Compton, T. (1999). Engagement of the cellular receptor for glycoprotein B of human cytomegalovirus activates the interferon-responsive pathway. Mol. Cell Bio. 19, 3607–3613.

Britt, W.J. (1984). Neutralizing antibodies detect a disulfide-linked glycoprotein complex within the envelope of human cytomegalovirus. Virology 135, 369–378.

Britt, W.J., and Auger, D. (1986). Human cytomegalovirus virion-associated protein with kinase activity. J. Virol. 59, 185–188.

Britt, W.J., Jarvis, M., Seo, J.Y., Drummond, D., and Nelson, J. (2004). Rapid genetic engineering of human cytomegalovirus by using a lambda phage linear recombination system: demonstration that pp28 (UL99) is essential for production of infectious virus. J. Virol. 78, 539–543.

Browne, E.P., Wing, B., Coleman, D., and Shenk, T. (2001). Altered cellular mRNA levels in human cytomegalovirus-infected fibroblasts: viral block to the accumulation of antiviral mRNAs. J. Virol. 75, 12319–12330.

Browne, H., Smith, G., Beck, S., and Minson, T. (1990). A complex between the MHC class I homologue encoded by human cytomegalovirus and beta 2 microglobulin. Nature 347, 770–772.

Carlson, C., Britt, W.J., and Compton, T. (1997). Expression, purification and characterization of a soluble form of human cytomegalovirus glycoprotein B. Virology 239, 198–205.

Cary, L.A., Han, D.C., and Guan, J.L. (1999). Integrin-mediated signal transduction pathways. Histol. Histopathol. 14, 1001–1009.

Cella, M., Jarrossay, D., Facchetti, F., Alebardi, O., Nakajima, H., Lanzavecchia, A., and Colonna, M. (1999). Plasmacytoid monocytes migrate to inflamed lymph nodes and produce large amounts of type I interferon. Nat. Med. 5, 919–923.

Chan, G., Nogalski, M.T., and Yurochko, A.D. (2009). Activation of EGFR on monocytes is required for human cytomegalovirus entry and mediates cellular motility. Proc. Natl. Acad. Sci. U.S.A. 106, 22369–22374.

Chin, K.C., and Cresswell, P. (2001). Viperin (cig5), an IFN-inducible antiviral protein directly induced by human cytomegalovirus. Proc. Natl. Acad. Sci. U.S.A. 98, 15125–15130.

Chuang, T., and Ulevitch, R.J. (2001). Identification of hTLR10: a novel human Toll-like receptor preferentially expressed in immune cells. Biochim. Biophys. Acta. 1518, 157–161.

Chuang, T.H., and Ulevitch, R.J. (2000). Cloning and characterization of a sub-family of human toll-like receptors: hTLR7, hTLR8 and hTLR9. Eur. Cyto. Netw. 11, 372–378.

Compton, T., Nowlin, D.M., and Cooper, N.R. (1993). Initiation of human cytomegalovirus infection requires initial interaction with cell surface heparan sulfate. Virology 193, 834–841.

Compton, T., Kurt-Jones, E.A., Boehme, K.W., Belko, J., Latz, E., Golenbock, D.T., and Finberg, R.W. (2003). Human cytomegalovirus activates inflammatory cytokine responses via CD14 and Toll-like receptor 2. J. Virol. 77, 4588–4596.

Conti, C., Cirone, M., Sgro, R., Altieri, F., Zompetta, C., and Faggioni, A. (2000). Early interactions of human herpesvirus 6 with lymphoid cells: role of membrane protein components and glycosaminoglycans in virus binding. J. Med. Virol. 62, 487–497.

Dai, J., Megjugorac, N.J., Amrute, S.B., and Fitzgerald-Bocarsly, P. (2004). Regulation of IFN regulatory factor-7 and IFN-alpha production by enveloped virus and lipopolysaccharide in human plasmacytoid dendritic cells. J. Immunol. *173*, 1535–1548.

Dalod, M., Salazar-Mather, T.P., Malmgaard, L., Lewis, C., Asselin-Paturel, C., Briere, F., Trinchieri, G., and Biron, C.A. (2002). Interferon alpha/beta and interleukin 12 responses to viral infections: pathways regulating dendritic cell cytokine expression in vivo. J. Exp. Med. *195*, 517–528.

Dalod, M., Hamilton, T., Salomon, R., Salazar-Mather, T.P., Henry, S.C., Hamilton, J.D., and Biron, C.A. (2003). Dendritic cell responses to early murine cytomegalovirus infection: subset functional specialization and differential regulation by interferon alpha/beta. J. Exp. Med. *197*, 885–898.

Diebold, S.S., Kaisho, T., Hemmi, H., Akira, S., and Reis, E.S.C. (2004). Innate antiviral responses by means of TLR7-mediated recognition of single-stranded RNA. Science *303*, 1529–1531.

Du, X., Poltorak, A., Wei, Y., and Beutler, B. (2000). Three novel mammalian toll-like receptors: gene structure, expression, and evolution. Eur. Cyto. Netw. *11*, 362–371.

Dunn, W., Chou, C., Li, H., Hai, R., Patterson, D., Stolc, V., Zhu, H., and Liu, F. (2003). Functional profiling of a human cytomegalovirus genome. Proc. Natl. Acad. Sci. U.S.A. *100*, 14223–14228.

Eto, K., Huet, C., Tarui, T., Kupriyanov, S., Liu, H.Z., Puzon-McLaughlin, W., Zhang, X.P., Sheppard, D., Engvall, E., and Takada, Y. (2002). Functional classification of ADAMs based on a conserved motif for binding to integrin alpha 9beta 1: implications for sperm-egg binding and other cell interactions. J. Biol. Chem. *277*, 17804–17810.

Evinger, M., Rubinstein, M., and Pestka, S. (1981). Antiproliferative and antiviral activities of human leukocyte interferons. Arch. Biochem. Biophys. *210*, 319–329.

Fairley, J.A., Baillie, J., Bain, M., and Sinclair, J.H. (2002). Human cytomegalovirus infection inhibits epidermal growth factor (EGF) signalling by targeting EGF receptors. J. Gen. Virol. *83*, 2803–2810.

Feire, A.L., Koss, H., and Compton, T. (2004). Cellular integrins function as entry receptors for human cytomegalovirus via a highly conserved disintegrin-like domain. Proc. Natl. Acad. Sci. U.S.A. *101*, 15470–15475.

Feire, A.L., Roy, R.M., Manley, K., and Compton, T. (2010). The glycoprotein B disintegrin-like domain binds beta 1 integrin to mediate cytomegalovirus entry. J. Virol. *84*, 10026–10037.

Feldman, S.B., Ferraro, M., Zheng, H.M., Patel, N., Gould-Fogerite, S., and Fitzgerald-Bocarsly, P. (1994). Viral induction of low frequency interferon-alpha producing cells. Virology *204*, 1–7.

Fitzgerald, K.A., McWhirter, S.M., Faia, K.L., Rowe, D.C., Latz, E., Golenbock, D.T., Coyle, A.J., Liao, S.M., and Maniatis, T. (2003). IKKepsilon and TBK1 are essential components of the IRF3 signaling pathway. Nat. Immunol. *4*, 491–496.

Fonteneau, J.F., Gilliet, M., Larsson, M., Dasilva, I., Munz, C., Liu, Y.J., and Bhardwaj, N. (2003). Activation of influenza virus-specific CD4+ and CD8+ T-cells: a new role for plasmacytoid dendritic cells in adaptive immunity. Blood *101*, 3520–3526.

Fortunato, E.A., McElroy, A.K., Sanchez, I., and Spector, D.H. (2000). Exploitation of cellular signaling and regulatory pathways by human cytomegalovirus. Trends Microbiol. *8*, 111–119.

Foster, G.R., Rodrigues, O., Ghouze, F., Schulte-Frohlinde, E., Testa, D., Liao, M.J., Stark, G.R., Leadbeater, L., and Thomas, H.C. (1996). Different relative activities of human cell-derived interferon-alpha subtypes: IFN-alpha 8 has very high antiviral potency. J. Interferon Cytokine Res. *16*, 1027–1033.

Giugni, T.D., Soderberg, C., Ham, D.J., Bautista, R.M., Hedlund, K.O., Moller, E., and Zaia, J.A. (1996). Neutralization of human cytomegalovirus by human CD13-specific antibodies. J. Infect. Dis. *173*, 1062–1071.

Gredmark, S., Britt, W.B., Xie, X., Lindbom, L., and Soderberg-Naucler, C. (2004). Human cytomegalovirus induces inhibition of macrophage differentiation by binding to human aminopeptidase N/CD13. J. Immunol. *173*, 4897–4907.

Gretch, D.R., Kari, B., Rasmussen, L., Gehrz, R.C., and Stinski, M.F. (1988). Identification and characterization of three distinct families of glycoprotein complexes in the envelopes of human cytomegalovirus. J. Virol. *62*, 875–881.

Grundy, J.E., McKeating, J.A., and Griffiths, P.D. (1987a). Cytomegalovirus strain AD169 binds beta 2 microglobulin in vitro after release from cells. J. Gen. Virol. *68*, 777–784.

Grundy, J.E., McKeating, J.A., Ward, P.J., Sanderson, A.R., and Griffiths, P.D. (1987b). Beta 2 microglobulin enhances the infectivity of cytomegalovirus and when bound to the virus enables class I HLA molecules to be used as a virus receptor. J. Gen. Virol. *68*, 793–803.

Grundy, J.E., McKeating, J.A., Sanderson, A.R., and Griffiths, P.D. (1988). Cytomegalovirus and beta 2 microglobulin in urine specimens. Reciprocal interference in their detection is responsible for artifactually high levels of urinary beta 2 microglobulin in infected transplant recipients. Transplantation *45*, 1075–1079.

Hahn, G., Revello, M.G., Patrone, M., Percivalle, E., Campanini, G., Sarasini, A., Wagner, M., Gallina, A., Milanesi, G., Koszinowski, U., et al. (2004). Human cytomegalovirus UL131-128 genes are indispensable for virus growth in endothelial cells and virus transfer to leukocytes. J. Virol. *78*, 10023–10033.

Heil, F., Hemmi, H., Hochrein, H., Ampenberger, F., Kirschning, C., Akira, S., Lipford, G., Wagner, H., and Bauer, S. (2004). Species-specific recognition of single-stranded RNA via toll-like receptor 7 and 8. Science *303*, 1526–1529.

Hemmi, H., Takeuchi, O., Kawai, T., Kaisho, T., Sato, S., Sanjo, H., Matsumoto, M., Hoshino, K., Wagner, H., Takeda, K., et al. (2000). A Toll-like receptor recognizes bacterial DNA. Nature *408*, 740–745.

Hobom, U., Brune, W., Messerle, M., Hahn., G., and Koszinowski, U. (2000). Fast screening procedures for random transposon libraries of cloned herpesvirus genomes: mutational analysis of human cytomegalovirus envelope glycoprotein genes. J. Virol. *74*, 7720–7729.

Honda, K., Yanai, H., Mizutani, T., Negishi, H., Shimada, N., Suzuki, N., Ohba, Y., Takaoka, A., Yeh, W.C., and Taniguchi, T. (2004). Role of a transductional-transcriptional processor complex involving MyD88 and IRF-7 in Toll-like receptor signaling. Proc. Natl. Acad. Sci. U.S.A. *101*, 15416–15421.

Hornung, V., Rothenfusser, S., Britsch, S., Krug, A., Jahrsdorfer, B., Giese, T., Endres, S., and Hartmann, G. (2002). Quantitative expression of toll-like receptor 1–10 mRNA in cellular subsets of human peripheral blood mononuclear cells and sensitivity to CpG oligodeoxynucleotides. J. Immunol. 168, 4531–4537.

Huber, M.T., and Compton, T. (1997). Characterization of a novel third member of the human cytomegalovirus glycoprotein H-glycoprotein L complex. J. Virol. 71, 5391–5398.

Huber, M.T., and Compton, T. (1998). The human cytomegalovirus UL74 gene encodes the third component of the glycoprotein H-glycoprotein L-containing envelope complex. J. Virol. 72, 8191–8197.

Ibanez, C.E., Schrier, R., Ghazal, P., Wiley, C., and Nelson, J.A. (1991). Human cytomegalovirus productively infects primary differentiated macrophages. J. Virol. 65, 6581–6588.

Ihara, S., Saito, S., and Watanabe, Y. (1982). Suppression of fibronectin synthesis by an early function(s) of human cytomegalovirus. J. Gen. Virol. 59, 409–413.

Isaacson, M.K., Feire, A.L., and Compton, T. (2007). Epidermal growth factor receptor is not required for human cytomegalovirus entry or signaling. J. Virol. 81, 6241–6247.

Jones, P.L., Crack, J., and Rabinovitch, M. (1997). Regulation of tenascin-C, a vascular smooth muscle cell survival factor that interacts with the alpha v beta 3 integrin to promote epidermal growth factor receptor phosphorylation and growth. J. Cell. Biol. 139, 279–293.

Jones, T.R., Lee, S.W., Johann, S.V., Razinkov, V., Visalli, R.J., Feld, B., Bloom, J.D., and O'Connell, J. (2004). Specific inhibition of human cytomegalovirus glycoprotein B-mediated fusion by a novel thiourea small molecule. J. Virol. 78, 1289–1300.

Kari, B., and Gehrz, R. (1992). A human cytomegalovirus glycoprotein complex designated gC-II is a major heparin-binding component of the envelope. J. Virol. 66, 1761–1764.

Kari, B., and Gehrz, R. (1993). Structure, composition and heparin binding properties of a human cytomegalovirus glycoprotein complex designated gC-II. J. Gen. Virol. 74, 255–264.

Kawai, T., Sato, S., Ishii, K.J., Coban, C., Hemmi, H., Yamamoto, M., Terai, K., Matsuda, M., Inoue, J., Uematsu, S., et al. (2004). Interferon-alpha induction through Toll-like receptors involves a direct interaction of IRF7 with MyD88 and TRAF6. Nat. Immunol. 5, 1061–1068.

Keay, S., and Baldwin, B. (1991). Anti-idiotype antibodies that mimic gp86 of human cytomegalovirus inhibit viral fusion but not attachment. J. Virol. 65, 5124–5128.

Keay, S., and Baldwin, B. (1992). The human fibroblast receptor for gp86 of human cytomegalovirus is a phosphorylated glycoprotein. J. Virol. 66, 4834–4838.

Keay, S., and Baldwin, B.R. (1996). Evidence for the role of cell protein phosphorylation in human cytomegalovirus/host cell fusion. J. Gen. Virol. 77, 2597–2604.

Keay, S., Rasmussen, L., and Merigan, T.C. (1988). Syngeneic monoclonal anti-idiotype antibodies that bear the internal image of a human cytomegalovirus neutralization epitope. J. Immunol. 140, 944–948.

Keay, S., Merigan, T.C., and Rasmussen, L. (1989). Identification of cell surface receptors for the 86-kilodalton glycoprotein of human cytomegalovirus. Proc. Natl. Acad. Sci. U.S.A. 86, 10100–10103.

Keay, S., Baldwin, B.R., Smith, M.W., Wasserman, S.S., and Goldman, W.F. (1995). Increases in [Ca2+]i mediated by the 92.5-kDa putative cell membrane receptor for HCMV gp86. Am. J. Physiol. 269, C11–21.

Kowalik, T.F., Wing, B., Haskill, J.S., Azizkhan, J.C., Baldwin, A.S., Jr., and Huang, E.S. (1993). Multiple mechanisms are implicated in the regulation of NF-kappa B activity during human cytomegalovirus infection. Proc. Natl. Acad. Sci. U.S.A. 90, 1107–1111.

Krug, A., French, A.R., Barchet, W., Fischer, J.A., Dzionek, A., Pingel, J.T., Orihuela, M.M., Akira, S., Yokoyama, W.M., and Colonna, M. (2004a). TLR9-dependent recognition of MCMV by IPC and DC generates coordinated cytokine responses that activate antiviral NK cell function. Immunity 21, 107–119.

Krug, A., Luker, G.D., Barchet, W., Leib, D.A., Akira, S., and Colonna, M. (2004b). Herpes simplex virus type 1 activates murine natural interferon-producing cells through toll-like receptor 9. Blood 103, 1433–1437.

Larsson, S., Soderberg-Naucler, C., Wang, F.Z., and Moller, E. (1998). Cytomegalovirus DNA can be detected in peripheral blood mononuclear cells from all seropositive and most seronegative healthy blood donors over time. Transfusion 38, 271–278.

Li, L., Nelson, J.A., and Britt, W.J. (1997). Glycoprotein H-related complexes of human cytomegalovirus: Identification of a third protein in the gCIII complex. J. Virol. 71, 3090–3097.

Lopper, M., and Compton, T. (2004). Functional coiled coils distributed among glycoprotein B and H in human cytomegalovirus membrane fusion. J. Virol. 78, 8333–8341.

Lund, J., Sato, A., Akira, S., Medzhitov, R., and Iwasaki, A. (2003). Toll-like receptor 9-mediated recognition of Herpes simplex virus-2 by plasmacytoid dendritic cells. J. Exp. Med. 198, 513–520.

Lund, J.M., Alexopoulou, L., Sato, A., Karow, M., Adams, N.C., Gale, N.W., Iwasaki, A., and Flavell, R.A. (2004). Recognition of single-stranded RNA viruses by Toll-like receptor 7. Proc. Natl. Acad. Sci. U.S.A. 101, 5598–5603.

Mach, M., Kropff, B., Dal Monte, P., and Britt, W. (2000). Complex formation by human cytomegalovirus glycoproteins M (gpUL100) and N (gpUL73). J. Virol. 74, 11881–11892.

McKeating, J.A., Grundy, J.E., Varghese, Z., and Griffiths, P.D. (1986). Detection of cytomegalovirus by ELISA in urine samples is inhibited by beta 2 microglobulin. J. Med. Virol. 18, 341–348.

McKeating, J.A., Griffiths, P.D., and Grundy, J.E. (1987). Cytomegalovirus in urine specimens has host beta 2 microglobulin bound to the viral envelope: a mechanism of evading the host immune response? J. Gen. Virol. 68, 785–792.

Medzhitov, R., Preston-Hurlburt, P., and Janeway, C.A., Jr. (1997). A human homologue of the Drosophila Toll protein signals activation of adaptive immunity. Nature 388, 394–397.

Miyamoto, S., Teramoto, H., Gutkind, J.S., and Yamada, K.M. (1996). Integrins can collaborate with growth factors for phosphorylation of receptor tyrosine kinases and MAP kinase activation: roles of integrin aggregation and occupancy of receptors. J. Cell Biol. 135, 1633–1642.

Moro, L., Venturino, M., Bozzo, C., Silengo, L., Altruda, F., Beguinot, L., Tarone, G., and Defilippi, P. (1998). Integrins induce activation of EGF receptor: role in MAP kinase induction and adhesion-dependent cell survival. EMBO J. *17*, 6622–6632.

Myerson, D., Hackman, R.C., Nelson, J.A., Ward, D.C., and McDougall, J.K. (1984). Widespread presence of histologically occult cytomegalovirus. Hum. Pathol. *15*, 430–439.

Navarro, L., Mowen, K., Rodems, S., Weaver, B., Reich, N., Spector, D., and David, M. (1998). Cytomegalovirus activates interferon immediate-early response gene expression and an interferon regulatory factor 3-containing interferon-stimulated response element-binding complex. Mol. Cell. Biol. *18*, 3796–3802.

Netterwald, J.R., Jones, T.R., Britt, W.J., Yang, S.J., McCrone, I.P., and Zhu, H. (2004). Postattachment events associated with viral entry are necessary for induction of interferon-stimulated genes by human cytomegalovirus. J. Virol. *78*, 6688–6691.

Nowlin, D.M., Cooper, N.R., and Compton, T. (1991). Expression of a human cytomegalovirus receptor correlates with infectibility of cells. J. Virol. *65*, 3114–3121.

O'Neill, L.A., Fitzgerald, K.A., and Bowie, A.G. (2003). The Toll-IL-1 receptor adaptor family grows to five members. Trends Immunol. *24*, 286–290.

Ogawa, T., Asai, Y., Hashimoto, M., and Uchida, H. (2002). Bacterial fimbriae activate human peripheral blood monocytes utilizing TLR2, CD14 and CD11a/CD18 as cellular receptors. Eur. J. Immunol. *32*, 2543–2550.

Perussia, B., Fanning, V., and Trinchieri, G. (1985). A leukocyte subset bearing HLA-DR antigens is responsible for in vitro alpha interferon production in response to viruses. Nat. Immun. Cell Growth Regul. *4*, 120–137.

Pietropaolo, R., and Compton, T. (1997). Direct interaction between human cytomegalovirus glycoprotein B and cellular annexin II. J. Virol. *71*, 9803–9807.

Pietropaolo, R., and Compton, T. (1999). Interference with annexin II has no effect on entry of human cytomegalovirus into fibroblast cells. J. Gen. Virol. *80*, 1807–1816.

Pleskoff, O., Treboute, C., Brelot, A., Heveker, N., Seman, M., and Alizon, M. (1997). Identification of a chemokine receptor encoded by human cytomegalovirus as a cofactor for HIV-1 entry. Science *276*, 1874–1878.

Pleskoff, O., Treboute, C., and Alizon, M. (1998). The cytomegalovirus-encoded chemokine receptor US28 can enhance cell–cell fusion mediated by different viral proteins. J. Virol. *72*, 6389–6397.

Polic, B., Jonjic, S., Pavic, I., Crnkovic, I., Zorica, I., Hengel, H., Lucin, P., and Koszinowski, U.H. (1996). Lack of MHC class I complex expression has no effect on spread and control of cytomegalovirus infection in vivo. J. Gen. Virol. *77*, 217–225.

Poltorak, A., He, X., Smirnova, I., Liu, M.Y., Huffel, C.V., Du, X., Birdwell, D., Alejos, E., Silva, M., Galanos, C., et al. (1998a). Defective LPS signaling in C3H/HeJ and C57BL/10ScCr mice: mutations in Tlr4 gene. Science *282*, 2085–2088.

Poltorak, A., Smirnova, I., He, X., Liu, M.Y., Van Huffel, C., McNally, O., Birdwell, D., Alejos, E., Silva, M., Du, X., et al. (1998b). Genetic and physical mapping of the Lps locus: identification of the toll-4 receptor as a candidate gene in the critical region. Blood Cells Mol. Dis. *24*, 340–355.

Preston, C.M., Harman, A.N., and Nicholl, M.J. (2001). Activation of interferon response factor-3 in human

cells infected with herpes simplex virus type 1 or human cytomegalovirus. J. Virol. *75*, 8909–8916.

Qureshi, S.T., Lariviere, L., Leveque, G., Clermont, S., Moore, K.J., Gros, P., and Malo, D. (1999). Endotoxin-tolerant mice have mutations in Toll-like receptor 4 (Tlr4). J. Exp. Med. *189*, 615–625.

Rassa, J.C., Meyers, J.L., Zhang, Y., Kudaravalli, R., and Ross, S.R. (2002). Murine retroviruses activate B cells via interaction with Toll-like receptor 4. Proc. Natl. Acad. Sci. U.S.A. *99*, 2281–2286.

Raynor, C.M., Wright, J.F., Waisman, D.M., and Pryzdial, E.L. (1999). Annexin II enhances cytomegalovirus binding and fusion to phospholipid membranes. Biochemistry *38*, 5089–5095.

Rock, F.L., Hardiman, G., Timans, J.C., Kastelein, R.A., and Bazan, J.F. (1998). A family of human receptors structurally related to Drosophila Toll. Proc. Natl. Acad. Sci. U.S.A. *95*, 588–593.

Ryckman, B.J., Jarvis, M.A., Drummond, D.D., Nelson, J.A., and Johnson, D.C. (2006). Human cytomegalovirus entry into epithelial and endothelial cells depends on genes UL128 to UL150 and occurs by endocytosis and low-pH fusion. J. Virol. *80*, 710–722.

Ryckman, B.J., Rainish, B.L., Chase, M.C., Borton, J.A., Nelson, J.A., Jarvis, M.A., and Johnson, D.C. (2008). Characterization of the human cytomegalovirus gH/gL/UL128–131 complex that mediates entry into epithelial and endothelial cells. J. Virol. *82*, 60–70.

Salazar-Mather, T.P., Lewis, C.A., and Biron, C.A. (2002). Type I interferons regulate inflammatory cell trafficking and macrophage inflammatory protein 1alpha delivery to the liver. J. Clin. Invest. *110*, 321–330.

Santoro, F., Kennedy, P.E., Locatelli, G., Malnati, M.S., Berger, E.A., and Lusso, P. (1999). CD46 is a cellular receptor for human herpesvirus 6. Cell *99*, 817–827.

Sato, M., Hata, N., Asagiri, M., Nakaya, T., Taniguchi, T., and Tanaka, N. (1998). Positive feedback regulation of type I IFN genes by the IFN-inducible transcription factor IRF-7. FEBS Lett. *441*, 106–110.

Sato, M., Suemori, H., Hata, N., Asagiri, M., Ogasawara, K., Nakao, K., Nakaya, T., Katsuki, M., Noguchi, S., Tanaka, N., et al. (2000). Distinct and essential roles of transcription factors IRF-3 and IRF-7 in response to viruses for IFN-alpha/beta gene induction. Immunity *13*, 539–548.

Servant, M.J., ten Oever, B., LePage, C., Conti, L., Gessani, S., Julkunen, I., Lin, R., and Hiscott, J. (2001). Identification of distinct signaling pathways leading to the phosphorylation of interferon regulatory factor 3. J. Biol. Chem. *276*, 355–363.

Sharma, S., TenOever, B.R., Grandvaux, N., Zhou, G.P., Lin, R., and Hiscott, J. (2003). Triggering the interferon antiviral response through an IKK-related pathway. Science *300*, 1148–52.

Siegal, F.P., Kadowaki, N., Shodell, M., Fitzgerald-Bocarsly, P.A., Shah, K., Ho, S., Antonenko, S., and Liu, Y.J. (1999). The nature of the principal type 1 interferon-producing cells in human blood. Science *284*, 1835–1837.

Simmen, K.A., Singh, J., Luukkonen, B.G., Lopper, M., Bittner, A., Miller, N.E., Jackson, M.R., Compton, T., and Fruh, K. (2001). Global modulation of cellular transcription by human cytomegalovirus is initiated by viral glycoprotein B. Proc. Natl. Acad. Sci. U.S.A. *98*, 7140–7145.

Sinzger, C., Kahl, M., Laib, K., Klingel, K., Rieger, P., Plachter, B., and Jahn, G. (2000). Tropism of human cytomegalovirus for endothelial cells is determined by a post-entry step dependent on efficient translocation to the nucleus. J. Gen. Virol. *81*, 3021–3035.

Soderberg, C., Giugni, T.D., Zaia, J.A., Larsson, S., Wahlberg, J.M., and Moller, E. (1993a). CD13 (human aminopeptidase-N) mediates human cytomegalovirus infection. J. Virol. *67*, 6576–6585.

Soderberg, C., Larsson, S., Bergstedtlindqvist, S., and Moller, E. (1993b). Definition of a subset of human peripheral blood mononuclear cells that are permissive to human cytomegalovirus infection. J. Virol. *67*, 3166–3175.

Soroceanu, L., Akhavan, A., and Cobbs, C.S. (2008). Platelet-derived growth factor-alpha receptor activation is required for human cytomegalovirus infection. Nature *455*, 391–395.

Spear, P.G., and Longnecker, R. (2003). Herpesvirus entry: an update. J. Virol. *77*, 10179–10185.

Stannard, L.M. (1989). Beta 2 microglobulin binds to the tegument of cytomegalovirus: an immunogold study. J. Gen. Virol. *70*, 2179–2184.

Stark, G.R., Kerr, I.M., Williams, B.R., Silverman, R.H., and Schreiber, R.D. (1998). How cells respond to interferons. Annu. Rev. Biochem. *67*, 227–264.

Stone, A.L., Kroeger, M., and Sang, Q.X. (1999). Structure-function analysis of the ADAM family of disintegrin-like and metalloproteinase-containing proteins (review). J. Protein Chem. *18*, 447–465.

Suhara, W., Yoneyama, M., Kitabayashi, I., and Fujita, T. (2002). Direct involvement of CREB-binding protein/p300 in sequence-specific DNA binding of virus-activated interferon regulatory factor-3 holocomplex. J. Biol. Chem. *277*, 22304–22313.

Tabeta, K., Georgel, P., Janssen, E., Du, X., Hoebe, K., Crozat, K., Mudd, S., Shamel, L., Sovath, S., Goode, J., et al. (2004). Toll-like receptors 9 and 3 as essential components of innate immune defense against mouse cytomegalovirus infection. Proc. Natl. Acad. Sci. U.S.A. *101*, 3516–3521.

Takeda, K., and Akira, S. (2003). Toll receptors and pathogen resistance. Cell. Microbiol. *5*, 143–153.

Takeuchi, O., and Akira, S. (2001). Toll-like receptors; their physiological role and signal transduction system. Int. Immunopharmacol. *1*, 625–635.

Takeuchi, O., Hoshino, K., Kawai, T., Sanjo, H., Takada, H., Ogawa, T., Takeda, K., and Akira, S. (1999). Differential roles of TLR2 and TLR4 in recognition of gram-negative and gram-positive bacterial cell wall components. Immunity *11*, 443–451.

Taniguchi, T., and Takaoka, A. (2002). The interferon-alpha/beta system in antiviral responses: a multimodal machinery of gene regulation by the IRF family of transcription factors. Curr. Opin. Immunol. *14*, 111–116.

Taniguchi, T., Ogasawara, K., Takaoka, A., and Tanaka, N. (2001). IRF family of transcription factors as regulators of host defense. Annu. Rev. Immunol. *19*, 623–655.

Taylor, H.P., and Cooper, N.R. (1990). The human cytomegalovirus receptor on fibroblasts is a 30-kilodalton membrane protein. J. Virol. *64*, 2484–2490.

Triantafilou, M., Miyake, K., Golenbock, D.T., and Triantafilou, K. (2002). Mediators of innate immune recognition of bacteria concentrate in lipid rafts and facilitate lipopolysaccharide-induced cell activation. J. Cell. Sci. *115*, 2603–2611.

Tugizov, S., Navarro, D., Paz, P., Wang, Y.L., Qadri, I., and Pereira, L. (1994). Function of human cytomegalovirus glycoprotein B: Syncytium formation in cells constitutively

expressing gB is blocked by virus-neutralizing antibodies. Virology *201*, 263–276.

Utz, U., Britt, W., Vugler, L., and Mach, M. (1989). Identification of a neutralizing epitope on glycoprotein gp58 of human cytomegalvirus. J. Virol. *63*, 1995–2001.

Wagner, H. (2004). The immunobiology of the TLR9 subfamily. Trends Immunol. *25*, 381–386.

Wang, D., and Shenk, T. (2005). Human cytomegalovirus virion protein complex required for epithelial and endothelial cell tropism. Proc. Natl. Acad. Sci. U.S.A. *102*, 18153–18158.

Warren, A.P., Owens, C.N., Borysiewicz, L.K., and Patel, K. (1994). Down-regulation of integrin alpha 1/beta 1 expression and association with cell rounding in human cytomegalovirus-infected fibroblasts. J. Gen. Virol. *75*, 3319–3325.

Weck, P.K., Apperson, S., May, L., and Stebbing, N. (1981). Comparison of the antiviral activities of various cloned human interferon-alpha subtypes in mammalian cell cultures. J. Gen. Virol. *57*, 233–237.

White, J.M. (2003). ADAMs: modulators of cell–cell and cell–matrix interactions. Curr. Opin. Cell Biol. *15*, 598–606.

Wolfsberg, T.G., Primakoff, P., Myles, D.G., and White, J.M. (1995). ADAM, a novel family of membrane proteins containing a disintegrin and metalloprotease domain: multipotential functions in cell–cell and cell–matrix interactions. J. Cell. Biol. *131*, 275–278.

Wright, J.F., Kurosky, A., and Wasi, S. (1994). An endothelial cell-surface form of annexin II binds human cytomegalovirus. Biochem. Biophys. Res. Commun. *198*, 983–989.

Wright, J.F., Kurosky, A., Pryzdial, E.L., and Wasi, S. (1995). Host cellular annexin II is associated with cytomegalovirus particles isolated from cultured human fibroblasts. J. Virol. *69*, 4784–4791.

Wright, R., Kurosky, A., and Wasi, S. (1993). Annexin II associated with human cytomegalovirus particles: possible implications for cell infectivity. FASEB J. *7*, A1301.

Wu, Q.H., Trymbulak, W., Tatake, R.J., Forman, S.J., Zeff, R.A., and Shanley, J.D. (1994). Replication of human cytomegalovirus in cells deficient in beta(2)-microglobulin gene expression. J. Gen. Virol. *75*, 2755–2759.

Yoneyama, M., Suhara, W., Fukuhara, Y., Fukuda, M., Nishida, E., and Fujita, T. (1998). Direct triggering of the type I interferon system by virus infection: activation of a transcription factor complex containing IRF-3 and CBP/p300. EMBO J. *17*, 1087–1095.

Yu, D., Smith, G.A., Enquist, L.W., and Shenk, T. (2002). Construction of a self-excisable bacterial artificial chromosome containing the human cytomegalovirus genome and mutagenesis of the diploid TRL/IRL13 gene. J. Virol. *76*, 2316–2328.

Yurochko, A.D., and Huang, E.S. (1999). Human cytomegalovirus binding to human monocytes induces immunoregulatory gene expression. J. Immunol. *162*, 4806–4816.

Yurochko, A.D., Kowalik, T.F., Huong, S.M., and Huang, E.S. (1995). Human cytomegalovirus up-regulates NF-kappa B activity by transactivating the NF-kappa B p105/p50 and p65 promoters. J. Virol. *69*, 5391–5400.

Yurochko, A.D., Hwang, E.S., Rasmussen, L., Keay, S., Pereira, L., and Huang, E.S. (1997a). The human cytomegalovirus

UL55 (gB) and UL75 (gH) glycoprotein ligands initiate the rapid activation of Sp1 and NF-kappaB during infection. J. Virol. *71*, 5051–5059.

Yurochko, A.D., Mayo, D.W., Poma, E.E., Baldwin, A.S., and Huang, E. (1997b). Inudction of the transcription factor Sp1 during human cytomegalovirus infection mediates up-regulation of the p65 and p105/p50 NF-kB promoters. J. Virol. *71*, 4638–4648.

Zarember, K.A., and Godowski, P.J. (2002). Tissue expression of human Toll-like receptors and differential regulation of Toll-like receptor mRNAs in leukocytes in response to microbes, their products, and cytokines. J. Immunol. *168*, 554–561.

Zhang, D., Zhang, G., Hayden, M.S., Greenblatt, M.B., Bussey, C., Flavell, R.A., and Ghosh, S. (2004). A toll-like receptor that prevents infection by uropathogenic bacteria. Science *303*, 1522–1526.

Zhu, H., Cong, J.P., and Shenk, T. (1997). Use of differential display analysis to assess the effect of human cytomegalovirus infection on the accumulation of cellular RNAs: induction of interferon-responsive RNAs. Proc. Natl. Acad. Sci. U.S.A. *94*, 13985–13990.

Zhu, H., Cong, J.P., Mamtora, G., Gingeras, T., and Shenk, T. (1998). Cellular gene expression altered by human cytomegalovirus: global monitoring with oligonucleotide arrays. Proc. Natl. Acad. Sci. U.S.A. *95*, 14470–14475.

Pre-Immediate Early Tegument Protein Functions

1.9

Robert F. Kalejta

Abstract

As virions disassemble during viral entry, they must expertly navigate and manage the complex and unwelcoming environments they encounter in order to successfully infect host cells. Herpesviruses incorporate proteins into their virions in a layer between the capsid and envelope termed the tegument to assist in this hostile takeover. When delivered to infected cells subsequent to membrane fusion, tegument proteins begin to facilitate viral infection after entry but before immediate-early (IE) gene expression (referred to as the pre-IE stage of infection). Tegument-delivered proteins mediate capsid migration through the cytoplasm to nuclear pore complexes and the transmission of the genome into the nucleus. Furthermore, they modulate viral transcription, and help infected cells avoid all three classes of immune function (intrinsic, innate and adaptive). While they are most often studied during lytic infections, a new appreciation for the role that the proper regulation of tegument-delivered protein function may play during viral latency is emerging. Here the pre-IE functions of tegument proteins during both lytic and latent infections are reviewed and analysed.

Introduction

Tegument proteins are those incorporated into herpesvirus virions between the genome-containing capsid and the glycoprotein-containing envelope. They are deposited into the cytoplasm of the infected cell upon fusion of the viral and cellular membranes, and as such are the very first viral proteins that act intracellularly during viral infections. As discussed below, they have important activities in priming cells for viral gene expression during a productive, lytic infection (a function that must be suppressed to allow for the establishment of latency) and in helping the infected cell avoiding immune surveillance.

Herpesviruses are prominent human pathogens, with human cytomegalovirus (HCMV) likely the most ubiquitous. Primary HCMV infection of the unborn causes birth defects, and reactivations of latent HCMV causes severe disease in immunocompromised or immunosuppressed populations, such as AIDS, cancer, and transplant patients (Mocarski et al., 2007). HCMV infection is also associated with chronic conditions such as immunosenescence (Pawelec and Derhovanessian, 2011), atherosclerosis, and restenosis (Caposio et al., 2011). The virus is also present in glioblastoma multiforme tumours (Ranganathan et al., 2012; Soroceanu and Cobbs, 2011), though the consequences of these infections for tumour biology remain unclear. In addition to the productive, lytic infection in which high levels of the majority of HCMV proteins are synthesized and infectious progeny virions are assembled and released, HCMV also establishes persistent and latent infections. Persistent infections are productive, however lower levels of proteins and infectious particles are produced, and the infected cell survives significantly longer than a lytically infected cell (Britt, 2008). During latency viral lytic gene expression, specifically the immediate-early (IE) genes, must be silenced, and the genome is maintained over time without the production of infectious virions (Sinclair, 2010). In response to certain stimuli, latent virus can reactivate and complete a productive, lytic infection.

This chapter briefly covers the genesis and organization of the tegument, but focuses on the biological effects of tegument proteins prior to the onset of viral gene expression. As the first viral genes to be expressed are the IE genes, the time period before their synthesis is called the pre-immediate-early (pre-IE) stage of infection. Please note that in addition to the pre-IE functions of the tegument-delivered proteins discussed below, de novo expressed tegument proteins also play substantial roles throughout all phases of a productive HCMV infection. For those seeking further information, readers are directed to a previous review of this topic (Kalejta, 2008a) as well as to a more thorough review of tegument protein function throughout the course of viral infection (Kalejta, 2008b).

Tegument composition

Approximately 35 viral proteins are consistently found in HCMV teguments (Kalejta, 2008b). Many are phosphorylated (and named for their apparent molecular mass with a 'pp' prefix that stands for phosphoprotein) although for most, any significance of their phosphorylation status to their function is unknown. Tegument proteins act throughout the infectious cycle, including during entry, gene expression, immune evasion, assembly, and egress. Along with viral proteins, cellular proteins and viral and cellular RNAs are also packaged within virions.

Tegument assembly

Mechanisms through which tegument proteins are packaged into virions are poorly understood. Bioinformatic and experimental searches for short, linear amino acid motifs that could mediate tegument localization, akin to a nuclear localization signal, have failed to reveal a universal packaging sequence. Sporadic examples of specific tegument proteins being underrepresented in the virions of recombinant viruses lacking a different tegument protein have been reported, but they fail to illuminate any conserved mechanism of tegument protein incorporation, calling into question whether or not such a mechanism exists. Interestingly, recent evidence appears to indicate that independent derivatives of the same viral strain (i.e. different isolates of AD169) can have substantially different tegument protein compositions (Reyda et al., 2011) further confounding issues dealing with tegument assembly.

In the absence of mechanistic data addressing tegument assembly, a stepwise protein–protein interaction cascade is the current working model for tegumentation. This model is based on the observations that some tegument proteins are nuclear while others are cytoplasmic, and that capsids form in the nucleus but are enveloped with their final membrane in the cytoplasm. In attempts to formulate a more defined model, high-throughput tegument protein interaction yeast two-hybrid screens have been employed to identify those tegument proteins that interact with each other, and those that interact with capsids or envelope glycoproteins (Phillips and Bresnahan, 2011; To et al., 2011). The two screens detected a core set of overlapping interactions, as well as unique binding pairs found in one screen but not the other. Both screens confirmed binary interactions between pp71, pp150, UL35 and UL94. It will be interesting to see if these proteins

form a higher order complex that might be essential for tegument assembly or for proper tegument protein function. One screen (To et al., 2011) detected a substantial number of interactions between UL24, UL25, UL89 and other virion proteins, suggesting that these three proteins may serve as organizing centres during tegument assembly. However, both UL24 and UL25 are non-essential for viral replication in vitro (Dunn et al., 2003; Yu et al., 2003) making them unlikely to be significant tegument organizers. Likewise, the UL89 protein that is part of the terminase complex required for genome cleavage and packaging is present in vanishingly small quantities within the tegument (Varnum et al. 2004) and thus unlikely to play a significant role in tegument organization. In summary these interaction screens have provided a wealth of data to fuel and inspire future research efforts, but by themselves do not provide any specific insight into the process of tegumentation.

Tegument structure

Visualization by cryo-electron tomography (cryoET) and biochemical fractionation has differentiated two general classes of tegument proteins, those that are densely packed and tightly associated with the capsid, and those that are loosely packed and not detectably associated with capsids (Yu et al., 2011). The tegument proteins found strongly associated with capsids were pp150 (UL32) and UL48, with UL47 also substantially capsid-associated, but apparently not as tightly bound as pp150 or UL48. CryoET experiments combined with pp150 antibody labelling detected at least some structurally ordered pp150 molecules directly adjacent to the capsid. Thus it appears that tegument proteins involved in capsid stability and movement (see below) are strongly capsid associated, whereas those that have other cytoplasmic and nuclear functions (see below) are weakly or not at all associated with capsids.

Tegument disassembly

Just as the tegument must assemble upon virion egress, it must also disassemble upon viral entry. How the tegument disassembles after entry is unknown, as is whether or not this process is similar in instances where HCMV enters by membrane fusion or by endocytosis (Ryckman et al., 2006). Specific proteins such as pp150, UL96 (Tandon and Mocarski, 2011) and UL26 (Munger et al., 2006) that appear to regulate virion stability may, in part, function through effects on the tegument, and thus may impact tegument disassembly. One clear observation is that upon infection of fibroblasts, some tegument proteins such as pp150 remain capsid associated and cytoplasmic, whereas others, such as pp65 and pp71 rapidly enter the nucleus. Interestingly, tegument-delivered pp65 and pp71 fail to enter nuclei upon infection of undifferentiated NT2 cells (Penkert and Kalejta, 2010) and CD34⁺ cells (Saffert *et al.*, 2010; E.R. Albright and R.F. Kalejta, unpublished observations) where HCMV establishes quiescent or latent infections, respectively. Whether this cytoplasmic localization of normally nuclear proteins represents a defect in trafficking of these two individual tegument components or a more global defect in tegument disassembly remains to be determined.

Capsid delivery to nuclear pores

Membrane fusion deposits the genome-containing capsid into the cytoplasm, where it, along with closely associated tegument proteins, makes its way along cellular microtubules to nuclear pore complexes (Ogawa-Goto *et al.*, 2003; Kalejta, 2008a), allowing for entry of the viral genome into the nucleus during the initiation of a lytic infection (Fig. I.9.1). If microtubules promote genome delivery to the nucleus during the establishment of latency, or if tegument proteins promote HCMV capsid movement during the entry process in any cell type has not been directly studied. However, during herpes simplex virus type 1 (HSV-1) lytic infections, the VP1/2 (UL36) tegument protein is responsible for trafficking capsids along microtubules and mediating the release of genomes into the nucleus (Roberts *et al.*, 2009; Abaitua *et al.*, 2011).

The HCMV orthologue of VP1/2 is the UL48 protein. UL48 is a deubiquitinating protease (Wang *et al.*, 2006) with both ubiquitin-specific carboxy-terminal hydrolase and isopeptidase activity (Kim *et al.*, 2009). Catalytic site mutants show only mild growth phenotypes at low multiplicities of infection (Wang *et al.*, 2006; Kim *et al.*, 2009), so the significance

of the deubiquitinating activity of this protein for viral replication is not currently appreciated. The protein itself however is critical for viral infection, as shown by independent mutational analysis (Dunn *et al.*, 2003; Yu *et al.*, 2003).

UL48 interacts with another tegument protein, UL47, as well as with the major capsid protein (Bechtel and Shenk, 2002). UL47-null mutant viruses replicate 100-fold less than wild type viruses and show delays in viral IE gene expression, implying a role for this protein during the pre-IE stage of virus infection. However, UL47-null viruses contain less UL48 within their teguments (Bechtel and Shenk, 2002), so it is premature to assign a pre-IE function to UL47 based on these experiments. Interestingly, recent studies with HSV-1 argue that only the UL48 orthologue (HSV-1 VP1/2 (UL36)), and not the UL47 orthologue (HSV-1 UL37) is required for capsid trafficking to nuclei during the entry process (Roberts *et al.*, 2009). Further experimentation is required to determine if the same holds true for HCMV, and if UL47 and/or UL48 play roles during the assembly and egress of infectious HCMV virions.

Recently, the UL77 protein has been implicated in pre-IE events. UL77 is orthologous to the HSV-1 UL25 protein, overexpression of which impairs HSV-1 IE gene expression without disrupting the targeting of incoming capsids to the nuclear pores (Rode *et al.*, 2011). Fewer HSV-1 genomes were found in the nuclei of cells overexpressing UL25 as compared to control cells, leading to the speculation that UL25 plays some role in the efficient delivery of viral genomes to the nucleus. Quantitative immunofluorescence studies concluded that overexpression of UL77 in HSV-1 infected cells reduced ICP8 expression, although the magnitude of inhibition appeared to be minor. In this experiment, effects on nuclear pore docking or genome delivery to the nucleus were not analysed (Rode *et*

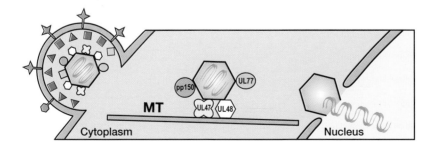

Figure I.9.1 Genome delivery to the nucleus. After being released from virions by membrane fusion, capsids and tightly associated tegument proteins travel along microtubules (MT) to nuclear pores, where the viral genome is released into the nucleus. Likely roles for tegument proteins during this process are capsid stabilization (pp150), interaction with MTs and cellular motor proteins (UL47 and UL48), and facilitating genome delivery (UL77).

al., 2011). Nevertheless, it is likely that HCMV UL77 participates in the efficient nuclear delivery of infecting viral genomes. Little more is known about UL77 (Kalejta, 2008b) except that it is an essential gene.

Activation of IE gene expression

Upon entry into the nucleus, the HCMV genome is immediately silenced by a cellular intrinsic immune defence (Fig. I.9.2A) mediated in large part by proteins that localize to PML nuclear bodies (Kalejta, 2008a; Tavalai and Stamminger, 2011). Intrinsic immune defences (Bieniasz, 2004; Neil and Bieniasz, 2009) are mediated by constitutively expressed proteins, and were originally discovered as retroviral restriction factors. HCMV was the first DNA virus found to be subject to intrinsic immunity (Saffert and Kalejta, 2006), but now it is appreciated that other DNA viruses are also controlled by such defences (Tavalai and Stamminger, 2008).

HCMV DNA is bereft of histones in virions, but becomes rapidly chromatinized upon entry into the nucleus (Nevels et al., 2011). Initially, the histones associated with incoming viral genomes bear post-translational modifications consistent with transcriptionally inactive heterochromatin (Sinclair, 2010; Reeves, 2011). The cellular intrinsic defence against HCMV plays a role in the formation of this repressive chromatin structure (Woodhall et al., 2006). However, an open chromatin structure indicative of active transcription soon replaces the initial repressive histone markings, and the tegument protein pp71 is responsible for initiating the transformation from silenced to active viral transcription (Kalejta, 2008a,b).

Upon viral entry into cells where a lytic viral replication cycle will initiate, tegument-delivered pp71 rapidly migrates to the nucleus where it interacts with (Hoffman et al., 2002) and degrades (Saffert and Kalejta, 2006) the cellular transcriptional co-repressor Daxx (Fig. I.9.2B). In the absence of sufficient Daxx neutralization (e.g. during Daxx overexpression, proteasome inhibition, or infection with a pp71-deficient virus), the histones associated with viral genomes

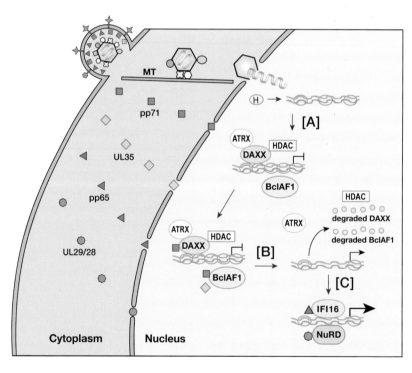

Figure I.9.2 De-repression and activation of IE gene expression. A. Nuclear genomes are chromatinized by their association with cellular histones (H) and then transcriptionally silenced by an intrinsic defence mediated by Daxx, histone deacetylases (HDACs), ATRX, and BclAF1 that induces a repressive chromatin structure at the major immediate-early locus (schematically illustrated here by the tight packing of histones). B. Tegument-delivered pp71 migrates to the nucleus and degrades Daxx leading to ATRX dispersal and relieving HDAC-mediated repression of IE gene expression. In cooperation with UL35, pp71 also degrades BclAF1 to stimulate IE gene expression. C. Tegument-delivered pp65 and UL29/28 migrate to the nucleus, associate with cellular proteins, and activate the transcription of viral IE genes (larger arrowhead) after the pp71-mediated neutralization of the intrinsic defence.

retain heterochromatic markings (Woodhall *et al.*, 2006), and viral IE gene expression is silenced (Cantrell and Bresnahan, 2006; Saffert and Kalejta, 2006). By degrading Daxx, pp71 inactivates a cellular intrinsic defence that would otherwise prevent viral IE gene expression and inhibit viral replication. It is important to note that, in addition to Daxx, HCMV IE gene expression is also inhibited by other intrinsic immune (Adler *et al.*, 2011) and cell cycle-related (Zydek *et al.*, 2011) mechanisms that do not appear to be targets of tegument-delivered proteins.

The mechanism through which pp71 induces protein degradation is unknown, but has been characterized as proteasome-dependent yet ubiquitin-independent (Kalejta and Shenk, 2003; Hwang and Kalejta, 2007). While pp71 induces the covalent addition of the small ubiquitin-like modifier (SUMO) protein to Daxx, this does not appear to be required for pp71-mediated Daxx degradation (Hwang and Kalejta, 2009). Recently, the 19S regulatory particle, a prominent proteasome activator, has been shown to be required for pp71-mediated protein degradation (Winkler and Kalejta, 2011), a unique finding among the characterized examples of virus protein-mediated proteasome-dependent, ubiquitin-independent protein degradation (Hwang *et al.*, 2011a).

In addition to Daxx, pp71 also induces the degradation of all three members of the retinoblastoma (Rb) family of tumour suppressors (Kalejta *et al.*, 2003; Hume *et al.*, 2008), as well as the BclAF1 protein (Lee *et al.*, 2012). Furthermore, pp71 disrupts the association of ATRX with Daxx (Lukashchuk *et al.*, 2008). While Rb family member degradation by HCMV does not appear to modulate IE gene expression nor be required for efficient HCMV replication in fibroblasts *in vitro* (Cantrell and Bresnahan, 2005), BcLAF1 degradation and ATRX dispersal by pp71 do stimulate IE gene expression (Fig. I.9.2B).

Both pp71 function and IE gene expression appear to be modulated by the viral UL35 proteins (Kalejta, 2008b). The UL35 gene is expressed as two isoforms, UL35 and UL35a (Liu and Biegalke, 2002), both of which interact with pp71 (Schierling *et al.*, 2004), but likely modulate its activity in different ways. The full-length protein (UL35) forms dot like structures within nuclei of transfected cells that recruit co-transfected pp71 and the cellular PML, Sp100 and Daxx proteins (Salsman *et al.*, 2011). In contrast, transfected UL35a (which consists of the C-terminal 193 amino acids of UL35) redirects co-transfected pp71 from the nucleus to the cytoplasm, which may be important during virion egress for the proper incorporation of pp71 into the tegument (Schierling *et al.*, 2005). UL35a is not a tegument protein (Liu

and Biegalke, 2002), and UL35 is only found in the tegument in very small quantities (Varnum *et al.*, 2004). Thus, how UL35 might affect the pre-IE stage of HCMV infection is currently unclear. However, recent experiments indicate that UL35 is required for the ability of pp71 to efficiently degrade the cellular BclAF1 protein (Fig. I.9.2B) that acts as a restriction factor to inhibit IE gene expression (Lee *et al.*, 2012). The mechanisms of pp71 and UL35 cooperation and the BclAF1-mediated repression of IE gene expression are unknown. Interestingly, pp71-mediated degradation of Daxx does not require UL35, indicating that pp71 has both UL35-dependent and -independent functions. Likewise, it is possible that UL35 has pp71-independent functions in addition to its role in the pp71-dependent degradation of BclAF1.

While pp71-dependent processes clearly de-repress the IE genes, other tegument proteins also appear to activate them (Fig. I.9.2C). The pp65 protein was recently shown to interact with the cellular IFI16 protein and recruit it to the major immediate-early locus where these proteins stimulated IE gene expression through an unknown mechanism (Cristea *et al.*, 2010). Knockdown of IFI16 inhibited IE gene expression and productive viral replication. Interestingly, IFI16 was also recently identified as an innate DNA sensor that induces the production of interferon-β, an antiviral cytokine (Unterholzner, *et al.*, 2010). As IFI16 is required for efficient IE gene expression and replication of not only HCMV (Cristea *et al.*, 2010) but also murine CMV (Hertel *et al.*, 1999; Rolle *et al.*, 2001), it appears that the positive effects of this protein may be more important for cytomegaloviruses than its presumptive negative effects as a DNA sensor and interferon inducer.

Another tegument protein, UL29/28, also activates immediate-early gene expression (Fig. I.9.2C). Exons produced from the UL29 and UL28 genes are spliced together in a transcript that is translated to generate a protein termed UL29/28 (Mitchell *et al.*, 2009). This protein interacts with the nucleosome remodelling and deacetylase (NuRD) complex (Terhune *et al.*, 2010). UL29/28 and NuRD complex components are found at the MIEP, and the loss of either UL29/28 or a functional NuRD complex inhibits IE gene expression. As the HDAC inhibitor trichostatin A (TSA) fails to rescue the growth defect of a UL29/28 null virus, it appears that this viral protein activates IE gene expression through an HDAC-independent mechanism (Terhune *et al.*, 2010).

Finally, while much of the regulation of IE protein production occurs at the level of transcriptional activation, later steps can also be modulated. For example, it is likely that enhancement of transcriptional elongation, probably through histone monoubiqui-tination by an Elongin B-containing ubiquitin ligase complex, also plays a role in IE gene expression during HCMV infection (Hwang *et al.*, 2011b). Furthermore, although most of the viral IE transcripts are spliced, the Us3 mRNA is not (Mocarski *et al.*, 2007), indicating perhaps that alterations to nuclear mRNA export might also facilitate the expression of at least this IE gene. To date, there have been no reports of translation control of IE gene expression. Whether or not tegument-delivered proteins regulate IE mRNA elongation and export during viral infection remains to be determined. However, pp71 has been shown to interact with the Elongin B protein (R.F. Kalejta, unpublished observations) that appears to promote transcriptional elongation of at least one HCMV late gene (Hwang *et al.*, 2011b). Furthermore, the UL69 tegument protein that binds RNA and shuttles between the nucleus and the cytoplasm may (Zielke *et al.*, 2011) or may not (Kronemann *et al.*, 2010) play a critical role during HCMV infection by facilitating the export of unspliced HCMV messenger RNAs.

Immune evasion

Avoiding detection by host immune defences is another likely activity of tegument proteins, with several having demonstrated immune-evasive activities. However, very little evidence exists to indicate whether or not virion-delivered (as opposed to *de novo* expressed) tegument proteins actually perform these functions at pre-IE times. Regardless, the known roles that tegument proteins play in preventing the infected cell from being detected and killed by host immune functions are listed below.

Host immunity is divided into intrinsic, innate, and adaptive/acquired branches. Intrinsic immunity is mediated by constitutively expressed proteins. As discussed above, it is clear that during lytic infection, tegument-delivered pp71 (and likely tegument-delivered UL35) inactivates intrinsic defences that would otherwise silence viral IE gene expression (Figs. I.9.2B and I.9.3A). Innate immunity is a rapid and varied response to viral infection. Multiple tegument proteins can thwart individual innate immune functions during lytic infection (Fig. I.9.3A). The viral proteins UL36 (McCormick *et al.*, 2010) and UL38 (Xuan *et al.*, 2009) inhibit apoptosis, which can be considered an arm of the innate immune system. The pp65 protein blocks natural killer cell-mediated cytotoxicity by interacting with the NKp30 activating receptor (Arnon *et al.*, 2005), and has been reported to diminish the interferon response to HCMV infection (Browne and Shenk, 2003; Abate *et al.*, 2004), although that result

has been challenged (Taylor and Bresnahan, 2006). Furthermore, two similar proteins, IRS1 and TRS1, bind double stranded RNA and inhibit the ability of PKR to shut down translation in HCMV infected cells (Marshall *et al.*, 2009). Finally, pp65 may modulate adaptive immune recognition of lytically infected cells (Fig. I.9.3A) by causing the degradation of the HLA-DR alpha chain by mediating the accumulation of HLA class II molecules in the lysosome (Odeberg *et al.*, 2003). The biological significance of the reported ability of pp65 to block the presentation of IE peptides through MHC class I molecules (Gilbert *et al.*, 1996) is debatable due to the strong IE-protein specific immune response mounted by infected individuals.

Tegument proteins and latency

Tegument proteins are also delivered to cells that will establish a latent infection (Penkert and Kalejta, 2011). Such cells are less differentiated (often referred to as 'undifferentiated') than the fibroblasts where lytic infection is most often studied. Several undifferentiated cell types are used to study quiescent (e.g. NT2 and THP-1 cells) and latent (e.g. CD34$^+$ cells) infections. Quiescent and latent infections are similar in all aspects examined to date except for one important difference, namely that quiescent infections have as yet been shown to animate (i.e. to initiate IE gene expression) or reactivate (i.e. to produce infectious progeny virions) efficiently in response to the proper stimulus. This separates them from true latent infections, for which animation and reactivation have been unequivocally demonstrated. The pre-IE stages of lytic and quiescent/latent infections have conserved similarities and important differences. While the process of viral entry and the delivery of viral genomes to the nucleus has not been studied during latency, it is expected that these events occur similarly as they do during lytic infection. For example, the efficiency of viral genome delivery to the nucleus of THP-1 cells where a quiescent infection is established was found to be similar to that in THP-1-derived macrophage where a lytic infection is initiated (Saffert and Kalejta, 2007). However, the next steps, de-repression and activation (Fig. I.9.2B,C) of IE gene expression do not take place in latently infected cells due to alternative subcellular localization of key tegument-delivered proteins.

Upon entry into NT2, THP-1, or CD34$^+$ cells, tegument-delivered pp71 and pp65 remain in the cytoplasm and do not enter the nucleus (Saffert and Kalejta, 2007; Saffert *et al.*, 2010; E.R. Albright and R.F. Kalejta, unpublished observations). In these cells, Daxx is not degraded and thus the intrinsic defence is not inactivated but allowed to silence viral IE gene

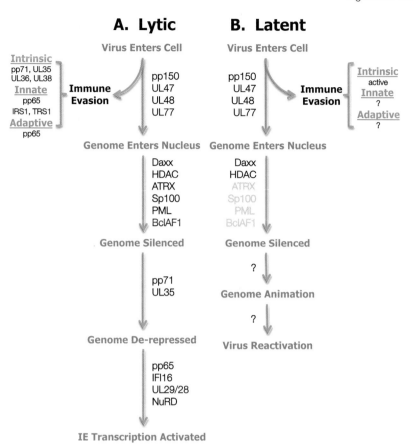

A. Lytic

Virus Enters Cell

Intrinsic
pp71, UL35
UL36, UL38
Innate
pp65
IRS1, TRS1
Adaptive
pp65

Immune Evasion

pp150
UL47
UL48
UL77

Genome Enters Nucleus

Daxx
HDAC
ATRX
Sp100
PML
BclAF1

Genome Silenced

pp71
UL35

Genome De-repressed

pp65
IFI16
UL29/28
NuRD

IE Transcription Activated

B. Latent

Virus Enters Cell

pp150
UL47
UL48
UL77

Immune Evasion

Intrinsic
active
Innate
?
Adaptive
?

Genome Enters Nucleus

Daxx
HDAC
ATRX
Sp100
PML
BclAF1

Genome Silenced

?

Genome Animation

?

Virus Reactivation

Figure I.9.3 Functions of tegument-delivered proteins during the pre-IE stage of lytic infection. A. Flowchart depicting the viral and cellular proteins that execute immune evasion and gene expression functions during the pre-IE stage of a lytic infection. See text and Fig. I.9.2 for details. B. Flowchart depicting the viral and cellular proteins that execute immune evasion and gene expression functions during the establishment of latency. Grey text indicates expected but unconfirmed protein functional roles. Question marks indicate processes that have yet to be examined. See text and Fig. I.9.4 for details.

expression (Fig. I.9.4). If Daxx levels are depleted with RNA interference, or if HDAC activity is inhibited with a small molecule, viral IE gene expression is initiated (Saffert and Kalejta 2007; Saffert *et al.*, 2010). Repressive effects of ATRX, BclAF1, PML or Sp100 have not been examined during quiescent/latent infections (Fig. I.9.4), although there is no reason to suspect that they play different roles in these cell types than they do in fibroblasts. The sequestering of tegument-delivered proteins in the cytoplasm of cells in which latency will be established not only inhibits the deactivation of the Daxx-mediated intrinsic defence, but should also prevent the activation of IE gene expression by pp65. While it is likely that other key tegument-delivered proteins (e.g. UL35 and UL29/28) also localize to the cytoplasm upon infection of undifferentiated cells, this has not yet been examined.

Mechanisms for the differentiation-dependent

subcellular localization of tegument-delivered proteins such as pp71 and pp65 remain unexplained. Interestingly, it appears the cytoplasm is the default localization for tegument-delivered proteins. Heterogeneous (NT2/fibroblast) cell fusion experiments indicate that tegument-delivered proteins found in the cytoplasm of undifferentiated cells can be driven into the nucleus by one or more factors expressed in differentiated cells (Penkert and Kalejta, 2010). Whether this unidentified factor or factors work by modulating tegument protein nuclear trafficking or simply facilitate tegument disassembly remains to be determined. The observation (Saffert and Kalejta, 2007; Saffert *et al.*, 2010) that *de novo* expressed pp71 and pp65 localize to the nucleus in the same undifferentiated cells where the tegument-delivered proteins remain cytoplasmic is provocative, but does not provide mechanistic details.

In addition to the known tegument proteins whose

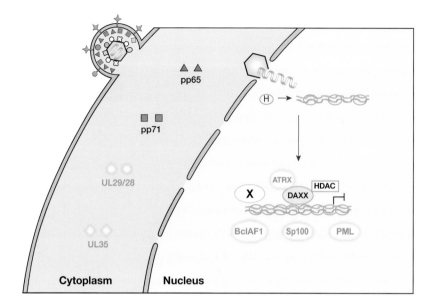

Figure I.9.4 IE gene de-repressing and activating tegument proteins are inactive during the establishment of latency due to their cytoplasmic localization. During the establishment of latency, tegument proteins that de-repress (pp71) or activate (pp65) IE gene expression are found in the cytoplasm and are thus unable to perform their normal activities. Other tegument proteins with similar functions may also be cytoplasmic, but their localization in latently infected cells has yet to be examined (depicted as blurred grey symbols to indicate expected but untested location). This allows the Daxx mediated intrinsic defence to silence IE gene expression. Other intrinsic defence proteins may also participate in this silencing, although their activities during latency have not been tested (depicted as blurred grey symbols to indicate expected but untested activity). Furthermore, a clinical strain-specific, HDAC-independent *trans*-dominant inhibition of IE gene expression (X) is also observed during the establishment of latency.

activities are muted during the establishment of latency by their cytoplasmic localization, an HDAC-independent, *trans*-dominant inhibition of IE gene expression (Fig. I.9.4) is present in cells infected with clinical viral strains, but not with the laboratory-adapted AD169 strain (Saffert *et al.*, 2010). The identity and mechanism of this clinical strain-specific restriction to IE gene expression during latency (that exists in addition to the host-mediated intrinsic defence that also silences IE gene expression) is not known, but could conceivably be mediated by a tegument-delivered protein. It is also unclear whether or not tegument proteins participate in the animation and/or reactivation of latent HCMV genomes. Depending on how long the genome remains latent, it might seem unlikely that tegument proteins delivered during viral entry would remain functional to allow their participation in these later processes. However, recent evidence indicates that proteins commonly incorporated into virion teguments could be produced *de novo* from latent genomes, and as such might initiate latent genome animation as the critical first step leading to productive reactivations (Penkert and Kalejta, 2011).

Finally, the potential immune evasion roles of tegument-delivered proteins during latency have generally not been examined (Fig. I.9.3B). One exception, of course, is the intrinsic defence that is not inactivated during latency because tegument-delivered pp71 remains cytoplasmic. Conceivably the ability of UL36 and UL38 to block apoptosis, IRS1 and TRS1 to promote translation, and pp65 to interact with the NKp30 activating receptor would remain in undifferentiated cells. Thus, tegument-delivered proteins could in fact inhibit some innate immune functions during latency. Whether or not pp65 could dampen the interferon response is questionable, but it might possibly modulate antigen presentation from the cytoplasm.

Conclusion

Activities and roles of individual tegument proteins at pre-IE times are coming into focus (Fig. I.9.3). They play roles in delivery of the capsid to the nuclear pores and genomes into the nucleus (Fig. I.9.1), de-repression and activation (Fig. I.9.2) of IE gene expression, and likely immune evasion as well. While information about binary tegument protein interactions has increased, mechanistic information about tegument assembly or disassembly is lacking. Finally, the previously unexplored roles of tegument proteins during the early events of latency are beginning to be spotlighted (Fig. I.9.4). Interventions that could modulate tegument protein function at the beginning of lytic replication, latency, or animation/reactivation could prevent replication cycles at very early (pre-IE) time points, and thus might prove to be effective antivirals. More mechanistic knowledge about how tegument proteins work after entry but prior to viral gene expression will be required to design such treatments.

Acknowledgements

I thank Leanne Olds for the illustrations, Rhiannon Penkert, Song Hee Lee, and Laura Winkler for reviewing the manuscript, and all the members of my laboratory for their helpful comments. This work was supported by NIH grant AI074984 and a Burroughs Wellcome Fund Investigator in the Pathogenesis of Infectious Disease Fellowship.

References

Abaitua, F., Daikoku, T., Crump, C.M., Bolstad, M., and O'Hare, P. (2011). A single mutation responsible for temperature-sensitive entry and assembly defects in the VP1–2 protein of herpes simplex virus. J. Virol. *85*, 2024–2036.

Abate, D.A., Watanabe, S., and Mocarski, E. (2004). Major human cytomegalovirus structural protein pp65 (ppUL83) prevents interferon response factor 3 activation in the interferon response. J. Virol. *78*, 10995–11006.

Adler, M., Tavalai, N., Muller, R., and Stamminger, T. (2011). Human cytomegalovirus immediate-early gene expression is restricted by the nuclear domain 10 component Sp100. J. Gen. Virol. *92*, 1532–1538.

Arnon, T.I., Achdout, H., Levi, O., Markel, G., Saleh, N., Katz, G., Gazit, R., Gonen-Gross, T., Hanna, J., Nahari, E., *et al.* (2005). Inhibition of the NKp30 activating receptor by pp65 of human cytomegalovirus. Nat. Immunol. *6*, 515–523.

Bechtel, J.T., and Shenk, T. (2002). Human cytomegalovirus UL47 tegument protein functions after entry and before immediate-early gene expression. J. Virol. *76*, 1043–1050.

Bieniasz, P.D. (2004). Intrinsic Immunity: a front-line defense against viral attack. Nat. Immunol. *5*, 1109–1115.

Britt, W. (2008). Manifestations of human cytomegalovirus infection: proposed mechanisms of acute and chronic disease. Curr. Top. Microbiol. Immunol. *325*, 417–470.

Browne, E.P., and Shenk, T. (2003). Human cytomegalovirus UL83-coded pp65 virion protein inhibits antiviral gene expression in infected cells. Proc. Natl. Acad. Sci. U.S.A. *100*, 11439–11444.

Cantrell, S.R., and Bresnahan, W.A. (2005). Interaction between the human cytomegalovirus UL82 gene product (pp71) and hDaxx regulates immediate-early gene expression and viral replication. J. Virol. *79*, 7792–7802.

Cantrell, S.R., and Bresnahan, W.A. (2006). Human cytomegalovirus (HCMV) UL82 gene product (pp71) relieves hDaxx-mediated repression of HCMV replication. J. Virol. *80*, 6188–6191.

Caposio, P., Orloff, S.L., and Streblow, D.N. (2011). The role of cytomegalovirus in angiogenesis. Virus Res. *157*, 204–211.

Cristea, I.M., Moorman, N.J., Terhune, S.S., Cuevas, C.D., O'Keefe, E.S., Rout, M.P., Chait, B.T., and Shenk, T. (2010). Human cytomegalovirus pUL83 stimulates activity of the viral immediate-early promoter through its interaction with the cellular IFI16 protein. J. Virol. *84*, 7803–7814.

Dunn, W., Chou, C., Li, H., Hai, R., Patterson, D., Stolc, V., Zhu, H., and Liu, F. (2003). Functional profiling of a human cytomegalovirus genome. Proc. Natl. Acad. Sci. U.S.A. *100*, 14223–14228.

Gilbert, M.J., Riddell, S.R., Plachter, B., and Greenberg, P.D. (1996). Cytomegalovirus selectively blocks antigen processing and presentation of its immediate-early gene product. Nature *383*, 720–722.

Hertel, L., De Andrea, M., Azzimonti, B., Rolle, A., Gariglio, M., and Landolfo, S. (1999). The interferon-inducible 204 gene, a member of the IFi 200 family, is not involved in the antiviral state induction by IFN-alpha, but is required by the mouse cytomegalovirus for its replication. Virology *262*, 1–8.

Hoffman, H., Sindre, H., and Stamminger, T. (2002). Functional interaction between the pp71 protein of human cytomegalovirus and the PML-interacting protein human Daxx. J. Virol. *76*, 5769–5783.

Hume, A.J., Finkel, J.S., Kamil, J.P., Coen, D.M., Culbertson, M.R., and Kalejta, R.F. (2008). Phosphorylation of retinoblastoma protein by viral protein with cyclin-dependent kinase function. Science *320*, 797–799.

Hwang, J., and Kalejta, R.F. (2007). Proteasome-dependent, ubiquitin-independent degradation of Daxx by the viral pp71 protein in human cytomegalovirus-infected cells. Virology *367*, 334–338.

Hwang, J., and Kalejta, R.F. (2009). Human cytomegalovirus pp71 induces Daxx SUMOylation. J. Virol. *83*, 6591–6598.

Hwang, J., Winkler, L., and Kalejta, R.F. (2011a). Ubiquitin-independent proteasomal degradation during oncogenic viral infections. Biochim. Biophys. Acta. *1816*, 147–157.

Hwang, J., Saffert, R.T., and Kalejta, R.F. (2011b). Elongin B-mediated epigenetic alteration of viral chromatin correlates with efficient human cytomegalovirus gene expression and replication. MBio *2*, e00023–11.

Kalejta, R.F. (2008a) Functions of human cytomegalovirus tegument proteins. Curr. Top. Microbiol. Immunol. *325*, 101–115.

Kalejta, R.F. (2008b) Tegument proteins of human cytomegalovirus. Microbiol. Mol. Biol. Rev. *72*, 249–265.

Kalejta, R.F., and Shenk, T. (2003). Proteasome-dependent, ubiquitin-independent degradation of the Rb family of tumor suppressors by the human cytomegalovirus pp71 protein. Proc. Natl. Acad. Sci. U.S.A. *100*, 3263–3268.

Kalejta, R.F., Bechtel, J.T., and Shenk, T. (2003). Human cytomegalovirus pp71 stimulates cell cycle progression by inducing the proteasome-dependent degradation of the retinoblastoma family of tumor suppressors. Mol. Cell. Biol. *23*, 1885–1895.

Kim, E.T., Oh, S.E., Lee, Y.O., Gibson, W., and Ahn, J.H. (2009). Cleavage specificity of the UL48 deubiquitinating protease activity of human cytomegalovirus and the growth of an active-site mutant virus in cultured cells. J. Virol. *83*, 12046–12056.

Kronemann, D., Hagemeier, S.R., Cygnar, D., Phillips, S., and Bresnahan, W.A. (2010). Binding of the human cytomegalovirus (HCMV) tegument protein UL69 to UAP56/URH49 is not required for efficient replication of HCMV. J. Virol. 84, 9649–9654.

Lee, S.H., Kalejta, R.F., Kerry, J., Semmes, O.J., O'Connor, C.M., Khan, Z., Garcia, B.A., Shenk, T., and Murphy, E. (2012). BclAF1 restriction factor is neutralized by proteasomal degradation and microRNA repression during human cytomegalovirus infection. Proc. Natl. Acad. Sci. USA, 109, 9575–9580.

Liu, Y., and Biegalke, B.J. (2002). B The human cytomegalovirus UL35 gene encodes two proteins with different functions. J. Virol. 76, 2460–2468.

Lukashchuk, V., McFarlane, S., Everett, R.D., and Preston, C.M. (2008). Human cytomegalovirus protein pp71 displaces that chromatin-associated factor ATRX from nuclear domain 10 at early stages of infection. J. Virol. 82, 12543–12544.

McCormick, A.L., Roback, L., Livingston-Rosanoff, D., and St. Clair, C. (2010). The human cytomegalovirus UL36 gene controls caspase-dependent and –independent cell death programs activated by infection of monocytes differentiating to macrophages. J. Virol. 84, 5108–5123.

Marshall, E.E., Bierle, C.J., Brune, W., and Geballe, A.P. (2009). Essential role for either TRS1 or IRS1 in human cytomegalovirus replication. J. Virol. 83, 4112–4120.

Mitchell, D.P., Savaryn, J.P., Moorman, N.J., Shenk, T., and Terhune, S.S. (2009). Human cytomegalovirus UL28 and UL29 open reading frames encode a spliced mRNA and stimulate accumulation of immediate-early RNAs. J. Virol. 83, 10187–10197.

Mocarski, E., Shenk, T., and Pass, R. (2007). Cytomegaloviruses, In Fields Virology, Knipe, D.M., and Howley. P.M., eds. (Lippincott Williams & Wilkins, Philadelphia, PA), pp. 2701–2772.

Munger, J., Yu, D., and Shenk, T. (2006). UL26-deficient human cytomegalovirus produces virions with hypophosphorylated pp28 tegument protein that is unstable within newly infected cells. J. Virol. 80, 3541–3548.

Neil, S., and Bieniasz, P. (2009). Human immunodeficiency virus, restriction factors, and interferon. J. Interferon Cytokine Res. 29, 569–580.

Nevels, M., Nitzsche, A., and Paulus, C. (2011). How to control an infectious bead string: nucleosome-based regulation and targeting of herpesvirus chromatin. Rev. Med. Virol. 21, 154–180.

Ogawa-Goto, K., Tanaka, K., Gibson, W., Moriishi, E., Miura, Y., Kurata, T., Irie, S., and Sata, T. (2003). Microtubule network facilitates nuclear targeting of human cytomegalovirus capsid. J. Virol. 77, 8541–8547.

Odeberg, J., Plachter, B., Branden, L., and Soderberg-Naucler, C. (2003). Human cytomegalovirus protein pp65 mediates accumulation of HLA-DR in lysosomes and destruction of the HLA-DR alpha-chain. Blood 101, 4870–4877.

Pawelec, G., and Derhovanessian, E. (2011). Role of CMV in immune senescence. Virus Res. 157, 175–179.

Penkert, R.R., and Kalejta, R.F. (2010). Nuclear localization of tegument-delivered pp71 in human cytomegalovirus-infected cells is facilitated by one or more factors present in terminally differentiated fibroblasts. J. Virol. 84, 9853–9863.

Penkert, R.R., and Kalejta, R.F. (2011). Tegument protein control of latent herpesvirus establishment and animation. Herpesviridae 2, 3.

Phillips, S.L., and Bresnahan, W.A. (2011). Identification of binary interactions between human cytomegalovirus virion proteins. J. Virol. 85, 440–447.

Ranganathan, P., Clark, P.A., Kuo, J.S., Salamat, M.S., and Kalejta, R.F. (2012). Significant association of multiple human cytomegalovirus genomic loci with glioblastoma multiforme samples. J. Virol. 86, 854–864.

Reeves, M.B. (2011). Chromatin-mediated regulation of cytomegalovirus gene expression. Virus Res. 157, 134–143.

Reyda, S., Aue, S., Buescher, N., Becke, S., Besold, K., Hesse, J., Tenzer, S., and Plachter, B. (2011). Mass Spectrometry discloses limited impact of pp65 on the protein composition of HCMV virions but reveals high variation of AD169-derived viruses. 36th International Herpesvirus Workshop, Gdansk, Poland, Abstract 4.08.

Roberts, A.P.E., Abaitua, F., O'Hare, P., McNab, D., Rixon, F.J and Pasdeloup, D. (2009). Differing roles of inner tegument proteins pUL36 and pUL37 during entry of herpes simplex virus type 1. J. Virol. 83, 105–116.

Rode, K., Dohner, K., Binz, A., Glass, M., Strive, T., Bauerfeind, R., and Sodeik, B. (2011). Uncoupling uncoating of herpes simplex virus genomes from their nuclear import and gene expression. J. Virol. 85, 4271–4283.

Rolle, S., De Andrea, M., Gioia, D., Lembo, D., Hertel, L., Landolfo, S., and Gariglio, M. (2001). The interferon-inducible 204 gene is transcriptionally activated by mouse cytomegalovirus and is required for its replication. Virology 286, 249–255.

Ryckman, B.J., Jarvis, M.A., Drummond, D.D., Nelson, J.A., and Johnson, D.C. (2006). Human cytomegalovirus entry into epithelial and endothelial cells depends on genes UL128 to UL150 and occurs by endocytosis and low-pH fusion. J. Virol. 80, 710–722.

Saffert, R.T., and Kalejta, R.F. (2006). Inactivating a cellular intrinsic immune defense mediated by Daxx is the mechanism through which the human cytomegalovirus pp71 protein stimulates viral immediate-early gene expression. J. Virol. 80, 3863–3871.

Saffert, R.T., and Kalejta, R.F. (2007). Human cytomegalovirus gene expression is silenced by Daxx-mediated intrinsic immune defense in model latent infections established in vitro. J. Virol. 81, 9109–9120.

Saffert, R.T., Penkert, R.R., and Kalejta, R.F. (2010). Cellular and viral control over the initial events of human cytomegalovirus experimental latency in CD34+ cells. J. Virol. 84, 5594–5604.

Salsman, J., Wang, X., and Frappier, L. (2011). Nuclear body formation and PML body remodeling by the human cytomegalovirus protein UL35. Virology 414, 119–129.

Schierling, K., Stamminger, T., Mertens, T., and Winkler, T. (2004). Human cytomegalovirus tegument proteins ppUL82 (pp71) and ppUL35 interact and cooperatively activate the major immediate-early enhancer. J. Virol. 78, 9512–9523.

Schierling, K., Buser, C., Mertens, T., and Winkler, T. (2005). Human cytomegalovirus tegument protein ppUL35 is important for viral replication and particle formation. J. Virol. 79, 3804–3096.

Sinclair, J. (2010). Chromatin structure regulates human cytomegalovirus gene expression during latency, reactivation and lytic infection. Biochim. Biophys. Acta *1799*, 286–295.

Soroceanu, L., and Cobbs, C.S. (2011). Is HCMV a tumor promoter? Virus Res. *157*, 193–203.

Tandon, R., and Mocarski, E.S. (2011). Cytomegalovirus pUL96 is critical for the stability of pp150-associated nucleocapsids. J. Virol. *85*, 7129–7141.

Tavalai, N., and Stamminger, T. (2008). New insights into the role of the subnuclear structure ND10 for viral infection. Biochim. Biophys. Acta *1783*, 2207–2221.

Tavalai, N., and Stamminger, T. (2011). Intrinsic cellular defense mechanisms targeting human cytomegalovirus. Virus Res. *157*, 128–133.

Taylor, R.T., and Bresnahan, W.A. (2006). Human cytomegalovirus immediate-early 2 protein IE86 blocks virus induced chemokine expression. J. Virol. *80*, 920–928.

Terhune, S.S., Moorman, N.J., Cristea, I.M., Savaryn, J.P., Cuevas-Bennett, C., Rout, M.P., Chait, B.T., and Shenk, T. (2010). Human cytomegalovirus UL29/28 protein interacts with components of the NuRD complex which promote accumulation of immediate-early RNA. PLoS Pathog. *6*, e1000965.

To, A., Bai, Y., Shen, A., Gong, H., Umamoto, S., Lu, S., and Liu, F. (2011). Yeats two hybrid analyses reveal novel binary interactions between human cytomegalovirus-encoded virion proteins. PLoS One *6*, e17796.

Unterholzner, L., Keating, S.E., Baran, M., Horan, K.A., Jensen, S.B., Sharma, S., Sirois, C.M., Jin, T., Latz, E., Xiao, T.S., et al. (2010). IFI16 is an innate immune sensor for intracellular DNA. Nat. Immunol. *11*, 997–1004.

Varnum, S.M., Streblow, D.N., Monroe, M.E., Smith, P., Auberry, K.J., Pasa-Tolic, L., Wang, D., Camp, D.G., Roland, K., Wiley, S., et al. (2004). Identification of proteins in human cytomegalovirus (HCMV) particles: the HCMV proteome. J. Virol. *78*, 10960–10966.

Wang, J., Loveland, A.N., Kattenhorn, L.M., Ploegh, H.L., and Gibson, W. (2006). High-molecular-weight protein (pUL48) of human cytomegalovirus is a competent deubiquitinating protease: mutant viruses altered in its active-site cysteine or histidine are viable. J. Virol. *80*, 6003–6012.

Winkler, L.L., and Kalejta, R.F. (2011). Ubiquitin-independent, proteasome-dependent degradation of Daxx by human cytomegalovirus pp71 requires the 19S regulatory particle of the proteasome. *36th International Herpesvirus Workshop, Gdansk, Poland,* Abstract 9.29.

Woodhall, D.L., Groves, I.J., Reeves, M.B., Wilkinson, G., and Sinclair, J.H. (2006). Human Daxx-mediated repression of human cytomegalovirus gene expression correlates with a repressive chromatin structure around the major immediate-early promoter. J. Biol. Chem. *281*, 37652–37660.

Xuan, B., Qian, Z., Torigoi, E., and Yu, D. (2009). Human cytomegalovirus protein pUL38 induces ATF4 expression, inhibits persistent JNK phosphorylation, and suppresses endoplasmic reticulum stress-induced cell death. J. Virol. *83*, 3463–3474.

Yu, D., Silva, M.C., and Shenk, T. (2003). Functional map of human cytomegalovirus AD169 defined by global mutational analysis. Proc. Natl. Acad. Sci. U.S.A. *100*, 12396–12401.

Yu, X., Shah, S., Lee, M., Dai, W., Lo, P., Britt, W., Zhu, H., Liu, F., and Zhou, Z.H. (2011). Biochemical and structural characterization of the capsid-bound tegument proteins of human cytomegalovirus. J. Struct. Biol. *174*, 451–460.

Zielke, B., Thomas, M., Giede-Jeppe, A., Muller, R., and Stamminger, T. (2011). Characterization of the betaherpesviral pUL69 family reveals binding of the cellular mRNA export factor UAP56 as a prerequisite for stimulation of nuclear mRNA export and for efficient viral replication. J. Virol. *85*, 1804–1819.

Zydek, M., Uecker, R., Tavalai, N., Stamminger, T., Hagemeier, C., and Wiebusch, L. (2011). General blockade of HCMV immediate-early mRNA expression in S/G2 by a nuclear, Daxx- and PML-independent mechanism. J. Gen. Virol. *92*, 2757–2769.

Major Immediate-Early Enhancer and Its Gene Products

I.10

Jeffery L. Meier and Mark F. Stinski

Abstract

CMV major immediate-early (MIE) gene expression activates the viral replicative cycle in both acute and reactivation infections and is greatly restricted in latent infection. Specific signalling cascades, transcriptional regulatory hierarchies, and *cis*-regulatory codes govern the initiation efficacy, magnitude, and sustainability of MIE gene transcription. The MIE enhancer/promoter, a major determinant in viral fitness, is at the heart of this control. It is equipped with complex regulatory circuitry that integrates diverse viral, cellular, and environmental cues. The MIE genes via differential RNA splicing produce a set of multifunctional proteins that function directly in advancing the viral life cycle. CMV-induced disease genesis is driven by the regulatory mechanisms underlying both the expression of the MIE genes and the actions of the MIE gene products. A better understanding of the MIE enhancer and its gene products could potentially spawn novel strategies for preventing CMV-related disease.

Introduction

CMV replicates in a wide range of differentiated cell types of endoderm, mesoderm, and ectoderm origin (Sinzger and Jahn, 1996). CMV targets select types of poorly differentiated cells for latent infection. Reactivation of the virus is brought about by cellular differentiation and/or stimulation (Sinclair and Sissons, 1996; Soderberg-Naucler and Nelson, 1999; see also Chapter I.19). Divergent infection outcomes partly arise from the regulatory control of transcription from the CMV major immediate-early (MIE) promoter, whose gene products initiate the viral lytic cycle and are rarely expressed during viral latency (Meier and Stinski, 1996). The enhancer of the MIE genes is at the core of the coordinate regulation of MIE promoter activity, viral fitness, and pathogenesis. Engineered enhancer alterations in the CMV genome have produced findings that support this view. Mutations of the MIE proteins

demonstrate that the viral proteins control intrinsic and innate cellular immune responses, transcription of cellular and viral genes, auto-regulation of MIE gene expression, and progression of the cell cycle. Here, we review the functions of the MIE enhancer/promoter and downstream viral gene products.

Enhancer requirement in viral replication and pathogenesis

Murine CMV recombinants lacking the entire MIE enhancer are greatly impaired in viral replication in cultured fibroblasts (Angulo *et al.*, 1998) and fail to actively infect and spread in young adult CB17 SCID mice inoculated via the intraperitoneal route (Ghazal *et al.*, 2003). In neonatal and severely immunodeficient adult BALB/c mice having defects in natural killer cells and adaptive immunity, the enhancerless virus is profoundly impaired in exponential viral growth and focal spread after intravenous infection, yet establishes and maintains a low-grade infection in spleens, livers, kidneys, and lungs of these mice (Podlech *et al.*, 2010). Coinfection with wild-type (WT) virus does not improve the abnormally low growth rate of the enhancerless virus.

The large replication defect of an enhancerless virus that is observed in cultured NIH 3T3 fibroblasts is corrected by replacing the missing enhancer with the comparable human CMV enhancer (Angulo *et al.*, 1998). This enhancer substitution also renders murine CMV fully fit for replication in mouse embryo fibroblasts, the endothelial cell line SVEC-4, and the liver-derived cell line MMH (Gustems *et al.*, 2008). In contrast, the same virus replicates less well (~ 10-fold) in the epithelial C1217 and macrophage RAW264.7 cell lines, compared with the parent and restored WT virus. A murine CMV recombinant having a human MIE enhancer and promoter in place of its own also exhibits normal replication kinetics in cultured fibroblasts (Grzimek *et al.*, 1999). After intravenous inoculation into immunodeficient BALB/c mice, the virus is less able to infect lungs, spleens, and adrenal glands, but efficiently seeds the livers of these mice (Grzimek et al., 1999). Focal spread of the virus proceeds efficiently in all organs. These findings are consistent with a defect in viral dissemination to organs other than liver. For the virus having the human enhancer substitution, titres of infectious viral progeny are reduced from 10- to 100-fold in spleens, livers, kidneys, lungs, and salivary glands of healthy adolescent BALB/c mice and immunocompromised CB17 SCID mice, after infection via the intraperitoneal route (Gustems et al., 2008). This enhancer-swap virus is less virulent in neonatal BALB/c mice, exhibiting an LD_{50} of 10^5 PFU/mouse versus 10^3 PFU/mouse for WT virus. Normalization of viral inoculums at 0.3 LD_{50} yields levels of viral production in spleens, livers, kidneys, and lungs of these mice that are equivalent between the enhancer-swap and control WT viruses, after intraperitoneal infection. However, salivary glands produce substantially less of the enhancer-swap virus at 14 days and this growth defect is not ameliorated by direct inoculation of the virus into the salivary gland. Spleen and lung explants from the latently infected BALB/c mice that are infected as neonates produce infectious virus with frequencies and kinetics that are equivalent between the enhancer-swap and WT viruses (Gustems et al., 2008). Replacing the enhancers in rat (Sanford et al., 2001) and human (Isomura and Stinski, 2003) CMVs with that of murine CMV reduces viral replication in cultured rat and human fibroblasts, respectively. The crippled rat CMV also falls short in production of infectious virus in rat salivary gland after intraperitoneal inoculation, despite replicating well in the spleen (Sanford et al., 2001). Thus, enhancer swaps between different CMV species result in aberrancies in efficacy of infection initiation, viral fitness, and pathogenesis.

The human CMV MIE enhancer is absolutely essential for MIE gene expression and viral replication in cultured cells. The 550-bp enhancer is composed of distal and proximal halves that differ in structural makeup yet function in unison to efficiently activate MIE promoter-dependent transcription and viral replication (Meier and Pruessner, 2000; Isomura and Stinski, 2003). Removal of the distal enhancer significantly decreases MIE promoter activity, viral replication, and viral plaquing efficiency (number of plaques and rate of their growth) in human fibroblasts (HF) at levels commensurate with decreasing multiplicity of infection (MOI) (Meier and Pruessner, 2000). These phenotypic defects are not observed at MOI of ≥ 1 PFU per cell or when the defective virus at MOI of < 0.1 is complemented with inactivated purified CMV particles (Meier and Pruessner, 2000;

Meier et al., 2002). This suggests that high amounts of one or more virion components compensate for the missing distal enhancer and partly function to increase MIE gene expression (see also Chapter I.9). Multiple cis-acting elements in the distal enhancer generate the MIE promoter activity (Meier et al., 2002). The proximal enhancer is also comprised of multiple positive cis-acting elements, and viruses with successively larger 5′-truncation deletions of this region are progressively more defective in MIE promoter activity and viral replication (Isomura et al., 2004). Retaining only the proximal enhancer GC-box nearest the core promoter in an otherwise enhancerless virus permits recovery of viable human CMV Towne, but the virus is severely defective in MIE gene expression and viral replication (Isomura et al., 2004). The functional abnormalities created by taking out either the distal or proximal enhancer are not completely remedied by inserting the murine CMV enhancer equivalent (Isomura and Stinski, 2003). Hence, the distal and proximal enhancers of human CMV contribute multiple cis-acting elements that govern the initiation and magnitude of MIE promoter-dependent transcription and viral replication. Interchange of MIE enhancers between different CMV species can result in loss of fitness of the virus in its natural host, suggesting that enhancer structure has co-evolved with its host.

Enhancer mechanics in activation

Constraints imposed by methods available in the early years of molecular virology focused investigation on CMV MIE enhancer/promoter segments contained in heterologous DNA vectors and assayed in cell-free, cell culture, or transgenic animal settings in order to understand the mechanisms underlying MIE enhancer/promoter regulation. Important insights gleaned from these studies recognized that MIE enhancer/promoter activity: (a) varies in accord with cell type, stage of cellular differentiation, and action of specific signalling pathways; (b) is dynamically regulated by interplay of a diverse array of specific cellular transcription factors that bind to cognate DNA sites in the MIE regulatory region and function in a condition-dependent and cooperative or synergistic manner; (c) is brought about by a basal transcriptional complex containing RNA polymerase II, the participation of transcriptional co-activators and additional regulators of transcription elongation, and the modification of chromatin components; and (d) is regulated by specific viral proteins that act on the MIE regulatory region (reviewed in Meier and Stinski, 1996).

This extensive body of early work also led to the initial characterization of many different types of cis-regulatory elements in the human CMV MIE regulatory region, schematically depicted in Fig. I.10.1. The human CMV MIE regulatory region is operationally partitioned into core promoter, enhancer, unique region, and modulator. The enhancer is composed of several different types of positive cis-acting elements that are bound by their respective cellular transcription factors, which include NF-κB, CREB/ATF, AP-1, SP1, serum response factor, ELK-1, and retinoic acid/retinoid X receptor (reviewed in Meier and Stinski, 1996). The specific transcription factors are regulated in quantity and activity by signalling pathways and cues coming from the host cell, the extracellular milieu, and the virus. Human CMV MIE enhancer/promoter activity is also increased by several different virion components, i.e. pp71 (ppUL82), UL69, TRS1/IRS1, and ppUL35 proteins (Liu and Stinski, 1992; Winkler and Stamminger, 1996; Romanowski et al., 1997; Homer et al., 1999; Bresnahan and Shenk, 2000; Schierling et al., 2004) and by positive feedback provided by the viral MIE gene product, IE1-72 protein (Cherrington and Mocarski, 1989). Positive cis-acting elements in the unique (e.g. PDX1) and modulator regions also add to MIE enhancer/promoter activity in some models (Nelson et al., 1987; Chao et al., 2004).

Subsequent innovations in technology have enabled delineation of the regulatory mechanisms that function in the context of the CMV genome in infected cells. This door of opportunity was opened by the development of technology for constructing recombinant CMV genomes (see also Chapter I.3), into which mutations are engineered in select viral DNA sequences. The rational design of mutations in MIE cis-regulatory sequences of the CMV genome is a mainstay for the determination of the function of these DNA elements in relation to various conditions of the infection. Findings produced by this approach have not always aligned with results of transient assays of CMV enhancer/promoter fragments. Apparent disconnect between approaches is illustrated for the following putative cis-regulatory sequences of the human CMV MIE regulatory region:

1   Removal of the unique region and/or modulator fails to reproduce results of transient assays done in the same cell type (Meier and Stinski, 1997).
2   Removal or substitution of the distal enhancer yields the opposite outcome from results of transient assays in HF (Meier and Pruessner, 2000).
3   In contrast to predictions based on transient assays, MIE gene expression and viral replication are minimally to negligibly changed in HF by placement of base-substitution mutations alone in the enhancer's five copies of the cyclic AMP

response elements (CRE) (Keller et al., 2003), four copies of NF-κB binding elements (κB) (Benedict et al., 2004; Gustems et al., 2006), or two copies of AP-1 binding sites (Isern et al., 2011).

Compensatory or cooperative functioning of specific types of enhancer cis-acting elements is a feature of both clinical-like (VR1814/FIX BAC) and laboratory (Ad169 or Towne BAC) human CMV strains. The combination of κB mutations with mutations in ELK-1/SRF (VR1814/FIX BAC) or AP1 (Ad169 BAC) binding sites decreases both MIE gene expression and viral replication in HF, whereas mutations limited to one type of site do not produce the phenotypic defects under normal growth conditions (Caposio et al., 2010; Isern et al., 2011). In contrast, the combination of CRE mutations with mutations in ELK-1 or κB (Towne BAC) does not appreciably lower MIE gene expression and viral replication in these cells (Yuan et al., 2009; Liu et al., 2010a). In the absence of virion-associated transactivators, a cooperative interaction between different types of cis-acting elements in the proximal enhancer confers optimal MIE promoter activity in human CMV genomes (Towne BAC) transfected into HF (Lashmit et al., 2009). Interplay between GC-box and CRE stands out among the possible cis-acting combinations in achieving greatest levels of proximal enhancer-dependent MIE gene expression, viral replication, and frequency of viral plaques (Lashmit et al., 2009). Deletion of a small segment of proximal enhancer containing the two GC-boxes (base positions, −55 and −75), as well as a CRE/AP-1 element, markedly decreases MIE RNA expression, viral replication, and viral plaque-size in HF (Isomura et al., 2005). In the distal enhancer, the targeting of two IFN-gamma-activated site (GAS)-like elements with mutations results in a decrease in human CMV Ad169 (Ad169 BAC) MIE enhancer/promoter activity at low MOI in acutely infected HF (Netterwald et al., 2005). However, the same or different GAS element mutations in human CMV Towne (Towne BAC) do not produce this abnormal phenotype (Meier laboratory, unpublished data). The cellular transcription factor peroxisome proliferator-activated receptor gamma (PPAR-gamma) relocates from cytoplasm to nucleus upon human CMV infection, binds in vitro to PPAR-gamma response elements located in the MIE distal enhancer, and attaches to the MIE enhancer in infected U373MG cells (Rauwel et al., 2010). Pharmacologic inhibition of PPAR-gamma decreases expression of viral IE2 RNA and protein, but does not alter expression the IE1 counterparts. This abnormality is linked to a large decrease in viral production. Recruitment of the

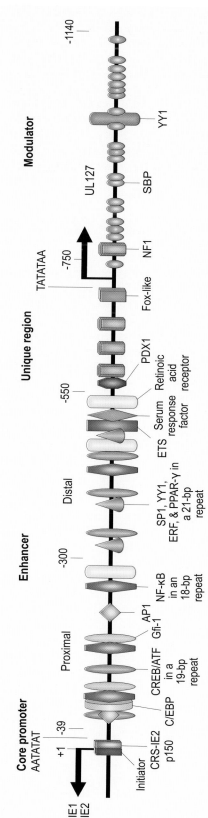

Figure I.10.1 The human CMV MIE regulatory region is composed of core promoter, enhancer, unique region, and modulator. The enhancer is divided into proximal (−39 to −299) and distal (−300 to −579) portions. Positions of putative binding sites for initiator complex, C/EBP, CREB/ATF, Gfi-1, AP1, NF-κB, SP1, YY1, ERF, ETS, serum response factor, retinoic acid receptor, PDX1, NF1, and SBP are shown relative to the +1 RNA start-site (leftward arrow) of the core promoter, as oriented in the viral genome. Binding sites for NF-κB, CREB/ATF, and the combination of SP1, YY1, ERF, and PPAR-gamma are nested in the 18-, 19-, and 21-bp repeats, respectively. IE2-86 binds to the *cis*-repression sequence (crs). The TGGGC^A/_G elements are depicted in Fig. I.10.2. Locations of AT-rich sequences in the unique region for binding of STAB1 and CDP (Lee *et al.*, 2007) are not shown.

PPAR-gamma coactivator 1 (PGC-1) to the enhancer by PPAR-gamma is postulated as a means for favouring the IE2 RNA splicing pattern, based on reports that PGC-1 can change pre-mRNA splicing by binding to PPAR-gamma that occupies upstream recognition elements in other promoters (Rauwel *et al.*, 2010). Taken together, the studies indicate that (1) viral replication fitness is driven by the synergistic actions of distinct sets of different types of cis-acting enhancer elements; (2) certain types of cis-acting elements are functionally interchangeable; and (3) cis-regulatory function may change as a consequence of virion-delivered transactivators, the cellular or environmental milieu, or genetic background of the CMV strain or BAC construct.

Cell cycle phase determines human CMV MIE enhancer activation state (see also Chapter I.14). Permissive cells infected in the G_0/G_1-phase of the cell cycle support MIE gene expression. The MIE genes are not expressed in primary cells infected in the S/G_2-phase of the cell cycle despite viral genome penetration into the cell nucleus (Fortunato *et al.*, 2002). MIE gene expression can be forced to occur in S/G_2 by treatment with chemical inhibitors of the cellular proteasome or cyclins (Fortunato *et al.*, 2002; Zydek *et al.*, 2010). The S/G_2-dependent block in viral gene expression extends to other human CMV IE-class genes, human CMV strains, and permissive cell types. In contrast, murine CMV expresses its MIE genes without restrain in the S/G_2-phase of the cell cycle in both mouse and human fibroblasts, whereas human CMV only expresses its MIE genes in murine fibroblasts in the G_0/G_1-phase of the cell cycle at time of infection (Zydek *et al.*, 2011). Neither the replacement of murine CMV's MIE enhancer with the human CMV MIE enhancer nor the addition of human CMV virion particles imparts cell cycle dependency to murine CMV MIE gene expression. Conversely, the addition of murine CMV virion particles does not counteract the S/G_2-dependent block to human CMV infection. The S/G_2-dependent block in human CMV MIE gene expression does not occur in transient assays of a human CMV genome fragment containing the MIE genes and regulatory region (Zydek *et al.*, 2011). The findings support the possibility that human CMV specifies a factor(s) that participates in bringing about the S/G_2-dependent gene expression silencing.

Serum factors influence human CMV MIE enhancer activity. In HF arrested in the G_0-phase by growth to confluence and serum-starvation, the addition of serum at the time of human CMV infection substantially increases the level of human CMV MIE RNA expression (Keller *et al.*, 2003). Difference in level of MIE RNA expression between serum-replete versus serum-deprived infections increases as MOI decreases. In serum-deprived non-dividing HF, the κB mutations in a clinical-like human CMV recombinant (VR1814/FIX BAC) result in a decrease in MIE RNA expression and viral replication (Caposio *et al.*, 2007). This abnormal viral phenotype is not observed in the same cells amply furnished with serum (Caposio *et al.*, 2007), a result that is mirrored by studies of laboratory CMV strains containing κB mutations (CMV Ad169 and Towne) (Gustems *et al.*, 2006; Liu *et al.*, 2010a). Mutations in either the ELK-1 or SRF binding site in a clinical-like human CMV recombinant (VR1814/FIX BAC) result in a decrease in viral replication in serum-deprived, non-dividing HF at MOI of 0.1 PFU per cell, while this defect is less evident in serum-replete proliferating HF (Caposio *et al.*, 2010). Chemical inhibition of the MEK1/2 signalling pathway or RNAi-mediated knockdown of ELK-1 or SRF attenuates MIE RNA expression from laboratory CMV Ad169 at high-MOI in serum-deprived non-dividing HF, but not in serum-nourished proliferating HF (Caposio *et al.*, 2010). The collective findings highlight the importance of cellular and environmental conditions in modulating the regulatory pathways that act on the enhancer.

The human CMV MIE enhancer contains GC-box repeats in both its proximal and distal segments. Some of the GC-boxes overlap with a TGGGC$^A/_G$ motif, which is repeated throughout the enhancer as an overlapping and non-overlapping element (Fig. I.10.2). The TGGGC$^A/_G$ sequence contributes to enhancer activity at low MOI in HF (C.S. Galle and J.L. Meier, unpublished data). It is bound *in vitro* by a yet unidentified cellular protein complex, which does not bind to the GC-box. Expectedly, Sp1 and Sp3 in nuclear extract of infected HF bind to the enhancer GC-box *in vitro* (Lang *et al.*, 1992; Isomura *et al.*, 2005). The specific transcription factors that bind to the GC-box, TGGGC$^A/_G$, and overlapping hybrid sequences in living infected cells remain to be determined. Incorporating base-substitution mutations in all of the TGGGC$^A/_G$, GC-box, and overlapping hybrid elements in the enhancer of human CMV Towne (Towne BAC) results in a decrease in IE1/IE2 gene expression at ≥ 6 hpi at low MOI in HF (C.S. Galle and Meier J.L., unpublished data). The post-IE deficiency in IE1/IE2 gene expression is not dependent on viral DNA synthesis and is not observed at high MOI. Complexity in arrangement of these repeats throughout the entire enhancer provides redundancy in structure and function, because the elimination of segments of this cis-regulatory network has little effect on functional outcome. The discovery that this cluster of cis-acting elements does not contribute to MIE gene expression during the IE timeframe of infection suggests that an enhancer cis-regulatory

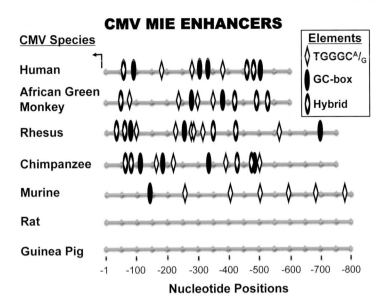

Figure I.10.2 Comparison of the quantity and spatial organization of TGGGC$^{A/}_{G}$ motifs, GC-boxes, and overlapping hybrids of these elements in the MIE enhancers of human, African green monkey, rhesus monkey, chimpanzee, murine, rat, and guinea pig CMVs. The TGGGC$^{A/}_{G}$ motifs (open diamonds), GC-boxes (black ovals), hybrid elements (open diamond within black oval) are depicted in the MIE enhancers of the indicated CMV species. The elements are aligned at nucleotide position −1 relative to the +1 RNA start-site of the MIE promoter (leftward arrow) and in an orientation that they naturally sit in their respective viral genomes. The 5′-extent of enhancers for rat (−840) and guinea pig (−1280) CMVs are not shown and are devoid of these elements.

code has evolved to sustain rather than to initiate expression of the human CMV IE1 and IE2 genes.

Human CMV IE1/IE2 gene expression in the post-IE timeframe is regulated at multiple levels. The repertoire of viral and cellular proteins involved in modulating the level of MIE enhancer activation in early/late stage of infection varies somewhat from the proteins which are involved in the IE stage of infection. Part of this early/late stage-modulation is linked to viral DNA synthesis. A decrease in human CMV IE1 RNA expression at ≥ 10 hpi can result from deletion of the UL29 gene (Ad169 BAC) (Terhune et al., 2010). The viral UL29/28 protein (pUL29/28) stimulates the MIE enhancer/promoter in transient assays and stimulates IE1 RNA accumulation, as well as accumulation of RNA from IE-class US3 and UL37 genes, through a mechanism that likely involves the nucleosome remodelling and histone deacetylase (NuRD) complex (Mitchell et al., 2009; Terhune et al., 2010). pUL29/28 binds to the chromatin-remodelling NuRD complex and knockdown of NuRD components CHD4 or RBB4 decreases IE1 RNA at 10 hpi (Terhune et al., 2010). The human CMV pUL21a protein increases IE-class gene expression in the late stage of the viral life cycle via a mechanism that is linked to viral DNA synthesis (Fehr and Yu, 2011). Interestingly, disruption of UL21a in the viral genome (Ad169 BAC) lowers

levels of IE2 RNA and protein, but not IE1 RNA and protein. The cis-repression sequence (crs) located between nucleotide positions −13 and −1 relative to the human CMV IE1/IE2 RNA start-site doubles as a binding site for IE2-86 protein and an unidentified cellular protein of ~ 150 kDa (Macias and Stinski, 1993, 1996). Binding of IE2-86 protein to the crs results in repression of transcription from MIE enhancer/promoter fragments in transient transfection and in vitro transcription assays. Swapping the MIE crs in the human CMV genome with the murine CMV crs, which differs in 11 of the 13 nucleotides, results in a large increase in the ratio of unspliced-to-spliced MIE RNA, which consequently impairs IE1/IE2 protein production and viral replication in HF (Isomura et al., 2008). The simian CMV crs differs in sequence by only one nucleotide and functions well as a substitute for the human CMV crs. However, changing as few as two or three additional nucleotides within the native crs of human CMV is enough to reproduce the abnormalities caused by a murine CMV crs swap. The abnormality develops in the absence of new protein synthesis (Isomura et al., 2008). A functionally intact crs is needed for IE2-86 protein-mediated generation of a repressive chromatin at the MIE enhancer/promoter at the post-IE time after infection (Cuevas-Bennett and Shenk, 2008). Other viral cis-regulatory elements and cellular factors, including those involved in innate antiviral mechanisms, probably vie for a negative regulatory role in the post-IE stage of enhancer/promoter activity. Post-transcriptional mechanisms of regulation also operate in governing balance in the amounts of IE1 and IE2 RNAs and proteins. The human CMV microRNA (miR-UL112-1) that targets IE1 message and decreases IE1-72 protein abundance at early/late stages of HF infection is one example of this form of regulation (Grey et al., 2007; Murphy et al., 2008; see also Chapter I.5).

The evolutionary divergence of the CMV species (see also Chapter I.1) has resulted in MIE enhancers that differ in type and arrangement of consensus binding sites for specific cellular transcription factors. Human and non-human primate CMV species have four copies of κB elements in their enhancers, whereas the rat CMV enhancer (Maastricht strain) has none. The TGGGC$^A/_G$, GC-box, and overlapping hybrid elements are densely clustered in human and non-human primate CMV enhancers and are not present in rat or guinea pig CMV MIE enhancers (Fig. I.10.2). Primate CMV enhancers also have five or more copies of CRE elements, and the murine CMV enhancer has only one CRE. The murine CMV MIE enhancer operates as a bipartite regulatory unit that activates the transcriptionally divergent IE1/IE3 and IE2 genes. Each part of the bipartite enhancer can function separately, but both parts together function synergistically in increasing levels of murine CMV MIE gene expression (Kropp et al., 2009) without increasing viral replication rate. By contrast, human CMV has the transcriptionally divergent UL127 promoter located immediately upstream of the enhancer and is silent in lytically infected HF. Removing a negative cis-acting element in the unique region allows UL127 promoter activation at early/late times of HF infection (Lashmit et al., 1998; Lundquist et al., 1999; Angulo et al., 2000). The structural differences between the enhancers of different CMV species suggest that cis-regulatory code has co-evolved with the host.

In nuclei of lytically infected cells, human CMV genomes take on a structure reminiscent of cellular chromatin (Kierszenbaum and Huang, 1978; St. Jeor et al., 1982), which is partly organized into an ordered nucleosomal array (Nitzshe et al., 2008; see also Chapter I.19). The innovation of the chromatin immunoprecipitation (ChIP) approach enabled use of the formaldehyde-cross-linking method to show that CMV genomes are coupled to cellular histones in lytically infected nuclei. The histones are attached to promoters, open reading frames, and non-coding segments of CMV genomes in infected cells, but not in virion particles (Nevels et al., 2004b; Reeves et al.,

2004, 2005, 2006; Ioudinkova et al., 2006; Yee et al., 2007; Cuevas-Bennett and Shenk, 2008). ChIP-based studies have not addressed the question about extent of heterogeneity and proportional distribution in superstructures formed by viral DNA–RNA-protein interactions that likely change in accord with time after infection, cell type, and environmental cues. Post-translational modification of histones is governed by the actions of signalling pathways and transcription factors and provides one means of epigenetic regulation of gene expression (see also Chapter I.19). Histone modifications that signify transcriptionally permissive chromatin, i.e. acetylation of histone H3 Lys9 and Lys14 and methylation of H3 Lys4, become concentrated at active CMV genes at times in the viral life cycle that mirror the temporal pattern of expression of these genes (Yee et al., 2007; Cuevas-Bennett and Shenk, 2008). The CMV MIE enhancer/promoter is no exception in this regard. A high level of pro-transcription histone modifications marks the human CMV MIE enhancer/promoter at 6 hpi in fibroblasts (Cuevas-Bennett and Shenk, 2008). A decrease in these marks at 12 hpi is prevented by treatment with an inhibitor of histone deacetylases (HDAC) or protein synthesis. Chromatin marks of repression, i.e. histone H3 Lys9 methylation and HP1, increase at ≥ 24 hpi. The accumulation of these repressive marks reflect the negative autoregulation conferred by the IE2-86 protein binding to the crs element (Cuevas-Bennett and Shenk, 2008) and interacting with HDAC1 and the histone methyltransferases G9a and Suvar(3–9) H1 (Reeves et al., 2006). In fibroblasts synchronized by starvation and infected at low MOI, the MIE promoter and other IE-class promoters bear chromatin marks of repression at 3 hpi (Groves et al., 2009). Shortly thereafter, the repressive chromatin marks decline in relation to the burst in IE RNA expression. During the early phase of infection, the IE1-72 protein is involved in alleviating the marks of repressive chromatin at the MIE enhancer/promoter (Nevels et al., 2004b). These findings are consistent with the idea that incoming human CMV genomes initially contend with an overarching mechanism of transcriptional repression and that at some of these viral genomes subsequently undergo de-repression.

Cellular nuclear domains 10 (ND10 domains; also named PML-associated nuclear bodies or oncogenic domains) play a role in the global repression of CMV genomes. ND10 are rich in chromatin modifying enzymes and other transcriptional regulators (e.g. hDAXX, PML, ATRX and HDAC). Specific virion proteins entering the cell (e.g. pp71), as well as viral proteins synthesized during infection (e.g. IE1-72), counteract the transcriptional repressive effects of the ND10 complex. This multifaceted interplay is discussed in more detail in Chapters I.9 and I.11 of this book. Notably, the cell cycle-dependent block in human CMV MIE gene expression does not appear to involve hDAXX, PML, or HDAC activity (Zydek *et al.*, 2011).

Recent findings of Kropp *et al.* (2009) suggest that murine CMV's MIE enhancer increases probability of transcriptional initiation to achieve a threshold level of MIE RNA expression that drives a fixed rate of viral replication *in vivo* and in acutely infected fibroblasts in culture. Exceeding this threshold does not accelerate viral replication. While the human CMV MIE enhancer also serves to increase probability of transcriptional initiation, there is only circumstantial evidence to suggest that exceeding a threshold level of MIE RNA expression is sufficient for clearing way to a fixed level of subsequent viral life cycle events (Du *et al.*, 2011).

Enhancer quiescence

Murine CMV latently infects sinusoidal endothelial cells, resting neuronal stems cells, and renal peritubular epithelial cells (Tsutsui *et al.*, 2002; Seckert *et al.*, 2009). Monocytes and dendritic cells do not appear to be major sites of murine CMV latency (Marquardt *et al.*, 2011). Transcription from the murine CMV MIE enhancer/promoter is greatly restricted during the latent infection, with the exception of sporadic blips in transcription that occur in some host conditions (Kurz *et al.*, 1999; Grzimek *et al.*, 2001; Simon *et al.*, 2005). These sporadic blips in CMV gene expression are discussed as a molecular basis of perpetual immune stimulation and T-cell activation during viral latency (Holtappels *et al.*, 2000; Simon *et al*, 2006; reviewed in Reddehase *et al.*, 2008; see Chapter I.22), an important finding of relevance for CMV-based vaccine vectors (see Chapter II.21). In kidneys of latently infected mice, the quiescent murine CMV MIE enhancer/promoter is bound by cellular YY1, CBF1/RBK-Jk, and CBF-associated co-repressor CIF, as determined by ChIP method (Liu *et al.*, 2010b). These factors likely help bring about the collection at this location of chromatin modifications that signify epigenetic silencing (Liu *et al.*, 2008; see Chapter I.22). Other cellular factors, such as ND10 components (Tang and Maul, 2003), are also likely involved in restricting murine MIE enhancer/promoter activity during viral latency, whereas CpG methylation is not found in the quiescent murine MIE enhancer/promoter *in vivo* (Hummel *et al.*, 2007).

In human CMV-seropositive healthy persons, the precursors of macrophage–dendritic cells carry latent human CMV genomes, in which the MIE promoter is transcriptionally inactive until reactivated by cellular differentiation and stimulation (Taylor-Wiedeman *et al.*, 1991, 1994; Maciejewski *et al*, 1992; Kondo *et al.*, 1994; Soderberg-Naucler *et al.*, 1997, 2001; Hahn *et al.*, 1998). Messenger RNAs mapping to both DNA strands of the human CMV MIE locus are detected in a small subset of latently infected monocytic–dendritic cell precursors, but not in peripheral blood monocytes (Kondo *et al.*, 1994, 1996; Hahn *et al.*, 1998; Slobedman and Mocarski, 1999; Lofgren-White *et al.*, 2000). Messages expressed from the same DNA strand as the MIE genes originate from within the enhancer at nucleotide positions -292 and -356, with respect to the +1 start site for lytic cycle transcription. The largest of the open reading frames in these mRNAs corresponds to UL126 and codes for the ORF94 protein (Kondo *et al.*, 1994, 1996). UL126's function is unknown and is dispensable for latent and lytic infections in cultured cells (Lofgren-White *et al.*, 2000). Transcription associated with human CMV latency is further discussed in Chapter I.20.

Studies performed years ago in conditionally permissive cells derived from undifferentiated monocytic and pluripotent NTera2 (NT2) lineages had determined that the human CMV MIE enhancer/promoter and subsequent lytic cycle events are inactive in the undifferentiated cells, but are active in the terminally differentiated cellular counterparts (Gonczol *et al.*, 1984). The establishment of viral quiescence in these cells simulates human CMV latency in primitive haematopoietic progenitors, thus providing a tractable model in which to characterize the regulatory pathways involved in silencing of human CMV MIE expression (Meier, 2001). Because NT2 have features of pluripotent stem cells, they are able to become neurons and astrocytes after treatment with retinoic acid or become epithelial and smooth muscle-like cells after treatment with bone morphogenetic protein-2. Results from these models have pointed to a shift in balance from positive to negative regulatory mechanisms that reduce enhancer-dependent transcription in quiescently infected cells (Meier and Stinski, 1996).

The negative cis-regulatory mechanisms that silence the human MIE enhancer/promoter in undifferentiated cells remain poorly understood. Three copies of the 21-bp repeat in the enhancer, as well as the upstream modulator, decrease MIE enhancer/promoter activity in reporter plasmids transfected into undifferentiated NT2 and THP-1 cells (Nelson *et al.*, 1987; Kothari *et al.*, 1991; Sinclair *et al.*, 1992). Cellular transcriptional repressors YY1, ETS2-repressor factor (ERF), and/or silencing binding protein (SBP) bind to these negative cis-acting elements *in vitro* (Fig. I.10.1) (Liu *et al.*, 1994; Huang *et al.*, 1996; Thrower *et al.*, 1996; Bain *et al.*, 2003). However, removal of the 21-bp repeats, modulator, or both 21-bp repeats and modulator from the viral genome does not alleviate MIE enhancer/promoter silencing in quiescently infected cells (Meier and Stinski, 1997; Meier, 2001). Transient assays also implicate cellular growth factor independence-1 (Gfi-1) (Zweidler-McKay et al., 1996) and CCAAT/enhancer binding proteins (Prösch et al., 2001) in enhancer/promoter silencing. However, mutation of the two Gfi-1 binding sites in the human CMV genome fails to relieve silencing in undifferentiated monocytic cells (R. Schnetzer and M. Stinski, unpublished data). Whether enhancer quiescence in the intact viral genome is brought about by extensive redundancy in previously recognized negative cis-acting sites or from undiscovered negative cis-regulatory mechanisms remains to be resolved.

Cellular ND10-associated regulators (e.g. hDAXX, PML, ATRX and HDAC) contribute to CMV genome-wide transcriptional repression in the different quiescently infected cell models examined. This mechanism of repression is addressed in Chapters I.9 and I.11. The MIE enhancer/promoter silence is further sponsored by heterochromatin components (e.g. hypoacetylated histones, methylated histones, heterochromatin protein 1, and HDAC) that are attracted to this site in the viral genomes in quiescently infected cell models (Murphy et al., 2002; Reeves et al., 2005; Ioudinkova et al., 2006) and in naturally infected dendritic cell precursors (Reeves et al., 2004; Reeves and Sinclair, 2010; see also Chapter I.19). Additional mechanisms of repression are likely operating in these systems and await future characterization. Virus-specified determinants also likely play a role in human MIE enhancer/promoter silencing. In this nonpermissive cellular environment, the virion particle components fail to activate MIE promoter-dependent transcription. The inability of viral pp71 to translocate from cytoplasm to nucleus in pluripotent NTera2 cells, monocytic THP-1 cells, and primary CD34+ haematopoietic progenitor cells correlates with lack of MIE gene expression in these cells (Saffert and Kalejta, 2007, 2010). Conversely, pp71 translocation to nuclei of the differentiated cellular counterparts correlates with MIE gene expression (Saffert and Kalejta, 2007, 2010; see also Chapter I.9).

Enhancer's role in reactivation

The frequency of detectable CMV reactivation events increases as host cellular immunity declines. The conditions and pathways that underlie CMV reactivation in the latently infected host are best detailed for the murine system. Allogeneic stimulation and acute inflammation reactivate MIE promoter-dependent transcription, which is first among multiple checkpoints in place to control the murine CMV reactivation cascade (Kurz and Reddehase, 1999; Simon et al., 2005; reviewed by Reddehase et al., 2008). The proinflammatory cytokine TNF-alpha induces murine CMV MIE promoter reactivation in latently infected mice (Hummel et al., 2001; Simon et al., 2005). In the absence of TNF-alpha receptor, other stimuli can reactive the CMV MIE promoter (Zhang et al., 2009). In latently infected BALB/c mice, either intra-abdominal sepsis or intraperitoneal infusion of sublethal doses of LPS (endotoxin), TNF-alpha, or IL-1 beta results in pulmonary reactivation of latent murine CMV (Cook et al., 2006). These findings suggest that pro-inflammatory stimuli are involved in inducing murine CMV enhancer/promoter reactivation in vivo. Signalling-mediated activation via NF-κB and AP-1 transcription factors that bind to the enhancer is proposed as a mechanism by which the pro-inflammatory stimuli operate (Hummel and Abecassis, 2002).

Latent human CMV in blood monocytes and CD34+ haematopoietic progenitor cells of healthy donors reactivates after induction of dendritic cell differentiation and subsequent stimulation with LPS or IL-6 (Reeves et al., 2004; Reeves and Compton, 2011). Reactivation by this approach is also achieved for macrophage-dendritic cell precursors that are experimentally infected in culture (Reeves et al., 2005; Hargett and Shenk, 2010; Reeves and Sinclair, 2010). In an experimental latency-myeloid cell system, the addition of TNF-α stimulation in conjunction with cellular differentiation promotes CMV reactivation (Hahn et al., 1998). In all studies, the induction of MIE promoter-dependent transcription marks the initiation of the CMV reactivation cascade. This transcription is accompanied by change from a repressive chromatin structure to a transcriptionally permissive chromatin structure at the MIE enhancer/promoter (Reeves et al., 2004, 2005; Reeves and Sinclair, 2010). Immature dendritic cells derived from naturally or experimentally infected blood CD14+ monocytes, via differentiation in the absence of serum, are highly susceptible to IL-6 stimulation-induced activation of human CMV MIE gene expression, which occurs within 24 hours of the IL-6 stimulation (Reeves and Compton, 2011). Chemical inhibitors of cellular ERK activity block the IL-6-activated expression of viral MIE RNA, suggesting that IL-6 might be acting on the MIE enhancer/promoter through the ERK/MAPK signalling pathway (Reeves and Compton, 2011).

The NT2 model has illuminated enhancer-based mechanisms that operate in reversing human CMV MIE gene silence during a quiescent infection. Stimulation of distinctly different signalling pathways can act separately on the enhancer to partially overcome the silence. MIE gene de-silencing is best achieved through the combined actions of multiple stimuli, signalling pathways, and cis-regulatory elements. For some types of stimuli, cellular differentiation is not required for the initial induction of MIE gene expression (Yuan *et al.*, 2009; Liu *et al.*, 2010a), though cellular differentiation may play a part in increasing efficiency of completion of the reactivation cascade. The molecular mechanisms by which enhancer-dependent transcription is activated are known for several conditions of stimulation:

1   Vasoactive intestinal peptide (VIP), a 28-amino acid immunosuppressive neuropeptide hormone, activates MIE gene expression via the enhancer CRE and the cellular PKA-CREB-TORC2 signalling cascade (Yuan *et al.*, 2009). This gene activation requires VIP-induced PKA-dependent TORC2 Ser171 dephosphorylation and TORC2 nuclear entry. Forskolin (FSK), an adenyl cyclase activator, produces the same outcome via the same signalling pathway (Keller *et al.*, 2007; Yuan *et al.*, 2009).

2   Phorbol 12-myristate 13-acetate (PMA) reverses MIE gene silence by stimulating a PKC-delta dependent, TORC2-independent signalling cascade that acts through cellular CREB and NF-κB (RelA and p50) transcription factors, as well as their cognate binding sites in the MIE enhancer (Liu *et al.*, 2010a).

3   VIP and PMA act together via enhancer CRE and κB to synergistically increase levels of expression of MIE RNA and protein, as well as the proportion of cells yielding this expression (J. Yuan and J.L. Meier, unpublished data).

4   Addition of retinoic acid or vorinostat to VIP and PMA sustains MIE gene expression at a higher level for a longer period of time than is produced by fewer of these stimuli. This durability of enhanced transcriptional output is conferred by the combination of CRE, κB, and other types of cis-regulatory elements located in the MIE enhancer (J. Yuan and J.L. Meier, unpublished data). Notably, retinoic acid alone has minimal effect on MIE gene expression, despite activation of MIE enhancer/promoter fragments in transient assays (Meier, 2001). Vorinostat, an inhibitor of HDAC 1, 2, 3, and 6, activates MIE gene expression through a mechanism that largely relies on enhancer CRE (Yuan J., and Meier, J.L., unpublished data), unlike the action of trichostatin A, an inhibitor of HDAC class I (HDAC 1, 2, 3, and 8) and class II (HDAC 4,

5, 6, 7 and 9) enzymes (Meier, 2001). Neither vorinostat nor trichostatin A activate MIE gene expression from recombinant viruses having mutations in both enhancer CRE and κB (J. Yuan and J.L. Meier, unpublished data).

Collectively, the findings in the NT2 model support the view that human CMV reactivation likely involves the concerted actions of multiple regulatory mechanisms that must first initiate and sustain MIE enhancer/promoter-dependent gene expression.

The MIE transcriptosome

Transcription from the MIE promoter occurs best in cells that are in the G_0/G_1 compartment of the cell cycle and are p53[+/+] WT. The MIE promoter of human CMVs, for reasons that are not completely understood, is inhibited in the presence of active cyclin-dependent kinase one and two (cdk1/2) (Zydek *et al.*, 2010). When the hyperacetylated histones marks on the MIE promoter are associated with activated transcription, the viral transcriptosome is also associated with cellular factors that favour the initiation of transcription such as RNA polymerase II (RNAP II), TATA box binding protein (TBP), TATA binding factor IIB (TFIIB), ckd7, cdk9, and cyclin T1 (Kapasi and Spector, 2008; Kapasi *et al.*, 2009). In addition, a cis-acting element positioned upstream of the transcription start site has an essential role in promoting the initial transcription and RNA splicing at the early stages after infection (Isomura *et al.*, 2008).

The two viral genes downstream of the MIE promoter, UL123 (IE1) and UL122 (IE2) in human CMV, are equivalent in function to the *ie1* and *ie3* genes of murine CMV, respectively. These viral gene products regulate subsequent expression of early and late viral genes, and a failure of robust expression of the MIE gene products can result in either latency or abortive infection. The IE1/*ie1*, as well as the human CMV UL29/28 gene products, contribute to opening the viral chromatin structure by dispersing the ND10s and, consequently, to the inactivation of HDACs that are initially drawn to the viral genome (Ahn *et al.*, 1998; Nevels *et al.*, 2004a; Terhune *et al.*, 2010; Cosme-Cruz *et al.*, 2011). The human CMV IE2-86 protein remains associated with the viral transcriptosome, where it recruits histone acetyltransferases (HATs) and facilitates viral promoter recognition (Sourvinos *et al.*, 2007). Therefore, the IE1/*ie1* and IE2/*ie3* proteins are transactivators of viral transcription. While other viral proteins become associated with the viral transcriptosome early after infection (e.g. human CMV UL112–113 and UL69) and at delayed early and late times after infection (e.g. human CMV UL79, 87 and 95), we will focus here only on the MIE gene products.

Major IE proteins

The CMV MIE genes and their functions have been reviewed previously (Meier and Stinski, 2006; Stinski and Meier, 2007; Stinski and Isomura, 2008). We will attempt to integrate primarily the newly published information for a more up-to-date view of the essential functions of the MIE proteins. The human CMV MIE primary transcript undergoes splicing and polyadenylation at alternative sites to produce multiple mRNAs as shown in Fig. I.10.3. The IE1 mRNAs consist of exons 1, 2, 3 and 4 (UL123). IE1 RNAs of 1.95 and 0.65 kb code viral proteins IE1-72 and IE1-38, respectively (Stenberg et al., 1984; Du et al., 2011). The 0.65 kb mRNA results from a rare splice within exon 4 in infected HF (Shirakata et al., 2002; Du et al., 2011). The IE2 mRNAs consist of exons 1, 2, 3 and 5 (UL122). Alternative splice sites within exon 5 are rarely used depending on the conditions and cell type. However, there are two late IE2 mRNAs that are driven by late promoters either upstream or within exon 5 that play a more dominant role in the viral infection at late

times (Stamminger et al., 1991; White et al., 2007; Fig. I.10.3).

Translation of the mRNAs initiates in exon 2 and consequently the IE1 (491 amino acids, 72 kDa) and IE2 (579 amino acids, 86 kDa) proteins have the first 85 amino acids in common (Stenberg et al., 1984, 1985). The IE2 mRNAs transcribed later in the infection are translated from initiator methionines within exon 5 and consequently these viral proteins, IE2-60 and IE2-40, have the carboxyl amino acids in common with the larger IE2-86 protein (White et al., 2007). Other primate and non-primate CMVs have similarities to human CMV. Since human CMV does not infect animals and murine CMV is the most frequently used animal model, we will frequently compare human and murine CMVs.

After translation of the mRNAs, the MIE proteins are transported to the nucleus where they are post-translationally modified by sumoylation (the covalent attachment of small ubiquitin-like modifiers or SUMO) and phosphorylation, which determines function of the viral proteins in the nucleus of the virus-infected cell (Ahn et al., 2001; Nevels et al., 2004a; Barrasa et al., 2005; Berndt et al., 2009; see also Chapter I.11). Covalent attachment of SUMO-1 or -2 to lysine 450 of IE1-72 gives an apparent molecular mass of 92 kDa.

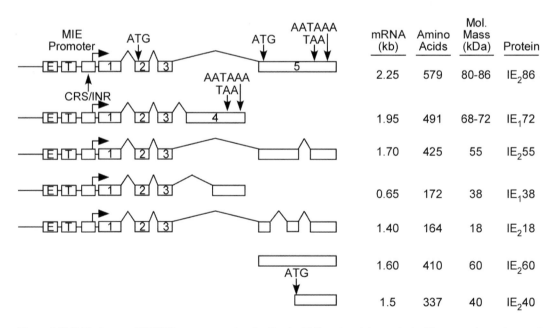

Figure I.10.3 The human CMV MIE genes encode a family of mRNA and protein products. The arrow downstream of the enhancer (E), TATA (T) box, *cis* repression sequence (crs) and initiator (INR) designates the start of transcription for the 5′ end of the viral RNA. AATAAA designates the polyadenylation signal for the 3′ end of the viral RNA. The mRNAs consist of exons 1, 2, and 3 spliced to either exon 4 or 5 and are indicated in kilobases (kb). Splices within exon 4 or 5 also occur. ATG designates the start site for protein translation. The proteins are indicated in number of amino acids. The MIE proteins are designated as IE1 or IE2 and according to their apparent molecular mass in kilodaltons (kDa).

The IE2-86 protein has a non-covalent SUMO interaction motif (SIM) that facilitates covalent sumoylation at lysine residues 175 and 180 to give an apparent molecular mass of 105 kDa. Mutation at these amino acids reduces viral protein and DNA accumulation and therefore, sumoylation in this case favours infectious virus replication (Nevels *et al.*, 2004a; Berndt *et al.*, 2009). The IE2-86 protein is also phosphorylated by ERK and MAPK at threonines (residues 27 and 233) and serines (residues 144 and 234), which positively affects viral replication. In contrast, the serine residues between 258 and 275 can positively or negatively affect viral replication depending on the location of the residue (Barrasa *et al.*, 2005).

Functions of IE1/*ie*1 proteins

The CMV IE1/*ie*1 proteins (acidic proteins) play an important role in the establishment of productive infection by augmenting the activity of the essential IE2/*ie*3 proteins. The most abundantly expressed protein of human CMV immediately after infection is the IE1-72 protein. The viral protein has a nuclear localization signal that targets it to the nucleus and a large central hydrophobic region that targets it to the ND10. The IE1/*ie*1 proteins, which induce dispersal of the structure by interfering with the sumoylation of PML, reduce HDAC activity at the viral genome. Since ND10 components contribute to silencing the viral genomes, dispersal of the ND10 is an important step in productive infection (Tavalai *et al.*, 2008). To insure this step, the human CMV also introduces into the cell a tegument protein designated pp71 that counteracts the repressive activity of the ND10 by reducing the activity of the HDAC associated with the MIE promoter (Kalejta, 2008; see also Chapter I.9). In cell culture, the human CMV IE1 protein is not essential after infection with high MOI, but it contributes to the efficiency of viral replication at low MOI. In addition, the *ie*1 gene of murine CMV is not required for the maintenance of latency or the reactivation from latency in the mouse (Busche *et al.*, 2009). However, deletion of the murine CMV *ie*1 gene attenuates viral growth in the mouse, but not in cell culture (Ghazal *et al.*, 2005; Wilhelmi *et al.*, 2008; Busche *et al.*, 2009). One of the important functions of CMV IE1/*ie*1 viral proteins is the binding to STAT2, which inhibits the type I interferon response. When IE1 protein is sumoylated, it no longer binds to STAT2, but the effect is negligible since sumoylation of IE1 protein is necessary to disperse the ND10, which is critical for neutralizing intrinsic cellular defence mechanisms (Huh *et al.*, 2008; Krauss *et al.*, 2009; Kim and Ahn, 2010). Sumoylation of IE1 protein increases IE2 protein expression, which promotes productive

viral replication (Nevels *et al.*, 2004a; Huh *et al.*, 2008; Nitzshe *et al.*, 2008; Krauss *et al.*, 2009). Recombinant human CMVs lacking IE1 accumulate low levels of early viral gene products and are attenuated for growth in fibroblast cells (Greaves and Mocarski, 1998). The IE1 protein of human CMV is also reported to prevent apoptosis, activate the MIE promoter, stimulate a variety of viral and cellular promoters, bind to metaphase chromatin, and down-regulate expression of LUNA, a transcript antisense to pp71 (UL84) (Meier and Stinski, 2006; Stinski and Meier, 2007; Reeves *et al.*, 2010). The IE1 protein of human CMV augments the activity of the essential IE2 protein by interacting with multiple transcriptional regulators (e.g. CTF-1, SP1, E2F1 to 5, $TAF_{II}130/TAF4$, p107, HDAC-2, PML, and Daxx) and plays an important role in the efficiency of viral replication. In addition, IE1 protein is reported to have kinase activity, which stimulates cellular kinase activities (e.g. JNK, cyclin E/cdk2) to promote viral replication (reviewed in Meier and Stinski, 2006; Stinski and Meier, 2007).

Functions of IE2/*ie*3 proteins

The IE2-86 protein of human CMV locates adjacent to the ND10 with the viral DNA where it interacts with the cellular basal transcription machinery (Sourvinos *et al.*, 2007). Both the IE1 and IE2 viral proteins increase the level of acetylated histones at early viral promoters, which favours productive viral replication (Yee *et al.*, 2007; Cuevas-Bennett and Shenk, 2008). In addition, the IE1-72 and IE2-86 proteins move the cell cycle from G_0/G_1 to the G_1/S interface, as described below in the section, 'Dysregulation of the cell cycle by the MIE proteins'. As the human CMV IE2 protein accumulates in the infected cell, it negatively autoregulates its own expression by binding to an AT-rich crs sequence (5'-CGTTTAGTGAACCG-3'). The crs is located between the MIE promoter's TATA box and mRNA start-site. Inhibition of transcription from the MIE promoter occurs by blocking RNA polymerase II and attracting HDACs to remodel the viral chromatin (reviewed in Stinski *et al.*, 1991, 1993; Stinski 1999; Reeves *et al.*, 2006). Concomitant with an increase in the precursors for DNA synthesis, the IE2 protein (Petrik *et al.*, 2006) and the early viral UL117 protein (Qian *et al.*, 2010) shut down cellular DNA synthesis and consequently, the virus-infected cells are devoted to viral DNA synthesis. Initiation of viral DNA synthesis from the viral origin of DNA replication (OriLyt) requires the IE2-86 protein to interact with the viral UL84 protein (Pari, 2008). The viral protein complex also interacts with viral proteins UL44 and UL112/113 (Kim and Ahn, 2010). As the viral infection transitions from early to late viral protein synthesis, the IE2-60 and -40 proteins are expressed and they also interact with the UL84 protein. Even though these IE2 protein isomers are dispensable, they are required for efficient viral replication (White *et al.* 2007).

The functional roles of the human CMV essential IE2 protein and the critical domains of the viral protein were previously reviewed (Stinski and Petrik, 2008). The human and murine CMV IE2/*ie*3 genes are essential for the cascade of viral gene expression and the production of infectious virus. The human CMV IE2 protein is a multifunctional viral regulatory protein that promotes as well as arrests cell cycle progression, inhibits cellular DNA synthesis, represses the MIE promoter, activates multiple viral and cellular promoters, and limits the expression of pro-inflammatory cytokines. The IE2/*ie*3 proteins must be expressed at sufficient levels to activate early viral gene expression and control the cell cycle, but overexpression is deleterious to the host cell. The IE2-86 protein of human CMV interacts with itself to form a homodimer (amino acids 388 to 542). The homodimer negatively regulates expression from the MIE promoter and binds (amino acids 346 to 579) to the crs between −14 and +1 relative to the +1 transcription start site. The carboxyl end of the viral protein is required for autoregulation. Amino acids within this region are conserved between all primate and non-primate CMV IE2-86 homologues (Petrik *et al.*, 2007). Homodimerization, viral DNA binding, negative autoregulation, and activation of early viral promoters have all been mapped to the carboxyl end of the viral protein (Asmar *et al.*, 2004). While the IE2 protein binding to the crs does not prevent the binding of TBP to the TATA box of the MIE promoter, it interferes with the recruitment of RNA polymerase II and recruits HDACs to remodel the viral chromatin to repress transcription at later stages of infection (reviewed in Stinski *et al.*, 1991, 1993). IE2-86 and IE2-40 binding to the crs are required for controlling the MIE promoter. Mutation of the crs in human CMV is deleterious to virus replication because overexpression of IE2-86 is cytotoxic to the host cell. With murine CMV, an early viral protein (M112/113) is proposed to abrogate auto-repression of the MIE promoter by *ie*3 and promote continued MIE expression (Maul, 2008). With murine and human CMVs, the IE2/ie3 and UL112/113 homologues are found associated with the viral replication centres (RCs) in the nucleus to promote viral DNA replication (Pari, 2008; Martinez *et al.*, 2010).

The human CMV IE2-86 protein also binds to sites upstream of early viral and cellular promoters through crs-like sites (e.g. viral UL4 and UL112; cellular cyclin E) (reviewed in Stinski and Petrik, 2008). In addition,

the IE2-86 protein interacts directly or indirectly with a long list of transcriptional regulatory proteins (TBP, TFIIB, TBP-associated factors (TAFs), TAF12, pRb, p53, p21, mdm2, HAT, CBP/p300, P/CAF, HDACs, methyltransferases, SUMOs, SUMO conjugating enzyme, SUMO ligase, Sp1, Ap-1, CREB, Tef-1, c-jun, c-myc, c-fos, JunB, ATF-2, Egr-1 and others) (reviewed in Meier and Stinski, 2006; Stinski and Petrik, 2008; Kim *et al.*, 2010a). The majority of these viral-host cell protein interactions were deduced from either transient transfection experiments or *in vitro* protein binding experiments. It has been difficult to demonstrate most of these viral-host cell protein interactions in the virus-infected cell. Since the IE2-86 protein interacts with AT-rich sequences in the minor groove of the DNA helix, it interacts in a relatively non-specific manner to assemble higher-order protein complexes. This may explain why the viral protein interacts with a long list of cellular transcription factors and why the viral protein can recognize an extended spectrum of DNA sequence. However, the IE2-86 protein, as well as the IE2-60 and IE2-40 proteins, bind strongly to the viral UL84 protein in the infected cell, which influences recognition of the viral OriLyt site for viral DNA replication initiation (Gao *et al.*, 2008; Pari, 2008).

The domains of the IE2-86 that affect activity were determined by transient transfection assays, *in vitro* protein-protein binding assays, and mutant recombinant virus, and have been reviewed previously (Stinski and Petrik, 2008). Mutations within the IE2 gene frequently cripple the virus or lead to non-viable virus. There is a core domain (amino acids 450 to 552) in the carboxyl end of the protein that affects all the major functions of IE2-86 (Asmar *et al.*, 2004). Within this region, there are conserved amino acid residues that influence the function of the viral protein. The amino acid residues at positions 446 and 452, influence regulation of the MIE promoter (Petrik *et al.*, 2007). The histidine residues at positions 446 and 452 may determine in part the structure of a zinc finger/LIM domain (Hwang, *et al.*, 2009). Both human CMV IE1-72 and IE2-86 have zinc finger/LIM-like domains. The orientation of the zinc finger of IE1-72 ($HX_2HXFX_3LX_2C_4C$) is opposite to that of IE2-86 ($CX_4CXL_5FX_3HX_4H$) (Hwang *et al.*, 2009). In IE2-86, the tyrosine residue at 544 may influence the homodimerization domain. The proline residue at 535 and the proline/tyrosine residues at 535/537 influence transcription from early viral promoters (Petrik *et al.*, 2007; Harris *et al.*, 2010). IE2-86 proteins with mutations at amino acid residues 535 and 537 are not recruited to early viral promoters (UL4 and UL112) (Petrik *et al.*, 2007). In contrast, a third conservative mutation at amino acid 534 retained all tested functions of the IE2-86 protein. These data suggest that given amino acid residues, such as 535, have a critical function in the viral protein, which requires further investigation. While the IE2-86 mutation at 535 failed to regulate the MIE promoter (Harris *et al.*, 2010), the mutation at 535 and 537 was reported to do so (Petrik *et al.*, 2007). These data are complicated because the mutant recombinant BACs cannot replicate and, consequently, one is limited to transient transfection experiments with the viral BAC DNA. However, when the mutation at amino acid residues 535 and 537 are in the IE2-86 protein expressed from a replication deficient adenovirus vector in the absence of other viral protein-binding partners, then the IE2-86 protein failed to regulate the MIE promoter (G. Du and M.F. Stinski, unpublished data). There is also an activation domain (amino acids 550 to 573) carboxyl to the core domain that is rich in acidic amino acids (Yeung *et al.*, 1993; Stinski, 2008).

The effects of the IE2-86 protein on viral and cellular promoters was reviewed previously (Stinski and Petrik, 2008) and are not discussed here. Early viral gene expression in a latently infected cell, such as human CMV-infected undifferentiated THP-1 cells, or in murine CMV-infected mice, requires the expression of both of the MIE proteins (Kurz *et al.*, 1999; Yee *et al.*, 2007). These MIE proteins activate early viral promoters by inactivating HDACs and attracting HATs and, consequently, re-modelling the viral chromatin. The MIE promoters of human and murine CMVs are repressed by IE2/ie3, whereas the early viral promoters are activated. The dynamic changes in the viral chromatin from IE to early and late viral gene expression have been described (Cuevas-Bennett and Shenk, 2008). How these events are triggered is possibly linked with the host cell cycle. Lastly, like IE1-72, the IE2-86 protein is reported to have antiapoptotic activity (Zhu *et al.*, 1995; Bai *et al.*, 2009).

Dysregulation of the cell cycle by the MIE proteins

The CMVs have a unique approach towards regulating the cell cycle (see also Chapter I.14). The virus initiates replication in a terminally differentiated cell in the G_0/G_1 component of the cell cycle. While the virus can infect differentiated cells during all phases of the cell cycle, expression from the human CMV MIE promoter occurs predominantly during the G_0/G_1 compartment of the cell cycle and not during the S phase (Fortunato *et al.*, 2002). High levels of cyclin A cdk1/cdk2 activity during the S phase are thought to inhibit transcription from the MIE promoter (Zydek *et al.*, 2010). While cyclin A is repressed early after human CMV infection, cyclins E and B are

subsequently induced. However, cyclin B/cdk1 activity must be controlled for productive virus replication or an abortive infection will occur (Du *et al.*, 2011). In addition, human CMV requires WT-cellular p53 for efficient initiation of viral transcription from the MIE promoter at the earliest stages of infection (Luo *et al.*, 2007; Hannemann *et al.*, 2009). p53 plays a pivotal role in human CMV infection. Cellular p53 influences viral IE and early gene expression and, consequently, the virus replicates less efficiently in a p53 null cell. As IE2-86 accumulates, it binds to p53 and inhibits p53 transactivation functions (Tsai *et al.*, 1996).

In vitro binding assays indicate that IE1-72 binds to the repressor p107 and IE2-86 binds to pRb. Infection with human CMV or expression of IE2-86 from a replication defective adenovirus vector activates an array of E2F responsive genes in HF cells (Song and Stinski, 2002). As a result, there is an activation of the cellular enzymes and initiation factors for cellular DNA synthesis. With murine CMV, stimulation of nucleotide metabolism for viral replication in cell culture is not necessary (Wilhelmi *et al.*, 2008). The human CMV also inhibits cyclin B/cdk1. A failure to control the levels of cyclin B/cdk1 results in an abortive viral infection (Du *et al.*, 2011). High levels of cyclin B/cdk1 can irreversibly commit the cell to mitosis (Hertel and Mocarski, 2004). The IE2-86 protein, and possibly the IE1-72 protein, plays an important role in stopping cell cycle progression and over- expression of cdk1 activity.

The IE1-72 and IE2-86 proteins of human CMV affect the levels of p53 in the virus-infected cells. IE2-86 binds to ubiquitin E3 ligase mdm-2, thus preventing the ubiquitination and degradation of p53 (Zhang *et al.*, 2006). This enhances phosphorylation of p53 at serine 15 by the ATM (ataxia telangiectasia mutated) serine/threonine protein kinase and, thereby, stabilizes the p53 protein in the infected cell (Zhang *et al.*, 2006). The cellular p53, like the viral IE2-86 protein, localizes to foci containing the viral DNA in the nucleus, which are referred to as pre-replication centres (pre-RCs). In addition, cellular kinase Chk1 is activated and associates with the pre-RCs, which may inactivate Cdc25 phosphatase and prevent the activation of cyclin B/cdk1. Transcription from the MIE promoter continues during the G_1 component of the cell cycle, but the virus stops the cell cycle progression at the G_1/S transition point. Human CMV replicates best in a p53 WT-cell where p53 can contribute to inhibition of cell cycle progression (Holger *et al.*, 2009). The IE2-86 protein of human CMV is both directly and indirectly involved in controlling the cell cycle. The IE2-p86 expressed from a replication defective adenovirus vector can independently stop cell cycle progression in either a p53$^{+/+}$ or a p53$^{-/-}$ cell (Murphy et al., 2000; Song and Stinski, 2004). A single mutation in IE2-86 that substituted arginine for glutamine at amino acid 548 altered the ability of the virus to activate expression from many of the early and late viral genes and to efficiently stop entry into the S phase (Petrik et al., 2006). In contrast, mutation at the same position that substituted alanine for arginine had no effect. These experiments are complicated by the potential for unknown compensatory second mutations that occur for replication of the most-fit virus. However, the level of the WT IE2-86 protein within the infected cell also influences the ability to stop cell cycle progression at G_1/S. Within exon 4 of the human CMV IE1 mRNA is a 24-nucleotide region that encodes conserved amino acids 412–419 of IE1-72. This region prevents an alternative splice in exon 4 in HF cells. Mutation of this site increases the alternate splice to exon 4 and the abundance of IE1-38 mRNA and protein (see Fig. I.10.3), which is accompanied by a slight decrease in IE1-72 and a significant decrease in IE2-86 mRNA and protein (Du et al., 2011). In this context, IE2-86 has all the properties of the WT viral protein, but the viral protein level and the timing and balance of expression are significantly altered. Cells infected with this mutant recombinant virus are delayed in early viral gene expression and produce lower levels of infectious virus. Many of the infected cells enter the S phase, and accumulate in the G_2/M compartment of the cell cycle with high relative levels of cyclin B and cdk1. This results in condensation and fragmentation of the cellular chromatin and a round-cell phenotype (Du et al. 2011). Under these circumstances, the level of the IE2-86 protein fails to control the level of cdk1, the cell progresses towards mitosis, the chromatin is condensed and fragmented, and the IE1 proteins are bound to the chromatin.

The IE2-86 protein of human CMV also controls indirectly the cell cycle. IE2-86 activates the expression of early viral genes, such as UL97 and UL117. The UL97 protein has cdk-like activity, which inactivates Rb-family repressor proteins and activates E2F-1 responsive cellular promoters (Hume et al., 2008). While this prepares the cell for DNA synthesis, UL117 prevents cellular mini-chromosome maintenance complex (MCM) loading at origins of cellular DNA replication and consequently, inhibits cellular DNA synthesis and progression into the S phase (Hume et al., 2008). Failure to arrest cell cycle progression at G_1/S interphase frequently results in an abortive human CMV infection. Cellular DNA synthesis is inhibited with human CMV and, consequently, the accumulation of the precursors for DNA synthesis is used for viral DNA synthesis. Elevated cdk activity and the S phase of the cell cycle prevent MIE gene expression from the human CMV genome. These properties appear to be unique to human CMV, since murine CMV lacks the G_1 dependence for MIE gene expression and, consequently, the virus can express viral genes in the S phase (Wiebusch et al., 2008). Murine CMV can arrest cell cycle progression predominantly in the late G_2 compartment.

Conclusion

The CMV MIE regulatory region and its gene products serve as a master switch in regulating the lytic and latent infection programmes. Silence of MIE gene expression precludes lytic cycle events, which are most likely a detriment to viral persistence in latently infected cells. Conversely, activated expression of these genes initiates the lytic cycle for production of viral progeny. The MIE enhancer governs these outcomes by integrating a diverse array of signals provided by the cell, the virus, and external environment. The enhancer's complex structure affords the regulatory means for effectively initiating and maintaining MIE gene transcription in a variety of permissive cell types. This functional resilience is a major determinant of viral fitness. The human CMV MIE genes encode the multifunctional IE1-72 and IE2-86 proteins, which are vital for advancing the viral lytic life cycle. These proteins directly activate the expression of other essential viral genes and vastly change host cell physiology and behaviour in ways that promote viral replication but may also harm the host. Hence, the mechanisms by which both the MIE genes are regulated and the MIE gene products function are at the core of the genesis of CMV-associated disease. While our knowledge about such mechanisms has advanced greatly, it is far from complete. A better understanding of the enhancer and its gene products has the potential to spawn novel strategies for combating CMV infection and disease.

Acknowledgements

We regret that other publications and contributions have not been included because of space constraints. We are grateful to members of the Stinski and Meier laboratories for their intellectual and scientific contributions.

Our work is supported by grants from the United States Departments of Health and Human Services and Veterans Affairs.

References

Ahn, J.-H., Xu, Y., Jang, W.-J., Matunis, M., and Hayward, G.S. (1998). Disruption of PML subnuclear domains by the acidic IE1 protein of human cytomegalovirus is mediated through interaction with PML and may modulate a RING finger-dependent cryptic transactivator function of PML. Mol. Cell Biol. *18*, 4899–4913.

Ahn, J.-H., Xu, Y., Jang, W.-J., Matunis, M.J., and Hayward, G.S. (2001). Evaluation of interactions of human cytomegalovirus immediate-early IE2 regulatory protein with small ubiquitin-like modifiers and their conjugation enzyme Ubc9. J. Virol. *75*, 3859–3872.

Angulo, A., Messerle, M., Koszinowski, U.H., and Ghazal, P. (1998). Enhancer requirment for murine cytomegalovirus growth and genetic complementation by the human cytomegalovirus enhancer. J. Virol. *72*, 8502–8509.

Angulo, A., Kerry, D., Haung, H., Borst, E.-M., Razinsky, A., Wu, J., Hobom, U., Messerle, M., and Ghazal, P. (2000). Identification of a boundry domain adjacent to the potent human cytomegalovirus enhancer that represses transcription of the divergent UL127 promoter. J. Virol. *74*, 2826–2839.

Asmar, J., Wiebusch, L., Truss, M., and Hagemeier, C. (2004). The putative zinc-finger of the human cytomegalovirus IE286-kilodalton protein is dispensable for DNA-binding and autorepression thereby demarcating a concise core domain in the C-terminus of the protein. J. Virol. *78*, 11853–11864.

Bai, Z., Ling, L., Wang, B., Liu, Z., Wang, H., Yan, Z., Qian, D., Ding, S., and Song, X. (2009). Effect of inducible expressed cytomegalovirus immediate-early 86 protein on cell apoptosis. Biosci. Biotec. Biochem. *73*, 1268–1273.

Bain, M., Mendelson, M., and Sinclair, J. (2003). Ets-2 Repressor Factor (ERF) mediates repression of the human cytomegalovirus major immediate-early promoter in undifferentiated non-permissive cells. J. Gen. Virol. *84*, 41–49.

Barrasa, M.I., Harel, N.Y., and Alwine, J.C. (2005). The phosphorylation status of the serine-rich region of the human cytomegalovirus 86kDa major immediate-early protein IE2/IE86 affects temporal viral gene expression. J. Virol. *79*, 1428–1437.

Benedict, C.A., Angulo, A., Patterson, G., Ha, S., Huang, H., Messerle, M., Ware, C.F., and Ghazal, P. (2004). Neutrality of the canonical NF-κB -dependent pathway for human and murine cytomegalovirus transcription and replication in vitro. J. Virol. *78*, 741–750.

Berndt, A., Hofmann-Winkler, H., Tavalai, N., Hahn, G., and Stamminger, T. (2009). Importance of covalent and noncovalent SUMO interactions with the major human cytomegalovirus transactivator IE2p86 for viral infection. J. Virol. *83*, 12881–12894.

Bresnahan, W.A., and Shenk, T.E. (2000). UL82 virion protein activates expression of immediate-early genes in human cytomegalovirus infected cells. Proc. Natl. Acad. Sci. U.S.A. *97*, 14506–14511.

Busche, A., Marquardt, A., Bleich, A., Ghazal, P., Angulo, A., and Messerle, M. (2009). The mouse cytomegalovirus immediate-early 1 gene is not required for establishment of latency or for reactivation in the lungs. J. Virol. *83*, 4030–4038.

Caposio, P., Luganini, A., Hahn, G., Landolfo, S., and Gribaudo, G. (2007). Activation of the virus-induced IKK/NF-kappaB signalling axis is critical for the replication of human cytomegalovirus in quiescent cells. Cell. Microbiol. *9*, 2040–2054.

Caposio, P., Luganini, A., Bronzini, M., Landolfo, S., and Gribaudo, G. (2010). The Elk-1 and Serum Response Factor binding sites in the major immediate-early promoter of the human cytomegalovirus are required for efficient viral replication in quiescent cells and compensate for inactivation of the NF-{kappa}B sites in proliferating cells. J. Virol. *84*, 4481–4493.

Chao, S.-H., Harada, J.N., Hyndman, F., Gao, X., Nelson, C.G., Chanda, S.K., and Caldwell, J.S. (2004). PDX-1, a cellular homeoprotein, binds to and regulates the activity of human cytomegalovirus immediate-early promoter. J. Biol. Chem. *279*, 16111–16120.

Cherrington, J.M., and Mocarski, E.S. (1989). Human cytomegalovirus ie1 transactivates the a promoter-enhancer via an 18-base-pair repeat element. J. Virol. *63*, 1435–1440.

Cook, C.H., Trgovcich, J., Zimmerman, P.D., Zhang, Y., and Sedmak, D.D. (2006). Lipopolysaccharide, tumor necrosis factor alpha, or interleukin-1b triggers reactivation of latent cytomegalovirus in immunocompetent mice. J. Virol. *80*, 9151–9158.

Cosme-Cruz, R., Martinez, F.P., Perez, K.J., and Tang, Q. (2011). H2B homology region of major immediate-early protein 1 is essential for murine cytomegalovirus to disrupt nuclear domain 10, but is not important for viral replication in cell culture. J. Gen. Virol. *92*, 2006–2019.

Cuevas-Bennett, C., and Shenk, T. (2008). Dynamic histone H3 acetylation and methylation at human cytomegalovirus promoters during replication in fibroblasts. J. Virol. *82*, 9525–9536.

Du, G., Dutta, N., Lashmit, P., and Stinski, M.F. (2011). Alternative splicing of the human cytomegalovirus major immediate-early genes affects infectious-virus replication and control of cellular cyclin-dependent kinase. J. Virol. *85*, 804–817.

Fehr, A.R., and Yu, D. (2011). Human cytomegalovirus early protein pUL21a promotes efficient viral DNA synthesis and the late accumulation of immediate-early transcripts. J. Virol. *85*, 663–674.

Fortunato, E.A., Sanchez, V., Yen, J.Y., and Spector, D.H. (2002). Infection of cells with human cytomegalovirus during S phase results in a blockade to immediate-early gene expression that can be overcome by inhibition of the proteosome. J. Virol. *76*, 5369–5379.

Gao, Y., Colletti, K., and Pari, G.S. (2008). Identification of human cytomegalovirus UL84 virus- and cell-encoded binding partners by using proteomics analysis. J. Virol. *82*, 96–104.

Ghazal, P., Messerle, M., Osborn, K., and Angulo, A. (2003). An essential role for the enhancer for murine cytomegalovirus in vivo growth and pathogenesis. J. Virol. *77*, 3217–3228.

Ghazal, P., Visser, A.E., Gustems, M., Garcia, R., Borst, E.M., Sullivan, K., Messerle, M., and Angulo, A. (2005). Elimination of ie1 significantly attenuates murine cytomegalovirus virulence but does not alter replicative capacity in cell culture. J. Virol. *79*, 7182–7194.

Gonczol, E., Andrews, P.W., and Plotkin, S.A. (1984). Cytomegalovirus replicates in differentiated but not in undifferentiated human embryonal carcinoma cells. Science 224, 159–161.

Greaves, R.F., and Mocarski, E.S. (1998). Defective growth correlates with reduced accumulation of a viral DNA replication protein after low-multiplicity infection by a human cytomegalovirus ie1 mutant. J. Virol. 72, 366–379.

Grey, F., Meyers, H., White, E.A., Spector, D.H., and Nelson, J. (2007). A human cytomegalovirus-encoded microRNA regulates expression of multiple viral genes involved in replication. PLoS Pathog. 3, e136.

Groves, I.J., Reeves, M.B., and Sinclair, J.H. (2009). Lytic infection of permissive cells with human cytomegalovirus is regulated by an intrinsic 'pre-immediate-early' repression of viral gene expression mediated by histone post-translational modification. J. Gen. Virol. 90, 2364–2374.

Grzimek, N.K.A., Podlech, J., Steffens, H.P., Holtappels, R., Schmalz, S., and Reddehase, M.J. (1999). In vivo replication of recombinant murine cytomegalovirus driven by the paralogous major immediate-early promoter-enhancer of human cytomegalovirus. J. Virol. 73, 5043–5055.

Grzimek, N.K.A., Dreis, D., Schmalz, S., and Reddehase, M.J. (2001). Random, asynchronous, and asymmetric transcriptional activity of enhancer-flanking major immediate-early genes IE1/3 and IE2 during murine cytomegalovirus latency in the lungs. J. Virol. 75, 2692–2705.

Gustems, M., Borst, E., Bendict, C.A., Perez, C., Messerle, M., Ghazal, P., and Angulo, A. (2006). Regulation of the transcription and replication cycle of human cytomegalovirus is insensitive to genetic elimination of the cognate NF-kappaB binding sites in the enhancer. J. Virol. 80, 9899–9904.

Gustems, M., Busche, A., Messerle, M., Ghazal, P., and Angulo, A. (2008). In vivo competence of murine cytomegalovirus under the control of the human cytomegalovirus major immediate-early enhancer in the establishment of latency and reactivation. J. Virol. 82, 10302–10307.

Hahn, G., Jores, R., and Mocarski, E.S. (1998). Cytomegalovirus remains latent in a common precursor of dendritic and myeloid cells. Proc. Natl. Acad. Sci. U.S.A. 95, 3937–3942.

Hannemann, H., Rosenke, K., O'Dowd, J.M., and Fortunato, E.A. (2009). The presence of p53 influences the expression of multiple human cytomegalovirus genes at early times postinfection. J. Virol. 83, 4316–4325.

Hargett, D., and Shenk, T.E. (2010). Experimental human cytomegalovirus latency in CD14+ monocytes. Proc. Natl. Acad. Sci. U.S.A. 107, 20039–20044.

Harris, S.M., Bullock, B., Westgard, E., Zhu, H., Stenberg, R.M., and Kerry, J.A. (2010). Functional properties of the human cytomegalovirus IE86 protein required for transcriptional regulation and virus replication. J. Virol. 84, 8839–8848.

Hertel, L., and Mocarski, E.S. (2004). Global analysis of host cell gene expression late during cytomegalovirus infection reveals extensive dysregulation of cell cycle expression and induction of pseudomitosis independent of US28 function. J. Virol. 78, 11988–12011.

Holger, H., Rosenke, K., O'Dowd, J.M., and Fortunato, E.A. (2009). The presence of p53 influences the expression

of multiple human cytomegalovirus genes at early times postinfection. J. Virol. 83, 4316–4325.

Homer, E.G., Rinaldi, A., Nicholl, M.J., and Preston, C.M. (1999). Activation of herpesvirus gene expression by the human cytomegalovirus protein pp71. J. Virol. 73, 8512–8518.

Holtappels, R., Pahl-Seibert, M.F., Thomas, D., and Reddehase, M.J. (2000). Enrichment of immediate-early 1 (m123/pp89) peptide-specific CD8 T-cells in a pulmonary CD62L(lo) memory-effector cell pool during latent murine cytomegalovirus infection of the lungs. J. Virol. 74, 11495–11503.

Huang, T.H., Oka, T., Asai, T., Okada, T., Merrills, B.W., Gerston, R.H., Witson, R.H., and Itakura, K. (1996). Repression by a differentiation-specific factor of the human cytomegalovirus enhancer. Nucleic Acids Res. 24, 1695–1701.

Huh, Y.H., Kim, Y.E., Kim, E.T., Park, J.J., Song, M.J., Zhu, H., Hayward, G.S., and Ahn, J.-H. (2008). Binding STAT2 by the acidic domain of human cytomegalovirus IE1 promotes viral growth and is negatively regulated by SUMO. J. Virol. 82, 10444–10454.

Hume, A.J., Finkel, J.S., Kamil, J.P., Coen, D.M., Culbertson, M.R., and Kalejta, R.F. (2008). Phosphorylation of retinoblastoma protein by viral protein with cyclin-dependent kinase function. Science 320, 797–799.

Hummel, M., Zhang, Z., Yan, S., Deplaen, I., Golia, P., Varghese, T., Thomas, G., and Abecassis, M.I. (2001). Allogeneic transplantation induces expression of cytomegalovirus immediate-early genes in vivo: a model for reactivation from latency. J. Virol. 75, 4814–4822.

Hummel, M., and Abecassis, M. (2002). A model for reactivation of CMV from latency. J. Clin. Virol. 25, S123–S136.

Hummel, M., Yan, S., Zhigao, L., Vaghese, T.K., and Abecassis, M. (2007). Transcriptional reactivation of murine cytomegalovirus ie gene expression by 5-aza-2'-deoxycytidine and trichostatin A in latently infected cells despite lack of methylation of the major immediate-early promoter. J. Gen. Virol. 88, 1097–1102.

Hwang, E.-S., Zhang, Z., Cai, H., Huang, D.Y., Huong, S.-M., Cha, C.-Y., and Huang, E.-S. (2009). Human cytomegalovirus IE1-72 protein interacts with p53 and inhibits p53-dependent transactivation by a mechanism different from that of IE2-86 protein. J. Virol. 83, 12388–12398.

Ioudinkova, E., Arcangeletti, M.C., Rynditch, A., De Conto, F., Motta, F., Covan, S., Pinardi, F., Razin, S.V., and Chezzi, C. (2006). Control of human cytomegalovirus gene expression by differential histone modifications during lytic and latent infection of a monocytic cell line. Gene 384, 120–128.

Isern, E., Gustems, M., Messerle, M., Borst, E., Ghazal, P., and Angulo, A. (2011). The activator protein 1 binding motifs within the human cytomegalovirus major immediate-early enhancer are functionally redundant and act in a cooperative manner with the NF-{kappa}B sites during acute infection. J. Virol. 85, 1732–1746.

Isomura, H., and Stinski, M.F. (2003). Effect of substitution of the human cytomegalovirus enhancer or promoter on replication in human fibroblasts. J. Virol. 77, 3602–3614.

Isomura, H., Tsurumi, T., and Stinski, M.F. (2004). Role of the proximal enhancer of the major immediate-early promoter in human cytomegalovirus replication. J. Virol. 78, 12788–12799.

Isomura, H., Stinski, M.F., Kudoh, A., Daikoku, T., Shirata, N., and Tsurumi, T. (2005). Two Sp1/Sp3 binding sites in the major immediate-early proximal enhancer of human cytomegalovirus have a significant role in viral replication. J. Virol. 79, 9597–9607.

Isomura, H., Stinski, M.F., Kudoh, A., Nakayama, S., Murata, T., Sato, Y., Iwahori, S., and Tsurumi, T. (2008). A cis element between the TATA Box and the transcription start site of the major immediate-early promoter of human cytomegalovirus determines efficiency of viral replication. J. Virol. 82, 849–858.

Kalejta, R.F. (2008). Functions of human cytomegalovirus tegument proteins prior to immediate-early gene expression. In Human Cytomegalovirus, Shenk, T., and Stinski, M.F., eds. (Springer-Verlag, Berlin and Heidelberg), pp. 101–116.

Kapasi, A.J., and Spector, D.J. (2008). Inhibition of the cyclin-dependent kinases at the beginning of human cytomegalovirus infection specifically alters the levels and localization of the RNA polymerase II carboxyl-terminal domain kinases cdk9 and cdk7 at the viral transcriptosome. J. Virol. 82, 394–407.

Kapasi, A.J., Clark, C.L., Tran, K., and Spector, D.J. (2009). Recruitment of cdk9 to the immediate-early viral transcriptosomes during human cytomegalovirus infection requires efficient binding to cyclin T1, a treshold level of IE2 86, and active transcription. J. Virol. 83, 5904–5917.

Keller, M.J., Wheeler, D.G., Cooper, E., and Meier, J.L. (2003). Role of the human cytomegalovirus major immediate-early promoter's 19-base-pair-repeat cAMP-response element in acutely infected cells. J. Virol. 77, 6666–6675.

Keller, M.J., Wu, A.W., Andrews, J.I., McGonagill, P.W., Tibesar, E.E., and Meier, J.L. (2007). Reversal of human cytomegalovirus major immediate-early enhancer/promoter silencing in quiescently infected cells via the cyclic-AMP signaling pathway. J. Virol. 81, 6669–6681.

Kierszenbaum, A.L., and Huang, E.-S. (1978). Chromatin pattern consisting of repeating bipartite structures in WI-38 cells infected with human cytomegalovirus. J. Virol. 28, 661–664.

Kim, E.T., Kim, Y.-E., Huh, Y.H., and Ahn, J.-H. (2010). Role of noncovalent SUMO binding by the human cytomegalvorius IE2 transactivaor in lytic growth. J. Virol. 84, 8111–8123.

Kim, Y.-E., and Ahn, J.-H. (2010). Role of the specific interaction of UL112–113 p84 with UL44 DNA polymerase processivity factor in promoting DNA replication of human cytomegalovirus. J. Virol. 84, 8409–8421.

Kondo, K., Kaneshima, H., and Mocarski, E.S. (1994). Human cytomegalovirus latent infection of granulocyte–macrophage progenitors. Proc. Natl. Acad. Sci. U.S.A. 91, 11879–11883.

Kondo, K.J.X., Xu, J., and Mocarski, E.S. (1996). Human cytomegalovirus latent gene expression in granulocyte–macrophage progenitors in culture and in seropositive individuals. Proc. Natl. Acad. Sci. U.S.A. 93, 11137–11142.

Kothari, S., Baillie, J., Sissons, J.G., and Sinclair, J.H. (1991). The 21bp repeat element of the human cytomegalovirus major immediate-early enhancer is a negative regulator

of gene expression in undifferentiated cells. Nucleic Acids Res. 19, 1767–1771.

Krauss, S., Kaps, J., Czech, N., Paulus, C., and Nevels, M. (2009). Physical requirements and functional consequences of complex formation between the cytomegalovirus IE1 protein and human STAT2. J. Virol. 83, 12854–12870.

Kropp, K.A., Simon, C.O., Fink, A., Renzaho, A., Kühnapfel, B., Podlech, J., Reddehase, M.J., and Grzimek, N.K. (2009). Synergism between the components of the bipartite major immediate-early transcriptional enhancer of murine cytomegalovirus does not accelerate virus replication in cell culture and host tissues. J. Gen. Virol. 90, 2395–2401.

Kurz, S.K., and Reddehase, M.J. (1999). Patchwork pattern of transcriptional reactivation in the lungs indicates sequential checkpoints in the transition from murine cytomegalovirus latency to recurrence. J. Virol. 73, 8612–8622.

Kurz, S.K., Rapp, M., Steffens, H.P., Grzimek, N.K.A., Schmalz, S., and Reddehase, M.J. (1999). Focal transcriptional activity of murine cytomegalovirus during latency in the lungs. J. Virol. 73, 482–494.

Lang, D., Fickenscher, H., and Stamminger, T. (1992). Analysis of proteins binding to the proximal promoter region of the human cytomegalovirus IE-1/2 enhancer/promoter reveals both consensus and aberrant recognition sequences for transcription factors SP1 and CREB. Nucleic Acids Res. 20, 3287–3295.

Lashmit, P.E., Stinski, M.F., Murphy, E.A., and Bullock, G.C. (1998). A cis repression sequence adjacent to the transcription start site of the human cytomegalovirus US3 gene is required to down regulate gene expression at early and late times after infection. J. Virol. 72, 9575–9584.

Lashmit, P., Wang, S., Li, H., Isomura, H., and Stinski, M.F. (2009). The CREB site in the proximal enhancer is critical for cooperative interaction with the other transcription factor binding sites to enhance transcription of the major intermediate-early genes in human cytomegalovirus-infected cells. J. Virol. 83, 8893–8904.

Lee, J., Klase, Z., Gao, X., Caldwell, J.S., Stinski, M.F., Kashanchi, F., and Chao, S.H. (2007). Cellular homeoproteins, SATB1 and CDP, bind to the unique region between the human cytomegalovirus UL127 and major immediate-early genes. Virology 366, 117–125.

Liu, B., and Stinski, M.F. (1992). Human cytomegalovirus contains a tegument protein that enhances transcription from promoters with upstream ATF and AP-1 cis-acting elements. J. Virol. 66, 4434–4444.

Liu, R., Baillie, J., Sissons, J.G., and Sinclair, J.H. (1994). The transcription factor YY1 binds to negative regulatory elements in the human cytomegalovirus major immediate-early enhancer/promoter and mediates repression in non-permissive cells. Nucleic Acids Res. 22, 2453–2459.

Liu, X., Yuan, J., Wu, A.W., McGonagill, P.W., Galle, C.S., and Meier, J.L. (2010a). Phorbol ester-induced human cytomegalovirus MIE enhancer activation through PKC-delta, CREB, and NF-κB de-silences MIE gene expression in quiescently infected human pluripotent NTera2 cells. J. Virol. 84, 8495–8508.

Liu, X.F., Yan, S., Abecassis, M., and Hummel, M. (2008). Establishment of murine cytomegalovirus latency in vivo is associated with changes in histone modifications and recruitment of transcriptional repressors to the major immediate-early promoter. J. Virol. 82, 10922–10931.

Liu, X.F., Yan, S., Abecassis, M., and Hummel, M. (2010b). Biphasic recruitment of transcriptional repressors to the murine cytomegalovirus major immediate-early promoter during the course of infection in vivo. J. Virol. 84, 3631–3643.

Lofgren-White, K., Slobedman, B., and Mocarski, E.S. (2000). Human cytomegalovirus latency-associated protein 94 is dispensable for productive and latent infection. J. Virol. 74, 9333–9337.

Lundquist, C.A., Meier, J.L., and Stinski, M.F. (1999). A strong transcriptional negative regulatory region between the human cytomegalovirus UL127 gene and the major immediate-early enhancer. J. Virol. 73, 9032–9052.

Luo, M.H., Rosenke, K., Czornak, K., and Fortunato, E.A. (2007). Human cytomegalovirus disrupts both ataxia telangiectasia mutated protein (ATM)- and ATM-Rad3-related kinase-mediated DNA damage responses during lytic infection. J. Virol. 81, 1934–1950.

Macias, M.P., and Stinski, M.F. (1993). An in vitro system for human cytomegalovirus immediate-early 2 protein (IE2)-mediated site-dependent repression of transcription and direct binding of IE2 to the major immediate-early promoter. Proc. Natl. Acad. Sci. U.S.A. 90, 707–711.

Macias, M.P., and Stinski, M.F. (1996). Cellular or viral protein binding to a cytomegalovirus promoter transcription initiation site: effects on transcription. J. Virol. 70, 3628–3635.

Maciejewski, J.P., Bruening, E.E., Donahue, R.E., Mocarski, E.S., Young, N.S., and St Jeor, S.C. (1992). Infection of haematopoietic progenitor cells by human cytomegalovirus. Blood 80, 170–178.

Marquardt, A., Halle, S., Seckert, C.K., Lemmermann, N.A., Veres, T.Z., Braun, A., Maus, U.A., Förster, R., Reddehase, M.J., Messerle, M., et al. (2011). Single cell detection of latent cytomegalovirus reactivation in host tissue. J. Gen. Virol. 92, 1279–1291.

Martinez, F.P., Cruz Cosme, R.S., and Tang, Q. (2010). Murine cytomegalovirus major immediate-early protein 3 interacts with cellular and viral proteins in viral DNA replication compartments and is important for early gene activation. J. Gen. Virol. 91, 2664–2676.

Maul, G.G. (2008). Initiation of cytomegalovirus infection at ND10. In Human Cytomegalovirus, Shenk, T., and Stinski, M.F., eds. (Springer-Verlag, Berlin and Heidelberg,), pp. 117–132.

Meier, J.L. (2001). Reactivation of the human cytomegalovirus major immediate-early regulatory region and viral replication in embryonal NTera2 cells: Role of trichostatin A, retinoic acid, and deletion of the 21-base-pair repeats and modulator. J. Virol. 75, 1581–1593.

Meier, J.L., and Pruessner, J. (2000). The human cytomegalovirus major immediate-early distal enhancer region is required for efficient viral replication and immediate-early gene expression. J. Virol. 74, 1602–1613.

Meier, J.L., and Stinski, M.F. (1996). Regulation of cytomegalovirus immediate-early genes. Intervirology 39, 331–342.

Meier, J.L., and Stinski, M.F. (1997). Effect of a modulator deletion on transcription of the human cytomegalovirus major immediate-early genes in infected undifferentiated and differentiated cells. J. Virol. 71, 1246–1255.

Meier, J.L., and Stinski, M.F. (2006). Major immediate-early enhancer and its gene products. In Cytomegaloviruses: Molecular Biology and Immunity, Reddehase, M.J., ed. (Caister Academic Press, Norfolk, UK), pp. 151–166.

Meier, J.L., Keller, M.J., and McCoy, J.J. (2002). Requirement of multiple cis-acting elements in the human cytomegalovirus major immediate-early distal enhancer for activation of viral gene expression and replication. J. Virol. 76, 313–320.

Mitchell, D.P., Savaryn, J.P., Moorman, N.J., Shenk, T., and Terhune, S.S. (2009). Human cytomegalovirus UL28 and UL29 open reading frames encode a spliced mRNA and stimulate accumulation of immediate-early RNAs. J. Virol. 83, 10187–10197.

Murphy, E., Vanícek, J., Robins, H., Shenk, T., and Levine, A.J. (2008). Suppression of immediate-early viral gene expression by herpesvirus-coded microRNAs: implications for latency. Proc. Natl. Acad. Sci. U.S.A. 105, 5453–5458.

Murphy, E.A., Streblow, D.N., Nelson, J.A., and Stinski, M.F. (2000). The human cytomegalovirus IE86 protein can block cell cycle progression after inducing transition into the S phase of permissive cells. J. Virol. 74, 7108–7118.

Murphy, J.C., Fischle, W., Verdin, E., and Sinclair, J.H. (2002). Control of cytomegalovirus lytic gene expression by histone acetylation. EMBO J. 21, 1112–1120.

Nelson, J.A., Reynolds-Kohler, C., and Smith, B. (1987). Negative and positive regulation by a short segment in the 5′-flanking region of the human cytomegalovirus major immediate-early gene. Mol. Cell. Biol. 7, 4125–4129.

Netterwald, J., Yang, S., Wang, W., Ghanny, S., Cody, M., Soteropoulas, P., Tian, B., Dunn, W., Liu, F., and Zhu, H. (2005). Two gamma interferon-activated site-like elements in the human cytomegalovirus major immediate-early promoter/enhancer are important for viral replication. J. Virol. 79, 5035–5046.

Nevels, M., Brune, W., and Shenk, T. (2004a). SUMOylation of the human cytomegalovirus 72-kilodalton protein facilitates expression of the 86-kilodalton IE2 protein and promotes viral replication. J. Virol. 78, 7803–7812.

Nevels, M., Paulus, C., and Shenk, T. (2004b). Human cytomegalovirus immediate-early 1 protein facilitates viral replication by antagonizing histone deacetylation. Proc. Natl. Acad. Sci. U.S.A. 101, 17234–17239.

Nitzshe, A., Paulus, C., and Nevels, M. (2008). Temporal dynamics of cytomegalovirus chromatin assembly in productively infected human cells. J. Virol. 82, 11167–11180.

Pari, G.S. (2008). Nuts and bolts of human cytomegalovirus lytic DNA replication. In 'Human Cytomegalovirus', Shenk, T., and Stinski, M.F., eds. (Springer-Verlag, Berlin and Heidelberg), pp 153–166.

Petrik, D.T., Schmitt, K.P., and Stinski, M.F. (2006). Inhibition of cellular DNA synthesis by the human cytomegalovirus IE86 protein is necessary for efficient virus replication. J. Virol. 80, 3872–3883.

Petrik, D.T., Schmitt, K.P., and Stinski, M.F. (2007). The autoregulatory and transactivating functions of the human cytomegalovirus IE86 protein use independent mechanisms for promoter binding. J. Virol. *81*, 5807–5818.

Podlech, J., Pintea, R., Kropp, K.A., Fink, A., Lemmermann, N.A., Erlach, K.C., Isern, E., Angulo, A., Ghazal, P., and Reddehase, M.J. (2010). Enhancerless cytomegalovirus is capable of establishing a low-level maintenance infection in severely immunodeficient host tissues but fails in exponential growth. J. Virol. *84*, 6254–6261.

Prösch, S., Heine, A.-K., Volk, H.D., and Kruger, D.H. (2001). CAAT/enhancer-binding proteins a and b negatively influence capacity of tumor necrosis factor a to up-regulate the human cytomegalovirus IE1/2 enhancer/promoter by nuclear factor kB during monocyte differentiation. J. Biol. Chem. *276*, 40712–40720.

Qian, Z., Leung-Pineda, V., Xuan, B., Piwnica-Worms, H., and Yu, D. (2010). Human cytomegalovirus protein pUL117 targets the mini-chromosome maintenance complex and suppresses cellar DNA synthesis. PLoS Pathogens 6, e1000814.

Rauwel, B., Mariamé, B., Martin, H., Nielsen, R., Allart, S., Pipy, B., Mandrup, S., Devignes, M.D., Evain-Brion, D., Fournier, T., et al. (2010). Activation of peroxisome proliferator-activated receptor gamma by human cytomegalovirus for de novo replication impairs migration and invasiveness of cytotrophoblasts from early placentas. J. Virol. *84*, 2946–2954.

Reddehase, M.J., Simon, C.O., Seckert, C.K., Lemmermann, N., and Grzimek, N.K. (2008). Murine model of cytomegalovirus latency and reactivation. In 'Human Cytomegalovirus', Shenk, T., and Stinski, M.F., eds. (Springer-Verlag, Berlin and Heidelberg), pp. 315–331.

Reeves, M., Murphy, J., Greaves, R., Fairley, J., Brehm, A., and Sinclair, J. (2006). Autorepression of the human cytomegalovirus major immediate-early promoter/enhancer at late times of infection is mediated by recruitment for chromatin remodeling enzymes by IE86. J. Virol. *80*, 9989–10009.

Reeves, M., Woodhall, D., Compton, T., and Sinclair, J. (2010). Human cytomegalovirus IE72 protein interacts with the transcriptional repressor hDaxx to regulate LUNA gene expression during lytic infection. J. Virol. *84*, 7185–7194.

Reeves, M.B., and Compton, T. (2011). Inhibition of inflammatory interleukin-6 activity via extracellular signal-regulated kinase-mitogen-activated protein kinase signaling antagonizes human cytomegalovirus reactivation from dendritic cells. J. Virol. *85*, 12750–12758.

Reeves, M.B., and Sinclair, J.H. (2010). Analysis of latent viral gene expression in natural and experimental latency models of human cytomegalovirus and its correlation with histone modifications at a latent promoter. J. Gen. Virol. *91*, 599–604.

Reeves, M.B., MacAry, P.A., Lehner, P.J., Sissons, J.G., and Sinclair, J.H. (2004). Latency, chromatin remodeling and reactivation of human cytomegalovirus in the dendritic cells of healthy carriers. Proc. Natl. Acad. Sci. U.S.A. *102*, 4140–4145.

Reeves, M.B., Lehner, P.J., Sissons, J.G., and Sinclair, J.H. (2005). An in vitro model for the regulation of human cytomegalovirus latency and reactivation in dendritic cells by chromatin remodeling. J. Gen. Virol. *86*, 2949–2954.

Romanowski, M.J., Garrido-Guerrero, E., and Shenk, T. (1997). pIRS1 and pTRS1 are present in human cytomegalovirus virions. J. Virol. *71*, 5703–5705.

Saffert, R.T., and Kalejta, R.F. (2007). Human cytomegalovirus gene expression is silenced by Daxx-mediated intrinsic immune defense in model latent infections established in vitro. J. Virol. *81*, 9109–9120.

Saffert, R.T., and Kalejta, R.F. (2010). Cellular and viral control over the initial events of human cytomegalovirus experimental latency in CD34+ cells. J. Virol. *84*, 5594–5604.

Sanford, G.R., Brock, L.E., Voight, S., Forester, C.M., and Burns, W.H. (2001). Rat cytomegalovirus major immediate-early enhancer switching results in altered growth characteristics. J. Virol. 75, 5076–5083.

Schierling, K., Stamminger, T., Mertens, T., and Winkler, M. (2004). Human cytomegalovirus tegument proteins ppUL82 (pp71) and ppUL35 interact and cooperatively activate the major immediate-early enhancer. J. Virol. *78*, 9512–9523.

Seckert, C.K., Renzaho, A., Tervo, H.M., Krause, C., Deegen, P., Kühnapfel, B., Reddehase, M.J., and Grzimek, N.K. (2009). Liver sinusoidal endothelial cells are a site of murine cytomegalovirus latency and reactivation. J. Virol. *83*, 8869–8884.

Shirakata, M., Terauchi, M., Ablikim, M., Imadone, K.-I., Hirai, K., Aso, T., and Yamanashi, Y. (2002). Novel immediate-early protein IE19 of human cytomegalovirus activates the origin recognition complex I promoter in a cooperative manner with IE72. J. Virol. *76*, 3158–3167.

Simon, C.O., Seckert, C.K., Dreis, D., Reddehase, M.J., and Grzimek, N.K. (2005). Role of tumor necrosis factor alpha in murine cytomegalovirus transcriptional reactivation in latently infected lungs. J. Virol. *79*, 326–340.

Simon, C.O., Holtappels, R., Tervo, H.M., Böhm, V., Däubner, T., Oehrlein-Karpi, S.A., Kühnapfel, B., Renzaho, A., Strand, D., Podlech, J., et al. (2006). CD8 T-cells control cytomegalovirus latency by epitope-specific sensing of transcriptional reactivation. J. Virol. *80*, 10436–10456.

Sinclair, J., and Sissons, P. (1996). Cytomegalovirus: Latent and persistent infection of monocytes and macrophages. Intervirology 39, 293–301.

Sinclair, J.H., Baillie, J., Bryant, L.A., Taylor-Wiedeman, J.A., and Sissons, J.G. (1992). Repression of human cytomegalovirus major immediate-early gene expression in a monocytic cell line. J. Gen. Virol. 73, 433–435.

Sinzger, C., and Jahn, G. (1996). Human cytomegalovirus cell tropism and pathogenesis. Intervirology *39*, 302–319.

Slobedman, B., and Mocarski, E. (1999). Quantitative analysis of latent human cytomegalovirus. J. Virol. 73, 4806–4812.

Soderberg-Naucler, C., and Nelson, J.A. (1999). Human cytomegalovirus latency and reactivation – a delicate balance between the virus and its host's immune system. Intervirology *42*, 314–321.

Soderberg-Naucler, N.C., Fish, K.N., and Nelson, J.A. (1997). Reactivation of latent human cytomegalovirus by allogenic stimulation of blood cells from healthy donors. Cell *91*, 119–126.

Soderberg-Naucler, C., Streblow, D.N., Fish, K.N., Allan-Yorke, J., Smith, P.P., and Nelson, J.A. (2001). Reactivation of latent human cytomegalovirus in CD14+ monocytes is differentiation dependent. J. Virol. *75*, 7543–7554.

Song, Y.-J., and Stinski, M.F. (2002). Effect of the human cytomegalovirus IE86 protein on expression of E2F-responsive genes: a DNA microarray analysis. Proc. Natl. Acad. Sci. U.S.A. *99*, 2836–2841.

Song, Y.-J., and Stinski, M.F. (2004). Inhibition of cell division by the human cytomegalovirus IE86 protein: Role of the p53 pathway or cdk1/cyclin B1. J. Virol. *79*, 2597–2603.

Sourvinos, G., Tavalai, N., Berndt, A., Spandidos, D.A., and Stamminger, T. (2007). Recruitment of human cytomegalovirus immeidate-early 2 protein onto parental viral genomes in association with ND10 in live-infected cells. J. Virol. *81*, 10123–10136.

St. Jeor, S., Hall, C., McGraw, C., and Hall, M. (1982). Analysis of human cytomegalovirus nucleoprotein complexes. J. Virol. *41*, 309–312.

Stamminger, T., Puchtler, E., and Fleckenstein, B. (1991). Discordant expression of the immediate-early 1 and 2 gene regions of human cytomegalovirus at early times after infection involves posttranscriptional processing events. J. Virol. *65*, 2273–2282.

Stenberg, R.M., Thomsen, D.R., and Stinski, M.F. (1984). Structural analysis of the major immediate-early gene of human cytomegalovirus. J. Virol. *49*, 190–199.

Stenberg, R.M., Witte, P.R., and Stinski, M.F. (1985). Multiple spliced and unspliced transcripts from human cytomegalovirus immediate-early region 2 and evidence for a common initiation site within immediate-early region 1. J. Virol. *56*, 665–675.

Stinski, M.F. (1999). Cytomegalovirus promoter for expression in mammalian cells. In Gene Expression Systems: Using Nature for the Art of Expression, Ferandez, J.M., and Hoeffler, J.P., eds. (Academic Press, San Diego, CA), pp. 211–233.

Stinski, M.F., and Isomura, H. (2008). Role of the cytomegalovirus major immediate-early enhancer in acute infection and reactivation from latency. Med. Microbiol. Immunol. *197*, 223–231.

Stinski, M.F., and Meier, J.L. (2007). Immediate-early viral gene regulation and function. In Human Herpesviruses Biology, Therapy, and Immunoprophylaxis, Arvin, A., Campadelli-Fiume, G., Mocarski, E., Moore, P.S., Roizman, B., Whitley, R., and Yamanishi, K., eds. (Cambridge University Press, New York), pp 241–263.

Stinski, M.F., and Petrik, D.T. (2008). Functional roles of the human cytomegalovirus essential IE86 protein. In Human Cytomegalovirus, Shenk, T., and Stinski, M.F., eds. (Springer-Verlag, Berlin and Heidelberg), pp. 133–152.

Stinski, M.F., Malone, C.L., Hermiston, T.W., and Liu, B. (1991). Regulation of human cytomegalovirus transcription. In Herpesvirus transcription and its control, Wagner, E.K., ed. (CRC Press, Boca Raton, FL), pp. 245–260.

Stinski, M.F., Macias, M.P., Malone, C.L., Thrower, A.R., and Huang, L. (1993). Regulation of transcription from the cytomegalovirus immediate-early promoter by cellular and viral proteins. In Multidisciplinary

Approach to Understanding Cytomegalovirus Disease, Michelson, S., and Plotkin, S.A., eds. (Elisver Science, New York), pp. 3–12.

Tang, Q., and Maul, G.G. (2003). Mouse cytomegalovirus immediate-early protein 1 binds with host cell repressors to relieve suppressive effects on viral transcription and replication during lytic infection. J. Virol. *77*, 1357–1367.

Tavalai, N., Papior, P., Rechter, S., and Stamminger, T. (2008). Nuclear domain 10 components promeylocytic leukemia protein and hDaxx independently contribute to an intrinsic antiviral defense against human cytomegalovirus infection. J. Virol. *82*, 126–137.

Taylor-Wiedeman, J., Sissons, J.G., Borysiewicz, L.K., and Sinclair, J.H. (1991). Monocytes are a major site of persistence of human cytomegalovirus in peripheral blood mononuclear cells. J. Gen. Virol. *72*, 2059–2064.

Taylor-Wiedeman, J.A., Sissons, J.G.P., and Sinclair, J.H. (1994). Induction of endogenous human cytomegalovirus gene expression after differentiation of monocytes from healthy carriers. J. Virol. *68*, 1597–1604.

Terhune, S.S., Moorman, N.J., Cristea, I.M., Savaryn, J.P., Cuevas-Bennett, C., Rout, M.P., Chait, B.T., and Shenk, T. (2010). Human cytomegalovirus UL29/28 protein interacts with components of the NuRD complex which promote accumulation of immediate-early RNA. PLoS Pathogens 6, e1000965.

Thrower, A.R., Bullock, G.C., Bissell, J.E., and Stinski, M.F. (1996). Regulation of a human cytomegalovirus immediate-early gene (US3) by a silencer/enhancer combination. J. Virol. *70*, 91–100.

Tsai, H.L., Kou, G.H., Chen, S.C., Wu, C.W., and Lin, Y.S. (1996). Human cytomegalovirus immediate-early protein IE2 tethers a transcriptional repression domain to p53. J. Biol. Chem. *271*, 3534–3540.

Tsutsui, Y., Kawasaki, H., and Kosugi, I. (2002). Reactivation of latent cytomegalovirus infection in mouse brain cells detected after transfer to brain slice cultures. J. Virol. *76*, 7247–7254.

White, E.A., Del Rosario, C.J., Sanders, R.L., and Spector, D.H. (2007). The IE2 60-kilodalton and 40-kilodalton proteins are dispensable for human cytomegalovirus replication but are required for efficient delayed early and late gene expression and production of infectious virus. J. Virol. *81*, 2573–2583.

Wiebusch, L., Neuwirth, A., Grabenhenrich, L., Voigt, S., and Hagemeier, C. (2008). Cell cycle-independent expression of immediate-early gene 3 results in G1 and G2 arrest in murine cytomegalovirus-infected cells. J. Virol. *82*, 10188–10198.

Wilhelmi, V., Simon, C.O., Podlech, J., Bohm, V., Däuber, T., Emde, S., Strand, D., Renzaho, A., Lemmermann, N.A., Seckert, C.K., et al. (2008). Transactivation of cellular genes involved in nucleotide metabolism by the regulatory IE1 protein of murine cytomegalovirus is not critical for viral replicative fitness in quiescent cells and host tissues. J. Virol. *82*, 9900–9916.

Winkler, M., and Stamminger, T. (1996). A specific subform of the human cytomegalovirus transactivator protein pUL69 is contained within the tegument of virus particles. J. Virol. *70*, 8984–8987.

Yee, L.-F., Lin, P.L., and Stinski, M.F. (2007). Ectopic expression of HCMV IE72 and IE86 proteins is sufficient to induce early gene expression but not production of infectious virus in undifferentiated promonocytic THP-1 cells. Virology 363, 174–188.

Yeung, K.C., Stoltzfus, C.M., and Stinski, M.F. (1993). Mutations of the cytomegalovirus immediate-early 2 protein defines regions and amino acid motifs important in transactivation of transcription from the HIV-1 LTR promoter. Virology 195, 786–792.

Yuan, J., Liu, X., Wu, A.W., McGonagill, P.W., Keller, M.J., Galle, C.S., and Meier, J.L. (2009). Breaking human cytomegalovirus major immediate-early gene silence by vasoactive intestinal peptide stimulation of the protein kinase A-CREB-TORC2 signaling cascade in human pluripotent embryonal NTera2 cells. J. Virol. 83, 6391–6401.

Zhang, Z., Evers, D.L., McCarville, J.F., Dantonel, J.C., Huong, S.M., and Huang, E.S. (2006). Evidence that the human cytomegalovirus IE2-86 protein binds mdm2 and facilitates mdm2 degradation. J. Virol. 80, 3833–3843.

Zhang, Z., Li, Z., Yan, S., Wang, X., and Abecassis, M. (2009). TNF-alpha signaling is not required for in vivo transcriptional reactivation of latent murine cytomegalovirus. Tranplantation 88, 640–645.

Zhu, H., Shen, Y., and Shenk, T. (1995). Human cytomegalovirus IE1 and IE2 proteins block apoptosis. J. Virol. 69, 7960–7970.

Zweidler-McKay, P.A., Grimes, H.L., Flubacher, M.M., and Tsichlis, P.N. (1996). Gfi-1 encodes a nuclear zinc finger protein that binds DNA and functions as a transcripitonal repressor. Mol. Cell. Biol. 16, 4024–4034.

Zydek, M., Hagemeier, C., and Wiebusch, L. (2010). Cyclin-dependent kinase activity controls the onset of the HCMV lytic cycle. PLoS Pathog. 6, e1001096.

Zydek, M., Uecker, R., Tavalai, N., Stamminger, T., Hagemeier, C., and Wiebusch, L. (2011). General blockade of HCMV immediate-early mRNA expression in S/G2 by a nuclear, Daxx- and PML-independent mechanism. J. Gen. Virol. 92, 2757–2769.

Multifaceted Regulation of Human Cytomegalovirus Gene Expression

I.11

Marco Thomas, Nina Reuter and Thomas Stamminger

Abstract

Research of the last two decades revealed that human cytomegalovirus (HCMV) developed a multitude of sophisticated mechanisms to usurp and manipulate the cellular gene expression machinery in order to achieve efficient viral protein synthesis. Furthermore, there is increasing evidence that the virus has to antagonize cellular restriction factors in order to avoid a silencing of viral transcription. This chapter summarizes our present knowledge on how viral regulatory proteins modulate chromatin structure, promoter activities, transcriptional elongation, RNA processing and mRNA export. Interestingly, as exemplified by the pleiotropic effector pUL69 of HCMV, specific viral regulatory proteins appear to be able to affect different cellular machineries, thus providing evidence for an extensive interconnection between transcriptional and post-transcriptional regulatory processes.

Introduction

Nuclear replicating DNA viruses like HCMV have evolved sophisticated mechanisms to exploit the machinery of the host cell for viral RNA polymerization, processing and transport. The compartmentalization of the host cell into nucleus and cytoplasm provides the opportunity for a spatial and temporal separation of transcription and translation offering multifaceted possibilities for the regulation of gene expression. Within the eukaryotic nucleus the cellular genome is organized in highly dynamic chromatin complexes and gene expression is controlled by the transformation of transcriptionally inactive heterochromatin into actively transcribed euchromatin. Gene expression starts with transcription and is accompanied by posttranscriptional processing of the pre-mRNA by 5′ capping, splicing and polyadenylation. Upon nucleocytoplasmic export of the fully processed mRNA, translation initiates within the cytoplasm. At each step of their biogenesis mRNAs are subjected to nuclear and cytoplasmic surveillance mechanisms that remove aberrantly processed and mutated mRNAs to minimize the production of potentially deleterious proteins. Notably, all of the aforementioned steps of eukaryotic gene expression are tightly interconnected and have profound effects on the fate of the respective transcript. The following chapter describes some of the mechanisms how HCMV utilizes and modulates the host cell machinery to achieve efficient viral gene expression (Fig. I.11.1).

Epigenetic regulation of viral transcription

Epigenetic processes, including DNA methylation, histone modification or various RNA-based mechanisms constitute an important regulatory mechanism of how gene activity can be modulated. Although all cells in an organism contain essentially the same copy of genomic DNA, the gene expression profile differs for each cell type due to chromatin-based regulatory mechanisms which control gene expression primarily on the level of transcription. Similarly, epigenetic processes like nucleosome occupancy or post-translational histone modifications are also significantly involved in the regulation of viral transcription and contribute to the outcome of infection. For instance, even though genetically largely identical HCMV genomes reach the nucleus of infected cells, viral gene expression can vary between fully active to almost completely restricted depending on the infected cell type (Gonczol et al., 1985).

Chromatinization of HCMV genomes

On the level of primary chromatin structure, HCMV genomes have been shown to follow a temporal pattern of histone occupancy. Distinct chromatin states can be distinguished which are characterized by the degree of nucleosome content (Paulus et al., 2010). While the encapsidated linear HCMV DNA is completely

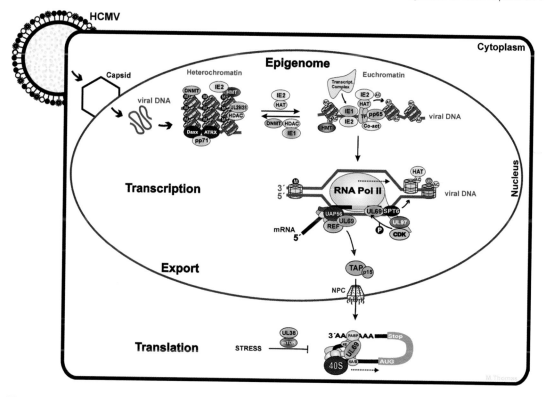

Figure I.11.1 Regulation of HCMV gene expression – an overview. Viral proteins that modulate the host gene expression machinery for efficient viral protein synthesis are colour-highlighted. Briefly: Upon infection, incoming viral genomes are transported into the nucleus where they are silenced by heterochromatinization. This repressive chromatin is relieved by the concerted action of the viral proteins IE1p72, IE2p86, pp71, and pUL29/28 (for a more detailed sequence of these events, see Fig. I.11.2). The viral proteins IE1p72, IE2p86 and pp65 (UL83) then transactivate viral E gene promoters in order to achieve productive cycle viral gene expression. Upon transcription, pUL69 is recruited to elongating mRNA by interacting with hSPT6. Next, pUL69 recruits UAP56- and REF in order to facilitate the nuclear mRNP export via the cellular TAP/p15 export pathway. After translocation of the mRNP into the cytoplasm pUL69 is involved in viral protein translation which is further sustained by the activity of UL38 which maintains mTOR in an active state. Ac, acetyl; CDK, cycline-dependent kinase; Co-act, co-activator; DNMT, DNA methyltransferase; HDAC, histone deacetylase; HAT, histone acetyltransferase; HMT, histone methyltransferase; M, methyl; mTOR, mammalian target of rapamycin; NPC, nuclear pore complex; P, phosphate; PABP, poly(A)-binding protein; REF, RNA export factor; RNA pol II, RNA polymerase II; TAP, tip-associated protein complexed by its cofactor p15; TF, transcription factor; Ub, ubiquitin; 4A/B and 4E, eukaryotic translation initiation factors A/B and E; 40S, small ribosomal subunit.

naked as it is not associated with any chromatin protein (Varnum *et al.*, 2004), already as early as 30 minutes after infection nucleosomes are assembled onto the circularizing HCMV genome within the host cell nucleus, which occurs in a DNA replication-independent manner (Nitzsche *et al.*, 2008). However, during productive infection the nuclear HCMV DNA appears to adopt a state of intermediate chromatin formation since the nucleosomes are irregularly spaced compared to the cellular chromatin (Nitzsche *et al.*, 2008). This low average viral genome chromatinization is maintained throughout the early (E) stage of infection (Nitzsche *et al.*, 2008). Thus, it is tempting to speculate that HCMV has evolved mechanisms

to antagonize chromatinization in order to facilitate efficient viral gene expression since this unique state of viral chromatin presumably allows unrestricted access for transcription factors. As an alternative model, the HCMV genome may be regularly chromatinized albeit with highly unstable nucleosomes that get easily lost by standard chromatin preparation techniques as recently proposed for HSV-1 DNA during lytic infection (Lacasse and Schang, 2010). Unstable nucleosomes are characterized by the incorporation of certain histone variants like H3.3, H3t or H2AZ, which are known to be predominantly assembled onto DNA during the process of DNA replication-independent nucleosome depositioning. Interestingly, the nuclear domain 10

(ND10) protein hDaxx (in collaboration with ATRX) has recently been shown to act as a histone chaperone for replication-independent deposition of H3.3 (Drane *et al.*, 2010). Since hDaxx is initially antagonized, but strongly up-regulated at early/late (E/L) times of HCMV infection (Tavalai *et al.*, 2011), it is tempting to speculate that this up-regulation at E/L times is associated with a function of hDaxx in the modulation of viral chromatinization. Finally, coupled with the process of viral DNA synthesis one can observe an extensive increase in nucleosome density during the late phase of infection (Nitzsche *et al.*, 2008). This is in clear contrast to the histone-free DNA, that can be found in HCMV virions and implies that the replicated, chromatinized viral genomes have to undergo a complete disassembly before they are packaged into capsids. However, the underlying mechanisms of this disassembly await further investigation.

Role of nuclear domain 10 (ND10) proteins for chromatin-mediated regulation of HCMV gene expression

In the meantime, it is well established that epigenetic-based processes play a crucial role in determining the outcome of HCMV infection. Especially the chromatin structure around the major immediate-early enhancer-promoter (MIEP) (see also Chapters I.10 and I.19) is of major importance since it provides an epigenetic switch that defines whether a lytic or a latent HCMV infection occurs (Sinclair, 2010; Nevels *et al.*, 2011; Reeves, 2011). Here, we focus on chromatin-mediated regulation of viral lytic gene expression since epigenetic regulation during HCMV latency and reactivation is covered by Chapter I.19. During lytic infection the chromatin state of the HCMV genome determines the efficiency of productive infection as all classes of lytic HCMV promoters seem to be subject to regulation by histone modifications. Immediately upon infection, HCMV genomes are targeted by an intrinsic defence mechanism of the cell which attempts to silence viral transcription by recruitment of transcriptionally repressive chromatin marks. Incoming viral DNA can be found in close proximity to a subnuclear structure known as ND10 (alternatively termed PML bodies based on the presence of promyelocytic leukaemia, PML, antigen) (Maul, 1998), which is generally accepted as a site of epigenetic regulation (Torok *et al.*, 2009). Our own studies and those of other laboratories have shown that ND10 bodies act as an antiviral environment for HCMV (Tavalai and Stamminger, 2008, 2009, 2011). All three major ND10 components, PML (Tavalai *et al.*, 2006, 2008), hDaxx (Cantrell and Bresnahan, 2006; Preston and Nicholl, 2006; Saffert

and Kalejta, 2006; Woodhall *et al.*, 2006; Tavalai *et al.*, 2008) and Sp100 (Adler *et al.*, 2011; Kim *et al.*, 2011; Tavalai *et al.*, 2011), along with the chromatin-remodelling factor ATRX (Lukashchuk *et al.*, 2008), were identified as host restriction factors that cooperate to induce a state of pre-immediate-early repression of viral IE gene expression (Groves *et al.*, 2009) (see Fig. I.11.2). The hDaxx protein was the first ND10 component for which evidence was provided that suppression is mediated on an epigenetic level. By recruitment of histone deacetylase (HDAC) activity, the hDaxx protein is capable of inducing a transcriptionally inactive chromatin structure around the MIEP which is characterized by hypoacetylated (H3/H4) as well as hypermethylated histones (H3K9me2) and the recruitment of HP-1β (Woodhall *et al.*, 2006; Groves *et al.*, 2009). Accordingly, knock-down of hDaxx prior to infection or pre-treatment of cells with the HDAC inhibitor trichostatin A (TSA) is clearly correlated with a loss of repressive histone modifications associated with the MIEP and results in transcription of the IE gene locus (Woodhall *et al.*, 2006). Only recently, Kim and colleagues could demonstrate that Sp100 contributes to the epigenetic control of the MIE promoter, since ablation of Sp100 expression similarly enhances the acetylation level of histones around the MIEP (Kim *et al.*, 2011). Whether the repressive effect of PML on HCMV IE gene expression is likewise based on epigenetic regulatory mechanisms still has to be determined. However, PML has also been shown to physically interact with a series of chromatin modifying enzymes that are all linked to transcriptional silencing (Wu *et al.*, 2001; Di *et al.*, 2002; Carbone *et al.*, 2006; Villa *et al.*, 2007; Cho *et al.*, 2011).

Viral antagonism of chromatin-mediated silencing

In order to efficiently initiate the viral gene expression programme, HCMV has evolved mechanisms to compromise the antiviral properties of PML nuclear bodies (Tavalai and Stamminger, 2011; see also Chapter I.9) (Fig. I.11.2). As a first line of defence to overcome ND10-based restriction, the imported HCMV structural protein pp71 induces the dissociation of ATRX from ND10 (Lukashchuk *et al.*, 2008). Thereafter, pp71 inactivates hDaxx's antagonistic function by promoting its proteasomal hydrolysis (Saffert and Kalejta, 2006). This converts the viral chromatin into a transcriptionally permissive state that facilitates viral IE gene expression as the MIEP becomes associated with transcriptionally active histone marks like acetylated H4, H3K9 or H3K14. In addition, the HCMV-encoded protein pUL29/28

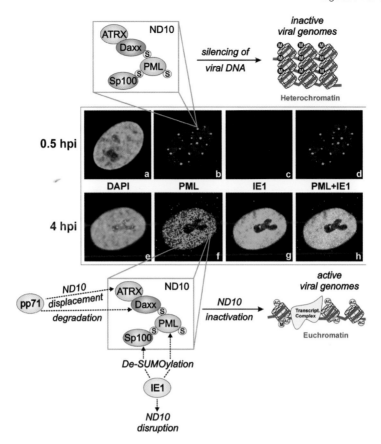

Figure I.11.2 ND10-based epigenetic regulation of viral IE transcription and its antagonization by HCMV. Directly upon infection, incoming HCMV DNA is subject to cellular epigenetic regulatory mechanisms which are instituted by the antiviral structure ND10. Already as early as 0.5 hours post infection, ND10-localized host restriction factors have converted HCMV genomes into a heterochromatic, transcriptionally inactive state through the recruitment of transcriptionally non-conducive histone marks. In order to efficiently initiate the lytic gene expression programme, the chromatin-mediated silencing of viral DNA by ND10 has to be overcome by HCMV. This is achieved by an HCMV-based sequential inactivation of ND10 which starts with the pp71-promoted displacement of ATRX from ND10 and its ability to induce the intranuclear proteasomal degradation of hDaxx. The concomitant changes in the chromatinization of HCMV DNA leading to a euchromatin formation, which becomes accessible for transcription factors, allows the initiation of viral IE transcription. This finally results in the expression of the viral effector protein IE1p72 which has the ability to compromise the post-translational SUMO-modification of PML and Sp100, thereby facilitating a complete disruption of ND10. Ac, acetyl; M, methyl; ND10, nuclear domain 10; PML, promyelocytic leukaemia protein; S, SUMO, small ubiquitin-like modifier.

has been shown to interact with components of the nucleosome remodelling and deacetylase (NuRD) complex in order to further enhance the accumulation of viral IE RNA (Terhune *et al.*, 2010) (Fig. I.11.1). The synthesis of high levels of the IE protein IE1p72, in a next step, overcomes PML- as well as Sp100-mediated repression which finally leads to a loss of ND10 integrity (Lee *et al.*, 2004; Tavalai *et al.*, 2011) (Fig. I.11.2). At the same time, however, E and L viral promoters still carry histone modifications being indicative of transcriptional repression like methylated H3K9 (Cuevas-Bennett and Shenk, 2008; Groves *et al.*, 2009). As infection proceeds, dynamic

changes in the chromatin structure around these promoters can be observed. According to their known temporal expression pattern, they gradually become associated with histone modification patterns that are conducive to transcription (acetylated H4, H3K9, and H3K14). This is most probably achieved by the viral transactivators IE1p72 and IE2p86 as for both an interaction with HDACs has been demonstrated (Nevels *et al.*, 2004; Park *et al.*, 2007). By sequestration of HDACs away from E and L promoters, they may act synergistically to drive the coordinated expression of E and L gene products. Hence, it is becoming increasingly clear that epigenetic processes play a marked role in the dynamic regulation of HCMV transcription throughout the course of lytic infection.

Viral promoter activation and transcription

The regulation of HCMV gene expression by *cis*-acting promoter elements and, consequently, the promoter-specific binding of transcription factors, which ultimately leads to the initiation of transcription by cellular RNA polymerases, has been the subject of numerous studies. The majority of HCMV genes is thought to be transcribed by cellular RNA polymerase II (Fig. I.11.1). In contrast to other herpesviruses such as Epstein–Barr virus, no confirmed RNA polymerase III transcription units are yet known for HCMV (Rosa *et al.*, 1981; Marschalek *et al.*, 1989). Viral promoter activation occurs in a temporally regulated cascade consisting of three sequential phases, termed immediate-early (IE), E and L (Demarchi, 1981; Wathen and Stinski, 1982; McDonough and Spector, 1983). Per definition, viral IE genes are transcribed in the absence of *de novo* protein synthesis, while the activation of E and L promoters depends on newly synthesized viral factors. Indeed, IE promoters exhibit a readily detectable constitutive activity which can be further stimulated either by structural proteins of the virus or by cellular signal transduction cascades. In contrast, the constitutive activity of E and L promoters is low in the absence of viral transactivators indicating that cellular transcription factors are not sufficient to stimulate those promoters.

Initiation of immediate-early transcription

Several recent studies demonstrated that the onset of HCMV IE transcription is strictly synchronized with the host cell cycle (see also Chapter I.14). While infected G0/G1 cells support viral IE gene expression, S/G2 cells can be infected but efficiently block IE gene expression (Salvant *et al.*, 1998; Fortunato *et al.*, 2002; Zydek *et al.*, 2010, 2011). Interestingly, this cell cycle-dependent block of IE transcription can be alleviated by cyclin-dependent kinase (cdk) inhibitors, indicating that cdk activity negatively controls the onset of HCMV gene expression. Since activated p53 can inhibit cdk activity, it has been postulated that stimulation of p53 by inflammatory stress could thus serve as a switch for reactivation from latency (Zydek *et al.*, 2010).

The main sites of IE transcription are the genomic loci corresponding to the open reading frames UL122–123 (alternatively termed IE1/IE2; major immediate-early, MIE), UL36–38, US3, and IRS1/ TRS1. Transcription ultimately leading to the synthesis of the most abundant IE proteins IE1p72 (UL123) and IE2p86 (UL122) initiates by the concerted action of viral tegument proteins and cellular transcription factors that interact either directly or indirectly with numerous *cis*-regulatory elements within the well-characterized MIEP, which is located upstream of the IE1/2 transcription start site (Thomsen *et al.*, 1984; Boshart *et al.*, 1985) (for further details on the architecture and regulation of the MIEP, see the preceding Chapter I.10). Upon infection, HCMV induces pro-inflammatory signalling, which in turn transactivates IE gene expression via NFκB, activator protein 1 (AP-1) and gamma interferon-activated site (GAS) elements within the MIEP (Netterwald *et al.*, 2005; Caposio *et al.*, 2007; Isaacson *et al.*, 2008; Isern *et al.*, 2011). MIEP activity is further stimulated by the viral tegument protein pp65 (pUL83), which has recently been shown to recruit the interferon-inducible p200-family protein IFI16 to the MIEP (Cristea *et al.*, 2010). Although the mechanism of stimulation has not been fully clarified, it was speculated that the pUL83–IFI16 complex at the MIEP might affect transcription via regulation of NFκB which binds at four distinct sites within the MIEP (Cristea *et al.*, 2010). In addition, pUL83 blocks the expression of interferon-responsive antiviral genes by mechanisms involving the regulation of both interferon response factor 3 (IRF3) and the NFκB subunit p65 (Browne and Shenk, 2003; Abate *et al.*, 2004). The second viral tegument protein with a well-documented effect on MIEP activity is the phosphoprotein pp71 (Liu and Stinski, 1992). Research of the last years, however, revealed that pp71 does not act as a direct transcriptional activator. In contrast, this protein was shown to antagonize a cellular repression mechanism of MIEP activity that is instituted by the ND10 factors ATRX and hDaxx (Saffert and Kalejta, 2006; Lukashchuk *et al.*, 2008; Tavalai *et al.*, 2008).

Modulation of RNA polymerase II activity by phosphorylation

MIEP-bound transcription factors then recruit the hypophosphorylated cellular RNA polymerase II (RNAPIIA) to the viral genome. There, the C-terminal domain (CTD) of RNAPII gets phosphorylated at serine 5 (Ser_5P) by the cdk 7/cyclin H/MAT1 complex, which is a component of the cellular transcription initiation complex TFIIH. The subsequent CTD-phosphorylation on serine 2 (Ser_2P) is mediated by the cdk 9/cyclin T1 complex (alternatively termed P-TEFb, positive transcription elongation factor b) and coincides with the transit from pausing to elongating

RNAPII (Price, 2000). In order to modulate the cellular environment for optimal viral transcription, HCMV infection induces elevated levels of the cdk-activating kinase (CAK) components cdk 7, cyclin H and MAT 1. Furthermore, it elevates the levels of cdk 9 and cyclin T1 of the P-TEFb-complex, which leads to an increase of hyperphosphorylated RNAPII CTD (Tamrakar et al., 2005). Interestingly, HCMV-infection also induces an RNAPII with intermediate electrophoretic mobility (RNAPIIi) appearing at approximately 24 hpi and becoming the most prominent form from 60 to 72 hpi. However, the functional relevance of this new phosphorylation pattern of RNAPII modification has yet to be determined (Baek et al., 2004). In addition to cellular kinases, RNAPII CTD is also phosphorylated by the HCMV-encoded serine/threonine kinase pUL97, which enters the host cell as a part of the virion tegument. This viral kinase predominantly modifies the CTD at serine 5 in vitro (Baek et al., 2004). The UL97 protein has been demonstrated to be of major importance for viral replication, as deletion of UL97 from the viral genome or pharmacological inhibition of pUL97 kinase activity significantly reduce viral replication (Prichard et al., 1999; Marschall et al., 2002; Krosky et al., 2003; Herget et al., 2004). However, pUL97-mediated CTD phosphorylation likely plays only a moderate role for RNAPII regulation in vivo and for the requirement of pUL97 for optimal viral replication (Baek et al., 2004). Rather, recent studies revealed that pUL97 phosphorylates and thereby inactivates the retinoblastoma tumour suppressor protein (Rb). Thus, this multifunctional viral protein kinase with structural similarity to human cdks plays a critical role for stimulating cell cycle progression in quiescent cells, however, in a manner that is insensitive to cellular cdk regulators, such as cellular p21, that normally attenuate cdk activity (Romaker et al., 2006; Hume et al., 2008; Prichard et al., 2008).

The immediate-early viral transcriptosome

Accumulating evidence indicates that HCMV transcription not only occurs in a temporally but also in a spatially regulated manner. Immunolocalization studies revealed the existence of subnuclear foci, termed viral transcriptosomes, that consist of several viral and cellular components and are thought to function as sites of IE transcription. Initial reports associated viral transcriptosomes with the cellular subnuclear structure ND10 since immunolocalization in many instances detected an apparent juxtaposition of these structures close to PML bodies (Korioth et al., 1996; Ahn and Hayward, 1997; Ahn et al., 1999). Moreover, viral DNA

and IE transcripts were also observed at the periphery of ND10. This gave rise to the hypothesis that the virus develops an immediate transcript environment constrained by pre-existing nuclear structures, suggesting that ND10 sites provide an optimal and even necessary environment for the initiation of viral transcription (Ishov and Maul, 1996; Ishov et al., 1997). However, the recent advent of siRNA technology facilitated the generation of primary human fibroblasts that are highly depleted of endogenous PML protein and thus lack genuine ND10 structures (Everett et al., 2006; Tavalai et al., 2006). HCMV infection of those cells clearly demonstrated that viral transcriptosomes are efficiently assembled in the absence of genuine ND10 (Sourvinos et al., 2007; Tavalai et al., 2008). Together with reports on the high exchange dynamics of ND10 components (Weidtkamp-Peters et al., 2008), on the dynamic association of ND10 with viral transcriptosomes as observed by live cell imaging (Sourvinos et al., 2007) and on the antiviral effects instituted by the ND10 proteins PML, Sp100, hDaxx and ATRX (reviewed in Tavalai and Stamminger, 2011), this supports the view that ND10 factors are transiently deposited at viral transcriptosomes in order to induce a silencing of viral gene expression. However, as suggested by a recent report on herpes simplex virus gene expression (Kalamvoki and Roizman, 2011), the possibility that HCMV usurps single ND10-associated factors for a function in viral transcription cannot be dismissed totally and requires further investigation.

As demonstrated by several studies, the IE protein IE2p86 can serve as a marker protein of viral transcriptosomes and it was shown that the DNA binding activity of this protein determines its supramolecular association with viral DNA to form intranuclear dot-like nucleoprotein complexes (Ishov et al., 1997; Ahn et al., 1999; Sourvinos et al., 2007). However, the IE transcriptosome was reported to contain numerous other viral and cellular factors which include the viral proteins IE1p72, pUL112–113, pUL69, the basal transcription factors TATA box binding protein and TATA binding factor IIB, the regulators of transcription DSIF (DRB sensitivity inducing factor, consisting of hSPT4/hSPT5) and cyclin T1, the chromatin modifying proteins Brd4, HDAC1 and HDAC2, the cdk proteins cdk7 and cdk9 as well as the transcriptionally active hyperphosphorylated RNAPII (RNAPIIO) (Ishov et al., 1997; Tamrakar et al., 2005; Kapasi and Spector, 2008; Kapasi et al., 2009). IE1p72 only transiently co-localizes with the transcriptosome, since this protein is responsible for the dispersal of ND10, thus antagonizing the repressive effect of ND10 (Korioth et al., 1996; Ahn and Hayward, 1997). IE2p86, however, persists at the viral transcription sites as infection progresses, and live cell imaging experiments demonstrated that the viral transcriptosomes develop into replication compartments where viral DNA synthesis occurs (Ahn *et al.*, 1999; Sourvinos *et al.*, 2007). Clearly, further studies will be necessary to define the full complement of viral transcriptosome factors and to elucidate the complex interplay between individual components.

Early and late promoter activation

Viral E gene expression per definition depends on the prior *de novo* synthesis of IE proteins and can thus be blocked by protein synthesis inhibitors such as cycloheximide. In contrast, viral true L gene expression additionally depends on the onset of viral DNA replication and is sensitive to inhibitors of the viral DNA polymerase. Furthermore, some transcription units, termed early-late or delayed-early, are activated at E times but show a further increase in expression at L times of the replicative cycle (for a detailed review on the different subgroups of E genes, see Fortunato and Spector, 1999). Numerous studies investigated the requirements for E and L promoter activation using the methodology of transient cotransfection of various cell types with reporter plasmids harbouring the corresponding viral promoter together with effector plasmids for viral transactivators. Regarding the *cis*-requirements for E and L promoter activation, these studies revealed the following salient features: while true L promoters are very simple in that they consist of a core element, in some cases corresponding to a non-canonical TATA-box (Puchtler and Stamminger, 1991; Depto and Stenberg, 1992; Kohler *et al.*, 1994; Isomura *et al.*, 2008), the architecture of E promoters exhibits a higher complexity. For instance, our group reported that activation of the prototypic E UL112/113 promoter depends on multiple *cis*-acting elements including a binding site for the cellular transcription factor CREB and direct binding sites for the IE protein IE2p86 (Arlt *et al.*, 1994; Lang *et al.*, 1995). Similar results were obtained for other E promoters (Huang *et al.*, 1994; Scully *et al.*, 1995; Kerry *et al.*, 1996, 1997), emphasizing that a binding of individual cellular transcription factors, interacting either directly or indirectly with viral transactivators, is necessary for viral E promoter activation. Transient transfection studies identified the IE protein IE2p86 as the most important viral transactivator of E promoters and this could be confirmed by the generation of a recombinant HCMV with IE2 (UL122) deleted, which fails to express E lytic genes (Pizzorno *et al.*, 1991; Marchini *et al.*, 2001). Furthermore, for a subclass of E promoters a cooperative transactivation together with IE1p72 was detected, while IE1p72 on its own was relatively weak (reviewed in White and Spector, 2007). However, it should be noted that, while transient transfection experiments proved to be very helpful for functional analyses, the interpretation of results requires great caution, since indirect effects might also contribute to increased reporter gene expression. Given the fact that IE1p72 has been identified as a powerful antagonist of intrinsic immunity counteracting a cellular repression mechanism (reviewed in Tavalai and Stamminger, 2011), it will be difficult to discern whether IE1p72 acts as a direct transcriptional activator or whether its observed effects on promoter activation are mediated in an indirect manner. A detailed overview of the mechanisms and functional domains of the two MIE gene products IE1p72 and IE2p86 in promoter activation is presented in Chapter I.10.

In contrast to viral E promoters, the HCMV IE proteins alone turned out to be not sufficient to activate true L promoters (Puchtler and Stamminger, 1991). Only recently, the E gene products encoded by open reading frames UL79, UL87 and UL95 were identified as essential transacting factors for the initiation of L gene expression (Isomura *et al.*, 2011; Perng *et al.*, 2011). These proteins exhibit homology to factors of murine herpesvirus 68 that are also required for true L gene expression, suggesting a general regulatory principle with conservation across the herpesvirus family. All three proteins were found to be recruited to viral replication compartments. Importantly, while mutation of the respective open reading frames abrogated L gene expression, viral DNA replication was not affected, thus discriminating between a function for viral replication control and L gene expression (Isomura *et al.*, 2011). However, the exact mechanisms of how these viral transactivators are able to control specific L genes have yet to be determined.

Regulation of transcriptional elongation

Since the detection of P-TEFb as a host cofactor for the HIV-1 transactivator Tat, there is increasing evidence that viruses extensively modulate the cellular transcriptional machinery at the level of transcriptional elongation (Ott *et al.*, 2011). For instance, as already described, HCMV infection induces the expression of cdk9 and cyclin T1, which are both components of P-TEFb (Tamrakar *et al.*, 2005). Then, a co-localization of cdk9, cyclin T1 and the viral tegument protein pUL69 at viral transcriptosomes was observed and this correlated with the demonstration of a direct protein–protein interaction between pUL69 and cyclin T1 (Kapasi and Spector, 2008; Feichtinger *et al.*, 2011). Furthermore, our group showed that pUL69 interacts with the cellular transcription elongation factor hSPT6 (Winkler *et al.*, 2000) (see Fig. I.11.1). Human hSPT6 along with hSPT5 and hSPT4 form the DSIF complex, that associates with RNA-polymerase II and histones in order to activate or repress transcriptional elongation (Bortvin and Winston, 1996; Hartzog *et al.*, 1998; Endoh *et al.*, 2004). Since a recent publication provided evidence that the interaction of pUL69 with hSPT6 is required for efficient HCMV replication (Cygnar *et al.*, 2012), these results are highly suggestive of modulation of transcriptional elongation by pUL69, although a formal proof for such an activity is still lacking. Additionally, it could be demonstrated that HCMV also subverts cellular Elongin B-mediated mono-ubiquitination of histone H2B, which is the epigenetic mark of actively transcribed genes, for enhancing transcriptional elongation and thus viral gene expression (Aso *et al.*, 1995; Pavri *et al.*, 2006; Hwang *et al.*, 2011). Given the fact that HCMV extensively exploits the cellular transcription machinery it is very likely that viral transcriptional termination as well as cotranscriptional mRNA processing (e.g. 5′ capping and 3′ polyadenylation) are also facilitated by a viral hijacking of cellular proteins (Stamminger *et al.*, 1991; Adair *et al.*, 2004). However, detailed mechanisms for this are still lacking.

Splicing regulation during HCMV infection

Before being transported to the cytoplasm, most pre-mRNAs of higher eukaryotes undergo a number of modifications that include the excision of introns by splicing. Nuclear pre-mRNA splicing takes place in multicomponent structures called spliceosomes, which are formed by an ordered assembly of five UsnRNPs (uridine-rich small nuclear ribonucleoproteins) and associated proteins. Mass spectrometry of spliceosome complexes has produced a catalogue of over 200 polypeptides emphasizing the complexity of regulatory events (Jurica and Moore, 2003). A typical mammalian gene contains nine introns and spans a genomic region of approximately 30 kbp. Owing to limitations in coding capacity, viruses have evolved sophisticated mechanisms for a more economical utilization of genetic information which includes extensive alternative splicing or the use of intronless genes. In fact, a large number of E and L genes of herpesviruses are intronless, whereas alternative splicing is a well known mechanism to increase the diversity of IE gene products. For some herpesviruses, this distinction between spliced IE genes and E as well as L intronless genes offers an attractive possibility for regulation: in herpes simplex virus, the IE protein ICP27 acts as an inhibitor of splicing, which is part of a viral host shut-off mechanism, while the expression of intronless genes is fostered, thus switching to the L phase of viral gene expression (Sandri-Goldin and Mendoza, 1992; Sciabica *et al.*, 2003). This is different for HCMV, where the homologous protein pUL69 does not act as a splicing inhibitor (reviewed in Toth and Stamminger, 2008).

Splicing of immediate-early genes

Extensive splicing is well documented for the two major sites of HCMV IE gene expression, the IE1/IE2 (UL123/UL122) and the UL36–38 gene loci (Stenberg *et al.*, 1985; Tenney and Colberg-Poley, 1991a). Interestingly, it was shown that splicing factors accumulate and co-localize with IE mRNA at and near the sites of active transcription and that this association is reversed by transcription inhibition (Dirks *et al.*, 1997). This indicates that RNA processing may occur independent of the position of the gene in the cell nucleus relative to speckle domains which represent intranuclear accumulations of mRNA splicing regulators. However, similar to the dynamic behaviour of ND10 proteins, active transcription attracts splicing factors, thus explaining the observed juxtaposition of viral transcriptosomes with accumulations of the spliceosome assembly factor SC35 (Dirks *et al.*, 1997; Ishov *et al.*, 1997). The predominant IE transcript IE1p72 (UL123) consists of four exons and specifies the 72-kDa nuclear protein IE1p72. The IE2 gene product, IE2p86 (UL122) is encoded by an alternatively spliced RNA, that contains the first three exons of IE1 plus the alternatively spliced terminal exon 5 (Stenberg *et al.*, 1985). The UL36–38 gene locus gives rise to at least five transcripts directed by three different promoters (Kouzarides *et al.*, 1988; Tenney and Colberg-Poley, 1991a,b; Adair *et al.*, 2003). One of these IE promoters directs the synthesis of several spliced 3.2–3.4 kb RNAs (UL37 and UL37M), that are present in small amounts at IE times. The same promoter also regulates the abundant 1.7 kb unspliced RNA encoding UL37 exon 1 (UL37X1/vMIA, the viral mitochondria-localized inhibitor of apoptosis), which is present from IE to L times of HCMV infection. The second IE promoter controls expression of the 1.65 kb spliced UL36 RNA that increases in abundance at E times after infection and codes for the viral inhibitor of caspase 8 activation (vICA) (see also Chapter I.15). Another promoter is located within UL37X1 and directs the synthesis of the abundant early UL38 transcript with a size of 1.35 kb.

The accuracy of alternative splicing and/or polyadenylation of the respective IE1/IE2 and UL37 transcripts relies on cellular RNA-processing factors binding to *cis* regulatory regions within their pre-mRNAs and is temporally regulated by cdk activity. The region determining expression of unspliced UL37X1 or spliced UL37 transcripts is very similar to the region between IE1 exon 4 and IE2 exon 5 in so far as in both sequences cleavage/polyadenylation signals overlap with signals for the downstream 3' splice acceptor site (Stenberg *et al.*, 1985; Adair *et al.*, 2003; Su *et al.*, 2003). Thus, competition between splicing and polyadenylation factors for binding to the UL37 pre-mRNA *cis* elements conclusively determines production of the UL37X1 unspliced RNA over that of spliced UL37 RNAs (Su *et al.*, 2003), a process that is tightly controlled by cdk 1, 2, 5, 7 and 9 activity (Sanchez *et al.*, 2004). Splicing of the IE1/IE2 genes is regulated analogously, as the potent splicing suppressor polypyrimidine tract-binding protein (PTB) competes with the U2 auxiliary factor 65 (U2AF65) for binding to the polypyrimidine tract in pre-mRNA (Cosme *et al.*, 2009). In line with these observations, HCMV infection not only increases the abundance of PTB, the essential polyadenylation cleavage stimulatory factor 64 (CstF-64), and the hypophosphorylated splicing factor 2, SF2 (Su *et al.*, 2003), but also temporally modulates their subcellular distribution for the regulated processing of UL37-transcripts (Gaddy *et al.*, 2010) and probably also IE1/2 pre-mRNAs. Besides IE1p72 and IE2p86, the MIE gene locus also gives rise to several minor gene products that are encoded by alternatively spliced RNAs, e.g. the transcriptional repressor IE1p19 (Shirakata *et al.*, 2002; Awasthi *et al.*, 2004), IE1p38, the expression of which is regulated by a 24-bp region in IE1 exon 4 (Du *et al.*, 2011), and the MIEP-transactivator protein IE2p55 (Baracchini *et al.*, 1992). The IE2p18 protein, encoded by an alternatively spliced 1.4 kb IE-mRNA, cannot be detected during normal infection but is very abundant in monocyte-derived macrophages or cycloheximide-treated HFFs, thus pointing to a cell type-specific regulation of IE mRNA-processing (Kerry *et al.*, 1995). At L times post infection IE2p60 and IE2p40 are generated from IE-transcripts, and although both are dispensable for HCMV replication they are required for optimal E and L gene expression and thus for productive infection (White *et al.*, 2007).

Other spliced genes of HCMV

Although splicing is extensively used to diversify the gene expression programme at IE times, it is not solely confined to this phase of the replicative cycle. In particular, a recent study aiming at a high resolution definition of the HCMV transcriptome by deep sequencing reported that RNA splicing is more common than recognized previously, affecting 58 protein-coding genes (Gatherer *et al.*, 2011). However, so far, little is known concerning the regulation of these splicing events.

Furthermore, there is some evidence that splicing regulation differs between lytic and latent infection. One example for this is the viral homologue of the potent immunosuppressor interleukin 10 (cmvIL-10), that is expressed from the alternatively spliced UL111.5A transcript (Kotenko *et al.*, 2000; Lockridge *et al.*, 2000). During productive lytic replication UL111.5A is processed to give rise to the 175 amino acid protein cmvIL-10 (Kotenko *et al.*, 2000), whereas in latently infected myeloid progenitor cells, the alternative splice variant, termed latency-associated (LA) cmvIL-10, is produced and translated to a 139 amino acid protein (Jenkins *et al.*, 2004). More recently, LAcmvIL-10 transcripts were also detected in productively infected HFFs, where they are expressed with E kinetics, while cmvIL-10 is expressed with L kinetics (Jenkins *et al.*, 2008b). Although LAcmvIL-10 and cmvIL-10 share their N-terminal 127 amino acids, they differ in their C-terminus due to the presence of an in-frame stop codon within the retained second intron of LAcmvIL-10 (Jenkins *et al.*, 2004). Conclusively, both proteins share some, but also differ in others, of their immunosuppressive capacities, e.g. LPS-activation of dendritic cells or down-regulation of MHC-II expression, and it was suggested that LAcmvIL-10-mediated down-regulation of MHC-II might contribute to the maintenance of HCMV-latency in myeloid progenitor cells (Jenkins *et al.*, 2008a) (see also Chapter I.20 on latency-associated transcripts of HCMV).

Since chromatin modification, transcriptional elongation and splicing are tightly coupled (Munoz *et al.*, 2010; Luco *et al.*, 2011), one might speculate that the so far underestimated splicing during HCMV gene expression (Gatherer *et al.*, 2011) may also be regulated by the modulation of the RNAPII elongation rate via Elongin B-mediated H2B mono-ubiquitination and pUL69-hSpt6 and/or pUL69-UAP56 interactions. In line with this idea, we observed that pUL69 localization and activity is dependent on cellular cdk- and/or viral pUL97-kinase activity (Rechter *et al.*, 2009; Thomas *et al.*, 2009). Therefore, pUL69 might represent the adjustable trigger that interconnects viral transcriptional elongation, splicing, and mRNA export.

Regulation of nuclear mRNA export

The compartmentalization of eukaryotic cells into nucleus and cytoplasm permits the spatial and temporal separation of transcription and translation and enables the tight regulation of protein synthesis. Besides cotranscriptional processing of eukaryotic mRNAs by 5' capping, splicing and 3' polyadenylation (Erkmann and Kutay, 2004; Vinciguerra and Stutz, 2004), intranuclear mRNAs are subjected to quality control during their biogenesis, e.g. by the exosome, which detects and retains partially spliced or unspliced mRNAs within the nucleus and leads to their destruction (Hilleren et al., 2001; Houseley et al., 2006). Cytoplasmic mRNAs, on the other hand, also underlie RNA surveillance mechanisms, e.g. by nonsense-mediated decay (NMD), which degrades mRNAs that contain preterminal termination codons. Although most cellular mRNAs derive from spliced genes, herpesviruses are dependent on the export of partially spliced or unspliced mRNAs during the course of infection and, therefore, had the necessity to evolve viral mRNA export factors in order to overcome cellular mRNA surveillance mechanisms and to facilitate the efficient export of viral mRNAs.

Cellular mRNA export receptors and adaptor proteins

Over the past decade several components of the cellular mRNA export pathway have been identified and they are quite well conserved from yeast to human. The functional mammalian cellular mRNA export receptor TAP (Tip-associated protein) – complexed by its cofactor p15 – directly interacts with the nuclear pore to transport messenger ribonucleoprotein particles (mRNPs) independently from the regulatory cofactor in nucleocytoplasmic transport RanGTP to the cytoplasm (Segref et al., 1997; Strasser et al., 2000; Erkmann and Kutay, 2004). Although TAP binds to RNA directly in vitro, additional adaptor proteins are required in vivo (Liker et al., 2000; Braun et al., 2001). Some of these adaptor proteins are members of the RNA export factor family, REF (named Aly or Yra1 in yeast). REF proteins exert nucleocytoplasmic shuttling activity, bind to RNA and TAP within the nucleus and thereby translocate mRNAs to the cytoplasm (Strasser and Hurt, 2000; Stutz et al., 2000; Zenklusen et al., 2001). Knock-down experiments in Drosophila melanogaster and Caenorhabditis elegans using siRNAs demonstrated that TAP is essential for mRNA export, whereas REF is dispensable (Gatfield and Izaurralde, 2002; Longman et al., 2003). This observation indicated that, besides REF, additional adaptor proteins are involved in eukaryotic mRNA export. Indeed, splicing regulatory serine/arginine-rich SR-proteins (e.g. SRp20 or 9G8) were identified that exert nucleocytoplasmic shuttling activity and

directly interact with TAP (Huang and Steitz, 2001; Huang et al., 2003).

Functional coupling of splicing and mRNA export

As a REF-interacting protein, UAP56 (alternatively named BAT1 in humans, HEL in Drosophila or Sub2p in yeast) has been identified. UAP56 is a member of the DExD/H-box family of RNA-helicases that are involved in various steps of RNA biogenesis (Linder and Stutz, 2001; Rocak and Linder, 2004). Initially, UAP56 has been identified as a U2AF65-associated protein, that is required during splicing, because it dissociates U2AF65 from the polypyrimidine tract within the intron of pre-mRNAs and thereby facilitates the binding of U2snRNP (Fleckner et al., 1997). It has been demonstrated by RNA interference in Drosophila that, besides its function during splicing, UAP56 fulfils an important function for cellular mRNA export by interacting with REF (Gatfield et al., 2001; Herold et al., 2003). In higher eukaryotes, pre-mRNA is loaded with the exon-junction complex (EJC) upstream of the splice site (Fig. I.11.3A). Proteins within this complex exert various functions during either mRNA export, translation, or NMD (Le Hir et al., 2001). Insofar as UAP56 and REF are both components of this complex, this led to the assumption that they are responsible for the export of spliced mRNAs via the TAP/p15 pathway and thereby connect splicing and mRNA export (Reed and Hurt, 2002; Reed, 2003).

Functional coupling of transcription and mRNA export

Interestingly, UAP56 and REF are also responsible for the export of intron-containing, unspliced mRNAs, suggesting the existence of an alternative pathway to load these export factors onto nascent mRNA (Strasser and Hurt, 2000; Gatfield et al., 2001; Jensen et al., 2001; Rodrigues et al., 2001; Kiesler et al., 2002; Lei and Silver, 2002). It was demonstrated in yeast that both UAP56 and REF are cotranscriptionally recruited to nascent mRNA (Lei et al., 2001; Zenklusen et al., 2002). Here, the yeast homologue Sub2p interacts directly with Hpr1p, a component of the THO complex, that functions in transcriptional elongation (Zenklusen et al., 2002; Abruzzi et al., 2004). Hence, the components of the THO complex in concert with Sub2p, Aly (REF) and most probably Mex67 (TAP) form a multiprotein complex, called TREX complex, that couples transcription–elongation with mRNA export (Fig. I.11.3B). Human UAP56 and REF can be found in complexes with various homologous components of the yeast

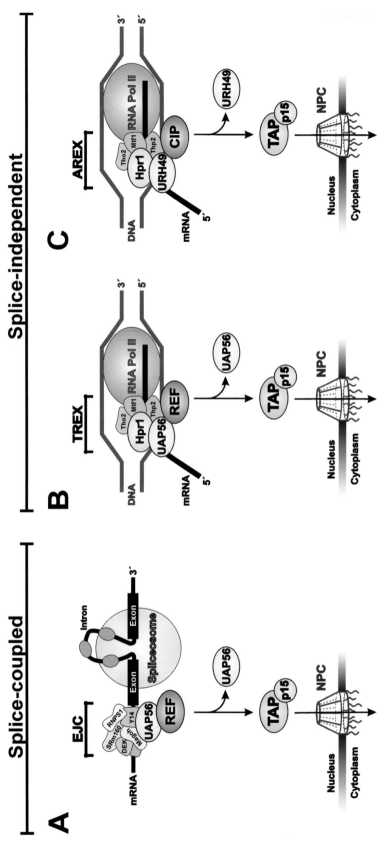

Figure I.11.3 Cellular mRNA export pathways involving UAP56 or URH49. (A) Splice-coupled mRNA-export: During splicing eukaryotic mRNA is loaded with the exon junction complex (EJC) 5′ upstream of the splice site. UAP56 and REF, amongst others, are components of this multi-protein complex. REF subsequently recruits the heterodimeric export receptor TAP/p15 to the mRNA, which is finally exported via the nuclear pore complex (NPC) into the cytoplasm. (B+C) Cotranscriptional, splice-independent mRNA-export: During transcription the THO complex (Hpr1, Tho2, Mtf1, Thp2) associates with RNA polymerase II (RNA Pol II) and acts mainly in transcriptional elongation. Additionally, the THO complex recruits UAP56 and REF to build up the TREX (transcription-elongation export) complex that translocates the mRNA via TAP/p15 to the cytoplasm. (C) In contrast to UAP56 forming the TREX complex, URH49 forms a URH49–CIP29 complex, termed the AREX (alternative mRNA export) complex that regulates a distinct subset of key mitotic regulatory mRNAs. In either case, UAP56/URH49 has to leave the mRNP-complex in order to facilitate binding of REF to TAP. CIP, cytokine-inducible protein 29kDa; NPC, nuclear pore complex; REF, RNA export factor; RNA pol II, RNA polymerase II; TAP, tip-associated protein complexed by its cofactor p15.

THO complex, implicating that the TREX complex is evolutionarily conserved (Strasser *et al.*, 2002). Interestingly, the human TREX complex is also recruited to intron-containing transcripts during splicing (Masuda *et al.*, 2005) and it has been suggested that UAP56 and REF are involved in the release of spliced mRNAs from nuclear speckles and their subsequent export to the cytoplasm (Dias *et al.*, 2010). Consistent with the idea that splicing is not an absolute requirement for mRNA export, UAP56 and REF have also been shown to interact with the cap-binding complex (CBP20 and CBP80) and this association was essential for the nuclear export of capped intronless mRNAs (Cheng *et al.*, 2006; Nojima *et al.*, 2007). In either case it is thought that UAP56 leaves the mRNP complex in the nucleus, since REF interacts with either TAP or UAP56 in a mutually exclusive manner (Strasser and Hurt, 2001). More recently, URH49 (UAP56-related helicase, 49 kDa; or DDX39) has been discovered (Pryor *et al.*, 2004). URH49 shares 90% amino acid sequence homology with UAP56 although it is 100% conserved within the DExD/H-box RNA -helicase motifs and shows the highest degree of divergence within its N-terminus. Even though UAP56 and URH49 are expressed in human and mouse, there is only one gene corresponding to UAP56 in *Saccharomyces cerevisiae* and *Drosophila melanogaster*. Since it has been demonstrated in yeast that both URH49 and UAP56 complement a Sub2p deletion and interact with REF, it has been proposed that both proteins exert similar or redundant functions in mammalian cells (Pryor *et al.*, 2004). The essential role of UAP56 and URH49 for cellular mRNA export was accordingly demonstrated, since UAP56/URH49 double knock-down cells were not viable due to a retention of poly(A)$^+$ mRNA within the nucleus (Kapadia *et al.*, 2006; M. Thomas, unpublished). Despite the high homology of UAP56 and URH49, both helicases seem to form distinct mRNA export machineries. While UAP56 preferentially forms the TREX complex, URH49 is a component of the AREX (alternative mRNA export) complex in order to export different mRNA substrates (Yamazaki *et al.*, 2010) (Fig. I.11.3C). In addition to the nuclear functions of UAP56/URH49, recent work by our group suggests an additional cytoplasmic function for both RNA helicases, as we could demonstrate that UAP56 and URH49 exhibit a CRM1-independent nucleocytoplasmic shuttling activity (Thomas *et al.*, 2011).

Herpesviral mRNA export

The *Herpesviridae* encode a conserved family of regulatory proteins, which are known to function as posttranscriptional activators facilitating the nuclear export of intronless mRNAs via recruitment of the cellular mRNA export machinery (Sandri-Goldin, 2001, 2008; Toth and Stamminger, 2008). Representatives of this so-called ICP27-protein family can be found in every mammalian or avian herpesvirus sequenced so far (e.g. HSV-1 ICP27, EBV EB2, or ORF57 of KSHV and HVS) and this underlines their functional importance. Although the overall amino acid identity between the ICP27-homologues of different herpesvirus subgroups is quite low (ranging from 16% to 37% amino acid identity) and specific motifs (e.g. RNA-binding motif, nuclear localization and export signals) diverged considerably during evolution, they share a conserved function as viral mRNA export factors.

The pleiotropic transactivator pUL69 of HCMV

The HCMV-encoded ICP27-homologue is pUL69, a nuclear phosphoprotein of 744 amino acids and a molecular mass of approximately 105–116 kDa, that is expressed during the early-late phase of viral replication (Winkler *et al.*, 1994). In contrast to its counterparts ICP27 and EB2 that are essential for replication of HSV-1 or EBV, respectively (Sacks *et al.*, 1985; Gruffat *et al.*, 2002), infection studies using pUL69 knock-out viruses revealed that these viruses yield severely reduced titres but are replication competent, suggesting that pUL69 is not absolutely essential for viral replication (Hayashi *et al.*, 2000; Zielke *et al.*, 2011). HCMV pUL69 enters the cell as part of the tegument and induces a cell-cycle arrest in the G1-phase (Lu and Shenk, 1999; Song *et al.*, 2001). Although three different isoforms of pUL69 with molecular masses of 105, 110 and 116 kDa are expressed during the viral replication cycle, only the 110 kDa isoform is incorporated into viral particles, suggesting that this protein is responsible for induction of the cell cycle arrest (Winkler and Stamminger, 1996). By analogy with its counterparts HSV-1 ICP27 and EBV EB2, HCMV pUL69 transactivates a broad spectrum of reporter constructs with different promoters (Rice and Knipe, 1988; Buisson *et al.*, 1989; Winkler *et al.*, 1994). Because pUL69 interacts via its conserved central domain (aa 269-574) with the transcription elongation factor hSPT6 and because hSPT6 binding-deficient mutants of pUL69 lost their transactivation capacity, it was assumed that this might represent the mechanism of pUL69-mediated transactivation (Winkler *et al.*, 2000) (Fig. I.11.4). With regard to posttranscriptional regulation of gene expression it is of note that RNAPII-bound hSPT6 interacts with the cellular SPT6-interacting protein IWS1, which in turn recruits the cellular RNA export factor REF to mRNAs (Yoh *et al.*, 2007). This cotranscriptional loading of

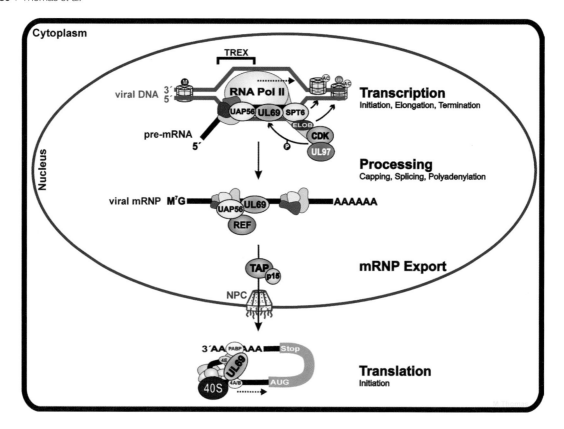

Figure I.11.4 Model of pUL69-mediated transactivation of viral gene expression. The pUL69 protein interacts with UAP56 and hSPT6, which are both associated with RNA polymerase II (RNA pol II) and regulate transcriptional elongation. As a component of the transcription–elongation export complex (TREX) UAP56 exerts an additional essential function by recruiting the cellular mRNA export adapter protein REF. Then, via its UAP56- and RNA-binding capacity, pUL69 forms an export-competent mRNP, which is translocated via the nuclear pore complex (NPC) into the cytoplasm. There, pUL69 binds to eIF4A and the poly(A)-binding protein (PABP) in order to promote translation. Viral (pUL97) and cellular kinases (cdk) are regulating pUL69-activity by phosphorylation. Ac, acetyl; CDK, cycline-dependent kinase; ELOB, Elongin-B; M, methyl; NPC, nuclear pore complex; P, phosphate; PABP, poly(A)-binding protein; REF, RNA export factor; RNA pol II, RNA polymerase II; TAP, tip-associated protein complexed by its cofactor p15; TREX, transcription–elongation export complex; Ub, ubiquitin; 4A/B and 4E, eukaryotic translation initiation factors A/B and E; 40S, small ribosomal subunit.

factors involved in mRNA processing, surveillance, and export could therefore additionally account for pUL69-mediated pleiotropic up-regulation of cellular and viral gene expression.

pUL69 – the HCMV-encoded mRNA export factor

Like its counterparts of the α- or γ-herpesviruses HCMV pUL69 possesses properties of a viral mRNA export factor, insofar as it binds RNA (Toth *et al.*, 2006), shuttles from the nucleus to the cytoplasm (Lischka *et al.*, 2001) and recruits the cellular mRNA export machinery via interaction with UAP56/URH49 (Lischka *et al.*, 2006). During the past decade we were able to map the domains within pUL69 that are required

for its respective functions. Thus, CRM1-independent nucleocytoplasmic shuttling is achieved by its bipartite nuclear localization signal (NLS) comprising the N-terminal residues 21–45 and its nuclear export signal (NES), that was mapped to the C-terminal amino acids 597–624. This nonconventional NES comprises in total 28 residues separated into an N-terminal proline/glutamine-rich cluster and a C-terminal acidic cluster of amino acids and is sufficient for nucleocytoplasmic shuttling, as it can be transferred to heterologous proteins (Lischka *et al.*, 2001). RNA-binding is facilitated by the N-terminus that contains two arginine-rich stretches comprising the amino acids 17–46 (R1/R2-motifs) and 123–136 (RS-motif) of pUL69 (Toth *et al.*, 2006). One defining characteristic of HCMV pUL69 is its capacity to interact directly with the cellular mRNA export factors UAP56 and URH49. The N-terminal UAP56/URH49-interaction motif comprises the amino acids 18–30 and thereby partially overlaps with the bipartite NLS and RNA-binding motif of pUL69 (Lischka *et al.*, 2006; Toth *et al.*, 2006). Importantly, loss of pUL69-mediated mRNA export occurs both when protein–protein interaction with UAP56 is inhibited or when its CRM1-independent nuclear export is blocked (Lischka *et al.*, 2001, 2006). In contrast, however, RNA-binding seems to be dispensable for pUL69-mediated accumulation of unspliced RNA in the cytoplasm, which is in contrast to EB2 or ICP27, where RNA-binding is essential for mRNA export (Sandri-Goldin, 1998; Hiriart *et al.*, 2003; Lischka *et al.*, 2006; Toth *et al.*, 2006). Nevertheless, HCMV-mutants expressing RNA-binding-deficient pUL69ΔR2ΔRS exhibit a slight but significant delay in the release of progeny virions, thereby suggesting functional importance of pUL69-RNA binding for HCMV replication *in vivo* (Zielke *et al.*, 2011). To date, there is no evidence for sequence-specific RNA binding of pUL69 *in vitro*, but *in vivo* pUL69 efficiently binds to CAT-reporter mRNA but not to GAPDH mRNA, suggesting selective RNA-binding *in vivo* (Toth *et al.*, 2006). It is currently under intense investigation whether, and if so, how cytomegaloviruses preferentially export viral rather than cellular mRNAs. One might speculate that either defined secondary structures within viral RNAs confer this specificity or that specific protein–protein interactions with cellular as well as viral proteins determine the selectivity of herpes-viral mRNA export.

Cytomegaloviral mRNA export

Functional analyses of the more closely related betaherpesviral pUL69-homologues pC69 of chimpanzee cytomegalovirus, pRh69 of rhesus cytomegalovirus, pM69 of murine cytomegalovirus, pU42 of human herpesvirus 6 and pU42 of elephant endotheliotropic herpesvirus revealed that nuclear localization and nucleocytoplasmic shuttling activity are well conserved throughout the subfamily of *Betaherpesvirinae* (see also Chapter I.1). However, in contrast, binding to the mRNA export factors UAP56 and/or URH49, that are evolutionary highly conserved, is confined to members of the genus *Cytomegalovirus* (pUL69, pC69 and pRh69) and correlates with the mRNA export activity of the respective proteins. The evolutionary conservation of the UAP56/URH49-binding site within the genus *Cytomegalovirus* suggests an important function of this protein interaction motif for viral replication. Concordantly, recombinant HCMV-derivatives that lack a functional UAP56-interaction motif (UL69ΔR1ΔRS and UL69mutUAP) show a severe replication defect compared to wild-type virus (Zielke *et al.*, 2011). In line with this observation we could demonstrate recently that a chimeric protein encoding the functional UAP56-interaction motif of HCMV pUL69 in fusion with the murine cytomegalovirus M69-protein acts as an active mRNA export factor, which substitutes for pUL69 during HCMV infection (Zielke *et al.*, 2012). Taken together, our data strongly suggest that interaction of the cytomegaloviral proteins pUL69, pC69 or pRh69 with UAP56/URH49 is crucial for efficient cytomegaloviral replication.

Interconnection between viral mRNA export and translational regulation

Initiation of mRNA translation is considered to be the rate limiting step for eukaryotic protein synthesis and therefore also represents an important target for viral host cell manipulation (Sonenberg and Hinnebusch, 2009). Although translational control mechanisms are not within the focus of this chapter, we would like to briefly discuss the emerging evidence for an interconnection between nuclear mRNA export and translational regulation. For efficient viral protein expression, cytomegaloviruses have to compete with the host cell for the limited pool of eukaryotic translation initiation factors (eIFs) in order to recruit ribosomes selectively to viral transcripts. Consequently, in addition to its effect on mRNA export, pUL69 was recently shown to facilitate the translation of mRNAs via interaction with eIF4A1 and the poly(A)-binding protein (PABP), thereby excluding the translation regulator eIF4E-binding protein 1 (4EBP1) from the cap-binding complex (Aoyagi *et al.*, 2010) (Fig. I.11.4). Hypophosphorylated 4EBP1 binds and sequesters eIF4E, thus inhibiting translation (Jackson and Wickens, 1997). Eukaryotic translation is further regulated via mammalian target of rapamycin (mTOR) containing complex 1 (mTORC1)-mediated phosphorylation of 4EBP1 which is concomitant with abolished 4EBP1-binding to the translation initiation factor eIF4E (Gingras *et al.*, 2001). The HCMV-encoded pUL38 sustains mTORC1-activity within infected cells, and thereby retains 4EBP1 in its hyperphosphorylated form in order to allow viral translation (Moorman *et al.*, 2008) (Fig. I.11.1). Notably, specific nuclear EJC proteins remain bound to mRNAs and enhance their translation within the cytoplasm (Lu and Cullen, 2003; Nott *et al.*, 2004; Lee *et al.*, 2009). Thus, one might speculate that the nucleocytoplasmic shuttling EJC-components UAP56 and URH49 (Thomas *et al.*, 2011) exert additional cytoplasmic functions during translation and are therefore targeted by the shuttling cytomegaloviral proteins pUL69, pC69 and pRh69 (Zielke *et al.*, 2011). Whether, and if so, how the UAP56/URH49-interaction is also important for the regulation of viral translation will be the subject of further studies.

Future perspectives

Research of the last two decades revealed exciting insights into the multifaceted regulation of HCMV gene expression. It is quite clear that HCMV has developed a multitude of sophisticated mechanisms to usurp and manipulate the transcriptional and post-transcriptional cellular machinery. One emerging theme is that a single viral protein, e.g. pUL69, apparently is able to affect different cellular machineries, however, this also provides evidence for an extensive crosstalk between transcriptional and post-transcriptional regulatory processes. Research into the underlying mechanisms will not only provide novel insights concerning the coupling of these processes, but may also reveal Achilles heels of the virus that could be used for therapeutic approaches. One major drawback, for sure, is that no structural information is available for HCMV regulatory proteins yet. This would be extremely helpful for a detailed understanding of regulatory processes on the molecular level and also for the design of inhibitory substances, thus posing an important challenge for the future. Furthermore, novel technologies like next generation sequencing and the RNAi technology open up new avenues to a broader understanding of how the virus and viral proteins manipulate cellular processes and how the cell restricts viral gene expression. Finally, since, up to now, research on HCMV gene regulation was mainly conducted using the permissive cell system of primary human fibroblasts, it will be important to address the question of cell type-specific regulatory mechanisms that HCMV utilizes in pathogenetically relevant cell types such as monocytes, dendritic cells, and epithelial cells.

Acknowledgements

The authors thank past and present members of the laboratory whose work has contributed to studies that are described in this review. Research in the Stamminger laboratory was supported by the DFG (SFB796), the IZKF Erlangen and the graduate school BIGSS.

References

Abate, D.A., Watanabe, S., and Mocarski, E.S. (2004). Major human cytomegalovirus structural protein pp65 (ppUL83) prevents interferon response factor 3 activation in the interferon response. J. Virol. *78*, 10995–11006.

Abruzzi, K.C., Lacadie, S., and Rosbash, M. (2004). Biochemical analysis of TREX complex recruitment to intronless and intron-containing yeast genes. EMBO J. *23*, 2620–2631.

Adair, R., Liebisch, G.W., and Colberg-Poley, A.M. (2003). Complex alternative processing of human cytomegalovirus UL37 pre-mRNA. J. Gen. Virol. *84*, 3353–3358.

Adair, R., Liebisch, G.W., Su, Y., and Colberg-Poley, A.M. (2004). Alteration of cellular RNA splicing and polyadenylation machineries during productive human cytomegalovirus infection. J. Gen. Virol. *85*, 3541–3553.

Adler, M., Tavalai, N., Muller, R., and Stamminger, T. (2011). Human cytomegalovirus immediate-early gene expression is restricted by the nuclear domain 10 component Sp100. J. Gen. Virol. *92*, 1532–1538.

Ahn, J.H., and Hayward, G.S. (1997). The major immediate-early proteins IE1 and IE2 of human cytomegalovirus colocalize with and disrupt PML-associated nuclear bodies at very early times in infected permissive cells. J. Virol. *71*, 4599–4613.

Ahn, J.H., Jang, W.J., and Hayward, G.S. (1999). The human cytomegalovirus IE2 and UL112–113 proteins accumulate in viral DNA replication compartments that initiate from the periphery of promyelocytic leukemia protein-associated nuclear bodies (PODs or ND10). J. Virol. *73*, 10458–10471.

Aoyagi, M., Gaspar, M., and Shenk, T.E. (2010). Human cytomegalovirus UL69 protein facilitates translation by associating with the mRNA cap-binding complex and excluding 4EBP1. Proc. Natl. Acad. Sci. U.S.A. *107*, 2640–2645.

Arlt, H., Lang, D., Gebert, S., and Stamminger, T. (1994). Identification of binding sites for the 86-kilodalton IE2 protein of human cytomegalovirus within an IE2-responsive viral early promoter. J. Virol. *68*, 4117–4125.

Aso, T., Lane, W.S., Conaway, J.W., and Conaway, R.C. (1995). Elongin (SIII): a multisubunit regulator of elongation by RNA polymerase II. Science *269*, 1439–1443.

Awasthi, S., Isler, J.A., and Alwine, J.C. (2004). Analysis of splice variants of the immediate-early 1 region of human cytomegalovirus. J. Virol. *78*, 8191–8200.

Baek, M.C., Krosky, P.M., Pearson, A., and Coen, D.M. (2004). Phosphorylation of the RNA polymerase II carboxyl-terminal domain in human cytomegalovirus-infected cells and in vitro by the viral UL97 protein kinase. Virology *324*, 184–193.

Baracchini, E., Glezer, E., Fish, K., Stenberg, R.M., Nelson, J.A., and Ghazal, P. (1992). An isoform variant of the cytomegalovirus immediate-early auto repressor functions as a transcriptional activator. Virology *188*, 518–529.

Bortvin, A., and Winston, F. (1996). Evidence that Spt6p controls chromatin structure by a direct interaction with histones. Science *272*, 1473–1476.

Boshart, M., Weber, F., Jahn, G., Dorsch-Hasler, K., Fleckenstein, B., and Schaffner, W. (1985). A very strong enhancer is located upstream of an immediate-early gene of human cytomegalovirus. Cell *41*, 521–530.

Braun, I.C., Herold, A., Rode, M., Conti, E., and Izaurralde, E. (2001). Overexpression of TAP/p15 heterodimers bypasses nuclear retention and stimulates nuclear mRNA export. J. Biol. Chem. *276*, 20536–20543.

Browne, E.P., and Shenk, T. (2003). Human cytomegalovirus UL83-coded pp65 virion protein inhibits antiviral gene expression in infected cells. Proc. Natl. Acad. Sci. U.S.A. *100*, 11439–11444.

Buisson, M., Manet, E., Trescol-Biemont, M.C., Gruffat, H., Durand, B., and Sergeant, A. (1989). The Epstein–Barr virus (EBV) early protein EB2 is a posttranscriptional activator expressed under the control of EBV transcription factors EB1 and R. J. Virol. *63*, 5276–5284.

Cantrell, S.R., and Bresnahan, W.A. (2006). Human cytomegalovirus (HCMV) UL82 gene product (pp71) relieves hDaxx-mediated repression of HCMV replication. J. Virol. *80*, 6188–6191.

Caposio, P., Luganini, A., Hahn, G., Landolfo, S., and Gribaudo, G. (2007). Activation of the virus-induced IKK/NF-kappaB signalling axis is critical for the replication of human cytomegalovirus in quiescent cells. Cell Microbiol. *9*, 2040–2054.

Carbone, R., Botrugno, O.A., Ronzoni, S., Insinga, A., Di, C.L., Pelicci, P.G., and Minucci, S. (2006). Recruitment of the histone methyltransferase SUV39H1 and its role in the oncogenic properties of the leukemia-associated PML-retinoic acid receptor fusion protein. Mol. Cell Biol. *26*, 1288–1296.

Cheng, H., Dufu, K., Lee, C.S., Hsu, J.L., Dias, A., and Reed, R. (2006). Human mRNA export machinery recruited to the 5′ end of mRNA. Cell *127*, 1389–1400.

Cho, S., Park, J.S., and Kang, Y.K. (2011). Dual functions of histone-lysine N-methyltransferase Setdb1 protein at promyelocytic leukemia-nuclear body (PML-NB): maintaining PML-NB structure and regulating the expression of its associated genes. J. Biol. Chem. *286*, 41115–41124.

Cosme, R.S., Yamamura, Y., and Tang, Q. (2009). Roles of polypyrimidine tract binding proteins in major immediate-early gene expression and viral replication of human cytomegalovirus. J. Virol. *83*, 2839–2850.

Cristea, I.M., Moorman, N.J., Terhune, S.S., Cuevas, C.D., O'Keefe, E.S., Rout, M.P., Chait, B.T., and Shenk, T. (2010). Human cytomegalovirus pUL83 stimulates activity of the viral immediate-early promoter through its interaction with the cellular IFI16 protein. J. Virol. *84*, 7803–7814.

Cuevas-Bennett, C., and Shenk, T. (2008). Dynamic histone H3 acetylation and methylation at human cytomegalovirus promoters during replication in fibroblasts. J. Virol. *82*, 9525–9536.

Cygnar, D., Hagemeier, S., Kronemann, D., and Bresnahan, W.A. (2012). The cellular protein SPT6 is required for efficient replication of human cytomegalovirus. J. Virol. *86*, 2011–2020.

Demarchi, J.M. (1981). Human cytomegalovirus DNA: restriction enzyme cleavage maps and map locations for immediate-early, early, and late RNAs. Virology *114*, 23–38.

Depto, A.S., and Stenberg, R.M. (1992). Functional analysis of the true late human cytomegalovirus pp28 upstream promoter: cis-acting elements and viral trans-acting proteins necessary for promoter activation. J. Virol. *66*, 3241–3246.

Di, C.L., Raker, V.A., Corsaro, M., Fazi, F., Fanelli, M., Faretta, M., Fuks, F., Lo, C.F., Kouzarides, T., Nervi, C., et al. (2002). Methyltransferase recruitment and DNA hypermethylation of target promoters by an oncogenic transcription factor. Science *295*, 1079–1082.

Dias, A.P., Dufu, K., Lei, H., and Reed, R. (2010). A role for TREX components in the release of spliced mRNA from nuclear speckle domains. Nat. Commun. *1*, 97.

Dirks, R.W., de Pauw, E.S., and Raap, A.K. (1997). Splicing factors associate with nuclear HCMV-IE transcripts after transcriptional activation of the gene, but dissociate upon transcription inhibition: evidence for a dynamic organization of splicing factors. J. Cell Sci. *110*, 515–522.

Drane, P., Ouararhni, K., Depaux, A., Shuaib, M., and Hamiche, A. (2010). The death-associated protein DAXX is a novel histone chaperone involved in the replication-independent deposition of H3.3. Genes Dev. *24*, 1253–1265.

Du, G., Dutta, N., Lashmit, P., and Stinski, M.F. (2011). Alternative splicing of the human cytomegalovirus major immediate-early genes affects infectious-virus replication and control of cellular cyclin-dependent kinase. J. Virol. *85*, 804–817.

Endoh, M., Zhu, W., Hasegawa, J., Watanabe, H., Kim, D.K., Aida, M., Inukai, N., Narita, T., Yamada, T., Furuya, A., et al. (2004). Human Spt6 stimulates transcription elongation by RNA polymerase II in vitro. Mol. Cell Biol. *24*, 3324–3336.

Erkmann, J.A., and Kutay, U. (2004). Nuclear export of mRNA: from the site of transcription to the cytoplasm. Exp. Cell Res. *296*, 12–20.

Everett, R.D., Rechter, S., Papior, P., Tavalai, N., Stamminger, T., and Orr, A. (2006). PML contributes to a cellular mechanism of repression of herpes simplex virus type 1 infection that is inactivated by ICP0. J. Virol. *80*, 7995–8005.

Feichtinger, S., Stamminger, T., Muller, R., Graf, L., Klebl, B., Eickhoff, J., and Marschall, M. (2011). Recruitment of cyclin-dependent kinase 9 to nuclear compartments during cytomegalovirus late replication: importance of an interaction between viral pUL69 and cyclin T1. J. Gen. Virol. *92*, 1519–1531.

Fleckner, J., Zhang, M., Valcarcel, J., and Green, M.R. (1997). U2AF65 recruits a novel human DEAD box protein required for the U2 snRNP-branchpoint interaction. Genes Dev. *11*, 1864–1872.

Fortunato, E.A., Sanchez, V., Yen, J.Y., and Spector, D.H. (2002). Infection of cells with human cytomegalovirus during S phase results in a blockade to immediate-early gene expression that can be overcome by inhibition of the proteasome. J. Virol. *76*, 5369–5379.

Fortunato, E.A., and Spector, D.H. (1999). Regulation of human cytomegalovirus gene expression. Adv. Virus Res. *54*, 61–128.

Gaddy, C.E., Wong, D.S., Markowitz-Shulman, A., and Colberg-Poley, A.M. (2010). Regulation of the subcellular distribution of key cellular RNA-processing factors during permissive human cytomegalovirus infection. J. Gen. Virol. *91*, 1547–1559.

Gatfield, D., and Izaurralde, E. (2002). REF1/Aly and the additional exon junction complex proteins are dispensable for nuclear mRNA export. J. Cell Biol. *159*, 579–588.

Gatfield, D., Le Hir, H., Schmitt, C., Braun, I.C., Kocher, T., Wilm, M., and Izaurralde, E. (2001). The DExH/D box protein HEL/UAP56 is essential for mRNA nuclear export in Drosophila. Curr. Biol. *11*, 1716–1721.

Gatherer, D., Seirafian, S., Cunningham, C., Holton, M., Dargan, D.J., Baluchova, K., Hector, R.D., Galbraith, J., Herzyk, P., Wilkinson, G.W., et al. (2011). High-resolution human cytomegalovirus transcriptome. Proc. Natl. Acad. Sci. U.S.A. *108*, 19755–19760.

Gingras, A.C., Raught, B., and Sonenberg, N. (2001). Regulation of translation initiation by FRAP/mTOR. Genes Dev. *15*, 807–826.

Gonczol, E., Andrews, P.W., and Plotkin, S.A. (1985). Cytomegalovirus infection of human teratocarcinoma cells in culture. J. Gen. Virol. *66*, 509–515.

Groves, I.J., Reeves, M.B., and Sinclair, J.H. (2009). Lytic infection of permissive cells with human cytomegalovirus is regulated by an intrinsic 'pre-immediate-early' repression of viral gene expression mediated by histone post-translational modification. J. Gen. Virol. *90*, 2364–2374.

Gruffat, H., Batisse, J., Pich, D., Neuhierl, B., Manet, E., Hammerschmidt, W., and Sergeant, A. (2002). Epstein–Barr virus mRNA export factor EB2 is essential for production of infectious virus. J. Virol. *76*, 9635–9644.

Hartzog, G.A., Wada, T., Handa, H., and Winston, F. (1998). Evidence that Spt4, Spt5, and Spt6 control transcription elongation by RNA polymerase II in Saccharomyces cerevisiae. Genes Dev. *12*, 357–369.

Hayashi, M.L., Blankenship, C., and Shenk, T. (2000). Human cytomegalovirus UL69 protein is required for efficient accumulation of infected cells in the G1 phase of the cell cycle. Proc. Natl. Acad. Sci. U.S.A. *97*, 2692–2696.

Herget, T., Freitag, M., Morbitzer, M., Kupfer, R., Stamminger, T., and Marschall, M. (2004). Novel chemical class of pUL97 protein kinase-specific inhibitors with strong anticytomegaloviral activity. Antimicrob. Agents Chemother. *48*, 4154–4162.

Herold, A., Teixeira, L., and Izaurralde, E. (2003). Genome-wide analysis of nuclear mRNA export pathways in *Drosophila*. EMBO J. *22*, 2472–2483.

Hilleren, P., McCarthy, T., Rosbash, M., Parker, R., and Jensen, T.H. (2001). Quality control of mRNA 3'-end processing is linked to the nuclear exosome. Nature *413*, 538–542.

Hiriart, E., Bardouillet, L., Manet, E., Gruffat, H., Penin, F., Montserret, R., Farjot, G., and Sergeant, A. (2003). A region of the Epstein–Barr virus (EBV) mRNA export factor EB2 containing an arginine-rich motif mediates direct binding to RNA. J. Biol. Chem. *278*, 37790–37798.

Houseley, J., LaCava, J., and Tollervey, D. (2006). RNA-quality control by the exosome. Nat. Rev. Mol. Cell Biol. *7*, 529–539.

Huang, L., Malone, C.L., and Stinski, M.F. (1994). A human cytomegalovirus early promoter with upstream negative and positive cis-acting elements: IE2 negates the effect of the negative element, and NF-Y binds to the positive element. J. Virol. *68*, 2108–2117.

Huang, Y., and Steitz, J.A. (2001). Splicing factors SRp20 and 9G8 promote the nucleocytoplasmic export of mRNA. Mol. Cell *7*, 899–905.

Huang, Y., Gattoni, R., Stevenin, J., and Steitz, J.A. (2003). SR splicing factors serve as adapter proteins for TAP-dependent mRNA export. Mol. Cell *11*, 837–843.

Hume, A.J., Finkel, J.S., Kamil, J.P., Coen, D.M., Culbertson, M.R., and Kalejta, R.F. (2008). Phosphorylation of retinoblastoma protein by viral protein with cyclin-dependent kinase function. Science *320*, 797–799.

Hwang, J., Saffert, R.T., and Kalejta, R.F. (2011). Elongin B-mediated epigenetic alteration of viral chromatin correlates with efficient human cytomegalovirus gene expression and replication. MBio. *2*, e00023–11.

Isaacson, M.K., Juckem, L.K., and Compton, T. (2008). Virus entry and innate immune activation. Curr. Top. Microbiol. Immunol. *325*, 85–100.

Isern, E., Gustems, M., Messerle, M., Borst, E., Ghazal, P., and Angulo, A. (2011). The activator protein 1 binding motifs within the human cytomegalovirus major immediate-early enhancer are functionally redundant and act in a cooperative manner with the NF-{kappa}B sites during acute infection. J. Virol. *85*, 1732–1746.

Ishov, A.M., and Maul, G.G. (1996). The periphery of nuclear domain 10 (ND10) as site of DNA virus deposition. J. Cell Biol. *134*, 815–826.

Ishov, A.M., Stenberg, R.M., and Maul, G.G. (1997). Human cytomegalovirus immediate-early interaction with host nuclear structures: definition of an immediate transcript environment. J. Cell Biol. *138*, 5–16.

Isomura, H., Stinski, M.F., Kudoh, A., Murata, T., Nakayama, S., Sato, Y., Iwahori, S., and Tsurumi, T. (2008). Noncanonical TATA sequence in the UL44 late promoter of human cytomegalovirus is required for the accumulation of late viral transcripts. J. Virol. *82*, 1638–1646.

Isomura, H., Stinski, M.F., Murata, T., Yamashita, Y., Kanda, T., Toyokuni, S., and Tsurumi, T. (2011). The human cytomegalovirus gene products essential for late viral gene expression assemble into prereplication complexes before viral DNA replication. J. Virol. *85*, 6629–6644.

Jackson, R.J., and Wickens, M. (1997). Translational controls impinging on the 5'-untranslated region and initiation factor proteins. Curr. Opin. Genet. Dev. *7*, 233–241.

Jenkins, C., Abendroth, A., and Slobedman, B. (2004). A novel viral transcript with homology to human interleukin-10 is expressed during latent human cytomegalovirus infection. J. Virol. *78*, 1440–1447.

Jenkins, C., Garcia, W., Abendroth, A., and Slobedman, B. (2008a). Expression of a human cytomegalovirus latency-associated homolog of interleukin-10 during the productive phase of infection. Virology *370*, 285–294.

Jenkins, C., Garcia, W., Godwin, M.J., Spencer, J.V., Stern, J.L., Abendroth, A., and Slobedman, B. (2008b). Immunomodulatory properties of a viral homolog of human interleukin-10 expressed by human cytomegalovirus during the latent phase of infection. J. Virol. *82*, 3736–3750.

Jensen, T.H., Boulay, J., Rosbash, M., and Libri, D. (2001). The DECD box putative ATPase Sub2p is an early mRNA export factor. Curr. Biol. *11*, 1711–1715.

Jurica, M.S., and Moore, M.J. (2003). Pre-mRNA splicing: awash in a sea of proteins. Mol. Cell *12*, 5–14.

Kalamvoki, M., and Roizman, B. (2011). The histone acetyltransferase CLOCK is an essential component of the herpes simplex virus 1 transcriptome that includes TFIID, ICP4, ICP27, and ICP22. J. Virol. *85*, 9472–9477.

Kapadia, F., Pryor, A., Chang, T.H., and Johnson, L.F. (2006). Nuclear localization of poly(A)+ mRNA following siRNA reduction of expression of the mammalian RNA helicases UAP56 and URH49. Gene *384*, 37–44.

Kapasi, A.J., and Spector, D.H. (2008). Inhibition of the cyclin-dependent kinases at the beginning of human cytomegalovirus infection specifically alters the levels and localization of the RNA polymerase II carboxyl-terminal domain kinases cdk9 and cdk7 at the viral transcriptosome. J. Virol. *82*, 394–407.

Kapasi, A.J., Clark, C L., Tran, K., and Spector, D.H. (2009). Recruitment of cdk9 to the immediate-early viral transcriptosomes during human cytomegalovirus infection requires efficient binding to cyclin T1, a threshold level of IE2 86, and active transcription. J. Virol. *83*, 5904–5917.

Kerry, J.A., Sehgal, A., Barlow, S.W., Cavanaugh, V.J., Fish, K., Nelson, J.A., and Stenberg, R.M. (1995). Isolation and characterization of a low-abundance splice variant from the human cytomegalovirus major immediate-early gene region. J. Virol. *69*, 3868–3872.

Kerry, J.A., Priddy, M.A., Jervey, T.Y., Kohler, C P., Staley, T.L., Vanson, C.D., Jones, T.R., Iskenderian, A.C., Anders, D.G., and Stenberg, R.M. (1996). Multiple regulatory events influence human cytomegalovirus DNA polymerase (UL54) expression during viral infection. J. Virol. *70*, 373–382.

Kerry, J.A., Priddy, M.A., Staley, T.L., Jones, T.R., and Stenberg, R.M. (1997). The role of ATF in regulating the human cytomegalovirus DNA polymerase (UL54) promoter during viral infection. J. Virol. *71*, 2120–2126.

Kiesler, E., Miralles, F., and Visa, N. (2002). HEL/UAP56 binds cotranscriptionally to the Balbiani ring pre-mRNA in an intron-independent manner and accompanies the BR mRNP to the nuclear pore. Curr. Biol. *12*, 859–862.

Kim, Y.E., Lee, J.H., Kim, E.T., Shin, H.J., Gu, S.Y., Seol, H.S., Ling, P.D., Lee, C.H., and Ahn, J.H. (2011). Human cytomegalovirus infection causes degradation of Sp100 proteins that suppress viral gene expression. J. Virol. *85*, 11928–11937.

Kohler, C.P., Kerry, J.A., Carter, M., Muzithras, V.P., Jones, T.R., and Stenberg, R.M. (1994). Use of recombinant virus to assess human cytomegalovirus early and late promoters in the context of the viral genome. J. Virol. *68*, 6589–6597.

Korioth, F., Maul, G.G., Plachter, B., Stamminger, T., and Frey, J. (1996). The nuclear domain 10 (ND10) is disrupted by the human cytomegalovirus gene product IE1. Exp. Cell Res. *229*, 155–158.

Kotenko, S.V., Saccani, S., Izotova, L.S., Mirochnitchenko, O.V., and Pestka, S. (2000). Human cytomegalovirus harbors its own unique IL-10 homolog (cmvIL-10). Proc. Natl. Acad. Sci. U.S.A. *97*, 1695–1700.

Kouzarides, T., Bankier, A.T., Satchwell, S.C., Preddy, E., and Barrell, B.G. (1988). An immediate-early gene of human cytomegalovirus encodes a potential membrane glycoprotein. Virology *165*, 151–164.

Krosky, P.M., Baek, M.C., and Coen, D.M. (2003). The human cytomegalovirus UL97 protein kinase, an antiviral drug target, is required at the stage of nuclear egress. J. Virol. *77*, 905–914.

Lacasse, J.J., and Schang, L.M. (2010). During lytic infections, herpes simplex virus type 1 DNA is in complexes with the properties of unstable nucleosomes. J. Virol. *84*, 1920–1933.

Lang, D., Gebert, S., Arlt, H., and Stamminger, T. (1995). Functional interaction between the human cytomegalovirus 86-kilodalton IE2 protein and the cellular transcription factor CREB. J. Virol. *69*, 6030–6037.

Le Hir, H., Gatfield, D., Izaurralde, E., and Moore, M.J. (2001). The exon–exon junction complex provides a binding platform for factors involved in mRNA export and nonsense-mediated mRNA decay. EMBO J. *20*, 4987–4997.

Lee, H.C., Choe, J., Chi, S.G., and Kim, Y.K. (2009). Exon junction complex enhances translation of spliced mRNAs

at multiple steps. Biochem. Biophys. Res. Commun. *384*, 334–340.

Lee, H.R., Kim, D.J., Lee, J.M., Choi, C.Y., Ahn, B.Y., Hayward, G.S., and Ahn, J.H. (2004). Ability of the human cytomegalovirus IE1 protein to modulate sumoylation of PML correlates with its functional activities in transcriptional regulation and infectivity in cultured fibroblast cells. J. Virol. *78*, 6527–6542.

Lei, E.P., and Silver, P.A. (2002). Intron status and 3′-end formation control cotranscriptional export of mRNA. Genes Dev. *16*, 2761–2766.

Lei, E.P., Krebber, H., and Silver, P.A. (2001). Messenger RNAs are recruited for nuclear export during transcription. Genes Dev. *15*, 1771–1782.

Liker, E., Fernandez, E., Izaurralde, E., and Conti, E. (2000). The structure of the mRNA export factor TAP reveals a cis arrangement of a non-canonical RNP domain and an LRR domain. EMBO J. *19*, 5587–5598.

Linder, P., and Stutz, F. (2001). mRNA export: travelling with DEAD box proteins. Curr. Biol. *11*, R961–R963.

Lischka, P., Rosorius, O., Trommer, E., and Stamminger, T. (2001). A novel transferable nuclear export signal mediates CRM1-independent nucleocytoplasmic shuttling of the human cytomegalovirus transactivator protein pUL69. EMBO J. *20*, 7271–7283.

Lischka, P., Toth, Z., Thomas, M., Mueller, R., and Stamminger, T. (2006). The UL69 transactivator protein of human cytomegalovirus interacts with DEXD/H-Box RNA helicase UAP56 to promote cytoplasmic accumulation of unspliced RNA. Mol. Cell Biol. *26*, 1631–1643.

Liu, B., and Stinski, M.F. (1992). Human cytomegalovirus contains a tegument protein that enhances transcription from promoters with upstream ATF and AP-1 cis-acting elements. J. Virol. *66*, 4434–4444.

Lockridge, K.M., Zhou, S.S., Kravitz, R.H., Johnson, J.L., Sawai, E.T., Blewett, E.L., and Barry, P.A. (2000). Primate cytomegaloviruses encode and express an IL-10-like protein. Virology *268*, 272–280.

Longman, D., Johnstone, I.L., and Caceres, J.F. (2003). The Ref/Aly proteins are dispensable for mRNA export and development in *Caenorhabditis elegans*. RNA *9*, 881–891.

Lu, M., and Shenk, T. (1999). Human cytomegalovirus UL69 protein induces cells to accumulate in G1 phase of the cell cycle. J. Virol. *73*, 676–683.

Lu, S., and Cullen, B.R. (2003). Analysis of the stimulatory effect of splicing on mRNA production and utilization in mammalian cells. RNA *9*, 618–630.

Luco, R.F., Allo, M., Schor, I.E., Kornblihtt, A.R., and Misteli, T. (2011). Epigenetics in alternative pre-mRNA splicing. Cell *144*, 16–26.

Lukashchuk, V., McFarlane, S., Everett, R.D., and Preston, C.M. (2008). Human cytomegalovirus protein pp71 displaces the chromatin-associated factor ATRX from nuclear domain 10 at early stages of infection. J. Virol. *82*, 12543–12554.

McDonough, S.H., and Spector, D.H. (1983). Transcription in human fibroblasts permissively infected by human cytomegalovirus strain AD169. Virology *125*, 31–46.

Marchini, A., Liu, H., and Zhu, H. (2001). Human cytomegalovirus with IE-2 (UL122) deleted fails to express early lytic genes. J. Virol. *75*, 1870–1878.

Marschalek, R., mon-Bohm, E., Stoerker, J., Klages, S., Fleckenstein, B., and Dingermann, T. (1989). CMER, an RNA encoded by human cytomegalovirus is most likely transcribed by RNA polymerase III. Nucleic Acids Res. 17, 631–643.

Marschall, M., Stein-Gerlach, M., Freitag, M., Kupfer, R., van den, B.M., and Stamminger, T. (2002). Direct targeting of human cytomegalovirus protein kinase pUL97 by kinase inhibitors is a novel principle for antiviral therapy. J. Gen. Virol. 83, 1013–1023.

Masuda, S., Das, R., Cheng, H., Hurt, E., Dorman, N., and Reed, R. (2005). Recruitment of the human TREX complex to mRNA during splicing. Genes Dev. 19, 1512–1517.

Maul, G.G. (1998). Nuclear domain 10, the site of DNA virus transcription and replication. Bioessays 20, 660–667.

Moorman, N.J., Cristea, I.M., Terhune, S.S., Rout, M.P., Chait, B.T., and Shenk, T. (2008). Human cytomegalovirus protein UL38 inhibits host cell stress responses by antagonizing the tuberous sclerosis protein complex. Cell Host Microbe 3, 253–262.

Munoz, M.J., de la Mata, M., and Kornblihtt, A.R. (2010). The carboxy terminal domain of RNA polymerase II and alternative splicing. Trends Biochem. Sci. 35, 497–504.

Netterwald, J., Yang, S., Wang, W., Ghanny, S., Cody, M., Soteropoulos, P., Tian, B., Dunn, W., Liu, F., and Zhu, H. (2005). Two gamma interferon-activated site-like elements in the human cytomegalovirus major immediate-early promoter/enhancer are important for viral replication. J. Virol. 79, 5035–5046.

Nevels, M., Paulus, C., and Shenk, T. (2004). Human cytomegalovirus immediate-early 1 protein facilitates viral replication by antagonizing histone deacetylation. Proc. Natl. Acad. Sci. U.S.A. 101, 17234–17239.

Nevels, M., Nitzsche, A., and Paulus, C. (2011). How to control an infectious bead string: nucleosome-based regulation and targeting of herpesvirus chromatin. Rev. Med. Virol. 21, 154–180.

Nitzsche, A., Paulus, C., and Nevels, M. (2008). Temporal dynamics of cytomegalovirus chromatin assembly in productively infected human cells. J. Virol. 82, 11167–11180.

Nojima, T., Hirose, T., Kimura, H., and Hagiwara, M. (2007). The interaction between cap-binding complex and RNA export factor is required for intronless mRNA export. J. Biol. Chem. 282, 15645–15651.

Nott, A., Le Hir, H., and Moore, M.J. (2004). Splicing enhances translation in mammalian cells: an additional function of the exon junction complex. Genes Dev. 18, 210–222.

Ott, M., Geyer, M., and Zhou, Q. (2011). The control of HIV transcription: keeping RNA polymerase II on track. Cell Host Microbe 10, 426–435.

Park, J.J., Kim, Y.E., Pham, H.T., Kim, E.T., Chung, Y.H., and Ahn, J.H. (2007). Functional interaction of the human cytomegalovirus IE2 protein with histone deacetylase 2 in infected human fibroblasts. J. Gen. Virol. 88, 3214–3223.

Paulus, C., Nitzsche, A., and Nevels, M. (2010). Chromatinisation of herpesvirus genomes. Rev. Med. Virol. 20, 34–50.

Pavri, R., Zhu, B., Li, G., Trojer, P., Mandal, S., Shilatifard, A., and Reinberg, D. (2006). Histone H2B monoubiquitination functions cooperatively with FACT to regulate elongation by RNA polymerase II. Cell 125, 703–717.

Perng, Y.C., Qian, Z., Fehr, A.R., Xuan, B., and Yu, D. (2011). The human cytomegalovirus gene UL79 is required for the accumulation of late viral transcripts. J. Virol. 85, 4841–4852.

Pizzorno, M.C., Mullen, M.A., Chang, Y.N., and Hayward, G.S. (1991). The functionally active IE2 immediate-early regulatory protein of human cytomegalovirus is an 80-kilodalton polypeptide that contains two distinct activator domains and a duplicated nuclear localization signal. J. Virol. 65, 3839–3852.

Preston, C.M., and Nicholl, M.J. (2006). Role of the cellular protein hDaxx in human cytomegalovirus immediate-early gene expression. J. Gen. Virol. 87, 1113–1121.

Price, D.H. (2000). P-TEFb, a cyclin-dependent kinase controlling elongation by RNA polymerase II. Mol. Cell Biol. 20, 2629–2634.

Prichard, M.N., Gao, N., Jairath, S., Mulamba, G., Krosky, P., Coen, D.M., Parker, B.O., and Pari, G.S. (1999). A recombinant human cytomegalovirus with a large deletion in UL97 has a severe replication deficiency. J. Virol. 73, 5663–5670.

Prichard, M.N., Sztul, E., Daily, S.L., Perry, A.L., Frederick, S.L., Gill, R.B., Hartline, C.B., Streblow, D.N., Varnum, S.M., Smith, R.D., et al. (2008). Human cytomegalovirus UL97 kinase activity is required for the hyperphosphorylation of retinoblastoma protein and inhibits the formation of nuclear aggresomes. J. Virol. 82, 5054–5067.

Pryor, A., Tung, L., Yang, Z., Kapadia, F., Chang, T.H., and Johnson, L.F. (2004). Growth-regulated expression and G0-specific turnover of the mRNA that encodes URH49, a mammalian DExH/D box protein that is highly related to the mRNA export protein UAP56. Nucleic Acids Res. 32, 1857–1865.

Puchtler, E., and Stamminger, T. (1991). An inducible promoter mediates abundant expression from the immediate-early 2 gene region of human cytomegalovirus at late times after infection. J. Virol. 65, 6301–6306.

Rechter, S., Scott, G.M., Eickhoff, J., Zielke, K., Auerochs, S., Muller, R., Stamminger, T., Rawlinson, W.D., and Marschall, M. (2009). Cyclin-dependent kinases phosphorylate the cytomegalovirus RNA export protein pUL69 and modulate its nuclear localization and activity. J. Biol. Chem. 284, 8605–8613.

Reed, R. (2003). Coupling transcription, splicing and mRNA export. Curr. Opin. Cell Biol. 15, 326–331.

Reed, R., and Hurt, E. (2002). A conserved mRNA export machinery coupled to pre-mRNA splicing. Cell 108, 523–531.

Reeves, M.B. (2011). Chromatin-mediated regulation of cytomegalovirus gene expression. Virus Res. 157, 134–143.

Rice, S.A., and Knipe, D.M. (1988). Gene-specific transactivation by herpes simplex virus type 1 alpha protein ICP27. J. Virol. 62, 3814–3823.

Rocak, S., and Linder, P. (2004). DEAD-box proteins: the driving forces behind RNA metabolism. Nat. Rev. Mol. Cell Biol. 5, 232–241.

Rodrigues, J.P., Rode, M., Gatfield, D., Blencowe, B., Carmo-Fonseca, M., and Izaurralde, E. (2001). REF proteins mediate the export of spliced and unspliced mRNAs from the nucleus. Proc. Natl. Acad. Sci. U.S.A. 98, 1030–1035.

Romaker, D., Schregel, V., Maurer, K., Auerochs, S., Marzi, A., Sticht, H., and Marschall, M. (2006). Analysis of the structure–activity relationship of four herpesviral UL97 subfamily protein kinases reveals partial but not full functional conservation. J. Med. Chem. 49, 7044–7053.

Rosa, M.D., Gottlieb, E., Lerner, M.R., and Steitz, J.A. (1981). Striking similarities are exhibited by two small Epstein–Barr virus-encoded ribonucleic acids and the adenovirus-associated ribonucleic acids VAI and VAII. Mol. Cell Biol. *1*, 785–796.

Sacks, W.R., Greene, C.C., Aschman, D.P., and Schaffer, P.A. (1985). Herpes simplex virus type 1 ICP27 is an essential regulatory protein. J. Virol. *55*, 796–805.

Saffert, R.T., and Kalejta, R.F. (2006). Inactivating a cellular intrinsic immune defense mediated by Daxx is the mechanism through which the human cytomegalovirus pp71 protein stimulates viral immediate-early gene expression. J. Virol. *80*, 3863–3871.

Salvant, B.S., Fortunato, E.A., and Spector, D.H. (1998). Cell cycle dysregulation by human cytomegalovirus: influence of the cell cycle phase at the time of infection and effects on cyclin transcription. J. Virol. *72*, 3729–3741.

Sanchez, V., McElroy, A.K., Yen, J., Tamrakar, S., Clark, C.L., Schwartz, R.A., and Spector, D.H. (2004). Cyclin-dependent kinase activity is required at early times for accurate processing and accumulation of the human cytomegalovirus UL122–123 and UL37 immediate-early transcripts and at later times for virus production. J. Virol. *78*, 11219–11232.

Sandri-Goldin, R.M. (1998). ICP27 mediates HSV RNA export by shuttling through a leucine-rich nuclear export signal and binding viral intronless RNAs through an RGG motif. Genes Dev. *12*, 868–879.

Sandri-Goldin, R.M. (2001). Nuclear export of herpes virus RNA. Curr. Top. Microbiol. Immunol. *259*, 2–23.

Sandri-Goldin, R.M. (2008). The many roles of the regulatory protein ICP27 during herpes simplex virus infection. Front. Biosci. *13*, 5241–5256.

Sandri-Goldin, R.M., and Mendoza, G.E. (1992). A herpesvirus regulatory protein appears to act post-transcriptionally by affecting mRNA processing. Genes Dev. *6*, 848–863.

Sciabica, K.S., Dai, Q.J., and Sandri-Goldin, R.M. (2003). ICP27 interacts with SRPK1 to mediate HSV splicing inhibition by altering SR protein phosphorylation. EMBO J. *22*, 1608–1619.

Scully, A.L., Sommer, M.H., Schwartz, R., and Spector, D.H. (1995). The human cytomegalovirus IE2 86-kilodalton protein interacts with an early gene promoter via site-specific DNA binding and protein–protein associations. J. Virol. *69*, 6533–6540.

Segref, A., Sharma, K., Doye, V., Hellwig, A., Huber, J., Luhrmann, R., and Hurt, E. (1997). Mex67p, a novel factor for nuclear mRNA export, binds to both poly(A)+ RNA and nuclear pores. EMBO J. *16*, 3256–3271.

Shirakata, M., Terauchi, M., Ablikim, M., Imadome, K., Hirai, K., Aso, T., and Yamanashi, Y. (2002). Novel immediate-early protein IE19 of human cytomegalovirus activates the origin recognition complex I promoter in a cooperative manner with IE72. J. Virol. *76*, 3158–3167.

Sinclair, J. (2010). Chromatin structure regulates human cytomegalovirus gene expression during latency, reactivation and lytic infection. Biochim. Biophys. Acta *1799*, 286–295.

Sonenberg, N., and Hinnebusch, A.G. (2009). Regulation of translation initiation in eukaryotes: mechanisms and biological targets. Cell *136*, 731–745.

Song, B., Yeh, K.C., Liu, J., and Knipe, D.M. (2001). Herpes simplex virus gene products required for viral inhibition of expression of G1-phase functions. Virology *290*, 320–328.

Sourvinos, G., Tavalai, N., Berndt, A., Spandidos, D.A., and Stamminger, T. (2007). Recruitment of human cytomegalovirus immediate-early 2 protein onto parental viral genomes in association with ND10 in live-infected cells. J. Virol. *81*, 10123–10136.

Stamminger, T., Puchtler, E., and Fleckenstein, B. (1991). Discordant expression of the immediate-early 1 and 2 gene regions of human cytomegalovirus at early times after infection involves posttranscriptional processing events [published erratum appeared in J. Virol. 1991, 5654]. J. Virol. *65*, 2273–2282.

Stenberg, R.M., Witte, P.R., and Stinski, M.F. (1985). Multiple spliced and unspliced transcripts from human cytomegalovirus immediate-early region 2 and evidence for a common initiation site within immediate-early region 1. J. Virol. *56*, 665–675.

Strasser, K., and Hurt, E. (2000). Yra1p, a conserved nuclear RNA-binding protein, interacts directly with Mex67p and is required for mRNA export. EMBO J. *19*, 410–420.

Strasser, K., and Hurt, E. (2001). Splicing factor Sub2p is required for nuclear mRNA export through its interaction with Yra1p. Nature *413*, 648–652.

Strasser, K., Bassler, J., and Hurt, E. (2000). Binding of the Mex67p/Mtr2p heterodimer to FXFG, GLFG, and FG repeat nucleoporins is essential for nuclear mRNA export. J. Cell Biol. *150*, 695–706.

Strasser, K., Masuda, S., Mason, P., Pfannstiel, J., Oppizzi, M., Rodriguez-Navarro, S., Rondon, A.G., Aguilera, A., Struhl, K., Reed, R., et al. (2002). TREX is a conserved complex coupling transcription with messenger RNA export. Nature *417*, 304–308.

Stutz, F., Bachi, A., Doerks, T., Braun, I.C., Seraphin, B., Wilm, M., Bork, P., and Izaurralde, E. (2000). REF, an evolutionary conserved family of hnRNP-like proteins, interacts with TAP/Mex67p and participates in mRNA nuclear export. RNA *6*, 638–650.

Su, Y., Adair, R., Davis, C.N., DiFronzo, N.L., and Colberg-Poley, A.M. (2003). Convergence of RNA cis elements and cellular polyadenylation factors in the regulation of human cytomegalovirus UL37 exon 1 unspliced RNA production. J. Virol. *77*, 12729–12741.

Tamrakar, S., Kapasi, A.J., and Spector, D.H. (2005). Human cytomegalovirus infection induces specific hyperphosphorylation of the carboxyl-terminal domain of the large subunit of RNA polymerase II that is associated with changes in the abundance, activity, and localization of cdk9 and cdk7. J. Virol. *79*, 15477–15493.

Tavalai, N., and Stamminger, T. (2008). New insights into the role of the subnuclear structure ND10 for viral infection. Biochim. Biophys. Acta *1783*, 2207–2221.

Tavalai, N., and Stamminger, T. (2009). Interplay between herpesvirus infection and host defense by PML nuclear bodies. Viruses *1*, 1240–1264.

Tavalai, N., and Stamminger, T. (2011). Intrinsic cellular defense mechanisms targeting human cytomegalovirus. Virus Res. *157*, 128–133.

Tavalai, N., Papior, P., Rechter, S., Leis, M., and Stamminger, T. (2006). Evidence for a role of the cellular ND10 protein PML in mediating intrinsic immunity against human cytomegalovirus infections. J. Virol. *80*, 8006–8018.

Tavalai, N., Papior, P., Rechter, S., and Stamminger, T. (2008). Nuclear domain 10 components promyelocytic leukemia protein and hDaxx independently contribute to an intrinsic antiviral defense against human cytomegalovirus infection. J. Virol. *82*, 126–137.

Tavalai, N., Adler, M., Scherer, M., Riedl, Y., and Stamminger, T. (2011). Evidence for a dual antiviral role of the major nuclear domain 10 component Sp100 during the immediate-early and late phases of the human cytomegalovirus replication cycle. J. Virol. *85*, 9447–9458.

Tenney, D.J., and Colberg-Poley, A.M. (1991a). Expression of the human cytomegalovirus UL36–38 immediate-early region during permissive infection. Virology *182*, 199–210.

Tenney, D.J., and Colberg-Poley, A.M. (1991b). Human cytomegalovirus UL36–38 and US3 immediate-early genes: temporally regulated expression of nuclear, cytoplasmic, and polysome-associated transcripts during infection. J. Virol. *65*, 6724–6734.

Terhune, S.S., Moorman, N.J., Cristea, I.M., Savaryn, J.P., Cuevas-Bennett, C., Rout, M.P., Chait, B.T., and Shenk, T. (2010). Human cytomegalovirus UL29/28 protein interacts with components of the NuRD complex which promote accumulation of immediate-early RNA. PLoS Pathog. *6*, e1000965.

Thomas, M., Lischka, P., Muller, R., and Stamminger, T. (2011). The cellular DExD/H-box RNA-helicases UAP56 and URH49 exhibit a CRM1-independent nucleocytoplasmic shuttling activity. PLoS One *6*, e22671.

Thomas, M., Rechter, S., Milbradt, J., Auerochs, S., Muller, R., Stamminger, T., and Marschall, M. (2009). Cytomegaloviral protein kinase pUL97 interacts with the nuclear mRNA export factor pUL69 to modulate its intranuclear localization and activity. J. Gen. Virol. *90*, 567–578.

Thomsen, D.R., Stenberg, R.M., Goins, W.F., and Stinski, M.F. (1984). Promoter-regulatory region of the major immediate-early gene of human cytomegalovirus. Proc. Natl. Acad. Sci. U.S.A. *81*, 659–663.

Torok, D., Ching, R.W., and Bazett-Jones, D.P. (2009). PML nuclear bodies as sites of epigenetic regulation. Front Biosci. *14*, 1325–1336.

Toth, Z., and Stamminger, T. (2008). The human cytomegalovirus regulatory protein UL69 and its effect on mRNA export. Front. Biosci. *13*, 2939–2949.

Toth, Z., Lischka, P., and Stamminger, T. (2006). RNA-binding of the human cytomegalovirus transactivator protein UL69, mediated by arginine-rich motifs, is not required for nuclear export of unspliced RNA. Nucleic Acids Res. *34*, 1237–1249.

Varnum, S.M., Streblow, D.N., Monroe, M.E., Smith, P., Auberry, K.J., Pasa-Tolic, L., Wang, D., Camp, D.G., Rodland, K., Wiley, S., et al. (2004). Identification of proteins in human cytomegalovirus (HCMV) particles: the HCMV proteome. J. Virol. *78*, 10960–10966.

Villa, R., Pasini, D., Gutierrez, A., Morey, L., Occhionorelli, M., Vire, E., Nomdedeu, J.F., Jenuwein, T., Pelicci, P.G., Minucci, S., et al. (2007). Role of the polycomb repressive complex 2 in acute promyelocytic leukemia. Cancer Cell *11*, 513–525.

Vinciguerra, P., and Stutz, F. (2004). mRNA export: an assembly line from genes to nuclear pores. Curr. Opin. Cell Biol. *16*, 285–292.

Wathen, M.W., and Stinski, M.F. (1982). Temporal patterns of human cytomegalovirus transcription: mapping the viral RNAs synthesized at immediate-early, early, and late times after infection. J. Virol. *41*, 462–477.

Weidtkamp-Peters, S., Lenser, T., Negorev, D., Gerstner, N., Hofmann, T.G., Schwanitz, G., Hoischen, C., Maul, G., Dittrich, P., and Hemmerich, P. (2008). Dynamics of component exchange at PML nuclear bodies. J. Cell Sci. *121*, 2731–2743.

White, E.A., and Spector, D.H. (2007). Early viral gene expression and function. In Human Herpesviruses: Biology, Therapy and Immunoprophylaxis, Arvin, A., Campadelli-Fiume, G., Mocarski, E., Moore, P.S., Roizman, B., Whitley, R., and Yamanishi, K., eds. (Cambridge University Press, Cambridge, UK), Chapter 18.

White, E.A., Del Rosario, C.J., Sanders, R.L., and Spector, D.H. (2007). The IE2 60-kilodalton and 40-kilodalton proteins are dispensable for human cytomegalovirus replication but are required for efficient delayed early and late gene expression and production of infectious virus. J. Virol. *81*, 2573–2583.

Winkler, M., and Stamminger, T. (1996). A specific subform of the human cytomegalovirus transactivator protein pUL69 is contained within the tegument of virus particles. J. Virol. *70*, 8984–8987.

Winkler, M., Rice, S.A., and Stamminger, T. (1994). UL69 of human cytomegalovirus, an open reading frame with homology to ICP27 of herpes simplex virus, encodes a transactivator of gene expression. J. Virol. *68*, 3943–3954.

Winkler, M., aus Dem Siepen, T., and Stamminger, T. (2000). Functional interaction between pleiotropic transactivator pUL69 of human cytomegalovirus and the human homolog of yeast chromatin regulatory protein SPT6. J. Virol. *74*, 8053–8064.

Woodhall, D.L., Groves, I.J., Reeves, M.B., Wilkinson, G., and Sinclair, J.H. (2006). Human Daxx-mediated repression of human cytomegalovirus gene expression correlates with a repressive chromatin structure around the major immediate-early promoter. J. Biol. Chem. *281*, 37652–37660.

Wu, W.S., Vallian, S., Seto, E., Yang, W.M., Edmondson, D., Roth, S., and Chang, K.S. (2001). The growth suppressor PML represses transcription by functionally and physically interacting with histone deacetylases. Mol. Cell Biol. *21*, 2259–2268.

Yamazaki, T., Fujiwara, N., Yukinaga, H., Ebisuya, M., Shiki, T., Kurihara, T., Kioka, N., Kambe, T., Nagao, M., Nishida, E., et al. (2010). The closely related RNA helicases, UAP56 and URH49, preferentially form distinct mRNA export machineries and coordinately regulate mitotic progression. Mol. Biol. Cell. *21*, 2953–2965.

Yoh, S.M., Cho, H., Pickle, L., Evans, R.M., and Jones, K.A. (2007). The Spt6 SH2 domain binds Ser2-P RNAPII to direct Iws1-dependent mRNA splicing and export. Genes Dev. *21*, 160–174.

Zenklusen, D., Vinciguerra, P., Strahm, Y., and Stutz, F. (2001). The yeast hnRNP-Like proteins Yra1p and Yra2p participate in mRNA export through interaction with Mex67p. Mol. Cell Biol. *21*, 4219–4232.

Zenklusen, D., Vinciguerra, P., Wyss, J.C., and Stutz, F. (2002). Stable mRNP formation and export require cotranscriptional recruitment of the mRNA export factors Yra1p and Sub2p by Hpr1p. Mol. Cell Biol. *22*, 8241–8253.

Zielke, B., Thomas, M., Giede-Jeppe, A., Muller, R., and Stamminger, T. (2011). Characterization of the betaherpesviral pUL69 protein family reveals binding of the cellular mRNA export factor UAP56 as a prerequisite for stimulation of nuclear mRNA export and for efficient viral replication. J. Virol. *85*, 1804–1819.

Zielke, B., Wagenknecht, N., Pfeifer, C., Zielke, K., Thomas, M., and Stamminger, T. (2012). Transfer of the UAP56-interaction motif of human cytomegalovirus pUL69 to its murine cytomegalovirus homolog converts the protein into a functional mRNA export factor that can substitute for pUL69 during viral infection. J. Virol. *86*, 7448–7453.

Zydek, M., Hagemeier, C., and Wiebusch, L. (2010). Cyclin-dependent kinase activity controls the onset of the HCMV lytic cycle. PLoS Pathog. *6*, e1001096.

Zydek, M., Uecker, R., Tavalai, N., Stamminger, T., Hagemeier, C., and Wiebusch, L. (2011). General blockade of HCMV immediate-early mRNA expression in S/G2 by a nuclear, Daxx- and PML-independent mechanism. J. Gen. Virol. *92*, 2757–2769.

Intracellular Sorting and Trafficking of Cytomegalovirus Proteins during Permissive Infection

I.12

Anamaris M. Colberg-Poley and Chad D. Williamson

Abstract

As with most DNA viruses, which require nuclear and cytoplasmic phases of virion maturation, proper and coordinated trafficking of viral proteins is crucial for the CMV life cycle. Trafficking of CMV proteins enables jumpstarting its infection, partly determines whether lytic or latent infection is established, promotes nuclear and cytoplasmic assembly of virions, and enhances their stability and egress. To allow complex processes including viral DNA replication, packaging, nuclear and cytoplasmic egress, trafficking of CMV proteins is temporally and spatially regulated by modifications, particularly phosphorylation, and by interactions between viral or cellular proteins. Thus, orchestrated recruitment and colocalization of necessary components to enable functions are assured. In addition to conventional nuclear and cytoplasmic trafficking of viral proteins, HCMV encodes an antiapoptotic UL37 exon 1 protein or viral mitochondria-localized inhibitor of apoptosis, which circuitously traffics from the ER to mitochondria and through ER subdomains known as mitochondria-associated membranes. By this unconventional trafficking, HCMV is able to commandeer ER–mitochondrial cross-talk as well as mitochondrial functions, metabolism, antiviral responses, and apoptosis. The importance of proper intracellular trafficking of some key HCMV proteins such as those required for its DNA replication and assembly is supported by the deleterious effects of their inhibition on HCMV permissive growth.

Introduction

CMV is a herpesvirus with a large DNA genome and two sites of virion maturation: the nucleus and the cytoplasmic viral assembly compartment (cVAC), where further tegumentation and envelopment of virions occurs prior to their release from the infected cell as infectious progeny. For the initial stages of growth, CMV uses cellular nuclear machinery, including

transcription and RNA processing machineries, for its replication. At later stages of CMV production, CMV capsids are assembled in the nucleus, package progeny DNA and undergo primary tegumentation prior to their primary envelopment and nuclear egress and transit to the cVAC. Thus, timed and orchestrated trafficking of key viral proteins to their correct subcellular compartment is essential for their functions and, ultimately, for progression of the viral life cycle. The purpose of this chapter is to integrate known trafficking patterns of CMV products with their functions during viral growth. We focus our review primarily on well documented intracellular trafficking of HCMV proteins during the viral life cycle (overviewed in Table I.12.1). When available, we include published information concerning the targeting sequences and their modification or interactions with viral or cellular proteins that affect their intracellular trafficking.

Nucleocytoplasmic trafficking of HCMV proteins

The nuclear envelope divides eukaryotic cells into the nucleus and cytoplasm. Translocation of soluble proteins > 45 kDa is generally mediated by the importin (IMP) superfamily of nuclear transporters. As HCMV has a nuclear phase of replication, nuclear translocation of many of its proteins is crucial for its life cycle. Nucleocytoplasmic trafficking of soluble proteins occurs through the nuclear pore complex and is mediated by selective mechanisms that are controlled by transport receptors and NLSs or nuclear export signals (NESs). A review of the regulation of nucleocytoplasmic trafficking of viral proteins was recently published (Fulcher and Jans, 2011). Classical NLSs, such as the SV40 large tumour antigen NLS, are composed of a single (monopartite) cluster of basic residues (Kalderon *et al.*, 1984a,b) or of two basic motifs separated by a ten residue spacer as in the bipartite NLS of *Xenopus laevis* nucleoplasmin (Robbins *et al.*, 1991). Classical

Table I.12.1 Overview of notable CMV protein trafficking

Protein	Targeting residues	Subdomain	Function	Comments	Selected references
Nuclear events: tegument proteins					
pp71	aa 188–300	ND10, Late cytoplasmic	Daxx degradation, MIE activation	Regulated by $223T$ phosphorylation, Trafficking affected by undifferentiated cell proteins	Hensel et al. (1996), Shen et al. (2008), Penkert and Kalejta (2010)
pp65	aa 415–438, aa 537–561	Early diffuse in nucleus, Nuclear lamina, Nucleoli, Late cytoplasmic	Modulate IFN response, Nuclear egress	Nucleoplasmic shuttling, Regulated by phosphorylation, CRM1-dependent export	Schmolke et al. (1995), Arcangeletti et al. (2003), Prichard et al. (2005a), Sanchez et al. (2000, 2007)
pp150	No predicted targeting signals	Early, INM, Late, cVAC	virion stability during transit to cVAC	interaction with Bicaudal D1, Affects its trafficking to cVAC, Interacts with HCMV capsids, Key role in capsid maturation	Baxter and Gibson (2001), Indran et al. (2010), Tandon and Mocarski (2011), Moorman et al. (2010)
pUL69	NLS, aa 21–45	Early, nuclear diffuse excluding nucleoli	Shuttling	Nucleocytoplasmic shuttling is modulated by phosphorylation, Export is CRM1 independent	Winkler et al. (1994), Lischka et al. (2001, 2006b), Rechter et al. (2009),
	NES, aa 597–624	Late, nVRC	Viral mRNA export		
	CCMV (pC69), RhCMV (pRh69)	Diffuse exclude nucleoli	Viral mRNA export	Nucleocytoplasmic shuttling	Zielke et al. (2011)
pUL96	14kDa, no NLS or NES		Virion stability	Associates with nucleocapsids	Tandon and Mocarski (2011)
MIE proteins					
IE1	aa 1–24, aa 326–327 and aa 340–342	ND10, associates with metaphase chromatin	Disperses ND10s, HCMV gene expression	Essential at low MOIs	Ahn et al. (1998), Wilkinson et al. (1998), Delmas et al. (2005), Lee et al. (2007)
IE2	aa 1–24, aa 145–156, aa 321–328	ND10	Essential for early and late gene expression	Essential for HCMV growth, Multiple isoforms	Pizzorno et al. (1991), Caswell et al. (1993), Ishov et al. (1997), White et al. (2007)
HCMV DNA replication proteins					
pUL54	aa 1153–1159, aa 1222–1227	nVRC	DNA Pol, catalytic subunit	Small inhibitors of UL54-UL44, Interaction block growth	Loregian et al. (2000), Alvisi et al. (2006b), Loregian and Coen (2006)

Table I.12.1 Continued

Protein	Targeting residues	Subdomain	Function	Comments	Selected references
ppUL44	aa 425–431	Initial, nuclear diffuse Early, pre-nVRC Later, nVRC	DNA Pol, processivity factor	Requires phosphorylation pUL97 and PKC/PKA substrate Recruited by pUL112–113 Negatively regulated by BRAP2	Penfold and Mocarski (1997), Marschall *et al.* (2003), Krosky *et al.* (2003b), Alvisi *et al.* (2006a), Fulcher *et al.* (2010)
ppUL57		pre-nVRC later, nVRC	ssDNA-binding protein		Kemble *et al.* (1987), Penfold and Mocarski (1997)
pUL84	aa 226–508, NLS with NES1, aa 228–237 NES2, aa 359–366		Essential noncore protein	Nucleocytoplasmic shuttling CRM1-dependent Shuttling is essential for HCMV growth	Xu *et al.* (2002), Lischka *et al.* (2003, 2006a), Kaiser *et al.* (2009), Gao *et al.* (2010)
pUL97	aa 6–35 aa 164–198, aa 190–213	Nucleus	Serine/threonine protein kinase	Phosphorylates pUL44, Nucleoside analogues Two isoforms	Krosky *et al.* (2003b), Marschall *et al.* (2003), Webel *et al.* (2011)
ppUL112–113 84 kDa isoform	aa 1–125	Early, pre-nVRC Late, nVRC	Non-core proteins	Organizer of nVRC Recruits core proteins to nVRCs	Penfold and Mocarski, (1997), Ahn *et al.* (1999), Park *et al.* (2006), Kim and Ahn, (2010)
Nuclear capsid assembly proteins					
MCP			Major capsid protein	Piggybacks on pAP or pNP1	Wood *et al.* (1997)
mCP			Triplex		
mC-BP	NLS		Triplex	Carries mCP and SCP to nucleus	Plafker and Gibson (1998)
SCP			Triplex		
pAP, pNP1			Assembly protein precursor, maturational protease precursor		

Protein	aa / motif	Localization	Type	Function	References
HCMV	aa 510–513, aa 537–543			Scaffold, carry MCP to Nucleus	Nguyen *et al.* (2008), Plafker and Gibson (1998)
SCMV	aa 175–178, aa 202–208				
pUL56	aa 816–827	Early pre-nVRC nVRC	terminase	IMP-dependent	Giesen *et al.* (2000) Thoma *et al.* (2006)
Nuclear glycoproteins					
gB	aa 885–890	Nucleus cVAC	nuclear Glycoprotein Virion glycoprotein		Bogner *et al.* (1997) Meyer and Radsak (2000)
gpUL16	C-terminal tail	PM, INM	Immune evasion		Vales-Gomez *et al.* (2006)
pUS17		TGN, nVRC			Das *et al.* (2006)
Nuclear egress complex					
pUL50		Nuclear rim	NEC		Milbradt *et al.* (2007) Sam *et al.* (2009)
pUL53		Patches in nuclear periphery	NEC	pUL50 relocalizes pUL53 to nuclear rim	Dal Monte *et al.* (2002), Milbradt *et al.* (2007), Sam *et al.* (2009)
RASCAL		nuclear rim	NEC	Requires pUL50 to target nuclear rim	Miller *et al.* (2010)
Cytoplasmic events: secretory apparatus and cVAC					
gB	aa 1–24	ER, cVAC		Requirement for PACS-1	Singh and Compton (2000)
		TGN		CK2 phosphorylation	Crump *et al.* (2003) Jarvis *et al.* (2004) Moorman *et al.* (2010)
		Endosomes Fused vesicles			
gM	aa 359–371, acidic cluster aa 329–333, Y-based motif	cVAC	Major envelope gp	FIP4 binding Trafficking is Arf5 independent	Krzyzaniak *et al.* (2007, 2009)
gO	aa 14–40	ER, Golgi		Soluble, anchored to membranes by complexing with gH/gL	Theiler and Compton (2001) Rychman *et al.* (2010)

Table I.12.1 Continued

Protein	Targeting residues	Subdomain	Function	Comments	Selected references
Dual ER and mitochondrial localization					
pUL37x1/vMIA pUL37$_M$ pUL37$_{NH2}$	aa 3–36	ER, MAM, OMM	antiapoptotic (Bax inhibition) Affects ATP synthesis Recruits viperin to IMM	Sequential ER to mitochondrial trafficking MAM lipid raft association	Mavinakere and Colberg-Poley (2004a), Williamson and Colberg-Poley (2010), Bozidis et al. (2010), Goldmacher et al. (1999), Seo et al. (2011)
			Promoter transactivation		Colberg-Poley et al. (1992, 1998)
pUS9	ER, aa 1–24 MTS, aa 198–247	ER/mitochondria			Mandic et al. (2009)
Mitochondrial proteins					
m38.5		OMM	Antiapoptotic (Bax inhibition)		Jurak et al. (2008) Norris and Youle (2008) Arnoult et al. (2008)
m41.1		Mitochondria	Antiapoptotic (Bak inhibition)		Cam et al. (2010)
Mitochondrial RNA					
beta 2.7 RNA		IMM	Antiapoptotic	Untranslated RNA	Reeves et al. (2007) Zhao et al. (2010)

NLSs are recognized by the IMPα/β heterodimer or the IMPβ homodimer (Jans *et al.*, 2000). The IMPα/β heterodimer functions in nuclear import while the IMPβ can mediate transport in either direction, with proteins mediating export referred to as exportins (Fulcher and Jans, 2011). The alpha subunit of the IMPα/β heterodimer recognizes the NLS; whereas, the beta subunit facilitates the IMPα interaction by causing a conformation change in IMPα. IMP-dependent passage of cargo proteins through the nuclear pore complex is achieved by transient interactions of IMPβ with FG (phenylalanine–glycine repeat)-nucleoporins. Release of the cargo protein at the nuclear face requires Ran, in its activated GTP-bound form, to bind to IMPβ and thereby dissociate the import complex (Fulcher and Jans, 2011). Proteins that contain NLSs and NESs or combined nucleocytoplasmic shuttling signals have the capacity to shuttle between the nucleus and the cytoplasm (Michael, 2000). The best-characterized NESs comprise three or four hydrophobic residues interspersed with non-hydrophobic residues, such as leucine-rich signals, and mediate export by interaction with CRM1 (exportin 1) (Fried and Kutay, 2003).

Jumpstarting HCMV life cycle with tegument proteins

The tegument of HCMV, the proteinaceous layer between the lipid envelope and capsid, contains 20–25 viral proteins (Varnum *et al.*, 2004; Kalejta, 2008), some of which have been characterized in terms of intracellular trafficking and functions. Tegument proteins can accelerate the initiation of lytic infection and play important roles in virus assembly, particle stability and particle disassembly (Baldick *et al.*, 1997; Bechtel and Shenk, 2002; Britt *et al.*, 2004; Munger *et al.*, 2006; Tandon and Mocarski, 2011). The functions of tegument proteins in the initiation of the lytic cycle are reviewed in Chapter I.9. Upon entry, some tegument proteins, including pp71 (UL82) and pp65 (UL83), translocate from incoming virions to the nucleus.

pp71

Nuclear trafficking is important for the ability of pp71 to initiate HCMV lytic infection (Marshall *et al.*, 2002; Shen *et al.*, 2008; Saffert *et al.*, 2010). Upon HCMV entry of permissive cells, pp71, encoded by UL82, counteracts nuclear innate defence by blocking Daxx and ATRX-mediated silencing of MIE gene expression (Ishov *et al.*, 2002; Cantrell and Bresnahan, 2006; Preston and Nicholl, 2006; Saffert and Kalejta, 2006). In addition, pp71 can

activate transcription at promoters containing AP-1 *cis*-elements and can bind to retinoblastoma protein, increasing progression through the cell cycle (Liu and Stinski, 1992; Kalejta *et al.*, 2003).

The subcellular localization of pp71 upon delivery by the virion into the cell depends upon cell differentiation and is facilitated by factors from differentiated cells such as permissive human fibroblasts (Penkert and Kalejta, 2010). In non-permissive cells, pp71 subcellular localization appears to be controlled at nuclear import and not at export. Nuclear import of pp71 in the non-permissive, undifferentiated monocyte-like THP-1 cell line is blocked and expression of the incoming HCMV genome is repressed by histone deacetylases. Therefore, latency, rather than lytic infection, is established (Saffert and Kalejta, 2007). In contrast, permissive HFFs produce factors that facilitate nuclear import of incoming pp71 and, thereby, initiate HCMV IE expression and lytic infection ensues. Thus, intracellular trafficking of some HCMV proteins can be regulated by interactions with host cell proteins and their proper trafficking of HCMV proteins can in part determine the outcome of HCMV infection.

The nuclear import of pp71 is dependent upon its mid-region, which spans an expansive domain, aa 188–300 (Shen *et al.*, 2008). This domain appears to contain a non-conventional NLS, whose tertiary structure is needed but is not sufficient to target heterologous proteins to the nucleus. At late times of infection, pp71 preferentially traffics to the cytoplasm in perinuclear structures (Hensel *et al.*, 1996). This differential trafficking of pp71 during progression of the virus life cycle suggests that pp71 contains multiple targeting signals or that its NLS is susceptible to negative regulation. Indeed, phosphorylation of pp71 may control its subcellular trafficking. The phosphorylation status of threonine-223 within its mid-region plays a critical role in pp71 subcellular trafficking (Shen *et al.*, 2008). If a phosphomimetic residue (aspartic acid) replaces threonine (T223D), pp71 traffics to punctate cytoplasmic sites, where it colocalizes with syntaxin 11 and partially overlaps with cVACs. It is thought that phosphorylation of threonine-223 masks the pp71 NLS or prevents its binding to a cellular protein required for its nuclear transport (Shen *et al.*, 2008). Surprisingly, a pp71 T223A mutant, which cannot be phosphorylated, traffics to the cytoplasm at late stages of HCMV infection, suggesting another level of the regulation of pp71 nucleocytoplasmic trafficking. Indeed, pp71 can be relocalized from the nucleus to the cytoplasm of transfected cells by HCMV pUL35a (the C-terminal isoform) but not by full-length pUL35 (Salsman *et al.*, 2011).

pp65

Upon HCMV entry, tegument-delivered pp65, encoded by UL83, is immediately targeted to the nuclear matrix particularly to the nuclear lamina and to nucleoli of permissive cells (Schmolke *et al.*, 1995; Gallina *et al.*, 1996; Hensel *et al.*, 1996; Sanchez *et al.*, 1998; Arcangeletti *et al.*, 2003). pp65 is produced from 12 h p.i. when it accumulates more diffusely in the nucleoplasm (Arcangeletti *et al.*, 2003). By late times, pp65 traffics to the cytoplasm while the nucleus becomes devoid of pp65 (Sanchez *et al.*, 2000). pp65 has two NLSs (aa 415–438, aa 537–561) in its C-terminus (Schmolke *et al.*, 1995; Gallina *et al.*, 1996). While both NLSs contribute to its nuclear translocation, only the latter appears to be sufficient to target heterologous proteins to the nucleus. Nonetheless, deletion of both NLSs leads to a greater reduction of its nuclear targeting (Schmolke *et al.*, 1995).

In contrast to pp71, pp65 appears to be a nucleocytoplasmic shuttling protein whose differential trafficking appears to result from the balance of its nuclear import and its export (Sanchez *et al.*, 2007). At early times of HCMV infection, pp65 nuclear import is greater than its export. At later times of infection, CRM1-dependent nuclear export of pp65 increases and results in its cytoplasmic accumulation (Sanchez *et al.*, 2007). pp65 has protein kinase activity (Yao *et al.*, 2001). Mutation of the pp65 NLSs and its phosphate binding site ([436]lysine) reduced its nuclear accumulation and its effects on the host cell cycle while retaining immunogenicity of the protein (Zaia *et al.*, 2009). Because of its immune-dominance and biological properties, it has been proposed that this pp65 mutant may provide a valuable vaccine candidate for HCMV disease prevention (Zaia *et al.*, 2009).

During HCMV infection, pp65 forms a protein complex with another tegument protein, pUL97, a protein serine/threonine kinase, which phosphorylates the antiviral drug ganciclovir (Littler *et al.*, 1992; He *et al.*, 1997; Kamil and Coen, 2007). pp65 and pUL97 may influence each other's localization and functions during virion assembly (Kamil and Coen, 2007). pUL97 kinase activity is required for proper intranuclear trafficking of pp65. pUL97 appears to influence the physical properties of pp65 by direct phosphorylation, which is important for virion assembly and, possibly, nuclear egress (Prichard *et al.*, 2005a). In absence of pUL97 kinase activity, pp65 becomes aggregated in the nucleus and cytoplasm. This aberrant aggregation of pp65 is reduced by pUL97 kinase activity. In addition to HCMV pUL97 kinase, the phosphorylation state and subcellular localization of pp65 can be modulated by cellular kinases or by phosphatases (Sanchez *et al.*, 2007).

While pp65 appears to be a substrate for cyclin-dependent kinases (CDKs) in the infected cell, phosphorylation of its consensus CDK sites do not alter pp65 localization, suggesting an indirect role of CDKs on pp65 localization (Sanchez *et al.*, 2007).

pp150

pp150 (also known as basic phosphoprotein, BPP or pUL32), encoded by UL32, is a major component of the HCMV tegument (Gibson, 1983). pp150 is detected at late times (48 h p.i.) at the INM where primary tegumentation of the nucleocapsids take place (Hensel *et al.*, 1995). pp150 interacts with intact HCMV capsids yet does not interact with individual capsid proteins (Baxter and Gibson, 2001). Binding of pp150 to the capsids occurs early, prior to encapsidation of HCMV DNA (Sampaio *et al.*, 2005; Tandon and Mocarski, 2011). Similar to pp65 and pp71, pp150 traffics to the cytoplasm at late times of infection. Surprisingly, pp150 lacks obvious signals for nuclear import or export. Its interaction with Bicaudal D1 influences trafficking of pp150 to the cVAC (Indran *et al.*, 2010).

Recently, it was found that the HCMV nucleocapsid and a large number of tegument proteins bind to pp150, suggesting that pp150 plays a key role in nucleocapsid/tegument maturation (Moorman *et al.*, 2010). Consistent with that suggestion is the finding that pp150 is essential for progeny production and for stabilization of virions during their translocation from the nucleus to the cVACs (Baxter and Gibson, 2001; AuCoin *et al.*, 2006; Tandon and Mocarski, 2011).

pUL47

Because IE1 transcription is delayed in cells infected with an HCMV UL47 mutant, it has been proposed that pUL47 complex is involved in the release of viral DNA and disassembly of the viral particle (Bechtel and Shenk, 2002). pUL47 forms complexes with tegument proteins, pUL48 and pUL69, as well as the major capsid protein, pUL86. In absence of pUL47, nuclear translocation of pUL69 is reduced suggesting that pUL69 is destabilized in absence of pUL47.

pUL69

The HCMV UL69 protein (pUL69) blocks progression in the G1 phase of the cell cycle (Lu and Shenk, 1999). Further, pUL69 is a viral mRNA export factor with nucleocytoplasmic shuttling capabilities and the ability to recruit cellular mRNA export machinery in order to promote cytoplasmic accumulation of viral unspliced RNAs (Lischka *et al.*, 2001, 2006b; Toth *et* al., 2006). pUL69 functions during the HCMV life cycle are reviewed in greater detail in Chapter I.11.

pUL69 is homologous to ICP27 of herpes simplex virus (HSV) and localizes to the nucleus initially in a diffuse pattern with the exclusion of nucleoli and at late times to nuclear viral replication compartments (nVRCs) (Winkler et al., 1994). pUL69 has a classical bipartite NLS (aa 21–45) in its N-terminus and a NES (aa 597–624) which includes a proline/glutamine cluster followed by an acidic domain in its C terminus (Lischka et al., 2001, 2006b). pUL69 export is CRM1 (or exportin 1)-independent (Lischka et al., 2001). pUL69 recruits cellular mRNA export machinery by its interaction with the cellular mRNA export factor UAP56, a cellular DExD/H-box RNA helicase, and thereby promotes nuclear export of unspliced mRNAs (Lischka et al., 2006b; Toth et al., 2006). pUL69 homologues from chimpanzee (pC69), rhesus (pRh69), and mouse CMV (pM69) also localize to the cell nucleus (Zielke et al., 2011). While they share nuclear localization, the ability of UL69 homologues to stimulate export of unspliced RNAs is conserved in pUL69, pC69, and pRh69, but not in pM69. The RNA export activity requires interaction of pUL69 and the chimpanzee and rhesus monkey homologues with UAP56/URH49.

The nuclear localization of pUL69 is modulated by its phosphorylation. CDKs, particularly CDK9-cyclin T1, play a crucial role for pUL69 nuclear localization and function of pUL69 during HCMV replication (Rechter et al., 2009). Upon treatment with CDK inhibitors, pUL69 aggregates in the nucleus in nVRCs and its nuclear RNA export activity is reduced. In addition, pUL97 interacts directly with pUL69, and can also phosphorylate it (Thomas et al., 2009). pUL97 thereby increases the ability of pUL69 to mediate nuclear export of unspliced mRNAs. The pUL97 N-terminal binding site for pUL69, aa 231–336, differs from the ppUL44 interaction site (aa 366–459) within its kinase domain (aa 337–651) (Marschall et al., 2003; Thomas et al., 2009).

pUL96

pUL96 is expressed from early through late times of infection and traffics to the nucleus and cytoplasm (Tandon and Mocarski, 2011). pUL96 is small (14 kDa) and can freely diffuse through nuclear pores and, therefore, does not require either a NLS or a NES to differentially traffic in the cell. Like pp150, pUL96 appears to stabilize capsids as they transit from the nucleus to the cVACs. UL96 viral mutants show a defect in the cytoplasmic stage of virus maturation. However, pUL96 does not associate with nuclear B capsids;

whereas pp150 does. An accumulation of pp150 in the nucleus during UL96 mutant infection suggested the involvement of pUL96 in the translocation of pp150-bound nucleocapsids out of the nucleus, possibly by influencing pp150 incorporation or orientation on capsids (Tandon and Mocarski, 2011).

Nuclear trafficking of HCMV MIE proteins

The MIE gene products, IE1 (UL123) and IE2 (UL122), play critical roles in subsequent HCMV gene expression and are rate-limiting for HCMV lytic growth (Meier and Stinski, 1996; Meier et al., 2002; Isomura and Stinski, 2003). Following their synthesis, IE1-72 and IE2-86 proteins are targeted to the nucleus, particularly to ND10s (Ishov et al., 1997; Arcangeletti et al., 2003). At initial times of infection, ND10s are the sites of IE transcription (Ishov et al., 1997) and, as infection progresses, some of these sites enlarge to form nVRCs (Sourvinos et al., 2007). IE1 is known to disrupt ND10s (Korioth et al., 1996; Ahn and Hayward, 1997; Ahn et al., 1998; Wilkinson et al., 1998b; Muller and Dejean, 1999; Lee, H.R., et al., 2004). IE2-86 localizes with input viral genomes to the periphery of ND10 domains (Ishov et al., 1997). To optimize HCMV gene expression during its growth, the intracellular localization and abundance of essential cellular RNA processing factors that affect alternative processing of HCMV pre-mRNAs are also temporally and spatially regulated during HCMV infection (Gaddy et al., 2010). Despite the importance of the MIE proteins and the intensive biochemical and genetic research into their regulatory functions (reviewed in Chapter I.10), there is relatively little mechanistic information underlying the nuclear targeting of HCMV MIE proteins.

IE1

Nuclear IE1-72 is the most abundant HCMV protein produced at IE times of infection. While it is essential only at low MOIs (Greaves and Mocarski, 1998), IE-72 plays key roles throughout HCMV infection. IE1-72 augments IE2-mediated regulation of HCMV promoters (Mocarski et al., 2007), disperses ND10s (Korioth et al., 1996; Ahn et al., 1998; Wilkinson et al., 1998a), is tethered to metaphase chromatin (Lafemina et al., 1989; Ahn et al., 1998; Wilkinson et al., 1998a; Nevels et al., 2004a), antagonizes histone deacetylation (Nevels et al., 2004b), and blocks apoptosis (Zhu et al., 1995). Two NLSs have been mapped in IE1: one NLS encoded by exon 2 (aa 1–24), shared by IE1 and IE2 (Wilkinson et al., 1998a; Lee et al., 2007). The second NLS is an IE1-specific bipartite signal encoded by its unique exon 4 (^{326}KR^{327} and ^{340}KRR^{342}) (Delmas *et al.*, 2005).

IE2

IE2-86 is a multifunctional nuclear regulatory protein that is essential for HCMV growth (Marchini *et al.*, 2001). The HCMV IE2 gene encodes multiple isoforms, which have differential effects on temporal expression of HCMV genes. IE2-86 activates early viral gene expression by interacting with cellular transcription machinery (Caswell *et al.*, 1993; Lukac *et al.*, 1997), while IE2-p60 and IE2-p40 enhance delayed early and late gene expression (White *et al.*, 2004, 2007). IE2 negatively regulates its own transcription (Cherrington *et al.*, 1991; Liu *et al.*, 1991). Besides the UL123 exon 2 NLS (aa 1–24), IE2 carries two additional NLSs encoded by its unique exon 5, ^{145}SPRKKPRKTTRP^{156} and ^{321}PRKKKSKR^{328} (Pizzorno *et al.*, 1991). Each NLS is sufficient to translocate a heterologous protein into the nucleus.

Nuclear trafficking of HCMV DNA replication proteins

The regulation and orchestration of nuclear trafficking of HCMV DNA replication proteins are schematically represented in Fig. I.12.1. HCMV DNA replication occurs within nVRCs that initially form in the periphery of ND10s in the nucleus of the infected cells (Ahn *et al.*, 1999). HCMV DNA replication appears more complex than that of other studied herpesviruses. Similar to other herpesviruses, it requires a conserved set of six core DNA replication proteins, which includes HCMV DNA polymerase (pUL54, ppUL44), a single-stranded DNA-binding protein (ssDBP) (pUL57), and the helicase–primase complex (pUL70, pUL105, pUL102) (Pari and Anders, 1993; Pari *et al.*, 1993; Ahn *et al.*, 1999). However, in contrast to other herpesvirus origins, HCMV *oriLyt* DNA replication requires a noncore protein, pUL84 (Pari and Anders, 1993; Sarisky and Hayward, 1996). Further, its replication in a transient DNA replication assay is enhanced by products from the noncore loci, including the UL112–113, UL36–38, IE1/2 and TRS1/IRS1 loci (Pari and Anders, 1993; Pari *et al.*, 1993; Sarisky and Hayward, 1996; Ahn *et al.*, 1999). IE1 and IE2, UL36–38 and TRS1/IRS1 products appear to transactivate the promoters of HCMV DNA replication genes (Iskenderian *et al.*, 1996; Colberg-Poley *et al.*, 1998). In addition, TRS1 and IRS1 have been recently found to competitively associate with ppUL44, potentially affecting its functions differentially (Strang *et al.*, 2010).

pUL84 interacts with IE2 as well as ppUL44 and pp65 (Spector and Tevethia, 1994; Gao *et al.*, 2008; Strang *et al.*, 2009). The pUL84–IE2 complex has been proposed to serve as an initiator of HCMV DNA synthesis (Xu *et al.*, 2004). The pUL84–IE2 complex interacts with *oriLyt* within the region that contains RNA/DNA hybrid and stem loop structures, and CCAAT/enhancer binding protein α binding motifs (Sarisky and Hayward, 1996; Colletti *et al.*, 2007; Gao *et al.*, 2010). Conversely, UL112–113 phosphoproteins appear to organize nVRCs by assembling and recruiting the DNA core replication machinery at the initial stages of pre-nVRC formation in the nucleus (Penfold and Mocarski, 1997; Ahn *et al.*, 1999; Park *et al.*, 2006; Kim and Ahn, 2010). Finally, pUL97, a serine/threonine protein kinase, affects HCMV DNA replication by phosphorylating the essential protein ppUL44 as well as the nucleoside analogues ganciclovir and acyclovir. Phosphorylation of ganciclovir by pUL97 is responsible for much of its antiviral selectivity (Sullivan *et al.*, 1992). Therefore, while not essential for HCMV *oriLyt* replication, pUL97 is an important determinant for HCMV DNA replication (Pari and Anders, 1993).

pUL54

The catalytic subunit of the HCMV DNA polymerase is encoded by the UL54 ORF (Heilbronn *et al.*, 1987; Ertl and Powell, 1992). pUL54 can be immunoprecipitated with its processivity factor, ppUL44, also known as ICP36 (Ertl and Powell, 1992). Importantly, small inhibitors that target the pUL54–ppUL44 interaction can impair HCMV growth (Loregian and Coen, 2006).

The C-terminus of pUL54 contains NLSs (Loregian *et al.*, 2000). pUL54 has two NLSs aa 1153–1159 (NLSA, ^{1153}PAKK$\underline{R}$A$\underline{R}$^{1159}), immediately upstream of its ppUL44 binding site (aa 1213–1242), and aa 1222–1227 (NLSB) (Alvisi *et al.*, 2006b). NLSA, of which the two underlined arginines are essential, is sufficient to translocate heterologous proteins to the nucleus by the IMPα/β heterocomplex (Alvisi *et al.*, 2006b). NLSB acts by binding ppUL44 and pUL54 and the complex can enter the nucleus.

ppUL44

ppUL44 has a monopartite basic NLS (^{425}PNTK-KQK^{431}) (Alvisi *et al.*, 2005). ppUL44 requires phosphorylation for efficient nuclear accumulation (Plachter *et al.*, 1992; Alvisi *et al.*, 2005). ppUL44 is phosphorylated at serine-413 by casein kinase 2 (CK2) to enable higher affinity recognition of the NLS by the IMPα/β heterocomplex and increased nuclear import (Alvisi *et al.*, 2005). ppUL44 appears to also

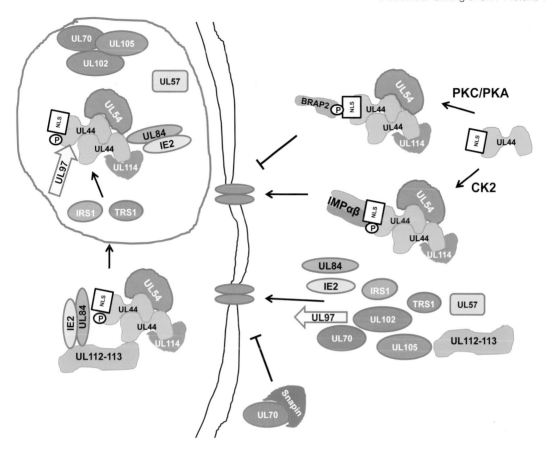

Figure I.12.1 Regulation and orchestration of trafficking of HCMV proteins involved in *ori*Lyt DNA replication. Shown is the increased nuclear import by IMPα/β of ppUL44 following its phosphorylation by CK2 at [413]S and upstream of its NLS (aa 425–431) (Alvisi *et al.*, 2005). Conversely, PKC/PKA-dependent phosphorylation of ppUL44 within its NLS ([427]T) augments BRAP2 binding, which inhibits ppUL44 nuclear trafficking (Fulcher *et al.*, 2010). pUL97, the HCMV serine/threonine protein kinase, interacts with and can phosphorylate ppUL44 (Krosky *et al.*, 2003b; Marschall *et al.*, 2003). The key role of ppUL44 as a scaffold for HCMV DNA synthesis is evidenced by its ability to bind and translocate pUL54 and pUL114 to the nucleus (Ertl and Powell, 1992; Alvisi *et al.*, 2006a;) and to mediate regulation of HCMV DNA polymerase by binding pUL84 and competitively binding IRS1 or TRS1 (Gao *et al.*, 2008; Strang *et al.*, 2010). Correct subnuclear trafficking of HCMV replication machinery is enhanced by ppUL112–113 p84-mediated recruitment of pUL84-IE2 and ppUL44 to nVRCs (Kim and Ahn, 2010). The nuclear trafficking and the NLSs required for some HCMV DNA replication factors, including its helicase-primase (pUL70, pUL102, and pUL105) and pUL57 (ssDBP) are less well defined. Nonetheless, binding of pUL70 by Snapin, a cellular vesicle-associated protein, was recently found to inhibit pUL70 trafficking to the nucleus (Shen *et al.*, 2011).

be a substrate for pUL97 kinase (Krosky *et al.*, 2003b; Marschall *et al.*, 2003).

ppUL44 dimerizes in the cytoplasm and is translocated to the nucleus as a dimer (Alvisi *et al.*, 2006a). The ability of ppUL44 to dimerize suggests that it can function as a scaffold for the recruitment of other factors to the HCMV DNA replication fork (Appleton *et al.*, 2004). Consistent with this model, the presence of one ppUL44 NLS on a subunit of the dimer is sufficient to allow nuclear transport of the assembled dimer. The presence of functional NLSs on both DNA polymerase subunits, pUL54

and ppUL44, suggests a mechanism to maximize their nuclear import (Alvisi *et al.*, 2006b). Further, ppUL44 appears to facilitate import of other HCMV replication proteins, including the cytoplasmic uracil DNA glycosylase pUL114, which assembles on the ppUL44 dimer before nuclear import (Prichard *et al.*, 2005b; Alvisi *et al.*, 2006a).

The nuclear import of ppUL44 is negatively regulated by binding to a cytoplasmic retention factor, BRCA-1-associated protein 2 (BRAP2) (Fulcher *et al.*, 2010). BRAP2 binds to ppUL44 when it is phosphorylated at [427]threonine in a PKC/PKA-dependent manner. BRAP2 is the first known cellular negative regulator of nuclear import that inhibits viral cargo in a phosphorylation-dependent manner (Fulcher and Jans, 2011).

ppUL57

ppUL57, also known as single stranded DNA-binding protein (ssDBP), localizes to pre-nVRCs with ppUL44 and ppUL112–113 by early times of HCMV infection (12–24 h p.i.) and then at late times to developing nVRCs (Kemble et al., 1987; Penfold and Mocarski, 1997).

pUL70

HCMV primase is a tight helicase–primase complex which consists of pUL70, pUL105 and pUL102 (Smith and Pari, 1995; Smith et al., 1996; McMahon and Anders, 2002). It is believed that pUL70 synthesizes RNA primers, which are extended by pUL54 (Urban et al., 2009). pUL70 interacts with Snapin, a cellular vesicle-associated protein that is localized predominantly in the cytoplasm (Shen et al., 2011). When Snapin is overexpressed, pUL70 translocation to the nucleus is reduced, and, importantly, HCMV DNA synthesis and progeny are decreased (Shen et al., 2011). Together these results suggest that Snapin plays a key role in regulating pUL70 subcellular location and thereby modulating viral DNA synthesis and productive infection.

pUL84

pUL84 is an early protein that colocalizes with ppUL44 in nVRCs (Xu et al., 2002). In contrast to core proteins required for HCMV oriLyt DNA synthesis, pUL84 shuttles from the nucleus to the cytoplasm (Lischka et al., 2006a). The complex, non-conventional NLS of pUL84 mediates its interaction with the IMPα/β pathway (Lischka et al., 2003). A large region (aa 226–508) including multiple basic residues and leucine-rich elements is required for the interaction with importin α via a domain that differs from the binding pocket for cellular proteins containing classical NLSs (Lischka et al., 2003). Two leucine-rich NESs (aa 228–237, aa 359–366) within the region appear to enable CRM1-dependent nucleocytoplasmic shuttling (Lischka et al., 2006a). Thus, pUL84, similar to pUL69, is a nucleocytoplasmic shuttling RNA binding protein.

pUL84 shuttling is essential for HCMV replication as demonstrated by findings that treatment of HCMV-infected cells with aptamers targeted to the UL84 NLS caused its mislocalization to the cytoplasm and inhibited HCMV replication (Kaiser et al., 2009). Secondly, inhibition of UL84 nucleocytoplasmic shuttling showed that its shuttling does not directly affect HCMV DNA replication but, nonetheless, plays an important role in HCMV growth and affects the cytoplasmic accumulation of at least one HCMV RNA encoding IRS1 (Gao et al., 2010).

pUL97

pUL97 is expressed with early/late kinetics and localizes to the nucleus (Michel et al., 1996; Webel et al., 2011). pUL97 phosphorylates a number of viral and cellular proteins including ppUL44 (Krosky et al., 2003b; Marschall et al., 2003), pUL69 (Thomas et al., 2009), and pp65 (Becke et al., 2010). In addition, pUL97 phosphorylation of lamins is involved in the disruption of the nuclear lamina during HCMV nuclear egress (see below) (Marschall et al., 2005; Milbradt et al., 2009).

It has recently been found that pUL97 is expressed as two isoforms, both of which are catalytically active (Webel et al., 2011). The pUL97 isoforms are initiated at two alternative sites of translation initiation (aa 1 and aa 74) and not by alternative splicing. A classical bipartite NLS of the large, full-length pUL97 isoform is localized to its extreme N-terminal sequences, aa 6–35. A second, less efficient NLS is present in both pUL97 isoforms and maps downstream of aa 74, including two overlapping sequences, aa 164–198 and aa 190–213 for optimal nuclear transport (Webel et al., 2011).

ppUL112-UL113 and pUL114

The HCMV UL112–113 locus encodes four phosphoproteins with common N-terminal sequences (Wright et al., 1988). The UL112–113 phosphoproteins appear to organize the nVRCs by recruiting the core replication machinery at initial stages of pre-nVRC formation (Penfold and Mocarski, 1997; Ahn et al., 1999). UL112–113 phosphoproteins co-localize with IE2 at the periphery of ND10s by early times (from 6 h p.i.) and, subsequently, in viral pre-nVRCs and mature nVRCs (Penfold and Mocarski, 1997; Yamamoto et al., 1998; Ahn et al., 1999). UL112–113 phosphoproteins interact with each other and recruit ppUL44 from diffuse nuclear distribution to viral pre-nVRCs (Park et al., 2006). A region spanning UL112–113 aa 1–125 is required for self-interaction and for localization to nuclear foci. At least ppUL112–113 84 kDa forms a complex with ppUL44, pUL84 and IE2 in HCMV-infected cells and enhances the efficiency of HCMV DNA replication (Kim and Ahn, 2010).

pUL79, pUL87, and pUL95

It has recently been found that HCMV early gene products, pUL79, pUL87, and pUL95, are required for expression of late genes (Isomura et al., 2011). These proteins colocalize with ppUL44 in pre-nVRCs, prior to HCMV DNA synthesis. These findings suggest that pre-nVRCs may regulate the timing of HCMV late gene expression.

Nuclear capsid assembly of virions and virion DNA cleavage-packaging

MCP, mCP, mC-BP, SCP, NP1, and pAP

As with all herpesviruses, HCMV capsid assembly and encapsidation of viral progeny DNA occur in the nucleus (Mocarski et al., 2007). The HCMV capsid consists predominantly of the MCP, encoded by UL86, which forms the capsomers. mCP, encoded by UL85, mCP binding protein (mC-BP), encoded by UL46, and the smallest capsid protein (SCP), encoded by UL48/49, together form the triplexes that interface with the capsomers. The assembly of capsids requires the proteinase precursor (pNP1, encoded by UL80a) and its genetically related assembly protein precursor (pAP), encoded by UL80.5. The assembly of CMV virions is reviewed comprehensively in Chapter I.13. We focus on their nuclear trafficking and NLSs.

Surprisingly, the CMV MCP does not appear to traffic to the nucleus alone. It is efficiently transported therein by pAP but not its cleavage product (Wood et al., 1997). Thus, CMV MCP seemingly lacks a nuclear localization signal and piggy-backs on pAP or pNP1 to localize to the nucleus. Simian CMV (SCMV) pAP contains two classical NLSs, each of which is sufficient to translocate heterologous proteins to the nucleus (Plafker and Gibson, 1998). Only SCMV NLS1 (^{175}KRRK^{178}) has an apparent counterpart in the pAP homologues of all herpesviruses, which may reflect more importance in familial function; whereas NLS2 (^{202}RARKRLK^{208}) is conserved in the β-herpesvirus homologues potentially fulfilling a group- specific function. The HCMV NLS1 (^{510}KRRK^{513}) and NLS2 (^{537}RARKRLK^{543}) share sequence identity with the SCMV sequences (Nguyen et al., 2008). Multimeric complexes form between MCP and the scaffolding proteins in the cytoplasm and their nuclear transport is mediated by the scaffolding proteins NLS1 and NLS2. The NLS in the SCMV UL80 ORF can act in trans through the pAP-MCP and pNP1–MCP complexes to enable nuclear translocation of HCMV MCP (Wood et al., 1997; Plafker and Gibson, 1998).

The importance of the HCMV UL80 NLS for nuclear translocation of MCP has been verified using HCMV viral mutants (Nguyen et al., 2008). Both UL80 NLSs are involved in translocation of HCMV MCP. Although NLS1 of SCMV was found to be more efficient at nuclear translocation of HCMV MCP in transfected cells (Plafker and Gibson, 1998), an HCMV NLS2-deficient mutant was more severely impeded for growth than the HCMV NLS1-deficient mutant. These findings suggest that NLS2 has functions in addition to its role as a nuclear localization signal (Nguyen et al., 2008). Further, inactivating both UL80 NLSs was lethal to HCMV growth. HCMV mC-BP, which contains one NLS, analogously appears to act in trans to translocate its triplex partners, mCP and SCP to the nucleus (Plafker and Gibson, 1998).

pUL56, pUL89, pUL104 and pUL52

pUL56 and pUL89 are the subunits of the HCMV terminase, which cleaves the HCMV genome carrying packaging motifs; whereas pUL104 is the portal protein (Bogner et al., 1993, 1998; Krosky et al., 1998; Underwood et al., 1998; Dittmer and Bogner, 2005; Dittmer et al., 2005). pUL56 binds the packaging motifs on HCMV DNA, provides energy for its translocation into capsids, and associates with the capsid allowing DNA entry into the capsids. pUL89 is required for cleavage of concatemeric progeny viral DNA into unit length genomes. Their nuclear translocation is thus essential for assembly of infectious HCMV virus.

pUL56, pUL89, and pUL104 localize in the nucleus at sites of cleavage-packaging of the HCMV genome, which overlap with nVRC (Giesen et al., 2000; Borst et al., 2008). By early times of HCMV infection (12–24h p.i.), pUL89 colocalizes with pUL56 in pre-nVRCs and by later times (after 48h p.i.) at nVRCs (Thoma et al., 2006). In contrast, pUL52 which is essential for HCMV DNA packaging is predominantly nuclear but its subnuclear distribution is different from that of pUL56, pUL89, and pUL104 (Borst et al., 2008).

Most is known about targeting of pUL56 to the nucleus. pUL56 carries a classical monopartite NLS (^{816}RRVRAT$\underline{R}$K$\underline{R}$PRR^{827}) of which the underlined basic residues are essential for its nuclear translocation (Giesen et al., 2000). The pUL56 NLS is sufficient to translocate a heterologous protein to the nucleus. pUL56 appears to be recognized by hSRP1α, a subunit of the IMPα complex, and is subsequently translocated into the nucleus in an IMP-dependent pathway. It has recently been found that a potent, non-nucleoside anti-HCMV drug, Letermovir or AIC246, targets pUL56, in a mechanism distinct from other inhibitors that target the terminase complex (Goldner et al., 2011).

Trafficking of HCMV proteins from the ER to the nucleus

As herpesviruses undergo the nuclear phase of virion maturation, some viral glycoproteins are translocated from their site of synthesis to the INM at sites of primary envelopment of virions. HCMV assembly requires gB translocation from the rough ER to the INM. The gB C-terminal sequence ([885]DRLRHR[890]) is required for its translocation from the ER to the INM (Radsak et al., 1990; Bogner et al., 1997; Meyer and Radsak, 2000). The gB NLS is sufficient to translocate a membrane-bound CD8 reporter protein to the INM and the underlined arginines are essential for this function. The consensus RxR motif is sufficient for localization of the CD8 reporter in the INM (Meyer et al., 2002). However, this NLS cannot translocate soluble β-galactosidase to the INM nor can the SV40 T NLS target gB to the INM (Meyer and Radsak, 2000).

Similarly, HCMV UL16, which encodes a glycoprotein that interferes with the immune response to HCMV-infected cells, localizes to the ER and to the TGN (Vales-Gomez et al., 2006). An Endoglycosidase H sensitive (EndoH[S]) form of gpUL16 localizes to the INM suggesting its trafficking thereto from the ER or cis-Golgi apparatus. gpUL16 contains two consensus RxR motifs (213QRLRIR218 and 224QRLRTED230) in its cytoplasmic tail that could be involved with its trafficking to the INM, as they are for HCMV gB (Meyer et al., 2002; Vales-Gomez et al., 2006). Together, these studies suggest that a small modular signal sequence mediates transport of TMD proteins from the ER to the INM and that their nuclear translocation machinery is distinct from the importin pathway used by soluble nuclear proteins.

pUS17

HCMV pUS17 is a member of the seven TMD G protein-coupled receptor-related proteins encoded by HCMV. pUS17 localizes predominantly to the TGN and to early endosomes (Das et al., 2006). In addition, a segment of pUS17 localizes to nVRCs of infected cells but not in transfected cells (Das et al., 2006). As pUS17 does not carry a conventional NLS, its nuclear import is likely based upon a non-canonical signal or its interaction with another protein that translocates to the nucleus.

Nuclear egress of virions

Following initial tegumentation, nucleocapsids must egress the nucleus to finalize assembly in the cytoplasm. Nuclear egress must be orchestrated because of the size of the CMV nucleocapsid (~ 130 nm) exceeds the opening of the nuclear pore (~ 40 nm). Before the virions can egress the nucleus, the nuclear lamina, a thick meshwork of intermediate filaments or lamins, which supports the INM, must be disassembled. PKC and CDK1 phosphorylate lamins during the pro-metaphase stage of the cell cycle leading to destabilization of the nuclear lamina (Collas et al., 1997; Likhacheva and Bogachev, 2001). MCMV was first documented to recruit PKC to the nuclear rim via M50, which is inserted in the INM, and to increase phosphorylation of the lamins and promote nuclear lamina dissolution (Muranyi et al., 2002). HCMV similarly encodes a nuclear egress complex (NEC) composed of pUL50 and pUL53, as well as pUL97 and RASCAL (Dal Monte et al., 2002; Krosky et al., 2003a; Milbradt et al., 2009; Sam et al., 2009; Miller et al., 2010). Direct interactions between pUL50 and pUL53 mediate the formation of the HCMV NEC (Camozzi et al., 2008; Milbradt et al., 2009; Sam et al., 2009). Cellular proteins, PKC, p32, and lamin B receptor, are recruited to destabilize the nuclear lamina (Marschall et al., 2005; Milbradt et al., 2007, 2009, 2010). pUL50 appears to have important recruitment functions as it recruits PKC and relocalizes pUL53 to the NEC (Milbradt et al., 2007). pUL97 and/or PKC phosphorylate [22]S of lamin A/C and thus generating a minimal binding site for the peptidyl-prolyl cis/trans-isomerase Pin1, which appears to result in conformational changes of proteins leading to localized depletion of the nuclear lamina (Milbradt et al., 2010). Interaction of BiP (GRP78) with pUL50 is required for nuclear lamina remodelling that enables egress of the HCMV nucleocapsids (Milbradt et al., 2009; Buchkovich et al., 2010). Thus, the combined activities of cellular and CMV protein kinases and chaperones actuate the formation of lamina-depleted areas through which HCMV particles readily traverse to egress the nucleus (Milbradt et al., 2010).

pUL50, pUL53, and pUL97

pUL53 has a nuclear localization signal in its N-terminus and localizes diffusely in the nucleus when pUL50 is not present (Milbradt et al., 2007; Camozzi et al., 2008; Sam et al., 2009). Conversely, when expressed alone, pUL50 localized mainly to the nuclear rim. When co-expressed, both pUL50 and pUL53 colocalize at the nuclear rim, forming aggregates, in a pattern similar to that in HCMV-infected cells. Thus, pUL50 appears to form a complex with pUL53 and targets it to the nuclear rim. Mutation of pUL53 L79A greatly reduced its ability to complex with pUL50 and reduced its colocalization to the

NEC (Sam *et al.*, 2009). In addition, pUL97 can be found concentrated at the nuclear lamina/INM by its interaction with p32 (Marschall *et al.*, 2005; Milbradt *et al.*, 2010).

RASCAL (nuclear r̲im-a̲s̲s̲ociated c̲ytomegalovir̲al protein)

The role of a newly identified HCMV protein, RASCAL, in the NEC has been recently suggested (Miller *et al.*, 2010). RASCAL, encoded by c-ORF29, whose sequence in primary strain TB40/E is extended compared to that in HCMV strains Toledo or Towne (Murphy *et al.*, 2003). Like pUL53, RASCAL requires pUL50 to localize to nuclear lamina (Miller *et al.*, 2010). Because RASCAL interacts with pUL50, it is likely to be a component of the NEC. At late times of HCMV infection, RASCAL localizes to cytoplasmic vesicles.

Subnuclear trafficking of other HCMV proteins

Using a genomic wide screen, it has been recently found that numerous HCMV proteins localize to specific compartments within the nucleus of transfected U2OS and 293T-cells (Salsman *et al.*, 2008). In particular, HCMV TRL5, TRL7, TRL9, UL29, UL31, UL76, UL108, and US33 targeted nucleoli; whereas, UL3, UL35, UL80a, and US32 trafficked to ND10s (Salsman *et al.*, 2008, 2011). Although UL30 and UL137 trafficked to specific subnuclear compartments, they did not colocalize with the markers examined in these studies. Further, HCMV UL30 and UL3 were found to disrupt Cajal bodies; while UL76, US25, UL68, TRL9, UL30, UL98 UL69, UL29 disrupted ND10s. These studies provide interesting clues as to the ability of HCMV proteins to traffic to specific subnuclear compartments. Nonetheless, caution is needed in extrapolating the results from transfected cells to subnuclear protein trafficking during HCMV infection as some discrepancies between those experimental systems have been found. HCMV pUL56, which does not normally localize to nucleoli during infection, displayed nucleolar localization in transfected cells (Giesen *et al.*, 2000). Another HCMV protein known to associate with nucleoli, pp65, during infection was not detected by transfection. Thus, examination of trafficking of these HCMV proteins to subnuclear compartments during infection is warranted and will predictably provide valuable information as to their authentic trafficking during the HCMV life cycle.

Targeting nascent HCMV proteins to the secretory apparatus

The ER is employed to synthesize, process, and distribute proteins encoded by nearly one-third of all human genes (Chen *et al.*, 2010). As such, the ER acts as an important quality control checkpoint for protein synthesis and folding. In addition, the ER functions as a site for steroid metabolism, intracellular calcium (Ca^{2+}) signalling, lipid and *de novo* ceramide synthesis, and xenobiotic metabolism (Bionda *et al.*, 2004; Lavoie and Paiement, 2008). By targeting viral proteins to the ER, HCMV can gain access to, and manipulate, these critical cellular machineries.

Nascent polypeptides destined for the secretory pathway are generally marked as such with N-terminal signal sequences and are directed to the ER immediately upon initiation of translation (Milstein *et al.*, 1972; Blobel and Dobberstein, 1975a). Alternatively, proteins can also be directed to the ER post-translationally, via a hydrophobic C-terminal domain or by acylation (Romisch, 1996). N-terminal signal peptides of secretory proteins are typically between 20–30 aa in length, with a centralized hydrophobic core of 6–15 aa being flanked on either side with residues providing specific cues for topology within the ER membrane and potential signal peptidase cleavage sites to allow signal sequence removal from mature proteins (Sabatini *et al.*, 1971; Milstein *et al.*, 1972; Blobel and Dobberstein, 1975a; von Heijne, 1988; Nothwehr and Gordon, 1990; Dalbey and Von Heijne, 1992).

Critical to the recognition and action of signal sequence peptides is a cytosolic ribonucleoprotein complex, called the signal recognition particle (SRP), comprised of six proteins (SRP9, SRP14, SRP19, SRP54, SRP68, and SRP72) and a 7S RNA transcript (Walter and Blobel, 1982, 1983). The SRP complex recognizes and binds signal peptides as they emerge from the ribosome (Walter *et al.*, 1981), and a heterodimer of its subunits SRP9 and SRP14 then mediates translation arrest of the nascent protein (Siegel and Walter, 1985, 1988). Translation inhibition early in protein synthesis ensures that preproteins do not undergo any folding, which would preclude a productive interaction with translocon components on the ER, and successful transfer into the secretory pathway (Hegde and Bernstein, 2006). Direct interaction of SRP with signal sequences occurs via a flexible, methionine-rich 'M domain' within the SRP54 subunit protein (Krieg *et al.*, 1986; Kurzchalia *et al.*, 1986; Zopf *et al.*, 1990; High and Dobberstein, 1991; Lutcke *et al.*, 1992). This interaction, driven predominantly by hydrophobic interactions, is fairly non-specific and allows for the recognition of diverse signal sequences.

Upon SRP interaction with the N-terminal signal

sequence on a nascent protein, and concomitant translation arrest, the entire SRP nascent chain–ribosome complex is then shuttled to the ER membrane to dock with an SRP receptor (SR) and the translocon machinery. This effectively creates a cellular partition between mRNA destined for ER compartments, from all other mRNA (Blobel and Dobberstein, 1975a,b; Lingappa and Blobel, 1980; Walter and Johnson, 1994; Rapoport et al., 1996). Large ribosomal subunits become stabilized upon ER association, with in vitro studies revealing exceptionally slow rates of release from the ER back into the cytosol, and subsequent separation into constituent components (Borgese et al., 1973, 1974; Kalies et al., 1994; Raden et al., 2000). Attached to these ER sites, stabilized ribosomes can further permit multiple rounds of mRNA loading and translation, and may act as privileged sites of protein synthesis during conditions of cell stress (Stephens et al., 2005; Lerner and Nicchitta, 2006).

Trafficking through the secretory pathway

Located between the ER and the Golgi apparatus in mammalian cells is an intermediate compartment of tubulovesicular membrane clusters, called the ER/Golgi intermediate compartment (ERGIC) (Appenzeller-Herzog and Hauri, 2006). The ERGIC is a complex membrane system that acts as the first post-ER bidirectional sorting station for anterograde and retrograde protein traffic (Aridor et al., 1995; Klumperman et al., 1998; Appenzeller et al., 1999; Martinez-Menarguez et al., 1999; Ben-Tekaya et al., 2005). The ERGIC was originally defined using a 53-kDa lectin (ERGIC-53), which displays predominant localization to these membranes (Schweizer et al., 1988; Hauri et al., 2000).

Anterograde transport through the ERGIC is often considered to occur by default, and is mediated by coat protein complex (COP)II-coated vesicles (Barlowe, 2003a,b; M.C. Lee et al., 2004b). Steady-state ER localization of proteins thus requires ER-retention or ER-retrieval sequences (Nilsson and Warren, 1994). Sorting into the retrograde pathway commonly utilizes a K/HDEL retrieval sequence found on the C-terminus of soluble, ER luminal proteins. This sequence binds to a receptor containing a dibasic cytosolic signal, in turn recognized by COPI, and brings about the recycling of proteins back to the ER from the Golgi apparatus, allowing their efficient concentration in the ER (Pelham, 1989). Several ER-resident membrane proteins similarly carry ER-retrieval sequences in their cytoplasmic domains. For type I membrane proteins this retrieval motif (KKXX or KXKXX) involves critical C-terminal lysine residues (Jackson et al., 1990; Pelham, 1995;

Teasdale and Jackson, 1996), whereas type II membrane proteins utilize N-terminal arginine motifs (RR) (Michelsen et al., 2005). In addition, the membrane-spanning domains of ER-resident transmembrane proteins can be sufficient to prevent movement of these proteins from the ER to the Golgi apparatus, serving thus as ER-retention signals (Smith and Blobel, 1993; Nilsson and Warren, 1994). In analogy, the TMDs of two CMV glycoproteins, MCMV gp36.5/m164 and HCMV gpUL142, are sufficient to mediate ER retention (Ashiru et al., 2009; Däubner et al., 2010).

The Golgi apparatus is critical for final maturation and sorting of secretory proteins, utilizing a number of sorting signals to direct protein cargo to appropriate destinations. Golgi-specific maturation of proteins includes complex carbohydrate modifications as well as proteolytic cleavage events (Nilsson et al., 2009; Dancourt and Barlowe, 2010; Wilson et al., 2010). As the main site of glycosylation, the Golgi apparatus obviously is an important site for viral glycoproteins, and particularly for HCMV as it contains at least 57 ORFs predicted to encode glycoproteins in the commonly used laboratory strain AD169 (Chee et al., 1990). Because of space constraints, we will focus on the trafficking of the best studied HCMV glycoproteins present in its envelope, gB, gM/gN, and gH/gL/gO, which play roles in HCMV cell attachment, entry and virus assembly (Simmen et al., 2001; Lopper and Compton, 2004; Netterwald et al., 2004; Shimamura et al., 2006) as well as two newly described HCMV glycoproteins, gpUL142 and gpUL144, that are present in clinical strains but not in laboratory strains of HCMV.

Final tegumentation and envelopment, including its glycoproteins, of HCMV virions occurs in the cVAC, which appears to be formed by rearrangement of the cellular secretory apparatus (Das et al., 2007; Das and Pellett, 2011). In the cVAC, vesicles from the early and recycling endosomes are surrounded by the Golgi and TGN (Das et al., 2007; Krzyzaniak et al., 2009; Cepeda et al., 2010).

gB

gB, encoded by the HCMV UL55 ORF, is an essential protein involved in HCMV attachment and entry. gB is a type I membrane protein that requires its hydrophobic signal sequence (aa 1–24) to target to the ER (Singh and Compton, 2000). Oligomerization of gB and accurate folding is critical for subsequent trafficking steps including cell surface expression (Yamashita et al., 1996; Zheng et al., 1996). Nonetheless, it appears that gB folding intermediates can traffic to the plasma membrane (PM), although at slower rates than wild-type gB (Singh and Compton, 2000). Phosphofurin acidic cluster sorting protein-1 (PACS-1) is required for correct gB localization to the TGN (Crump *et al.*, 2003; Jarvis *et al.*, 2004). PACS-1 binds to the gB acidic cluster, in a CK2 phosphorylation-dependent manner, connects gB to cellular adaptor proteins (AP1 and AP3), and retrieves it from endosomes to the TGN. However, phosphorylation of the gB acidic cluster is not the only requirement for TGN localization (Jarvis *et al.*, 2004). In contrast to pp28 (TGN/ESCRT) and pp150 (clathrin-coated vesicles), gB traffics to endosomes at initial times of cytoplasmic packaging (72 h p.i.); whereas, at later times (120 h p.i.), multiple vesicles of distinct origins merge into one site of virion assembly (Moorman *et al.*, 2010).

gM/gN

The gM/gN complex is the most abundant component of the HCMV envelope (Varnum *et al.*, 2004). gM is an essential, type III glycoprotein with seven TMDs, encoded by the UL100 ORF. The complex of gM and gN, encoded by UL73 ORF, is necessary for their trafficking from the ER to the TGN and endosomal compartment (Mach *et al.*, 2007). The C-terminal long acidic cluster (aa 359–371) of gM influences the rate of gM/gN trafficking to the cVAC (Krzyzaniak *et al.*, 2007). The C-terminus of gM interacts with Rab11 effector protein FIP4 (Krzyzaniak *et al.*, 2009). Rab11 GTPase activity appears to play a central role in HCMV assembly. In addition, gM has a tyrosine-based motif (aa 329–333), an intracellular sorting signal, which may impact its interactions with adaptor proteins (Krzyzaniak *et al.*, 2007).

gN is a small type I glycoprotein, which is extensively glycosylated (Mach *et al.*, 2000). gN is palmitoylated either at ^{125}C or ^{126}C, but this modification did not alter intracellular trafficking from the ER to more distal compartments of the secretory pathway (Mach *et al.*, 2007).

gH/gL/gO

gO is encoded by the UL74 ORF. gO is a soluble luminal glycoprotein with a single hydrophobic signal (aa 14–40) which is cleaved and absent in mature gO (Theiler and Compton, 2001). While remaining soluble in transiently transfected cells, mature gO, which lacks a membrane anchor, is associated with membranes in infected cells due to its interactions with other HCMV proteins within the ER and Golgi (Theiler and Compton, 2001).

gO processing and trafficking appears to differ between HCMV strains, due to sequence variation within the UL74 gene as well as differences in the coding capacity of HCMV strains for other genes

unessential for growth in cell culture. In HCMV (strain AD169)-infected HFFs, gO covalently binds to gH/gL heterodimers in the ER (Huber and Compton, 1998, 1999; Kaye *et al.*, 1992). Subsequently, the gH/gL/gO complex traffics to the Golgi apparatus where gO becomes phosphorylated, O-glycosylated, and its N-linked oligosaccharides are modified to EndoH resistant moieties (Huber and Compton, 1998, 1999; Kinzler *et al.*, 2002). This mature gH/gL/gO complex is then incorporated into HCMV cVAC, where gO plays a role in secondary envelopment of virions and is incorporated into viral progeny (Huber and Compton, 1998; Jiang *et al.*, 2008). Not surprisingly, in HCMV (AD169)-infected HFFs, gO is mainly detected within the Golgi apparatus (Theiler and Compton, 2002). In contrast, in HCMV (TR)-infected HFFs, gO is detected predominantly within the ER (Ryckman *et al.*, 2010). Here, gO still functions as a chaperone promoting gH/gL export from the ER, trafficking to the Golgi apparatus, and incorporation into virions. However, gO was not found to be incorporated into virions of the TR strain (Ryckman *et al.*, 2010).

gpUL142

The glycoprotein encoded by UL142 in clinical isolates and low-passage HCMV strains inhibits NK cell-mediated lysis of infected host cells (Wills *et al.*, 2005). Sequential mutations in the region encoding UL142 and UL144 (see below) occurred in some HCMV primary strains during their adaptation to cell culture (Dargan *et al.*, 2010). gpUL142, which lacks ER localization signals such as KDEL or K(X)KXX, unconventionally employs its TMD to localize predominantly to the ER and its luminal domain to localize to the *cis*-Golgi apparatus (Ashiru *et al.*, 2009). gpUL142 inhibits NK cell-mediated lysis by down-modulating surface expression of MHC class I-related chain A (MICA) proteins (Chalupny *et al.*, 2006). gpUL142 interacts with nascent MICA and sequesters full-length MICA in the *cis*-Golgi apparatus (Ashiru *et al.*, 2009). Thus, the ER/*cis*-Golgi localization of gpUL142 appears to be important for its immune evasion function.

gpUL144

The UL144 ORF found in clinical isolates of HCMV encodes a TNF receptor structural homologue (Benedict *et al.*, 1999), and is one of 19 viral genes missing from attenuated laboratory strains of HCMV (Cha *et al.*, 1996). gpUL144 potently activates, in a TRAF-6-dependent manner, NFκB-induced transcription, and as a result induces expression of the cellular chemokine CCL22 in infected cells (Poole et al., 2006). Specific genotypic polymorphisms of UL144, within clinically isolated HCMV, have recently been associated with high plasma viral loads and poor developmental outcomes in congenitally infected infants (Waters et al., 2010). The YRTL sequence in the conserved cytoplasmic tail of gpUL144 acts as a sorting motif to block movement of gpUL144 from the Golgi to the PM. Mutation of the critical tyrosine residue within this motif to an alanine allowed significant levels of mutant UL144 to accumulate on the PM (Benedict et al., 1999). The YRTL sequence resembles a YXXZ motif (where Z represents a bulky hydrophobic residue), which has been described to directly interact with clathrin-mediated receptor internalization adaptor complexes at the PM and TGN (Marks et al., 1997).

Dual trafficking of HCMV UL37 proteins to the ER and to mitochondria

In contrast to cellular mitochondrial proteins, HCMV UL37 proteins, including UL37 exon 1 protein (pUL37x1), also known as viral mitochondria-localized inhibitor of apoptosis (vMIA), UL37 medium protein (pUL37$_M$), and the N-terminal unglycosylated fragment (pUL37$_{NH2}$) of the full-length UL37 glycoprotein (gpUL37), localize dually to the ER and mitochondria (Al-Barazi and Colberg-Poley, 1996; Goldmacher et al., 1999; Colberg-Poley et al., 2000; Hayajneh et al., 2001; Mavinakere and Colberg-Poley, 2004a,b). The pUL37x1/vMIA is the predominant UL37 isoform produced during HCMV infection (Adair et al., 2004; Mavinakere and Colberg-Poley, 2004a) and inhibits apoptosis by inactivating the proapoptotic activity of Bax at the outer mitochondrial membrane (OMM) (Arnoult et al., 2004; Poncet et al., 2004, 2006; Pauleau et al., 2007; Norris and Youle, 2008). pUL37x1/vMIA rapidly causes mitochondrial fragmentation in transfected cells (Fig. I.12.2) and during infection (McCormick et al., 2003b; Castanier et al., 2010). pUL37x1/vMIA also controls HtrA2/Omi activity (McCormick et al., 2008), affects mitochondrial ATP synthesis (Poncet et al., 2006; Seo et al., 2011), and disrupts the F-actin cytoskeleton (Poncet et al., 2006; Sharon-Friling et al., 2006) by relocalizing the antiviral protein, viperin from the ER to mitochondria (Seo et al., 2011).

Mitochondrial targeting

The vast majority of cellular mitochondrial proteins are encoded in the nucleus and synthesized in the cytosol. Import of mitochondrial proteins and their

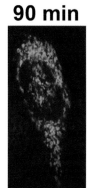

Figure I.12.2 Live cell imaging of mitochondrial fragmentation by pUL37x1-monomeric EGFP (mEGFP). HFFs were transiently transfected with a vector expressing pUL37x1-mEGFP and translation was blocked with anisomycin (14 hours). Following reversal of inhibition, mitochondrial fragmentation was rapidly visualized (within 90 minutes) in live cells expressing pUL37x1-mEGFP.

targeting sequences have been recently reviewed (Neupert and Herrmann, 2007; Dukanovic and Rapaport, 2011). Virtually all mitochondrial proteins are imported by the translocase of the outer membrane of mitochondria (TOM) complex (Rapaport, 2003; Neupert and Herrmann, 2007). The TOM translocase is a macromolecular complex of 490–600 kDa (Kunkele et al., 1998; Model et al., 2002). The diameter of the pore is ~2 nm, sufficient to accommodate two α helices (Kunkele et al., 1998). The pre-protein receptors, Tom20 and Tom22, localized on the OMM, are involved in translocation of mitochondrial protein precursors, particularly those with N-terminal mitochondrial targeting signal (MTS) (Lithgow et al., 1995). The other TOM receptor, Tom70, recognizes polytopic inner mitochondrial membrane (IMM) proteins (Brix et al., 1999). The pore protein, Tom40, appears to act in a chaperone-like manner supporting partial unfolding of substrate proteins during their translocation (Esaki et al., 2003).

One class of mitochondrial proteins that are targeted to the OMM, typified by Tom20 and Tom70, contain a single TMD at their N-termini (Rapaport, 2003). The MTS for Tom20 has been defined as a TMD of moderate hydrophobicity and a net positive charge within five residues immediately downstream of the leader sequence (Kanaji et al., 2000). Upon insertion into the OMM, the bulk of Tom20 is exposed to the cytosol.

Sequential ER to mitochondrial trafficking of HCMV UL37 proteins

Similar to Tom20, HCMV UL37 proteins, which traffic to the OMM, are N-terminally anchored at the ER membrane and the downstream sequences are localized in the cytosol (Mavinakere *et al.*, 2006). The pUL37x1/vMIA intracellular sorting signals are similar in layout to those of Tom20 with analogously positioned TMD leader and downstream basic residues (Williamson and Colberg-Poley, 2010). Despite these similarities, UL37 proteins traffic sequentially from the ER to mitochondria as a gpUL37 cleavage site mutant was *N*-glycosylated prior to its mitochondrial import (Mavinakere *et al.*, 2006). Most cellular proteins with dual ER and mitochondrial trafficking undergo competitive trafficking to ER or to mitochondria (Fig. I.12.3). This differential trafficking is often determined by post-translational processing (Colombo *et al.*, 2005; Boopathi *et al.*, 2008). Nonetheless, sequential ER to mitochondrial trafficking of cellular proteins has been documented for a cellular EndoH^S glycoprotein, which associates with two major mitochondrial complexes

(Complex I and Complex V) in the IMM (Chandra *et al.*, 1998).

Not surprisingly, ER to mitochondrial trafficking of HCMV UL37 proteins is genetically distinguishable from direct mitochondrial import of Tom20 (Kanaji *et al.*, 2000; Williamson and Colberg-Poley, 2010). The HCMV UL37 MTS is bipartite and includes a moderately hydrophobic leader (aa 3–22) that is required for ER association and mitochondrial import as well as juxtaposed downstream sequences (aa 23–36) that are additionally required for OMM import (Mavinakere and Colberg-Poley, 2004a). Three features affect pUL37x1/vMIA dual trafficking: (i) the moderate hydrophobicity of its leader (aa 3–22), (ii) a consensus phosphorylation site (^{21}SY), and (iii) a proline rich domain (^{33}PLPP) (Williamson and Colberg-Poley, 2010). The strength of these trafficking signals is evidenced by the finding that the HCMV UL37 downstream MTS (aa 23–36) overruled the Tom20 MTS and retargeted the Tom20 leader to the sequential ER to mitochondrial pathway (Williamson and Colberg-Poley, 2010). The biological significance

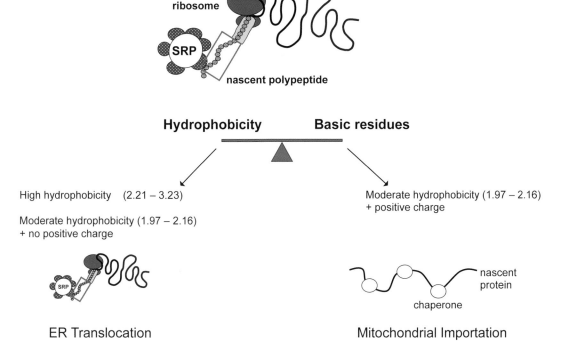

Figure I.12.3 Leader sequences targeting the ER or mitochondria. The salient features of hydrophobic leaders and downstream sequences used to target cellular proteins to the ER (left) or mitochondria (right) are schematically represented. ER-associated translation is favoured by moderate to high hydrophobicity of the leader and the absence of neighbouring basic residues; whereas, mitochondrial import is favoured by the presence of a leader of moderate hydrophobicity with juxtaposed basic residues (Kanaji *et al.*, 2000).

of UL37 ER to mitochondrial trafficking for primate CMV growth is suggested by the conservation of the UL37x1 MTS in all sequenced primate CMVs, including rhesus macaque, chimpanzee, and African green monkey CMVs (McCormick *et al.*, 2003a; Williamson and Colberg-Poley, 2010).

Trafficking to the mitochondria associated membranes (MAM)

Following its synthesis in the ER, pUL37x1 is rapidly localized in close proximity to the OMM (Figs. I.12.4 and I.12.5). An ER subdomain, known as the MAM, provides direct ER-mitochondrial contact at sites where the IMM and OMM meet (Ardail *et al.*, 1993). The MAM plays a pivotal role in regulating ER–mitochondrial cross-talk (Hajnoczky *et al.*, 2006; Rizzuto and Pozzan, 2006; Hayashi *et al.*, 2009). Ca^{2+}-rich microdomains generated by the MAM enable the function of the mitochondrial uniporter without increasing bulk cytosolic Ca^{2+} levels (Rizzuto and Pozzan, 2006). These Ca^{2+}-microdomains are generated by a macromolecular complex, composed of inositol 1,4,5 trisphosphate receptors (IP3R), glucose response protein 75 (GRP75) and voltage dependent anion channel (VDAC). The MAM IP3R–GRP75–VDAC complex spans ER–mitochondrial junctions facilitating efficient mitochondrial Ca^{2+} uptake (Szabadkai *et al.*, 2006). Sigma 1 receptors (Sig-1Rs) are MAM localized, ligand-operated chaperones that stabilize IP3R thus regulating Ca^{2+} influx from the ER to mitochondria (Hayashi and Su, 2007). Constitutive low level IP3R-mediated Ca^{2+} release is essential for efficient mitochondrial respiration and maintenance of normal cellular bioenergetics (Cardenas *et al.*, 2010). Increased mitochondrial Ca^{2+}

drives the adaptive metabolic phase of early ER stress (Bravo *et al.*, 2011).

The MAM is also enriched in lipid synthetic enzymes, such as phosphatidylserine synthase type 1 (PSS-1) and acyl-CoA:diacylglycerol acyltransferase 2 (DGAT2), and is capable of accumulating high levels of cholesterol and bioactive sphingolipids, including ceramide (Rusiñol *et al.*, 1994; Stone and Vance, 2000; Bionda *et al.*, 2004; Stone *et al.*, 2009; Hayashi and Fujimoto, 2010). These lipids, including phospholipids, cholesterol, and ceramide, are transferred from the ER to mitochondria via the MAM (Stiban *et al.*, 2008; Hayashi *et al.*, 2009). In addition, the MAM is enriched in internal lipid rafts (Hayashi and Fujimoto, 2010), which could serve to connect extrinsic and intrinsic apoptotic pathways (Gajate *et al.*, 2009). Finally, mitochondrial antiviral responses have also been recently linked to the MAM (Castanier and Arnoult, 2010; Castanier *et al.*, 2010; Horner *et al.*, 2011). Mitochondrial antiviral signalling (MAVS) protein, an OMM protein, triggers production of type I interferons and pro-inflammatory cytokines (Seth *et al.*, 2005). Upon activation, MAVS protein binds to stimulator of interferon genes (STING), an ER protein, at the MAM (Castanier *et al.*, 2010). Binding of MAVS protein to STING is critical for type I interferon production and NFκB activation after viral production (Ishikawa and Barber, 2008; Zhong *et al.*, 2008; Castanier and Arnoult, 2010). Mitochondrial elongation enhances ER-mitochondrial tethering and favours further association of MAVS-STING (Castanier and Arnoult, 2010). Because it promotes mitochondrial fragmentation, pUL37x1/vMIA disrupts MAVS–STING association and downstream signalling (Castanier *et al.*, 2010).

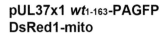

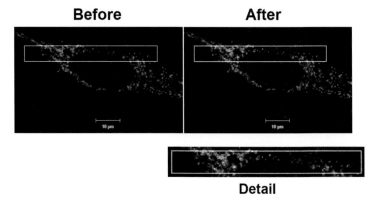

Figure I.12.4 Photoactivation of pUL37x1-photoactivatable GFP (PAGFP). HFFs transiently expressing pUL37x1-PAGFP and DsRed1-mito were irradiated with 405 nm (boxed area) and 488 nm lasers. Shown is the region before (left) and after (right) photoactivation. A detailed view is shown in the lower panel. Photoactivated pUL37x1-PAGFP predominantly colocalized with the mitochondrial marker.

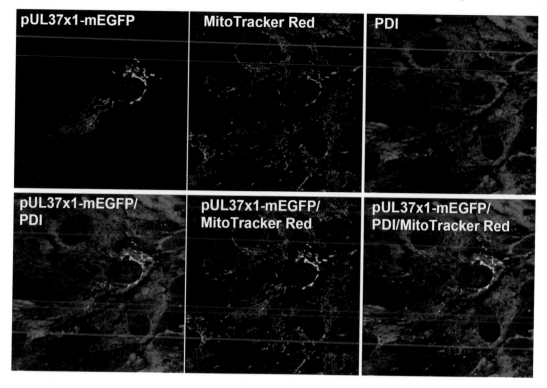

Figure I.12.5 Live cell imaging of pUL37x1-mEGFP synthesis and trafficking to the ER and mitochondria. HFFs transiently transfected with a vector expressing pUL37x1-mEGFP were treated with anisomycin. When the inhibitor was removed, fluorescence was monitored at 1.25 hours after reversal. Confocal images of the individual fluorophores (pUL37x1-mEGFP, green, an ER marker, PDI, blue, and a mitochondrial marker, MitoTracker Red, red) are shown on the top. Overlay of pUL37x1-mEGFP/PDI and pUL37x1-mEGFP/MitoTracker Red are shown in the lower left and lower middle panels, respectively. The overlay image of all three fluorophores is shown on the lower right hand panel.

Targeting of cellular proteins to the MAM

While some cellular MAM resident proteins have been identified, only few have been genetically dissected to determine their targeting signals for MAM localization. PSS-1 contains two lysines at its C-terminus (aa 472, 473) and this motif is similar to that reported for ER targeting. However, deletion of these residues did not affect PSS-1 import into the MAM (Stone and Vance, 2000).

The final reaction of triacylglycerol synthesis is cata-lysed by the major enzyme DGAT2, which localizes to the ER and particularly in the MAM, near the surface of lipid droplets (Wakimoto et al., 2003; Stone et al., 2009). From the MAM, DGAT2 appears to associate peripherally to the OMM via its N-terminus (aa 1–67), possibly to promote efficient synthesis of lipids (Stone et al., 2009).

Sig-1R partially localizes within cytoskeleton associated MAM lipid rafts that differ from GM1-con-taining lipid rafts on the PM (Hayashi and Su, 2003).

Sig-1R has functional cholesterol binding sites (CBDs), [171]VEYGR[175] (single CBD) and [199]LFYTLRSYAR[208] (two overlapping CBDs) in its C-terminus close to its TMDs (Palmer et al., 2007). Cholesterol and ceramide are important membrane components that anchor Sig-1R in the MAM (Hayashi and Fujimoto, 2010).

pUL37x1 and Sig-1R appear to be in very close proximity in the MAM (Williamson and Colberg-Poley, 2009). Similar to Sig-1R, HCMV UL37 proteins have a CBD ([14]LLAFWYFSYR[23]) in their N-terminal TMDs (Williamson et al., 2011b). The functional importance of UL37x1 CBDs is suggested by their conservation in primate CMV UL37x1 sequences including those of rhesus, chimpanzee, and African green monkey CMVs (McCormick et al., 2003a; Wil-liamson et al., 2011b). Association of pUL37x1/vMIA with lipid rafts is cholesterol-dependent (Williamson et al., 2011b). However, in contrast to Sig-1R, mutation of the UL37x1 CBD and the consequent reduction in lipid raft association did not block trafficking to the MAM or to mitochondria (Williamson et al., 2011b).

In contrast to mitochondrial import sequences, evidence for a consensus MAM targeting sequence is still lacking. Rather, a moderately hydrophobic signal sequence targets HCMV UL37 proteins to the MAM. Because of the presence of ribosomes at the MAM between the ER and mitochondrial membranes (Csordas et al., 2006), it is possible that targeted translation of HCMV UL37 proteins at the MAM underlies their importation therein. Moreover, HCMV infection has been found to increase the abundance of translation machinery associated with the MAM subcompartment (Zhang et al., 2011).

We have begun to use Förster (fluorescence) resonance energy transfer (FRET) to examine pUL37x1 interactions during its sequential trafficking (Fig. I.12.6). FRET detects transfer of energy from an excited fluorophore to an acceptor molecule. This transfer results in a decrease in donor emission paralleled by an increase in acceptor emission. The efficiency of transfer is dependent upon the distance between the molecules and the orientation angle between donor and acceptor molecules. For intermolecular and intramolecular FRET, CFP and YFP can be monitored as the YFP/CFP emission intensity. FRET efficiency measures the increased donor (CFP) and decreased acceptor (YFP) intensity following photobleaching of the acceptor (YFP). The consistently high efficiencies of FRET observed with the pUL37x1-CFP/pUL37x1-YFP

donor–acceptor pair are consistent with the finding that pUL37x1/vMIA forms homo-oligomers (Pauleau et al., 2007). In contrast, variable FRET efficiencies between pUL37x1-CFP and Sig-1R-YFP are consistent with our previous findings that pUL37x1 partially colocalizes with Sig-1R in the MAM and associates with MAM lipid rafts (Williamson and Colberg-Poley, 2009; Williamson et al., 2011b).

Dual trafficking of other HCMV products to ER and mitochondria

pUS9

pUS9, a member of the US6 family with hydrophobic sequences at its N- and C-termini, localizes to the ER (Maidji et al., 1998; Huber et al., 2002; Mandic et al., 2009). pUS9 is not detectable by immunofluorescence in HCMV-infected normal human fibroblasts, as it is expressed at low levels (Huber et al., 2002). Therefore, localization of transfected tagged pUS9 was performed (Mandic et al., 2009). pUS9 was found to also localize to mitochondria. Residues 1–24, which form a hydrophobic leader sequence, serve to target US9 to the ER; whereas, US9 aa 198–247 are sufficient to target heterologous GFP to mitochondria. As its ER signal sequence is N-terminal, nascent pUS9 may become ER-associated and then translocated to mitochondria

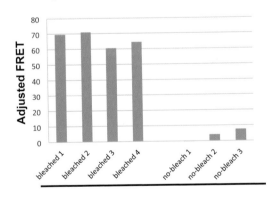

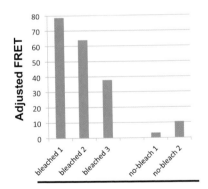

Figure I.12.6 FRET efficiencies of pUL37x1 oligomers and of colocalized pUL37x1 and Sig-1R during trafficking. HFFs co-expressing pUL37x1-CFP/pUL37x1-YFP (left panel) or pUL37x1-CFP/Sig-1R-YFP (right panel) were imaged on an Olympus Fluoview 1000. The fluorescence intensities in different areas of the cell (regions of interest 1–4) were measured prior to and following photobleaching the acceptor fluorophore (YFP). The plots show the adjusted FRET efficiencies for bleached regions of interest compared to different, unbleached regions on the same slides. The consistently high FRET efficiencies of pUL37x1-CFP/pUL37x1-YFP reflect its ability to homo-oligomerize; whereas, the disparate FRET efficiencies of pUL37x1-CFP/Sig-1R-YFP likely represent their partial colocalization in MAM lipid rafts or transient interactions during calcium efflux from the ER (Sharon-Friling et al., 2006).

similarly to pUL37x1/vMIA. Nonetheless, it still remains to be determined whether pUS9 dually targets the ER and mitochondria during HCMV infection.

Trafficking of other HCMV products to mitochondria

m38.5 and m41.1

The MCMV genome encodes m38.5 with an analogous function of inhibiting Bax but with little sequence similarity to pUL37x1/vMIA (Arnoult et al., 2008; Jurak et al., 2008; Norris and Youle, 2008). Another MCMV protein, m41.1, inhibits Bak oligomerization and synergizes with m38.5 (Cam et al., 2010). Both m38.5 and 41.1 localize to mitochondria but there is no evidence yet that they dually traffic to the ER and to mitochondria as HCMV UL37 proteins do.

beta 2.7 RNA

Lastly, an HCMV untranslated RNA has been found to target mitochondria (Reeves et al., 2007). HCMV β2.7 RNA targets mitochondrial complex I (reduced nicotinamide adenine dinucleotide-ubiquinone oxidoreductase) in the IMM and prevents the relocalization of the essential subunit genes associated with retinoid/interferon–induced mortality-19, in response to apoptotic stimuli. This interaction stabilizes mitochondrial membrane potential and continued production of ATP. Import of RNA into mitochondria is less well understood than import of mitochondrial proteins. Almost every organism with mitochondria imports small non-coding tRNAs (Alfonzo and Soll, 2009). Polynucleotide phosphorylase plays a central role in augmenting import of small RNAs required for DNA replication and RNA processing into mitochondrial matrix (Wang et al., 2010). For larger coding mRNAs, the MTS and 3′ untranslated regions play a role in targeting these transcripts to the OMM, for the purpose of increasing efficiency of mitochondria import of encoded proteins (Corral-Debrinski et al., 2000; Margeot et al., 2002; Sylvestre et al., 2003; Herpers and Rabouille, 2004). Puf3, a member of the eukaryotic family of RNA-binding proteins termed PUF (Pumilio and FBF), appears to play a role in mRNA targeting to the mitochondria (Quenault et al., 2011). Puf3 may assist in the interaction of Tom20 with the newly emerging MTS to facilitate association of the coding mRNA with mitochondria (Eliyahu et al., 2010). However, the HCMV β2.7 mRNA is unique, in that it is long (2.7 kb) and is not translated, and yet targets the IMM to play a role in protecting infected cells from apoptosis (Reeves et al., 2007; Zhao et al., 2010). The

RNA sequences which selectively target the HCMV β2.7 RNA to mitochondria and translocate it across mitochondrial membranes remain to be determined.

Why do HCMV UL37 proteins traffic circuitously to mitochondria through the MAM?

In understanding why HCMV UL37 proteins traffic indirectly to mitochondria through the MAM rather than directly from the cytosol, it is helpful to consider the emerging functions of the MAM. The MAM is increasingly being recognized as a hub for ER and mitochondrial cross-talk (Hayashi et al., 2009). By targeting the MAM through its UL37 proteins, HCMV infection acquires the ability to regulate this crucial conversation.

The MAM has roles in Ca^{2+} signalling, which is necessary for normal mitochondrial metabolism and in adaptive metabolism during early ER stress; however, when mitochondrial Ca^{2+} is overloaded, apoptosis can be induced. HCMV pUL37x1/vMIA causes Ca^{2+} efflux from the ER (Sharon-Friling et al., 2006). HCMV infection reprogrammes cellular metabolism to aerobic glycolysis and anaplerotic use of the tricarboxylic acid cycle (Munger et al., 2006, 2008; Chambers et al., 2010; Yu et al., 2010). Regulation of ER Ca^{2+} signalling may play a role in reprogramming of cellular metabolism. Further, regulation of Ca^{2+}-rich microdomains in the MAM to avoid mitochondrial Ca^{2+} overload may increase the ability of HCMV to inhibit apoptosis at yet another mechanism.

Mitochondrial antiviral responses, potentiated by a mitochondrial protein (MAVS), require binding of an ER protein (STING). Because MAVS–STING interaction is favoured by mitochondrial elongation, mitochondrial fragmentation promoted by pUL37x1/vMIA inhibits mitochondrial antiviral responses and thereby reduces amplification propagation of antiviral signalling (McCormick et al., 2003b; Castanier et al., 2010). pUL37x1/vMIA also causes degradation of retinoic acid-inducible gene I (RIG-I) antiviral protein (Scott, 2009). Recently, pUL37x1/vMIA has been found to recruit the cellular antiviral protein, viperin, from the ER and translocate it to mitochondria where its function is subverted to favour HCMV growth (Seo et al., 2011). Thus, by its sequential trafficking from the ER through the MAM and to mitochondria, pUL37x1/vMIA acquires the ability to modulate innate immune responses against HCMV infection.

HCMV infection is known to modulate ER stress responses (Alwine, 2008) and Sig-1R, a MAM protein, translocates during ER stress signalling from the MAM to bulk ER (Hayashi and Su, 2007). Localization of pUL37x1 in close proximity to Sig-1R (Williamson and Colberg-Poley, 2009) may affect the localization of Sig-1R and its response to ER stress. We recently found that HCMV infection relocalizes Sig-1R from a nuclear location to a predominantly cytoplasmic (MAM) location (Zhang *et al.*, 2011).

Furthermore, pUL37x1 associates with lipid rafts localized in the MAM (Williamson *et al.*, 2011b). Internal lipid rafts are enriched in the MAM (Hayashi and Fujimoto, 2010) and lipid rafts have recently been found to serve as apoptosome assembly sites and to connect extrinsic to intrinsic apoptotic signalling (Gajate *et al.*, 2009). Thus, HCMV may affect apoptosome assembly in MAM lipid rafts.

The MAM is enriched in lipid synthetic enzymes. Lipid synthesis is a complex anabolic process whose synthetic enzymes are found in the cytoplasm, ER membranes, and mitochondria.

HCMV growth is highly dependent upon lipid biosynthesis (Munger *et al.*, 2008; Spencer *et al.*, 2011). This enrichment of lipid synthetic enzymes in the MAM may enable HCMV to regulate synthesis of key lipids required for its envelopment and growth.

The MAM contacts mitochondria at sites of junction between the OMM and IMM. By targeting the MAM, pUL37x1 may access internal mitochondrial proteins, whose activities it affects, such as HtrA2/Omi and mitochondrial phosphate carrier (Poncet *et al.*, 2006; McCormick *et al.*, 2008).

The crucial role of the MAM in cellular metabolism, lipid synthesis, mitochondrial Ca^{2+} signalling, antiviral and stress responses, as well as survival suggest that HCMV, by trafficking of UL37 proteins circuitously from the ER to mitochondria through the MAM, acquires the ability to affect these functions. Evidence to the importance of the MAM for HCMV growth is our recent finding the HCMV infection restructures the MAM proteome by late times (Zhang *et al.*, 2011). Chaperones, including GRP78 (BiP), which is known to be required for HCMV growth (Buchkovich *et al.*, 2008), and components of the calcium signalling complex are greatly increased in the MAM, as are lipid synthetic enzymes and mitochondrial metabolic enzymes (Zhang *et al.*, 2011).

Final comments

Proper and coordinated trafficking of viral products is essential for the concerted function of viral machines and progression of the viral life cycle. We have gained considerable insight into how CMV products mimic organelle-specific targeting signals and taps into existing cellular trafficking pathways to enable trafficking of its proteins. Moreover, trafficking of many of its proteins is coordinated to enable assembly of its DNA replication, virion packaging, and egress complexes. By this coordinated trafficking of its proteins, CMV is able to produce highly complex, infectious progeny for subsequent infection of susceptible cells. HCMV also uses the less well-understood protein trafficking pathway from the ER to mitochondria. This dominant targeting of pUL37x1/vMIA confers to HCMV the ability to regulate the functions of the MAM and the exchange of ER and mitochondrial information, which could have deleterious effects on its growth as they impact cellular metabolism, mitochondrial antiviral responses or mitochondrial-mediated apoptosis.

Another intriguing aspect of intracellular CMV protein trafficking is its regulation by cellular proteins (such as BRAP2 and Snapin) and by undifferentiated cells, which as a result are non-permissive to infection. Thus, the interplay between the virus and its host cell can dictate viral protein trafficking and the location of viral proteins can affect its ability to grow. These regulatory mechanisms may ultimately serve as a foundation for new classes of antiviral drugs targeting HCMV protein trafficking (Williamson *et al.*, 2011a) and thereby viral growth.

Acknowledgements

Our work was supported by grants from the National Institutes of Health (R01 AI057906, R21 AI081957 and National Center for Research Resources UL1RR031988) and by Children's Research Institute funds (to A.C.P.). Its contents are solely the responsibility of the authors and do not necessarily represent the official views of the National Center for Research Resources or of the National Institutes of Health. Confocal microscopy imaging was supported by a core grant (5P30HD40677) to the Children's Intellectual and Developmental Disabilities Research Center.

References

Adair, R., Liebisch, G.W., Su, Y., and Colberg-Poley, A.M. (2004). Alteration of cellular RNA splicing and polyadenylation machineries during productive human cytomegalovirus infection. J. Gen. Virol. *85*, 3541–3553.

Ahn, J.H., Brignole, E.J., 3rd, and Hayward, G.S. (1998). Disruption of PML subnuclear domains by the acidic IE1 protein of human cytomegalovirus is mediated through interaction with PML and may modulate a RING finger-dependent cryptic transactivator function of PML. Mol. Cell Biol. *18*, 4899–4913.

Ahn, J.H., and Hayward, G.S. (1997). The major immediate-early proteins IE1 and IE2 of human cytomegalovirus colocalize with and disrupt PML-associated nuclear bodies at very early times in infected permissive cells. J. Virol. *71*, 4599–4613.

Ahn, J.H., Jang, W.J., and Hayward, G.S. (1999). The human cytomegalovirus IE2 and UL112–113 proteins accumulate in viral DNA replication compartments that initiate from the periphery of promyelocytic leukemia protein-associated nuclear bodies (PODs or ND10). J. Virol. 73, 10458–10471.

Al-Barazi, H.O., and Colberg-Poley, A.M. (1996). The human cytomegalovirus UL37 immediate-early regulatory protein is an integral membrane N-glycoprotein which traffics through the endoplasmic reticulum and Golgi apparatus. J. Virol. 70, 7198–7208.

Alfonzo, J.D., and Soll, D. (2009). Mitochondrial tRNA import-the challenge to understand has just begun. Biol. Chem. 390, 717–722.

Alvisi, G., Jans, D.A., Guo, J., Pinna, L.A., and Ripalti, A. (2005). A protein kinase CK2 site flanking the nuclear targeting signal enhances nuclear transport of human cytomegalovirus ppUL44. Traffic 6, 1002–1013.

Alvisi, G., Jans, D.A., and Ripalti, A. (2006a). Human cytomegalovirus (HCMV) DNA polymerase processivity factor ppUL44 dimerizes in the cytosol before translocation to the nucleus. Biochemistry 45, 6866–6872.

Alvisi, G., Ripalti, A., Ngankeu, A., Giannandrea, M., Caraffi, S.G., Dias, M.M., and Jans, D.A. (2006b). Human cytomegalovirus DNA polymerase catalytic subunit pUL54 possesses independently acting nuclear localization and ppUL44 binding motifs. Traffic 7, 1322–1332.

Alwine, J.C. (2008). Modulation of host cell stress responses by human cytomegalovirus. Curr. Top. Microbiol. Immunol. 325, 263–279.

Appenzeller, C., Andersson, H., Kappeler, F., and Hauri, H.P. (1999). The lectin ERGIC-53 is a cargo transport receptor for glycoproteins. Nat. Cell Biol. 1, 330–334.

Appenzeller-Herzog, C., and Hauri, H.P. (2006). The ER-Golgi intermediate compartment (ERGIC): in search of its identity and function. J. Cell. Sci. 119, 2173–2183.

Appleton, B.A., Loregian, A., Filman, D.J., Coen, D.M., and Hogle, J.M. (2004). The cytomegalovirus DNA polymerase subunit UL44 forms a C clamp-shaped dimer. Mol. Cell. 15, 233–244.

Arcangeletti, M.C., De Conto, F., Ferraglia, F., Pinardi, F., Gatti, R., Orlandini, G., Calderaro, A., Motta, F., Medici, M.C., Martinelli, M., et al. (2003). Human cytomegalovirus proteins PP65 and IEP72 are targeted to distinct compartments in nuclei and nuclear matrices of infected human embryo fibroblasts. J. Cell Biochem. 90, 1056–1067.

Ardail, D., Gasnier, F., Lerme, F., Simonot, C., Louisot, P., and Gateau-Roesch, O. (1993). Involvement of mitochondrial contact sites in the subcellular compartmentalization of phospholipid biosynthetic enzymes. J. Biol. Chem. 268, 25985–25992.

Aridor, M., Bannykh, S.I., Rowe, T., and Balch, W.E. (1995). Sequential coupling between COPII and COPI vesicle coats in endoplasmic reticulum to Golgi transport. J. Cell Biol. 131, 875–893.

Arnoult, D., Bartle, L.M., Skaletskaya, A., Poncet, D., Zamzami, N., Park, P.U., Sharpe, J., Youle, R.J., and Goldmacher, V.S. (2004). Cytomegalovirus cell death suppressor vMIA blocks Bax- but not Bak-mediated apoptosis by binding and sequestering Bax at mitochondria. Proc. Natl. Acad. Sci. U.S.A. 101, 7988–7993.

Arnoult, D., Skaletskaya, A., Estaquier, J., Dufour, C., and Goldmacher, V.S. (2008). The murine cytomegalovirus cell death suppressor m38.5 binds Bax and blocks Bax-mediated mitochondrial outer membrane permeabilization. Apoptosis 13, 1100–1110.

Ashiru, O., Bennett, N.J., Boyle, L.H., Thomas, M., Trowsdale, J., and Wills, M.R. (2009). NKG2D ligand MICA is retained in the cis-Golgi apparatus by human cytomegalovirus protein UL142. J. Virol. 83, 12345–12354.

AuCoin, D.P., Smith, G.B., Meiering, C.D., and Mocarski, E.S. (2006). Betaherpesvirus-conserved cytomegalovirus tegument protein ppUL50 (pp150) controls cytoplasmic events during virion maturation. J. Virol. 80, 8199–8210.

Baldick, C.J., Jr., Marchini, A., Patterson, C.E., and Shenk, T. (1997). Human cytomegalovirus tegument protein pp71 (ppUL82) enhances the infectivity of viral DNA and accelerates the infectious cycle. J. Virol. 71, 4400–4408.

Barlowe, C. (2003a). Molecular recognition of cargo by the COPII complex: a most accommodating coat. Cell 114, 395–397.

Barlowe, C. (2003b). Signals for COPII-dependent export from the ER: what's the ticket out? Trends Cell Biol. 13, 295–300.

Baxter, M.K., and Gibson, W. (2001). Cytomegalovirus basic phosphoprotein (pUL32) binds to capsids in vitro through its amino one-third. J. Virol. 75, 6865–6873.

Bechtel, J.T., and Shenk, T. (2002). Human cytomegalovirus UL47 tegument protein functions after entry and before immediate-early gene expression. J. Virol. 76, 1043–1050.

Becke, S., Fabre-Messeman, V., Aue, S., Auerochs, S., Sedmak, T., Wolfrum, U., Strand, D., Marschall, M., Plachter, B., and Reyda, S. (2010). Modification of the major tegument protein pp65 of human cytomegalovirus inhibits virus growth and leads to the enhancement of a protein complex with pUL69 and pUL97 in infected cells. J. Gen. Virol. 91, 2531–2541.

Ben-Tekaya, H., Miura, K., Pepperkok, R., and Hauri, H.P. (2005). Live imaging of bidirectional traffic from the ERGIC. J. Cell Sci. 118, 357–367.

Benedict, C.A., Butrovich, K.D., Lurain, N.S., Corbeil, J., Rooney, I., Schneider, P., Tschopp, J., and Ware, C.F. (1999). Cutting edge: a novel viral TNF receptor superfamily member in virulent strains of human cytomegalovirus. J. Immunol. 162, 6967–6970.

Bionda, C., Portoukalian, J., Schmitt, D., Rodriguez-Lafrasse, C., and Ardail, D. (2004). Subcellular compartmentalization of ceramide metabolism: MAM (mitochondria-associated membrane) and/or mitochondria? Biochem. J. 382, 527–533.

Blobel, G., and Dobberstein, B. (1975a). Transfer of proteins across membranes. I. Presence of proteolytically processed and unprocessed nascent immunoglobulin light chains on membrane-bound ribosomes of murine myeloma. J. Cell Biol. 67, 835–851.

Blobel, G., and Dobberstein, B. (1975b). Transfer of proteins across membranes. II. Reconstitution of functional rough microsomes from heterologous components. J. Cell Biol. 67, 852–862.

Bogner, E., Reschke, M., Reis, B., Mockenhaupt, T., and Radsak, K. (1993). Identification of the gene product encoded by ORF UL56 of the human cytomegalovirus genome. Virology 196, 290–293.

Bogner, E., Anheier, B., Offner, F., Smuda, C., Reschke, M., Eickmann, M., and Radsak, K. (1997). Nuclear translocation of mutagenized forms of human cytomegalovirus glycoprotein B (gpUL55). J. Gen. Virol. 78, 1647–1651.

Bogner, E., Radsak, K., and Stinski, M.F. (1998). The gene product of human cytomegalovirus open reading frame UL56 binds the pac motif and has specific nuclease activity. J. Virol. 72, 2259–2264.

Boopathi, E., Srinivasan, S., Fang, J.K., and Avadhani, N.G. (2008). Bimodal protein targeting through activation of cryptic mitochondrial targeting signals by an inducible cytosolic endoprotease. Mol. Cell 32, 32–42.

Borgese, D., Blobel, G., and Sabatini, D.D. (1973). In vitro exchange of ribosomal subunits between free and membrane-bound ribosomes. J. Mol. Biol. 74, 415–438.

Borgese, N., Mok, W., Kreibich, G., and Sabatini, D.D. (1974). Ribosomal–membrane interaction: in vitro binding of ribosomes to microsomal membranes. J. Mol. Biol. 88, 559–580.

Borst, E.M., Wagner, K., Binz, A., Sodeik, B., and Messerle, M. (2008). The essential human cytomegalovirus gene UL52 is required for cleavage-packaging of the viral genome. J. Virol. 82, 2065–2078.

Bravo, R., Vicencio, J.M., Parra, V., Troncoso, R., Munoz, J.P., Bui, M., Quiroga, C., Rodriguez, A.E., Verdejo, H.E., Ferreira, J., et al. (2011). Increased ER-mitochondrial coupling promotes mitochondrial respiration and bioenergetics during early phases of ER stress. J. Cell Sci. 124, 2143–2152.

Britt, W.J., Jarvis, M., Seo, J.Y., Drummond, D., and Nelson, J. (2004). Rapid genetic engineering of human cytomegalovirus by using a lambda phage linear recombination system: demonstration that pp28 (UL99) is essential for production of infectious virus. J. Virol. 78, 539–543.

Brix, J., Rudiger, S., Bukau, B., Schneider-Mergener, J., and Pfanner, N. (1999). Distribution of binding sequences for the mitochondrial import receptors Tom20, Tom22, and Tom70 in a presequence-carrying preprotein and a non-cleavable preprotein. J. Biol. Chem. 274, 16522–16530.

Buchkovich, N.J., Maguire, T.G., Yu, Y., Paton, A.W., Paton, J.C., and Alwine, J.C. (2008). Human cytomegalovirus specifically controls the levels of the endoplasmic reticulum chaperone BiP/GRP78, which is required for virion assembly. J. Virol. 82, 31–39.

Buchkovich, N.J., Maguire, T.G., and Alwine, J.C. (2010). Role of the endoplasmic reticulum chaperone BiP, SUN domain proteins, and dynein in altering nuclear morphology during human cytomegalovirus infection. J. Virol. 84, 7005–7017.

Cam, M., Handke, W., Picard-Maureau, M., and Brune, W. (2010). Cytomegaloviruses inhibit Bak- and Bax-mediated apoptosis with two separate viral proteins. Cell. Death. Differ. 17, 655–665.

Camozzi, D., Pignatelli, S., Valvo, C., Lattanzi, G., Capanni, C., Dal Monte, P., and Landini, M.P. (2008). Remodelling of the nuclear lamina during human cytomegalovirus infection: role of the viral proteins pUL50 and pUL53. J. Gen. Virol. 89, 731–740.

Cantrell, S.R., and Bresnahan, W.A. (2006). Human cytomegalovirus (HCMV) UL82 gene product (pp71) relieves hDaxx-mediated repression of HCMV replication. J. Virol. 80, 6188–6191.

Cardenas, C., Miller, R.A., Smith, I., Bui, T., Molgo, J., Muller, M., Vais, H., Cheung, K.H., Yang, J., Parker, I., et al. (2010). Essential regulation of cell bioenergetics by constitutive

InsP3 receptor Ca2+ transfer to mitochondria. Cell 142, 270–283.

Castanier, C., and Arnoult, D. (2010). Mitochondrial localization of viral proteins as a means to subvert host defense. Biochim. Biophys. Acta 1813, 575–583.

Castanier, C., Garcin, D., Vazquez, A., and Arnoult, D. (2010). Mitochondrial dynamics regulate the RIG-I-like receptor antiviral pathway. EMBO Rep. 11, 133–138.

Caswell, R., Hagemeier, C., Chiou, C.J., Hayward, G., Kouzarides, T., and Sinclair, J. (1993). The human cytomegalovirus 86K immediate-early (IE) 2 protein requires the basic region of the TATA-box binding protein (TBP) for binding, and interacts with TBP and transcription factor TFIIB via regions of IE2 required for transcriptional regulation. J. Gen. Virol. 74, 2691–2698.

Cepeda, V., Esteban, M., and Fraile-Ramos, A. (2010). Human cytomegalovirus final envelopment on membranes containing both trans-Golgi network and endosomal markers. Cell. Microbiol. 12, 386–404.

Cha, T.A., Tom, E., Kemble, G.W., Duke, G.M., Mocarski, E.S., and Spaete, R.R. (1996). Human cytomegalovirus clinical isolates carry at least 19 genes not found in laboratory strains. J. Virol. 70, 78–83.

Chalupny, N.J., Rein-Weston, A., Dosch, S., and Cosman, D. (2006). Down-regulation of the NKG2D ligand MICA by the human cytomegalovirus glycoprotein UL142. Biochem. Biophys. Res. Commun. 346, 175–181.

Chambers, J.W., Maguire, T.G., and Alwine, J.C. (2010). Glutamine metabolism is essential for human cytomegalovirus infection. J. Virol. 84, 1867–1873.

Chandra, N.C., Spiro, M.J., and Spiro, R.G. (1998). Identification of a glycoprotein from rat liver mitochondrial inner membrane and demonstration of its origin in the endoplasmic reticulum. J. Biol. Chem. 273, 19715–19721.

Chee, M.S., Bankier, A.T., Beck, S., Bohni, R., Brown, C.M., Cerny, R., Horsnell, T., Hutchison, C.A., 3rd, Kouzarides, T., Martignetti, J.A., et al. (1990). Analysis of the protein-coding content of the sequence of human cytomegalovirus strain AD169. Curr. Top. Microbiol. Immunol. 154, 125–169.

Chen, X., Karnovsky, A., Sans, M.D., Andrews, P.C., and Williams, J.A. (2010). Molecular characterization of the endoplasmic reticulum: insights from proteomic studies. Proteomics 10, 4040–4052.

Cherrington, J.M., Khoury, E.L., and Mocarski, E.S. (1991). Human cytomegalovirus ie2 negatively regulates alpha gene expression via a short target sequence near the transcription start site. J. Virol. 65, 887–896.

Colberg-Poley, A.M., Huang, L., Soltero, V.E., Iskenderian, A.C., Schumacher, R.F., and Anders, D.G. (1998). The acidic domain of pUL37x1 and gpUL37 plays a key role in transactivation of HCMV DNA replication gene promoter constructions. Virology 246, 400–408.

Colberg-Poley, A.M., Patel, M.B., Erezo, D.P., and Slater, J.E. (2000). Human cytomegalovirus UL37 immediate-early regulatory proteins traffic through the secretory apparatus and to mitochondria. J. Gen. Virol. 81, 1779–1789.

Collas, P., Thompson, L., Fields, A.P., Poccia, D.L., and Courvalin, J.C. (1997). Protein kinase C-mediated interphase lamin B phosphorylation and solubilization. J. Biol. Chem. 272, 21274–21280.

Colletti, K.S., Smallenburg, K.E., Xu, Y., and Pari, G.S. (2007). Human cytomegalovirus UL84 interacts with an RNA stem–loop sequence found within the RNA/DNA hybrid region of oriLyt. J. Virol. *81*, 7077–7085.

Colombo, S., Longhi, R., Alcaro, S., Ortuso, F., Sprocati, T., Flora, A., and Borgese, N. (2005). N-myristoylation determines dual targeting of mammalian NADH-cytochrome b5 reductase to ER and mitochondrial outer membranes by a mechanism of kinetic partitioning. J. Cell Biol. *168*, 735–745.

Corral-Debrinski, M., Blugeon, C., and Jacq, C. (2000). In yeast, the 3′ untranslated region or the presequence of ATM1 is required for the exclusive localization of its mRNA to the vicinity of mitochondria. Mol. Cell Biol. *20*, 7881–7892.

Crump, C.M., Hung, C.H., Thomas, L., Wan, L., and Thomas, G. (2003). Role of PACS-1 in trafficking of human cytomegalovirus glycoprotein B and virus production. J. Virol. *77*, 11105–11113.

Csordas, G., Renken, C., Varnai, P., Walter, L., Weaver, D., Buttle, K.F., Balla, T., Mannella, C.A., and Hajnoczky, G. (2006). Structural and functional features and significance of the physical linkage between ER and mitochondria. J. Cell Biol. *174*, 915–921.

Dal Monte, P., Pignatelli, S., Zini, N., Maraldi, N.M., Perret, E., Prevost, M.C., and Landini, M.P. (2002). Analysis of intracellular and intraviral localization of the human cytomegalovirus UL53 protein. J. Gen. Virol. *83*, 1005–1012.

Dalbey, R.E., and Von Heijne, G. (1992). Signal peptidases in prokaryotes and eukaryotes – a new protease family. Trends Biochem. Sci. *17*, 474–478.

Dancourt, J., and Barlowe, C. (2010). Protein sorting receptors in the early secretory pathway. Annu. Rev. Biochem. *79*, 777–802.

Dargan, D.J., Douglas, E., Cunningham, C., Jamieson, F., Stanton, R.J., Baluchova, K., McSharry, B.P., Tomasec, P., Emery, V.C., Percivalle, E., et al. (2010). Sequential mutations associated with adaptation of human cytomegalovirus to growth in cell culture. J. Gen. Virol. *91*, 1535–1546.

Das, S., and Pellett, P.E. (2011). Spatial relationships between markers for secretory and endosomal machinery in human cytomegalovirus-infected cells versus those in uninfected cells. J. Virol. *85*, 5864–5879.

Das, S., Skomorovska-Prokvolit, Y., Wang, F.Z., and Pellett, P.E. (2006). Infection-dependent nuclear localization of US17, a member of the US12 family of human cytomegalovirus-encoded seven-transmembrane proteins. J. Virol. *80*, 1191–1203.

Das, S., Vasanji, A., and Pellett, P.E. (2007). Three-dimensional structure of the human cytomegalovirus cytoplasmic virion assembly complex includes a reoriented secretory apparatus. J. Virol. *81*, 11861–11869.

Daubner, T., Fink, A., Seitz, A., Tenzer, S., Muller, J., Strand, D., Seckert, C.K., Janssen, C., Renzaho, A., Grzimek, N.K., et al. (2010). A novel transmembrane domain mediating retention of a highly motile herpesvirus glycoprotein in the endoplasmic reticulum. J. Gen. Virol. *91*, 1524–1534.

Delmas, S., Martin, L., Baron, M., Nelson, J.A., Streblow, D.N., and Davignon, J.L. (2005). Optimization of CD4+ T lymphocyte response to human cytomegalovirus nuclear IE1 protein through modifications of both size and cellular localization. J. Immunol. *175*, 6812–6819.

Dittmer, A., and Bogner, E. (2005). Analysis of the quaternary structure of the putative HCMV portal protein PUL104. Biochemistry *44*, 759–765.

Dittmer, A., Drach, J.C., Townsend, L.B., Fischer, A., and Bogner, E. (2005). Interaction of the putative human cytomegalovirus portal protein pUL104 with the large terminase subunit pUL56 and its inhibition by benzimidazole-D-ribonucleosides. J. Virol. *79*, 14660–14667.

Dukanovic, J., and Rapaport, D. (2011). Multiple pathways in the integration of proteins into the mitochondrial outer membrane. Biochim. Biophys. Acta *1808*, 971–980.

Eliyahu, E., Pnueli, L., Melamed, D., Scherrer, T., Gerber, A.P., Pines, O., Rapaport, D., and Arava, Y. (2010). Tom20 mediates localization of mRNAs to mitochondria in a translation-dependent manner. Mol. Cell Biol. *30*, 284–294.

Ertl, P.F., and Powell, K.L. (1992). Physical and functional interaction of human cytomegalovirus DNA polymerase and its accessory protein (ICP36) expressed in insect cells. J. Virol. *66*, 4126–4133.

Esaki, M., Kanamori, T., Nishikawa, S., Shin, I., Schultz, P.G., and Endo, T. (2003). Tom40 protein import channel binds to non-native proteins and prevents their aggregation. Nat. Struct. Biol. *10*, 988–994.

Fried, H., and Kutay, U. (2003). Nucleocytoplasmic transport: taking an inventory. Cell. Mol. Life Sci. *60*, 1659–1688.

Fulcher, A.J., and Jans, D.A. (2011). Regulation of nucleocytoplasmic trafficking of viral proteins: An integral role in pathogenesis? Biochim. Biophys. Acta *1813*, 2176–2190.

Fulcher, A.J., Roth, D.M., Fatima, S., Alvisi, G., and Jans, D.A. (2010). The BRCA-1 binding protein BRAP2 is a novel, negative regulator of nuclear import of viral proteins, dependent on phosphorylation flanking the nuclear localization signal. FASEB J. *24*, 1454–1466.

Gaddy, C.E., Wong, D.S., Markowitz-Shulman, A., and Colberg-Poley, A.M. (2010). Regulation of the subcellular distribution of key cellular RNA-processing factors during permissive human cytomegalovirus infection. J. Gen. Virol. *91*, 1547–1559.

Gajate, C., Gonzalez-Camacho, F., and Mollinedo, F. (2009). Lipid raft connection between extrinsic and intrinsic apoptotic pathways. Biochem. Biophys. Res. Commun. *380*, 780–784.

Gallina, A., Percivalle, E., Simoncini, L., Revello, M.G., Gerna, G., and Milanesi, G. (1996). Human cytomegalovirus pp65 lower matrix phosphoprotein harbours two transplantable nuclear localization signals. J. Gen. Virol. *77*, 1151–1157.

Gao, Y., Colletti, K., and Pari, G.S. (2008). Identification of human cytomegalovirus UL84 virus- and cell-encoded binding partners by using proteomics analysis. J. Virol. *82*, 96–104.

Gao, Y., Kagele, D., Smallenberg, K., and Pari, G.S. (2010). Nucleocytoplasmic shuttling of human cytomegalovirus UL84 is essential for virus growth. J. Virol. *84*, 8484–8494.

Gibson, W. (1983). Protein counterparts of human and simian cytomegaloviruses. Virology *128*, 391–406.

Giesen, K., Radsak, K., and Bogner, E. (2000). The potential terminase subunit of human cytomegalovirus, pUL56, is translocated into the nucleus by its own nuclear localization signal and interacts with importin alpha. J. Gen. Virol. *81*, 2231–2244.

Goldmacher, V.S., Bartle, L.M., Skaletskaya, A., Dionne, C.A., Kedersha, N.L., Vater, C.A., Han, J.W., Lutz, R.J., Watanabe, S., Cahir McFarland, E.D., et al. (1999). A cytomegalovirus-encoded mitochondria-localized inhibitor of apoptosis structurally unrelated to Bcl-2. Proc. Natl. Acad. Sci. U.S.A. 96, 12536–12541.

Goldner, T., Hewlett, G., Ettischer, N., Ruebsamen-Schaeff, H., Zimmermann, H., and Lischka, P. (2011). The novel anticytomegalovirus compound AIC246 (Letermovir) inhibits human cytomegalovirus replication through a specific antiviral mechanism that involves the viral terminase. J. Virol. 85, 10884–10893.

Greaves, R.F., and Mocarski, E.S. (1998). Defective growth correlates with reduced accumulation of a viral DNA replication protein after low-multiplicity infection by a human cytomegalovirus ie1 mutant. J. Virol. 72, 366–379.

Hajnoczky, G., Csordas, G., Das, S., Garcia-Perez, C., Saotome, M., Sinha Roy, S., and Yi, M. (2006). Mitochondrial calcium signalling and cell death: approaches for assessing the role of mitochondrial Ca2+ uptake in apoptosis. Cell. Calcium 40, 553–560.

Hauri, H.P., Kappeler, F., Andersson, H., and Appenzeller, C. (2000). ERGIC-53 and traffic in the secretory pathway. J. Cell Sci. 113, 587–596.

Hayajneh, W.A., Colberg-Poley, A.M., Skaletskaya, A., Bartle, L.M., Lesperance, M.M., Contopoulos-Ioannidis, D.G., Kedersha, N.L., and Goldmacher, V.S. (2001). The sequence and antiapoptotic functional domains of the human cytomegalovirus UL37 exon 1 immediate-early protein are conserved in multiple primary strains. Virology 279, 233–240.

Hayashi, T., and Fujimoto, M. (2010). Detergent-resistant microdomains determine the localization of sigma-1 receptors to the endoplasmic reticulum–mitochondria junction. Mol. Pharmacol. 77, 517–528.

Hayashi, T., and Su, T.P. (2003). Sigma-1 receptors (sigma(1) binding sites) form raft-like microdomains and target lipid droplets on the endoplasmic reticulum: roles in endoplasmic reticulum lipid compartmentalization and export. J. Pharmacol. Exp. Ther. 306, 718–725.

Hayashi, T., and Su, T.P. (2007). Sigma-1 receptor chaperones at the ER–mitochondrion interface regulate Ca(2+) signaling and cell survival. Cell 131, 596–610.

Hayashi, T., Rizzuto, R., Hajnoczky, G., and Su, T.P. (2009). MAM: more than just a housekeeper. Trends Cell Biol. 19, 81–88.

He, Z., He, Y.S., Kim, Y., Chu, L., Ohmstede, C., Biron, K.K., and Coen, D.M. (1997). The human cytomegalovirus UL97 protein is a protein kinase that autophosphorylates on serines and threonines. J. Virol. 71, 405–411.

Hegde, R.S., and Bernstein, H.D. (2006). The surprising complexity of signal sequences. Trends Biochem. Sci. 31, 563–571.

von Heijne, G. (1988). Transcending the impenetrable: how proteins come to terms with membranes. Biochim. Biophys. Acta 947, 307–333.

Heilbronn, R., Jahn, G., Burkle, A., Freese, U.K., Fleckenstein, B., and zur Hausen, H. (1987). Genomic localization, sequence analysis, and transcription of the putative human cytomegalovirus DNA polymerase gene. J. Virol. 61, 119–124.

Hensel, G., Meyer, H., Gartner, S., Brand, G., and Kern, H.F. (1995). Nuclear localization of the human cytomegalovirus tegument protein pp150 (ppUL32). J. Gen. Virol. 76, 1591–1601.

Hensel, G.M., Meyer, H.H., Buchmann, I., Pommerehne, D., Schmolke, S., Plachter, B., Radsak, K., and Kern, H.F. (1996). Intracellular localization and expression of the human cytomegalovirus matrix phosphoprotein pp71 (ppUL82): evidence for its translocation into the nucleus. J. Gen. Virol. 77, 3087–3097.

Herpers, B., and Rabouille, C. (2004). mRNA localization and ER-based protein sorting mechanisms dictate the use of transitional endoplasmic reticulum-golgi units involved in gurken transport in Drosophila oocytes. Mol. Biol. Cell 15, 5306–5317.

High, S., and Dobberstein, B. (1991). The signal sequence interacts with the methionine-rich domain of the 54-kD protein of signal recognition particle. J. Cell Biol. 113, 229–233.

Horner, S.M., Liu, H.M., Park, H.S., Briley, J., and Gale, M., Jr. (2011). Mitochondrial-associated endoplasmic reticulum membranes (MAM) form innate immune synapses and are targeted by hepatitis C virus. Proc. Natl. Acad. Sci. U.S.A. 108, 14590–14595.

Huber, M.T., and Compton, T. (1998). The human cytomegalovirus UL74 gene encodes the third component of the glycoprotein H-glycoprotein L-containing envelope complex. J. Virol. 72, 8191–8197.

Huber, M.T., and Compton, T. (1999). Intracellular formation and processing of the heterotrimeric gH-gL-gO (gCIII) glycoprotein envelope complex of human cytomegalovirus. J. Virol. 73, 3886–3892.

Huber, M.T., Tomazin, R., Wisner, T., Boname, J., and Johnson, D.C. (2002). Human cytomegalovirus US7, US8, US9, and US10 are cytoplasmic glycoproteins, not found at cell surfaces, and US9 does not mediate cell-to-cell spread. J. Virol. 76, 5748–5758.

Indran, S.V., Ballestas, M.E., and Britt, W.J. (2010). Bicaudal D1-dependent trafficking of human cytomegalovirus tegument protein pp150 in virus-infected cells. J. Virol. 84, 3162–3177.

Ishikawa, H., and Barber, G.N. (2008). STING is an endoplasmic reticulum adaptor that facilitates innate immune signalling. Nature 455, 674–678.

Ishov, A.M., Stenberg, R.M., and Maul, G.G. (1997). Human cytomegalovirus immediate-early interaction with host nuclear structures: definition of an immediate transcript environment. J. Cell Biol. 138, 5–16.

Ishov, A.M., Vladimirova, O.V., and Maul, G.G. (2002). Daxx-mediated accumulation of human cytomegalovirus tegument protein pp71 at ND10 facilitates initiation of viral infection at these nuclear domains. J. Virol. 76, 7705–7712.

Iskenderian, A.C., Huang, L., Reilly, A., Stenberg, R.M., and Anders, D.G. (1996). Four of eleven loci required for transient complementation of human cytomegalovirus DNA replication cooperate to activate expression of replication genes. J. Virol. 70, 383–392.

Isomura, H., and Stinski, M.F. (2003). The human cytomegalovirus major immediate-early enhancer determines the efficiency of immediate-early gene transcription and viral replication in permissive cells at low multiplicity of infection. J. Virol. 77, 3602–3614.

Isomura, H., Stinski, M.F., Murata, T., Yamashita, Y., Kanda, T., Toyokuni, S., and Tsurumi, T. (2011). The human cytomegalovirus gene products essential for late viral gene expression assemble into prereplication complexes before viral DNA replication. J. Virol. 85, 6629–6644.

Jackson, M.R., Nilsson, T., and Peterson, P.A. (1990). Identification of a consensus motif for retention of transmembrane proteins in the endoplasmic reticulum. EMBO J. 9, 3153–3162.

Jans, D.A., Xiao, C.Y., and Lam, M.H. (2000). Nuclear targeting signal recognition: a key control point in nuclear transport? Bioessays 22, 532–544.

Jarvis, M.A., Jones, T.R., Drummond, D.D., Smith, P.P., Britt, W.J., Nelson, J.A., and Baldick, C.J. (2004). Phosphorylation of human cytomegalovirus glycoprotein B (gB) at the acidic cluster casein kinase 2 site (Ser900) is required for localization of gB to the trans-Golgi network and efficient virus replication. J. Virol. 78, 285–293.

Jiang, X.J., Adler, B., Sampaio, K.L., Digel, M., Jahn, G., Ettischer, N., Stierhof, Y.D., Scrivano, L., Koszinowski, U., Mach, M., et al. (2008). UL74 of human cytomegalovirus contributes to virus release by promoting secondary envelopment of virions. J. Virol. 82, 2802–2812.

Jurak, I., Schumacher, U., Simic, H., Voigt, S., and Brune, W. (2008). Murine cytomegalovirus m38.5 protein inhibits Bax-mediated cell death. J. Virol. 82, 4812–4822.

Kaiser, N., Lischka, P., Wagenknecht, N., and Stamminger, T. (2009). Inhibition of human cytomegalovirus replication via peptide aptamers directed against the nonconventional nuclear localization signal of the essential viral replication factor pUL84. J. Virol. 83, 11902–11913.

Kalderon, D., Richardson, W.D., Markham, A.F., and Smith, A.E. (1984a). Sequence requirements for nuclear location of simian virus 40 large-T antigen. Nature 311, 33–38.

Kalderon, D., Roberts, B.L., Richardson, W.D., and Smith, A.E. (1984b). A short amino acid sequence able to specify nuclear location. Cell 39, 499–509.

Kalejta, R.F. (2008). Tegument proteins of human cytomegalovirus. Microbiol. Mol. Biol. Rev. 72, 249–265.

Kalejta, R.F., Bechtel, J.T., and Shenk, T. (2003). Human cytomegalovirus pp71 stimulates cell cycle progression by inducing the proteasome-dependent degradation of the retinoblastoma family of tumor suppressors. Mol. Cell Biol. 23, 1885–1895.

Kalies, K.U., Gorlich, D., and Rapoport, T.A. (1994). Binding of ribosomes to the rough endoplasmic reticulum mediated by the Sec61p-complex. J. Cell Biol. 126, 925–934.

Kamil, J.P., and Coen, D.M. (2007). Human cytomegalovirus protein kinase UL97 forms a complex with the tegument phosphoprotein pp65. J. Virol. 81, 10659–10668.

Kanaji, S., Iwahashi, J., Kida, Y., Sakaguchi, M., and Mihara, K. (2000). Characterization of the signal that directs Tom20 to the mitochondrial outer membrane. J. Cell Biol. 151, 277–288.

Kaye, J.F., Gompels, U.A., and Minson, A.C. (1992). Glycoprotein H of human cytomegalovirus (HCMV) forms a stable complex with the HCMV UL115 gene product. J. Gen. Virol. 73, 2693–2698.

Kemble, G.W., McCormick, A.L., Pereira, L., and Mocarski, E.S. (1987). A cytomegalovirus protein with properties of herpes simplex virus ICP8: partial purification of the polypeptide and map position of the gene. J. Virol. 61, 3143–3151.

Kim, Y.E., and Ahn, J.H. (2010). Role of the specific interaction of UL112–113 p84 with UL44 DNA polymerase processivity factor in promoting DNA replication of human cytomegalovirus. J. Virol. 84, 8409–8421.

Kinzler, E.R., Theiler, R.N., and Compton, T. (2002). Expression and reconstitution of the gH/gL/gO complex of human cytomegalovirus. J. Clin. Virol. 25, 87–95.

Klumperman, J., Schweizer, A., Clausen, H., Tang, B.L., Hong, W., Oorschot, V., and Hauri, H.P. (1998). The recycling pathway of protein ERGIC-53 and dynamics of the ER-Golgi intermediate compartment. J. Cell Sci. 111, 3411–3425.

Korioth, F., Maul, G.G., Plachter, B., Stamminger, T., and Frey, J. (1996). The nuclear domain 10 (ND10) is disrupted by the human cytomegalovirus gene product IE1. Exp. Cell Res. 229, 155–158.

Krieg, U.C., Walter, P., and Johnson, A.E. (1986). Photocrosslinking of the signal sequence of nascent preprolactin to the 54-kilodalton polypeptide of the signal recognition particle. Proc. Natl. Acad. Sci. U.S.A. 83, 8604–8608.

Krosky, P.M., Underwood, M.R., Turk, S.R., Feng, K.W., Jain, R.K., Ptak, R.G., Westerman, A.C., Biron, K.K., Townsend, L.B., and Drach, J.C. (1998). Resistance of human cytomegalovirus to benzimidazole ribonucleosides maps to two open reading frames: UL89 and UL56. J. Virol. 72, 4721–4728.

Krosky, P.M., Baek, M.C., and Coen, D.M. (2003a). The human cytomegalovirus UL97 protein kinase, an antiviral drug target, is required at the stage of nuclear egress. J. Virol. 77, 905–914.

Krosky, P.M., Baek, M.C., Jahng, W.J., Barrera, I., Harvey, R.J., Biron, K.K., Coen, D.M., and Sethna, P.B. (2003b). The human cytomegalovirus UL44 protein is a substrate for the UL97 protein kinase. J. Virol. 77, 7720–7727.

Krzyzaniak, M., Mach, M., and Britt, W.J. (2007). The cytoplasmic tail of glycoprotein M (gpUL100) expresses trafficking signals required for human cytomegalovirus assembly and replication. J. Virol. 81, 10316–10328.

Krzyzaniak, M.A., Mach, M., and Britt, W.J. (2009). HCMV-encoded glycoprotein M (UL100) interacts with Rab11 effector protein FIP4. Traffic 10, 1439–1457.

Kunkele, K.P., Heins, S., Dembowski, M., Nargang, F.E., Benz, R., Thieffry, M., Walz, J., Lill, R., Nussberger, S., and Neupert, W. (1998). The preprotein translocation channel of the outer membrane of mitochondria. Cell 93, 1009–1019.

Kurzchalia, T.V., Wiedmann, M., Girshovich, A.S., Bochkareva, E.S., Bielka, H., and Rapoport, T.A. (1986). The signal sequence of nascent preprolactin interacts with the 54K polypeptide of the signal recognition particle. Nature 320, 634–636.

Lafemina, R.L., Pizzorno, M.C., Mosca, J.D., and Hayward, G.S. (1989). Expression of the acidic nuclear immediate-early protein (IE1) of human cytomegalovirus in stable cell lines and its preferential association with metaphase chromosomes. Virology 172, 584–600.

Lavoie, C., and Paiement, J. (2008). Topology of molecular machines of the endoplasmic reticulum: a compilation of proteomics and cytological data. Histochem. Cell Biol. 129, 117–128.

Lee, H.R., Kim, D.J., Lee, J.M., Choi, C.Y., Ahn, B.Y., Hayward, G.S., and Ahn, J.H. (2004). Ability of the human cytomegalovirus IE1 protein to modulate sumoylation of PML correlates with its functional activities in transcriptional regulation and infectivity in cultured fibroblast cells. J. Virol. 78, 6527–6542.

Lee, H.R., Huh, Y.H., Kim, Y.E., Lee, K., Kim, S., and Ahn, J.H. (2007). N-terminal determinants of human cytomegalovirus IE1 protein in nuclear targeting and disrupting PML-associated subnuclear structures. Biochem. Biophys. Res. Commun. 356, 499–504.

Lee, M.C., Miller, E.A., Goldberg, J., Orci, L., and Schekman, R. (2004). Bi-directional protein transport between the ER and Golgi. Annu. Rev. Cell Dev. Biol. 20, 87–123.

Lerner, R.S., and Nicchitta, C.V. (2006). mRNA translation is compartmentalized to the endoplasmic reticulum following physiological inhibition of cap-dependent translation. RNA *12*, 775–789.

Likhacheva, E.V., and Bogachev, S.S. (2001). Lamins and their functions in cell cycle. Membr. Cell Biol. *14*, 565–577.

Lingappa, V.R., and Blobel, G. (1980). Early events in the biosynthesis of secretory and membrane proteins: the signal hypothesis. Recent Prog. Horm. Res. *36*, 451–475.

Lischka, P., Rosorius, O., Trommer, E., and Stamminger, T. (2001). A novel transferable nuclear export signal mediates CRM1-independent nucleocytoplasmic shuttling of the human cytomegalovirus transactivator protein pUL69. EMBO J. *20*, 7271–7283.

Lischka, P., Sorg, G., Kann, M., Winkler, M., and Stamminger, T. (2003). A nonconventional nuclear localization signal within the UL84 protein of human cytomegalovirus mediates nuclear import via the importin alpha/beta pathway. J. Virol. *77*, 3734–3748.

Lischka, P., Rauh, C., Mueller, R., and Stamminger, T. (2006a). Human cytomegalovirus UL84 protein contains two nuclear export signals and shuttles between the nucleus and the cytoplasm. J. Virol. *80*, 10274–10280.

Lischka, P., Toth, Z., Thomas, M., Mueller, R., and Stamminger, T. (2006b). The UL69 transactivator protein of human cytomegalovirus interacts with DEXD/H-Box RNA helicase UAP56 to promote cytoplasmic accumulation of unspliced RNA. Mol. Cell Biol. *26*, 1631–1643.

Lithgow, T., Glick, B.S., and Schatz, G. (1995). The protein import receptor of mitochondria. Trends Biochem. Sci. *20*, 98–101.

Littler, E., Stuart, A.D., and Chee, M.S. (1992). Human cytomegalovirus UL97 open reading frame encodes a protein that phosphorylates the antiviral nucleoside analogue ganciclovir. Nature *358*, 160–162.

Liu, B., and Stinski, M.F. (1992). Human cytomegalovirus contains a tegument protein that enhances transcription from promoters with upstream ATF and AP-1 cis-acting elements. J. Virol. *66*, 4434–4444.

Liu, B., Hermiston, T.W., and Stinski, M.F. (1991). A cis-acting element in the major immediate-early (IE) promoter of human cytomegalovirus is required for negative regulation by IE2. J. Virol. *65*, 897–903.

Lopper, M., and Compton, T. (2004). Coiled coil domains in glycoproteins B and H are involved in human cytomegalovirus membrane fusion. J. Virol. *78*, 8333–8341.

Loregian, A., and Coen, D.M. (2006). Selective anti-cytomegalovirus compounds discovered by screening for inhibitors of subunit interactions of the viral polymerase. Chem. Biol. *13*, 191–200.

Loregian, A., Piaia, E., Cancellotti, E., Papini, E., Marsden, H.S., and Palu, G. (2000). The catalytic subunit of herpes simplex virus type 1 DNA polymerase contains a nuclear localization signal in the UL42-binding region. Virology *273*, 139–148.

Lu, M., and Shenk, T. (1999). Human cytomegalovirus UL69 protein induces cells to accumulate in G1 phase of the cell cycle. J. Virol. *73*, 676–683.

Lukac, D.M., Harel, N.Y., Tanese, N., and Alwine, J.C. (1997). TAF-like functions of human cytomegalovirus immediate-early proteins. J. Virol. *71*, 7227–7239.

Lutcke, H., High, S., Romisch, K., Ashford, A.J., and Dobberstein, B. (1992). The methionine-rich domain of the 54 kDa subunit of signal recognition particle is sufficient for the interaction with signal sequences. EMBO J. *11*, 1543–1551.

McCormick, A.L., Skaletskaya, A., Barry, P.A., Mocarski, E.S., and Goldmacher, V.S. (2003a). Differential function and expression of the viral inhibitor of caspase 8-induced apoptosis (vICA) and the viral mitochondria-localized inhibitor of apoptosis (vMIA) cell death suppressors conserved in primate and rodent cytomegaloviruses. Virology *316*, 221–233.

McCormick, A.L., Smith, V.L., Chow, D., and Mocarski, E.S. (2003b). Disruption of mitochondrial networks by the human cytomegalovirus UL37 gene product viral mitochondrion-localized inhibitor of apoptosis. J. Virol. *77*, 631–641.

McCormick, A.L., Roback, L., and Mocarski, E.S. (2008). HtrA2/Omi terminates cytomegalovirus infection and is controlled by the viral mitochondrial inhibitor of apoptosis (vMIA). PLoS Pathog. *4*, e1000063.

Mach, M., Kropff, B., Dal Monte, P., and Britt, W. (2000). Complex formation by human cytomegalovirus glycoproteins M (gpUL100) and N (gpUL73). J. Virol. *74*, 11881–11892.

Mach, M., Osinski, K., Kropff, B., Schloetzer-Schrehardt, U., Krzyzaniak, M., and Britt, W. (2007). The carboxy-terminal domain of glycoprotein N of human cytomegalovirus is required for virion morphogenesis. J. Virol. *81*, 5212–5224.

McMahon, T.P., and Anders, D.G. (2002). Interactions between human cytomegalovirus helicase-primase proteins. Virus Res. *86*, 39–52.

Maidji, E., Tugizov, S., Abenes, G., Jones, T., and Pereira, L. (1998). A novel human cytomegalovirus glycoprotein, gpUS9, which promotes cell-to-cell spread in polarized epithelial cells, colocalizes with the cytoskeletal proteins E-cadherin and F-actin. J. Virol. *72*, 5717–5727.

Mandic, L., Miller, M.S., Coulter, C., Munshaw, B., and Hertel, L. (2009). Human cytomegalovirus US9 protein contains an N-terminal signal sequence and a C-terminal mitochondrial localization domain, and does not alter cellular sensitivity to apoptosis. J. Gen Virol. *90*, 1172–1182.

Marchini, A., Liu, H., and Zhu, H. (2001). Human cytomegalovirus with IE-2 (UL122) deleted fails to express early lytic genes. J. Virol. *75*, 1870–1878.

Margeot, A., Blugeon, C., Sylvestre, J., Vialette, S., Jacq, C., and Corral-Debrinski, M. (2002). In Saccharomyces cerevisiae, ATP2 mRNA sorting to the vicinity of mitochondria is essential for respiratory function. EMBO J. *21*, 6893–6904.

Marks, M.S., Ohno, H., Kirchnausen, T., and Bonracino, J.S. (1997). Protein sorting by tyrosine-based signals: adapting to the Ys and wherefores. Trends Cell Biol. *7*, 124–128.

Marschall, M., Freitag, M., Suchy, P., Romaker, D., Kupfer, R., Hanke, M., and Stamminger, T. (2003). The protein kinase pUL97 of human cytomegalovirus interacts with and phosphorylates the DNA polymerase processivity factor pUL44. Virology *311*, 60–71.

Marschall, M., Marzi, A., aus dem Siepen, P., Jochmann, R., Kalmer, M., Auerochs, S., Lischka, P., Leis, M., and Stamminger, T. (2005). Cellular p32 recruits cytomegalovirus kinase pUL97 to redistribute the nuclear lamina. J. Biol. Chem. *280*, 33357–33367.

Marshall, K.R., Rowley, K.V., Rinaldi, A., Nicholson, I.P., Ishov, A.M., Maul, G.G., and Preston, C.M. (2002). Activity and intracellular localization of the human cytomegalovirus protein pp71. J. Gen. Virol. 83, 1601–1612.

Martinez-Menarguez, J.A., Geuze, H.J., Slot, J.W., and Klumperman, J. (1999). Vesicular tubular clusters between the ER and Golgi mediate concentration of soluble secretory proteins by exclusion from COPI-coated vesicles. Cell 98, 81–90.

Mavinakere, M.S., and Colberg-Poley, A.M. (2004a). Dual targeting of the human cytomegalovirus UL37 exon 1 protein during permissive infection. J. Gen Virol. 85, 323–329.

Mavinakere, M.S., and Colberg-Poley, A.M. (2004b). Internal cleavage of the human cytomegalovirus UL37 immediate-early glycoprotein and divergent trafficking of its proteolytic fragments. J. Gen Virol. 85, 1989–1994.

Mavinakere, M.S., Williamson, C.D., Goldmacher, V.S., and Colberg-Poley, A.M. (2006). Processing of human cytomegalovirus UL37 mutant glycoproteins in the endoplasmic reticulum lumen prior to mitochondrial importation. J. Virol. 80, 6771–6783.

Meier, J.L., and Stinski, M.F. (1996). Regulation of human cytomegalovirus immediate-early gene expression. Intervirology 39, 331–342.

Meier, J.L., Keller, M.J., and McCoy, J.J. (2002). Requirement of multiple cis-acting elements in the human cytomegalovirus major immediate-early distal enhancer for viral gene expression and replication. J. Virol. 76, 313–326.

Meyer, G., Gicklhorn, D., Strive, T., Radsak, K., and Eickmann, M. (2002). A three-residue signal confers localization of a reporter protein in the inner nuclear membrane. Biochem Biophys. Res. Commun. 291, 966–971.

Meyer, G.A., and Radsak, K.D. (2000). Identification of a novel signal sequence that targets transmembrane proteins to the nuclear envelope inner membrane. J. Biol. Chem. 275, 3857–3866.

Michel, D., Pavic, I., Zimmermann, A., Haupt, E., Wunderlich, K., Heuschmid, M., and Mertens, T. (1996). The UL97 gene product of human cytomegalovirus is an early-late protein with a nuclear localization but is not a nucleoside kinase. J. Virol. 70, 6340–6346.

Michelsen, K., Yuan, H., and Schwappach, B. (2005). Hide and run. Arginine-based endoplasmic-reticulum-sorting motifs in the assembly of heteromultimeric membrane proteins. EMBO Rep. 6, 717–722.

Milbradt, J., Auerochs, S., and Marschall, M. (2007). Cytomegaloviral proteins pUL50 and pUL53 are associated with the nuclear lamina and interact with cellular protein kinase C. J. Gen. Virol. 88, 2642–2650.

Milbradt, J., Auerochs, S., Sticht, H., and Marschall, M. (2009). Cytomegaloviral proteins that associate with the nuclear lamina: components of a postulated nuclear egress complex. J. Gen Virol. 90, 579–590.

Milbradt, J., Webel, R., Auerochs, S., Sticht, H., and Marschall, M. (2010). Novel mode of phosphorylation-triggered reorganization of the nuclear lamina during nuclear egress of human cytomegalovirus. J. Biol. Chem. 285, 13979–13989.

Miller, M.S., Furlong, W.E., Pennell, L., Geadah, M., and Hertel, L. (2010). RASCAL is a new human cytomegalovirus-encoded protein that localizes to the nuclear lamina and in cytoplasmic vesicles at late times postinfection. J. Virol. 84, 6483–6496.

Milstein, C., Brownlee, G.G., Harrison, T.M., and Mathews, M.B. (1972). A possible precursor of immunoglobulin light chains. Nat. New. Biol. 239, 117–120.

Mocarski, E.S., Shenk, T., and Pass, R.F. (2007). Cytomegaloviruses. In Fields Virology, D.M. Knipe, and P.M. Howley, eds. (Wolters Kluwer Health, Lippincott Williams & Wilkins, Philadelphia, PA), pp. 2701–2772.

Model, K., Prinz, T., Ruiz, T., Radermacher, M., Krimmer, T., Kuhlbrandt, W., Pfanner, N., and Meisinger, C. (2002). Protein translocase of the outer mitochondrial membrane: role of import receptors in the structural organization of the TOM complex. J. Mol. Biol. 316, 657–666.

Moorman, N.J., Sharon-Friling, R., Shenk, T., and Cristea, I.M. (2010). A targeted spatial-temporal proteomics approach implicates multiple cellular trafficking pathways in human cytomegalovirus virion maturation. Mol Cell Proteomics 9, 851–860.

Muller, S., and Dejean, A. (1999). Viral immediate-early proteins abrogate the modification by SUMO-1 of PML and Sp100 proteins, correlating with nuclear body disruption. J. Virol. 73, 5137–5143.

Munger, J., Yu, D., and Shenk, T. (2006). UL26-deficient human cytomegalovirus produces virions with hypophosphorylated pp28 tegument protein that is unstable within newly infected cells. J. Virol. 80, 3541–3548.

Munger, J., Bennett, B.D., Parikh, A., Feng, X.J., McArdle, J., Rabitz, H.A., Shenk, T., and Rabinowitz, J.D. (2008). Systems-level metabolic flux profiling identifies fatty acid synthesis as a target for antiviral therapy. Nat. Biotechnol. 26, 1179–1186.

Muranyi, W., Haas, J., Wagner, M., Krohne, G., and Koszinowski, U.H. (2002). Cytomegalovirus recruitment of cellular kinases to dissolve the nuclear lamina. Science 297, 854–857.

Murphy, E., Yu, D., Grimwood, J., Schmutz, J., Dickson, M., Jarvis, M.A., Hahn, G., Nelson, J.A., Myers, R.M., and Shenk, T.E. (2003). Coding potential of laboratory and clinical strains of human cytomegalovirus. Proc. Natl. Acad. Sci. U.S.A. 100, 14976–14981.

Netterwald, J.R., Jones, T.R., Britt, W.J., Yang, S.J., McCrone, I.P., and Zhu, H. (2004). Postattachment events associated with viral entry are necessary for induction of interferon-stimulated genes by human cytomegalovirus. J. Virol. 78, 6688–6691.

Neupert, W., and Herrmann, J.M. (2007). Translocation of proteins into mitochondria. Annu. Rev. Biochem. 76, 723–749.

Nevels, M., Brune, W., and Shenk, T. (2004a). SUMOylation of the human cytomegalovirus 72-kilodalton IE1 protein facilitates expression of the 86-kilodalton IE2 protein and promotes viral replication. J. Virol. 78, 7803–7812.

Nevels, M., Paulus, C., and Shenk, T. (2004b). Human cytomegalovirus immediate-early 1 protein facilitates viral replication by antagonizing histone deacetylation. Proc. Natl. Acad. Sci. U.S.A. 101, 17234–17239.

Nguyen, N.L., Loveland, A.N., and Gibson, W. (2008). Nuclear localization sequences in cytomegalovirus capsid assembly proteins (UL80 proteins) are required for virus production: inactivating NLS1, NLS2, or both affects replication to strikingly different extents. J. Virol. 82, 5381–5389.

Nilsson, T., Au, C.E., and Bergeron, J.J. (2009). Sorting out glycosylation enzymes in the Golgi apparatus. FEBS Lett. 583, 3764–3769.

Nilsson, T., and Warren, G. (1994). Retention and retrieval in the endoplasmic reticulum and the Golgi apparatus. Curr. Opin. Cell Biol. 6, 517–521.

Norris, K.L., and Youle, R.J. (2008). Cytomegalovirus proteins vMIA and m38.5 link mitochondrial morphogenesis to Bcl-2 family proteins. J. Virol. 82, 6232–6243.

Nothwehr, S.F., and Gordon, J.I. (1990). Targeting of proteins into the eukaryotic secretory pathway: signal peptide structure/function relationships. Bioessays 12, 479–484.

Palmer, C.P., Mahen, R., Schnell, E., Djamgoz, M.B., and Aydar, E. (2007). Sigma-1 receptors bind cholesterol and remodel lipid rafts in breast cancer cell lines. Cancer Res. 67, 11166–11175.

Pari, G.S., and Anders, D.G. (1993). Eleven loci encoding trans-acting factors are required for transient complementation of human cytomegalovirus oriLyt-dependent DNA replication. J. Virol. 67, 6979–6988.

Pari, G.S., Kacica, M.A., and Anders, D.G. (1993). Open reading frames UL44, IRS1/TRS1, and UL36–38 are required for transient complementation of human cytomegalovirus oriLyt-dependent DNA synthesis. J. Virol. 67, 2575–2582.

Park, M.Y., Kim, Y.E., Seo, M.R., Lee, J.R., Lee, C.H., and Ahn, J.H. (2006). Interactions among four proteins encoded by the human cytomegalovirus UL112–113 region regulate their intranuclear targeting and the recruitment of UL44 to prereplication foci. J. Virol. 80, 2718–2727.

Pauleau, A.-L., Larochette, N., Giordanetto, F., Scholz, S.R., Poncet, D., Zamzami, N., Golmacher, V.S., and Koemer., G. (2007). Structure-function analysis of the interaction between Bax and the cytomegalovirus-encoded protein vMIA. Oncogene 26, 7067–7080.

Pelham, H.R. (1989). Control of protein exit from the endoplasmic reticulum. Annu. Rev. Cell Biol. 5, 1–23.

Pelham, H.R. (1995). Sorting and retrieval between the endoplasmic reticulum and Golgi apparatus. Curr. Opin. Cell Biol. 7, 530–535.

Penfold, M.E., and Mocarski, E.S. (1997). Formation of cytomegalovirus DNA replication compartments defined by localization of viral proteins and DNA synthesis. Virology 239, 46–61.

Penkert, R.R., and Kalejta, R.F. (2010). Nuclear localization of tegument-delivered pp71 in human cytomegalovirus-infected cells is facilitated by one or more factors present in terminally differentiated fibroblasts. J. Virol. 84, 9853–9863.

Pizzorno, M.C., Mullen, M.A., Chang, Y.N., and Hayward, G.S. (1991). The functionally active IE2 immediate-early regulatory protein of human cytomegalovirus is an 80-kilodalton polypeptide that contains two distinct activator domains and a duplicated nuclear localization signal. J. Virol. 65, 3839–3852.

Plachter, B., Nordin, M., Wirgart, B.Z., Mach, M., Stein, H., Grillner, L., and Jahn, G. (1992). The DNA-binding protein P52 of human cytomegalovirus reacts with monoclonal antibody CCH2 and associates with the nuclear membrane at late times after infection. Virus Res. 24, 265–276.

Plafker, S.M., and Gibson, W. (1998). Cytomegalovirus assembly protein precursor and proteinase precursor contain two nuclear localization signals that mediate their own nuclear translocation and that of the major capsid protein. J. Virol. 72, 7722–7732.

Poncet, D., Larochette, N., Pauleau, A.L., Boya, P., Jalil, A.A., Cartron, P.F., Vallette, F., Schnebelen, C., Bartle, L.M.,

Skaletskaya, A., et al. (2004). An antiapoptotic viral protein that recruits Bax to mitochondria. J. Biol Chem 279, 22605–22614.

Poncet, D., Pauleau, A.L., Szabadkai, G., Vozza, A., Scholz, S.R., Le Bras, M., Briere, J.J., Jalil, A., Le Moigne, R., Brenner, C., et al. (2006). Cytopathic effects of the cytomegalovirus-encoded apoptosis inhibitory protein vMIA. J. Cell Biol. 174, 985–996.

Poole, E., King, C.A., Sinclair, J.H., and Alcami, A. (2006). The UL144 gene product of human cytomegalovirus activates NFkappaB via a TRAF6-dependent mechanism. EMBO J. 25, 4390–4399.

Preston, C.M., and Nicholl, M.J. (2006). Role of the cellular protein hDaxx in human cytomegalovirus immediate-early gene expression. J. Gen. Virol. 87, 1113–1121.

Prichard, M.N., Britt, W.J., Daily, S.L., Hartline, C.B., and Kern, E.R. (2005a). Human cytomegalovirus UL97 Kinase is required for the normal intranuclear distribution of pp65 and virion morphogenesis. J. Virol. 79, 15494–15502.

Prichard, M.N., Lawlor, H., Duke, G.M., Mo, C., Wang, Z., Dixon, M., Kemble, G., and Kern, E.R. (2005b). Human cytomegalovirus uracil DNA glycosylase associates with ppUL44 and accelerates the accumulation of viral DNA. Virol. J. 2, 55.

Quenault, T., Lithgow, T., and Traven, A. (2011). PUF proteins: repression, activation and mRNA localization. Trends Cell Biol. 21, 104–112.

Raden, D., Song, W., and Gilmore, R. (2000). Role of the cytoplasmic segments of Sec61alpha in the ribosome-binding and translocation-promoting activities of the Sec61 complex. J. Cell Biol. 150, 53–64.

Radsak, K., Brucher, K.H., Britt, W., Shiou, H., Schneider, D., and Kollert, A. (1990). Nuclear compartmentation of glycoprotein B of human cytomegalovirus. Virology 177, 515–522.

Rapaport, D. (2003). Finding the right organelle. Targeting signals in mitochondrial outer-membrane proteins. EMBO Rep. 4, 948–952.

Rapoport, T.A., Rolls, M.M., and Jungnickel, B. (1996). Approaching the mechanism of protein transport across the ER membrane. Curr. Opin. Cell Biol. 8, 499–504.

Rechter, S., Scott, G.M., Eickhoff, J., Zielke, K., Auerochs, S., Muller, R., Stamminger, T., Rawlinson, W.D., and Marschall, M. (2009). Cyclin-dependent kinases phosphorylate the cytomegalovirus RNA export protein pUL69 and modulate its nuclear localization and activity. J. Biol. Chem. 284, 8605–8613.

Reeves, M.B., Davies, A.A., McSharry, B.P., Wilkinson, G.W., and Sinclair, J.H. (2007). Complex I binding by a virally encoded RNA regulates mitochondria-induced cell death. Science 316, 1345–1348.

Rizzuto, R., and Pozzan, T. (2006). Microdomains of intracellular Ca2+: molecular determinants and functional consequences. Physiol. Rev. 86, 369–408.

Robbins, J., Dilworth, S.M., Laskey, R.A., and Dingwall, C. (1991). Two interdependent basic domains in nucleoplasmin nuclear targeting sequence: identification of a class of bipartite nuclear targeting sequence. Cell 64, 615–623.

Romisch, K., and Corsi, A. (1996). Protein translocation into the endoplasmic reticulum. In Protein Targeting, Hurtley, S.M., ed. (Oxford University Press, Oxford, UK), pp. 101–122.

Rusinol, A.E., Cui, Z., Chen, M.H., and Vance, J.E. (1994). A unique mitochondria-associated membrane fraction from rat liver has a high capacity for lipid synthesis and contains pre-Golgi secretory proteins including nascent lipoproteins. J. Biol. Chem. *269*, 27494–27502.

Ryckman, B.J., Chase, M.C., and Johnson, D.C. (2010). Human cytomegalovirus TR strain glycoprotein O acts as a chaperone promoting gH/gL incorporation into virions but is not present in virions. J. Virol. *84*, 2597–2609.

Sabatini, D.D., Blobel, G., Nonomura, Y., and Adelman, M.R. (1971). Ribosome–membrane interaction: Structural aspects and functional implications. Adv. Cytopharmacol. *1*, 119–129.

Saffert, R.T., and Kalejta, R.F. (2006). Inactivating a cellular intrinsic immune defense mediated by Daxx is the mechanism through which the human cytomegalovirus pp71 protein stimulates viral immediate-early gene expression. J. Virol. *80*, 3863–3871.

Saffert, R.T., and Kalejta, R.F. (2007). Human cytomegalovirus gene expression is silenced by Daxx-mediated intrinsic immune defense in model latent infections established in vitro. J. Virol. *81*, 9109–9120.

Saffert, R.T., Penkert, R.R., and Kalejta, R.F. (2010). Cellular and viral control over the initial events of human cytomegalovirus experimental latency in CD34+ cells. J. Virol. *84*, 5594–5604.

Salsman, J., Wang, X., and Frappier, L. (2011). Nuclear body formation and PML body remodeling by the human cytomegalovirus protein UL35. Virology *414*, 119–129.

Salsman, J., Zimmerman, N., Chen, T., Domagala, M., and Frappier, L. (2008). Genome-wide screen of three herpesviruses for protein subcellular localization and alteration of PML nuclear bodies. PLoS Pathog. *4*, e1000100.

Sam, M.D., Evans, B.T., Coen, D.M., and Hogle, J.M. (2009). Biochemical, biophysical, and mutational analyses of subunit interactions of the human cytomegalovirus nuclear egress complex. J. Virol. *83*, 2996–3006.

Sampaio, K.L., Cavignac, Y., Stierhof, Y.D., and Sinzger, C. (2005). Human cytomegalovirus labeled with green fluorescent protein for live analysis of intracellular particle movements. J. Virol. *79*, 2754–2767.

Sanchez, V., Angeletti, P.C., Engler, J.A., and Britt, W.J. (1998). Localization of human cytomegalovirus structural proteins to the nuclear matrix of infected human fibroblasts. J. Virol. *72*, 3321–3329.

Sanchez, V., Greis, K.D., Sztul, E., and Britt, W.J. (2000). Accumulation of virion tegument and envelope proteins in a stable cytoplasmic compartment during human cytomegalovirus replication: characterization of a potential site of virus assembly. J. Virol. *74*, 975–986.

Sanchez, V., Mahr, J.A., Orazio, N.I., and Spector, D.H. (2007). Nuclear export of the human cytomegalovirus tegument protein pp65 requires cyclin-dependent kinase activity and the Crm1 exporter. J. Virol. *81*, 11730–11736.

Sarisky, R.T., and Hayward, G.S. (1996). Evidence that the UL84 gene product of human cytomegalovirus is essential for promoting oriLyt-dependent DNA replication and formation of replication compartments in cotransfection assays. J. Virol. *70*, 7398–7413.

Schmolke, S., Drescher, P., Jahn, G., and Plachter, B. (1995). Nuclear targeting of the tegument protein pp65 (UL83) of human cytomegalovirus: an unusual bipartite nuclear localization signal functions with other portions of the protein to mediate its efficient nuclear transport. J. Virol. *69*, 1071–1078.

Schweizer, A., Fransen, J.A., Bachi, T., Ginsel, L., and Hauri, H.P. (1988). Identification, by a monoclonal antibody, of a 53-kD protein associated with a tubulo-vesicular compartment at the cis-side of the Golgi apparatus. J. Cell Biol. *107*, 1643–1653.

Scott, I. (2009). Degradation of RIG-I following cytomegalovirus infection is independent of apoptosis. Microbes. Infect. *11*, 973–979.

Seo, J.Y., Yaneva, R., Hinson, E.R., and Cresswell, P. (2011). Human cytomegalovirus directly induces the antiviral protein viperin to enhance infectivity. Science *332*, 1093–1097.

Seth, R.B., Sun, L., Ea, C.K., and Chen, Z.J. (2005). Identification and characterization of MAVS, a mitochondrial antiviral signaling protein that activates NF-kappaB and IRF 3. Cell *122*, 669–682.

Sharon-Friling, R., Goodhouse, J., Colberg-Poley, A.M., and Shenk, T. (2006). Human cytomegalovirus pUL37x1 induces the release of endoplasmic reticulum calcium stores. Proc. Natl. Acad. Sci. U.S.A. *103*, 19117–19122.

Shen, A., Lei, J., Yang, E., Pei, Y., Chen, Y.C., Gong, H., Xiao, G., and Liu, F. (2011). Human cytomegalovirus primase UL70 specifically interacts with cellular factor snapin. J. Virol. *85*, 11732–11741.

Shen, W., Westgard, E., Huang, L., Ward, M.D., Osborn, J.L., Chau, N.H., Collins, L., Marcum, B., Koach, M.A., Bibbs, J., *et al.* (2008). Nuclear trafficking of the human cytomegalovirus pp71 (ppUL82) tegument protein. Virology *376*, 42–52.

Shimamura, M., Mach, M., and Britt, W.J. (2006). Human cytomegalovirus infection elicits a glycoprotein M (gM)/gN-specific virus-neutralizing antibody response. J. Virol. *80*, 4591–4600.

Siegel, V., and Walter, P. (1985). Elongation arrest is not a prerequisite for secretory protein translocation across the microsomal membrane. J. Cell Biol. *100*, 1913–1921.

Siegel, V., and Walter, P. (1988). Each of the activities of signal recognition particle (SRP) is contained within a distinct domain: analysis of biochemical mutants of SRP. Cell *52*, 39–49.

Simmen, K.A., Singh, J., Luukkonen, B.G., Lopper, M., Bittner, A., Miller, N.E., Jackson, M.R., Compton, T., and Fruh, K. (2001). Global modulation of cellular transcription by human cytomegalovirus is initiated by viral glycoprotein B. Proc. Natl. Acad. Sci. U.S.A. *98*, 7140–7145.

Singh, J., and Compton, T. (2000). Characterization of a panel of insertion mutants in human cytomegalovirus glycoprotein B. J. Virol. *74*, 1383–1392.

Smith, J.A., and Pari, G.S. (1995). Human cytomegalovirus UL102 gene. J. Virol. *69*, 1734–1740.

Smith, J.A., Jairath, S., Crute, J.J., and Pari, G.S. (1996). Characterization of the human cytomegalovirus UL105 gene and identification of the putative helicase protein. Virology *220*, 251–255.

Smith, S., and Blobel, G. (1993). The first membrane spanning region of the lamin B receptor is sufficient for sorting to the inner nuclear membrane. J. Cell Biol. *120*, 631–637.

Sourvinos, G., Tavalai, N., Berndt, A., Spandidos, D.A., and Stamminger, T. (2007). Recruitment of human cytomegalovirus immediate-early 2 protein onto parental viral genomes in association with ND10 in live-infected cells. J. Virol. *81*, 10123–10136.

Spector, D.J., and Tevethia, M.J. (1994). Protein–protein interactions between human cytomegalovirus IE2-580aa and pUL84 in lytically infected cells. J. Virol. 68, 7549–7553.

Spencer, C.M., Schafer, X.L., Moorman, N.J., and Munger, J. (2011). Human cytomegalovirus induces the activity and expression of acetyl-coenzyme A carboxylase, a fatty acid biosynthetic enzyme whose inhibition attenuates viral replication. J. Virol. 85, 5814–5824.

Stephens, S.B., Dodd, R.D., Brewer, J.W., Lager, P.J., Keene, J.D., and Nicchitta, C.V. (2005). Stable ribosome binding to the endoplasmic reticulum enables compartment-specific regulation of mRNA translation. Mol. Biol. Cell. 16, 5819–5831.

Stiban, J., Caputo, L., and Colombini, M. (2008). Ceramide synthesis in the endoplasmic reticulum can permeabilize mitochondria to proapoptotic proteins. J. Lipid Res. 49, 625–634.

Stone, S.J., and Vance, J.E. (2000). Phosphatidylserine synthase-1 and -2 are localized to mitochondria-associated membranes. J. Biol. Chem. 275, 34534–34540.

Stone, S.J., Levin, M.C., Zhou, P., Han, J., Walther, T.C., and Farese, R.V., Jr. (2009). The endoplasmic reticulum enzyme DGAT2 is found in mitochondria-associated membranes and has a mitochondrial targeting signal that promotes its association with mitochondria. J. Biol. Chem. 284, 5352–5361.

Strang, B.L., Sinigalia, E., Silva, L.A., Coen, D.M., and Loregian, A. (2009). Analysis of the association of the human cytomegalovirus DNA polymerase subunit UL44 with the viral DNA replication factor UL84. J. Virol. 83, 7581–7589.

Strang, B.L., Geballe, A.P., and Coen, D.M. (2010). Association of human cytomegalovirus proteins IRS1 and TRS1 with the viral DNA polymerase accessory subunit UL44. J. Gen. Virol. 91, 2167–2175.

Sullivan, V., Talarico, C.L., Stanat, S.C., Davis, M., Coen, D.M., and Biron, K.K. (1992). A protein kinase homologue controls phosphorylation of ganciclovir in human cytomegalovirus-infected cells. Nature 358, 162–164.

Sylvestre, J., Margeot, A., Jacq, C., Dujardin, G., and Corral-Debrinski, M. (2003). The role of the 3′ untranslated region in mRNA sorting to the vicinity of mitochondria is conserved from yeast to human cells. Mol. Biol. Cell 14, 3848–3856.

Szabadkai, G., Bianchi, K., Varnai, P., De Stefani, D., Wieckowski, M.R., Cavagna, D., Nagy, A.I., Balla, T., and Rizzuto, R. (2006). Chaperone-mediated coupling of endoplasmic reticulum and mitochondrial Ca2+ channels. J. Cell Biol. 175, 901–911.

Tandon, R., and Mocarski, E.S. (2011). Cytomegalovirus pUL96 is critical for the stability of pp150-associated nucleocapsids. J. Virol. 85, 7129–7141.

Teasdale, R.D., and Jackson, M.R. (1996). Signal-mediated sorting of membrane proteins between the endoplasmic reticulum and the golgi apparatus. Annu. Rev. Cell. Dev. Biol. 12, 27–54.

Theiler, R.N., and Compton, T. (2001). Characterization of the signal peptide processing and membrane association of human cytomegalovirus glycoprotein O. J. Biol. Chem. 276, 39226–39231.

Theiler, R.N., and Compton, T. (2002). Distinct glycoprotein O complexes arise in a post-Golgi compartment of cytomegalovirus-infected cells. J. Virol. 76, 2890–2898.

Thoma, C., Borst, E., Messerle, M., Rieger, M., Hwang, J.S., and Bogner, E. (2006). Identification of the interaction domain of the small terminase subunit pUL89 with the large subunit pUL56 of human cytomegalovirus. Biochemistry 45, 8855–8863.

Thomas, M., Rechter, S., Milbradt, J., Auerochs, S., Muller, R., Stamminger, T., and Marschall, M. (2009). Cytomegaloviral protein kinase pUL97 interacts with the nuclear mRNA export factor pUL69 to modulate its intranuclear localization and activity. J. Gen Virol. 90, 567–578.

Toth, Z., Lischka, P., and Stamminger, T. (2006). RNA-binding of the human cytomegalovirus transactivator protein UL69, mediated by arginine-rich motifs, is not required for nuclear export of unspliced RNA. Nucleic Acids Res. 34, 1237–1249.

Underwood, M.R., Harvey, R.J., Stanat, S.C., Hemphill, M.L., Miller, T., Drach, J.C., Townsend, L.B., and Biron, K.K. (1998). Inhibition of human cytomegalovirus DNA maturation by a benzimidazole ribonucleoside is mediated through the UL89 gene product. J. Virol. 72, 717–725.

Urban, M., Joubert, N., Hocek, M., Alexander, R.E., and Kuchta, R.D. (2009). Herpes simplex virus-1 DNA primase: a remarkably inaccurate yet selective polymerase. Biochemistry 48, 10866–10881.

Vales-Gomez, M., Winterhalter, A., Roda-Navarro, P., Zimmermann, A., Boyle, L., Hengel, H., Brooks, A., and Reyburn, H.T. (2006). The human cytomegalovirus glycoprotein UL16 traffics through the plasma membrane and the nuclear envelope. Cell Microbiol. 8, 581–590.

Varnum, S.M., Streblow, D.N., Monroe, M.E., Smith, P., Auberry, K.J., Pasa-Tolic, L., Wang, D., Camp, D.G., 2nd, Rodland, K., Wiley, S., et al. (2004). Identification of proteins in human cytomegalovirus (HCMV) particles: the HCMV proteome. J. Virol. 78, 10960–10966.

Wakimoto, K., Chiba, H., Michibata, H., Seishima, M., Kawasaki, S., Okubo, K., Mitsui, H., Torii, H., and Imai, Y. (2003). A novel diacylglycerol acyltransferase (DGAT2) is decreased in human psoriatic skin and increased in diabetic mice. Biochem. Biophys. Res. Commun. 310, 296–302.

Walter, P., and Blobel, G. (1982). Signal recognition particle contains a 7S RNA essential for protein translocation across the endoplasmic reticulum. Nature 299, 691–698.

Walter, P., and Blobel, G. (1983). Disassembly and reconstitution of signal recognition particle. Cell 34, 525–533.

Walter, P., and Johnson, A.E. (1994). Signal sequence recognition and protein targeting to the endoplasmic reticulum membrane. Annu. Rev. Cell Biol. 10, 87–119.

Walter, P., Ibrahimi, I., and Blobel, G. (1981). Translocation of proteins across the endoplasmic reticulum. I. Signal recognition protein (SRP) binds to in-vitro-assembled polysomes synthesizing secretory protein. J. Cell Biol. 91, 545–550.

Wang, G., Chen, H.W., Oktay, Y., Zhang, J., Allen, E.L., Smith, G.M., Fan, K.C., Hong, J.S., French, S.W., McCaffery, J.M., et al. (2010). PNPASE regulates RNA import into mitochondria. Cell 142, 456–467.

Waters, A., Hassan, J., De Gascun, C., Kissoon, G., Knowles, S., Molloy, E., Connell, J., and Hall, W.W. (2010). Human cytomegalovirus UL144 is associated with viremia and infant development sequelae in congenital infection. J. Clin. Microbiol. 48, 3956–3962.

Webel, R., Milbradt, J., Auerochs, S., Schregel, V., Held, C., Nobauer, K., Razzazi-Fazeli, E., Jardin, C., Wittenberg, T., Sticht, H., et al. (2011). Two isoforms of the protein kinase pUL97 of human cytomegalovirus are differentially regulated in their nuclear translocation. J. Gen. Virol. 92, 638–649.

White, E.A., Clark, C.L., Sanchez, V., and Spector, D.H. (2004). Small internal deletions in the human cytomegalovirus IE2 gene result in nonviable recombinant viruses with differential defects in viral gene expression. J. Virol. 78, 1817–1830.

White, E.A., Del Rosario, C.J., Sanders, R.L., and Spector, D.H. (2007). The IE2 60-kilodalton and 40-kilodalton proteins are dispensable for human cytomegalovirus replication but are required for efficient delayed early and late gene expression and production of infectious virus. J. Virol. 81, 2573–2583.

Wilkinson, G.W., Kelly, C., Sinclair, J.H., and Rickards, C. (1998a). Disruption of PML-associated nuclear bodies mediated by the human cytomegalovirus major immediate-early gene product. J. Gen. Virol. 79, 1233–1245.

Wilkinson, G.W., Kelly, C., Sinclair, J.H., and Rickards, C. (1998b). Disruption of PML-associated nuclear bodies mediated by the human cytomegalovirus major immediate-early gene product. J. Gen. Virol. 79, 1233–1245.

Williamson, C.D., and Colberg-Poley, A.M. (2009). Access of viral proteins to mitochondria via mitochondria-associated membranes. Rev. Med. Virol. 19, 147–164.

Williamson, C.D., and Colberg-Poley, A.M. (2010). Intracellular sorting signals for sequential trafficking of human cytomegalovirus UL37 proteins to the endoplasmic reticulum and mitochondria. J. Virol. 84, 6400–6409.

Williamson, C.D., De Biasi, R.L., and Colberg-Poley, A.M. (2011a). Viral product trafficking to mitochondria, mechanisms and roles in pathogenesis. Infect Disord. Drug Targets 12, 18–37.

Williamson, C.D., Zhang, A., and Colberg-Poley, A.M. (2011b). The human cytomegalovirus protein UL37 exon 1 associates with internal lipid rafts. J. Virol. 85, 2100–2111.

Wills, M.R., Ashiru, O., Reeves, M.B., Okecha, G., Trowsdale, J., Tomasec, P., Wilkinson, G.W., Sinclair, J., and Sissons, J.G. (2005). Human cytomegalovirus encodes an MHC class I-like molecule (UL142) that functions to inhibit NK cell lysis. J. Immunol. 175, 7457–7465.

Wilson, C., Venditti, R., Rega, L.R., Colanzi, A., D'Angelo, G., and De Matteis, M.A. (2010). The Golgi apparatus: an organelle with multiple complex functions. Biochem. J. 433, 1–9.

Winkler, M., Rice, S.A., and Stamminger, T. (1994). UL69 of human cytomegalovirus, an open reading frame with homology to ICP27 of herpes simplex virus, encodes a transactivator of gene expression. J. Virol. 68, 3943–3954.

Wood, L.J., Baxter, M.K., Plafker, S.M., and Gibson, W. (1997). Human cytomegalovirus capsid assembly protein precursor (pUL80.5) interacts with itself and with the major capsid protein (pUL86) through two different domains. J. Virol. 71, 179–190.

Wright, D.A., Staprans, S.I., and Spector, D.H. (1988). Four phosphoproteins with common amino termini are encoded by human cytomegalovirus AD169. J. Virol. 62, 331–340.

Xu, Y., Colletti, K.S., and Pari, G.S. (2002). Human cytomegalovirus UL84 localizes to the cell nucleus via a nuclear localization signal and is a component of viral replication compartments. J. Virol. 76, 8931–8938.

Xu, Y., Cei, S.A., Rodriguez Huete, A., Colletti, K.S., and Pari, G.S. (2004). Human cytomegalovirus DNA replication requires transcriptional activation via an IE2- and UL84-responsive bidirectional promoter element within oriLyt. J. Virol. 78, 11664–11677.

Yamamoto, T., Suzuki, S., Radsak, K., and Hirai, K. (1998). The UL112/113 gene products of human cytomegalovirus which colocalize with viral DNA in infected cell nuclei are related to efficient viral DNA replication. Virus Res. 56, 107–114.

Yamashita, Y., Shimokata, K., Mizuno, S., Daikoku, T., Tsurumi, T., and Nishiyama, Y. (1996). Calnexin acts as a molecular chaperone during the folding of glycoprotein B of human cytomegalovirus. J. Virol. 70, 2237–2246.

Yao, Z.Q., Gallez-Hawkins, G., Lomeli, N.A., Li, X., Molinder, K.M., Diamond, D.J., and Zaia, J.A. (2001). Site-directed mutation in a conserved kinase domain of human cytomegalovirus-pp65 with preservation of cytotoxic T lymphocyte targeting. Vaccine 19, 1628–1635.

Yu, Y., Maguire, T.G., and Alwine, J.C. (2010). Human cytomegalovirus activates glucose transporter 4 expression to increase glucose uptake during infection. J. Virol. 85, 1573–1580.

Zaia, J.A., Li, X., Franck, A.E., Wu, X., Thao, L., and Gallez-Hawkins, G. (2009). Biologic and immunologic effects of knockout of human cytomegalovirus pp65 nuclear localization signal. Clin. Vaccine Immunol. 16, 935–943.

Zhang, A., Williamson, C.D., Wong, D.S., Bullough, M.D., Brown, K.J., Hathout, Y., and Colberg-Poley, A.M. (2011). Quantitative proteomic analyses of human cytomegalovirus-induced restructuring of endoplasmic reticulum-mitochondrial contacts at late times of infection. Mol. Cell Proteomics 10, M111.009936.

Zhao, J., Sinclair, J., Houghton, J., Bolton, E., Bradley, A., and Lever, A. (2010). Cytomegalovirus beta2.7 RNA transcript protects endothelial cells against apoptosis during ischemia/reperfusion injury. J. Heart Lung Transplant. 29, 342–345.

Zheng, Z., Maidji, E., Tugizov, S., and Pereira, L. (1996). Mutations in the carboxyl-terminal hydrophobic sequence of human cytomegalovirus glycoprotein B alter transport and protein chaperone binding. J. Virol. 70, 8029–8040.

Zhong, B., Yang, Y., Li, S., Wang, Y.Y., Li, Y., Diao, F., Lei, C., He, X., Zhang, L., Tien, P., et al. (2008). The adaptor protein MITA links virus-sensing receptors to IRF3 transcription factor activation. Immunity 29, 538–550.

Zhu, H., Shen, Y., and Shenk, T. (1995). Human cytomegalovirus IE1 and IE2 proteins block apoptosis. J. Virol. 69, 7960–7970.

Zielke, B., Thomas, M., Giede-Jeppe, A., Muller, R., and Stamminger, T. (2011). Characterization of the betaherpesviral pUL69 protein family reveals binding of the cellular mRNA export factor UAP56 as a prerequisite for stimulation of nuclear mRNA export and for efficient viral replication. J. Virol. 85, 1804–1819.

Zopf, D., Bernstein, H.D., Johnson, A.E., and Walter, P. (1990). The methionine-rich domain of the 54 kd protein subunit of the signal recognition particle contains an RNA binding site and can be crosslinked to a signal sequence. EMBO J. 9, 4511–4517.

Morphogenesis of the Cytomegalovirus Virion and Subviral Particles

I.13

Wade Gibson and Elke Bogner

Abstract

Formation and maturation of the cytomegalovirus capsid is reviewed. Recent information about protein–protein interactions involved in capsid assembly, molecular interactions relating to the mechanism of DNA packaging, and the sequence of events in primary envelopment is considered as it establishes similarities and reveals differences between CMV and other herpesviruses.

Introduction

Lytic replication of herpesviruses culminates in the production of infectious virus, enhancing transmission to a new host and spread of the pathogen. Evolution and specialization between and within the subfamilies of *Herpesviridae* has resulted in differences thought to confer biological niche-specific advantages. Among these are significant variations, if not important differences, in the structure and formation of infectious virus.

We focus here on early steps of virus production in cytomegalovirus-infected cells: capsid assembly and associated proteolytic cleavage, DNA packaging and the proteins involved, and primary envelopment of the nucleocapsid. We note similarities in these steps between CMV and other herpesviruses, primarily the alpha-herpesviruses herpes simplex virus (HSV) and pseudorabies virus (PRV), and some notable differences that are of particular interest. The composition, structure, and formation of CMV, HSV, and PRV has been discussed more comprehensively in recent reviews, which this chapter is intended to complement and to which the reader is directed for additional information and details (Steven and Spear, 1997; Brown *et al.*, 2002; Holzenburg and Bogner, 2002; Britt and Boppana, 2004; Mettenleiter, 2004; Gibson, 2006, 2008; Liu and Zhou, 2007; Mettenleiter *et al.*, 2009; Conway and Homa, 2011; Johnson and Baines, 2011).

Structure and composition of the virion

Virions of cytomegalovirus and the other herpesviruses are composed of a *capsid shell* containing the linear double-strand DNA genome, embedded in a complex of proteins organized into the *tegument* layer, which interfaces the capsid with the *envelope* (Roizman and Pellett, 2001).

The capsid shell is an icosahedral lattice, approximately 100 nm in diameter, composed of three abundant protein species organized into 162 capsomers and 320 triplexes. One hundred and fifty of the capsomers (the hexons) are situated in the twenty triangular faces of the shell, and 60 (the pentons) are located at its 12 vertices. Capsomeres are hexamers or pentamers of major capsid protein (MCP, pUL86, 154 kDa in HCMV). Triplexes are distributed between the capsomers and are composed of two copies of the minor capsid protein (mCP, pUL85, triplex dimer protein or TDP, 35 kDa in HCMV) and one copy of the mCP-binding protein (mCP-BP, pUL46, triplex monomer protein or TMP, 33 kDa in HCMV). One vertex of the shell is modified or replaced by a symmetrical portal complex composed of 12 copies of the portal protein (pUL104, 79 kDa in HCMV), through which the viral DNA enters and leaves the capsid (Newcomb *et al.*, 2009) (e.g. Fig. I.13.1).

Herpesvirus capsids also have a protein tightly bound to the exterior face of their capsomers (Booy *et al.*, 1994; Wingfield *et al.*, 1997; Trus *et al.*, 1999; Yu *et al.*, 2005). It is the smallest capsid protein (SCP) detected and is encoded in CMV by an ORF between UL48 and UL49, provisionally called UL48/49 or UL48.5 (Baldick and Shenk, 1996; Gibson *et al.*, 1996). Six copies bind to each hexon, and in HSV but not CMV five copies also bind to each penton. Its function is essential to produce infectious CMV (Borst *et al.*, 2001), unlike that of its more dispensable homologue in HSV (Desai *et al.*, 1998).

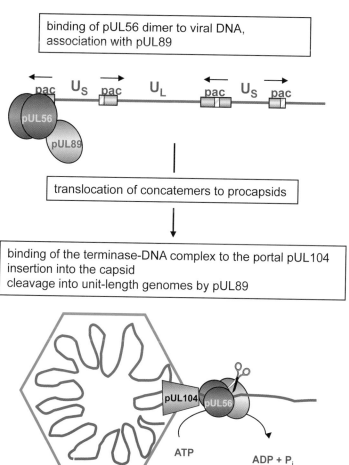

binding of pUL56 dimer to viral DNA, association with pUL89

translocation of concatemers to procapsids

binding of the terminase-DNA complex to the portal pUL104
insertion into the capsid
cleavage into unit-length genomes by pUL89

ATP   ADP + P$_i$

Figure I.13.1 Packaging of DNA into preformed nascent capsids. The large terminase subunit pUL56 dimerize and associates with the small terminase subunit pUL89 (top drawing). Binding to DNA is mediated via pUL56, followed by cleavage of excess *a* sequences. The terminase–DNA complex is translocated to preformed nascent capsids and interacts with the portal (bottom drawing; pUL89 behind pUL56 dimer). ATP hydrolysis mediates the insertion of one unit-length genome into the capsid. After packaging of one genome the concatenated DNA is cleaved.

The tegument layer contains most of the virion phosphoproteins, several enzymatic activities, and components that modulate cellular and viral gene expression (Irmiere and Gibson, 1983; Roby and Gibson, 1986; Kalejta, 2008; also see Chapter I.9). It is ~ 50 nm thick and by mass is predominated by pUL32 (basic phosphoprotein, pp150, 149 kDa), pUL82 (upper matrix protein, pp71, 74 kDa), and pUL83 (lower matrix protein, pp65, 69 kDa). All three of these and two other tegument proteins ('80k', pUL97 and pp28, pUL28) are phosphorylated during infection and can serve as substrates *in vitro* for a virion-associated protein kinase(s) (Mar *et al.*, 1981; Roby and Gibson, 1986; Gallina *et al.*, 1999; Kamil and Coen, 2007).

Basis phosphoprotein (BPP) is distinguished as being the tegument protein most tightly bound to the capsid, based on its resistance to detergent stripping from the virion nucleocapsid (Gibson, 1996) and its binding to capsids *in vitro* (Baxter and Gibson, 2001). It is also the predominant or only virion protein bearing O-linked N-acetylglucosamine monosaccharides (Benko and Gibson, 1988; Benko *et al.*, 1988; Greis *et al.*, 1994), a modification associated with regulatory and targeting mechanisms (Slawson and Hart, 2011). BPP has recognized counterparts only among other beta-herpesviruses and, although its role during virion formation remains undetermined, altering its structure through mutagenesis corrupts the process and reduces virus production (AuCoin *et al.*, 2006).

The high molecular weight protein (HMWP, pUL48, 212 kDa) is also closely associated with the capsid surface (Chen *et al.*, 1999), but is stripped away more easily than BPP by detergent treatment and remains selectively complexed with the HMWP-binding protein (hmwBP, pUL47, 115 kDa) (Gibson, 1996; Harmon and Gibson, 1996). The amino end of HMWP is a ubiquitin-specific cysteine protease, whose activity is conserved throughout the herpesvirus family (Kattenhorn et al., 2005; Schlieker et al., 2005; Wang et al., 2006; Jarosinski et al., 2007; Kim et al., 2009). The enzyme is active as part of the full-length HWMP within the mature virion, but is not absolutely required for production of infectious virus in cell culture (Wang et al., 2006). Although its biological target(s) and purpose during infection remains undefined, its presence in the virion suggests a role during virus formation and egress and/or during entry and nucleocapsid uncoating (Wang et al., 2006). The reduced pathogenicity in animals observed for mutants of PRV and Marek's disease virus (MDV) that encode enzymatically inactive pUL48 homologues (Jarosinski et al., 2007; Bottcher et al., 2008; Lee et al., 2009) is also fully compatible with impaired entry or egress functions, but underscores the possible involvement of substrates more critical for natural infection than cell culture replication.

The tegument interfaces the capsid with the virion envelope, which is ~ 10 nm thick and contains at least ten abundant protein species, including G-protein coupled receptors (Margulies et al., 1996; Margulies and Gibson, 2007; O'Connor and Shenk, 2011), immunogenic CMV glycoprotein complexes (Mach, 2006), and cellular proteins (Varnum et al., 2004). Interactions between specific tegument and envelope proteins remain elusive, but the pUL83 upper matrix protein has been proposed as one candidate tegument anchor, based on findings that HCMV dense bodies are essentially enveloped spherical aggregates of that protein (Irmiere and Gibson, 1983). Additional or alternate tegument proteins are likely to be involved, however, since mutant viruses unable to express pUL83 do produce infectious virus, though no dense bodies (Schmolke et al., 1995). The tegument and envelope contain additional less abundant virus-encoded and host cell proteins (Varnum et al., 2004), as well as phospholipids (Roby and Gibson, 1996; Liu et al., 2011), polyamines (Gibson et al., 1984), and small RNAs (Terhune et al., 2004).

Aberrant enveloped particles

In addition to virions, HCMV-infected cells produce and release non-infectious enveloped particles (NIEP) and dense bodies (DB). Comparisons of infectivity, structure, DNA content, and protein composition established that: (i) NIEPs are similar in structure and protein composition to virions, but have no DNA

and retain a set of capsid scaffolding proteins (UL80 proteins, see below) absent from virions. And, (ii) DBs are simpler in structure and composition than virions and NIEPs, but several times larger and more heterogeneous in their size distribution. They are essentially enveloped spheroidal aggregates of tegument protein pUL83, which contain no DNA, capsid, or intact capsid proteins (Irmiere and Gibson, 1983; Roby and Gibson, 1986). As reported and discussed before, such measurements and comparisons are significantly influenced by the strain of HCMV used and the methods used to prepare and analyse the particles (Irmiere and Gibson, 1983; Roby and Gibson, 1986).

Formation of the capsid

Herpesvirus capsids assemble and receive their DNA genome in the nucleus, progressing from nascent scaffold-containing procapsids, to DNA-containing nucleocapsids competent for primary envelopment. Three stable intracellular capsid forms are routinely recovered from herpesvirus-infected cells: A-, B-, and C-capsids.

Including procapsids, these four characterized capsid forms are suggested to be closely linked through a hypothetical transient nuclear intermediate called the transition B-capsid (B_t), as illustrated in Figs. I.13.2 and I.13.3. Stable accumulating B-capsids are speculated to be B_t-capsids that underwent scaffold cleavage but failed to engage a DNA/terminase complex to incorporate DNA during a critical window of opportunity. Stable accumulating A-capsids are thought to be B_t-capsids that completed scaffold cleavage and elimination, and initiated but failed to complete and stabilize DNA incorporation, resulting in a ~ 7% larger-diameter shell devoid of scaffolding proteins and DNA (Lee et al., 1988). Successful maturation of B_t-capsids yields intranuclear nucleocapsids (NC), distinguished in electron micrographs from A- and B-capsids by the comparatively higher density of their heavy-metal-binding DNA centres.

Different NC forms are present (i) within primary enveloped particles between the inner and outer nuclear membranes (primary enveloped NC, NCpe), (ii) in the cytoplasm following de-envelopment of primary enveloped particles (cytoplasmic NC, NCcy), and (iii) within virions following secondary envelopment (secondary enveloped NC, NCse). Depending upon cell type and the method of cell disruption and fractionation used, these different NC forms are expected to remain compartmentalized to a greater or lesser extent.

Initial interactions of the capsid subunit proteins occur in the cytoplasm (Fig. I.13.3). MCP is unable to

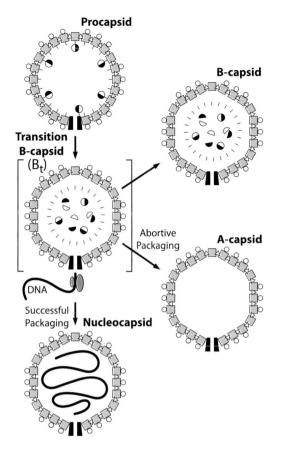

Figure I.13.2 Model depicting inter-relatedness of CMV capsid forms. Spherical procapsids (top left), containing the internal scaffolding proteins pAP (short radial lines) and pPR (short radial lines with half-filled circles), mature to DNA-containing nucleocapsids (bottom left) through a hypothetical intermediate called the transition B-capsid (B$_t$-capsid, bracketed). Expected to be heterogeneous in composition and shape, B$_t$-capsids undergo cleavage of the scaffolding proteins (indicated by altered distribution and integrity of internal line and circle symbols), conformational changes of the shell (depicted as more angular appearance), and incorporation of the ds-DNA genome (thick irregular line beneath B$_t$-capsid) through the portal complex (dark opposing trapezoids at bottom of capsid). Stable A- and B-capsids are typically abundant homogeneous particles thought to arise as by-products of nucleocapsid formation, accumulating prior to the cleaved scaffolding proteins being eliminated and DNA incorporated (B-capsids), or following unsuccessful DNA incorporation (A-capsids). Grey-filled squares represent capsomeres, empty triangles represent triplexes, small empty circles represent the smallest capsid protein, and grey ovals on DNA proximal to the portal represent the large and small subunits of the DNA cleavage/packaging complex, terminase (see below and Fig. I.13.4).

move into the nucleus on its own, and is translocated by associating with the assembly protein precursor (pAP, pUL80.5, 38 kDa in HCMV) (Wood et al., 1997; Plafker and Gibson, 1998). This interaction is mediated by the carboxy-terminal amino acids of pAP (Thomsen et al., 1995; Beaudet-Miller et al., 1996; Desai and Person, 1996; Pelletier et al., 1997; Wood et al., 1997), and potentiated or stabilized by the self-association of pAP through a conserved domain near its amino end (Wood et al., 1997; Loveland et al., 2007). The region of MCP participating in this interaction in HSV has been localized to its amino end (Desai and Person, 1999), which is exposed on the interior face of the capsid shell (Bowman et al., 2003) (Fig. I.13.3, step 1).

The relatively slow movement of MCP into the nucleus, as measured by pulse-chase radiolabeling experiments (Gibson, 1981; de Bruyn Kops et al., 1998), may reflect the cytoplasmic organization of relatively large capsomere substructures or protomers (Fig. I.13.3, step 2). Translocation of these complexes into the nucleus depends on nuclear localization sequences (NLS) within pAP (Plafker and Gibson, 1998; Nguyen et al., 2008) (Fig. I.13.3, step 3). Unlike the scaffolding protein homologues of other herpesviruses, HCMV pAP contains two NLS. NLS1 has a positional counterpart in all herpesviruses, just upstream of the group-conserved Pro-Gly-Glu sequence, but NLS2 has recognized counterparts only among the beta-herpesviruses (Plafker and Gibson, 1998). Mutating either NLS alone reduces production of infectious virus, but mutating both is lethal (Nguyen et al., 2008). The phenotype of the NLS2 mutation is more profound that that of NLS1 and may reflect an involvement of this sequence in critical interactions between pAP and the portal protein, extrapolating from studies done with the homologous HSV proteins (Singer et al., 2005; Nguyen et al., 2008; Yang and Baines, 2008, 2009).

Dimers of the minor capsid protein pUL85 (mCP) likewise require an escort protein for nuclear translocation – a function of their triplex partner pUL46 (mC-BP), which is thought to provide NLS for the complex (Baxter and Gibson, 1997; Spencer et al., 1998) (Fig. I.13.3, step 4).

Within the nucleus the triplex, capsomere, and portal complexes coalesce to form procapsids (Fig. I.13.3, step 5). This process requires and is driven by the scaffolding proteins. In their absence or when their self-interaction is disrupted, aberrant structures are formed and production of infectious virus is interrupted (Person and Desai, 1998; Loveland et al., 2007). The portal complex may provide a nucleation site from which capsid formation extends, as happens in some bacteriophage (Murialdo and Becker, 1978; Moore and Prevelige, 2002), and also a means of

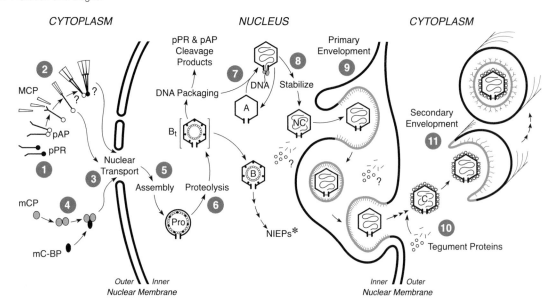

Figure I.13.3 Formation the CMV virion. Shown here is a schematic representation of assembly and maturational steps described in the text. (1) Oligomerization of pUL80.5 (pAP, empty circles) and pUL80a (pPR, filled circles); (2) interaction of pAP oligomers with MCP to form protomers of the capsid pentons and hexons; (3) nuclear translocation of pAP-MCP hetero-oligomers; (4) association of the two triplex proteins to form trimers that translocate into the nucleus; (5) interaction of the integral capsid protomers (pUL86/pUL80.5, pUL85/pUL46, and pUL104) to form the procapsid (Pro); (6 and 7) activation of pUL80a to cleave itself and pAP, resulting in transition B-capsids (B$_t$) that either mature no further and accumulate as stable B-capsids (B), or loose all remnants of the pUL80 scaffolding proteins and receive DNA to become nucleocapsids (NC) – A-capsids (A) are represented as products resulting from abortively, incompletely, or unstably packaged DNA (also see Fig. I.13.1); (8) stabilization of the nucleocapsid through interaction with initial tegument proteins; (9) primary envelopment of the nucleocapsid involving interactions of the capsid with pUL53, and with pUL50 located in the inner lamina of the nuclear membrane (amino portion of pUL50 indicated as black lines on nucleoplasmic side of membrane), resulting in transiently enveloped particles that become de-enveloped as they pass through the outer nuclear membrane (all membrane and envelope bilayers are represented here by single black or grey lines); (10) addition of more tegument proteins [e.g. empty circles and short lines on outside of cytoplasmic nucleocapsid (C)] in the virus assembly compartment adjacent to the nucleus; and (11) secondary envelopment of fully tegumented capsids at endosomal membranes to give enveloped virions within tubules and vesicles (luminal wavy lines and Y-shaped lines represent immature and mature glycoproteins, respectively). Asterisk in nucleus denotes progression of HCMV B-capsids through primary and secondary envelopment pathways, producing NIEPs (described in Irmiere and Gibson, 1983).

localizing capsid assembly and insuring a single portal per capsid. However, the formation of herpesvirus capsids in recombinant baculovirus expression systems containing no portal protein demonstrates that it is not absolutely required to form an icosahedral shell closely resembling those in herpesvirus-infected cells (Tatman et al., 1994; Thomsen et al., 1994; Newcomb et al., 2005; Perkins et al., 2008; Henson et al., 2009).

Procapsid formation is followed by proteolytic cleavage and elimination of the internal UL80 scaffolding proteins to make r3oom for the DNA genome. These steps are represented by the B$_t$-capsid in Figs. I.13.2 and I.13.3. Interaction of the scaffolding protein with the MCP is broken by a cleavage that removes the carboxyl end (6-kDa 'Tail') of pAP (Fig. I.13.3, step 6). The resulting fragments (AP + Tail) are eliminated from the nascent capsid, as evidenced by their absence

from A-capsids and virions (Irmiere and Gibson, 1983, 1985; Chan et al., 2002; Loveland et al., 2005). pAP is also phosphorylated at multiple sites and this modification weakens its self-interaction (Plafker et al., 1999; Casaday et al., 2004), suggesting a mechanism whereby phosphorylation promotes both dissociation of the cleaved pAP oligomers and their elimination from the capsid due to electrostatic repulsion by the incoming viral DNA (McClelland et al., 2002; Gibson, 2006).

The protease that cleaves pAP is encoded by a longer ORF (UL80a) that is in frame and 3′ co-terminal with ORF UL80.5 encoding pAP (Liu and Roizman, 1991; Welch et al., 1991a). The catalytic activity of the UL80a maturational serine protease (pPR, 74 kDa) resides in its 28-kDa amino end (Welch et al., 1991b, 1993; Liu and Roizman, 1992). This genetic organization functionally fuses the proteolytic domain to the pAP-scaffolding domain (via a short 'linker' sequence), and insures its delivery as a pAP mimic to the interior of the nascent capsid where its activity is required. Upon activation within B_t-capsids by an undefined mechanism, HCMV pPR undergoes self-cleavage at five sites (Brignole and Gibson, 2007). Two of these, the maturational (M) site and the release (R) site, are present in all herpesvirus pPR homologues (Welch et al., 1991b; Gibson et al., 1995). M-site cleavage removes the 6-kDa Tail sequence from both pPR and pAP, breaking their interaction with MCP. R-site cleavage releases the amino proteolytic domain of pPR, called assemblin, from the carboxyl scaffolding sequence.

In both HCMV and HSV the scaffolding portion of pPR is eliminated from the capsid, along with pAP, but the fate of their assemblin domains differs. That of HSV (VP21) remains in the capsid and is present in the mature virion (Gibson and Roizman, 1972; Spear and Roizman, 1972; Person et al., 1993), whereas that of HCMV is cleaved twice more – once at the internal (I) site and once at the cryptic (C) site (Baum et al., 1993; Burck et al., 1994), and is absent from virions (Chan et al., 2002; Loveland et al., 2005). Blocking these two cleavages reduces production of infectious virus by 90%, suggesting a role in facilitating the complete elimination of all pPR remnants from HCMV capsids to accommodate their longer DNA (Loveland et al., 2005). Additional information about the structure, self-cleavage, and substrate specificity of HCMV pPR and assemblin have been presented and reviewed recently (Brignole and Gibson, 2007; Fernandes et al., 2011; Gibson, 2012).

DNA packaging

HCMV has a linear double-stranded DNA that circularizes immediately after infection (Stinski, 1991; Garber et al., 1993; Roizman, 1996). During infection, HCMV DNA is replicated in the nuclei of infected cells. Replication of the covalently closed circular DNA is thought to occur via a rolling circle mechanism leading to head-to-tail linked genomes, called concatamers (Ben-Porat, 1983; McVoy and Adler, 1994). As an essential step in the formation of infectious virus, these concatamers have to be cleaved into single-genome lengths and packaged into preassembled capsids (Fig. I.13.3, step 7). The enzymes catalysing this process form a DNA cleavage/packaging complex called terminase.

Genome structure

The HCMV genome is a linear double-stranded DNA of approximately 235 kb, containing repetitive and unique components and having a complex class E

organization like that of herpes simplex virus (HSV-1 and HSV-2) (Roizman and Pellett, 2001). The genome has a unique long (U_L) and unique short (U_S) segment, separated by two inverted repeats. U_L is flanked by two copies of a *b* sequence and U_S is flanked by two copies of a *c* sequence (Stinski, 1991). In addition, the genome contains multiple *a* sequences that are located within the internal (IR) and terminal repeats (TR; Fig. I.13.4). U_S termini contain one or no copies of *a* adjacent to the *c* sequence. In contrast, U_L termini and the U/L junction contain a variable number of *a* sequence copies (Tamashiro et al., 1984; Mocarski et al., 1987).

During HCMV DNA replication, U_L and U_S can invert relative to each other, resulting in four genomic isomers in equimolar amounts (Stinski, 1991; Fig. I.13.4B). These inversions result from recombination of concatameric DNA or an event occurring during early rounds of replication (Zhang et al., 1994), but are not absolutely required for herpesvirus DNA replication since not all herpesvirus genomes isomerize (Roizman and Pellett, 2001). Recently it was shown by using an HCMV mutant with deleted internal repeats that efficient replication in fibroblasts is not dependent on isomerization (Sauer et al., 2010). Nevertheless, isomerization may confer an evolutionary advantage, perhaps more evident during replication in different cell types or in the host, such as maintaining or correcting genome length during replication (Cui et al., 2009).

Packaging sites

Since only concatamers with accessible U_S termini have been reported, *cis*-acting packaging and cleavage sequences must exist at the termini. The two *cis*-acting motifs *pac1* and *pac2* are located in the *a* sequence, and are necessary for packaging unit-length genomes (Stow et al., 1983; Deiss et al., 1986). In general, *pac1* motifs consist of 3- to 7-bp A- or T-rich sequences flanked on each side by a short *c* sequence (Tamashiro et al., 1984, Deiss et al., 1986). In contrast, *pac2* motifs are characterized by 5- to 10-bp A-rich sequences. The *pac1* and *pac2* motifs are on opposite sides of the DNA cleavage site (Deiss et al., 1986). There is evidence from guinea pig CMV (GPCMV) that *pac2* may be involved in mediating docking of the concatamer/terminase complex with capsids (McVoy et al., 2000), and evidence from HSV that *pac1* may be important for terminal cleavage (Tong and Stow, 2010). In the case of HCMV the situation is more complicated. While a typical *pac2* ATAAA motif was identified (Kemble and Mocarski, 1989), its considerable separation from the *pac1* motif is not consistent with that of other herpesviruses. McVoy and colleagues (2000) showed that guinea pig cytomegalovirus uses cryptic *pac2* motifs

A

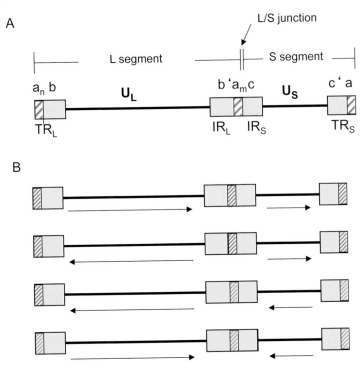

B

Figure I.13.4 Structure of the HCMV genome. (A) The L (long) and S (short) segments are indicated above the genome. The unique long (U_L) and the unique short (U_S) are flanked by terminal (T_R) and internal repeats (I_R). The a sequence is located within the I_R and T_R. While in the S segment only a single copy of the a is found, the L segments contains a variable number of a sequence copies (a_n, a_m). (B) Four genomic isoforms that are found in equimolar amounts.

without A-rich elements, and more recently Wang and McVoy (2011) demonstrated that HCMV uses a similar cryptic, cis-acting pac2 motif that together with pac1 is sufficient to enable packaging and cleavage of concatameric DNA. In addition, cleavage in HCMV is augmented by an element located within a sequence located 41 to 100 nucleotides from the predicted cleavage site (Wang and McVoy, 2011).

The players in DNA cleavage and packaging

At least three viral proteins participate in DNA cleavage and packaging.

The HCMV terminase

Terminases are enzymatic protein complexes that were first described in tailed dsDNA bacteriophage and have multiple activities, all of which are required for DNA packaging (Feiss and Becker, 1983; Black, 1988; Bhattacharyya and Rao, 1993). In these bacteriophage, DNA packaging involves (i) recognition of viral DNA by a specific protein, (ii) binding of that protein to

specific nucleotide sequences (packaging signals, e.g. pac1 and pac2), (iii) initial cleavage of the DNA to generate a concatamer end, (iv) targeting the DNA–protein complex to the nascent capsid, (v) interaction of the DNA–protein complex with the capsid portal complex, (vi) translocation of DNA into the capsid, requiring ATP hydrolysis by a subunit of the terminase complex, and (vii) completion of the packaging process by DNA cleavage (two-strand nicking; Fig. I.13.1). Some bacteriophage terminases (e.g. λ, T3 and T7) cut DNA at specific sites, yielding unit-length genomes (Becker and Murialdo, 1990; Fujisawa, 1994). It has been calculated for bacteriophage that DNA translocation requires one molecule of ATP per 2 to 2.5 base pairs (Guo et al., 1987; Shibata et al., 1987; Moffitt et al., 2009), and that the packaging motor generates a force of 57 pN, making it one of the strongest biological nanomotor (Smith et al., 2001; Guo and Lee, 2007). In most terminase complexes the larger of two subunits catalyses the ATP-dependent translocation step (Rao and Black, 1988; Becker and Murialdo, 1990; Morita et al., 1993; Bhattacharyya and Rao, 1994; Catalano et al., 1995; Hang et al., 2000). The HCMV terminase also has two subunits, pUL56 and pUL89 (Bogner, 1993, 2002;

Krosky et al., 1998; Underwood et al., 1998; Scheffczik et al., 2002), each with a different function. Both terminase subunits are essential for virus replication and are highly conserved throughout the herpesvirus family (Dunn et al., 2003; Yu et al., 2003).

Terminase subunit pUL56

The terminase subunit pUL56 is translocated into replication centres by its own nuclear localization signal, via an importin alpha dependent pathway (Giesen et al., 2000). In this compartment pUL56 drives DNA translocation into capsids by providing energy through its ATPase activity (Hwang and Bogner, 2002). DNA encapsidation is one of the most intriguing steps in virus formation. Even though pUL56 lacks a typical Walker box (glycine-rich motifs; G/AXXXXGKT/S), amino acids 709–716 (YNETFGKQ) constitute an ATP-binding site, with Gly714-Lys conserved (Walker et al., 1982; Scholz et al., 2003). The homologous murine CMV subunit, M56, has a similar sequence that is essential for virus replication (Wang and McVoy, 2008). The HCMV pUL56 subunit is also a DNA-binding protein, with specificity for the packaging motifs, pac1 and pac2 (Bogner et al., 1998; Scheffczik et al., 2002). Comparable results were reported from experiments done with the HSV-1 homologue of pUL56 (pUL28), interacting with a pac1 motif (Adelman et al., 2001). It has been suggested that HSV pUL28 serves as a sequence-specific DNA-binding protein during HSV DNA packaging (Adelman et al., 2001).

pUL56 is a dimer, formed by the association of two open ring-like structures positioned on top of each other and connected by a pronounced density on one side. Each ring is about 9 nm in diameter, with a central protein deficit or groove approximately 3.5 by 2.5 nm across that is sufficient for DNA-binding (Savva et al., 2004). This structure is typical of DNA metabolizing proteins.

Terminase subunit pUL89

The HCMV terminase subunit pUL89 and its HSV counterpart pUL15 (Baines et al., 1994) are homologues of one of the most highly conserved herpesvirus proteins (McGeoch et al., 2006). Consistent with its function in DNA packaging, pUL89 moves to viral replication centres of infected cells (Thoma et al., 2006), where it forms hetero-multimers with pUL56. The interaction is mediated by the pUL89 amino acid sequence G_{580}RDKALAVEQFISR-FNSGYIK$_{600}$ (Hwang and Bogner, 2002; Thoma et al., 2006), which is proposed to form an amphipathic alpha helix (Couvreux et al., 2010). The main function

of pUL89 appears to be cleaving the concatameric DNA into unit-length genomes (Scheffczik et al., 2002). Although it has not been verified that cleavage is by site-specific duplex nicking, this would seem necessary for unit-length genome packaging. HCMV pUL89, like pUL56, is a toroidal DNA-metabolizing protein. Its rings are approximately 8 nm in diameter and the central protein deficit or groove is approximately 2 nm across (Scheffczik et al., 2002). Therefore all structural requirements for a concatamer-cutting endonuclease are fulfilled.

The portal protein

A key component for DNA translocation into capsids is the portal protein (HCMV pUL104) (Dittmer et al., 2005). By oligomerization of twelve radially organized pUL104 monomers, a channel is formed (Holzenburg et al., 2009). Although the portal proteins of different DNA viruses have little or no sequence homology, the channels they form have a similar structure and function and are located at a single vertex of the capsid (Casjens and Huang, 1982; Kochan et al., 1984; Bazinet et al., 1988; Trus et al., 2004). Both bacteriophage and HSV portals are composed of 12 copies of a single protein, resulting in a 6- and 12-fold symmetrical ring (Guasch et al., 2002; Agirrezabala et al., 2005; Guo and Lee, 2007). A structure-based model for DNA translocation through the portal proposes a cooperation between the ATPase activity of terminase and a 'valve' or stepping mechanism provided by the portal (Jing et al., 2010).

Antivirals that target DNA packaging

For antivirals active against HCMV, in general, see Chapter II.19. There are several classes of compounds that specifically target HCMV DNA packaging. One is the benzimidazole D-ribonucleosides (Drach et al., 1992; Townsend et al., 1995; Krosky et al., 1998; Underwood et al., 1998), another is the dihydroquinazoline-acetic acid derivates (Lischka et al., 2010), and a third is the compound, BAY-384766 (Reefschlaeger et al., 2001).

The problematic instability of the initial benzimidazole ribonucleosides BDCRB and TCRB (Good et al., 1994) has been overcome in newer derivatives 2-bromo-4,5,6-trichloro-1-(2,3,5-tri-O-acetyl-β-D-ribofuranosyl) benzimidazole (BTCRB) and 2,4,5,6-tetrachloro-1-(2,3,5-tri-O-acetyl-β-D-ribofuranosyl) benzimidazole (Cl_4RB) (Fig. I.13.5), both of which inhibit DNA cleavage but may have different mechanisms of action (Hwang et al., 2007,

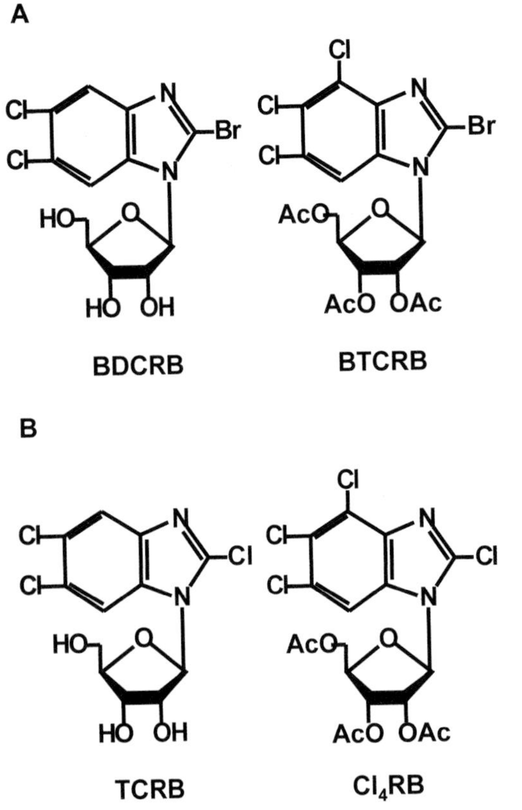

A

BDCRB

BTCRB

AIC246 (Letermovir)

B

TCRB

Cl$_4$RB

Figure **I.13.6** Chemical structure of AIC246 (Letermovir).

Figure **I.13.5** Structures of the benzimidazole D-ribonucleosides. (A) Parental BDCRB and the tetra-halogenated derivate BTCRB. (B) Parental TCRB and the tetra-halogenated derivate Cl$_4$RB.

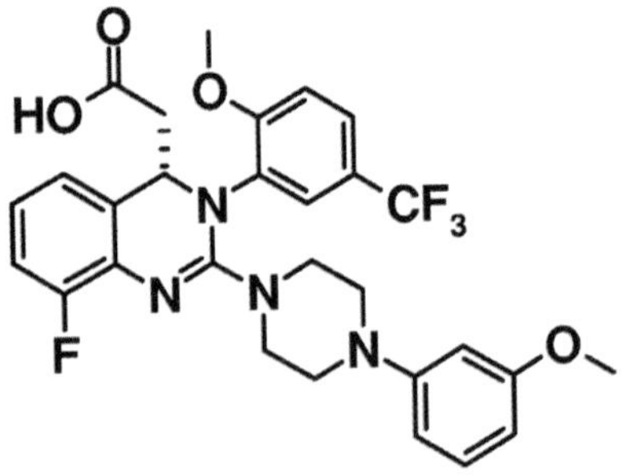

2009). Whereas Cl$_4$RB interferes with interaction of terminase pUL56 and the portal, BTCRB does not (Dittmer and Bogner, 2005).

More recently, Lischka and colleagues (2010) reported that a 3,4-dihydro-quinazoline-4-yl-acetic acid (AIC246, Letermovir, Fig. I.13.6) inhibits HCMV replication with efficacy in a mouse model. Genotypic mapping of AIC246-resistant mutants implicates terminase subunit pUL56 as the drug target (Goldner et al., 2011), but HCMV DNA packaging appears less affected by AIC246 than by the benzimidazole D-ribonucleosides. Its mechanism of action is not yet understood.

BAY-384766 (Tomeglovir) is a non-nucleosidic phenylenediamine sulfonamide that prevents the cleavage of high molecular weight viral DNA concatamers to unit-length genomes. Phase I trials were conducted and demonstrated a good tolerability in humans but its development as a drug was discontinued (Lischka and Zimmermann, 2008). It nevertheless represents a novel class of terminase inhibitors.

Capsid maturation and envelopment

Several proteins in addition to capsomere-bound SCP associate with the capsid prior to its envelopment (Fig. I.13.3, step 8). In HSV a pUL17–pUL25 complex is proposed to bind selectively to DNA-containing capsids (C-capsid selective complex; capsid vertex-specific complex), localizing to the vertices and providing stabilization to the nucleocapsid (McNab et al., 1998; Salmon et al., 1998; Conway et al., 2010). Considering the apparent instability of the HCMV capsid (Irmiere and Gibson, 1985), and the observation that HCMV B-capsids containing no DNA can be enveloped (e.g. NIEPs; Irmiere and Gibson, 1983), the functional equivalency of the HCMV homologues, pUL93 and pUL77 (Meissner et al., 2011) is of particular interest.

DNA-containing nucleocapsids leave the nucleus and acquire an envelope in their final maturation steps to becoming infectious virus. They are too large to pass through intact nuclear pores and share an egress mechanism that is a distinguishing feature of herpesvirus replication, and without a recognized equivalent in non-infected cells. Although variations and exceptions in specific aspects of the pathway have been reported among the different herpesviruses (e.g. Buser et al., 2007; Klupp et al., 2011; O'Connor and Shenk, 2011), common features are recognized. Most notably, all herpesviruses share the requirement for a pair of proteins that mediate budding through the inner nuclear membrane – primary envelopment (Fig. I.13.3, step 9). In CMV these are pUL50 and pUL53 (Muranyi et al., 2002), which interact with each other through a proline-rich domain at the amino end of pUL50 distal to its carboxyl membrane anchor (Rupp et al., 2007), and an alpha-helical domain at the amino end of pUL53 (Sam *et al.*, 2009), to form the nuclear envelopment complex (NEC). The HSV and PRV homologues are pUL31 and pUL34, respectively (Klupp *et al.*, 2000, 2001; Roller *et al.*, 2000; Reynolds *et al.*, 2001, 2002; Fuchs *et al.*, 2002).

pUL53 links the capsid to pUL50 in the inner leaflet of the nuclear membrane, but it is unsettled whether it associates with the capsid before interacting with pUL50 or whether it first forms a membrane-bound pUL50-pUL53 NEC, which then captures the capsid. In HSV and PRV evidence supports pUL31 (CMV pUL53 homologue) binding to capsids (Leelawong *et al.*, 2011; Yang and Baines, 2011). Cellular protein kinase C is recruited to help dissociate the nuclear lamins and promote formation of the capsid-NEC, in CMV-infected cells (Muranyi *et al.*, 2002). UL97 also participates in this critical process and recruits host peptidyl-prolyl *cis–trans* isomerase (PIN1) to the nuclear rim (Krosky *et al.*, 2003; Milbradt *et al.*, 2010). HSV encodes its own protein kinase (pUS3) for this purpose (Klupp *et al.*, 2001; Reynolds *et al.*, 2001, 2002).

Productive interaction of the capsid, NEC, and protein kinase results in the capsid budding through the inner leaflet of the nuclear membrane into the perinuclear space, and acquiring its primary envelope. During this process several tegument proteins and glycoproteins are thought to become transiently or permanently associated with the capsid (Eickmann *et al.*, 2006). Analyses of these particles recovered from HSV-infected cells showed they contain only some of the tegument and envelope proteins present in mature virions (Reynolds *et al.*, 2002; Naldinho-Souto *et al.*, 2006; Padula *et al.*, 2009).

In the generalized primary envelopment–de-envelopment–secondary envelopment model, immature particles in the perinuclear space are proposed to bind and bud through the outer leaflet of the nuclear membrane, shedding their primary envelope and entering the cytoplasm as non-enveloped particles – the membrane-associated pUL50 and glycoprotein homologues being lost with the primary envelope (Mettenleiter *et al.*, 2009; Johnson and Baines, 2011). In HCMV-infected cells the de-enveloped particles move into a large cytoplasmic inclusion adjacent to the nucleus called the assembly compartment (Sanchez *et al.*, 2000a,b). It is proposed that the bulk of HCMV tegument proteins are added to the capsid in this compartment (Fig. I.13.3, step 10), which is organized as nested cylinders composed of vesicles derived from Golgi bodies, the trans-Golgi network, and early endosomes (Das *et al.*, 2007; Das and Pellett, 2011). Virion tegument protein pUL71 appears to have a key role in this restructuring

and be intimately involved in final envelopment as evidenced by the severely reduced envelopment efficiency of mutations in that gene (Womack and Shenk, 2010; Schauflinger *et al.*, 2011; Meissner *et al.*, 2012). Consistent with the final virion tegument layer being added outside the nucleus, non-enveloped cytoplasmic HCMV capsids have long been recognized and distinguished by having a thick fibrillar coating that is absent from nuclear capsids, and appears to be retained as those particles bud through endosomal membranes to complete their secondary envelopment (e.g. Severi *et al.*, 1988; Tooze *et al.*, 1993; Silvia *et al.*, 2003; Gibson, 2008; Liu *et al.*, 2011). In the absence of tegument phosphoprotein pUL99, final envelopment fails and heavily tegumented capsids accumulate in the cytoplasm (Silvia *et al.*, 2003). The reader is referred to Chapter 13 in the first edition of this book for additional details and discussion of HCMV envelopment (Eickmann *et al.*, 2006; see also Chapter I.12).

Egress of mature virus particles from the cell following maturational secondary envelopment (Fig. I.13.2, step 11) (Eickmann *et al.*, 2006) results in the release of infectious virus, along with two enveloped but incomplete and non-infectious particles (NIEPs and DBs; see, Craighead *et al.*, 1972; Irmiere and Gibson, 1983; Severi *et al.*, 1988) of undetermined biological significance but noted potential for vaccine development (Gibson and Irmiere, 1984).

Summary and perspectives

Herpesviruses commit a large percentage of their genome to encoding functions that insure production and transmission of infectious progeny. Understanding the mechanisms involved in (i) overcoming the barriers to delivering their genome to the nucleus, (ii) packaging their concatenated DNA into preformed capsids filled with obstructing scaffolding material, and (iii) enabling the DNA-containing capsids to escape their nuclear assembly site will advance understanding of CMV replication and provide leads to inhibiting these essential processes with new antiviral strategies.

Acknowledgements

We thank members of our research groups whose work contributed to this review, express our appreciation to colleagues for helpful discussions and advice, and are grateful for support from the German Research Foundation (BO 1214/15–1 to E.B.), the Wilhelm Sander Foundation (no. 2004.031.2 to E.B.), and the United States Public Health Service (AI082246 and AI092374 to W.G)..

References

Adelman, K., Salmon, B., and Baines, J.D. (2001). Herpes simplex virus DNA packaging sequences adopt novel structures that are specifically recognized by a component of the cleavage and packaging machinery. Proc. Natl. Acad. Sci. U.S.A. *98*, 3086–3091.

Agirrezabala, X., Martín-Benito, J., Valle, M., González, J.M., Valencia, A., Valpuesta, J.M., and Carrascosa, J.L. (2005). Structure of the connector of bacteriophage T7 at 8 Å resolution: structural homologies of a basic component of a DNA translocating machinery. J. Mol. Biol. *347*, 895–902.

AuCoin, D.P., Smith, G.B., Meiering, C.D., and Mocarski, E.S. (2006). Betaherpesvirus-conserved cytomegalovirus tegument protein ppUL32 (pp150) controls cytoplasmic events during virion maturation. J. Virol. *80*, 8199–8210.

Baines, J., Poon, A., Rovnak, J., and Roizman, B. (1994). The herpes virus 1 UL15 gene encodes two proteins and is required for cleavage of genomic viral DNA. J. Virol. *68*, 8118–8124.

Baldick, C.J., Jr., and Shenk, T. (1996). Proteins associated with purified human cytomegalovirus particles. J. Virol. *70*, 6097–6105.

Baum, E.Z., Bebernitz, G.A., Hulmes, J.D., Muzithras, V.P., Jones, T.R., and Gluzman, Y. (1993). Expression and analysis of the human cytomegalovirus UL80-encoded protease: identification of autoproteolytic sites. J. Virol. *67*, 497–506.

Baxter, M.K., and Gibson, W. (1997). The putative human cytomegalovirus triplex proteins, minor capsid protein (mCP) and mCP-binding protein (mC-BP), form a heterotrimeric complex that localizes to the cell nucleus in the absence of other viral proteins. Paper presented at: 22nd International Herpesvirus Workshop (La Jolla, CA).

Baxter, M.K., and Gibson, W. (2001). Cytomegalovirus basic phosphoprotein (pUL32) binds to capsids in vitro through its amino one-third. J. Virol. *75*, 6865–6873.

Bazinet, C., Benbasat, J., King, J., Carazo, J.M., and Carrascosa, J.L. (1988). Purification and organization of the gene 1 portal protein required for phage P22 DNA packaging. Biochemistry *27*, 1849–1856.

Beaudet-Miller, M., Zhang, R., Durkin, J., Gibson, W., Kwong, A.D., and Hong, Z. (1996). Viral specific interaction between the human cytomegalovirus major capsid protein and the C-terminus of the precursor assembly protein. J. Virol. *70*, 8081–8088.

Becker, A., and Murialdo, H. (1990). Bacteriophage lambde DNA: The beginning of the end. J. Bacteriol. *172*, 2819–2824.

Ben-Porat, T. (1983). Replication of herpesvirus DNA. In The Herpesviruses, Roizman, R., ed. (Plenum Press, New York), pp. 81–106.

Benko, D.M., and Gibson, W. (1988). A major tegument phosphoprotein of human cytomegalovirus contains O-linked GlcNAc. Paper presented at: The UCLA symposia on molecular and cellular biology, Taos, NM, 24–30 January (Alan R. Liss, Inc., New York).

Benko, D.M., Haltiwanger, R.S., Hart, G.W., and Gibson, W. (1988). Virion basic phosphoprotein from human cytomegalovirus contains O-linked N-acetylglucosamine. Proc. Natl. Acad. Sci. U.S.A. *85*, 2573–2577.

Bhattacharyya, S.P., and Rao, V.B. (1993). A novel terminase activity associated with the DNA packaging protein gp17 of bacteriophage T4. Virology *196*, 34–44.

Bhattacharyya, S.P., and Rao, V.B. (1994). Structural analysis of DNA cleaved in vivo by bacteriophage T4 terminase. Gene *146*, 67–72.

Black, L.W. (1988). DNA packaging in dsDNA bacteriophages. In The Bacteriophages Vol 2, Calendar, R., ed. (Plenum Press, New York), pp. 321–373.

Bogner, E. (2002). Human cytomegalovirus terminase as a target for antiviral chemotherapy. Rev. Med. Virol. *12*, 115–127.

Bogner, E., Reschke, M., Reis, B., Richter, A., Mockenhaupt, T., and Radsak, K. (1993). Identification of the gene product encoded by ORF UL56 of th human cytomegalovirus. Virology *196*, 290–293.

Bogner, E., Radsak, K., and Stinski, M.F. (1998). The gene product of human cytomegalovirus open reading frame UL56 binds the pac motif and has specific nuclease activity. J. Virol. *72*, 2259–2264.

Booy, F.P., Trus, B.L., Newcomb, W.W., Brown, J.C., Conway, J.F., and Steven, A.C. (1994). Finding a needle in a haystack: detection of a small protein (the 12-kDa VP26) in a large complex (the 200-MDa capsid of herpes simplex virus). Proc. Natl. Acad. Sci. U.S.A. *91*, 5652–5656.

Borst, E.M., Mathys, S., Wagner, M., Muranyi, W., and Messerle, M. (2001). Genetic evidence of an essential role for cytomegalovirus small capsid protein in viral growth. J. Virol. *75*, 1450–1458.

Bottcher, S., Maresch, C., Granzow, H., Klupp, B.G., Teifke, J.P., and Mettenleiter, T.C. (2008). Mutagenesis of the active site cysteine in the ubiquitin-specific protease contained in the large tegument protein pUL36 of pseudorabies virus impairs viral replication in vitro and neuroinvasion in vivo. J. Virol. *82*, 6009–6016.

Bowman, B.R., Baker, M.L., Rixon, F.J., Chiu, W., and Quiocho, F.A. (2003). Structure of the herpesvirus major capsid protein. EMBO J. *22*, 757–765.

Brignole, E.J., and Gibson, W. (2007). Enzymatic activities of human cytomegalovirus maturational protease assemblin and its precursor (pPR, pUL80a) are comparable: maximal activity of pPR requires self-interaction through its scaffolding domain. J. Virol. *81*, 4091–4103.

Britt, W.J., and Boppana, S. (2004). Human cytomegalovirus virion proteins. Hum. Immunol. *65*, 395–402.

Brown, J.C., McVoy, M.A., and Homa, F.L. (2002). Packaging DNA into herpesvirus capsids. In Structure–Function Relationships of Human Pathogenic Viruses, Holzenburg, A., and Bogner, E., eds. (Kluwer Academic/Plenum Publishers, New York), pp. 111–153.

de Bruyn Kops, A., Uprichard, S.L., Chen, M., and Knipe, D.M. (1998). Comparison of the intranuclear distributions of herpes simplex virus proteins involved in various viral functions. Virology *252*, 162–178.

Burck, P.J., Berg, D.H., Luk, T.P., Sassmannshausen, L.M., Wakulchik, M., Smith, D.P., Hsiung, H.M., Becker, G.W., Gibson, W., and Villarreal, E.C. (1994). Human cytomegalovirus maturational proteinase: expression in *Escherichia coli*, purification, and enzymatic characterization by using peptide substrate mimics of natural cleavage sites. J. Virol. *68*, 2937–2946.

Buser, C., Walther, P., Mertens, T., and Michel, D. (2007). Cytomegalovirus primary envelopment occurs at large infoldings of the inner nuclear membrane. J. Virol. *81*, 3042–3048.

Casaday, R.J., Bailey, J.R., Kalb, S.R., Brignole, E.J., Loveland, A.N., Cotter, R.J., and Gibson, W. (2004). Assembly protein precursor (pUL80.5 Homolog) of simian cytomegalovirus is phosphorylated at a glycogen synthase kinase 3 site and its downstream 'priming' site: phosphorylation affects interactions of protein with itself and with major capsid protein. J. Virol. *78*, 13501–13511.

Casjens, S., and Huang, W.M. (1982). Initiation of sequential packaging of bacteriophage-P22 DNA. J. Mol. Biol. *157*, 287–298.

Catalano, C.E., Cue, D., and Feiss, M. (1995). Virus-DNA packaging – the strategy used by phage-lambda. Mol. Microbiol. *16*, 1075–1086.

Chan, C.K., Brignole, E.J., and Gibson, W. (2002). Cytomegalovirus assemblin (pUL80a): cleavage at internal site not essential for virus growth; proteinase absent from virions. J. Virol. *76*, 8667–8674.

Chen, D.H., Jiang, H., Lee, M., Liu, F., and Zhou, Z.H. (1999). Three-dimensional visualization of tegument/capsid interactions in the intact human cytomegalovirus. Virology *260*, 10–16.

Conway, J.F., and Homa, F.L. (2011). Nucleocapsid structure, assembly and DNA packaging of herpes simplex virus. In Alphaheresviruses, Weller, S.K., ed. (Caister Academic Press, Norfolk, UK), pp. 175–194.

Conway, J.F., Cockrell, S.K., Copeland, A.M., Newcomb, W.W., Brown, J.C., and Homa, F.L. (2010). Labeling and localization of the herpes simplex virus capsid protein UL25 and its interaction with the two triplexes closest to the penton. J. Mol. Biol. *397*, 575–586.

Couvreux, A., Hantz, S., Marquant, R., Champier, G., Alain, S., Morellet, N., and Bouaziz, S. (2010). Insight into the structure of the pUL89 C-terminal domain of the human cytomegalovirus terminase complex. Proteins *78*, 1520–1530.

Craighead, J.E., Kanich, R.E., and Almeida, J.D. (1972). Nonviral microbodies with viral antigenicity produced in cytomegalovirus-infected cells. J. Virol. *10*, 766–775.

Cui, X., McGregor, A., Schleiss, M.R., and McVoy, M.A. (2009). The impact of genome length on replication and genome stability of the herpesvirus guinea pig cytomegalovirus. Virology *386*, 132–138.

Das, S., and Pellett, P.E. (2011). Spatial relationships between markers for secretory and endosomal machinery in human cytomegalovirus-infected cells versus those in uninfected cells. J. Virol. *85*, 5864–5879.

Das, S., Vasanji, A., and Pellett, P.E. (2007). Three-dimensional structure of the human cytomegalovirus cytoplasmic virion assembly complex includes a reoriented secretory apparatus. J. Virol. *81*, 11861–11869.

Deiss, L.P., Chou, J., and Frenkel, N. (1986). Functional domains within the a sequence involved in the cleavage-packaging of herpes simplex virus DNA. J. Virol. *59*, 605–618.

Desai, P., and Person, S. (1996). Molecular interactions between the HSV-1 capsid proteins as measured by the yeast two-hybrid system. Virology *220*, 516–521.

Desai, P., and Person, S. (1999). Second site mutations in the N-terminus of the major capsid protein (VP5) overcome a block at the maturation cleavage site of the capsid scaffold proteins of herpes simplex virus type 1. Virology *261*, 357–366.

Desai, P., DeLuca, N.A., and Person, S. (1998). Herpes simplex virus type 1 VP26 is not essential for replication in cell culture but influences production of infectious virus in the nervous system of infected mice. Virology *247*, 115–124.

Dittmer, A., and Bogner, E. (2005). Analysis of the quaternary structure of the putative HCMV portal protein PUL104. Biochemistry *44*, 759–765.

Dittmer, A., Drach, J.C., Townsend, L.B., Fischer, A., and Bogner, E. (2005). Interaction of the putative human cytomegalovirus portal protein pUL104 with the large terminase subunit pUL56 and its inhibition by benzimidazole-D-ribonucleosides. J. Virol. *79*, 14660–14667.

Drach, J.C., Townsend, L.B., Nassiri, M.R., Turk, S.R., Coleman, L.A., Devivar, R.V., Genzlinger, G., Kreske, E.D., Renau, T.E., Westerman, A.C., *et al.* (1992). Benzimidazole ribonucleosides: a new class of antivirals with potent and selective activity against human cytomegalovirus. Antiviral Res. *17(Suppl. 1)*, 49.

Dunn, W., Chou, C., Li, H., Hai, R., Patterson, D., Stolc, V., Zhu, H., and Liu, F. (2003). Functional profiling of a human cytomegalovirus genome. Proc. Natl. Acad. Sci. U.S.A. *100*, 14223–14228.

Eickmann, M., Gicklhorn, D., and Radsak, K. (2006). Glycoprotein trafficking in virion morphogenesis. In Cytomegaloviruses Molecular Biology and Immunology, Reddehase, M.J., ed. (Caister Academic Press, Norfolk, UK), pp. 245–264.

Feiss, M., and Becker, A. (1983). DNA packaging and cutting. In Lambda II, Hendrix, R.W., Roberts, J.W., Stahl, F.W., and Weisberg, R.A., eds. (Cold Spring Harbor Press, Cold Spring Harbor, NY), pp. 305–330.

Fernandes, S.M., Brignole, E.J., and Gibson, W. (2011). Cytomegalovirus capsid protease: biological subsrtes cleaved more efficiently by full-length enzyme (pUL80) than by catalytic domain (assemblin). J. Virol. *85*, 3526–3534.

Fuchs, W., Klupp, B.G., Granzow, H., Osterrieder, N., and Mettenleiter, T.C. (2002). The interacting UL31 and UL34 gene products of pseudorabies virus are involved in egress from the host cell nucleus and represent components of primary enveloped but not mature virions. J. Virol. *76*, 364–378.

Fujisawa, H., and Hearing, P. (1994). Structure, function and specificty of the DNA packaging signals in double-stranded DNA viruses. Sem. Virol. *5*, 5–13.

Gallina, A., Simoncini, L., Garbelli, S., Percivalle, E., Pedrali-Noy, G., Lee, K.S., Erikson, R.L., Plachter, B., Gerna, G., and Milanesi, G. (1999). Polo-like kinase 1 as a target for human cytomegalovirus pp65 lower matrix protein. J. Virol. *73*, 1468–1478.

Garber, D.A., Beverley, S.M., and Coen, D.M. (1993). Demonstration of circularization of herpes simplex virus DNA following infection using pulsed field gel electrophoresis. Virology *197*, 459–462.

Gibson, W. (1981). Structural and nonstructural proteins of strain Colburn cytomegalovirus. Virology *111*, 516–537.

Gibson, W. (1996). Structure and assembly of the virion. Intervirology *39*, 389–400.

Gibson, W. (2006). Assembly and maturation of the capsid. In Cytomegaloviruses: Molecular Biology and Immunology, Reddehase, M.J., ed. (Caister Academic Press, Norfolk, UK), pp. 231–244.

Gibson, W. (2008). Structure and function of the cytomegalovirus virion. Curr. Top. Microbiol. Immunol. *325*, 187–204.

Gibson, W. (2012). Cytomegalovirus assemblin and precursor. In Handbook of Proteolytic Enzymes, Barrett,

A., Rawlings, N., and Woessner, J., eds. (Elsevier, New York).

Gibson, W., and Irmiere, A. (1984). Selection of particles and proteins for use as human cytomegalovirus subunit vaccines. Birth Defects 20, 305–324.

Gibson, W., and Roizman, B. (1972). Proteins specified by herpes simplex virus. VIII. characterization and composition of multiple capsid forms of subtypes 1 and 2. J. Virol. 10, 1044–1052.

Gibson, W., Clopper, K.S., Britt, W.J., and Baxter, M.K. (1996). Human cytomegalovirus smallest capsid protein identified as product of short open reading frame located between HCMV UL48 and UL49. J. Virol. 70, 5680–5683.

Gibson, W., VanBreemen, R., Fields, A., LaFemina, R., and Irmiere, A. (1984). D,L-alpha-difluoromethylornithine inhibits human cytomegalovirus replication. J. Virol. 50, 145–154.

Gibson, W., Welch, A.R., and Hall, M.R. (1995). Assemblin, a herpes virus serine maturational proteinase and new molecular target for antivirals. Perspect. Drug. Discov. Design 2, 413–426.

Giesen, K., Radsak, K., and Bogner, E. (2000). The potential terminase subunit of human cytomegalovirus, pUL56, is translocated into the nucleus by its own nuclear localization signal and interacts with importin alpha. J. Gen. Virol. 81, 2231–2244.

Goldner, T., Hewlett, G., Ettischer, N., Ruebsamen-Schaeff, H., Zimmermann, H., and Lischka, P. (2011). The novel anti-cytomegaloviruscCompound AIC246 inhibits HCMV replication through a specific antiviral mechanism that involves the viral terminase. J. Virol. 85, 10884–10893.

Good, S.S., Owens, B.S., Townsend, L.B., and Drach, J.C. (1994). The disposition in rats and monkey of 2-Bromo-5,6-dichloro-1-(β-D-ribofuranosyl) -benzimidazole (BDCRB) and its 2,5,6-trichloro congener (TCRB). Antiviral Res. 23, 103.

Greis, K.D., Gibson, W., and Hart, G.W. (1994). Site-specific glycosylation of the human cytomegalovirus tegument basic phosphoprotein (UL32) at serine 921 and serine 952. J. Virol. 68, 8339–8349.

Guasch, A., Pous, J., Ibarra, B., Gomis-Ruth, F.X., Valpuesta, J.M., Sousa, N., Carrascosa, J.L., and Coll, M. (2002). Detailed architecture of a DNA translocating machine: the high-resolution structure of the bacteriophage phi29 connector particle. J. Mol. Biol. 315, 663–676.

Guo, P., and Lee, T.J. (2007). Viral nanomotors for packaging of dsDNA and dsRNA. Mol. Microbiol. 64, 886–903.

Guo, P., Peterson, C., and Anderson, D. (1987). Prohead and DNA-gp3-dependent ATPase activity of the DNA packaging protein gp16 of bacteriophage phi 29. J. Mol. Biol. 197, 229–236.

Hang, J.Q., Tack, B.F., and Feiss, M. (2000). ATPase center of bacteriophage lambda terminase involved in post-cleavage stages of DNA packaging: identification of ATP-interactive amino acids. J. Mol. Biol. 302, 777–795.

Harmon, M.-E., and Gibson, W. (1996). High molecular weight virion protein of human cytomegalovirus forms complex with product of adjacent open reading frame. Proc. Am. Soc. Virol. Abstract. W35–4, 144.

Henson, B.W., Perkins, E.M., Cothran, J.E., and Desai, P. (2009). Self-assembly of Epstein–Barr virus capsids. J. Virol. 83, 3877–3890.

Holzenburg, A., and Bogner, E. (2002). From concatemeric DNA into unit-length genomes – a miracle or clever

genes? In Structure-Function Relationships of Human Pathogenic Viruses, Holzenburg, A., and Bogner, E., eds. (Kluwer Academic/Plenum Publishers, New York), pp. 155–173.

Holzenburg, A., Dittmer, A., and Bogner, E. (2009). Assembly of monomeric human cytomegalovirus pUL104 into portal structures. J. Gen. Virol. 90, 2381–2385.

Hwang, J.S., and Bogner, E. (2002). ATPase activity of the terminase subunit pUL56 of human cytomegalovirus. J. Biol. Chem. 277, 6943–6948.

Hwang, J.-S., Kregler, O., Schilf, R., Bannert, N., Drach, J.C., Townsend, L.B., and Bogner, E. (2007). Identification of acetylated, tetrahalogenated benzimidazole D-ribonucleotides with enhanced activity against human cytomegalovirus. J. Virol. 81, 11604–11611.

Hwang, J.-S., Schilf, R., Drach, J.C., Townsend, L.B., and Bogner, E. (2009). Susceptibilities of HCMV clinical isolates and other herpesviruses to new acetylated, tetrahalogenated benzimidazole D-ribonucleosides. Antimicrob. Agents. Chemother. 53, 5095–5101.

Irmiere, A., and Gibson, W. (1983). Isolation and characterization of a noninfectious virion-like particle released from cells infected with human strains of cytomegalovirus. Virology 130, 118–133.

Irmiere, A., and Gibson, W. (1985). Isolation of human cytomegalovirus intranuclear capsids, characterization of their protein constituents, and demonstration that the B-capsid assembly protein is also abundant in noninfectious enveloped particles. J. Virol. 56, 277–283.

Jarosinski, K., Kattenhorn, L.M., Kaufer, B., Ploegh, H.L., and Osterrieder, N. (2007). A herperpesvirus ubiquitin-specific protease is critical for efficient T-cell lymphoma formation. Proc. Natl. Acad. Sci. U.S.A. 104, 20025–20030.

Jing, P., Haque, F., Shu, D., Montemagno, C., and Guo, P. (2010). One-way traffic of a viral motor channel for double-stranded DNA translocation. Nano Lett. 10, 3620–3627.

Johnson, D.C., and Baines, J.D. (2011). Herpesviruses remodel host membranes for virus egress. Nat. Rev. Microbiol. 9, 382–394.

Kalejta, R.F. (2008). Fuctions of human cytomegalovirus tegument proteins prior to immediate-early gene expression. In Human Cytomegalovirus, Shenk, T., and Stinski, M.F., eds. (Springer, Heidelberg, Germany), pp. 101–115.

Kamil, J.P., and Coen, D.M. (2007). Human cytomegalovirus protein kinase UL97 forms a complex with the tegument phosphoprotein pp65. J. Virol. 81, 10659–10668.

Kattenhorn, L.M., Korbel, G.A., Kessler, B.M., Spooner, E., and Ploegh, H.L. (2005). A deubiquitinating enzyme encoded by HSV-1 belongs to a family of cysteine proteases that is conserved across the family Herpesviridae. Mol. Cell. 19, 547–557.

Kemble, G.W., and Mocarski, E.S. (1989). A host cell protein binds to a highly conserved sequence element (pac-2) within the cytomegalovirus a sequence. J. Virol. 63, 4715–4728.

Kim, E.T., Oh, S.E., Lee, Y.O., Gibson, W., and Ahn, J.H. (2009). Cleavage specificity of the UL48 deubiquitinating protease activity of human cytomegalovirus and the growth of an active-site mutant virus in cultured cells. J. Virol. 83, 12046–12056.

Klupp, B.G., Granzow, H., and Mettenleiter, T.C. (2000). Primary envelopment of pseudorabies virus at the nuclear membrane requires the UL34 gene product. 25th International Herpesvirus Workshop, Abstract 7.04.

Klupp, B.G., Granzow, H., and Mettenleiter, T.C. (2001). Effect of the pseudorabies virus US3 protein on nuclear membrane localization of the UL34 protein and virus egress from the nucleus. J. Gen. Virol. 82, 2363–2371.

Klupp, B.G., Granzow, H., and Mettenleiter, T.C. (2011). Nuclear envelope breakdown can substitute for primary envelopment-mediated nuclear egress of herpesviruses. J. Virol. 85, 8285–8292.

Kochan, J., Carrascosa, J.L., and Murialdo, H. (1984). Bacteriophage lambda preconnectors. Purification and structure. J. Mol. Biol. 174, 433–447.

Krosky, P.M., Underwood, M.R., Turk, S.R., Feng, K.W., Jain, R.K., Ptak, R.G., Westerman, A.C., Biron, K.K., Townsend, L.B., and Drach, J.C. (1998). Resistance of human cytomegalovirus to benzimidazole ribonucleosides maps to two open reading frames: UL89 and UL56. J. Virol. 72, 4721–4728.

Krosky, P.M., Baek, M.C., and Coen, D.M. (2003). The human cytomegalovirus UL97 protein kinase, an antiviral drug target, is required at the stage of nuclear egress. J. Virol. 77, 905–914.

Lee, J.I., Sollars, P.J., Baver, S.B., Pickard, G.E., Leelawong, M., and Smith, G.A. (2009). A herpesvirus encoded deubiquitinase is a novel neuroinvasive determinant. PLoS Pathog. 5, e1000387.

Lee, J.Y., Irmiere, A., and Gibson, W. (1988). Primate cytomegalovirus assembly: evidence that DNA packaging occurs subsequent to B capsid assembly. Virology 167, 87–96.

Leelawong, M., Guo, D., and Smith, G.A. (2011). A physical link between the Pseudorabies Virus and the nuclear egress complex. J. Virol. 85, 11675–11684.

Lischka, P., and Zimmermann, H. (2008). Antiviral strategies to combat cytomegalovirus infections in transplant recipients. Curr. Opin. Pharmacol. 8, 541–548.

Lischka, P., Hewlett, G., Wunberg, T., Baumeister, J., Paulsen, D., Goldner, T., Ruebsamen-Schaeff, H., and Zimmermann, H. (2010). In vitro and in vivo activities of the novel anticytomegalovirus compound AIC246. Antimicrob. Agents Chemother. 54, 1290–1297.

Liu, F., and Roizman, B. (1991). The promoter, transcriptional unit, and coding sequence of herpes simplex virus 1 family 35 proteins are contained within and in frame with the UL26 open reading frame. J. Virol. 65, 206–212.

Liu, F., and Roizman, B. (1992). Differentiation of multiple domains in the herpes simplex virus 1 protease encoded by the UL26 gene. Proc. Natl. Acad. Sci. U.S.A. 89, 2076–2080.

Liu, F., and Zhou, Z.H. (2007). Comparative virion structures of human herpesviruses. In Human Herpesviruses: Biology, Therapy, and Immunoprophylaxis, Arvin, A., Campadelli-Fiume, G., Mocarski, E., Moore, P.S., Roizman, B., Whitley, R., and Yamanishi, K., eds. (Cambridge University Press, Cambridge, UK), pp. 27–43.

Liu, S.T., Sharon-Friling, R., Ivanova, P., Milne, S.B., Myers, D.S., Rabinowitz, J.D., Brown, H.A., and Shenk, T. (2011). Synaptic vesicle-like lipidome of human cytomegalovirus virions reveals a role for SNARE machinery in virion egress. Proc. Natl. Acad. Sci. U.S.A. 108, 12869–12874.

Loveland, A.N., Chan, C.K., Brignole, E.J., and Gibson, W. (2005). Cleavage of human cytomegalovirus protease pUL80a at internal and cryptic sites is not essential but enhances infectivity. J. Virol. 79, 12961–12968.

Loveland, A.N., Nguyen, N.L., Brignole, E.J., and Gibson, W. (2007). The amino-conserved domain of human cytomegalovirus UL80a proteins is required for key interactions during early stages of capsid formation and virus production. J. Virol. 81, 620–628.

McClelland, D.A., Aitken, J.D., Bhella, D., McNab, D., Mitchell, J., Kelly, S.M., Price, N.C., and Rixon, F.J. (2002). pH reduction as a trigger for dissociation of herpes simplex virus type 1 scaffolds. J. Virol. 76, 7407–7417.

McGeoch, D.J., Rixon, F.J., and Davison, A.J. (2006). Topics in herpesvirus genomics and evolution. Virus Res. 117, 90–104.

Mach, M. (2006). Antibody-mediated neutralization of infectivity, In Cytomegaloviruses Molecular Biology and Immunology, Reddehase, M.J., ed. (Caister Academic Press, Norfolk, UK), pp. 265–284.

McNab, A.R., Desai, P., Person, S., Roof, L.L., Thomsen, D.R., Newcomb, W.W., Brown, J.C., and Homa, F.L. (1998). The product of the herpes simplex virus type 1 UL25 gene is required for encapsidation but not for cleavage of replicated viral DNA. J. Virol. 72, 1060–1070.

McVoy, M.A., and Adler, S.P. (1994). Human cytomegalovirus DNA replicates after early circularization by concatemer formation, and inversion occurs within the concatemer. J. Virol. 68, 1040–1051.

McVoy, M.A., Nixon, D.E., Hur, J.K., and Adler, S.P. (2000). The ends on herpesvirus DNA replicative concatemers contain pac2 cis cleavage/packaging elements and their formation is controlled by terminal cis sequences. J. Virol. 74, 1587–1592.

Mar, E.-C., Patel, P.C., and Huang, E.-S. (1981). Human cytomegalovirus-associated DNA polymerase and protein kinase activities. J. Gen. Virol. 57, 149–156.

Margulies, B.J., and Gibson, W. (2007). The chemokine receptor homologue encoded by US27 of human cytomegalovirus is heavily glycosylated and is present in infected human foreskin fibroblasts and enveloped virus particles. Virus Res. 123, 57–71.

Margulies, B.J., Browne, H., and Gibson, W. (1996). Identification of the human cytomegalovirus G protein-coupled receptor homologue encoded by UL33 in infected cells and enveloped virus particles. Virology 225, 111–126.

Meissner, C.S., Koppen-Rung, P., Dittmer, A., Lapp, S., and Bogner, E. (2011). A 'coiled coil' motif is important for oligomerization and DNA binding properties of human cytomegalovirus protein UL77. PLoS One 6, e25115.

Meissner, C.S., Suffner, S., Schauflinger, M., von Einem, J., and Bogner, E. (2012). A leucine zipper motif of a tegument protein triggers final envelopment of human cytomegalovirus. J. Virol. 86, 3370–3382.

Mettenleiter, T.C. (2004). Budding events in herpesvirus morphogenesis. Virus Res. 106, 167–180.

Mettenleiter, T.C., Klupp, B.G., and Granzow, H. (2009). Herpesvirus assembly: an update. Virus Res. 143, 222–234.

Milbradt, J., Webel, R., Auerochs, S., Sticht, H., and Marschall, M. (2010). Novel mode of phosphorylation-triggered reorganization of the nuclear lamina during nuclear egress of human cytomegalovirus. J. Biol. Chem. 285, 13979–13989.

Mocarski, E.S., Liu, A.C., and Spaete, R.R. (1987). Structure and variability of the a sequence in the genome of human cytomegalovirus (Towne strain). J. Gen. Virol. 68, 2223–2230.

Moffitt, J.R., Chemla, Y.R., Aathavan, K., Grimes, S., Jardine, P.J., Anderson, D.L., and Bustamante, C. (2009). Intersubunit coordination in a homomeric ring ATPase. Nature 457, 446–450.

Moore, S.D., and Prevelige, P.E., Jr. (2002). Bacteriophage p22 portal vertex formation in vivo. J. Mol. Biol. 315, 975–994.

Morita, M., Tasaka, M., and Fujisawa, H. (1993). DNA packaging ATPase of bacteriophage T3. Virology 193, 748–752.

Muranyi, W., Haas, J., Wagner, M., Krohne, G., and Koszinowski, U.H. (2002). Cytomegalovirus recruitment of cellular kinases to dissolve the nuclear lamina. Science 297, 854–857.

Murialdo, H., and Becker, A. (1978). Head morphogenesis of complex double-stranded deoxyribonucleic acid bacteriophages. Microbiol. Rev. 42, 529–576.

Naldinho-Souto, R., Browne, H., and Minson, T. (2006). Herpes simplex virus tegument protein VP16 is a component of primary enveloped virions. J. Virol. 80, 2582–2584.

Newcomb, W.W., Homa, F.L., and Brown, J.C. (2005). Involvement of the portal at an early step in herpes simplex virus capsid assembly. J. Virol. 79, 10540–10546.

Newcomb, W.W., Cockrell, S.K., Homa, F.L., and Brown, J.C. (2009). Polarized DNA ejection from the herpesvirus capsid. J. Mol. Biol. 392, 885–894.

Nguyen, N.N., Loveland, A.N., and Gibson, W. (2008). Nuclear localization sequences in cytomegalovirus capsid assembly proteins (UL80 proteins) are required for virus production: Inactivating NLS1, NLA2, or both affects replication to striningly different extents. J. Virol. 82, 5381–5389.

O'Connor, C.M., and Shenk, T. (2011). Human cytomegalovirus pUS27 G protein-coupled receptor homologue is required for efficient spread by the extracellular route but not for direct cell-to-cell spread. J. Virol. 85, 3700–3707.

Padula, M.E., Sydnor, M.L., and Wilson, D.W. (2009). Isolation and preliminary characterization of herpes simplex virus 1 primary enveloped virions from the perinuclear space. J. Virol. 83, 4757–4765.

Pelletier, A., Do, F., Brisebois, J.J., Lagace, L., and Cordingley, M.G. (1997). Self-association of herpes simplex virus type 1 ICP35 is via coiled coil interactions and promotes stable interaction with the major capsid protein. J. Virol. 71, 5197–5208.

Perkins, E.M., Anacker, D., Davis, A., Sankar, V., Ambinder, R.F., and Desai, P. (2008). Small capsid protein pORF65 is essential for assembly of Kaposi's sarcoma-associated herpesvirus capsids. J. Virol. 82, 7201–7211.

Person, S., and Desai, P. (1998). Capsids are formed in a mutant virus blocked at the maturation site of the UL26 and UL26.5 open reading frames of herpes simplex virus type 1 but are not formed in a null mutant of UL38 (VP19C). Virology 242, 193–203.

Person, S., Laquerre, S., Desai, P., and Hempel, J. (1993). Herpes simplex virus type 1 capsid protein VP21 originates within the UL26 open reading frame. J. Gen. Virol. 74, 2269–2273.

Plafker, S.M., and Gibson, W. (1998). Cytomegalovirus assembly protein precursor and proteinase precursor contain two nuclear localization signals that mediate their

own nuclear translocation and that of the major capsid protein. J. Virol. 72, 7722–7732.

Plafker, S.M., Woods, A.S., and Gibson, W. (1999). Phosphorylation of simian cytomegalovirus assembly protein precursor (pAPNG.5) and proteinase precursor (pAPNG1): multiple attachment sites identified, including two adjacent serines in a casein kinase II consensus sequence. J. Virol. 73, 9053–9062.

Rao, V.B., and Black, L.W. (1988). Cloning, overexpression and purification of the terminase proteins gp16 and gp17 of bacteriophage T4. Construction of a defined in-vitro DNA packaging system using purified terminase proteins. J. Mol. Biol. 200, 475–488.

Reefschlaeger, J., Bender, W., Hallenberger, S., Weber, O., Eckenberg, P., Goldmann, S., Haerter, M., Buerger, I., Trappe, J., Herrington, J.A., et al. (2001). Novel non-nucleoside inhibitors of cytomegaloviruses (BAY 38-4766): in vitro and in vivo antiviral activity and mechanism of action. J. Antimicrob. Chemother. 48, 757–767.

Reynolds, A.E., Ryckman, B.J., Baines, J.D., Zhou, Y., Liang, L., and Roller, R.J. (2001). U(L)31 and U(L)34 proteins of herpes simplex virus type 1 form a complex that accumulates at the nuclear rim and is required for envelopment of nucleocapsids. J. Virol. 75, 8803–8817.

Reynolds, A.E., Wills, E.G., Roller, R.J., Ryckman, B.J., and Baines, J.D. (2002). Ultrastructural localization of the herpes simplex virus type 1 UL31, UL34, and US3 proteins suggests specific roles in primary envelopment and egress of nucleocapsids. J. Virol. 76, 8939–8952.

Roby, C., and Gibson, W. (1986). Characterization of phosphoproteins and protein kinase activity of virions, noninfectious enveloped particles, and dense bodies of human cytomegalovirus. J. Virol. 59, 714–727.

Roizman, B. (1996). Herpes simplex viruses and their replication. In Fields Virology, Fields, B.N., Knipe, D.M., and Howley, P.M., eds. (Lippincott-Raven, Philadelphia, PA), pp. 2231–2295.

Roizman, B., and Pellett, P.E. (2001). The family herpesviridae: A brief introduction, Vol 2, 4th edn (Lippincott Williams & Wilkins, Baltimore, MD).

Roller, R.J., Zhou, Y., Schnetzer, R., Ferguson, J., and DeSalvo, D. (2000). Herpes simplex virus type 1 U(L)34 gene product is required for viral envelopment. J. Virol. 74, 117–129.

Rupp, B., Ruzsics, Z., Buser, C., Adler, B., Walther, P., and Koszinowski, U.H. (2007). Random screening for dominant-negative mutants of the cytomegalovirus nuclear egress protein M50. J. Virol. 81, 5508–5517.

Salmon, B., Cunningham, C., Davison, A.J., Harris, W.J., and Baines, J.D. (1998). The herpes simplex virus type 1 U(L)17 gene encodes virion tegument proteins that are required for cleavage and packaging of viral DNA. J. Virol. 72, 3779–3788.

Sam, M.D., Evans, B.T., Coen, D.M., and Hogle, J.M. (2009). Biochemical, biophysical, and mutational analyses of subunit interactions of the human cytomegalovirus nuclear egress complex. J. Virol. 83, 2996–3006.

Sanchez, V., Greis, K.D., Sztul, E., and Britt, W.J. (2000a). Accumulation of virion tegument and envelope proteins in a stable cytoplasmic compartment during human cytomegalovirus replication: characterization of a potential site of virus assembly. J. Virol. 74, 975–986.

Sanchez, V., Sztul, E., and Britt, W.J. (2000b). Human cytomegalovirus pp28 (UL99) localizes to a cytoplasmic compartment which overlaps the endoplasmic reticulum-golgi-intermediate compartment. J. Virol. 74, 3842–3851.

Sauer, A., Wang, J.B., Hahn, G., and McVoy, M.A. (2010). A human cytomegalovirus deleted of internal repeats replicates with near wild type efficiency but fails to undergo genome isomerization. Virology 401, 90–95.

Savva, C.G., Holzenburg, A., and Bogner, E. (2004). Insights into the structure of human cytomegalovirus large terminase subunit pUL56. FEBS Lett. 563, 135–140.

Schauflinger, M., Fischer, D., Schreiber, A., Chevillotte, M., Walther, P., Mertens, T., and von Einem, J. (2011). The tegument protein UL71 of human cytomegalovirus is involved in late envelopment and affects multivesicular bodies. J. Virol. 85, 3821–3832.

Scheffczik, H., Savva, C.G., Holzenburg, A., Kolesnikova, L., and Bogner, E. (2002). The terminase subunits pUL56 and pUL89 of human cytomegalovirus are DNA-metabolizing proteins with toroidal structure. Nucleic Acids Res. 30, 1695–1703.

Schlieker, C., Korbel, G.A., Kattenhorn, L.M., and Ploegh, H.L. (2005). A deubiquitinating activity is conserved in the large tegument protein of the herpesviridae. J. Virol. 79, 15582–15585.

Schmolke, S., Kern, H.F., Drescher, P., Jahn, G., and Plachter, B. (1995). The dominant phosphoprotein pp65 (UL83) of human cytomegalovirus is dispensable for growth in cell culture. J. Virol. 69, 5959–5968.

Scholz, B., Rechter, S., Drach, J.C., Townsend, L.B., and Bogner, E. (2003). Identification of the ATP-binding site in the terminase subunit pUL56 of human cytomegalovirus. Nucleic Acids Res. 31, 1426–1433.

Severi, B., Landini, M.P., and Govoni, E. (1988). Human cytomegalovirus morphogenesis: an ultrastructural study of the late cytoplasmic phases. Arch. Virol. 98, 51–64.

Shibata, H., Fujisawa, H., and Minagawa, T. (1987). Characterization of the bacteriophage T3 DNA packaging reaction in vitro in a defined system. J. Mol. Biol. 196, 845–851.

Silvia, M., Yu, Q.C., Enquist, L.W., and Shenk, T. (2003). pp28 tegument protein of HCMV function in assembly of the virions. Paper presented at: Cytomegalovirus Workshop (Mastrich, Netherlands).

Singer, G.P., Newcomb, W.W., Thomsen, D.R., Homa, F.L., and Brown, J.C. (2005). Identification of a region in the herpes simplex virus scaffolding protein required for interaction with the portal. J. Virol. 79, 132–139.

Slawson, C., and Hart, G.W. (2011). O-GlcNAc signalling: implications for cancer. Nat. Rev. Cancer 11, 678–684.

Smith, D.E., Tans, S.J., Smith, S.B., Grimes, S., Anderson, D.L., and Bustamante, C. (2001). The bacteriophage straight phi29 portal motor can package DNA against a large internal force. Nature 413, 748–752.

Spear, P.G., and Roizman, B. (1972). Proteins specified by herpes simplex virus. V. Purification and structural proteins of the herpesvirion. J. Virol. 9, 143–159.

Spencer, J.V., Newcomb, W.W., Thomsen, D.R., Homa, F.L., and Brown, J.C. (1998). Assembly of the herpes simplex virus capsid: preformed triplexes bind to the nascent capsid. J. Virol. 72, 3944–3951.

Steven, A.C., and Spear, P.G. (1997). Herpesvirus capsid assembly and envelopment. In Structural Biology of Viruses, Chiu, W., Burnett, R.M., and Garcea, R.L., eds. (Oxford University Press, New York), pp. 312–351.

Stinski, M.F. (1991). Cytomegalovirus and its replication. In Fundamental Virology, Fields, B.N., and Knipe, D.M., eds. (Raven Press, New York), pp. 929–950.

Stow, N.D., McMonagle, E.C., and Davison, A.J. (1983). Fragments from both termini of the herpes simplex virus type 1 genome contains signals required for the encapsidation of viral DNA. Nucl. Acids Res. 11, 8205–8220.

Tamashiro, J.C., Filpula, D., Friedmann, T., and Spector, D.H. (1984). Structure of the heterogeneous L–S junction region of human cytomegalovirus AS169 DNA. J. Virol. 52, 541–548.

Tatman, J.D., Preston, V.G., Nicholson, P., Elliott, R.M., and Rixon, F.J. (1994). Assembly of herpes simplex virus type 1 capsids using a panel of recombinant baculoviruses. J. Gen. Virol. 75, 1101–1113.

Terhune, S.S., Schroer, J., and Shenk, T. (2004). RNAs are packaged into human cytomegalovirus virions in proportion to their intracellular concentration. J. Virol. 78, 10390–10398.

Thoma, C., Borst, E., Messerle, M., Rieger, M., Hwang, J.S., and Bogner, E. (2006). Identification of the interaction domain of the small terminase subunit pUL89 with the large subunit pUL56 of human cytomegalovirus. Biochemistry 45, 8855–8863.

Thomsen, D.R., Roof, L.L., and Homa, F.L. (1994). Assembly of herpes simplex virus (HSV) intermediate capsids in insect cells infected with recombinant baculoviruses expressing HSV capsid proteins. J. Virol. 68, 2442–2457.

Thomsen, D.R., Newcomb, W.W., Brown, J.C., and Homa, F.L. (1995). Assembly of the herpes simplex virus capsid: requirement for the carboxyl terminal twenty five amino acids of the proteins encoded by the UL26 and UL26.5 genes. J. Virol. 69, 3690–3703.

Tong, L., and Stow, N.D. (2010). Analysis of herpes simplex virus type 1 DNA packaging signal mutations in the context of the viral genome. J. Virol. 84, 321–329.

Tooze, J., Hollinshead, M., Reis, K., and Kern, H. (1993). Progeny vaccinia and human cytomegalovisur particles utilize early endosomal cisternae for their envelopes. Eur. J. Cell. Biol. 60, 163–178.

Townsend, L.B., Devivar, R.V., Turk, S.R., Nassiri, M.R., and Drach, J.C. (1995). Design, synthesis, and antiviral activity of certain 2,5,6-trihalo-1-(β-D-ribofuranosyl) benzimidazoles. J. Med. Chem. 38, 4098–4105.

Trus, B.L., Gibson, W., Cheng, N., and Steven, A.C. (1999). Capsid structure of simian cytomegalovirus from cryoelectron microscopy: evidence for tegument attachment sites [published erratum appears in J. Virol. 1999 May;73(5):4530]. J. Virol. 73, 2181–2192.

Trus, B.L., Cheng, N., Newcomb, W.W., Homa, F.L., Brown, J.C., and Steven, A.C. (2004). Structure and polymorphism of the UL6 portal protein of herpes simplex virus type 1. J. Virol. 78, 12668–12671.

Underwood, M.R., Harvey, R.J., Stanat, S.C., Hemphill, M.L., Miller, T., Drach, J.C., Townsend, L.B., and Biron, K.K. (1998). Inhibition of human cytomegalovirus DNA maturation by a benzimidazole ribonucleoside is mediated through the UL89 gene product. J. Virol. 72, 717–725.

Varnum, S.M., Streblow, D.N., Monroe, M.E., Smith, P., Auberry, K.J., Pasa-Tolic, L., Wang, D., Camp, D.G., 2nd, Rodland, K., Wiley, S., et al. (2004). Identification of proteins in human cytomegalovirus (HCMV) particles: the HCMV proteome. J. Virol. 78, 10960–10966.

Walker, J.E., Saraste, M., Runswick, M.J., and Gay, N.J. (1982). Distantly related sequences in the alpha- and beta-subunits of ATP synthase, myosin, kinases and other ATP-requiring enzymes and a common nucleotide binding fold. EMBO J. *1*, 945–951.

Wang, J., Loveland, A.N., Kattenhorn, L.M., Ploegh, H.L., and Gibson, W. (2006). High-molecular-weight protein (pUL48) of human cytomegalovirus is a competent deubiquitinating protease: mutant viruses altered in its active-site cysteine or histidine are viable. J. Virol. *80*, 6003–6012.

Wang, J.B., and McVoy, M.A. (2011). A 128-base-pair sequence containing the pac1 and a presumed cryptic pac2 sequence includes cis elements sufficient to mediate efficient genome maturation of human cytomegalovirus. J. Virol. *85*, 4432–4439.

Welch, A.R., McNally, L.M., and Gibson, W. (1991a). Cytomegalovirus assembly protein nested gene family: four 3'-coterminal transcripts encode four in-frame, overlapping proteins. J. Virol. *65*, 4091–4100.

Welch, A.R., Woods, A.S., McNally, L.M., Cotter, R.J., and Gibson, W. (1991b). A herpesvirus maturational protease, *assemblin*: identification of its gene, putative active site domain, and cleavage site. Proc. Natl. Acad. Sci. U.S.A. *88*, 10792–10796.

Welch, A.R., McNally, L.M., Hall, M.R., and Gibson, W. (1993). Herpesvirus proteinase: site-directed mutagenesis used to study maturational, release, and inactivation cleavage sites of precursor and to identify a possible catalytic site serine and histidine. J. Virol. *67*, 7360–7372.

Wingfield, P.T., Stahl, S.J., Thomsen, D.R., Homa, F.L., Booy, F.P., Trus, B.L., and Steven, A.C. (1997). Hexon-only binding of VP26 reflects differences between the hexon and penton conformations of VP5, the major capsid protein of herpes simplex virus. J. Virol. *71*, 8955–8961.

Womack, A., and Shenk, T. (2010). Human cytomegalovirus tegument protein pUL71 is required for efficient virion egress. MBio 1 pii: e00282–10.

Wood, L.J., Baxter, M.K., Plafker, S.M., and Gibson, W. (1997). Human cytomegalovirus capsid assembly protein precursor (pUL80.5) interacts with itself and with the major capsid protein (pUL86) through two different domains. J. Virol. *71*, 179–190.

Yang, K., and Baines, J.D. (2008). Domain within Herpes Simplex Virus 1 scaffold proteins required for interaction with portal protein in infected cells and incorporation of the portal vertex into capsids. J. Virol. *82*, 5021–5030.

Yang, K., and Baines, J.D. (2009). Proline and tyrosine residues in scaffold proteins of Herpes Simplex Virus 1 critical to the interaction with portal protein and its incorporation into capsids. J. Virol. *83*, 8076–8081.

Yang, K., and Baines, J.D. (2011). Selection of HSV capsids for envelopment involves interaction between capsid surface components pUL31, pUL17, and pUL25. Proc. Natl. Acad. Sci. U.S.A. *108*, 14276–14281.

Yu, D., Silva, M.C., and Shenk, T. (2003). Functional map of human cytomegalovirus AD169 defined by global mutational analysis. Proc. Natl. Acad. Sci. U.S.A. *100*, 12396–12401.

Yu, X., Shah, S., Atanasov, I., Lo, P., Liu, F., Britt, W.J., and Zhou, Z.H. (2005). Three-dimensional localization of the smallest capsid protein in the human cytomegalovirus capsid. J. Virol. *79*, 1327–1332.

Zhang, X., Efstathiou, S., and Simmns, A. (1994). Identification of novel herpes simplex virus replicative intermediates by field inversion gel electrophoresis: implication for viral DNA amplification strategies. Virology *202*, 530–539.

Exploitation of Host Cell Cycle Regulatory Pathways by HCMV

I.14

Deborah H. Spector

Abstract

Successful replication of HCMV requires the deployment of multiple approaches to commandeer the host cell machinery and create a cellular milieu that is optimal for viral gene expression, DNA replication, and formation of infectious progeny. The complex regulatory network that drives cell cycle progression provides a rich source of factors that can be co-opted and combined in different ways to tailor the host cell's environment to meet the needs of the virus for productive infection. To this end, HCMV dramatically alters cell cycle regulatory pathways, leading to cell cycle arrest. These alterations begin as soon as the viral particle enters the cell and continue throughout the entire replicative cycle. The molecular mechanisms underlying the viral-mediated effects operate at multiple levels, including altered RNA transcription, inhibition of cell DNA synthesis, changes in the levels and activity of cyclin-dependent kinases as well as other cellular kinases involved in cell cycle control, modulation of protein stability through targeted effects on the ubiquitin-proteasome degradation pathway, and movement of proteins to different cellular locations. This chapter will focus on the interplay between the viral and cellular factors and the mechanisms utilized to effect these changes as they relate to the cell cycle.

Introduction – overview of the mammalian cell cycle

An understanding of how HCMV exploits and subverts the regulatory pathways used to control the host cell cycle first requires a synopsis of the multifaceted and highly ordered sequence of events that occur every time the cell divides (for review, see Murray, 2004; Satyanarayana and Kaldis, 2009). The cell cycle is primarily regulated by heterodimeric kinases that consist of a regulatory cyclin subunit and a catalytic subunit, the cyclin-dependent kinase (cdk). Multisubunit E3 ubiquitin ligases that ubiquitinate proteins and target

them for degradation by the proteasome also contribute significantly to cell cycle progression. One of the most important E3 ubiquitin ligases is the multisubunit anaphase-promoting complex (APC).

The cell cycle consists of four phases, G_1, S, G_2, and M. Progression into the G_1 phase, either from a resting state (G_0 phase) or mitosis is associated with expression of the D-type cyclins that form kinase complexes with cdk4 or cdk6 to phosphorylate their substrates. The INK4 family of inhibitors and p27 bind to cdk4 and cdk6 to prevent D-type cyclin activity. During the G_1 phase, multiple transcription factors and genes encoding proteins involved in nucleotide metabolism and DNA replication are induced to promote commitment to DNA synthesis and cell division if the necessary requirements are met. Prior to S phase, pre-replication complexes (pre-RC) assemble at the origins of DNA replication. The multisubunit origin recognition complex (ORC) is the first to bind to the DNA and serves as a platform for the recruitment of other factors. Cdc6 and Cdt1 bind to the complex and facilitate the loading of the family of at least six minichromosome maintenance (MCM) proteins (Blow and Dutta, 2005). Cyclin E1 (referred to as cyclin E in the text) expression is induced, and the active cdk2/cyclin E kinase promotes the transition into S phase.

At the beginning of S phase, cyclin A2 (referred to in the text as cyclin A) accumulates and forms an active kinase complex with cdk2. Regulation of cyclin A occurs at both the protein and mRNA levels (Glotzer et al., 1991; Henglein et al., 1994; Desdouets et al., 1995; Schulze et al., 1995; Zwicker et al., 1995; Zwicker and Muller, 1997; Bottazzi et al., 2001; Tessari et al., 2003). The cdk2/cyclin A complex and another kinase, cdc7/Dbf4, activate the firing of DNA origins of replication and initiation of DNA replication (for review, see Nishitani and Lygerou, 2002; Machida et al., 2005). During S and G2, cdt1 activity is blocked both by degradation and binding to geminin, which accumulates during S, G2, and M phases. This ensures that DNA replication initiates at each origin only once and prevents polyploidy.

After the DNA has been completely replicated, cells enter G_2, and there is an accumulation of the proteins required for separation of the chromosomes and mitosis. Cyclin B1 (referred to in the text as cyclin B) levels increase, and the cyclin associates with cdk1. The cdk1/cyclin B complex is initially non-functional, and dephosphorylation of cdk1 by cdc25 phosphatase at the G_2/M transition is required for the activation of the kinase (Millar and Russell, 1992). Multiple proteins are phosphorylated by cdk1/cyclin B and cdk1/cyclin A kinases to make sure that there is an ordered and even segregation of the chromosomes to the daughter cells. Inactivation of the cdk1 complexes occurs during mitosis by degradation of cyclins A and B through the ubiquitin-dependent proteolytic pathway involving the APC E3 ubiquitin ligase and the proteasome (Glotzer et al., 1991). The APC also ubiquitinates geminin and targets it for proteasome degradation as the cells exit mitosis, thereby allowing loading of the pre-RCs onto the chromatin during G_1 phase (McGarry and Kirschner, 1998). This degradation of the cyclins and geminin continues until the onset of S phase (Brandeis and Hunt, 1996).

Checkpoints exist at the G_1/S boundary, during S phase, at G_2/M, and during mitosis to stop cell cycle progression if there is DNA damage or aberrant spindle formation (for review, see Lukas et al., 2004). Two of the most important proteins that provide this checkpoint function are the tumour suppressors p53 and the Retinoblastoma (Rb) family of pocket proteins (Rb, p107, and p130). The Rb proteins, in their hypophosphorylated forms, bind to the E2F family of transcription factors, thus repressing RNA synthesis from promoters that are targets of these factors. Phosphorylation of the Rb proteins in late G_1 dissociates these complexes, allowing the E2F factors to activate transcription of multiple genes, many of which encode proteins required for DNA replication (Dyson, 1998). p53 coordinates multiple cellular processes through its activity as a transcriptional activator and repressor in response to stress and growth factors (Vousden and Lu, 2002; Slee et al., 2004). Phosphorylation of p53 controls its association with the murine double minute (MDM2) protein, which targets p53 for degradation by the proteasome (for review, see Lavin and Gueven, 2006). In response to DNA damage, nutrient deprivation, and other insults to the cell, p53 levels are stabilized. This can lead to the expression of p21, an inhibitor of most cdks, as well as induction of several pro-apoptotic genes. At mitosis, APC activity is tightly controlled through interaction with the mitotic checkpoint to ensure that chromosomes are properly aligned on the spindle before protein degradation allows their segregation to daughter cells.

Cell cycle arrest in infected cells is multifactorial

In the mid 1990s, it was recognized that cells infected with HCMV in the G_0/G_1 phase of the cycle do not proceed through mitosis. Surprisingly, the cells arrest in a pseudo-G_1 state that is distinguished by the expression of selected G_1-phase, S-phase and M-phase gene products and a block in cellular DNA synthesis (see Fig. I.14.1) (Jault et al., 1995; Bresnahan et al., 1996b; Lu and Shenk, 1996; Dittmer and Mocarski, 1997; Salvant et al., 1998; Wiebusch and Hagemeier, 1999, 2001; Challacombe et al., 2004; Hertel and Mocarski, 2004). Regulatory controls at multiple levels are affected, including accumulation, degradation, post-translational modifications, activation, and deactivation of numerous cell cycle proteins. For example, the tumour suppressors pRb, p130 and p107, which inhibit the expression of E2F-responsive genes, are maintained in a phosphorylated inactive state (Jault et al., 1995; McElroy et al., 2000), and p53 is stabilized and sequestered in viral replication centres (Muganda et al., 1994; Bresnahan et al., 1996b; Fortunato and Spector, 1998; Chen et al., 2001; Casavant et al., 2006). It would seem that these effects on Rb and p53 alone should inactivate the G_1/S checkpoint and promote progression through S phase, but as will be seen below, the modulation of many other cell cycle regulatory proteins impacts significantly on the final outcome.

Cdk activity is highly dysregulated in infected cells. The G_1/S- and G_2/M-phase cyclins E and B1, respectively, accumulate in infected cells with a concomitant increase in associated kinase activity (Jault et al., 1995; Wiebusch and Hagemeier, 2001; Sanchez et al., 2003). In contrast, the steady-state levels of the S-phase cyclin A and G_1-phase cyclin D1 are reduced (Jault et al., 1995; Bresnahan et al., 1996b; Salvant et al., 1998; Wiebusch and Hagemeier, 2001). The effects of the virus on cyclins E, A, and D1 are primarily at the level of transcription, while accumulation of cyclin B1 is associated with increased stability of the protein (Salvant et al., 1998; Sanchez et al., 2003).

Importance of cell cycle arrest for the viral infection

Initiation of HCMV gene expression requires that the cells be in G_0 or G_1 at the time of infection (Salvant et al., 1998; Fortunato et al., 2002). When cells are infected near or during S phase, major IE gene expression cannot begin, and the cells must pass through S

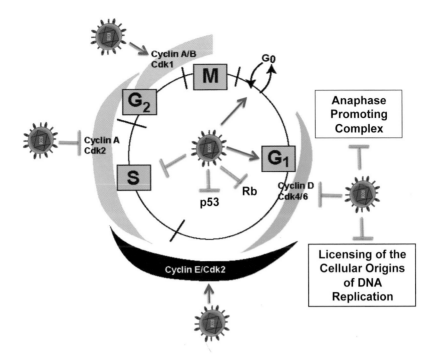

Figure I.14.1 Cell cycle arrest in HCMV infected cells is multifactorial. The cell cycle consists of four phases – G_1, S, G_2 and M. Cells in the G_0 resting state are stimulated through growth signals to express cyclin D and enter G_1 phase. In uninfected cells, the anaphase-promoting complex (APC), an E3 ubiquitin ligase, remains active and targets proteins for degradation by the proteasome. A pre-replication complex (licensing) is established at cellular origins of DNA replication and cyclin E is induced. During S phase, cyclin A accumulates and the cellular DNA is replicated. G_2 phase marks the transition prior to cell division in M phase. Both cyclin A and cyclin B are required during the G_2/M period. The major guardians of the cell cycle are Rb and p53. HCMV infection of cells during G_0/G_1 phase induces progression through G_1. However, the cell cycle is blocked before the replication of cellular DNA, and the expression of the cyclins and cyclin-dependent kinases is disrupted. HCMV inhibits the expression of cyclin D and cyclin A, but promotes accumulation of high levels of cyclin E and cyclin B. Licensing of the cellular origins of DNA replication is inhibited, and activity of the APC is blocked. p53 is stabilized, but it cannot activate its target promoters, and there is relief of the inhibitory activity of Rb on the E2F/DP transcription factors.

phase and undergo mitosis prior to the synthesis of IE and early gene products. The mechanisms governing these events have yet to be fully determined, but some clues have recently emerged. An important observation is that this block to IE gene expression during the S/G_2 phase is at the level of transcription, but is not specific for the major IE promoter driving expression of IE1-72 and IE2-86, and affects other regions of IE gene expression, including the US3 and the UL36–38 loci (Zydek et al., 2010, 2011). Moreover, several other known obstacles to viral IE gene expression are not involved. There is not a defect at the level of translocation of the viral genome and matrix proteins to the nucleus, and the block is independent of the intrinsic early inhibition of viral transcription at ND10 domains involving Daxx, PML, and HDAC (see Chapter I.9).

One clue to the mechanism is that prolonged exposure of the cells to replicative stress or DNA damaging agents, such as aphidicolin or doxorubicin treatment,

respectively, prior to and during the infection not only relieved the block to IE gene expression, but also allowed the viral infection to proceed to late gene expression in the S/G_2 cells (Zydek et al., 2010). The DNA damage-induced expression and accumulation of the checkpoint protein p53 plays a key role, as relief of repression does not occur in cells that have stable knockdown of p53. This finding is consistent with the earlier study by Fortunato et al. (2002) that showed that inhibition of the proteasome, which would stabilize p53, allowed IE gene expression. It also supports prior work showing that HCMV replication and formation of infectious virus is greater in cells that express WT p53 (Casavant et al., 2006; Hannemann et al., 2009). It is likely that other pathways are involved, since the use of a specific inhibitor of cdks (cdk1, cdk2, cdk5, cdk7, and cdk9) for a brief period at the beginning of the infection, can provide some relief of the block to IE gene expression (Zydek et al., 2010), but this does not occur if the inhibitor is present for longer periods (Fortunato et al., 2002). A likely target of the cdk inhibitor is Cyclin A2/cdk2, which normally increases in abundance during the S/G$_2$ phase and is specifically inhibited at the transcriptional level during productive infection (Salvant et al., 1998; Shlapobersky et al., 2006). In accord with this, inhibition of the proteasome should also allow the accumulation of the cdk inhibitors p27 and p21, which are degraded during S phase. Clearly, further studies will be needed to elucidate the basis of this restriction to viral replication in S phase. It is noteworthy, however, that MCMV does not strictly inhibit or depend on the phase of the cell cycle for replication (Maul and Negorev, 2008; Wiebusch et al., 2008). Infection of cells, either human or mouse, with MCMV can lead to a cell cycle block in either G$_1$ or G$_2$, and viral IE gene expression is not restricted when the cells are in S or G$_2$ phase (Wiebusch et al., 2008). In accord with this, my lab has found that expression of cyclin A is not inhibited in cells infected with MCMV (Spector, D., unpublished).

Involvement of proteins in the virion – pUL69 and pUL82

Two virion proteins, pUL69 and pp71 (pUL82), have been shown to modulate cell cycle progression, although as we begin to elucidate the function of viral structural proteins in different types of cells, others will likely be identified. As components of the virion tegument, pUL69 and pp71 can function as soon as the virus enters the cells (see Chapter I.9). In the case of pUL69, overexpression of this protein stimulates accumulation of cells in G$_1$ phase of the cell cycle (Lu and Shenk, 1999). In addition, cells infected with a mutant virus lacking functional pUL69 do not efficiently undergo cell cycle arrest (Hayashi et al., 2000). This mutant does not replicate to WT levels, but the growth defect may be attributed to other functions of pUL69.

The deletion of pp71 also creates a virus that is severely impaired for growth. The growth defect can be complemented by expression of the protein in trans (Bresnahan et al., 2000; Dunn et al., 2003). pp71 has been shown to interact with the cell growth suppressors Rb, p107, and p130 (Kalejta et al., 2003) and to target hypophosphorylated forms of these pocket proteins for ubiquitin-independent degradation by the proteasome (Kalejta et al., 2003). Consistent with this activity, pp71 expression in uninfected cells accelerates their progression through G$_1$ phase, but does not change the overall doubling time (Kalejta et al., 2003). These results suggest that pp71 delivered to cells as part of the incoming virus particles may stimulate the cell cycle at the beginning of the infection before IE gene expression. The finding that pp71 also interacts with ND10-associated transcription repressor Daxx and promotes its ubiquitin-independent and proteasome-mediated degradation, as well as dissociation of its binding partner ATRX, suggests that a significant function for pp71 is initiating viral transcription at ND10 sites (Hofmann et al., 2002; Ishov et al., 2002; Marshall et al., 2002; Cantrell and Bresnahan, 2005; Preston and Nicholl, 2006; Saffert and Kalejta, 2006; Hwang and Kalejta, 2007; Lukashchuk et al., 2008).

Role of the major IE proteins IE1 and IE2

The major region of IE transcription is described in detail in another chapter. This region includes two genes, IE1 and IE2. IE1 RNA has four exons; a single ORF (UL123) initiates in exon 2 and specifies a 72-kDa protein (IE1-72). The IE2 gene product, IE2-86 (ORF UL122), is encoded by an alternatively spliced RNA with the first three exons of IE1 and a different terminal exon. IE2 also encodes abundant late unspliced RNAs that specify 60-kDa and 40-kDa proteins corresponding to the C-terminus. Studies with mutant viruses have demonstrated that IE1-72 is required at low but not high MOI, while IE2-86 is essential regardless of the MOI. Both IE1-72 and IE2-86 have been shown to activate, as well as block, cell cycle progression in heterologous systems in the absence of infection, but as will be described below, there is still only minimal evidence that they have these functions in the context of the infection.

IE protein 1

The initial evidence suggesting that IE1-72 had an effect on the cell cycle, was that transient expression of IE1-72 in asynchronously cycling cells results in the accumulation of the cells in the S and G$_2$/M phases (Castillo et al., 2000). One proposed explanation for this result is that it is due to the interaction of IE1-72 with the pocket protein p107 (Margolis et al., 1995; Poma et al., 1996; Woo et al., 1997; Castano et al., 1998; Hansen et al., 2001; Zhang et al., 2003). IE1-72 alleviates p107-mediated repression of E2F-responsive promoters in transient transfection assays and thus may stimulate S-phase entry. Additionally, IE1-72 can reverse the inhibitory effects of p107 on cdk2/cyclin E kinase activity, which may also facilitate the G$_1$/S transition. It has been suggested that the formation of an IE1-72/p107 complex mediates these effects (Poma et al., 1996; Johnson et al., 1999; Zhang et al., 2003). One surprising finding is that IE1-72 was unable to induce quiescent cells to enter S phase in cells that express p53 (Castillo *et al.*, 2000). Based on transient expression assays that show that IE1-72 can induce the nuclear accumulation of p53, which activates p21, it has been suggested the kinase inhibitory activity of p21 may account for the inability of IE1-72 to push quiescent cells into S phase (Castillo *et al.*, 2005). However, others have argued that the interaction of IE1-72 with p53 inhibits binding to the p21 promoter and its transcription, but most of these experiments were performed in *in vitro* binding assays and transient expression assays (Hwang *et al.*, 2009).

IE protein 2

IE2-86 is a potent transactivator of transcription from HCMV and heterologous promoters and is essential for viral replication. Although its major binding partner is the viral protein UL84, it has been found to bind to multiple cellular proteins, including those involved in cell cycle control. Most notable are its interactions with Rb and p53 (Hagemeier *et al.*, 1994; Sommer *et al.*, 1994; Speir *et al.*, 1994; Bonin and McDougall, 1997; Fortunato *et al.*, 1997). In transient expression and *in vitro* systems, IE2-86 interacts with the C-terminus of p53, and the binding of p53 to target promoters is inhibited (Speir *et al.*, 1994; Bonin and McDougall, 1997; Hsu *et al.*, 2004). IE2-86 expression also blocks the acetylation of p53 and of histones in proximity to p53-dependent promoters (Hsu *et al.*, 2004). Thus IE2-86 may regulate expression of p53 target genes by multiple mechanisms. These effects on protein acetylation may result from down-regulation of p300 and cAMP-responsive element-binding protein-binding protein (CBP) histone acetyl transferase (HAT) activities, which have been detected in a complex with p53 and IE2-86 (Hsu *et al.*, 2004). It is possible that the inhibition of p300/CBP HAT activity is p53-promoter specific, as IE2-86 did not suppress histone acetylation globally. The biological relevance of these experiments must be considered with caution, given that none were performed in the context of the viral infection.

The early phase of HCMV infection is associated with the activation of the expression of many genes that are required for host cell DNA synthesis and proliferation, and IE2-86 has been implicated as playing an important role in their induction (Hirai and Watanabe, 1976; Estes and Huang, 1977; Isom, 1979; Boldogh *et al.*, 1991; Wade *et al.*, 1992; Browne *et al.*, 2001). Many of the up-regulated genes are controlled by the E2F/DP transcription factors, which are inhibited by complex formation with the Rb family of proteins. The observation that steady-state levels of RNA from several E2F responsive genes increase when uninfected fibroblasts are infected with an adenovirus expressing

IE2-86 suggests that this may be one of the functions of IE2-86 (Song and Stinski, 2002). The key question, however, is what are the underlying mechanisms for the activation of these growth regulatory genes in the context of the infection?

Several studies have shown that transient expression of IE2-86 alters cell cycle progression, with a block at the G_1/S boundary in a p53$^{+/+}$ cell or after entry into S phase in a p53 mutant cell (Murphy *et al.*, 2000; Wiebusch and Hagemeier, 2001; Noris *et al.*, 2002; Wiebusch *et al.*, 2003a; Song and Stinski, 2005). In transient transfection assays, deletion of aa 451–579 abolished the ability of IE2-86 to induce G_1 arrest in transient assays in U373 cells, but aa 25–85, aa 544–579 and aa 136–290 were not necessary (Wiebusch and Hagemeier, 1999). How this G_1 arrest is achieved, however, is yet to be determined. A report by Petrik *et al.* (2006) presented data showing that a recombinant mutant virus in which aa 548 of IE2-86 was changed from Q to R results in a growth-impaired virus that does not inhibit cellular DNA synthesis or the cell cycle. However, a recent study from my lab (Burgdorf *et al.*, 2011) with a virus that has the same mutation in the IE2 gene has provided evidence that this is not the case. The major difference between these studies is that we propagated the mutant virus in cells that could be induced to express IE2 proteins and complementing function, and thus were able to produce virus in high enough titres to be able to study it more thoroughly following infection of non-complementing cells. Although this mutant virus was severely debilitated at late times in the infection, it was able to block the cell cycle as efficiently as WT virus. We suspect that the major reason for these differing results is that in the study by Petrik *et al.* (2006), the mutant virus was not grown on complementing cells. It took many weeks for non-complementing cells transfected with the IE2 mutant virus to reach 100% CPE, and this protracted growth time increases the possibility of mutations in other regions of the genome. Preparing stocks of severely debilitated IE2 mutants in the complementing cells helps to alleviate the pressure on these mutants to develop secondary mutations that facilitate replication. This highlights the importance of propagating debilitated mutant viruses on complementing cells to study the function of specific genes.

Regulation of cdk2/cyclin E

Cyclin E transcription is regulated by E2F, and the potential role of IE2-86 in its accumulation during the infection has been the focus of several studies (Bresnahan *et al.*, 1998; McElroy *et al.*, 2000; Wiebusch and Hagemeier, 2001; Wiebusch *et al.*, 2003a). The majority of the experiments have used transient expression assays to examine the regulation of the cyclin E promoter driving a reporter gene. In one study, it was shown that IE2-86 could bind to sequences in the cyclin E promoter *in vitro* and could activate expression of a cyclin E promoter-driven reporter construct (Bresnahan *et al.*, 1998). The work demonstrating that cyclin E was one of the E2F responsive genes that showed induced transcription when IE2-86 was expressed from an adenovirus vector reinforced the notion that IE2-86 expression up-regulates cyclin E in infected cells (Song and Stinski, 2002). A role for IE2-86 is further suggested by the finding that a recombinant virus with a deletion of aa 30–77 in the N-terminal region shared by both IE1-72 and IE2-86 is deficient in up-regulating cyclin E (White and Spector, 2005). This is likely not due to the defect in IE1-72 as cyclin E is efficiently up-regulated by a recombinant virus with a deletion in IE1-72, and the recombinant virus lacking aa 30–77 is still unable to increase the levels of cyclin E even in complementing cells that express IE1-72 (McElroy *et al.*, 2000; White and Spector, 2005). However, this latter result may be due in part to an effect on early gene expression, as McElroy *et al.* (2000) reported that early viral gene expression, not IE2-86 expression, is necessary for accumulation of cyclin E protein. These conflicting results suggest that cyclin E accumulation in infected cells is controlled by several pathways and the increase in cyclin E might be regulated at both the mRNA and protein level.

In uninfected cells, the kinase cdk2 is activated through phosphorylation by the CDK-activating kinase (CAK), cdk7/cyclin H. The abundance of cdk7 and cyclin H and the kinase activity of the complex significantly increase during the infection (Tamrakar *et al.*, 2005). Cdk2/cyclin E complexes can also be inhibited by the binding of p21 and the pocket proteins, p107 and p130. Changes in each of these pathways in infected cells results in the maintenance of high levels of active cdk2/cyclin E. In addition to the large increase in cyclin E mRNA described above, there is a decrease in p21 during the infection (Bresnahan *et al.*, 1996a; Chen *et al.*, 2001), and high levels of active cdk2/cyclin E are found in complexes with p107 and p130 (Spector, D., unpublished data). It has been suggested that the binding of IE1-72 to p107 might relieve the repression of the cdk2/cyclin E kinase (Johnson *et al.*, 1999; Zhang *et al.*, 2003). However, because high levels of cdk2/cyclin E activity are also observed in cells infected with an HCMV recombinant that does not encode a functional IE1-72 (Spector, D., unpublished), it seems likely that other viral functions are also involved in overriding the inhibition imposed by the pocket proteins in the context of the infection. It is possible that the utilization of

multiple mechanisms to achieve the same goal assures that the virus can replicate efficiently in different cellular environments.

Cdk1/cyclin B regulation

In uninfected cells, the levels of the cyclin B subunit fluctuate during the cell cycle. The protein accumulates during S/G$_2$/M. It is then degraded as the cell progresses through mitosis, and this degradation continues through G$_1$. Regulated phosphorylation and dephosphorylation of cdk1 is the major mediator of the kinase activity of the complex. In uninfected cells, inhibitory phosphates are added to Thr14 and Tyr15 of cdk1 by two kinases, Wee1 and Myt1, whose expression and activity are also cell-cycle regulated (Heald *et al.*, 1993; Mueller *et al.*, 1995; Watanabe *et al.*, 1995; Booher *et al.*, 1997; Liu *et al.*, 1997, 1999). Dephosphorylation of these sites by members of the cdc25 family of dual-specificity protein phosphatases relieves this inhibition. The initial activation of cdk1 during S/G$_2$ phase is likely due to cdc25B-mediated dephosphorylation (Hoffmann *et al.*, 1993; Lammer *et al.*, 1998; Karlsson *et al.*, 1999). Cdk1/cyclin B can then phosphorylate and activate cdc25C. This ignites a feedback loop that increases the cdk-dependent kinase activity and leads to phosphorylation of substrates necessary for mitosis. Full activation of cdk1 is achieved through phosphorylation of Thr161 by cdk7/cyclin H (Harper and Elledge, 1998).

Although the infected cells do not progress through S phase, cyclin B accumulates to high levels due to stabilization of the protein (Sanchez *et al.*, 2003). As will be discussed below, this is a result of the viral-mediated inhibition of the APC E3 ubiquitin ligase that targets cyclin B for proteasome-mediated degradation. There are also several mechanisms to ensure that the cdk1/cyclin B complex is maintained in its active form in infected cells (Sanchez *et al.*, 2003). The activating cdc25 protein phosphatases accumulate. In contrast, the abundance and activity of the inhibitory kinases Myt1 and Wee1 are significantly reduced. Wee1 is targeted for proteasome degradation following its ubiquitination by the Skp, Cullin, F-box containing complex (SCF) E3 ubiquitin ligase. The observation that Wee1 is degraded by the proteasome in the infected cells is significant in that it showed that the accumulation of cyclin B was not likely due to some global inhibition of the proteasome.

The localization of cyclin B is also altered in infected cells. By immunofluorescence, it was found that the level of cyclin B in individual infected cells is low, but is evenly distributed between the nucleus and cytoplasm, with some concentration at the centrosome

(Sanchez *et al.*, 2003). By biochemical fractionation of HCMV-infected cells, it can be shown that the active form of cdk1/cyclin B is also higher in the cytoplasm, consistent with the increased levels of the protein at the centrosome.

Regulation of cyclin A

In contrast to Cyclin B and E, the expression of cyclin A is down-regulated, and this occurs at the transcriptional level (Salvant *et al.*, 1998). Based on a report that the high-mobility group AT-hook 2 (HMGA2) transcription factor regulates cyclin A transcription (Tessari *et al.*, 2003), my lab investigated whether the mechanism of inhibition involves HMGA2. HMGA proteins are architectural transcription factors that enhance or repress transcription by participating in the organization and assembly of nucleoprotein structures (enhanceosomes) at the promoter. HMGA2 can activate the expression of cyclin A by relieving repression of the promoter. In Shlapobersky *et al.* (2006), we demonstrated that HMGA2 transcription is repressed during the infection. To determine if repression of HMGA2 is directly related to cyclin A inhibition and impacts on the infection, we constructed an HCMV recombinant virus that expressed HMGA2. In cells infected with the recombinant virus, cyclin A mRNA and protein were induced, and there was a significant delay in viral early gene expression and DNA replication, indicating that the repression of HMGA2 is important for viral replication. With recombinant viruses that had mutations in either IE1-72 or IE2-86, it was also shown that IE2 is involved in the regulation of HMGA2 expression and resulting inhibition of cyclin A transcription. The mechanism involving the down-regulation of HMGA2 RNA expression by IE2-86 has yet to be determined. One possibility is that it is related to IE2 interaction with HDAC, as HMGA2 is an example of a gene that requires HDAC activity for its expression (Ferguson *et al.*, 2003). However, given all of the reported interactions of IE2-86 with other transcription factors, it is likely that this is not the only mechanism.

Checkpoint control and DNA damage pathways

Mechanisms to respond to different types of DNA damage are poised for activation during cell cycle progression to ensure that damaged DNA is repaired before the DNA is replicated and chromosomes with deleterious mutations are transferred to daughter cells. The alternative is the initiation of apoptosis. The observation that HCMV infection induces breaks in chromosome 1 (Fortunato *et al.*, 2000) prompted

investigations into the effects of the virus on DNA damage pathways and how they are exploited to benefit viral replication. During the infection, there are significant changes in proteins involved in G_1 and S phase checkpoint control. As noted above, the tumour suppressors Rb, p130, and p107, which inhibit the expression of E2F-responsive genes and the activity of cdk2 kinase complexes, are maintained in a phosphorylated inactivate state (Jault *et al.*, 1995; McElroy *et al.*, 2000). This hyperphosphorylation of Rb is mediated primarily by the viral-encoded kinase UL97, which performs functions similar to cdk1 and when expressed in serum starved rat fibroblasts, can induce the cells to enter S phase (Hume *et al.*, 2008; Prichard *et al.*, 2008).

Early studies reported that although p53 accumulated to high levels in HCMV-infected cells, the protein was sequestered into viral replication centres, thus contributing, at least in part, to its inability to activate the expression of some downstream cellular target genes such as p21 (Muganda *et al.*, 1994; Bresnahan *et al.*, 1996a; Fortunato and Spector, 1998; Chen and Fang, 2001). Subsequently, it was shown by Rosenke *et al.* (2006) that p53 binds to viral promoters and the DNA binding activity of the protein is required for this localization to the viral replication centres. It is likely that this relocalization of p53 commandeers some of p53 functions for specific use by the virus, as HCMV replication and formation of infectious virus is greater in cells that express WT p53 (Casavant *et al.*, 2006; Hannemann *et al.*, 2009). The reduction in virus production has been attributed to delays in viral DNA replication. Consistent with the binding of p53 to viral promoters, delays in the expression of early and late proteins are also observed in p53 null cells. Reintroduction of wild type but not mutant p53 partially rescues the phenotype, establishing a role for p53 in the viral lytic cycle.

Activation of the ataxia telangiectasia mutated (ATM) and ataxia telangiectasia and rad-3-related (ATR) kinases, which play a role in the sensing and response to the presence of double-stranded breaks or single-stranded breaks and lesions, respectively, occurs early in the infection (Shen *et al.*, 2004; Castillo *et al.*, 2005; Gaspar and Shenk, 2006; Luo *et al.*, 2007). However, a fully functional damage response does not occur due to inefficient relocalization of all of the required proteins (Gaspar and Shenk, 2006; Luo *et al.*, 2007). Although many of the proteins localize to the viral replication centres, several that are involved in the non-homologous end-joining (NHEJ) response appear to be completely excluded (Luo *et al.*, 2007). This may be necessary to prevent the rejoining of ends that are generated during the replication of the virus. It is unclear, however, which of the many pathways controlled by ATM and ATR are needed by the virus. Some have found that HCMV replicates normally in cells deficient in ATM (Luo *et al.*, 2007), but others report that replication of the virus in ATM minus or deficient cells is significantly inhibited (Pickering *et al.*, 2011). One difference between these studies is that in the case where ATM was not required, cells were synchronized in G_0 and released into G_1 at the time of the infection (Luo *et al.*, 2007). In contrast, ATM appeared to be required in the case where asynchronous cells were infected (Pickering *et al.*, 2011). Given that infection by HCMV is critically dependent on the cells being in G_0/G_1, it is possible that additional restrictions were placed on the infection in the absence of ATM if the cells were not in G_0/G_1 at the time of infection. Another difference is that different strains of HCMV are used, with the replication of strain AD169, but not strain Towne, showing dependence on ATM. Both are attenuated laboratory strains, and to date, this question has not been addressed with clinical isolates.

Inhibition of cellular DNA replication

The replication of cellular DNA is inhibited by several mechanisms, including the down-regulation of cyclin A discussed above and deleterious effects on the formation of pre-RCs at the origins of DNA replication (Biswas *et al.*, 2003; Wiebusch *et al.*, 2003a). This latter effect is due to decreased expression of several of the MCM proteins, and inhibition of the loading of these proteins onto chromatin (Biswas *et al.*, 2003; Wiebusch *et al.*, 2003a). This may be a result of the premature accumulation of geminin, which, as discussed above, normally accumulates in S phase to block cdt1 activity and ensure that there is no refiring of origins as the cells proceed through S and G_2/M. An alternative possibility is that there is a direct effect on the MCM proteins. Supporting this latter hypothesis is the recent finding that a recombinant virus deficient in the expression of the viral protein UL117 is unable to inhibit host cell DNA synthesis as efficiently as WT virus (Qian *et al.*, 2010). This effect is associated with increased levels and loading of the MCM proteins onto the chromatin in mutant infected cells, but the increased accumulation of geminin is comparable.

Regulation of RNA transcription

Microarrays have greatly facilitated analysis of the global effects of HCMV infection on the accumulation of cellular RNAs (Zhu *et al.*, 1997; Browne *et al.*, 2001; Challacombe *et al.*, 2004; Hertel and Mocarski, 2004),

and the advent of next generation deep sequencing techniques now makes it possible to look at the effects not only on mRNAs, but also noncoding RNAs, miRNAs, and other small RNAs (Stark *et al.*, 2012). While increases in the levels of mRNA do not necessarily correlate with changes in the steady-state levels of protein, these analyses provide a foundation for defining infection-associated changes in transcription. In general, it appears that the RNA levels of multiples genes involved in cell cycle regulation are altered, including kinases, cyclins, transcription factors, and regulators of proteasome activity.

It has become increasingly evident that miRNAs, both viral- and cell-encoded, play an important role in the regulation of gene expression, and thus it is not surprising that HCMV encodes at least 22 mature miRNAs (Dunn *et al.*, 2005; Grey *et al.*, 2005; Pfeffer *et al.*, 2005; Stern-Ginossar *et al.*, 2009; Stark *et al.*, 2012). A detailed review of the viral miRNAs is presented in Chapter I.5. Of relevance to the cell cycle is a recent study by Grey *et al.* (2010). In general, miRNAs primarily target the 3' UTRs of mRNAs, although it has been reported that the cellular miR-10a can induce protein expression by binding to the 5' UTRs of cellular ribosomal protein mRNAs (Ørom *et al.*, 2008), and the liver specific miRNA-122 facilitates the replication of Hepatitis C virus by binding to the 5' UTR of the viral RNA genome (Roberts *et al.*, 2011). Grey *et al.* (2010) have made the interesting finding that one of the HCMV viral miRNAs, miR-US25-1, also binds to the 5' UTR of several cell mRNAs, including a few involved in the cell cycle. The technology used was immunoprecipitation of the RNA-induced silencing complex (RISC) and analysis of the associated miRNAs and target mRNA sequences. A major gene identified was cyclin E2, which is different from the cyclin E1 that is the main cyclin studied in association with the cell cycle. Less is known about cyclin E2, which appears to have both overlapping and distinct functions with cyclin E1, but is regulated differently (Caldon and Musgrove, 2010). Cyclin E2 is up-regulated in WT-infected cells, and there are slightly higher levels of the cyclin E2 RNA and approximately 2-fold higher levels of the cyclin E2 protein in cells infected with recombinant virus lacking the miR-US25-1, suggesting that the miRNA might decrease translation or stability of the protein (Grey *et al.*, 2010). The mutant virus, however, replicates comparable to the WT virus, indicating that at least in fibroblasts, this viral miRNA is not essential for the infection. The extension of these types of studies to HCMV replication in other cell types relevant to the *in vivo* infection, however, may reveal important functions for this miRNA.

Regulation of protein stability

In the past few years, there has been increasing recognition that a major mechanism used by HCMV to control the cell cycle is regulation of protein stability. The proteasome degradation pathway plays a major role in the regulation of multiple cellular functions and its activity is essential for DNA replication and progression of the cell cycle (Ciechanover, 1994; Hershko and Ciechanover, 1998). Most proteins targeted for degradation by the 26S proteasome require prior attachment of ubiquitin. The enzyme E1 forms a thioester bond between itself and ubiquitin, which is then transferred to an E2 ubiquitin-conjugating enzyme. E2 then transfers the ubiquitin to a ubiquitin protein ligase E3, which catalyses the attachment of ubiquitin to a lysine on the target protein. The transfer of additional ubiquitin proteins generates a polyubiquitin chain. The specificity of the degradation occurs primarily at the step of substrate recognition and ubiquitination by the E3 ubiquitin ligases. The protein is translocated into the proteasome, where it is degraded and ubiquitin chains are removed and processed to ubiquitin monomers (Hadar *et al.*, 1992; Gupta *et al.*, 1993).

One of the best-studied multisubunit E3 ubiquitin ligase complexes is the APC (Castro *et al.*, 2005; Peters, 2006; Thornton and Toczyski, 2006). The APC ubiquitinates many cell cycle proteins, including cyclins B and A, securin, Plk1, CENP-E, Aurora A kinase, Aurora B kinase, thymidine kinase, Cdc20, cdc6, and geminin, and its periodic activation and inactivation is critical for accurate cell cycle progression. Specifically, it triggers exit from mitosis, prevents early onset of DNA replication in G_1, and blocks more than one round of DNA replication from a given origin.

The APC contains at least 11 core proteins. Two other subunits, Cdh1 and Cdc20 serve as specificity factors by binding to both the APC and the target protein, thus activating ubiquitination of the target (Eytan *et al.*, 2006). Cdc20 is the major regulatory protein for the APC at the beginning of mitosis, while Cdh1 association with the APC maintains its activity through mitosis and G_1 phase. Based on cryo-EM structures, the human APC can be viewed as two subcomplexes, APC3/APC6/APC7/APC8/APC13 and APC2/APC11/APC10, which are bound to an APC1/APC4/APC5 scaffold (Dube *et al.*, 2005). The binding between APC1, APC4, APC5, and APC8 is also interdependent, such that the loss of one subunit decreases the association of the other three (Thornton *et al.*, 2006).

The early observations from our laboratory and others that several substrates of the APC (e.g. cyclin B, Cdc6, and geminin) abnormally accumulate early in the HCMV infection led to the hypothesis that APC activity is down-regulated during the infection (Biswas *et al.*, 2003; Wiebusch *et al.*, 2003b, 2005; Sanchez and Spector, 2006). Subsequently, we and others showed that the APC is dysregulated at multiple levels (see Fig. I.14.2) (Wiebusch *et al.*, 2005; Tran *et al.*, 2008, 2010). The regulatory subunit Cdh1 is hyperphosphorylated in a cdk-independent manner and the APC becomes destabilized, as evidenced by the dissociation of not only Cdh1 but also APC1, the largest subunit of the APC. In contrast, subunits that contain the tetratricopeptide repeat (TPR) motif (APC3, APC6, APC7, and APC8) remain in a complex. This disassembly is also associated with specific proteasome-dependent degradation of the APC4 and APC5 subunits, although it is not clear whether the disassembly occurs before or after degradation of the proteins. In addition to the disassembly and degradation of specific subunits, there is also relocalization of the subunits with the retention of APC1 in the nucleus and redistribution of the TPR subunits to the cytoplasm.

Since hyperphosphorylation of Cdh1 by cdks occurs in uninfected cells during S phase and is associated with inactivation of the APC, it was initially hypothesized that this might be the major mechanism by which the APC was inactivated. Cdh1 becomes phosphorylated during HCMV infection beginning 8–12 h p.i., and this phosphorylation still occurs in the presence of the cdk inhibitor roscovitine. Based on work showing that the HCMV early protein kinase UL97 acts as a cdk mimic that is involved in the hyperphosphorylation of Rb during the infection, my lab reasoned that it might be involved in the phosphorylation of Cdh1. With the use of a UL97 deletion virus, it was determined that Cdh1 was not phosphorylated, but the APC subunits still dissociated, APC4 and APC5 were degraded, and the APC was still inactivated, albeit with delayed kinetics (Tran *et al.*, 2010). However, when expressed by itself in uninfected cells, it appears that this kinase is able to decrease APC activity (Spector, D., unpublished).

Based on the above studies, attention has turned to the degradation of the APC4 and APC5 subunits. The kinetics of the degradation during the infection indicated that the potential viral protein(s) involved were likely expressed at IE or early times of the infection or brought in with the viral tegument. The results of several experiments, however, have shown that input virion proteins and IE gene expression are not sufficient and that a viral early gene product or cellular gene induced at early times targets APC4 and APC5 (Tran *et al.*, 2010). Recently, it was reported that a small HCMV protein encoded by UL21A is likely to be responsible for the degradation (Fehr *et al.*, 2012). My lab has confirmed this result, and studies are currently in progress to determine the mechanism by which UL21A is able

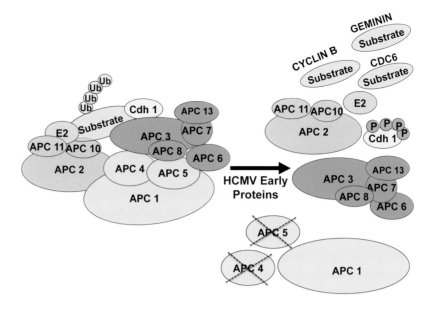

Figure I.14.2 The anaphase–promoting complex (APC) is disabled during the early phase of the HCMV infection. This inhibition is associated with multiple changes in the composition of the APC, including: degradation of the APC4 and APC5 subunits, hyperphosphorylation of the Cdh1 regulatory subunit, and disassembly of the complex. As a result, there is accumulation of proteins such as cyclin B, geminin, and cdc6 that would normally be ubiquitinated and targeted for degradation by the proteasome.

to degrade the proteins. Taken together, the data suggest that UL97 and UL21A may function together to ensure complete inhibition of the APC.

Importance of cyclin-dependent kinases for viral replication

Although limited, there have been some attempts to try to decipher the contributions of specific cdks during the infection by expressing dominant-negative forms of the catalytic subunits in infected cells. Work by Bresnahan and colleagues (1997) showed a decrease in the levels of some capsid proteins in infected cells transfected with a dominant-negative cdk2, suggesting that the activity of cdk2/cyclin E complexes is essential for viral replication. In contrast, work by Hertel et al. (2007) suggests that cdk1 activity is dispensable for the infection, given that virus titre was not markedly reduced in cell cultures transduced with a retrovirus expressing a GFP-tagged DNcdk1. The only effect that the DNcdk1 had was to inhibit a pseudomitosis phenotype that they observed in some cells infected with a variant of the HCMV strain AD169 (Hertel and Mocarski, 2004).

As an alternative approach to establish the importance of cdk activity during infection, we and others have used specific cdk inhibitors, such as the drug Roscovitine, to treat infected cells (Bresnahan et al.,

1997; Sanchez et al., 2004). This drug has high specificity for cdks 1, 2, 5, 7, and 9 and reversibly inhibits their activity by competing for binding of ATP (De Azevedo et al., 1997). Based on the inhibitor studies, there are two primary time intervals during which cdk activity appears to be required for the infection: one during the first 8 hours, and the second after the onset of viral DNA replication (Bresnahan et al., 1997; Sanchez et al., 2004; Sanchez and Spector, 2006). When the drug is added at the beginning of the infection, there is altered processing of IE transcripts, a significant decrease in the expression of select early and late genes, and a reduction in viral DNA replication. However, if the drug is added after 6 to 8 h p.i., these specific effects are not observed. Yet even if the drug is added as late as 48 h p.i., there is still a reduction in virus titre.

Function of cdks at the beginning of the infection

One of the most interesting results when Roscovitine was added at the beginning of the infection was that there was a change in the splicing pattern of the major IE transcripts, IE1-72 and IE2-86, which are alternatively spliced from the same primary transcript (Sanchez et al., 2004). Their first three exons are shared and they differ in the fourth exon, with the fourth exon of IE1-72 closer to the start site of transcription. IE2-86 transcripts are made initially, and the splicing pattern then changes so that IE1-72 RNA accumulates preferentially. At later times, IE2-86 transcripts increase, and there is only a modest further increase in the levels of the IE1-72 RNA. When Roscovitine is present from the initiation of the infection, IE2-86 RNA is preferentially made, and the switch in the splicing pattern to favour IE1-72 does not occur. The polyadenylation signals for IE1-72 and splicing signals for IE2-86 are juxtaposed, and thus it seems when the cdks are inhibited, splicing to yield IE2-86 is favoured over the cleavage/polyadenylation necessary to produce the IE1-72 transcript. This differential processing of transcripts when the polyadenylation and splicing signals are close to each other is seen at another region of IE gene expression that yield the family of UL37 RNAs, and it is affected in a similar way in the presence of the cdk inhibitor.

It was suspected that the altered processing of the UL122–123 and UL37 IE RNAs in the presence of Roscovitine might be associated with changes in the phosphorylation of the C-terminal domain (CTD) of the large subunit of RNA polymerase (RNAP) II by cdk7/cyclin H/and cdk9/cyclin T (Tamrakar et al., 2005). The CTD consists of 52 repeats of the heptapeptide YSPTSPS. Cdk7/cyclin H phosphorylates the Ser 5 of the CTD, and cdk9/cyclin T phosphorylates the CTD on Ser 2. The current consensus is that transcription and RNA processing are integrated events whereby the differential phosphorylation of the CTD repeats at Ser 2 and Ser 5 defines its affinity for various transcription factors, kinases, and RNA processing factors. When the transcription initiation complex is formed on a promoter, the CTD of RNAP II is unphosphorylated. Cdk7/cyclin H primarily phosphorylates Ser 5 on the CTD, which leads to the recruitment of the RNA capping enzymes. Further phosphorylation of the CTD Ser 2 residues by cdk9/cyclin T occurs upon entry to elongation and is associated with recruitment of the cleavage/polyadenylation and splicing machinery.

The results of studies investigating the abundance and activity of the cdk7/cyclin H and cdk9/cyclin T complexes and phosphorylation pattern of RNAP II during the infection supported the above hypothesis. Tamrakar et al. (2005) showed that by 24 h p.i., the infected cell contains more cdk7, cdk9, and cyclin T1 than the uninfected cell, and the abundance of the proteins and the kinase activity of the complexes continue to increase as the infection progresses. Associated with this increase in the cdk7/cyclin H and cdk9/cyclin T cdk activity is an enhanced phosphorylation of the RNAP II CTD on the Ser 2 and Ser 5 residues. As expected, the amount of CTD phosphorylation decreased in infected cells when Roscovitine was added

at the beginning of the infection, but not after 8 h p.i., in accord with the effects on the IE1/IE2 and UL37 IE transcripts.

A key question was why is cdk activity only required during the first 8 hours for accurate processing and expression of IE and early genes and normal viral DNA synthesis? One clue to explain this restricted interval in which cdk activity is required can be found in the differential localization of cdk9 and cdk7 at the beginning of the infection. Upon cell entry, incoming HCMV genomes localize near ND10, where viral IE transcription begins (Ishov and Maul, 1996; Ishov et al., 1997). Following translation, the IE1-72 and IE2-86 proteins return to the nucleus and concentrate near the ND10. IE1-72 mediates the dispersal of ND10 associated proteins and it also disperses, while IE2-86 persists at the site (Kelly et al., 1995; Korioth et al., 1996; Ahn and Hayward, 1997; Ishov et al., 1997; Ahn et al., 1998). It was found that as early as 4 h p.i., several proteins involved in RNA transcription, including cdk9, cyclin T1, cdk7, Brd4, HDAC1 and HDAC2 colocalize with IE2-86 in distinct aggregates (referred to as viral transcriptosomes) adjacent to the ND10 (Tamrakar et al., 2005; Kapasi and Spector, 2008). However, the addition of Roscovitine at the beginning of the infection specifically prevents the recruitment of cdk7 and cdk9, but not the other proteins, to the transcriptosome (Kapasi and Spector, 2008). In contrast, both cdks colocalize with IE2-86 in the transcriptosome if Roscovitine is added after 8 h p.i. These results suggest that during the first few hours of the infection, IE2-86 and other proteins required for viral RNA synthesis are recruited to form the viral transcriptosome, and one or more of the cdks targeted by Roscovitine must remain active for cdk7 and cdk9 to be recruited. Accordingly, the required level of phosphorylation of the RNAP II CTD at these sites is established within the first 8 hours. Interestingly, with additional inhibitors and cdk9 mutants, it was found that the cdk9 kinase activity is not necessary for its recruitment to the transcriptosome, but the cdk9 must be able to form a complex with cyclin T1 (Kapasi et al., 2009).

Function of cdks at later times in the infection

Sanchez and Spector (2006) showed that addition of Roscovitine to infected cells at 24 or 48 h p.i. results in a 1–2 log drop in virus titre, but viral DNA synthesis is reduced only 2- to 3-fold. The drop in titre is primarily due to inhibition of the maturation or release of virions. Consistent with these data, changes in the levels, post-translational modification, and localization of specific proteins have been observed in Roscovitine-treated cells at these later times. For example, the levels of the IE2-86 and pp150 (UL32) proteins are significantly reduced, but there is not a corresponding decrease in the mRNAs. In contrast, there is an increase in intranuclear aggregates of the pUL69 matrix protein, primarily in its hyperphosphorylated form, which may be due to enhanced phosphorylation by the viral UL97 kinase as a result of cdk inhibition (Thomas et al., 2009). These intranuclear aggregates also contain cdk9 and cyclin T1, and UL69 can interact with cyclin T1, but not cdk9 (Rechter et al., 2009; Feichtinger et al., 2011). However, there is no direct evidence that cdk9/cyclin T1 phosphorylates UL69 during the infection. Another effect of Roscovitine is the altered phosphorylation and nuclear retention of pp65 (Sanchez and Spector, 2006; Sanchez et al., 2007). This protein is a major component of the virion matrix, and its transfer from the nucleus to the cytoplasmic assembly compartment occurs during the late phase of the infection. Interestingly, mutation of the potential cdk phosphorylation sites does not affect the ability of pp65 to localize to the nucleus or re-localize to the cytoplasm late in the infection, and thus there may be another cdk target that modulates the localization of pp65. The above examples are only a small subset of the virion proteins that are phosphorylated. Whether the phosphorylation and function of other virion proteins is mediated by the cdks is an important question, and recent advances in whole cell proteomic analysis should facilitate studies to address this.

Perspectives

In this chapter, I have highlighted the many ways that HCMV alters cell cycle regulatory pathways. A key question is which changes are actually important for viral replication? For example, why does the virus inhibit the APC? Is inactivation of the APC solely needed to allow accumulation of proteins to promote cell cycle arrest, or are there other target proteins whose accumulation impacts on different cellular regulatory pathways to promote the infection. Future experiments involving knockdown of gene expression through the use of siRNAs or induced overexpression with lentiviral vectors should provide some insight into these questions. To date, only a few viral genes have been implicated in the modulation of these pathways, and it is important to identify the other viral genes, as they might be good targets for new antiviral therapies. It also should be recognized that most of the studies to date have been done in fibroblasts. To understand how exploitation of the host cell cycle regulatory pathways impact on HCMV pathogenesis, it is necessary to extend these studies to other relevant target cells such as EC, smooth muscle cells, cells of the neural lineage, and monocytes.

Acknowledgements

Research in the Spector laboratory was supported by NIH grants CA073490, CA037429, and AI0883991. I thank the members of my lab for their helpful comments.

References

Ahn, J.H., and Hayward, G.S. (1997). The major immediate-early proteins IE1 and IE2 of human cytomegalovirus colocalize with and disrupt PML-associated nuclear bodies at very early times in infected permissive cells. J. Virol. 71, 4599–4613.

Ahn, J.H., Brignole, E.R., and Hayward, G.S. (1998). Disruption of PML subnuclear domains by the acidic IE1 protein of human cytomegalovirus is mediated through interaction with PML and may modulate a RING finger-dependent cryptic transactivator function of PML. Mol. Cell. Biol. 18, 4899–4913.

Biswas, N., Sanchez, V., and Spector, D.H. (2003). Human cytomegalovirus infection leads to accumulation of geminin and inhibition of the licensing of cellular DNA replication. J. Virol. 77, 2369–2376.

Blow, J.J., and Dutta, A. (2005). Preventing re-replication of chromosomal DNA. Nat. Rev. Mol. Cell. Biol. 6, 476–486.

Boldogh, I., AbuBakar, S., Deng, C.Z., and Albrecht, T. (1991). Transcriptional activation of cellular oncogenes fos, jun and myc by human cytomegalovirus. J. Virol. 65, 1568–1571.

Bonin, L.R., and McDougall, J.K. (1997). Human cytomegalovirus IE2 86-kilodalton protein binds p53 but does not abrogate G1 checkpoint function. J. Virol. 71, 5831–5870.

Booher, R.N., Holman, P.S., and Fattaey, A. (1997). Human Myt1 is a cell cycle-regulated kinase that inhibits Cdc2 but not Cdk2 activity. J. Biol. Chem. 272, 22300–22306.

Bottazzi, M.E., Buzzai, M., Zhu, X., Desdouets, C., Brechot, C., and Assoian, R.K. (2001). Distinct effects of mitogens and actin cytoskeleton on CREB and pocket protein phosphorylation control the extent and timing of cyclin A promoter activity. Mol. Cell. Biol. 21, 7607–7616.

Brandeis, M., and Hunt, T. (1996). The proteolysis of mitotic cyclins in mammalian cells persists from the end of mitosis until the onset of S phase. EMBO J. 15, 5280–5289.

Bresnahan, W.A., Boldogh, I., Ma, T., Albrecht, T., and Thompson, E.A. (1996a). Cyclin E/CDK2 activity is controlled by different mechanisms in the G0 and G1 phases of the cell cycle. Cell Growth Differ. 7, 1283–1290.

Bresnahan, W.A., Boldogh, I., Thompson, E.A., and Albrecht, T. (1996b). Human cytomegalovirus inhibits cellular DNA synthesis and arrests productively infected cells in late G1. Virology 224, 156–160.

Bresnahan, W.A., Boldogh, I., Chi, P., Thompson, E.A., and Albrecht, T. (1997). Inhibition of cellular CDK2 activity blocks human cytomegalovirus replication. Virology 231, 239–247.

Bresnahan, W.A., Albrecht, T., and Thompson, E.A. (1998). The cyclin E promoter is activated by human cytomegalovirus 86-kDa immediate-early protein. J. Biol. Chem. 273, 22075–22082.

Bresnahan, W.A., Hultman, G.E., and Shenk, T. (2000). Replication of wild-type and mutant human cytomegalovirus in life-extended human diploid fibroblasts. J. Virol. 74, 10816–10818.

Browne, E.P., Wing, B., Coleman, D., and Shenk, T. (2001). Altered cellular mRNA levels in human cytomegalovirus-infected fibroblasts: viral block to the accumulation of antiviral mRNAs. J. Virol. 75, 12319–12330.

Burgdorf, S.W., Clark, C.L., Burgdorf, J.R., and Spector, D.H. (2011). Mutation of glutamine to arginine at position 548 of IE2 86 in human cytomegalovirus leads to decreased expression of IE2 40, IE2 60, UL83, and UL84 and increased transcription of US8–9 and US29–32. J. Virol. 85, 11098–11110.

Caldon, C.E., and Musgrove, E.A. (2010). Distinct and redundant functions of cyclin E1 and cyclin E2 in development and cancer. Cell Div. 5, 2.

Cantrell, S.R., and Bresnahan, W.A. (2005). Interaction between the human cytomegalovirus UL82 gene product (pp71) and hDaxx regulates immediate-early gene expression and viral replication. J. Virol. 79, 7792–7802.

Casavant, N.C., Luo, M.H., Rosenke, K., Winegardner, T., Zurawska, A., and Fortunato, E.A. (2006). Potential role for p53 in the permissive life cycle of human cytomegalovirus. J. Virol. 80, 8390–8401.

Castano, E., Kleyner, Y., and Dynlacht, B.D. (1998). Dual cyclin-binding domains are required for p107 to function as a kinase inhibitor. Mol. Cell. Biol. 18, 5380–5391.

Castillo, J.P., Yurochko, A., and Kowalik, T.F. (2000). Role of human cytomegalovirus immediate-early proteins in cell growth control. J. Virol. 74, 8028–8037.

Castillo, J.P., Frame, F.M., Rogoff, H.A., Pickering, M.T., Yurochko, A.D., and Kowalik, T.F. (2005). Human cytomegalovirus IE1-72 activates ataxia telangiectasia mutated kinase and a p53/p21-mediated growth arrest response. J. Virol. 79, 11467–11475.

Castro, A., Bernis, C., Vigneron, S., Labbe, J.C., and Lorca, T. (2005). The anaphase-promoting complex: a key factor in the regulation of cell cycle. Oncogene 24, 314–325.

Challacombe, J.F., Rechtsteiner, A., Gottardo, R., Rocha, L.M., Browne, E.P., Shenk, T., Altherr, M.R., and Brettin, T.S. (2004). Evaluation of the host transcriptional response to human cytomegalovirus infection. Physiol. Genomics 18, 51–62.

Chen, J., and Fang, G. (2001). MAD2B is an inhibitor of the anaphase-promoting complex. Genes Dev. 15, 1765–1770.

Chen, Z., Knutson, E., Kurosky, A., and Albrecht, T. (2001). Degradation of p21cip1 in cells productively infected with human cytomegalovirus. J. Virol. 75, 3613–3625.

Ciechanover, A. (1994). The ubiquitin-proteasome proteolytic pathway. Cell 79, 13–21.

De Azevedo, W.F., Leclerc, S., Meijer, L., Havlicek, L., Strnad, M., and Kim, S.H. (1997). Inhibition of cyclin-dependent kinases by purine analogues: crystal structure of human cdk2 complexed with roscovitine. Eur. J. Biochem. 243, 518–526.

Desdouets, C., Matesic, G., Molina, C.A., Foulkes, N.S., Sassone-Corsi, P., Bréchot, C., and Sobzak-Thepot, J. (1995). Cell cycle regulation of cyclin A gene expression by the cyclic AMP transcription factors CREB and CREM. Mol. Cell. Biol. 15, 3301–3309.

Dittmer, D., and Mocarski, E.S. (1997). Human cytomegalovirus infection inhibits G1/S transition. J. Virol. 71, 1629–1634.

Dube, P., Herzog, F., Gieffers, C., Sander, B., Riedel, D., Muller, S.A., Engel, A., Peters, J.M., and Stark, H. (2005). Localization of the coactivator Cdh1 and the cullin subunit Apc2 in a cryo-electron microscopy model of vertebrate APC/C. Mol. Cell 20, 867–879.

Dunn, C., Chalupny, N.J., Sutherland, C.L., Dosch, S., Sivakumar, P.V., Johnson, D.C., and Cosman, D. (2003). Human cytomegalovirus glycoprotein UL16 causes intracellular sequestration of NKG2D ligands, protecting against natural killer cell cytotoxicity. J. Exp. Med. 197, 1427–1439.

Dunn, W., Trang, P., Zhong, Q., Yang, E., Belle, C.v., and Liu, F. (2005). Human cytomegalovirus expresses novel microRNAs during productive viral infection. Cell. Microbiol. 7, 1684–1695.

Dyson, N. (1998). The regulation of E2F by pRB-family proteins. Genes Dev. 12, 2245–2262.

Estes, J.E., and Huang, E.-S. (1977). Stimulation of cellular thymidine kinases by human cytomegalovirus. J. Virol. 24, 13–21.

Eytan, E., Moshe, Y., Braunstein, I., and Hershko, A. (2006). Roles of the anaphase-promoting complex/cyclosome and of its activator Cdc20 in functional substrate binding. Proc. Natl. Acad. Sci. U.S.A. 103, 2081–2086.

Fehr, A.R., Gualberto, N.C., Savaryn, J.P., Terhune, S.S., and Yu, D. (2012). Proteasome-dependent disruption of the E3 ubiquitin ligase anaphase-promoting complex by HCMV protein pUL21a. PLoS Pathog. 8, e1002789.

Feichtinger, S., Stamminger, T., Müller, R., Graf, L., Klebl, B., Eickhoff, J., and Marschall, M. (2011). Recruitment of cyclin-dependent kinase 9 to nuclear compartments during cytomegalovirus late replication: importance of an interaction between viral pUL69 and cyclin T1. J. Gen. Virol. 92, 1519–1531.

Ferguson, M., Henry, P.A., and Currie, R.A. (2003). Histone deacetylase inhibition is associated with transcriptional repression of the Hmga2 gene. Nucl. Acids Res. 31, 3123–3133.

Fortunato, E.A., and Spector, D.H. (1998). p53 and RPA are sequestered in viral replication centers in the nuclei of cells infected with human cytomegalovirus. J. Virol. 72, 2033–2039.

Fortunato, E.A., Sommer, M.H., Yoder, K., and Spector, D.H. (1997). Identification of domains within the human cytomegalovirus major immediate-early 86-kilodalton protein and the retinoblastoma protein required for physical and functional interaction with each other. J. Virol. 71, 8176–8185.

Fortunato, E.A., Dell'Aquila, M.L., and Spector, D.H. (2000). Specific chromosome 1 breaks induced by human cytomegalovirus. Proc. Natl. Acad. Sci. U.S.A. 97, 853–858.

Fortunato, E.A., Sanchez, V., Yen, J.Y., and Spector, D.H. (2002). Infection of cells with human cytomegalovirus during S phase results in a blockade to immediate-early gene expression that can be overcome by inhibition of the proteasome. J. Virol. 76, 5369–5379.

Gaspar, M., and Shenk, T. (2006). Human cytomegalovirus inhibits a DNA damage response by mislocalizing checkpoint proteins. Proc. Natl. Acad. Sci. U.S.A. 103, 2821–2826.

Glotzer, M., Murray, A.W., and Kirschner, M.W. (1991). Cyclin is degraded by the ubiquitin pathway. Nature *349*, 132–138.

Grey, F., Antoniewicz, A., Allen, E., Saugstad, J., McShea, A., Carrington, J.C., and Nelson, J. (2005). Identification and characterization of human cytomegalovirus-encoded microRNAs. J. Virol. *79*, 12095–12099.

Grey, F., Tirabassi, R., Meyers, H., Wu, G., McWeeney, S., Hook, L., and Nelson, J.A. (2010). A viral microRNA down-regulates multiple cell cycle genes through mRNA 5'UTRs. PLoS Pathog. *6*, e1000967.

Gupta, K., Copeland, N.G., Gilbert, D.J., Jenkins, N.A., and Gray, D.A. (1993). Unp, a mouse gene related to the tre oncogene. Oncogene *8*, 2307–2310.

Hadar, T., Warms, J.V.B., Rose, I.A., and Hershko, A. (1992). A ubiquitin C-terminal isopeptidase that acts on polyubiquitin chains – role in protein degradation. J. Biol. Chem. *267*, 719–727.

Hagemeier, C., Caswell, R., Hayhurst, G., Sinclair, J., and Kouzarides, T. (1994). Functional interaction between the HCMV IE2 transactivator and the retinoblastoma protein. EMBO J. *13*, 2897–2903.

Hannemann, H., Rosenke, K., O'Dowd, J.M., and Fortunato, E.A. (2009). The presence of p53 influences the expression of multiple human cytomegalovirus genes at early times postinfection. J. Virol. *83*, 4316–4325.

Hansen, K., Farkas, T., Lukas, J., Holm, K., Roonstrand, L., and Bartek, J. (2001). Phosphorylation-dependent and -independent functions of p130 cooperate to evoke a sustained G1 block. EMBO J. *20*, 422–432.

Harper, J.W., and Elledge, S.J. (1998). The role of Cdk7 in CAK function, a retro-retrospective. Genes Dev. *12*, 285–289.

Hayashi, M.L., Blankenship, C., and Shenk, T. (2000). Human cytomegalovirus UL69 is required for efficient accumulation of infected cells in the G1 phase of the cell cycle. Proc. Natl. Acad. Sci. U.S.A. *97*, 2692–2696.

Heald, R., McLoughlin, M., and McKeon, F. (1993). Human Wee1 maintains mitotic timing by protecting the nucleus from cytoplasmically activated cdc2 kinase. Cell *74*, 463–474.

Henglein, B., Chenivesse, X., Wang, J., Eick, D., and Bréchot, C. (1994). Structure and cell cycle-regulated transcriptiion of the human cyclin A gene. Proc. Natl. Acad. Sci. U.S.A. *91*, 5490–5494.

Hershko, A., and Ciechanover, A. (1998). The ubiquitin system. Ann. Rev. Biochem. *67*, 425–479.

Hertel, L., and Mocarski, E.S. (2004). Global analysis of host cell gene expression late during cytomegalovirus infection reveals extensive dysregulation of cell cycle gene expression and induction of pseudomitosis independent of US28 function. J. Virol. *78*, 11988–12011.

Hertel, L., Chou, S., and Mocarski, E.S. (2007). Viral and cell cycle-regulated kinases in cytomegalovirus-induced pseudomitosis and replication. PLoS Pathog. *3*, e6.

Hirai, K., and Watanabe, Y. (1976). Induction of a-type DNA polymerases in human cytomegalovirus-infected WI-38 cells. Biochim. Biophys. Acta *447*, 328–339.

Hoffmann, I., Clarke, P.R., Marcote, M.J., Karsenti, E., and Draetta, G. (1993). Phosphorylation and activation of human cdc25-C by cdc2-cyclin B and its involvement in the self-amplification of MPF at mitosis. EMBO J. *12*, 53–63.

Hofmann, H., Sindre, H., and Stamminger, T. (2002). Functional interaction between the pp71 protein of human cytomegalovirus and the PML-interacting protein human Daxx. J. Virol. *76*, 5769–5783.

Hsu, C.-H., Chang, M.D.T., Tai, K.-Y., Yang, Y.-T., Wang, P.-S., Chen, C.-J., Wang, Y.-H., Lee, S.-C., Wu, C.-W., and Juan, L.-J. (2004). HCMV IE2-mediated inhibition of HAT activity down-regulates p53 function. EMBO J. *23*, 2269–2280.

Hume, A.J., Finkel, J.S., Kamil, J.P., Coen, D.M., Culbertson, M.R., and Kalejta, R.F. (2008). Phosphorylation of retinoblastoma protein by viral protein with cyclin-dependent kinase function. Science *320*, 797–799.

Hwang, J., and Kalejta, R.F. (2007). Proteasome-dependent, ubiquitin-independent degradation of Daxx by the viral pp71 protein in human cytomegalovirus-infected cells. Virology *367*, 334–338.

Hwang, E.S., Zhang, Z., Cai, H., Huang, D.Y., Huong, S.M., Cha, C.Y., and Huang, E.S. (2009). Human cytomegalovirus IE1-72 protein interacts with p53 and inhibits p53-dependent transactivation by a mechanism different from that of IE2-86 protein. J. Virol. *83*, 12388–11239.

Ishov, A.M., and Maul, G.G. (1996). The periphery of nuclear domain 10 (ND10) as site of DNA virus deposition. J. Cell Biol. *134*, 815–826.

Ishov, A.M., Stenberg, R.M., and Maul, G.G. (1997). Human cytomegalovirus immediate-early interaction with host nuclear structures: Definition of an immediate transcript environment. J. Cell Biol. *138*, 5–16.

Ishov, A.M., Vladimirova, O.V., and Maul, G.G. (2002). Daxx-mediated accumulation of human cytomegalovirus tegument protein pp71 at ND10 facilitates initiation of viral infection at these nuclear domains. J. Virol. *76*, 7705–7712.

Isom, H.C. (1979). Stimulation of ornithine carboxylase by human cytomegalovirus. J. Gen. Virol. *42*, 265–278.

Jault, F.M., Jault, J.-M., Ruchti, F., Fortunato, E.A., Clark, C., Corbeil, J., Richman, D.D., and Spector, D.H. (1995). Cytomegalovirus infection induces high levels of cyclins, phosphorylated RB, and p53, leading to cell cycle arrest. J. Virol. *69*, 6697–6704.

Johnson, R.A., Yurochko, A.D., Poma, E.E., Zhu, L., and Huang, E.-S. (1999). Domain mapping of the human cytomegalovirus IE1-72 and cellular p107 protein–protein interaction and the possible functional consequences. J. Gen. Virol. *80*, 1293–1303.

Kalejta, R.F., Bechtel, J.T., and Shenk, T. (2003). Human cytomegalovirus pp71 stimulates cell cycle progression by inducing the proteasome-dependent degradation of the retinoblastoma family of tumor suppressors. Mol. Cell. Biol. *23*, 1885–1895.

Kapasi, A.J., and Spector, D.H. (2008). Inhibition of the cyclin-dependent kinases at the beginning of the human cytomegalovirus infection specifically alters the levels and localization of the RNA polymerase II carboxyl-terminal domain kinases cdk9 and cdk7 at the viral transcriptosome. J. Virol. *82*, 394–407.

Kapasi, A.J., Clark, C.L., Tran, K., and Spector, D.H. (2009). Recruitment of cdk9 to the immediate-early viral transcriptosomes during human cytomegalovirus infection requires efficient binding to cyclin T1, a threshold level of IE2 86, and active transcription. J. Virol. *83*, 5904–5917.

Karlsson, C., Katich, S., Hagting, A., Hoffmann, I., and Pines, J. (1999). Cdc25B and cdc25C differ markedly in their properties as initiators of mitosis. J. Cell Biol. *146*, 573–584.

Kelly, C., Driel, R.V., and Wilkinson, G.W. (1995). Disruption of PML-associated nuclear bodies during human cytomegalovirus infection. J. Gen. Virol. *76*, 2887–2893.

Korioth, F., Maul, G.G., Plachter, B., Stamminger, T., and Frey, J. (1996). The nuclear domain 10 (ND10) is disrupted by the human cytomegalovirus gene product IE1. Exp. Cell Res. *229*, 155–158.

Lammer, C., Wagerer, S., Saffrich, R., Mertens, D., Ansorge, W., and Hoffmann, I. (1998). The cdc25B phosphatase is essential for the G2/M phase transition in human cells. J. Cell Sci. *111*, 2445–2453.

Lavin, M.F., and Gueven, N. (2006). The complexity of p53 stabilization and activation. Cell Death Differ. *13*, 941–950.

Liu, F., Stanton, J.J., Wu, Z., and Piwnica-Worms, H. (1997). The human MYT1 kinase preferentially phosphorylates CDC2 on threonine 14 and localizes to the endoplasmic reticulum and Golgi complex. Mol. Cell. Biol. *17*, 571–583.

Liu, F., Rothblum-Oviatt, C., Ryan, C.E., and Piwnica-Worms, H. (1999). Overproduction of human Myt1 kinase Induces a G2 cell cycle delay by interfering with the intracellular trafficking of cdc2-cyclin B1 complexes. Mol. Cell. Biol. *19*, 5113–5123.

Lu, M., and Shenk, T. (1996). Human cytomegalovirus infection inhibits cell cycle progression at multiple points, including the transition from G1 to S. J. Virol. *70*, 8850–8857.

Lu, M., and Shenk, T. (1999). Human cytomegalovirus UL69 protein induces cells to accumulate in G_1 phase of the cell cycle. J. Virol. *73*, 676–683.

Lukas, J., Lukas, C., and Bartek, J. (2004). Mammalian cell cycle checkpoints: signalling pathways and their organization in space and time. DNA Repair *3*, 997–1007.

Lukashchuk, V., McFarlane, S., Everett, R.D., and Preston, C.M. (2008). Human cytomegalovirus protein pp71 displaces the chromatin-associated factor ATRX from nuclear domain 10 at early stages of infection. J. Virol. *82*, 12543–12554.

Luo, M.H., Rosenke, K., Czornak, K., and Fortunato, E.A. (2007). Human cytomegalovirus disrupts both ataxia telangiectasia mutated protein (ATM)- and ATM-Rad3-related kinase-mediated DNA damage responses during lytic infection. J. Virol. *81*, 1934–1950.

McElroy, A.K., Dwarakanath, R.S., and Spector, D.H. (2000). Dysregulation of cyclin E gene expression in human cytomegalovirus-infected cells requires viral early gene expression and is associated with changes in the Rb-related protein p130. J. Virol. *74*, 4192–4206.

McGarry, T.J., and Kirschner, M.W. (1998). Geminin, an inhibitor of DNA replication, is degraded during mitosis. Cell *93*, 1043–1053.

Machida, Y.J., Hamlin, J.L., and Dutta, A. (2005). Right place, right time, and only once: replication initiation in metazoans. Cell *123*, 13–24.

Margolis, M.J., Panjovic, S., Wong, E.L., Wade, M., Jupp, R., Nelson, J.A., and Azizkhan, J.C. (1995). Interaction of the 72-kilodalton human cytomegalovirus IE1 gene product with E2F1 coincides with E2F-dependent activation of dihydrofolate reductase transcription. J. Virol. *69*, 7759–7767.

Marshall, K.R., Rowley, K.V., Rinaldi, A., Nicholson, I.P., Ishov, A.M., Maul, G.G., and Preston, C.M. (2002). Activity and

intracellular localization of the human cytomegalovirus protein pp71. J. Gen. Virol. *83*, 1601–1612.

Maul, G.G., and Negorev, D. (2008). Differences between mouse and human cytomegalovirus interactions with their respective hosts at immediate-early times of the replication cycle. Med. Microbiol. Immunol. *197*, 241–249.

Millar, J.B.A., and Russell, P. (1992). The cdc25 M-phase inducer: an unconventional protein phosphatase. Cell *68*, 407–410.

Mueller, P.R., Coleman, T.R., Kumagai, A., and Dunphy, W.G. (1995). Myt1: a membrane-associated inhibitory kinase that phosphorylates cdc2 on both threonine-14 and tyrosine-15. Science *270*, 86–90.

Muganda, P., Mendoza, O., Hernandez, J., and Qian, Q. (1994). Human cytomegalovirus elevates levels of the cellular protein p53 in infected fibroblasts. J. Virol. *68*, 8028–8034.

Murphy, E.A., Streblow, D.N., Nelson, J.A., and Stinski, M.F. (2000). The human cytomegalovirus IE86 protein can block cell cycle progression after inducing transition into the S phase of the cell cycle. J. Virol. *74*, 7108–7118.

Murray, A.W. (2004). Recycling the cell cycle: cyclins revisited. Cell *116*, 221–234.

Nishitani, H., and Lygerou, Z. (2002). Control of DNA Replication. Genes Cells *7*, 523–534.

Noris, E., Zannetti, C., Demurtas, A., Sinclair, J., De Andrea, M., Gariglio, M., and Landolfo, S. (2002). Cell cycle arrest by human cytomegalovirus 86-kDa IE2 protein resembles premature senescence. J. Virol. *76*, 12135–12148.

Ørom, U.A., Nielsen, F.C., and Lund, A.H. (2008). MicroRNA-10a binds the 5′UTR of ribosomal protein mRNAs and enhances their translation. Mol. Cell *23*, 460–471.

Peters, J.M. (2006). The anaphase promoting complex/cyclosome: a machine designed to destroy. Nat. Rev. Mol. Cell. Biol. *7*, 644–656.

Petrik, D.T., Schmitt, K.P., and Stinski, M.F. (2006). Inhibition of cellular DNA synthesis by the human cytomegalovirus IE86 protein is necessary for efficient virus replication. J. Virol. *80*, 3872–3883.

Pfeffer, S., Sewer, A., Lagos-Quintana, M., Sheridan, R., Sander, C., Grasser, F.A., Dyk, L.F.v., Ho, C.K., Shuman, S., Chien, M., et al. (2005). Identification of microRNAs of the herpesvirus family. Nat. Methods *2*, 269–276.

Pickering, M.T., Debatis, M., Castillo, J., Lagadinos, A., Wang, S., Lu, S., and Kowalik, T.F. (2011). An E2F1-mediated DNA damage response contributes to the replication of human cytomegalovirus. PLoS Pathog. *7*, e1001342.

Poma, E.E., Kowalik, T.F., Zhu, L., Sinclair, J.H., and Huang, E.-S. (1996). The human cytomegalovirus IE1-72 protein interacts with the cellular p107 protein and relieves p107-mediated transcriptional repression of an E2F-responsive promoter. J. Virol. *70*, 7867–7877.

Preston, C.M., and Nicholl, M.J. (2006). Role of the cellular protein hDaxx in human cytomegalovirus immediate-early gene expression. J. Gen. Virol. *87*, 1113–1121.

Prichard, M.N., Sztul, E., Daily, S.L., Perry, A.L., Frederick, S.L., Gill, R.B., Hartline, C.B., Streblow, D.N., Varnum, S.M., Smith, R.D., et al. (2008). Human cytomegalovirus UL97 kinase activity is required for the hyperphosphorylation of retinoblastoma protein and inhibits the formation of nuclear aggresomes. J. Virol. *82*, 5054–5067.

Qian, Z., Leung-Pineda, V., Xuan, B., Piwnica-Worms, H., and Yu, D. (2010). Human cytomegalovirus protein pUL117 targets the mini-chromosome maintenance complex and suppresses cellular DNA synthesis. PLoS Pathog. *6*, e1000814.

Rechter, S., Scott, G.M., Eickhoff, J., Zielke, K., Auerochs, S., Müller, R., Stamminger, T., Rawlinson, W.D., and Marschall, M. (2009). Cyclin-dependent kinases phosphorylate the cytomegalovirus RNA export protein pUL69 and modulate its nuclear localization and activity. J. Biol. Chem. *284*, 8605–8613.

Roberts, A.P., Lewis, A.P., and Jopling, C.L. (2011). miR-122 activates hepatitis C virus translation by a specialized mechanism requiring particular RNA components. Nucl. Acids Res. *39*, 7716–7729.

Rosenke, K., Samuel, M.A., McDowell, E.T., Toerne, M.A., and Fortunato, E.A. (2006). An intact sequence-specific DNA-binding domain is required for human cytomegalovirus-mediated sequestration of p53 and may promote in vivo binding to the viral genome during infection. Virology *348*, 19–34.

Saffert, R.T., and Kalejta, R.F. (2006). Inactivating a cellular intrinsic immune defense mediated by Daxx is the mechanism through which the human cytomegalovirus pp71 protein stimulates viral immediate-early gene expression. J. Virol. *80*, 3863–3871.

Salvant, B.S., Fortunato, E.A., and Spector, D.H. (1998). Cell cycle dysregulation by human cytomegalovirus: Influence of the cell cycle phase at the time of infection and effects on cyclin transcription. J. Virol. *72*, 3729–3741.

Sanchez, V., and Spector, D.H. (2006). Cyclin-dependent kinase activity is required for efficient expression and post-translational modification of human cytomegalovirus proteins and for production of extracellular particles. J. Virol. *80*, 5886–5896.

Sanchez, V., McElroy, A.K., and Spector, D.H. (2003). Mechanisms governing maintenance of cdk1/cyclin B1 kinase activity in cells infected with human cytomegalovirus. J. Virol. *77*, 13214–13224.

Sanchez, V., McElroy, A.K., Yen, J., Tamrakar, S., Clark, C.L., Schwartz, R.A., and Spector, D.H. (2004). Cyclin-dependent kinase activity is required at early times for accurate processing and accumulation of the human cytomegalovirus UL122–123 and UL37 immediate-early transcripts and at later times for virus production. J. Virol. *78*, 11219–11232.

Sanchez, V., Mahr, J.A., Orazio, N., and Spector, D.H. (2007). Nuclear export of the human cytomegalovirus tegument protein pp65 requires cyclin-dependent kinase activity and the Crm1 exporter. J. Virol. *81*, 11730–11736.

Satyanarayana, A., and Kaldis, P. (2009). Mammalian cell-cycle regulation: several Cdks, numerous cyclins and diverse compensatory mechanisms. Oncogene *28*, 2925–2939.

Schulze, A., Zerfass, K., Spitkovsky, D., Middendorp, S., Berges, J., Helin, K., Jansen-Dürr, P., and Henglein, B. (1995). Cell cycle regulation of the cyclin A gene promoter is mediated by a variant E2F site. Proc. Natl. Acad. Sci. U.S.A. *92*, 11264–11268.

Shen, Y.H., Utama, B., Wang, J., Raveendran, M., Senthil, D., Waldman, W.J., Belcher, J.D., Vercellotti, G., Martin, D., Mitchelle, B.M., *et al.* (2004). Human cytomegalovirus causes endothelial injury through the ataxia telangiectasia mutant and p53 DNA damage signaling pathways. Circ. Res. *94*, 1310–1317.

Shlapobersky, M., Sanders, R., Clark, C., and Spector, D.H. (2006). Repression of HMGA2 gene expression by

human cytomegalovirus involves the IE2 86-kilodalton protein and is necessary for efficient viral replication and inhibition of cyclin A transcription. J. Virol. *80*, 9951–9961.

Slee, E.A., O'Connor, D.J., and Lu, X. (2004). To die or not to die: how does p53 decide? Oncogene *23*, 2809–2818.

Sommer, M.H., Scully, A.L., and Spector, D.H. (1994). Trans-activation by the human cytomegalovirus IE2 86 kDa protein requires a domain that binds to both TBP and RB. J. Virol. *68*, 6223–6231.

Song, Y.-J., and Stinski, M.F. (2002). Effect of the human cytomegalovirus IE86 protein on expression of E2F responsive genes: a DNA microarray analysis. Proc. Natl. Acad. Sci. U.S.A. *99*, 2836–2841.

Song, Y.J., and Stinski, M.F. (2005). Inhibition of cell division by the human cytomegalovirus IE86 protein: role of the p53 pathway or cyclin-dependent kinase 1/cyclin B1. J. Virol. *79*, 2597–2603.

Speir, E., Modali, R., Huang, E.-S., Leon, M.B., Sahwl, F., Finkel, T., and Epstein, S.E. (1994). Potential role of human cytomegalovirus and p53 interaction in coronary restenosis. Science *265*, 391–394.

Stark, T.J., Arnold, J.D., Spector, D.H., and Yeo, G.W. (2012). High-Resolution Profiling and Analysis of Viral and Host Small RNAs during Human Cytomegalovirus Infection. J. Virol. *86*, 226–235.

Stern-Ginossar, N., Saleh, N., Goldberg, M.D., Prichard, M., Wolf, D.G., and Mandelboim, O. (2009). Analysis of human cytomegalovirus-encoded microRNA activity during infection. J. Virol. *83*, 10684–10693.

Tamrakar, S., Kapasi, A.J., and Spector, D.H. (2005). Human cytomegalovirus infection induces specific hyperphosphorylation of the carboxyl-terminal domain of the large subunit of RNA polymerase II that is associated with changes in the abundance, activity, and localization of cdk9 and cdk7. J. Virol. *79*, 15477–15493.

Tessari, M.A., Gostissa, M., Altamura, S., Sgarra, R., Rustighi, A., Salvagno, C., Caretti, G., Imbriano, C., Mantovani, R., Sal, G.D., *et al.* (2003). Transcriptional activation of the cyclin A gene by the architectural transcription factor HMGA2. Mol. Cell. Biol. *23*, 9104–9116.

Thomas, M., Rechter, S., Milbradt, J., Auerochs, S., Müller, R., Stamminger, T., and Marschall, M. (2009). Cytomegaloviral protein kinase pUL97 interacts with the nuclear mRNA export factor pUL69 to modulate its intranuclear localization and activity. J. Gen. Virol. *90*, 567–578.

Thornton, B.R., and Toczyski, D.P. (2006). Precise destruction: an emerging picture of the APC. Genes Dev. *20*, 3069–3078.

Thornton, B.R., Ng, T.M., Matyskiela, M.E., Carroll, C.W., Morgan, D.O., and Toczyski, D.P. (2006). An architectural map of the anaphase-promoting complex. Genes Dev. *20*, 449–460.

Tran, K., Mahr, J.A., Choi, J., Teodoro, J.G., Green, M.R., and Spector, D.H. (2008). Accumulation of substrates of the anaphase-promoting complex (APC) during human cytomegalovirus infection is associated with the phosphorylation of Cdh1 and the dissociation and relocalization of the APC subunits. J. Virol. *82*, 529–537.

Tran, K., Kamil, J.P., Coen, D.M., and Spector, D.H. (2010). Inactivation and disassembly of the anaphase-promoting complex during human cytomegalovirus infection is associated with degradation of the APC5 and

APC4 subunits and does not require UL97-mediated phosphorylation of Cdh1. J. Virol. *84*, 10832–10843.

Vousden, K.H., and Lu, X. (2002). Live or let die: the cell's response to p53. Nature Rev. Cancer *2*, 594–604.

Wade, M., Kowalik, T.F., Mudryj, M., Huang, E.S., and Azizkhan, J.C. (1992). E2F mediates dihydrofolate reductase promoter activation and multiprotein complex formation in human cytomegalovirus infection. Mol. Cell. Biol. *12*, 4364–4374.

Watanabe, N., Broome, M., and Hunter, T. (1995). Regulation of the human WEE1Hu CDK tyrosine 15-kinase during the cell cycle. EMBO J. *14*, 1878–1891.

White, E.A., and Spector, D.H. (2005). Exon 3 of the human cytomegalovirus major immediate-early region is required for efficient viral gene expression and for cellular cyclin modulation. J. Virol. *79*, 7438–7452.

Wiebusch, L., and Hagemeier, C. (1999). Human cytomegalovirus 86-kilodalton IE2 protein blocks cell cycle progression in G_1. J. Virol. *73*, 9274–9283.

Wiebusch, L., and Hagemeier, C. (2001). The human cytomegalovirus immediate-early 2 protein dissociates cellular DNA synthesis from cyclin-dependent kinase activation. EMBO J. *20*, 1086–1098.

Wiebusch, L., Asmar, J., Uecker, R., and Hagemeier, C. (2003a). Human cytomegalovirus immediate-early protein 2 (IE2)-mediated activation of cyclin E is cell-cycle-independent and forces S-phase entry in IE2-arrested cells. J. Gen. Virol. *84*, 51–60.

Wiebusch, L., Uecker, R., and Hagemeier, C. (2003b). Human cytomegalovirus prevents replication licensing by inhibiting MCM loading onto chromatin. EMBO Rep. *4*, 42–46.

Wiebusch, L., Bach, M., Uecker, R., and Hagemeier, C. (2005). Human cytomegalovirus inactivates the G0/G1-APC/C ubiquitin ligase by Cdh1 dissociation. Cell Cycle *4*, 1435–1439.

Wiebusch, L., Neuwirth, A., Grabenhenrich, L., Voigt, S., and Hagemeier, C. (2008). Cell cycle-independent expression of immediate-early gene 3 results in G1 and G2 arrest in murine cytomegalovirus-infected cells. J. Virol. *82*, 10188–10198.

Woo, M.S., Sanchez, I., and Dynlacht, B.D. (1997). p130 and p107 use a conserved domain to inhibit cellular cyclin-dependent kinase activity. Mol. Cell. Biol. *17*, 3566–3579.

Zhang, Z., Huong, S.-M., Wang, X., Huang, D.Y., and Huang, E.-S. (2003). Interactions between human cytomegalovirus IE1-72 and cellular p107: functional domains and mechanisms of up-regulation of cyclin E/cdk2 kinase activity. J. Virol. *77*, 12660–12670.

Zhu, H., Cong, J.P., and Shenk, T. (1997). Use of differential display analysis to assess the effect of human cytomegalovirus infection on the accumulation of cellular RNAs: induction of interferon-responsive RNAs. Proc. Natl. Acad. Sci. U.S.A. *94*, 13985–13990.

Zwicker, J., and Muller, R. (1997). Cell-cycle regulation of gene expression by transcriptional repression. Trends Genet. *13*, 3–6.

Zwicker, J., Lucibello, F.C., Wolfraim, L.A., Gross, C., Truss, M., Engeland, K., and Muller, R. (1995). Cell cycle regulation of the cyclin A, cdc25C and cdc2 genes is based on a common mechanism of transcriptional repression. EMBO J. *15*, 4514–4522.

Zydek, M., Hagemeier, C., and Wiebusch, L. (2010). Cyclin-dependent kinase activity controls the onset of the HCMV lytic cycle. PLoS Pathog. *6*, e1001096.

Zydek, M., Uecker, R., Tavalai, N., Stamminger, T., Hagemeier, C., and Wiebusch, L. (2011). General blockade of human cytomegalovirus immediate-early mRNA expression in the S/G2 phase by a nuclear, Daxx- and PML-independent mechanism. J. Gen. Virol. *92*, 2757–2769.

Cell Death Pathways Controlled by Cytomegaloviruses

I.15

A. Louise McCormick and Edward S. Mocarski

Abstract

Cytomegalovirus (CMV) deploys multiple strategies to overcome host intrinsic, innate, and adaptive responses that limit infection by triggering cell death. Multiple cell death suppressors are encoded by cytomegaloviruses infecting humans, monkeys and rodents. The viral inhibitor of caspase activation (vICA) and even the viral mitochondrial-localized inhibitor of apoptosis (vMIA) represent evolutionarily conserved strategies, whereas viral inhibitor of receptor-interacting protein kinase (RIP) activation (vIRA), the mitochondrial complex I-associated β2.7 RNA and other viral gene products whose mechanisms are not fully understood, appear to have evolved independently in primate and rodent CMVs. Initiators, effectors and interactions between CMV-induced cell death pathways have begun to emerge, through studies in cell culture and intact animals. It has become very clear that many cell death pathways are targeted by CMV-encoded cell death suppressors. Indeed, this subfamily of viruses has provided fundamental insights into pathogen-triggered regulated cell death pathways in mammals. Whereas features of the control and consequences of both intrinsic and extrinsic caspase-dependent apoptosis were already well-established, studies in CMV brought serine protease-dependent programmed cell death and RIP3 programmed necrotic death pathways to the fore. Virus-encoded cell death suppressors contribute resistance to cell stress as well as resistance to disruption of critical metabolic and aerobic respiration activities. Challenges for investigators focused on human CMV (HCMV) continue to be integration of findings using diverse viral strains that impact metabolic and stress pathways with seemingly subtle differences.

Introduction

Efforts to categorize regulated cell death pathways rely on comparisons to the morphological changes described for apoptosis and necrosis (Kerr *et al.*, 1972) together with biochemical evidence for defined processes that mediate death. Apoptosis has become uniquely identified by its dependence on caspases, a family of cysteine-aspartic acid proteases responsible for initiation and execution of this pathway (Festjens *et al.*, 2006) that orchestrate specific events to promote cell death. Various non-apoptotic, caspase-independent, regulated pathways have also been described (Leist and Jaattela, 2001; Lockshin and Zakeri, 2002, 2004; Jaattela, 2004; Vandenabeele *et al.*, 2006; Golstein and Kroemer, 2007) and a classification system has been established to distinguish related from unrelated processes (Kroemer *et al.*, 2009). HCMV, rhesus macaque CMV (RhCMV), and murine CMV (MCMV) inhibit caspase-dependent as well as caspase-independent pathways through the elaboration of cell death suppressors (McCormick, 2008; Brune, 2011; Mocarski *et al.*, 2011) and viral mutants inactivating individual cell death suppressor genes have played key roles in unveiling virus-induced death pathways and the cell type(s) where they occur. Investigations of CMV cell death suppressor mutants have provided a distinction between conventional, caspase-dependent apoptosis and two caspase-independent pathways, one initiated via mitochondrial serine protease HtrA2, and apparently mediated by additional serine proteases, and the other initiated by the pathogen recognition receptor, DNA activator of interferon (DAI) in complex with receptor interacting protein kinase (RIP)3 that together mediate programmed necrosis (Menard *et al.*, 2003; Reboredo *et al.*, 2004; Sharon-Friling *et al.*, 2006; Cicin-Sain *et al.*, 2008; McCormick *et al.*, 2008, 2010; Mack *et al.*, 2008; Upton *et al.*, 2010, 2012). Caspase-independent pathways are resistant to the caspase inhibitor zVAD. In addition, CMV also encodes genes that regulate cell death by dampening the impact of metabolic stress due to the unfolded protein response and to inhibitors of aerobic respiration (Reeves *et al.*, 2007; Terhune *et al.*, 2007; Moorman *et al.*, 2008; Xuan *et al.*, 2009; Qian *et al.*, 2011). Importantly, these cell death and cell stress suppressors appear to overlap to ensure protracted viability and metabolic activity of the infected cell. In this chapter, we summarize current knowledge of direct and indirect mechanisms employed to prolong infected cell viability and the impact on CMV pathogenesis.

Caspase-dependent regulated cell death

As a programmed pathway central to development and host defence, apoptosis is initiated by signalling pathways that activate caspases in a limited proteolytic cascade that drives cell disassembly into apoptotic bodies (Hengartner, 2000). Across the animal world, this pathway eliminates excess cells during development and during tissue homeostasis, settings where apoptotic debris is cleared by professional phagocytes without causing inflammation. Two general pathways activate apoptosis in host defence, one initiated from

within cells (intrinsic) and the other (extrinsic) mediated via trimeric cell surface receptors (death receptors) that bind immune mediators in the tumour necrosis factor (TNF) superfamily (see Fig. I.15.1). There are four TNF family receptors that mediate extrinsic death in human cells: Fas/CD95, TNF receptor 1 (TNFR1) and TNF-related apoptosis-inducing ligand (TRAIL) receptors (R)1 (also known as DR4) and TRAIL-R2/DR5. Once engaged by the appropriate ligand, protein–protein interactions mediated by the cytosolic portion of a death receptor drive recruitment of adapter proteins and proteolytic activation of initiator caspase 8 in a death domain (DD)-dependent complex with Fas-associated adaptor with a DD (FADD). The activation process is naturally regulated by cellular FADD-like interleukin 1β–converting enzyme-inhibitory protein (cFLIP), an enzymatically inactive paralogue of caspase 8 (Barnhart *et al.*, 2003). Intrinsic apoptosis is triggered by diverse cellular insults that damage cellular

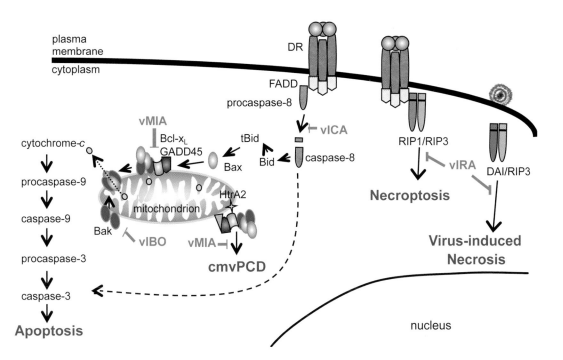

Figure I.15.1 Mechanisms of cytomegalovirus suppression of regulated cell death pathways. The virus-encoded viral inhibitor of mitochondrial apoptosis (vMIA), viral inhibitor of Bak activation (vIBO), viral inhibitor of caspase 8 activation (vICA) and viral inhibitor of RIP activation (vIRA) are shown. vMIA, vIBO and vICA inhibit death receptor (DR) induced caspase 8-dependent apoptosis and vIRA inhibits RIP1/RIP3-dependent necroptosis following engagement of DR ligand (DRL) through signals transduced via FADD to caspase 8 or RIP1. Depending on the cell type, caspase 8-dependent extrinsic apoptosis is either mediated by direct cleavage/activation of caspase 3 (dashed arrow) or by cleavage of Bid and subsequent amplification of caspase 9 and caspase 3 following cytochrome c release from mitochondria through a Bax/Bak-mediated permeability transitional. Following virus infection, vMIA is crucial for suppressing caspase-independent mitochondrial HtrA2 serine protease-dependent CMV-associated programmed cell death (cmvPCD) and vIRA is crucial for suppressing caspase-independent DAI/RIP3-dependent virus-induced necrosis, whereas, vIBO blocks Bak oligomerization and vICA blocks caspase 8 to prevent caspase-dependent apoptosis described in the text.

DNA, alter endoplasmic reticulum (ER) translation, increase cytosolic Ca^{2+} levels, or disrupt mitochondria membrane permeability or respiration.

A key step critical for intrinsic apoptosis that can also contribute to extrinsic apoptosis is the release of cytochrome c from mitochondria together with activation of prodeath factors such as APAF1 to promote activation of initiator caspase 9, that then activates effector/executioner caspase 3 (Festjens et al., 2006; Bratton and Salvesen, 2010). The release of cytochrome c is naturally regulated by B-cell lymphoma 2 (Bcl2) family members (Festjens et al., 2004). Both antiapoptotic as well as proapoptotic Bcl2 family proteins dictate mitochondrial membrane permeabilization. Bcl2 family proteins have from one to four signature Bcl2 homology (BH) domains (BH1 to BH4) important to interaction and activity. Bax and/or Bak oligomerization play central roles in the release (Kuwana and Newmeyer, 2003; Arnoult et al., 2004; Green and Kroemer, 2004; Antignani and Youle, 2006; Cassidy-Stone et al., 2008). BH domain interactions drive proapoptotic Bax and/or Bak oligomerization (Petros et al., 2004) and also mediate the antiapoptotic activity of Bcl2 and Bclx$_L$. In addition, a number of Bcl2 family members with a single BH3 domain (BH3 only proteins) promote the activation of proapoptotic Bax and/or Bak oligomerization. Overall, the balance of proapoptotic and antiapoptotic Bcl2 proteins in any cell determines the threshold for apoptosis. Prodeath signals may also be balanced by more global prosurvival signals. In many mammalian cell types, extrinsic apoptosis that follows caspase 8 activation proceeds through mitochondrial amplification events. When deployed, caspase 8 activation of the mitochondrial pathway occurs through the Bcl2 family proteins that promote release of prodeath factors from mitochondria. In particular, caspase 8 directly activates the prodeath Bcl2 family member Bid by proteolytic cleavage. Truncated (t)Bid in turn, is a direct activator of the proapoptotic Bcl2 family members, Bax and Bak (see Fig. I.15.1). Whereas intrinsic apoptosis directly triggers mitochondrial release of cytochrome c, in most cell types, extrinsic apoptosis makes use of a mitochondrial amplification step following death receptor activation. In some lymphoid and myeloid cell types, caspase 8 levels are high and target caspase 3 directly (Barnhart et al., 2003), independent of mitochondrial amplification (see Fig. I.15.1). Thus, intrinsic and extrinsic apoptosis converge on executioner caspase 3, which in turn cleaves and activates caspases 6 and 7. Depending on the particular setting, these caspases may all cleave downstream targets that control the morphological events associated with apoptosis.

The contribution of three key components (APAF1,

a caspase and an antiapoptotic Bcl2 family member) to regulation of developmental apoptosis are recognized as evolutionarily ancient (Hengartner, 2000). The unexpected viability as well as immune competence of mice that lack germline caspase 8 (once RIP3 was also eliminated) revealed that extrinsic apoptosis is surprisingly dispensable for development and gross tissue homeostasis (Kaiser et al., 2011; Mocarski et al., 2011). Further, core innate and adaptive immunity is retained in the absence of caspase 8 sufficient to control MCMV infection. Thus, the derivation of these mice demote caspase 8 as well as RIP3-dependent pathways from a central role in development, implicating intrinsic apoptosis as sufficient for elimination of excess cells during mammalian development as earlier recognized for invertebrates (Hengartner, 2000). Also as described in primitive organisms, intrinsic apoptosis must also contribute substantially to mammalian host defence, in agreement with virus-encoded modulators targeting mitochondria. The caspase 8 and RIP3 pathways are, however, implicated as major players in inflammatory tissue damage and disease pathogenesis that afflict wild type animals under a range of infection, autoimmune, tissue damage, and transplantation settings given that mice deficient in caspase 8 and RIP3 are remarkably resistant to inflammatory disease despite retaining intact cytokine activation and inflammatory responses (Kaiser et al., 2011).

Many viruses encode cell death suppressors that keep cells alive in order to produce progeny and facilitate spread in the infected host. The viral cell death suppressors that counteract caspase-dependent apoptosis appear to have been stolen from host cells over evolution, and thus are often sequence homologues of key cellular regulatory proteins such as the cellular FLICE inhibitory protein (cFLIP) that regulates caspase 8 activity or the Bcl2-related proteins that regulate mitochondrial steps in apoptosis (Irusta et al., 2003; Polster et al., 2004). Other viruses, such as the CMVs encode gene products that interfere with these steps but that lack sequence homology with cellular proteins. Whether related in sequence or not, functional analogues of the cFLIP and Bcl-2 proteins have provided novel insights into features of apoptosis including the roles of caspase 8 in programmed necrosis and Bax-dependent disruption of the mitochondrial network, as discussed in more depth in the remainder of the chapter.

vMIA, the viral mitochondria-localized inhibitor of apoptosis

Despite the lack of any overall sequence similarity to the antiapoptotic Bcl2 family members, Bcl2 and Bclx$_L$, or an obvious BH domain, vMIA, encoded by the UL37x1 ORF of HCMV, suppresses release of prodeath factors from the mitochondria by a mechanism that suggests evolutionary relatedness to Bcl2 (Goldmacher *et al.*, 1999). Consistent with this suggestion, vMIA prevents cell death induced by diverse stimuli that promote intrinsic and extrinsic apoptosis. vMIA is the most broadly antiapoptotic CMV cell death suppressor identified (Goldmacher *et al.*, 1999; Belzacq *et al.*, 2001; Jan *et al.*, 2002; Roumier *et al.*, 2002; Boya *et al.*, 2003; Andreau *et al.*, 2004; Arnoult *et al.*, 2004; Boya *et al.*, 2005; McCormick *et al.*, 2005). The function of vMIA is reconstituted by a 63-aa derivative composed of an amino terminal mitochondrial-targeting domain (aa 2–34) and a carboxyl-terminal antiapoptotic domain (AAD, aa 118–147). These together are both necessary and sufficient to block intrinsic and extrinsic apoptosis (Hayajneh *et al.*, 2001). The antiapoptotic function of vMIA requires mitochondrial localization (aa 2–34) and antiapoptotic (AAD, aa 118–147) domains that together are both necessary and sufficient to provide cell death protection (Hayajneh *et al.*, 2001). The mitochondrial-targeting domain, comprised of the amino-terminal hydrophobic signal and adjacent highly conserved basic residues, directs vMIA transfer from the ER to the mitochondrial outer membrane via mitochondrion-associated membrane (MAM) contact points (Mavinakere and Colberg-Poley, 2004; Mavinakere *et al.*, 2006; Bozidis *et al.*, 2008; Williamson and Colberg-Poley, 2009; Williamson *et al.*, 2011) (see also Chapter I.12). On transfer to the mitochondria, the targeting domain spans the mitochondrial membrane leaving the AAD exposed to the cytoplasm (Mavinakere *et al.*, 2006). vMIA function is dependent on a critical amphipathic α-helix (aa 126–140) within the AAD (Smith and Mocarski, 2005). Point mutations that disrupt AAD structure, alter amphipathicity, or add charge to the hydrophobic face of the predicted α-helix, completely abrogate vMIA function. In contrast, the predicted hydrophilic face of the AAD α-helix tolerates as many as five or six amino acid substitutions before function is disrupted. Overall, the absence of sequence homology to the Bcl2 family (Goldmacher *et al.*, 1999) is counterbalanced by these features of the functional domain and suggests the AAD α-helix provides a structural role that may be analogous to the α-helical BH domains in antiapoptotic Bcl2 family members (Smith and Mocarski, 2005; Pauleau *et al.*, 2007).

Proapoptotic mitochondrial membrane permeability is controlled by Bax and Bak and prevented by Bcl2 proteins which interact as homodimers and heterodimers (Kuwana and Newmeyer, 2003; Green and Kroemer, 2004; Antignani and Youle, 2006; Cassidy-Stone *et al.*, 2008). vMIA follows a similar interaction

mechanism and binds the antiapoptotic Bcl2 family protein $Bclx_L$ (Smith and Mocarski, 2005) as well as the proapoptotic Bcl-2 family proteins Bax (Arnoult *et al.*, 2004) and Bak (Karbowski *et al.*, 2006). Of these, the interaction with Bax has been most extensively investigated. Bax oligomerization requires a conformational change that exposes a membrane-insertion domain (Antonsson *et al.*, 2001; Kuwana and Newmeyer, 2003; Arnoult *et al.*, 2004; Green and Kroemer, 2004; Antignani and Youle, 2006; Cassidy-Stone *et al.*, 2008). vMIA interaction with Bax prevents the membrane insertion and this inactivating interaction is mediated by the carboxy-terminal AAD (Arnoult *et al.*, 2004; Poncet *et al.*, 2004). These observations underlie the hypothesis that vMIA sequesters Bax at mitochondria. In the presence of vMIA, oligomerized Bax at mitochondria fails to promote apoptosis, suggesting that sequestration is a component of the antiapoptotic mechanism while distinguishing the vMIA-dependent antiapoptotic mechanism from that of cellular and viral Bcl2 proteins that prevent Bax relocalization and oligomerization at mitochondria indirectly (see Fig. I.15.1). It remains likely that vMIA also interacts with Bak in specific settings (Karbowski *et al.*, 2006).

Evidence suggests vMIA inhibits Bax within a complex that includes additional cell fate regulatory proteins. A specific interaction has been demonstrated for growth arrest and DNA damage 45 alpha (GADD45α) that binds vMIA in both yeast and mammalian cells. Targeted knockdown of GADD45α, as well as GADD45β or GADD45γ, inhibits vMIA function. As also indicated for the interaction with Bax, binding GADD45 proteins requires an intact AAD α-helix (Smith and Mocarski, 2005). Conversely, overexpression of each GADD45 family protein enhances vMIA activity. These data indicate a direct link exists between the DNA damage response pathway and vMIA-mediated cell death suppression. Interestingly, GADD45 also enhances the function of the vMIA-binding partner $Bclx_L$ (Smith and Mocarski, 2005). Thus, vMIA likely sequesters Bax to control proapoptotic function and also functions together with $Bclx_L$ and GADD45α as an antiapoptotic complex.

vMIA function is dispensable for replication, but contributes to the resistance of infected cells to apoptosis when purposefully induced (Reboredo *et al.*, 2004; McCormick *et al.*, 2005). An abundant, unspliced immediate-early transcript encodes the 163 aa UL37x1 gene product protein, which, when expressed exogenously, is sufficient for all cell death protection functions identified to date. In addition, exogenous vMIA prevents infection-induced cell death during infection with UL37x1 mutant virus (McCormick *et al.*, 2005). During infection, a number of 5′ coterminal viral transcripts include UL37x1 and produce a family of antiapoptotic proteins (Tenney and Colberg-Poley, 1990, 1991b; Goldmacher *et al.*, 1999). Splicing to UL37x2 and UL37x3 yields larger antiapoptotic glycoproteins gpUL37 and pUL37$_M$ (Goldmacher *et al.*, 1999). In addition, many less abundant spliced transcripts are also predicted to encode vMIA-related antiapoptotic proteins (Adair *et al.*, 2003), although functional analyses of these variants have not been reported. Thus, identification of any potential for additional/novel roles by splice variants in regulating cell death requires additional attention.

For reasons that remain unclear, similar mutations of UL37x1 when introduced into different viral strains produce variable impacts on replication, indicating genetic differences potentially influence the outcome (Reboredo *et al.*, 2004; McCormick *et al.*, 2005; Sharon-Friling *et al.*, 2006; Kaarbo *et al.*, 2011). For example, ΔUL37x1, a deletion mutant made in Towne-BAC, a cloned derivative of the short form Towne strain genome lacking approximately 15 genes, also called TownevarATCC (Marchini *et al.*, 2001), replicates to levels as high as parental virus (McCormick *et al.*, 2005), whereas mutants derived from the AD169 strain, also called AD169varATCC, exhibit pronounced replication defects (Yu *et al.*, 2003; Reboredo *et al.*, 2004; Sharon-Friling *et al.*, 2006). Although it is clear that vMIA is dispensable when studied in the context of Towne strain infection, variations in experimental methodologies, especially with regard to direct evaluation of vMIA function, prevent conclusions regarding the underlying pathways contributing to the impact of this gene product in other viral strains. Nonetheless, these differences underscore the potential for combinatorial impacts from prosurvival genes sufficient to replace vMIA function within the right genetic context.

Primate and rodent CMVs retain genes that encode the vMIA function (McCormick *et al.*, 2003a, 2005). Chimpanzee, rhesus macaque, and African green monkey CMV orthologues each retain sequence similarity with the mitochondrial-targeting and AAD domains of vMIA. In contrast, vMIA orthologues encoded by ORF m38.5 in murine CMV (MCMV) and r38.5 in rat CMV (RCMV) retain limited amino acid sequence homology to primate CMV (McCormick *et al.*, 2003a, 2005; Brocchieri *et al.*, 2005). MCMV m38.5 encodes an antiapoptotic protein that localizes to mitochondria (McCormick *et al.*, 2005) and recruits Bax to mitochondria during infection (Arnoult *et al.*, 2008), which are both properties of primate CMV vMIA. In contrast to primate CMVs, rodent CMVs encode a second mitochondrial antiapoptotic protein that functions at mitochondria along with m38.5. Based on the characterization of m41.1 (MCMV),

r41.1 (RCMV Maastricht isolate), and e41.1 (RCMV English isolate) (Brocchieri *et al.*, 2005; Cam *et al.*, 2010) are all predicted to possess viral inhibitor of Bak oligomerization (vBO) function during infection (Cam *et al.*, 2010) (see Fig. I.15.1). Whether primate CMVs encode a second mitochondrial protein to interfere with Bak is not yet resolved, although reports that indicate vMIA fails to protect from Bak-dependent apoptosis or interact physically with Bak (Arnoult *et al.*, 2004, 2008; Cam *et al.*, 2010) are counterbalanced by evidence that HCMV interacts with both Bax and Bak (Karbowski *et al.*, 2006).

Mutant Δm38.5 virus replicates efficiently in cultured fibroblasts, but is impaired in endothelial cells, dendritic cells, and macrophages. When studied *in vivo*, this mutant fails to disseminate to salivary glands at WT levels in a pattern that suggests a failure to sustain dissemination in leucocytes (Manzur *et al.*, 2009). Overall, the positional conservation of the rodent CMV ORFs relative to UL37x1, along with functional and mechanistic studies, indicate that rodent and primate CMVs encode vMIA homologues with comparable function (McCormick *et al.*, 2003a; Brocchieri *et al.*, 2005; Arnoult *et al.*, 2008), and that vMIA and vBO in rodent CMVs constitute the activity of the single vMIA in primate CMVs. These data indicate CMV vMIAs are important in suppressing cell death pathways and promoting survival of the virus in the host.

vICA, the viral inhibitor of caspase 8 activation

vICA, the product of the UL36 gene, binds procaspase 8 to prevent proteolytic activation and caspase 8-dependent apoptosis (Skaletskaya *et al.*, 2001) (see Fig. I.15.1). As caspase 8 is an initiator caspase in the extrinsic, immune-regulated apoptosis pathway, this inhibitory function can be predicted to promote survival from proapoptotic signals originating from death receptor signalling. Because a UL36 homologue is conserved in all mammalian β-herpesviruses, this function is predicted to play an important, evolutionarily conserved role (McCormick *et al.*, 2003a; Menard *et al.*, 2003). Indeed, this gene appears to be the most highly conserved of all CMV-encoded modulatory functions, even though its function is dispensable for HCMV or MCMV replication in cultured fibroblasts (Patterson and Shenk, 1999; Skaletskaya *et al.*, 1999; Hayajneh *et al.*, 2001; Menard *et al.*, 2003). Consistent with this, routine passage of HCMV in cultured fibroblasts (HFs) can be associated with the emergence of adventitious mutations within UL36 (Skaletskaya *et al.*, 2001; McCormick, 2010). When evaluated UL36-deficient viruses are less resistant to extrinsic apoptosis induced via TNF receptor or Fas (McCormick *et al.*, 2010), based on comparisons of laboratory strains that carry mutations in this gene (Skaletskaya *et al.*, 2001). Evidence for the required biologic role for the caspase inhibitor comes from evaluation of M36 (Menard *et al.*, 2003). In cultured cells, MCMV infection of macrophages but not endothelial cells in the absence of M36 induces elevated caspase 8 activity that triggers apoptosis (Menard *et al.*, 2003; Cicin-Sain *et al.*, 2008). Consistent with this pattern, M36-mutant virus is severely attenuated *in vivo*, induces apoptosis within infected tissue macrophages and hepatocytes, and can be rescued by expression of dominant negative FADD from the virus (Cicin-Sain *et al.*, 2005, 2008). Infection with HCMV UL36 viruses also reveals cell-specific impacts due to elevated caspase 8 activity (Skaletskaya *et al.*, 2001; McCormick *et al.*, 2010). Thus, infection of THP-1 monocytic cells early in differentiation towards macrophages activates caspase 8 and promotes apoptosis in the absence of vICA. Combined, these observations are consistent with the hypothesis that vICA serves to promote replication in specific cells and to increase viral resistance to apoptosis induced by extrinsic signals in a wide range of cells.

Caspase-independent cell death

In some settings, broad-spectrum caspase inhibitors fail to prevent premature infected cell death, an observation that has been employed to explore activation and specificity of alternative non-apoptotic programmed cell death pathways (McCormick *et al.*, 2005; McCormick, 2008; Upton *et al.*, 2010). Cell death that fails to be suppressed by inhibition of caspase activity is generally referred to as caspase-independent, although additional supporting evidence is required for such an assignment. Two CMV-encoded cell-death suppressors have helped reveal the crucial role that non-apoptotic pathways play in the viral life cycle. Studies on MCMV mutant virus in gene M45-encoded viral inhibitor of RIP activation (vIRA) unveiled CMV-induced programmed necrosis and mutant HCMV in gene UL37x1-encoded vMIA revealed serine protease-dependent CMV-infected cell-specific programmed cell death (cmvPCD). Each of these pathways is supported by a considerable body of evidence showing virus-induced programmed necrosis and cmvPCD pathways are patterns of cell death completely distinct from apoptosis (see Fig. I.15.1). By expressing vIRA, MCMV inhibits programmed necrosis in cells with elevated levels of RIP3 (Mack *et al.*, 2008; Upton *et al.*, 2010). This death is triggered by a complex between DNA-dependent activator of interferon (DAI) in complex with RIP3 (Upton *et al.*, 2012). Although vMIA is a potent suppressor of

apoptosis, when mutated during HCMV infection, a serine protease-dependent cell death pathway that appears to be triggered via mitochondrial HtrA2/Omi emerges to prematurely kill infected cells (McCormick *et al.*, 2005). Initially characterized as antiapoptotic proteins (Goldmacher *et al.*, 1999; Brune *et al.*, 2001), vIRA and vMIA highlight the multifunctional role of CMV-encoded cell death suppressors, and the mechanistically complex interplay between cell death pathways during viral infection.

Programmed necrosis controlled by M45, the underline{v}iral underline{i}nhibitor of RIP underline{a}ctivation vIRA

Herpesviruses, including HCMV, encode genes related to ribonucleotide reductases (Chee *et al.*, 1990); however, neither the CMV UL45 gene product nor sequence homologues retain ribonucleotide reductase function (Chee *et al.*, 1990; Patrone *et al.*, 2003; Lembo *et al.*, 2004). Nonetheless, M45-deficient MCMV is severely attenuated *in vivo* as the result of cell death suppressor function provided by the gene product (Lembo *et al.*, 2004; Upton *et al.*, 2008). Initial evaluation of viral M45 mutants revealed an important role in preventing premature endothelial cell death (Brune *et al.*, 2001). Subsequently, when studies were carried out in additional cell lines, susceptibility to death was shown to be conferred by RIP3 (Upton *et al.*, 2010) and DAI (Upton *et al.*, 2012) rather than cell type and M45 was shown to block caspase-independent death (Mack *et al.*, 2008) as the result of impact on RIP1 and RIP3 interactions (Upton *et al.*, 2008). Thus, NIH3T3 fibroblasts resist to M45 mutant virus induced death due to the low levels of RIP3 and apparent lack of DAI, whereas 3T3-SA fibroblasts are fully susceptible (Upton *et al.*, 2008) due to the presence of both RIP3 and DAI (Upton *et al.*, 2012). Cell death suppression is associated with the modulation of RIP homotypic interaction motif (RHIM)-mediated interactions (Upton *et al.*, 2008, 2010; Rebsamen *et al.*, 2009). In general, RIP1 and RIP3 kinases are critical to cell fate decisions that control how the cell will die (Declercq *et al.*, 2009) and DAI is a RHIM adaptor that interacts with RIP1 and RIP3 (Kaiser *et al.*, 2008). RIP1 regulates the choice between two regulated cell death pathways induced by TNF: apoptosis and necroptosis (Denecker *et al.*, 2001; Degterev *et al.*, 2008). Necroptosis (necrostatin-sensitive programmed necrosis) differs from apoptosis in that this pathway is dependent on RIP1–RIP3 complexes that become activated specifically when caspase 8 is inhibited or absent in cells (Degterev *et al.*, 2008). In other experimental settings, i.e. following overexpression, RIP1 can promote apoptosis (Kaiser and Offermann, 2005). In addition to modulating interactions between RIP1, RIP3, and DAI, vIRA blocks interactions between RIP1 and RIP3 (Upton *et al.*, 2010), as well as interactions between either kinase and either DAI (Kaiser *et al.*, 2008; Rebsamen *et al.*, 2009) or Toll/interleukin-1 receptor (TIR) domain containing adaptor-inducing IFNβ (TRIF), an adapter that is activated through TLR3 or TLR4 (Kaiser, in preparation). During viral infection, the RHIM of M45 is required to prevent virus-induced programmed necrosis (Upton *et al.*, 2008, 2010) that is mediated by a DAI–RIP3 complex (Upton *et al.*, 2012). Apparently, any RHIM-dependent protein–protein interaction may be blocked by vIRA, whether the outcome is cell death (Upton *et al.*, 2012) or activation of NF-κB or IRF3 (Rebsamen *et al.*, 2009). These data highlight the potential significance of RHIM-containing proteins, including RIP kinases, in critical cell fate decisions impacting infected cell survival.

Consistent with initial observations (Brune *et al.*, 2001), M45 is sufficient to suppress induced apoptosis (Mack *et al.*, 2008; Upton *et al.*, 2008). As cellular factors and steps associated with death of the infected cell have been identified, however, it has become clear that M45-deficient viruses induce programmed necrosis, completely independent of death receptor activation, TRIF and RIP1 kinase, yet dependent on RIP3 kinase (Upton *et al.*, 2010). Thus, premature death induced by M45-deficient virus is morphologically and biochemically identified as necrosis (Kroemer *et al.*, 2009; Upton *et al.*, 2010). Further, the expression levels of RIP3 kinase and DAI appear to dictate whether a cell dies (Upton *et al.*, 2012). Given that a single cell death stimulus such as TNF can induce apoptosis or programmed necrosis (Declercq *et al.*, 2009; He *et al.*, 2009; Zhang *et al.*, 2009), the evaluation of the necrotic pathway during MCMV infection employed experimental manipulation of RIP3 and RIP1 levels as well as genetically deficient mouse fibroblasts. These mechanistic studies revealed RIP3 kinase levels, rather than the cell type, *per se*, determine cell susceptibility to the virus-induced programmed necrosis that limits replication. Interestingly, M45-deficient virus infection of cells that are insensitive to direct virus-induced programmed necrosis nevertheless remain susceptible to caspase-independent cell death induced by TNF in the presence of cycloheximide (Mack *et al.*, 2008) implying control of RIP1-dependent necroptosis, although the specific contributions of RIP1 and RIP3 have not been evaluated in this setting. In sum, vIRA inhibits RIP1-dependent apoptosis induced by overexpression of the TLR3/TLR4 adaptor protein TRIF, RIP1-RIP3 dependent necroptosis (Mack *et al.*, 2008; Upton *et al.*, 2010), and DAI-RIP3-dependent virus-induced

necrosis (Upton *et al.*, 2010, 2012). Overall, these data imply similarity between the programmed necrosis pathway activated by viral infection and the pattern of death following death receptor engagement in the presence of caspase inhibitors (Mocarski *et al.*, 2011).

Unlike M45-deficient MCMV, UL45-deficient HCMV replicates in endothelial cells (Hahn *et al.*, 2002). HCMV UL45 lacks an amino-terminal with homology to the M45 RHIM. Interestingly, HSV-2 RR1 includes a homologous region and controls cell death. Additionally, UL45 mutants in the laboratory strain AD169varATCC exhibit reduced replication in fibroblasts following a low multiplicity infection (Patrone *et al.*, 2003) and UL45 mutant in the low passage level strain TB40/E exhibits striking growth defects in endothelial cells (Sinzger, personal communication). Thus, the contribution of HCMV UL45 to cell death suppression remains to be fully elaborated.

cmvPCD, a vMIA-controlled, HtrA2/Omi-dependent death that follows replication

Mechanistic studies of premature cell death induced by the vMIA-deficient virus ΔUL37x1, made in TowneBAC, identified the role of caspase-independent, HtrA2/Omi-dependent, cmvPCD in the termination of HCMV infection (McCormick *et al.*, 2005, 2012; McCormick, 2008). Despite nearly equivalent yields, mutant virus-infected cultures had increased levels of infected cell death. Surprisingly, the mechanism of mutant and WT virus-induced death was similar whether the mitochondrial cell death inhibitor was present or absent, revealing that the novel HtrA2/Omi pathway controlled demise of both mutant and parent virus-infected cells, albeit with different kinetics. The serine protease HtrA2/Omi functions during apoptosis as well, and more specifically, is released from mitochondria prior to promoting apoptosis from within the cytoplasm (Suzuki *et al.*, 2001; Hegde *et al.*, 2002; Jesenberger and Jentsch, 2002; Festjens *et al.*, 2004; Jaattela, 2004). In contrast, the protease apparently remains mitochondrial during cmvPCD as localization of the protease to mitochondria continues throughout infection (McCormick, 2008). An evaluation that considered both the morphological changes that occur during CMV replication and those related to apoptosis revealed that cmvPCD initiates well after the general enlargement of the cell and formation of nuclear and cytoplasmic inclusions typical of late-stage infected cells. Initiation of cmvPCD begins with a process of cell fragmentation that appears distinct from the membrane blebbing of apoptosis besides being caspase independent. This is followed by continued progression until the entire cell body becomes fragmented. Morphological evidence of apoptosis, such as decrease in cell size and membrane blebbing, nuclear collapse, and disintegration of DNA without chromatin condensation are absent. In combination, these morphological characteristics distinguish cmvPCD from apoptosis. cmvPCD is dependent on serine protease activity, and as such, is resistant to the broad-spectrum caspase inhibitor zVAD, but inhibited by the broad-spectrum serine protease inhibitor, N-alpha-p-tosyl-L-lysine chloromethyl ketone (TLCK) or the HtrA2/Omi-specific inhibitor, UCF-101. Consistent with a late infection-induced cell death programme, each of these inhibitors can be added at late times of infection. A direct role for HtrA2/Omi has been suggested from premature overexpression that promotes premature cell death, yet, naturally, vMIA controls HtrA2/Omi function without altering expression levels. Mitochondrial respiration, altered early during apoptosis, apparently continues late through cmvPCD, as mitochondria function continues prior to and during cell fragmentation, consistent with continued mitochondrial localization of cytochrome c and HtrA2/Omi. Since cmvPCD can be identified within cultures infected by WT or mutant virus, the role of vMIA in this context is to delay the process that naturally terminates infection in fibroblasts.

Caspase-independent cell death controlled by vICA

In the presence of RIP3, caspase 8 has a critical role in promoting monocyte to macrophage differentiation while macrophages develop in mice that lack caspase 8 and RIP3 (Sordet et al., 2002; Schwerk and Schulze-Osthoff, 2003; Garrido and Kroemer, 2004; Kang et al., 2004; Gordon and Taylor, 2005; Launay et al., 2005; Cathelin et al., 2006; Lamkanfi et al., 2007; Droin et al., 2008; Kaiser et al., 2011; Oberst et al., 2011). Modifications of the protease and/or functions of antiapoptotic proteins have previously been suggested as important to caspase 8 regulation when RIP3 is present (Perlman et al., 1999; Ben Moshe et al., 2008). CMV replication is dependent on macrophage differentiation. Thus, differentiated macrophages are hosts for productive infection, while monocytes, although permissive, do not produce virus following CMV infection without first entering the differentiation programme. In an effort to evaluate how all these factors contribute to cell death pathways activated by HCMV during differentiation, THP-1 cells were infected with a UL36-deficient virus at various stages of differentiation in the presence or absence of the pan-caspase inhibitor zVAD. Although significant inhibitory activity was demonstrated at early stages of differentiation, zVAD neither inhibited

nor contributed to cell death of THP-1 cells infected late in differentiation. Whether these observations indicate a multifunctional role for vICA remains to be determined.

Cell death following stress

Proapoptotic stress can originate within any organelle. Sensors within the nucleus, endoplasmic reticulum, lysosomes, and Golgi apparatus are all actively involved in the signalling that promotes apoptosis (Ferri and Kroemer, 2001). The functions of the viral gene UL38 and the β2.7 RNA are examples of indirect control of cell death mediated by CMV as the result of direct modulations of cell stress.

pUL38 controls unfolded protein response (UPR)-induced apoptosis

Growth properties of UL38-deficient viruses revealed an important role in controlling proapoptotic stress as yields are reduced by approximately 100-fold (Dunn et al., 2003; Yu et al., 2003; Terhune et al., 2007). This phenotype is reversed in cells expressing UL38 or by using the pan-caspase inhibitor zVAD (Terhune et al., 2007). In the absence of pUL38, premature infected cell death is associated with morphological and biochemical patterns of apoptosis that initiates very early (24 hours) and reaches > 50% by 72 hours (Terhune et al., 2007). Consistent with these kinetics, pUL38 is produced from a unique, early transcript (Tenney and Colberg-Poley, 1990, 1991a,b). Further, pUL38 is sufficient to prevent apoptosis induced by infection with E1B-19K deficient adenovirus or treatment with thapsigargin or tunicamycin, both ER-stress inducers (Terhune et al., 2007; Xuan et al., 2009). The molecular mechanism of cell stress suppression is currently ill-defined, however, evidence suggests ER stress is induced in the absence of pUL38. Thus, in the absence of pUL38, c-Jun N-terminal kinase (JNK) phosphorylation becomes elevated by infection within 24 hours, a JNK inhibitor is sufficient to partially restore infected cell viability, and pUL38 expression is sufficient to decrease stress-induced JNK phosphorylation (Xuan et al., 2009). In other settings, consequences of JNK phosphorylation and activation have included both direct effects on proapoptotic Bcl-2 proteins and indirect effects through increased transcription of proapoptotic genes (Dhanasekaran and Reddy, 2008). In addition to these effects on the apoptosis pathway, additional roles of UL38 in the control of cell stress and growth have been revealed by UL38-dependent elevation in ATF4 expression (Xuan et al., 2009), a transcription factor required for antioxidant responses that increase protein folding in the endoplasmic reticulum following UPR. Although pUL38 has been reported to bind > 30 cellular proteins, none have yet been directly connected to cell death suppression (Moorman et al., 2008). One interaction, pUL38 binding of TSC2, a component of the tuberous sclerosis tumour suppressor protein complex (TSC1/2), is expected to activate the mammalian target of rapamycin complex 1 (mTORC1) and as a result provide viral control of cell stress related to metabolic stress (Moorman et al., 2008). Genetic experiments evaluating pUL38-dependent control of cell death and mTORC1 activation indicate the amino-terminal 239 aa of pUL38 are sufficient to restore viral replication and provide protection from ER-stress induced apoptosis while additional carboxy-terminal aa are required for mTORC1 activation (Qian et al., 2011). These data highlight the potential for future identification of cellular proteins that bind pUL38 and are specifically required for ER-stress control during infection. Whether these impacts will be directly or indirectly connected to suppression of cell death requires additional evaluation of pUL38, although the protein appears to have brought attention to the interface of cell death and stress.

β2.7 RNA controls stress induced by interference with mitochondrial respiration

The β2.7 noncoding RNA is highly conserved and accounts for as much as 60% of RNA in infected cells (McDonough and Spector, 1983; McDonough et al., 1985; McSharry et al., 2003; Gatherer et al., 2011). Consistent with the proposal that the virus benefits from the noncoding RNA, the RL4 ORF within the β2.7 RNA gene is maintained in a strain-specific pattern. The function of the β2.7 RNA is suggested from the physical interaction of the RNA with proteins of the nicotinamide adenine dinucleotide-ubiquinone oxidoreductase complex (mitochondrial respiration complex I) (Reeves et al., 2007). Consistent with the suggested control of complex I, the β2.7 RNA increases survival from the mitochondrial respiration poison rotenone (Reeves et al., 2007) and ischaemia/reperfusion (I/R) injury (Zhao et al., 2010). During infection, higher ATP production is sustained late in infection when the β2.7 RNA is present (McSharry et al., 2003; Reeves et al., 2007). However, β2.7 RNA-deficient virus shows no growth defects in culture preventing comprehensive conclusions. Continued mitochondrial function during CMV infection is certainly suggested from several studies evaluating mitochondrial DNA synthesis, mitochondrial protein expression profiles,

and ATP production (Furukawa et al., 1976; Hertel and Mocarski, 2004; Reeves et al., 2007), consistent with the suggestion that β2.7 RNA may be one of many factors that contribute to viral production by maintaining ATP production. In contrast, the cell-death suppressor role may be more specifically tied to intrinsic stresses associated with decreased ATP (Reeves et al., 2007). Thus, functional overlap may preclude a simple phenotypic observation following focused evaluation of virus replication. Alternatively the β2.7 RNA may be critical only under certain conditions of stress or within a specific cell type.

Additional functions of pUL37x1/vMIA

In addition to cell death protection, UL37x1/vMIA has been ascribed functions at the mitochondria and endoplasmic reticulum with as yet uncertain connections to cell death. At mitochondria, vMIA modulates energetic function and mitochondrial morphology (McCormick et al., 2003b, 2008; Poncet et al., 2006; Kaarbo et al., 2011; Seo et al., 2011). A role in down-modulating cellular oxidative phosphorylation and ATP levels was revealed from two studies that also identified physical interactions of vMIA with viperin, an interferon-inducible protein, and with the mitochondrial phosphate carrier, PiC, a component of the ATP synthasome (Poncet et al., 2006; Seo et al., 2011). In contrast to these studies, vMIA has also been reported to drive infection-related increases in mitochondrial biogenesis and activity (Kaarbo et al., 2011). All of these studies have employed vMIA-deficient mutant viruses made in the same viral strain, thus, additional evaluation may be needed to clarify mechanisms that determine vMIA-dependent control of ATP levels. vMIA is known to disrupt the reticular mitochondrial network, both during infection and following exogenous expression (McCormick et al., 2003b, 2008). The Bcl2 proteins Bax and Bak that bind vMIA and control mitochondria membrane permeabilization also control mitochondrial morphogenesis (Karbowski et al., 2006). Overexpression of Bax is sufficient to overcome vMIA-dependent disruption of the network and Bax-deficient cells are resistant to vMIA-dependent disruption (Karbowski et al., 2006; Poncet et al., 2006). Thus, the mechanism for vMIA-dependent disruption may be related to sequestration of Bax both in exogenous settings and within CMV infected cells (McCormick et al., 2003b; Andoniou et al., 2004; Arnoult et al., 2004; Poncet et al., 2006; McCormick, 2008) while any benefit to infection remains unclear. Finally, vMIA has been reported to promote calcium release from the endoplasmic reticulum (Sharon-Friling et al., 2006).

Conclusion

Knowledge of the intricate balance between cellular metabolism, intrinsic and extrinsic cell death pathways, and impacts on virus clearance continues to expand. The multiple CMV cell death suppression functions collectively fall within this broad range. The array of cell types and replication settings relevant to CMV (Mocarski *et al.*, 2006) likely contribute to selective pressures. While the adventitious mutations that arise naturally during propagation of virus in cultured cells may confound initial conclusions, the gain may ultimately be in a greater appreciation for the balance between prosurvival and prodeath signals confronted by CMV. Conserved and non-conserved functions in model viruses continue to provide relevant settings to identify pathway controls shared across CMVs and host species as well as unique mechanisms. Critical decisions regarding cell fate have been revealed at intersections of cell death pathways, metabolism and cell death, and pathogen control and cell death, and each of these control points apparently present prime targets for multifunctional viral protein, a common strategy to minimize the genetic burden required to confront multiple antiviral sensors.

Acknowledgements

This work was supported by Public Health Service Grants R01 AI030363 and AI020211.

References

Adair, R., Liebisch, G.W., and Colberg-Poley, A.M. (2003). Complex alternative processing human cytomegalovirus UL37 pre-mRNA. J. Gen. Virol. *84*, 3353–3358.

Andoniou, C.E., Andrews, D.M., Manzur, M., Ricciardi-Castagnoli, P., and Degli-Esposti, M.A. (2004). A novel checkpoint in the Bcl-2-regulated apoptotic pathway revealed by murine cytomegalovirus infection of dendritic cells. J. Cell Biol. *166*, 827–837.

Andreau, K., Castedo, M., Perfettini, J.L., Roumier, T., Pichart, E., Souquere, S., Vivet, S., Larochette, N., and Kroemer, G. (2004). Preapoptotic chromatin condensation upstream of the mitochondrial checkpoint. J. Biol. Chem. *279*, 55937–55945.

Antignani, A., and Youle, R.J. (2006). How do Bax and Bak lead to permeabilization of the outer mitochondrial membrane? Curr. Opin. Cell. Biol. *18*, 685–689.

Antonsson, B., Montessuit, S., Sanchez, B., and Martinou, J.C. (2001). Bax is present as a high molecular weight oligomer/complex in the mitochondrial membrane of apoptotic cells. J. Biol. Chem. *276*, 11615–11623.

Arnoult, D., Bartle, L.M., Skaletskaya, A., Poncet, D., Zamzami, N., Park, P.U., Sharpe, J., Youle, R.J., and Goldmacher, V.S. (2004). Cytomegalovirus cell death suppressor vMIA blocks Bax- but not Bak-mediated apoptosis by binding and sequestering Bax at mitochondria. Proc. Natl. Acad. Sci. U.S.A. *101*, 7988–7993.

Arnoult, D., Skaletskaya, A., Estaquier, J., Dufour, C., and Goldmacher, V.S. (2008). The murine cytomegalovirus cell death suppressor m38.5 binds Bax and blocks Bax-mediated mitochondrial outer membrane permeabilization. Apoptosis *13*, 1100–1110.

Barnhart, B.C., Alappat, E.C., and Peter, M.E. (2003). The CD95 type I/type II model. Sem. Immunol. *15*, 185–193.

Belzacq, A.S., El Hamel, C., Vieira, H.L., Cohen, I., Haouzi, D., Metivier, D., Marchetti, P., Brenner, C., and Kroemer, G. (2001). Adenine nucleotide translocator mediates the mitochondrial membrane permeabilization induced by lonidamine, arsenite and CD437. Oncogene *20*, 7579–7587.

Ben Moshe, T., Kang, T.B., Kovalenko, A., Barash, H., Abramovitch, R., Galun, E., and Wallach, D. (2008). Cell-autonomous and non-cell-autonomous functions of caspase-8. Cytokine Growth Factor Rev. *19*, 209–217.

Boya, P., Gonzalez-Polo, R.A., Poncet, D., Andreau, K., Vieira, H.L., Roumier, T., Perfettini, J.L., and Kroemer, G. (2003). Mitochondrial membrane permeabilization is a critical step of lysosome-initiated apoptosis induced by hydroxychloroquine. Oncogene *22*, 3927–3936.

Boya, P., Gonzalez-Polo, R.A., Casares, N., Perfettini, J.L., Dessen, P., Larochette, N., Metivier, D., Meley, D., Souquere, S., Yoshimori, T., *et al.* (2005). Inhibition of macroautophagy triggers apoptosis. Mol. Cell. Biol. *25*, 1025–1040.

Bozidis, P., Williamson, C.D., and Colberg-Poley, A.M. (2008). Mitochondrial and secretory human cytomegalovirus UL37 proteins traffic into mitochondrion-associated membranes of human cells. J. Virol. *82*, 2715–2726.

Bratton, S.B., and Salvesen, G.S. (2010). Regulation of the Apaf-1-caspase-9 apoptosome. J. Cell. Sci. *123*, 3209–3214.

Brocchieri, L., Kledal, T.N., Karlin, S., and Mocarski, E.S. (2005). Predicting coding potential from genome sequence: application to betaherpesviruses infecting rats and mice. J. Virol. *79*, 7570–7596.

Brune, W. (2011). Inhibition of programmed cell death by cytomegaloviruses. Virus Res. *157*, 144–150.

Brune, W., Menard, C., Heesemann, J., and Koszinowski, U.H. (2001). A ribonucleotide reductase homolog of cytomegalovirus and endothelial cell tropism. Science *291*, 303–305.

Cam, M., Handke, W., Picard-Maureau, M., and Brune, W. (2010). Cytomegaloviruses inhibit Bak- and Bax-mediated apoptosis with two separate viral proteins. Cell Death Differ. *17*, 655–665.

Cassidy-Stone, A., Chipuk, J.E., Ingerman, E., Song, C., Yoo, C., Kuwana, T., Kurth, M.J., Shaw, J.T., Hinshaw, J.E., Green, D.R., *et al.* (2008). Chemical inhibition of the mitochondrial division dynamin reveals its role in Bax/Bak-dependent mitochondrial outer membrane permeabilization. Dev. Cell *14*, 193–204.

Cathelin, S., Rebe, C., Haddaoui, L., Simioni, N., Verdier, F., Fontenay, M., Launay, S., Mayeux, P., and Solary, E. (2006). Identification of proteins cleaved downstream of caspase activation in monocytes undergoing macrophage differentiation. J. Biol. Chem. *281*, 17779–17788.

Chee, M.S., Bankier, A.T., Beck, S., Bohni, R., Brown, C.M., Cerny, R., Horsnell, T., Hutchison, C.A., 3rd, Kouzarides, T., Martignetti, J.A., et al. (1990). Analysis of the protein-coding content of the sequence of human cytomegalovirus strain AD169. Curr. Top. Microbiol. Immunol. 154, 125–169.

Cicin-Sain, L., Podlech, J., Messerle, M., Reddehase, M.J., and Koszinowski, U.H. (2005). Frequent coinfection of cells explains functional in vivo complementation between cytomegalovirus variants in the multiply infected host. J. Virol. 79, 9492–9502.

Cicin-Sain, L., Ruzsics, Z., Podlech, J., Bubic, I., Menard, C., Jonjic, S., Reddehase, M.J., and Koszinowski, U.H. (2008). Dominant-negative FADD rescues the in vivo fitness of a cytomegalovirus lacking an antiapoptotic viral gene. J. Virol. 82, 2056–2064.

Declercq, W., Vanden Berghe, T., and Vandenabeele, P. (2009). RIP kinases at the crossroads of cell death and survival. Cell 138, 229–232.

Degterev, A., Hitomi, J., Germscheid, M., Ch'en, I.L., Korkina, O., Teng, X., Abbott, D., Cuny, G.D., Yuan, C., Wagner, G., et al. (2008). Identification of RIP1 kinase as a specific cellular target of necrostatins. Nat. Chem. Biol. 4, 313–321.

Denecker, G., Vercammen, D., Declercq, W., and Vandenabeele, P. (2001). Apoptotic and necrotic cell death induced by death domain receptors. Cell. Mol. Life Sci. 58, 356–370.

Dhanasekaran, D.N., and Reddy, E.P. (2008). JNK signaling in apoptosis. Oncogene 27, 6245–6251.

Droin, N., Cathelin, S., Jacquel, A., Guery, L., Garrido, C., Fontenay, M., Hermine, O., and Solary, E. (2008). A role for caspases in the differentiation of erythroid cells and macrophages. Biochimie 90, 416–422.

Dunn, W., Chou, C., Li, H., Hai, R., Patterson, D., Stolc, V., Zhu, H., and Liu, F. (2003). Functional profiling of a human cytomegalovirus genome. Proc. Natl. Acad. Sci. U.S.A. 100, 14223–14228.

Ferri, K.F., and Kroemer, G. (2001). Organelle-specific initiation of cell death pathways. Nat. Cell Biol. 3, E255–E263.

Festjens, N., Cornelis, S., Lamkanfi, M., and Vandenabeele, P. (2006). Caspase-containing complexes in the regulation of cell death and inflammation. Biol. Chem. 387, 1005–1016.

Festjens, N., van Gurp, M., van Loo, G., Saelens, X., and Vandenabeele, P. (2004). Bcl-2 family members as sentinels of cellular integrity and role of mitochondrial intermembrane space proteins in apoptotic cell death. Acta Haematol. 111, 7–27.

Furukawa, T., Sakuma, S., and Plotkin, S.A. (1976). Human cytomegalovirus infection of WI-38 cells stimulates mitochondrial DNA synthesis. Nature 262, 414–416.

Garrido, C., and Kroemer, G. (2004). Life's smile, death's grin: vital functions of apoptosis-executing proteins. Curr. Opin. Cell Biol. 16, 639–646.

Gatherer, D., Seirafian, S., Cunningham, C., Holton, M., Dargan, D.J., Baluchova, K., Hector, R.D., Galbraith, J., Herzyk, P., Wilkinson, G.W., et al. (2011). High-resolution human cytomegalovirus transcriptome. Proc. Natl. Acad. Sci. U.S.A. 108, 19755–19760.

Goldmacher, V.S., Bartle, L.M., Skaletskaya, A., Dionne, C.A., Kedersha, N.L., Vater, C.A., Han, J.W., Lutz, R.J., Watanabe, S., Cahir McFarland, E.D., et al. (1999). A cytomegalovirus-encoded mitochondria-localized inhibitor of apoptosis structurally unrelated to Bcl-2. Proc. Natl. Acad. Sci. U.S.A. 96, 12536–12541.

Golstein, P., and Kroemer, G. (2007). Cell death by necrosis: towards a molecular definition. TIBS 32, 37–43.

Gordon, S., and Taylor, P.R. (2005). Monocyte and macrophage heterogeneity. Nat. Rev. Immunol. 5, 953–964.

Green, D.R., and Kroemer, G. (2004). The pathophysiology of mitochondrial cell death. Science 305, 626–629.

Hahn, G., Khan, H., Baldanti, F., Koszinowski, U.H., Revello, M.G., and Gerna, G. (2002). The human cytomegalovirus ribonucleotide reductase homolog UL45 is dispensable for growth in endothelial cells, as determined by a BAC-cloned clinical isolate of human cytomegalovirus with preserved wild-type characteristics. J. Virol. 76, 9551–9555.

Hayajneh, W.A., Colberg-Poley, A.M., Skaletskaya, A., Bartle, L.M., Lesperance, M.M., Contopoulos-Ioannidis, D.G., Kedersha, N.L., and Goldmacher, V.S. (2001). The sequence and antiapoptotic functional domains of the human cytomegalovirus UL37 exon 1 immediate-early protein are conserved in multiple primary strains. Virology 279, 233–240.

He, S., Wang, L., Miao, L., Wang, T., Du, F., Zhao, L., and Wang, X. (2009). Receptor interacting protein kinase-3 determines cellular necrotic response to TNF-alpha. Cell 137, 1100–1111.

Hegde, R., Srinivasula, S.M., Zhang, Z., Wassell, R., Mukattash, R., Cilenti, L., DuBois, G., Lazebnik, Y., Zervos, A.S., Fernandes-Alnemri, T., et al. (2002). Identification of Omi/HtrA2 as a mitochondrial apoptotic serine protease that disrupts inhibitor of apoptosis protein–caspase interaction. J. Biol. Chem. 277, 432–438.

Hengartner, M.O. (2000). The biochemistry of apoptosis. Nature 407, 770–776.

Hertel, L., and Mocarski, E.S. (2004). Global analysis of host cell gene expression late during cytomegalovirus infection reveals extensive dysregulation of cell cycle gene expression and induction of Pseudomitosis independent of US28 function. J. Virol. 78, 11988–12011.

Irusta, P.M., Chen, Y.B., and Hardwick, J.M. (2003). Viral modulators of cell death provide new links to old pathways. Curr. Opin. Cell. Biol. 15, 700–705.

Jaattela, M. (2004). Multiple cell death pathways as regulators of tumour initiation and progression. Oncogene 23, 2746–2756.

Jan, G., Belzacq, A.S., Haouzi, D., Rouault, A., Metivier, D., Kroemer, G., and Brenner, C. (2002). Propionibacteria induce apoptosis of colorectal carcinoma cells via short-chain fatty acids acting on mitochondria. Cell Death Differ. 9, 179–188.

Jesenberger, V., and Jentsch, S. (2002). Deadly encounter: ubiquitin meets apoptosis. Nat. Rev. Mol. Cell. Biol. 3, 112–121.

Kaarbo, M., Ager-Wick, E., Osenbroch, P.O., Kilander, A., Skinnes, R., Muller, F., and Eide, L. (2011). Human cytomegalovirus infection increases mitochondrial biogenesis. Mitochondrion 11, 935–945.

Kaiser, W.J., and Offermann, M.K. (2005). Apoptosis induced by the toll-like receptor adaptor TRIF is dependent on its receptor interacting protein homotypic interaction motif. J. Immunol. 174, 4942–4952.

Kaiser, W.J., Upton, J.W., Long, A.B., Livingston-Rosanoff, D., Daley-Bauer, L.P., Hakem, R., Caspary, T., and Mocarski, E.S. (2011). RIP3 mediates the embryonic lethality of caspase-8-deficient mice. Nature *471*, 368–372.

Kaiser, W.J., Upton, J.W., and Mocarski, E.S. (2008). Receptor-interacting protein homotypic interaction motif-dependent control of NF-kappa B activation via the DNA-dependent activator of IFN regulatory factors. J. Immunol. *181*, 6427-6434

Kang, T.B., Ben-Moshe, T., Varfolomeev, E.E., Pewzner-Jung, Y., Yogev, N., Jurewicz, A., Waisman, A., Brenner, O., Haffner, R., Gustafsson, E., *et al.* (2004). Caspase-8 serves both apoptotic and nonapoptotic roles. J. Immunol. *173*, 2976–2984.

Karbowski, M., Norris, K.L., Cleland, M.M., Jeong, S.Y., and Youle, R.J. (2006). Role of Bax and Bak in mitochondrial morphogenesis. Nature *443*, 658–662.

Kerr, J.F., Wyllie, A.H., and Currie, A.R. (1972). Apoptosis: a basic biological phenomenon with wide-ranging implications in tissue kinetics. Br. J. Cancer. *26*, 239–257.

Kroemer, G., Galluzzi, L., Vandenabeele, P., Abrams, J., Alnemri, E.S., Baehrecke, E.H., Blagosklonny, M.V., El-Deiry, W.S., Golstein, P., Green, D.R., *et al.* (2009). Classification of cell death: recommendations of the Nomenclature Committee on Cell Death 2009. Cell Death Differ. *16*, 3–11.

Kuwana, T., and Newmeyer, D.D. (2003). Bcl-2-family proteins and the role of mitochondria in apoptosis. Curr. Opin. Cell. Biol. *15*, 691–699.

Lamkanfi, M., Festjens, N., Declercq, W., Vanden Berghe, T., and Vandenabeele, P. (2007). Caspases in cell survival, proliferation and differentiation. Cell Death Differ. *14*, 44–55.

Launay, S., Hermine, O., Fontenay, M., Kroemer, G., Solary, E., and Garrido, C. (2005). Vital functions for lethal caspases. Oncogene *24*, 5137–5148.

Leist, M., and Jaattela, M. (2001). Four deaths and a funeral: from caspases to alternative mechanisms. Nat. Rev. Mol. Cell. Biol. *2*, 589–598.

Lembo, D., Donalisio, M., Hofer, A., Cornaglia, M., Brune, W., Koszinowski, U., Thelander, L., and Landolfo, S. (2004). The ribonucleotide reductase R1 homolog of murine cytomegalovirus is not a functional enzyme subunit but is required for pathogenesis. J. Virol. *78*, 4278–4288.

Lockshin, R.A., and Zakeri, Z. (2002). Caspase-independent cell deaths. Curr. Opin. Cell Biol. *14*, 727–733.

Lockshin, R.A., and Zakeri, Z. (2004). Apoptosis, autophagy, and more. Int. J. Biochem. Cell. Biol. *36*, 2405–2419.

McCormick, A.L. (2008). Control of apoptosis by human cytomegalovirus. Curr. Top. Microbiol. Immunol. *325*, 281–295.

McCormick, A.L., Skaletskaya, A., Barry, P.A., Mocarski, E.S., and Goldmacher, V.S. (2003a). Differential function and expression of the viral inhibitor of caspase 8-induced apoptosis (vICA) and the viral mitochondria-localized inhibitor of apoptosis (vMIA) cell death suppressors conserved in primate and rodent cytomegaloviruses. Virology *316*, 221–233.

McCormick, A.L., Smith, V.L., Chow, D., and Mocarski, E.S. (2003b). Disruption of mitochondrial networks by the human cytomegalovirus UL37 gene product viral mitochondrion-localized inhibitor of apoptosis. J. Virol. *77*, 631–641.

McCormick, A.L., Meiering, C.D., Smith, G.B., and Mocarski, E.S. (2005). Mitochondrial cell death suppressors carried by human and murine cytomegalovirus confer resistance to proteasome inhibitor-induced apoptosis. J. Virol. *79*, 12205–12217.

McCormick, A.L., Roback, L., and Mocarski, E.S. (2008). HtrA2/Omi terminates cytomegalovirus infection and is controlled by the viral mitochondrial inhibitor of apoptosis (vMIA). PLoS Pathog. *4*, e1000063.

McCormick, A.L., Roback, L., Livingston-Rosanoff, D., and St Clair, C. (2010). The human cytomegalovirus UL36 gene controls caspase-dependent and -independent cell death programs activated by infection of monocytes differentiating to macrophages. J. Virol. *84*, 5108–5123.

McCormick, A.L., Roback, L., Wynn, G., and Mocarski, E.S. (2012). Multiplicity-dependent activation of a serine protease-dependent cytomegalovirus-associated programmed cell death pathway. Virology, S0042-6822(12)00423-0.

McDonough, S.H., and Spector, D.H. (1983). Transcription in human fibroblasts permissively infected by human cytomegalovirus strain AD169. Virology *125*, 31–46.

McDonough, S.H., Staprans, S.I., and Spector, D.H. (1985). Analysis of the major transcripts encoded by the long repeat of human cytomegalovirus strain AD169. J. Virol. *53*, 711–718.

Mack, C., Sickmann, A., Lembo, D., and Brune, W. (2008). Inhibition of proinflammatory and innate immune signaling pathways by a cytomegalovirus RIP1-interacting protein. Proc. Natl. Acad. Sci. U.S.A. *105*, 3094–3099.

McSharry, B.P., Tomasec, P., Neale, M.L., and Wilkinson, G.W. (2003). The most abundantly transcribed human cytomegalovirus gene (beta 2.7) is non-essential for growth in vitro. J. Gen. Virol. *84*, 2511–2516.

Manzur, M., Fleming, P., Huang, D.C., Degli-Esposti, M.A., and Andoniou, C.E. (2009). Virally mediated inhibition of Bax in leukocytes promotes dissemination of murine cytomegalovirus. Cell Death Differ. *16*, 312–320.

Marchini, A., Liu, H., and Zhu, H. (2001). Human cytomegalovirus with IE-2 (UL122) deleted fails to express early lytic genes. J. Virol. *75*, 1870–1878.

Mavinakere, M.S., and Colberg-Poley, A.M. (2004). Dual targeting of the human cytomegalovirus UL37 exon 1 protein during permissive infection. J. Gen. Virol. *85*, 323–329.

Mavinakere, M.S., Williamson, C.D., Goldmacher, V.S., and Colberg-Poley, A.M. (2006). Processing of human cytomegalovirus UL37 mutant glycoproteins in the endoplasmic reticulum lumen prior to mitochondrial importation. J. Virol. *80*, 6771–6783.

Menard, C., Wagner, M., Ruzsics, Z., Holak, K., Brune, W., Campbell, A.E., and Koszinowski, U.H. (2003). Role of murine cytomegalovirus US22 gene family members in replication in macrophages. J. Virol. *77*, 5557–5570.

Mocarski, E.S., Jr., Shenk, T., and Pass, R.F. (2006). Cytomegaloviruses, 5th ed (Lippincott Williams & Wilkins, Philadelphia, PA).

Mocarski, E.S., Upton, J.W., and Kaiser, W.J. (2011). Viral infection and the evolution of caspase 8-regulated apoptotic and necrotic death pathways. Nat. Rev. Immunol. 12, 79–88.

Moorman, N.J., Cristea, I.M., Terhune, S.S., Rout, M.P., Chait, B.T., and Shenk, T. (2008). Human cytomegalovirus protein UL38 inhibits host cell stress responses by antagonizing the tuberous sclerosis protein complex. Cell Host Microbe 3, 253–262.

Oberst, A., Dillon, C.P., Weinlich, R., McCormick, L.L., Fitzgerald, P., Pop, C., Hakem, R., Salvesen, G.S., and Green, D.R. (2011). Catalytic activity of the caspase-8-FLIP(L) complex inhibits RIPK3-dependent necrosis. Nature 471, 363–367.

Patrone, M., Percivalle, E., Secchi, M., Fiorina, L., Pedrali-Noy, G., Zoppe, M., Baldanti, F., Hahn, G., Koszinowski, U.H., Milanesi, G., et al. (2003). The human cytomegalovirus UL45 gene product is a late, virion-associated protein and influences virus growth at low multiplicities of infection. J. Gen. Virol. 84, 3359–3370.

Patterson, C.E., and Shenk, T. (1999). Human cytomegalovirus UL36 protein is dispensable for viral replication in cultured cells. J. Virol. 73, 7126–7131.

Pauleau, A.L., Larochette, N., Giordanetto, F., Scholz, S.R., Poncet, D., Zamzami, N., Goldmacher, V.S., and Kroemer, G. (2007). Structure-function analysis of the interaction between Bax and the cytomegalovirus-encoded protein vMIA. Oncogene 26, 7067–7080.

Perlman, H., Pagliari, L.J., Georganas, C., Mano, T., Walsh, K., and Pope, R.M. (1999). FLICE-inhibitory protein expression during macrophage differentiation confers resistance to fas-mediated apoptosis. J. Exp. Med. 190, 1679–1688.

Petros, A.M., Olejniczak, E.T., and Fesik, S.W. (2004). Structural biology of the Bcl-2 family of proteins. Biochim. Biophys. Acta 1644, 83–94.

Polster, B.M., Pevsner, J., and Hardwick, J.M. (2004). Viral Bcl-2 homologs and their role in virus replication and associated diseases. Biochim. Biophys. Acta 1644, 211–227.

Poncet, D., Larochette, N., Pauleau, A.L., Boya, P., Jalil, A.A., Cartron, P.F., Vallette, F., Schnebelen, C., Bartle, L.M., Skaletskaya, A., et al. (2004). An antiapoptotic viral protein that recruits Bax to mitochondria. J. Biol. Chem. 279, 22605–22614.

Poncet, D., Pauleau, A.L., Szabadkai, G., Vozza, A., Scholz, S.R., Le Bras, M., Briere, J.J., Jalil, A., Le Moigne, R., Brenner, C., et al. (2006). Cytopathic effects of the cytomegalovirus-encoded apoptosis inhibitory protein vMIA. J. Cell Biol. 174, 985–996.

Qian, Z., Xuan, B., Gualberto, N., and Yu, D. (2011). The human cytomegalovirus protein pUL38 suppresses endoplasmic reticulum stress-mediated cell death independently of its ability to induce mTORC1 activation. J. Virol. 85, 9103–9113.

Reboredo, M., Greaves, R.F., and Hahn, G. (2004). Human cytomegalovirus proteins encoded by UL37 exon 1 protect infected fibroblasts against virus-induced apoptosis and are required for efficient virus replication. J. Gen. Virol. 85, 3555–3567.

Rebsamen, M., Heinz, L.X., Meylan, E., Michallet, M.C., Schroder, K., Hofmann, K., Vazquez, J., Benedict, C.A., and Tschopp, J. (2009). DAI/ZBP1 recruits RIP1 and RIP3 through RIP homotypic interaction motifs to activate NF-kappaB. EMBO Rep. 10, 916–922.

Reeves, M.B., Davies, A.A., McSharry, B.P., Wilkinson, G.W., and Sinclair, J.H. (2007). Complex I binding by a virally encoded RNA regulates mitochondria-induced cell death. Science 316, 1345–1348.

Roumier, T., Vieira, H.L., Castedo, M., Ferri, K.F., Boya, P., Andreau, K., Druillennec, S., Joza, N., Penninger, J.M., Roques, B., et al. (2002). The C-terminal moiety of HIV-1 Vpr induces cell death via a caspase-independent mitochondrial pathway. Cell Death Differ. 9, 1212–1219.

Schwerk, C., and Schulze-Osthoff, K. (2003). Non-apoptotic functions of caspases in cellular proliferation and differentiation. Biochem. Pharmacol. 66, 1453–1458.

Seo, J.Y., Yaneva, R., Hinson, E.R., and Cresswell, P. (2011). Human cytomegalovirus directly induces the antiviral protein viperin to enhance infectivity. Science 332, 1093–1097.

Sharon-Friling, R., Goodhouse, J., Colberg-Poley, A.M., and Shenk, T. (2006). Human cytomegalovirus pUL37x1 induces the release of endoplasmic reticulum calcium stores. Proc. Natl. Acad. Sci. U.S.A. 103, 19117–19122.

Skaletskaya, A., Bartle, L.M., Chittenden, T., McCormick, A.L., Mocarski, E.S., and Goldmacher, V.S. (2001). A cytomegalovirus-encoded inhibitor of apoptosis that suppresses caspase-8 activation. Proc. Natl. Acad. Sci. U.S.A. 98, 7829–7834.

Smith, G.B., and Mocarski, E.S. (2005). Contribution of GADD45 family members to cell death suppression by cellular Bcl-xL and cytomegalovirus vMIA. J. Virol. 79, 14923–14932.

Sordet, O., Rebe, C., Dubrez-Daloz, L., Boudard, D., and Solary, E. (2002). Intracellular redistribution of procaspases during TPA-induced differentiation of U937 human leukemic cells. Leukemia 16, 1569–1570.

Suzuki, Y., Imai, Y., Nakayama, H., Takahashi, K., Takio, K., and Takahashi, R. (2001). A serine protease, HtrA2, is released from the mitochondria and interacts with XIAP, inducing cell death. Mol. Cell 8, 613–621.

Tenney, D.J., and Colberg-Poley, A.M. (1990). RNA analysis and isolation of cDNAs derived from the human cytomegalovirus immediate-early region at 0.24 map units. Intervirology 31, 203–214.

Tenney, D.J., and Colberg-Poley, A.M. (1991a). Expression of the human cytomegalovirus UL36–38 immediate-early region during permissive infection. Virology 182, 199–210.

Tenney, D.J., and Colberg-Poley, A.M. (1991b). Human cytomegalovirus UL36–38 and US3 immediate-early genes: temporally regulated expression of nuclear, cytoplasmic, and polysome-associated transcripts during infection. J. Virol. 65, 6724–6734.

Terhune, S., Torigoi, E., Moorman, N., Silva, M., Qian, Z., Shenk, T., and Yu, D. (2007). Human cytomegalovirus UL38 protein blocks apoptosis. J. Virol. 81, 3109–3123.

Upton, J.W., Kaiser, W.J., and Mocarski, E.S. (2008). Cytomegalovirus M45 cell death suppression requires receptor-interacting protein (RIP) homotypic interaction motif (RHIM)-dependent interaction with RIP1. J. Biol. Chem. 283, 16966–16970.

Upton, J.W., Kaiser, W.J., and Mocarski, E.S. (2010). Virus inhibition of RIP3-dependent necrosis. Cell Host Microbe 7, 302–313.

Upton, J.W., Kaiser, W.J., and Mocarski, E.S. (2012). DAI/ZBP1/DLM-1 complexes with RIP3 to mediate virus-induced programmed necrosis that is targeted by murine cytomegalovirus vIRA. Cell Host Microbe *11*, 290–297.

Vandenabeele, P., Declercq, W., Van Herreweghe, F., and Vanden Berghe, T. (2010). The role of the kinases RIP1 and RIP3 in TNF-induced necrosis. Sci. Signal. *3*, re4.

Williamson, C.D., and Colberg-Poley, A.M. (2009). Access of viral proteins to mitochondria via mitochondria-associated membranes. Rev. Med. Virol. *19*, 147–164.

Williamson, C.D., Zhang, A., and Colberg-Poley, A.M. (2011). The human cytomegalovirus protein UL37 exon 1 associates with internal lipid rafts. J. Virol. *85*, 2100–2111.

Xuan, B., Qian, Z., Torigoi, E., and Yu, D. (2009). Human cytomegalovirus protein pUL38 induces ATF4 expression, inhibits persistent JNK phosphorylation, and suppresses endoplasmic reticulum stress-induced cell death. J. Virol. *83*, 3463–3474.

Yu, D., Silva, M.C., and Shenk, T. (2003). Functional map of human cytomegalovirus AD169 defined by global mutational analysis. Proc. Natl. Acad. Sci. U.S.A. *100*, 12396–12401.

Zhang, D.W., Shao, J., Lin, J., Zhang, N., Lu, B.J., Lin, S.C., Dong, M.Q., and Han, J. (2009). RIP3, an energy metabolism regulator that switches TNF-induced cell death from apoptosis to necrosis. Science *325*, 332–336.

Zhao, J., Sinclair, J., Houghton, J., Bolton, E., Bradley, A., and Lever, A. (2010). Cytomegalovirus beta2.7 RNA transcript protects endothelial cells against apoptosis during ischemia/reperfusion injury. J. Heart Lung Transplant *29*, 342–345.

Cytomegaloviruses and Interferons

I.16

Mirko Trilling and Hartmut Hengel

Abstract

Interferons (IFNs) comprise a family of three different subtypes (I, II and III) of related cytokines which share their potent immuno-stimulatory and antiviral function. IFN secretion is initiated by synchronous activation of distinct classes of transcription factors (ATF/c-Jun, IRFs, NF-κB) upon recognition of conserved pathogen-associated molecular patterns (PAMPs) by germ-line-encoded pathogen recognition receptors (PRRs). Binding of the transcription factors to the *ifn-β* promoter/enhancer assembles the IFN enhanceosome, leading to IFN transcription. Secreted IFNs signal in an autocrine and paracrine manner via Jak-STAT signal transduction pathways stimulating a far-reaching transcriptional programme of >300 differentially expressed genes to orchestrate intrinsic, innate and adaptive immunity. The intimate co-adaptation of cytomegaloviruses with their host species led to the evolution of multiple viral countermeasures which mitigate the antiviral effect of IFNs. The number of identified HCMV- and MCMV-encoded gene products interfering with IFN induction, IFN receptor signalling or IFN effector functions, is steadily growing. This review aims to provide a snapshot of our current understanding of the balance of power between pro- and antiviral measures positioned between CMV and the host IFN system. Given the immense selective pressure elicited by IFNs, it is tempting to speculate that IFNs have driven CMV to evolve a high number of antagonistic genes ensuring the complex counterbalance with IFNs and promoting CMV replication in an IFN containing environment. Counterintuitively, CMV appears also to exploit IFN induced transcription to enhance its gene expression under appropriate conditions.

Introduction: interferon compendium

Since their discovery and initial description in 1957 (Isaacs and Lindenmann, 1957), IFNs started their triumph. Described as factors which interfere (hence the name) with viral replication, the unique antiviral activity of IFNs was documented in numerous settings *in vitro* and *in vivo*. IFNs are structurally and functionally related pleiotropic cytokines which are secreted upon pathogen encounter eliciting an extremely effective anti-pathogenic response by reinforcing intrinsic resistance, invoking innate immunity and recruitment and stimulation of adaptive immunity. Consistently, spontaneous mutations or targeted deletions in the induction or signal transduction components lead to severe immuno-deficiencies and infections with opportunistic agents or live attenuated vaccines become perilous (reviewed by Zhang *et al.*, 2008). Mice lacking IFN receptors are hyper-susceptible towards a great variety of pathogens (reviewed by van den Broek *et al.*, 1995). Conversely, recombinant IFNs are clinically approved as drugs and constitute an effective component of the standard treatment regime against e.g. hepatitis B and C virus.

Type I, type II and type III IFNs

IFNs are subdivided, based on structural homology and receptor usage, into type I IFNs (all IFN-α subtypes and IFN-β), type II IFNs (IFN-γ) and the recently discovered type III IFNs. Different IFN types have partly overlapping but also distinct biologic effects. Although type I and type III IFNs engage different receptor complexes, they share a similar intracellular signalling pathway and induce similar transcriptional responses (Kotenko *et al.*, 2003).

IFN-β was initially described as 'fibroblast IFN' but in fact most cells are capable to secrete IFN-β. Upon pathogen encounter, IFN-β is the first IFN subtype being produced – in the mouse together with IFN-α4 (Marie *et al.*, 1998). Other type I IFNs require a positive feedback loop, which depends on functional IFN signal transduction (Marie *et al.*, 1998). Nevertheless, the 'initiator' IFN-β itself is also under the control of this positive feedback loop, since IFNAR-deficient cells are known to produce less IFN-α and IFN-β (Dai et al., 2011).

Type II IFN, IFN-γ, is secreted by T lymphocytes ('immune IFN') and NK cells. Whether macrophages secrete relevant amounts of IFN-γ under certain circumstances is a matter of debate (Bogdan and Schleicher, 2006).

Type I IFN induction

IFN induction requires specific sensors dedicated to the perception of pathogens. In contrast to the receptors of adaptive immunity, the receptors of the innate immune system are germline-encoded and fully functional at first pathogen encounter. This is achieved by recognition of conserved structures associated with infections. Receptors of innate immunity respond to so called PAMPs and are thus called PRRs (Janeway and Medzhitov, 2002). PAMPs like e.g. LPS of Gram-negative bacteria are molecules or alterations absent from 'healthy' cells. PAMPs constitute essential structural components of pathogens or necessary consequences

of their life cycle, reducing the likelihood that pathogens can avoid recognition. Mislocated molecules like the cytoplasmic presence of dsDNA can also serve as PAMP. PAMPs frequently associated with viral infections are dsDNA in the cytoplasm and dsRNA in the cytoplasm or the endosome.

Three major classes of PRRs are known: Toll-like receptors (TLRs), RIG-I-like receptors (RLRs) and NOD-like receptors (NLRs). Recently, the families of AIM2-like receptors and PARP-like receptors have joined this fast-growing list. Engagement of PRRs leads to activation of IRFs, NF-κB and ATF/c-Jun transcription factors, which are the major constituents of the IFN-β enhanceosome and synergistically initiate the type I IFN transcription (see Fig. I.16.1) by binding to an enhancer element upstream of the *ifn-β* gene (reviewed by Panne, 2008). Initially, IRF-1 was implicated in IFN induction, but now it is clear that IRF-3 plays a more important role. After IFN expression, IRF-7 becomes an important part of the positive feedback loop to maintain sustained IFN responses.

A complex and highly interconnected system of adaptor proteins couples PAMP receptors and the

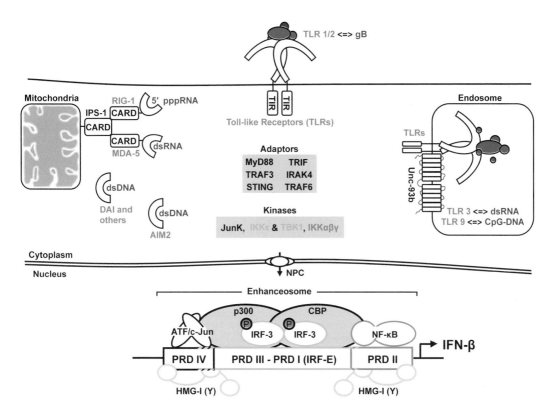

Figure I.16.1 Simplified overview over the IFN induction pathway following recognition of cytomegaloviral PAMPs by PRRs. PRRs are written in red and their cognate CMV ligands are written in blue. The ATF/c-Jun system is depicted in black, the IRF system in orange and the NF-κB system in green colour. NPC, nuclear pore complex.

transcription factors. TLRs signal either via MyD88 (most TLRs), or via TRIF (almost exclusively in the case of TLR3) or a combination of both. Downstream of these adaptor proteins IRAKs, TRIFs and TAK1 (or RIP1) activate the IKK complex, leading to activation of NF-κB by phosphorylating inhibitory proteins like IκBα, inducing their subsequent proteasomal degradation and releasing the transcriptionally active p50:p65 NF-κB complex. IRF-3 phosphorylation is achieved by a TRIF-dependent activation of IKKε or TBK kinases.

IFN signal transduction

IFNs signal via Jak-STAT signal transduction pathways: Upon binding of the IFN to its receptor complex, receptor-associated kinases from the Janus kinase family (Jak1, Jak2 or Tyk2) become activated. These kinases phosphorylate the receptor chains and create docking sites for STATs. STATs bind and become tyrosine phosphorylated themselves. This modification leads to a reciprocal interaction of the src-homology 2 (SH2) domain of one STAT with the phospho-tyrosine residue of another STAT, inducing dimerization. STAT

dimers translocate into the nucleus, bind to specific DNA-consensus elements, recruit the transcriptional machinery and induce gene expression. Dephosphorylation and return to the cytoplasm close the cycle of activation and deactivation (see Fig. I.16.2).

Recently, this simple scheme has become more complex: It was found that STATs are latently dimerized even in their inactive state and that the homodimers (and heterodimers?) change the orientation from an anti-parallel to a parallel conformation (Mao et al., 2005; Zhong et al., 2005). Although the phosphorylation itself is dispensable for DNA-binding, the phospho-Tyr-SH2 interaction instructs a conformation which drastically enhances DNA-binding (Wenta et al., 2008). The phosphorylation is also critically important for the nuclear accumulation, since phosphorylated STATs can not leave the nucleus until they become dephosphorylated. Sequence-specific DNA-binding and dephosphorylation are interrelated since binding to the cognate DNA consensus motif protects STAT molecules more efficiently from dephosphorylation than binding to other sequences, thereby prolonging the retention time on specific promoters (Meyer et al.,

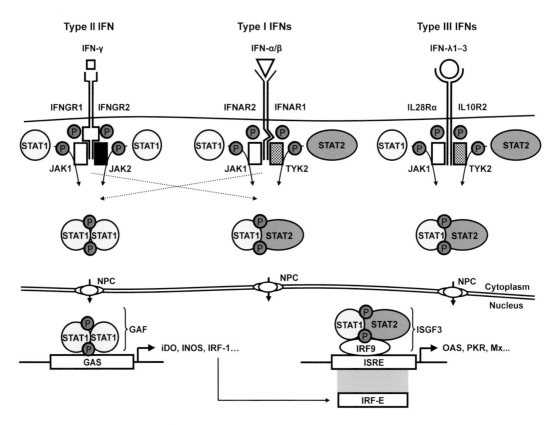

Figure I.16.2 Schematic overview over the IFN-Jak-STAT signal transduction pathway. For details see the text. The dashed lines indicate cross-talk events, whereas the straight lines indicate canonical signal transduction. NPC: nuclear pore complex.

2003). Thus, the fundamental scenario whereby STATs translocate into the nucleus only upon activation and return into the cytoplasm when their task is accomplished is oversimplified. Rather, STATs continuously shuttle between nucleus and cytoplasm, and activation stimulates nuclear retention rather than the rate of import (Meyer et al., 2003; Vinkemeier, 2004).

Type I IFNs initiate gene expression upon binding to a complex composed of the two receptor chains IFNAR1 and IFNAR2, which are associated with the kinases Tyk2 and Jak1, respectively, which are both essential for type I IFN signalling (Fig. I.16.2) (Velazquez et al., 1992; Muller et al., 1993). Upon ligand binding, the Janus kinases tyrosine-phosphorylate the receptor chains to create docking sites for STAT molecules. The binding of STAT molecules occurs in an ordered and sequential manner and STAT2 binding is required for subsequent STAT1 binding and phosphorylation (Li et al., 1997) and thus STAT1 phosphorylation is diminished in STAT2-deficient cells (Leung et al., 1995). Therefore, mechanisms that deprive STAT2 from IFNAR should also negatively influence IFN-α-dependent STAT1 phosphorylation. After binding of the STATs to the phosphorylated receptor, the Janus kinases phosphorylate the STATs. Type I IFNs mainly phosphorylate STAT1 and STAT2 but to a lesser extent also STAT3, STAT4, STAT5A/B and STAT6.

Phosphorylated STAT1 and STAT2 molecules heterodimerize via a reciprocal interaction of the phosphorylated tyrosine residue (STAT1-Y701 and STAT2-Y690, respectively) with the SH2-domain of the other STAT. Additionally, IRF-9 is recruited (Veals et al., 1992) and the heterotrimer, called ISGF3, translocates into the nucleus (Fig. I.16.2) (Levy et al., 1989). IRF-9 seems to be essential for the nuclear translocation, which is mediated by Importin-α3, Importin-α4 and to a lesser extent by Importin-α7. The nuclear export is Leptomycin B sensitive and therefore regarded as CRM1-dependent (Banninger and Reich, 2004). The heterotrimer composed of STAT1α or STAT1β, STAT2 and IRF-9 binds to so-called IFN-stimulated response element (ISRE) consensus elements and activates gene expression of ISGs. Consistent with the fact that IRF-9 is a constituent of this heterotrimeric complex, ISRE and IRF-binding sites, denominated IRF-E elements, are partly overlapping (see Fig. I.16.2).

Similarly, IFN-γ induces phosphorylation and homodimerization of STAT1 (Shuai et al., 1992). IFN-γ binds to a complex composed of the two IFN-γ receptor chains IFNGR1 and IFNGR2, which are associated with Jak1 (Muller et al., 1993; Shuai et al., 1993; Silvennoinen et al., 1993) and Jak2 (Fig. I.16.2) (Watling et al., 1993), respectively. Upon binding of

IFN-γ to the receptor, Jak kinases, the receptor itself and STAT1 become phosphorylated and thus activated (Igarashi et al., 1994). Active STAT1 molecules homodimerize, constituting the gamma activated factor (GAF) (Decker et al., 1991; Shuai et al., 1992, 1993). GAF translocates into the nucleus and binds specific promoter elements called gamma activated sequence (GAS) to induce transcription of downstream genes.

Type III IFNs were discovered in 2003 and termed IFN-λ1-3 or IL-28A, IL-28B and IL-29 (Kotenko et al., 2003; Sheppard et al., 2003). Although structurally related to the IL-10 family and to IL-22 (Gad et al., 2009), IFN-λs evoke an IFN-like response, induce STAT2 phosphorylation and limit replication of viruses (Robek et al., 2005; Gad et al., 2009), including HSV-2 (Ank et al., 2006). An important difference between the type I and the type III IFN system is the distribution of their receptor complexes. While IFNAR is broadly expressed, the presence of IFN-λ receptors seems to be limited to certain tissues. IFN-λ signals via a heterodimer composed of IL-10R2 and IFN-λR1 (Fig. I.16.2) the expression of which is prominently found on cells of epithelial origin (Sommereyns et al., 2008), thus playing an important role in combating e.g. rotavirus infection of the intestine (Pott et al., 2011).

IFN-induced effector mechanisms

Depending on the cell type and the duration of IFN exposure between 120 and 300 host genes are directly regulated by IFN (Der et al., 1998; de Veer et al., 2001). An online database can be used to learn about the IFN inducibility for a given gene (http://www.lerner.ccf. org/labs/williams/xchip-html.cgi). Although a definitive function is not yet known for all gene products, many ISGs elicit profound antiviral effects against various types of viruses (Schoggins et al., 2011).

PKR is among the most prominent IFN induced antiviral effectors representing a dsRNA-binding Ser/Thr kinase which phosphorylates Ser-51 of the eukaryotic translation initiation factor eIF-2α and thereby blocks its recycling, shutting down translation (Kimchi et al., 1979; Wu and Kaufman, 1996).

Another mainly IFN-α-inducible factor with antiviral activity is the OAS – RNase L system. In presence of dsRNA, OAS induces 2′–5′ linkage of ATP (Zilberstein et al., 1978). This unstable 2′–5′-oligoadenylate induces a local dimerization and activation of the latently monomeric RNase L (Clemens and Williams, 1978; Ratner et al., 1978; Jacobs and Langland, 1996). Dimerized RNase L in turn degrades mRNA and rRNA.

Moreover, IFN stimulates the expression of a distinct set of dynamin-related GTPases clustered into three groups: Mx, 47-kDa and 65-kDa GTPases. Mx

GTPases confer resistance to many RNA viruses like influenza and some members of the 65-kDa GTPases also have been shown to act antivirally against viruses including VSV, HCV or influenza (Carter *et al.*, 2005; Itsui *et al.*, 2009; Nordmann *et al.*, 2011). Although 47-kDa GTPases are induced upon MCMV infection *in vivo*, viral plaque numbers remain unchanged in animals individually deficient for the prominent family members of the 47-kDa GTPase family LRG-47, IRG-47 or IGTP (Taylor *et al.*, 2000; Collazo *et al.*, 2001).

A mainly IFN-γ-induced gene known to inhibit poxviruses (Karupiah *et al.*, 1993) and herpesviruses (Croen, 1993) is iNOS. iNOS is an enzyme which catalyses the synthesis of reactive NO species. Although HCMV seemingly interferes with IFN-induced NO production by iNOS, the addition of an exogenous NO donor compound reduces HCMV replication (Bodaghi *et al.*, 1999). Conversely, iNOS-deficient mice, are more susceptible to MCMV and generate higher virus titres (Fernandez *et al.*, 2000; Noda *et al.*, 2001).

The IFN-γ-inducible IDO enzyme consumes the essential amino acid L-tryptophan. In monocytes, HCMV infection induces IDO expression (Furset *et al.*, 2008) and IDO has been shown to reduce HCMV replication in retinal pigment epithelium cells (Bodaghi *et al.*, 1999) and in astrocytes (Suh *et al.*, 2007).

IFN induction upon CMV infection

MCMV infection via the intraperitoneal route leads to type I IFN production detectable in mice already few hours post infection (h p.i.) and reaching peak titres ca. 36 h p.i., while 4 days post infection (d p.i.) IFN is no longer detectable (Kelsey *et al.*, 1977 – see also related Chapters II.9, II.11 and II.12). Interestingly, a super-infection fails to induce normal IFN secretion during acute MCMV replication (Kelsey *et al.*, 1977), suggesting the establishment of a system-wide interference of MCMV infection with IFN secretion. RT-PCR experiments showed that MCMV induces IFN-β and different IFN-α subtypes in infected mice (Yeow *et al.*, 1997). The induction of type I IFNs in the serum was found to be biphasic with a first peak ca. 8 h p.i. and a second peak ca. 2 d p.i. – contemporaneous with the initial infection and the starting of viral replication (Schneider *et al.*, 2008). In humans IFN-γ was found in the serum during the course of acute HCMV infection (Rhodes-Feuillette *et al.*, 1983). Type III IFN (IL-28A) mRNA induction was also documented in the intestine upon MCMV infection (Brand *et al.*, 2005).

MCMV establishes productive replication in macrophages albeit the replication cycle is protracted in comparison to fibroblasts. IFN-α/β induction is increased and prolonged in macrophages reaching peak titres at 3 d p.i. (Yamaguchi *et al.*, 1988). High titres of IFN-β are also produced by HCMV-infected endothelial cells in cell culture (Sedmak *et al.*, 1995). Surprisingly, even IFN-γ was found to become secreted by fibroblasts during late stages of HCMV replication (Boldogh *et al.*, 1997).

Consistent with the IFN induction upon CMV infection, a large fraction of CMV-induced genes (*cigs*) were identified as ISGs, for instance *cig1* is ISG54 and *cig5* is Viperin (Zhu *et al.*, 1997; Browne *et al.*, 2001; Chin and Cresswell, 2001). This ISG induction requires the fusion between virion envelope and host membranes and can be blocked by fusion inhibitors and neutralizing antibodies (Netterwald *et al.*, 2004), a fact which strongly suggests that IFN induction upon CMV infection constitutes a natural process and is not just a reaction to cellular and viral debris contained in the virus stock preparation. Consistent with this IFN induction, HCMV activates IRF-3 (Navarro *et al.*, 1998; Preston *et al.*, 2001; Gravel and Servant, 2005), which is essential for ISG and IFN-β induction (DeFilippis *et al.*, 2006). Even HCMV-gB alone suffices to induce ISGs (Boyle *et al.*, 1999; Simmen *et al.*, 2001; Boehme *et al.*, 2004) and to activate IRF-3 (Boehme *et al.*, 2004 – see also Chapter I.8). While gB-induced ISG expression is IRF-3-dependent, it does seemingly not represent the native ISG induction elicited by regular HCMV infection, which is dsDNA- and ZBP1/DAI-dependent (DeFilippis *et al.*, 2010b).

TNF-α and lymphotoxin induce an NF-κB-dependent induction of IFN-β upon CMV infection, restricting viral replication *in vitro* and *in vivo* (Benedict *et al.*, 2001). The LT-LTR-IFN-β-IFNAR1 axis is crucial to protect lymphocytes from apoptosis during MCMV infection (Banks *et al.*, 2005) and TNFR1 and the LTβR are important for the induction of IFN-γ by NK cells and to induce IFN-β secretion in target cells (Iversen *et al.*, 2005). Especially, the first peak of type I IFN was found to critically require LTβ-expressing B cells and the expression of LTβR in splenic stroma cells (Schneider *et al.*, 2008).

The main IFN-producing cells during CMV infections

The main producers of IFN-α upon MCMV infection were found to be a subset of immature antigen-presenting DCs with plasmacytoid morphology (Asselin-Paturel *et al.*, 2001). IFN producing cells exhibit a CD8α+, Ly6G/C+ and CD11b− DC phenotype (Dalod *et al.*, 2002). These IFN-producing pDCs mature upon MCMV infection and acquire the potential to stimulate NK cells and to present antigens to activate T-cells (Dalod et al., 2003). Interestingly, pDCs are rather non-permissive for MCMV, whereas cDC become productively infected by MCMV (Andrews et al., 2001; Dalod et al., 2002). Usage of an IFN-β reporter mouse revealed that IFN-β-producing cells are activated GR-1^{int}, B220$^+$, CD11b$^-$, CD8α$^+$ pDCs positioned mainly at the T–/B-cell interface of the splenic white pulp and occasionally in the red pulp or the marginal zone (Scheu et al., 2008). Consistently, ablation of pDC reduces type I IFN production and augments the viral burden (Swiecki et al., 2010). CD11b$^+$ cDCs infected by MCMV secrete IFN-α and activate NK cells which subsequently produce IFN-γ (Andoniou et al., 2005).

Molecular sensors for CMV

CMVs possess a large dsDNA genome and therefore a comprehensible sensor is TLR9 – a receptor known to sense CpG-dsDNA (Hemmi et al., 2000). A mutant mouse line harbouring a mutation in TLR9 is highly susceptible to MCMV, reaching ca. 1000-fold higher viral titres in the spleen and a higher likelihood to succumb to infection (Tabeta et al., 2004). DCs from TLR9-deficient mice fail to induce IFN-α upon MCMV infection and to control MCMV replication (Krug et al., 2004). TLR9 signals via the adaptor protein MyD88. Consistently, DCs from MyD88-deficient mice fail to produce IFN-α upon MCMV infection and are hypersensitive towards MCMV in vivo (Krug et al., 2004; Tabeta et al., 2004; Delale et al., 2005), demonstrating the important role of the TLR9–MyD88 axis for sensing MCMV.

Nevertheless, the comparison of MyD88- and TLR9-deficient mice revealed that TLR9-independent but MyD88-dependent TLRs contribute to type I IFN induction during MCMV infection and that hepatic IFN-α induction is particularly TLR9-independent (Hokeness-Antonelli et al., 2007).

TLR2 becomes activated by HCMV virions and activates down-stream NF-κB signalling (Compton et al., 2003 – see also Chapter I.8). MCMV infections in TLR2-deficient mice revealed decreased type I IFN concentrations in conjunction with increased viral titres and a more pronounced liver pathology (Szomolanyi-Tsuda et al., 2006). TLR2 (together with TLR1) senses HCMV virions by an interaction with gB and gH (Boehme et al., 2006). Interestingly, a correlation between the risk of CMV disease in liver transplant recipients and a homozygocity for a polymorphism in TLR2 (R753Q) – found in 5% of the studied patients – has been reported (Kijpittayarit et al., 2007). Consistently, TLR2-R753Q has lost the potency to induce IL-8

and TNF-α upon recognition of HCMV gB (Brown et al., 2009).

Since large parts of the transcribed CMV genome simultaneously express sense as well as antisense transcripts (Zhang et al., 2007), dsRNA is found in CMV infected cells (Budt et al., 2009; Marshall et al., 2009). Despite the presence of dsRNA, IPS-1, an adaptor for the dsRNA sensors RIG-I and MDA-5, is not essential for type I IFN induction in HCMV infected fibroblasts (DeFilippis et al., 2010a), which might either indicate that IPS-1 is dispensable for the recognition of cytomegaloviral dsRNA or that this pathway is under stringent cytomegaloviral control. The findings that HCMV induces degradation of RIG-I (Scott, 2009) and that HCMV and MCMV actively inhibit Sendai virus (a known RIG-1 stimulus) induced IFN-β induction (Taylor and Bresnahan, 2005; Le et al., 2008b) favour the latter interpretation.

TLR3 is activated by dsRNA. In contrast to most other TLRs which signal via MyD88, TLR3 signals via TRIF (Yamamoto et al., 2003). Mice harbouring a TRIF-mutation are impaired in producing type I IFN upon MCMV infection and are hyper-susceptible to MCMV (Hoebe et al., 2003). TLR3-deficient animals are also more susceptible to MCMV (Tabeta et al., 2004). The fact that the mortality of TLR3$^{-/-}$ animals upon MCMV infection is not significantly altered (in contrast to TLR9-deficient mice) (Tabeta et al., 2004) indicates that TLR9-mediated dsDNA sensing is more important than TLR3-dependent dsRNA sensing.

From the intracellular nucleic acid sensors, DAI/ZBP1 has been shown to recognize HCMV and to signal via STING and DDX3 (DeFilippis et al., 2010a) initiating type I IFN transcription. Ablation of the NOD-like protein NLRC5 reduces IFN-α mRNA expression upon HCMV infection, indicating that NOD-like receptors contribute to the induction of type I IFN by CMVs (Kuenzel et al., 2010). Likewise, absence of AIM2 reduces IFN-γ induction upon MCMV infection (Rathinam et al., 2010).

Downstream of both TLRs, RLRs and receptors like DAI, the kinases TBK1 and IKKε are activated to phosphorylate IRF-3 which subsequently translocates into the nucleus, binds the IFN-β promoter and stimulates target genes. HCMV infection stimulates phosphotransferase activity of TBK1 towards IRF-3 (Gravel and Servant, 2005). Double deficiency of IRF-3 and IRF-7 abrogates IFN-β induction by MCMV (Rathinam et al., 2010). Interestingly, IRF-7, but not IRF-3, was found to be required for the systemic IFN-α response to MCMV (Steinberg et al., 2009). In DCs, IRF-8 additionally contributes to IFN-α and IFN-β induction (Tailor et al., 2007).

IFN possesses antiviral activity against CMV

The interplay of IFN and CMVs has been studied for almost five decades (Henson and Smith, 1964). Already the first studies reported that CMV is relatively resistant to IFN (Glasgow et al., 1967). Nevertheless, infection with MCMV produces a species-specific, proteinase-sensitive but acid-stable inhibitor of viral infection, which turned out to be IFN (Oie et al., 1975). In vitro, incubation with IFN modestly reduced MCMV replication (ca. 1 $\log_{10}$ inhibition by 1000 U/ml). Interestingly, the potency of IFN is inversely correlated with starting infectious doses of CMV (Oie et al., 1975; Postic and Dowling, 1977). IFN-α reduces viral replication in fibroblasts in vitro and is capable to decrease IE gene expression (Gribaudo et al., 1993; Martinotti et al., 1993), an effect which was attributed to the inhibition of NF-κB by IFN-α (Gribaudo et al., 1995). Conversely, neutralization of type I IFN increased viral titres in vivo (Chong et al., 1983). Clinical isolates of HCMV were also shown to be moderately sensitive to IFN (Postic and Dowling, 1977). Interestingly, the production of cell-free virus was found to be more sensitive towards IFN than the intracellular virus progeny (Holmes et al., 1978).

In vivo, MCMV replication is confined by the concerted action of innate and adaptive immune functions (Polic et al., 1998). Consistent with the moderate antiviral effect of IFN-γ in cell culture, IFN-γ is essential but not sufficient for the control of MCMV replication in the salivary gland (Lucin et al., 1992). The restricted antiviral effect of IFN-γ can be potentiated upon addition of TNF-α and blocks viral late (L) gene expression (Lucin et al., 1994). Ablation of TNF-α, IFN-α/β or IFN-γ by neutralizing antibodies during MCMV infection in vivo increased virus production in SCID mice – albeit with organ specific impact (Heise and Virgin, 1995).

Treatment of cells with recombinant IFN-γ inhibits HCMV replication (Yamamoto et al., 1987; Bodaghi et al., 1999; Sainz, et al., 2005), especially in combination with type I IFNs (Yamamoto et al., 1987; Sainz, et al., 2005). Ectopic expression of type I IFNs in vivo by gene therapy reduces MCMV replication in an IFNAR1-dependent manner. In this context, IFN-α1 was found to be more effective than IFN-α4 and IFN-α9 (Yeow et al., 1998).

An important insight into the relative impact of IFNs for immune control of MCMV arises from studies with gene-deficient mice lacking defined components of the IFN signal transduction machinery. The LD_{50} of MCMV in IFNGR1$^{-/-}$ and IFN-γ$^{-/-}$ mice is ca. 5-fold lower and in IFNAR1$^{-/-}$ even > 100-fold lower compared with WT mice. IFNAR1/IFNGR1 double

deficient mice succumb upon an infection with literally a few PFU MCMV (Presti et al., 1998; Gil et al., 2001; Loh et al., 2005). IFN-γ-deficient mice exhibit chronic vascular pathologies and fail to terminate productive MCMV infection (Presti et al., 1998). STAT1-deficient animals are ~ 100-fold more susceptible to MCMV, but clearly less susceptible than IFNAR1/IFNGR1 double deficient mice, indicating the existence of STAT1-dependent as well as STAT1-independent MCMV control mechanisms induced by IFN-γ (Gil et al., 2001; Ramana et al., 2001). An unbiased indication for the importance of Jak-STAT signal transduction for MCMV control was provided by Crozat et al. (2006): An in vivo 'MCMV resistome' screen in a large cohort of ENU-induced mutant mice yielded a highly MCMV-susceptible mouse line. The causative mutation resided in the DNA-binding domain of STAT1.

Tyk2-deficient mice suffer from hyper-susceptibility to MCMV infection. Especially in macrophages, the functionality of the type I IFN signal transduction pathway is required to limit viral replication. Noteworthy, MCMV replication was increased by > 4 orders of magnitude in IFNAR1-deficient macrophages and was also significantly increased in Tyk2$^{-/-}$ and IFN-β$^{-/-}$ cells (Strobl et al., 2005).

IFN-competent macrophages possess cell type specific IFN-γ-induced antiviral mechanisms efficiently controlling MCMV replication (Presti et al., 2001; Kropp et al., 2011). In this context, IFN-γ acts by priming of the type I IFN network which in turn controls MCMV immediate-early (IE) gene expression (Kropp et al., 2011).

Two case reports of patients either harbouring a deleterious mutation in the IFNGR1 or in STAT1 document an HCMV viraemia and a severe HCMV pneumonia, respectively, revealing the importance of IFNs for HCMV control in humans in vivo (Cunningham et al., 2000; Vairo et al., 2011).

CMV interferes with the host system interfering with viral replication: to square complexity

In principle, the IFN system can elicit a tremendous antiviral impact and therefore constitutes a prime selection factor for viral fitness. With this perception it is an evolutionary imperative that natural selection favours viruses which can circumvent the detrimental effects of IFNs. Basically all viruses subvert either IFN induction, IFN signalling and/or individual ISGs (reviewed by Hengel et al., 2005). Their adaptation is often so potent that the net effect of IFN is hardly measurable under experimental circumstances. Such findings should not be accounted as lack of antiviral potency of IFNs. Instead, it is an exciting result of perfect viral adaptation leading to evasion from innate immunity. Unfortunately, using the WT virus, it is hard to distinguish between 'IFN ineffectiveness' and 'viral evasion', unless a virus mutant is generated which lacks the respective IFN antagonists. The following sections will focus on individual HCMV and MCMV gene products, which exert a direct or an indirect effect on IFN expression, signalling or ISG function – ordered by ascending UL or M/m gene numbers (for an overview see Fig. I.16.3).

UL37x1/vMIA

Identified as potent inhibitor of mitochondrial apoptosis (Goldmacher et al., 1999), viral mitochondria-localized inhibitor of apoptosis (vMIA) is known to disrupt the mitochondrial network (McCormick et al., 2003 – see also Chapter I.15). RLR-dependent signal transduction is locally and physically tethered to mitochondria and influenced by mitochondrial dynamics. Expression of vMIA dampens signalling downstream of the adaptor protein MAVS by reducing the MAVS–STING association and subsequent IFN-β induction by Sendai virus (Castanier et al., 2010).

pp65/pUL83

Despite the high abundance of pp65/pUL83 as tegumentuos part of HCMV virions and its prominent role as immunogen, the function(s) of pp65/pUL83 for virus replication is not fully understood. ΔUL83-HCMV infected cells transcribe more ISGs than WT-HCMV infected cells and ectopic pp65/pUL83 expression interferes with IFN-α-induced ISG transcription (Browne and Shenk, 2003), indicating that pp65/pUL83 is essential and sufficient to inhibit IFN-α signalling. An analysis of the IFN-induced signal transduction pathways revealed that ΔUL83-HCMV fails to prevent DNA-binding of NF–κB complexes and IRF-1 induction (Browne and Shenk, 2003).

Abate et al. compared the transcriptomes of WT-HCMV and ΔUL83-HCMV infected cells and found differences especially in ISGs. An analysis of the underlying transcription factors revealed phosphorylation and nuclear translocation of IRF-3 upon ΔUL83-HCMV but not upon WT-HCMV infection while NF-κB translocation was comparable. Ectopic expression of pUL83 interfered with nuclear accumulation of IRF-3 upon dsDNA transfection, again indicating that pp65/pUL83 suffices

to antagonize IFN induction (Abate et al., 2004). However the direct contribution of pp65/pUL83 to the inhibition of IFN responses has been challenged when studying pp86/pUL122 since it was found that ΔUL83-HCMV expresses significantly less pp86/pUL122 and pp71/pUL82 proteins. A virus mutant harbouring stop codon insertions within UL83 showed unaltered pp71/pUL82 and pp86/pUL122 expression, but failed to reproduce the previous ΔUL83-HCMV phenotype in terms of IFN-β induction (Taylor and Bresnahan, 2006b), indicating that at least a part of the observed phenotype of the ΔUL83-HCMV mutant is due to a diminished pp86/pUL122 expression (for pre-IE functions of virion tegument proteins, see also Chapter I.9).

A recent study revealed a co-precipitation of pp65/pUL83 with the HIN-200 family members IFI16, IFI16b and IFIX. IFI16 and pp65/pUL83 bind to the HCMV MIE promoter and IFI16 is required for efficient IE gene expression and viral replication under low MOI infection conditions (Cristea et al., 2010). Interestingly, HIN-200 family members (e.g. AIM2) have been shown to be DNA sensor proteins for cytosolic dsDNA inducing the inflammasome (Fernandes-Alnemri et al., 2009; Hornung et al., 2009; Burckstummer et al., 2009) and to contribute to IFN induction by dsDNA-viruses including MCMV (Rathinam et al., 2010). IFI16 itself is a nuclear dsDNA sensor, inducing a STING-dependent activation of IRF-3 and NF-κB upon recognition of dsDNA and upon HSV-1 infection (Unterholzner et al., 2010), potentially explaining the effect of pp65/pUL83 on IFN and ISG induction.

vIL-10/pUL111a

Like its cellular homologue, the virus-encoded IL-10 acts immuno-suppressive and dampens type I IFN production by pDCs (Chang et al., 2009 – see also Chapter I.20).

IE2-pp86/pUL122

IE2-pp86 acts both as transactivator as well as transrepressor. Although IE2-pp86/pUL122 transactivates a co-transfected reporter gene under the control of the IFN-β promoter (Pizzorno et al., 1988), it inhibits transcription of endogenous IFN-β induced by UV-inactivated HCMV and Sendai virus infection (Taylor and Bresnahan, 2005, 2006a) by interfering with DNA-binding of NF-κB without antagonizing IκB-α Ser-32 phosphorylation, proteasomal degradation of IκB-α or nuclear translocation of p50 (Taylor and Bresnahan, 2006a).

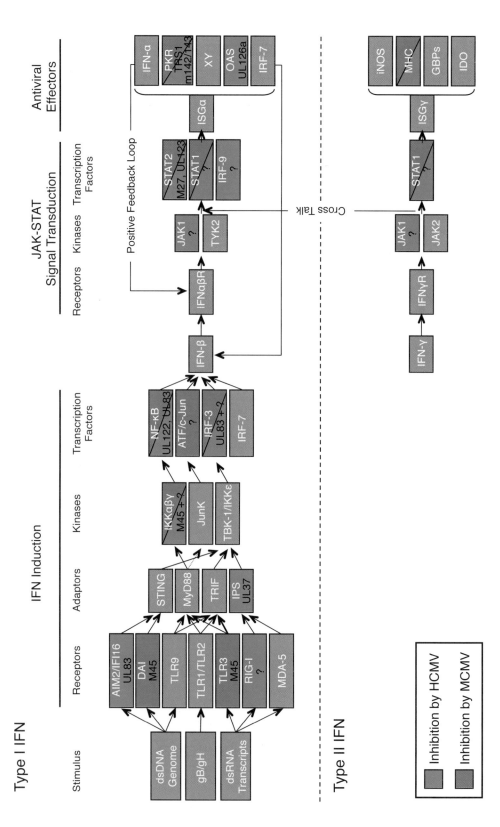

Figure I.16.3 Cytomegaloviral inhibition of the IFN system. The upper panel depicts the type I IFN system and the lower panel the type II IFN system. Inhibition of the respective building blocks by HCMV is indicated in green and inhibition by MCMV in blue colour. The inhibitors (written in black) and their functions are described in the text. The arrows indicate signal transduction pathways.

IE1-pp72/pUL123

Among other functions, IE1-pp72/pUL123 interferes with signalling downstream of the IFNAR (Paulus *et al.*, 2006). An HCMV mutant lacking the ie1-specific exon 4 of the IE region exhibits a severe growth deficit under low MOI infection conditions (Greaves and Mocarski, 1998). This ie1-deficient Towne strain mutant (CR208) is more susceptible towards IFN-α, and neutralization of IFN-β partly reconstitutes viral replication (Paulus *et al.*, 2006). Additionally, ISG induction is increased and prolonged in CR208-infected cells. Conversely, IE1-pp72/pUL123-positive cells express decreased amounts of ISGs (Paulus *et al.*, 2006), indicating that IE1-pp72/pUL123 is essential and sufficient to block IFN signal transduction. Amounts and phosphorylation of STAT1, STAT2 and IRF-9 as well as nuclear translocation of ISGF3 remain unaffected by IE1-pp72/pUL123. Nevertheless, the DNA-binding of ISGF3 to ISRE elements is diminished in presence of IE1-pp72/pUL123 (Paulus *et al.*, 2006). IE1-pp72/pUL123 co-localizes with ISGF3 and co-precipitates with STAT2, weakly with STAT1, but not with IRF-9 (Paulus *et al.*, 2006). The STAT2:IE1–pp72/pUL123 interaction was confirmed and found to require an acidic stretch (aa 421–475) (Huh *et al.*, 2008). A virus mutant lacking this stretch shows decreased replication under low MOI conditions. Upon IFN-β pre-incubation, the mutant hardly replicates (Huh *et al.*, 2008). IE1-pp72/pUL123 re-localizes STAT2 to nuclear ND10 domains and to metaphase chromosomes, requiring a core (aa 373–420) and an ancillary region (aa 421–445) (Krauss *et al.*, 2009). Surprisingly, ectopic expression of IE1-pp72/pUL123 induced an IFN-γ-like subset of ISGs by a STAT1-dependent mechanism (Knoblach *et al.*, 2011), suggesting that IE1-pp72/pUL123 is a janus-faced factor exerting antagonistic and agonistic modulation of IFN signalling.

ORF94/pUL126a

pUL126a is a nuclear localized latency-associated protein encoded by the *ie1/ie2* gene region. It is dispensable for productive HCMV replication and latent infection (Kondo *et al.*, 1996; White *et al.*, 2000) but sufficed to antagonize OAS expression in presence and absence of type I or type II IFN (Tan *et al.*, 2011 – see also Chapter I.20).

TRS/IRS1 m142/m143

Both MCMV and HCMV express inhibitors of PKR. Superinfection with HCMV trans-complements the growth deficit of viruses lacking known PKR antagonists due to the expression of TRS/IRS1 (Child *et al.*, 2004; Cassady, 2005). IRS1 and TRS1 are dsRNA-binding proteins (Hakki and Geballe, 2005) which directly interact with PKR to sequester it in the nucleus away from cytoplasmic dsRNA and its substrate eIF2-α (Hakki *et al.*, 2006). MCMV expresses the proteins m142 and m143 which are essential for MCMV replication but can be complemented by HCMV TRS1 (Valchanova *et al.*, 2006). Both MCMV proteins interact with PKR in an RNA-independent manner (Budt *et al.*, 2009). The essential nature of m142 and m143 depends on PKR but not on RNase L (Budt *et al.*, 2009).

pM27

pM27 is an MCMV *early/late* encoded 79-kDa protein dispensable for viral replication in fibroblasts *in vitro*, but essential for efficient viral replication *in vivo* (Abenes *et al.*, 2001; Zimmermann *et al.*, 2005). ΔM27-MCMV replicates almost WT-like *in vitro*, unless cells are treated with IFN (Zimmermann *et al.*, 2005). Specifically, ΔM27-MCMV exhibits an increased susceptibility towards type I IFN and hardly replicates in IFN-γ-conditioned cells. pM27 is both essential and sufficient to reduce cellular STAT2 amounts by reducing the STAT2 half-life in a proteasome-dependent manner. pM27 induces STAT2 poly-ubiquitination by the recruitment of a specific multiprotein E3 ubiquitin ligase complex, the DDB1-Cullin 4A complex (Trilling *et al.*, 2011). The inability of ΔM27-MCMV to interfere with Jak-STAT signalling leads to an increased ISG gene expression, exemplified by the IFN-inducible subunits of the immune proteasome (Khan *et al.*, 2004).

pM45

Initially described as a nucleotide reductase sequence homologue (Rawlinson *et al.*, 1996) involved in endothelial cell tropism (Brune *et al.*, 2001), it was shown to lack nucleotide reductase activity (Lembo *et al.*, 2004) but to bind and inactivate RIP1 and to blunt TNFR1 and TLR3 signalling (Mack *et al.*, 2008). pM45 harbours a RHIM motif which is crucial for the interaction with RIP1 and RIP3 (Upton *et al.*, 2008). Furthermore, RIP is involved in DAI-dependent signal transduction. Consequently, pM45 interferes with the recruitment of RIP to DAI and subsequent NF-κB activation (Rebsamen *et al.*, 2009). The fact that pM45 blocks TLR and DAI signal transduction places pM45 into the class of inhibitors of IFN induction. Interestingly, pM45 interferes with NF-κB signalling by a second mechanism, targeting NEMO to autophagosomes for subsequent degradation in the lysosomes (Fliss *et al.*, 2012).

Phenotypically defined HCMV-encoded IFN-antagonizing functions not yet assigned to a responsible HCMV gene product

Miller and colleagues investigated cytomegaloviral interference with IFN signal transduction and its impact on IFN-γ induced MHC-II expression (Miller *et al.*, 1998, 1999). HCMV abolishes IFN-γ-dependent induction of the transcription factors IRF-1 and CIITA, the latter being a master regulator of MHC-II. Consistently, the IFN-γ-induced surface disposition of HLA-DR is abrogated by HCMV (Miller *et al.*, 1998). An analysis of the upstream signalling revealed that HCMV reduces Jak1 amounts, blocking downstream signalling events. The loss of Jak1 is sensitive to inhibitors of the proteasome, but resistant to ganciclovir/phosphonoacetic acid, suggesting proteasomal degradation by an IE or early (E) gene product (Miller *et al.*, 1998). Beside this pronounced effect on the IFN-γ-Jak1-STAT1 axis, HCMV targets IRF-1 and CIITA-dependent HLA-DR gene expression by IFN-γ, even at time points when STAT1 phosphorylation is still evident (Le Roy *et al.*, 1999), demonstrating several levels of redundant interference.

HCMV also interferes with IFN-α-dependent induction of ISGs. Since Jak1 is essential for IFN-α signal transduction, the Tyr-phosphorylation of Jak2, IFNAR1, Tyk2, STAT2 and STAT1 become abrogated. Additionally, HCMV induces a post-transcriptional loss of IRF-9 by a phosphonoformic acid resistant mechanism (Miller *et al.*, 1999). On top of that, HCMV induces proteasomal degradation of STAT2 – albeit in a strain-specific manner (Le *et al.*, 2008a). Preceding these degradation events, HCMV also induces the cellular phosphatase SHP2, which dephosphorylates STAT1, thereby blocking induction of IDO and HLA-DRα (Baron and Davignon, 2008).

Phenotypically defined MCMV-encoded IFN-antagonizing functions not yet assigned to a causative MCMV gene product

Unlike HCMV, MCMV retains a large fraction of the receptor proximal Jak-STAT signal transduction machinery intact. STAT1 is phosphorylated, translocates into the nucleus and binds to GAS-elements, indicating that the function of the IFN receptors and Jaks is not impaired (Popkin *et al.*, 2003; Zimmermann *et al.*, 2005; Trilling *et al.*, 2011). Nevertheless, IFN-γ fails to up-regulate MHC-II presentation in infected macrophages (Heise *et al.*, 1998b) due to a blockade of MHC-II gene induction (Heise *et al.*, 1998a) through interference with IFN-γ-induced promoter assembly

by inhibiting IRF-1 and RNA polymerase II recruitment to the CIITA promoter (Popkin *et al.*, 2003). MCMV also blocks IFN-β induction by targeting NF-κB, ATF/c-Jun and IRF-3, the relevant set of transcription factors binding to the IFN-β enhanceosome (Le *et al.*, 2008b).

Conclusions

First come – first served

How to reconcile the abundance of viral antagonists devoted to evade from innate immunity with the importance of the IFN system for CMV control? One might argue that if something is efficiently inhibited it should not be restrictive. In the absence of IFN antagonists, CMV is susceptible towards IFNs (Zimmermann *et al.*, 2005; Paulus *et al.*, 2006; Budt *et al.*, 2009; Trilling *et al.*, 2011). Even in WT-MCMV infection, IFN is required to confine the infection, as seen in IFN signalling incompetent mice, such as IFNAR1- or STAT1-deficient mice. We think the answer to the problem is timing and location: CMV can efficiently control IFN induction, IFN signal transduction, and the action of several ISGs during its productive replication in permissive cells. Yet, this control is cell-restricted and is not (or only to a limited extent) transmitted to bystander cells (Fig. I.16.4). Additionally, it takes some time to express these antagonists. In the case of MCMV, it requires 6 hours to block IFN-β induction in permissive fibroblasts and even longer in macrophages (Le *et al.*, 2008b). The same holds true for ISRE IFN signal transduction – it takes MCMV 8–12 hours to inhibit ISRE signalling (Zimmermann *et al.*, 2005). IFNs act *in vivo* on infected cells as well as on bystander cells which represent the potential next target cells. These neighbouring cells start to express antiviral gene products and turn themselves less permissive, thereby further decelerating viral gene expression upon infection, including IFN antagonistic gene products. Under these circumstances CMV suffers from the action of these antiviral proteins during a viral 'life' phase in which CMV is vulnerable due to its restricted gene expression. This distinction is crucial and confirmed by the finding that IFN efficiency is much higher after pre-incubation of cells than incubation at (or after) the time of infection.

From immune evasion to immune exploitation

As outlined above, immune evasion is an evolutionary imperative. Although this conceptually represents

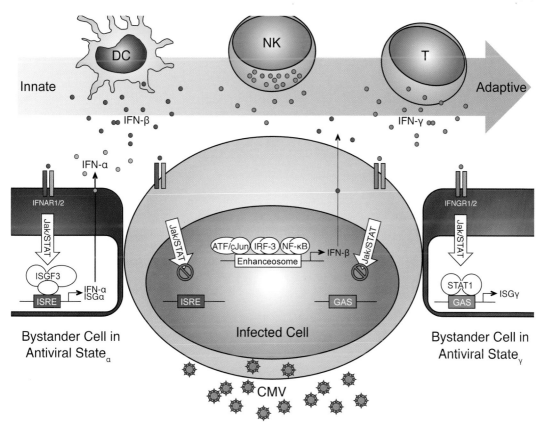

Figure I.16.4 CMV surrounded by IFN competent cells. The schema shows the interplay between a CMV infected cell (enlarged grey cell in the centre) and its natural and immunologic bystander cells. The type I IFN system is depicted in blue and the type II IFN system in red colours. Early events are depicted on the left side ('innate' phase) of the figure and later events on the right-hand side ('adaptive' phase). Please note that bystander cells are receptive for both types of IFN. DC, dendritic cell; NK, natural killer cell; T, T-lymphocyte.

a major leap forward for understanding virus–host interactions, it goes not far enough. Trivial to say that the host cell is the major constituent of the ecological niche of a virus. Despite of the above mentioned countermeasures, CMV infection does always lead to the immediate induction of IFN. Therefore, the transcription factors activated by pathogen recognition receptors, IFNs or cytokines as well as the induced gene products are a usual feature encountered by CMV and consequently CMV learned to exploit them: The HCMV MIE enhancer (see Chapter I.10) contains two so called HCMV response sites (VRS) elements which are reminiscent of IFN responsive GAS elements, making the IE gene expression IFN inducible. An AD169 BAC-derived virus carrying mutations in these GAS-like sites shows a diminished growth replication at low MOI infection conditions (Netterwald *et al.*, 2005). In this context it is noteworthy that the transactivator IE1-pp72 interacts with STAT molecules and induces an IFN-independent,

STAT1-dependent IFN type II like gene expression profile (Knoblach *et al.*, 2011).

The facts that the major tegument component pp65-pUL83 interacts with the DNA sensor IFI16 (which normally induces IFN-β) and that both proteins together stimulate the MIE promoter (Cristea *et al.*, 2010) further strengthen the argument of active innate immune exploitation by CMV. Similarly, MCMV requires p204, an IFN-inducible murine homologue of IFI16, for efficient replication (Hertel *et al.*, 1999).

IFN-γ treatment was sufficient to induce differentiation from not permissive monocytes to HCMV-permissive macrophages, thereby potentiating viral replication (Soderberg-Naucler *et al.*, 1997). HCMV remains in a state of latency in granulocyte–macrophage progenitor cells but IE gene expression, indicative for reactivation, can be substantially induced by IFN-γ (Hahn *et al.*, 1998).

Even downstream of the level of transcription factors acting on viral promoters, an active exploitation of

IFN-induced gene products has been shown. Viperin is induced upon HCMV infection and sought to suppress viral replication (Chin and Cresswell, 2001). Nevertheless, Viperin and pUL37x1-vMIA interact with each other, redistributing Viperin to mitochondria where it acts to the favour of virus replication by stimulating ATP generation (Seo *et al.*, 2011).

Acknowledgements

We apologize to our colleagues whose work could not be cited adequately due to word count constraints. We thank Vu Thuy Khanh Le for help and insightful discussions. The authors receive funding by the DFG (SFB 974 project A09 and GK1045 to HH) and an intramural research grant by the Medical Faculty, Heinrich-Heine-University Düsseldorf No. 9772473 (MT).

References

Abate, D.A., Watanabe, S., and Mocarski, E.S. (2004). Major human cytomegalovirus structural protein pp65 (ppUL83) prevents interferon response factor 3 activation in the interferon response. J. Virol. 78, 10995–11006.

Abenes, G., Lee, M., Haghjoo, E., Tong, T., Zhan, X., and Liu, F. (2001). Murine cytomegalovirus open reading frame M27 plays an important role in growth and virulence in mice. J. Virol. 75, 1697–1707.

Andoniou, C.E., van Dommelen, S.L., Voigt, V., Andrews, D.M., Brizard, G., Asselin- Paturel, C., Delale, T., Stacey, K.J., Trinchieri, G., and Degli-Esposti, M.A. (2005). Interaction between conventional dendritic cells and natural killer cells is integral to the activation of effective antiviral immunity. Nat. Immunol. 6, 1011–1019.

Andrews, D.M., Andoniou, C.E., Granucci, F., Ricciardi-Castagnoli, P., and Degli-Esposti, M.A. (2001). Infection of dendritic cells by murine cytomegalovirus induces functional paralysis. Nat. Immunol. 2, 1077–1084.

Ank, N., West, H., Bartholdy, C., Eriksson, K., Thomsen, A.R., and Paludan, S.R. (2006). Lambda interferon (IFN-lambda), a type III IFN, is induced by viruses and IFNs and displays potent antiviral activity against select virus infections in vivo. J. Virol. 80, 4501–4509.

Asselin-Paturel, C., Boonstra, A., Dalod, M., Durand, I., Yessaad, N., Dezutter-Dambuyant, C., Vicari, A., O'Garra, A., Biron, C., Briere, F., et al. (2001). Mouse type I IFN-producing cells are immature APCs with plasmacytoid morphology. Nat. Immunol. 2, 1144–1150.

Banks, T.A., Rickert, S., Benedict, C.A., Ma, L., Ko, M., Meier, J., Ha, W., Schneider, K., Granger, S.W., Turovskaya, O., et al. (2005). A lymphotoxin-IFN-beta axis essential for lymphocyte survival revealed during cytomegalovirus infection. J. Immunol. 174, 7217–7225.

Banninger, G., and Reich, N.C. (2004). STAT2 nuclear trafficking. J. Biol. Chem. 279, 39199–39206.

Baron, M., and Davignon, J.L. (2008). Inhibition of IFN-gamma-induced STAT1 tyrosine phosphorylation by human CMV is mediated by SHP2. J. Immunol. 181, 5530–5536.

Benedict, C.A., Banks, T.A., Senderowicz, L., Ko, M., Britt, W.J., Angulo, A., Ghazal, P., and Ware, C.F. (2001).

Lymphotoxins and cytomegalovirus cooperatively induce interferon-beta, establishing host-virus detente. Immunity 15, 617–626.

Bodaghi, B., Goureau, O., Zipeto, D., Laurent, L., Virelizier, J.L., and Michelson, S. (1999). Role of IFN-gamma-induced indoleamine 2,3 dioxygenase and inducible nitric oxide synthase in the replication of human cytomegalovirus in retinal pigment epithelial cells. J. Immunol. 162, 957–964.

Boehme, K.W., Singh, J., Perry, S.T., and Compton, T. (2004). Human cytomegalovirus elicits a coordinated cellular antiviral response via envelope glycoprotein B. J. Virol. 78, 1202–1211.

Boehme, K.W., Guerrero, M., and Compton, T. (2006). Human cytomegalovirus envelope glycoproteins B and H are necessary for TLR2 activation in permissive cells. J. Immunol. 177, 7094–7102.

Bogdan, C., and Schleicher, U. (2006). Production of interferon-gamma by myeloid cells – fact or fancy? Trends Immunol. 27, 282–290.

Boldogh, I., Bui, T.K., Szaniszlo, P., Bresnahan, W.A., Albrecht, T., and Hughes, T.K. (1997). Novel activation of gamma-interferon in nonimmune cells during human cytomegalovirus replication. Proc. Soc. Exp. Biol. Med. 215, 66–73.

Boyle, K.A., Pietropaolo, R.L., and Compton, T. (1999). Engagement of the cellular receptor for glycoprotein B of human cytomegalovirus activates the interferon-responsive pathway. Mol. Cell Biol. 19, 3607–3613.

Brand, S., Beigel, F., Olszak, T., Zitzmann, K., Eichhorst, S.T., Otte, J.M., Diebold, J., Diepolder, H., Adler, B., Auernhammer, C.J. et al. (2005). IL-28A and IL-29 mediate antiproliferative and antiviral signals in intestinal epithelial cells and murine CMV infection increases colonic IL-28A expression. Am. J. Physiol Gastrointest. Liver Physiol. 289, 960–968.

van den Broek, M.F., Muller, U., Huang, S., Zinkernagel, R.M., and Aguet, M. (1995). Immune defence in mice lacking type I and/or type II interferon receptors. Immunol. Rev. 148, 5–18.

Brown, R.A., Gralewski, J.H., and Razonable, R.R. (2009). The R753Q polymorphism abrogates toll-like receptor 2 signaling in response to human cytomegalovirus. Clin. Infect. Dis. 49, e96–e99.

Browne, E.P., and Shenk, T. (2003). Human cytomegalovirus UL83-coded pp65 virion protein inhibits antiviral gene expression in infected cells. Proc. Natl. Acad. Sci. U.S.A. 100, 11439–11444.

Browne, E.P., Wing, B., Coleman, D., and Shenk, T. (2001). Altered cellular mRNA levels in human cytomegalovirus-infected fibroblasts: viral block to the accumulation of antiviral mRNAs. J. Virol. 75, 12319–12330.

Brune, W., Menard, C., Heesemann, J., and Koszinowski, U.H. (2001). A ribonucleotide reductase homolog of cytomegalovirus and endothelial cell tropism. Science 291, 303–305.

Budt, M., Niederstadt, L., Valchanova, R.S., Jonjic, S., and Brune, W. (2009). Specific inhibition of the PKR-mediated antiviral response by the murine cytomegalovirus proteins m142 and m143. J. Virol. 83, 1260–1270.

Burckstummer, T., Baumann, C., Bluml, S., Dixit, E., Durnberger, G., Jahn, H., Planyavsky, M., Bilban, M., Colinge, J., Bennett, K.L., et al. (2009). An orthogonal proteomic-genomic screen identifies AIM2 as a cytoplasmic DNA sensor for the inflammasome. Nat. Immunol. 10, 266–272.

Carter, C.C., Gorbacheva, V.Y., and Vestal, D.J. (2005). Inhibition of VSV and EMCV replication by the interferon-induced GTPase, mGBP-2: differential requirement for wild- type GTP binding domain. Arch. Virol. *150*, 1213–1220.

Cassady, K.A. (2005). Human cytomegalovirus TRS1 and IRS1 gene products block the double-stranded-RNA-activated host protein shutoff response induced by herpes simplex virus type 1 infection. J. Virol. *79*, 8707–8715.

Castanier, C., Garcin, D., Vazquez, A., and Arnoult, D. (2010). Mitochondrial dynamics regulate the RIG-I-like receptor antiviral pathway. EMBO Rep. *11*, 133–138.

Chang, W.L., Barry, P.A., Szubin, R., Wang, D., and Baumgarth, N. (2009). Human cytomegalovirus suppresses type I interferon secretion by plasmacytoid dendritic cells through its interleukin 10 homolog. Virology *390*, 330–337.

Child, S.J., Hakki, M., De Niro, K.L., and Geballe, A.P. (2004). Evasion of cellular antiviral responses by human cytomegalovirus TRS1 and IRS1. J. Virol. *78*, 197–205.

Chin, K.C., and Cresswell, P. (2001). Viperin (cig5), an IFN-inducible antiviral protein directly induced by human cytomegalovirus. Proc. Natl. Acad. Sci. U.S.A. *98*, 15125–15130.

Chong, K.T., Gresser, I., and Mims, C.A. (1983). Interferon as a defence mechanism in mouse cytomegalovirus infection. J. Gen. Virol. *64*, 461–464.

Clemens, M.J., and Williams, B.R. (1978). Inhibition of cell-free protein synthesis by pppA2′p5′A2′p5′A: a novel oligonucleotide synthesized by interferon-treated L cell extracts. Cell *13*, 565–572.

Collazo, C.M., Yap, G.S., Sempowski, G.D., Lusby, K.C., Tessarollo, L., Woude, G.F., Sher, A., and Taylor, G.A. (2001). Inactivation of LRG-47 and IRG-47 reveals a family of interferon gamma-inducible genes with essential, pathogen-specific roles in resistance to infection. J. Exp. Med. *194*, 181–188.

Compton, T., Kurt-Jones, E.A., Boehme, K.W., Belko, J., Latz, E., Golenbock, D.T., and Finberg, R.W. (2003). Human cytomegalovirus activates inflammatory cytokine responses via CD14 and Toll-like receptor 2. J. Virol. *77*, 4588–4596.

Cristea, I.M., Moorman, N.J., Terhune, S.S., Cuevas, C.D., O'Keefe, E.S., Rout, M.P., Chait, B.T., and Shenk, T. (2010). Human cytomegalovirus pUL83 stimulates activity of the viral immediate-early promoter through its interaction with the cellular IFI16 protein. J. Virol. *84*, 7803–7814.

Croen, K.D. (1993). Evidence for antiviral effect of nitric oxide. Inhibition of herpes simplex virus type 1 replication. J. Clin. Invest. *91*, 2446–2452.

Crozat, K., Georgel, P., Rutschmann, S., Mann, N., Du, X., Hoebe, K., and Beutler, B. (2006). Analysis of the MCMV resistome by ENU mutagenesis. Mamm. Genome *17*, 398–406.

Cunningham, J.A., Kellner, J.D., Bridge, P.J., Trevenen, C.L., Mcleod, D.R., and Davies, H.D. (2000). Disseminated bacille Calmette–Guerin infection in an infant with a novel deletion in the interferon-gamma receptor gene. Int. J. Tuberc. Lung Dis. *4*, 791–794.

Dai, P., Cao, H., Merghoub, T., Avogadri, F., Wang, W., Parikh, T., Fang, C.M., Pitha, P.M., Fitzgerald, K.A., Rahman, M.M. et al. (2011). Myxoma virus induces type I interferon production in murine plasmacytoid dendritic cells via a TLR9/MyD88-, IRF5/IRF7-, and IFNAR-dependent pathway. J. Virol. *85*, 10814–10825.

Dalod, M., Salazar-Mather, T.P., Malmgaard, L., Lewis, C., Asselin-Paturel, C., Briere, F., Trinchieri, G., and Biron, C.A. (2002). Interferon alpha/beta and interleukin 12 responses to viral infections: pathways regulating dendritic cell cytokine expression in vivo. J. Exp. Med. *195*, 517–528.

Dalod, M., Hamilton, T., Salomon, R., Salazar-Mather, T.P., Henry, S.C., Hamilton, J.D., and Biron, C.A. (2003). Dendritic cell responses to early murine cytomegalovirus infection: subset functional specialization and differential regulation by interferon alpha/beta. J. Exp. Med. *197*, 885–898.

Decker, T., Lew, D.J., Mirkovitch, J., and Darnell, J.E., Jr. (1991). Cytoplasmic activation of GAF, an IFN-gamma-regulated DNA-binding factor. EMBO J. *10*, 927–932.

DeFilippis, V.R., Robinson, B., Keck, T.M., Hansen, S.G., Nelson, J.A., and Fruh, K.J. (2006). Interferon regulatory factor 3 is necessary for induction of antiviral genes during human cytomegalovirus infection. J. Virol. *80*, 1032–1037.

DeFilippis, V.R., Alvarado, D., Sali, T., Rothenburg, S., and Fruh, K. (2010a). Human cytomegalovirus induces the interferon response via the DNA sensor ZBP1. J. Virol. *84*, 585–598.

DeFilippis, V.R., Sali, T., Alvarado, D., White, L., Bresnahan, W., and Fruh, K.J. (2010b). Activation of the interferon response by human cytomegalovirus occurs via cytoplasmic double-stranded DNA but not glycoprotein B. J. Virol. *84*, 8913–8925.

Delale, T., Paquin, A., Asselin-Paturel, C., Dalod, M., Brizard, G., Bates, E.E., Kastner, P., Chan, S., Akira, S., Vicari, A. et al. (2005). MyD88-dependent and -independent murine cytomegalovirus sensing for IFN-alpha release and initiation of immune responses in vivo. J. Immunol. *175*, 6723–6732.

Der, S.D., Zhou, A., Williams, B.R., and Silverman, R.H. (1998). Identification of genes differentially regulated by interferon alpha, beta, or gamma using oligonucleotide arrays. Proc. Natl. Acad. Sci. U.S.A. *95*, 15623–15628.

Fernandes-Alnemri, T., Yu, J.W., Datta, P., Wu, J., and Alnemri, E.S. (2009). AIM2 activates the inflammasome and cell death in response to cytoplasmic DNA. Nature *458*, 509–513.

Fernandez, J.A., Rodrigues, E.G., and Tsuji, M. (2000). Multifactorial protective mechanisms to limit viral replication in the lung of mice during primary murine cytomegalovirus infection. Viral Immunol. *13*, 287–295.

Fliss, P., Jowers, T., Brinkmann, M., Holstermann, B., Mack, C., Dickinson, P., Hohenberg, H., Ghazal, P., and Brune, W. (2012). Viral mediated redirection of NEMO/IKKc to autophagosomes curtails the inflammatory cascade. PLoS. Pathog. *8*, e1002517.

Furset, G., Floisand, Y., and Sioud, M. (2008). Impaired expression of indoleamine 2, 3- dioxygenase in monocyte-derived dendritic cells in response to Toll-like receptor-7/8 ligands. Immunology *123*, 263–271.

Gad, H.H., Dellgren, C., Hamming, O.J., Vends, S., Paludan, S.R., and Hartmann, R. (2009). Interferon-lambda is functionally an interferon but structurally related to the interleukin-10 family. J. Biol. Chem. *284*, 20869–20875.

Gil, M.P., Bohn, E., O'Guin, A.K., Ramana, C.V., Levine, B., Stark, G.R., Virgin, H.W., and Schreiber, R.D. (2001).

Biologic consequences of Stat1-independent IFN signaling. Proc. Natl. Acad. Sci. U.S.A. 98, 6680–6685.

Glasgow, L.A., Hanshaw, J.B., Merigan, T.C., and Petralli, J.K. (1967). Interferon and cytomegalovirus in vivo and in vitro. Proc. Soc. Exp. Biol. Med. 125, 843–849.

Goldmacher, V.S., Bartle, L.M., Skaletskaya, A., Dionne, C.A., Kedersha, N.L., Vater, C.A., Han, J.W., Lutz, R.J., Watanabe, S., Cahir McFarland, E.D., et al. (1999). A cytomegalovirus-encoded mitochondria-localized inhibitor of apoptosis structurally unrelated to Bcl-2. Proc. Natl. Acad. Sci. U.S.A. 96, 12536–12541.

Gravel, S.P., and Servant, M.J. (2005). Roles of an IkappaB kinase-related pathway in human cytomegalovirus-infected vascular smooth muscle cells: a molecular link in pathogen- induced proatherosclerotic conditions. J. Biol. Chem. 280, 7477–7486.

Greaves, R.F., and Mocarski, E.S. (1998). Defective growth correlates with reduced accumulation of a viral DNA replication protein after low-multiplicity infection by a human cytomegalovirus ie1 mutant. J. Virol. 72, 366–379.

Gribaudo, G., Ravaglia, S., Caliendo, A., Cavallo, R., Gariglio, M., Martinotti, M.G., and Landolfo, S. (1993). Interferons inhibit onset of murine cytomegalovirus immediate-early gene transcription. Virology 197, 303–311.

Gribaudo, G., Ravaglia, S., Gaboli, M., Gariglio, M., Cavallo, R., and Landolfo, S. (1995). Interferon-alpha inhibits the murine cytomegalovirus immediate-early gene expression by down-regulating NF-kappa B activity. Virology 211, 251–260.

Hahn, G., Jores, R., and Mocarski, E.S. (1998). Cytomegalovirus remains latent in a common precursor of dendritic and myeloid cells. Proc. Natl. Acad. Sci. U.S.A. 95, 3937–3942.

Hakki, M., and Geballe, A.P. (2005). Double-stranded RNA binding by human cytomegalovirus pTRS1. J. Virol. 79, 7311–7318.

Hakki, M., Marshall, E.E., De Niro, K.L., and Geballe, A.P. (2006). Binding and nuclear relocalization of protein kinase R by human cytomegalovirus TRS1. J. Virol. 80, 11817- 11826.

Heise, M.T., and Virgin, H.W. (1995). The T-cell-independent role of gamma interferon and tumor necrosis factor alpha in macrophage activation during murine cytomegalovirus and herpes simplex virus infections. J. Virol. 69, 904–909.

Heise, M.T., Connick, M., and Virgin, H.W. (1998a). Murine cytomegalovirus inhibits interferon gamma-induced antigen presentation to CD4 T-cells by macrophages via regulation of expression of major histocompatibility complex class II-associated genes. J. Exp. Med. 187, 1037–1046.

Heise, M.T., Pollock, J.L., O'Guin, A., Barkon, M.L., Bormley, S., and Virgin, H.W. (1998b). Murine cytomegalovirus infection inhibits IFN gamma-induced MHC class II expression on macrophages: the role of type I interferon. Virology 241, 331–344.

Hemmi, H., Takeuchi, O., Kawai, T., Kaisho, T., Sato, S., Sanjo, H., Matsumoto, M., Hoshino, K., Wagner, H., Takeda, K., et al. (2000). A Toll-like receptor recognizes bacterial DNA. Nature 408, 740–745.

Hengel, H., Koszinowski, U.H., and Conzelmann, K.K. (2005). Viruses know it all: new insights into IFN networks. Trends Immunol. 26, 396–401.

Henson, D., and Smith, R.D. (1964). Interferon Production in vitro by cells infected with the murine salivary gland virus. Proc. Soc. Exp. Biol. Med. 117, 517–520.

Hertel, L., De Andrea, M., Azzimonti, B., Rolle, A., Gariglio, M., and Landolfo, S. (1999). The interferon-inducible 20 gene, a member of the Ifi 200 family, is not involved in the antiviral state induction by IFN-alpha, but is required by the mouse cytomegalovirus for its replication. Virology 262, 1–8.

Hoebe, K., Du, X., Georgel, P., Janssen, E., Tabeta, K., Kim, S.O., Goode, J., Lin, P., Mann, N., Mudd, S., et al. (2003). Identification of Lps2 as a key transducer of MyD88 independent TIR signalling. Nature 424, 743–748.

Hokeness-Antonelli, K.L., Crane, M.J., Dragoi, A.M., Chu, W.M., and Salazar-Mather, T.P. (2007). IFN-alphabeta mediated inflammatory responses and antiviral defense in liver is TLR9-independent but MyD88-dependent during murine cytomegalovirus infection. J. Immunol. 179, 6176–6183.

Holmes, A.R., Rasmussen, L., and Merigan, T.C. (1978). Factors affecting the interferon sensitivity of human cytomegalovirus. Intervirology 9, 48–55.

Hornung, V., Ablasser, A., Charrel-Dennis, M., Bauernfeind, F., Horvath, G., Caffrey, D.R., Latz, E., and Fitzgerald, K.A. (2009). AIM2 recognizes cytosolic dsDNA and forms a caspase-1-activating inflammasome with ASC. Nature 458, 514–518.

Huh, Y.H., Kim, Y.E., Kim, E.T., Park, J.J., Song, M.J., Zhu, H., Hayward, G.S., and Ahn, J.H. (2008). Binding STAT2 by the acidic domain of human cytomegalovirus IE1 promotes viral growth and is negatively regulated by SUMO. J. Virol. 82, 10444–10454.

Igarashi, K., Garotta, G., Ozmen, L., Ziemiecki, A., Wilks, A.F., Harpur, A.G., Larner, A.C., and Finbloom, D.S. (1994). Interferon-gamma induces tyrosine phosphorylation of interferon-gamma receptor and regulated association of protein tyrosine kinases, Jak1 and Jak2, with its receptor. J. Biol. Chem. 269, 14333–14336.

Isaacs, A., and Lindenmann, J. (1957). Virus interference. I. The interferon. Proc. R. Soc. Lond B Biol. Sci. 147, 258–267.

Itsui, Y., Sakamoto, N., Kakinuma, S., Nakagawa, M., Sekine-Osajima, Y., Tasaka-Fujita, M., Nishimura-Sakurai, Y., Suda, G., Karakama, Y., Mishima, K., et al. (2009). Antiviral effects of the interferon-induced protein guanylate binding protein 1 and its interaction with the hepatitis C virus NS5B protein. Hepatology 50, 1727–1737.

Iversen, A.C., Norris, P.S., Ware, C.F., and Benedict, C.A. (2005). Human NK cells inhibit cytomegalovirus replication through a noncytolytic mechanism involving lymphotoxin- dependent induction of IFN-beta. J. Immunol. 175, 7568–7574.

Jacobs, B.L., and Langland, J.O. (1996). When two strands are better than one: the mediators and modulators of the cellular responses to double-stranded RNA. Virology 219, 339–349.

Janeway, C.A., Jr., and Medzhitov, R. (2002). Innate immune recognition. Annu. Rev. Immunol. 20, 197–216.

Karupiah, G., Xie, Q.W., Buller, R.M., Nathan, C., Duarte, C., and MacMicking, J.D. (1993). Inhibition of viral replication by interferon-gamma-induced nitric oxide synthase. Science 261, 1445–1448.

Kelsey, D.K., Olsen, G.A., Overall, J.C., Jr., and Glasgow, L.A. (1977). Alteration of host defense mechanisms by murine cytomegalovirus infection. Infect. Immun. 18, 754–760.

Khan, S., Zimmermann, A., Basler, M., Groettrup, M., and Hengel, H. (2004). A cytomegalovirus inhibitor gamma interferon signaling controls immunoproteasome induction. J. Virol. 78, 1831–1842.

Kijpittayarit, S., Eid, A.J., Brown, R.A., Paya, C.V., and Razonable, R.R. (2007). Relationship between Toll-like receptor 2 polymorphism and cytomegalovirus disease after liver transplantation. Clin. Infect. Dis. 44, 1315–1320.

Kimchi, A., Zilberstein, A., Schmidt, A., Shulman, L., and Revel, M. (1979). The interferon- induced protein kinase PK-i from mouse L cells. J. Biol. Chem. 254, 9846–9853.

Knoblach, T., Grandel, B., Seiler, J., Nevels, M., and Paulus, C. (2011). Human cytomegalovirus IE1 protein elicits a type II interferon-like host cell response that depends on activated STAT1 but not interferon-gamma. PLoS. Pathog. 7, e1002016.

Kondo, K., Xu, J., and Mocarski, E.S. (1996). Human cytomegalovirus latent gene expression in granulocyte–macrophage progenitors in culture and in seropositive individuals. Proc. Natl. Acad. Sci. U.S.A. 93, 11137–11142.

Kotenko, S.V., Gallagher, G., Baurin, V.V., Lewis-Antes, A., Shen, M., Shah, N.K., Langer, J.A., Sheikh, F., Dickensheets, H., and Donnelly, R.P. (2003). IFN-lambdas mediate antiviral protection through a distinct class II cytokine receptor complex. Nat. Immunol. 4, 69–77.

Krauss, S., Kaps, J., Czech, N., Paulus, C., and Nevels, M. (2009). Physical requirements and functional consequences of complex formation between the cytomegalovirus IE1 protein and human STAT2. J. Virol. 83, 12854–12870.

Kropp, K.A., Robertson, K.A., Sing, G., Rodriguez-Martin, S., Blanc, M., Lacaze, P., Hassim, M.F., Khondoker, M.R., Busche, A., Dickinson, P., et al. (2011). Reversible inhibition of murine cytomegalovirus replication by gamma interferon (IFN-gamma) in primary macrophages involves a primed type I IFN-signaling subnetwork for full establishment of an immediate-early antiviral state. J. Virol. 85, 10286–10299.

Krug, A., French, A.R., Barchet, W., Fischer, J.A., Dzionek, A., Pingel, J.T., Orihuela, M.M., Akira, S., Yokoyama, W.M., and Colonna, M. (2004). TLR9-dependent recognition of MCMV by IPC and DC generates coordinated cytokine responses that activate antiviral NK cell function. Immunity 21, 107–119.

Kuenzel, S., Till, A., Winkler, M., Hasler, R., Lipinski, S., Jung, S., Grotzinger, J., Fickenscher, H., Schreiber, S., and Rosenstiel, P. (2010). The nucleotide-binding oligomerization domain-like receptor NLRC5 is involved in IFN-dependent antiviral immune responses. J. Immunol. 184, 1990–2000.

Le, V.T.K., Trilling, M., Wilborn, M., Hengel, H., and Zimmermann, A. (2008a). Human cytomegalovirus interferes with signal transducer and activator of transcription (STAT) 2 protein stability and tyrosine phosphorylation. J. Gen. Virol. 89, 2416–2426.

Le, V.T.K., Trilling, M., Zimmermann, A., and Hengel, H. (2008b). Mouse cytomegalovirus inhibits beta interferon (IFN-beta) gene expression and controls activation pathways of the IFN-beta enhanceosome. J. Gen. Virol. 89, 1131–1141.

Le Roy, E., Muhlethaler-Mottet, A., Davrinche, C., Mach, B., and Davignon, J.L. (1999). Escape of human cytomegalovirus from HLA-DR-restricted CD4(+) T-cell response is mediated by repression of gamma

interferon-induced class II transactivator expression. J. Virol. 73, 6582–6589.

Lembo, D., Donalisio, M., Hofer, A., Cornaglia, M., Brune, W., Koszinowski, U., Thelander, L., and Landolfo, S. (2004). The ribonucleotide reductase R1 homolog of murine cytomegalovirus is not a functional enzyme subunit but is required for pathogenesis. J. Virol. 78, 4278–4288.

Leung, S., Qureshi, S.A., Kerr, I.M., Darnell, J.E., Jr., and Stark, G.R. (1995). Role of STAT2 in the alpha interferon signaling pathway. Mol. Cell Biol. 15, 1312–1317.

Levy, D.E., Kessler, D.S., Pine, R., and Darnell, J.E., Jr. (1989). Cytoplasmic activation of ISGF3, the positive regulator of interferon-alpha-stimulated transcription, reconstituted in vitro. Genes Dev. 3, 1362–1371.

Li, X., Leung, S., Kerr, I.M., and Stark, G.R. (1997). Functional subdomains of STAT2 required for preassociation with the alpha interferon receptor and for signaling. Mol. Cell Biol. 17, 2048–2056.

Loh, J., Chu, D.T., O'Guin, A.K., Yokoyama, W.M., and Virgin, H.W. (2005). Natural killer cells utilize both perforin and gamma interferon to regulate murine cytomegalovirus infection in the spleen and liver. J. Virol. 79, 661–667.

Lucin, P., Pavic, I., Polic, B., Jonjic, S., and Koszinowski, U.H. (1992). Gamma interferon-dependent clearance of cytomegalovirus infection in salivary glands. J. Virol. 66, 1977- 1984.

Lucin, P., Jonjic, S., Messerle, M., Polic, B., Hengel, H., and Koszinowski, U.H. (1994). Late phase inhibition of murine cytomegalovirus replication by synergistic action of interferon-gamma and tumour necrosis factor. J. Gen. Virol. 75, 101–110.

McCormick, A.L., Skaletskaya, A., Barry, P.A., Mocarski, E.S., and Goldmacher, V.S. (2003). Differential function and expression of the viral inhibitor of caspase 8-induced apoptosis (vICA) and the viral mitochondria-localized inhibitor of apoptosis (vMIA) cell death suppressors conserved in primate and rodent cytomegaloviruses. Virology 316, 221–233.

Mack, C., Sickmann, A., Lembo, D., and Brune, W. (2008). Inhibition of proinflammatory and innate immune signaling pathways by a cytomegalovirus RIP1-interacting protein. Proc. Natl. Acad. Sci. U.S.A. 105, 3094–3099.

Mao, X., Ren, Z., Parker, G.N., Sondermann, H., Pastorello, M.A., Wang, W., McMurray, J.S., Demeler, B., Darnell, J.E., Jr., and Chen, X. (2005). Structural bases of unphosphorylated STAT1 association and receptor binding. Mol. Cell 17, 761–771.

Marie, I., Durbin, J.E., and Levy, D.E. (1998). Differential viral induction of distinct interferon-alpha genes by positive feedback through interferon regulatory factor-7. EMBO J. 17, 6660–6669.

Marshall, E.E., Bierle, C.J., Brune, W., and Geballe, A.P. (2009). Essential role for either TRS1 or IRS1 in human cytomegalovirus replication. J. Virol. 83, 4112–4120.

Martinotti, M.G., Gribaudo, G., Gariglio, M., Caliendo, A., Lembo, D., Angeretti, A., Cavallo, R., and Landolfo, S. (1993). Effect of interferon-alpha on immediate-early gene expression of murine cytomegalovirus. J. Interferon Res. 13, 105–109.

Meyer, T., Marg, A., Lemke, P., Wiesner, B., and Vinkemeier, U. (2003). DNA binding controls inactivation and nuclear accumulation of the transcription factor Stat1. Genes Dev. 17, 1992–2005.

Miller, D.M., Rahill, B.M., Boss, J.M., Lairmore, M.D., Durbin, J.E., Waldman, J.W., and Sedmak, D.D. (1998). Human cytomegalovirus inhibits major histocompatibility complex class II expression by disruption of the Jak/Stat pathway. J. Exp. Med. *187*, 675–683.

Miller, D.M., Zhang, Y., Rahill, B.M., Waldman, W.J., and Sedmak, D.D. (1999). Human cytomegalovirus inhibits IFN-alpha-stimulated antiviral and immunoregulatory responses by blocking multiple levels of IFN-alpha signal transduction. J. Immunol. *162*, 6107–6113.

Muller, M., Briscoe, J., Laxton, C., Guschin, D., Ziemiecki, A., Silvennoinen, O., Harpur, A.G., Barbieri, G., Witthuhn, B.A., Schindler, C., et al.. (1993). The protein tyrosine kinase JAK1 complements defects in interferon-alpha/beta and -gamma signal transduction. Nature *366*, 129–135.

Navarro, L., Mowen, K., Rodems, S., Weaver, B., Reich, N., Spector, D., and David, M. (1998). Cytomegalovirus activates interferon immediate-early response gene expression and an interferon regulatory factor 3-containing interferon-stimulated response element-binding complex. Mol. Cell Biol. *18*, 3796–3802.

Netterwald, J.R., Jones, T.R., Britt, W.J., Yang, S.J., McCrone, I.P., and Zhu, H. (2004). Postattachment events associated with viral entry are necessary for induction of interferon-stimulated genes by human cytomegalovirus. J. Virol. *78*, 6688–6691.

Netterwald, J., Yang, S., Wang, W., Ghanny, S., Cody, M., Soteropoulos, P., Tian, B., Dunn, W., Liu, F., and Zhu, H. (2005). Two gamma interferon-activated site-like elements in the human cytomegalovirus major immediate-early promoter/enhancer are important for viral replication. J. Virol. *79*, 5035–5046.

Noda, S., Tanaka, K., Sawamura, S., Sasaki, M., Matsumoto, T., Mikami, K., Aiba, Y., Hasegawa, H., Kawabe, N., and Koga, Y. (2001). Role of nitric oxide synthase type 2 in acute infection with murine cytomegalovirus. J. Immunol. *166*, 3533–3541.

Nordmann, A., Wixler, L., Boergeling, Y., Wixler, V., and Ludwig, S. (2011). A new splice variant of the human guanylate-binding protein 3 mediates anti-influenza activity through inhibition of viral transcription and replication. FASEB J. *26*, 1290–1300.

Oie, H.K., Easton, J.M., Ablashi, D.V., and Baron, S. (1975). Murine cytomegalovirus: induction of and sensitivity to interferon in vitro. Infect. Immun. *12*, 1012–1017.

Panne, D. (2008). The enhanceosome. Curr. Opin. Struct. Biol. *18*, 236–242.

Paulus, C., Krauss, S., and Nevels, M. (2006). A human cytomegalovirus antagonist of type I IFN-dependent signal transducer and activator of transcription signaling. Proc. Natl. Acad. Sci. U.S.A. *103*, 3840–3845.

Pizzorno, M.C., O'Hare, P., Sha, L., LaFemina, R.L., and Hayward, G.S. (1988). Trans- activation and autoregulation of gene expression by the immediate-early region 2 gene products of human cytomegalovirus. J. Virol. *62*, 1167–1179.

Polic, B., Hengel, H., Krmpotic, A., Trgovcich, J., Pavic, I., Luccaronin, P., Jonjic, S., and Koszinowski, U.H. (1998). Hierarchical and redundant lymphocyte subset control precludes cytomegalovirus replication during latent infection. J. Exp. Med. *188*, 1047–1054.

Popkin, D.L., Watson, M.A., Karaskov, E., Dunn, G.P., Bremner, R., and Virgin, H.W. (2003). Murine cytomegalovirus paralyzes macrophages by blocking IFN gamma-induced promoter assembly. Proc. Natl. Acad. Sci. U.S.A. *100*, 14309–14314.

Postic, B., and Dowling, J.N. (1977). Susceptibility clinical isolates of cytomegalovirus to human interferon Antimicrob. Agents Chemother. *11*, 656–660.

Pott, J., Mahlakoiv, T., Mordstein, M., Duerr, C.U., Michiel T., Stockinger, S., Staeheli, P., and Hornef, M.W. (2011 IFN-lambda determines the intestinal epithelial antivir host defense. Proc. Natl. Acad. Sci. U.S.A. *108*, 7944 7949.

Presti, R.M., Pollock, J.L., Dal Canto, A.J., O'Guin, A.K., an Virgin, H.W. (1998). Interferon gamma regulates acu and latent murine cytomegalovirus infection and chron disease of the great vessels. J. Exp. Med. *188*, 577–588.

Presti, R.M., Popkin, D.L., Connick, M., Paetzold, S., an Virgin, H.W. (2001). Novel cell type- specific antivir mechanism of interferon gamma action in macrophage J. Exp. Med. *193*, 483–496.

Preston, C.M., Harman, A.N., and Nicholl, M.J. (2001 Activation of interferon response factor-3 in huma cells infected with herpes simplex virus type 1 or huma cytomegalovirus. J. Virol. *75*, 8909–8916.

Ramana, C.V., Gil, M.P., Han, Y., Ransohoff, R.M., Schreibe R.D., and Stark, G.R. (2001). Stat1-independer regulation of gene expression in response to IFN-gamm Proc. Natl. Acad. Sci. U.S.A. *98*, 6674–6679.

Rathinam, V.A., Jiang, Z., Waggoner, S.N., Sharma, S., Col L.E., Waggoner, L., Vanaja, S.K., Monks, B.G., Ganesa S., Latz, E., et al. (2010). The AIM2 inflammasome essential for host defense against cytosolic bacteria an DNA viruses. Nat. Immunol. *11*, 395–402.

Ratner, L., Wiegand, R.C., Farrell, P.J., Sen, G.C., Cabrer, B., an Lengyel, P. (1978). Interferon, double-stranded RNA an RNA degradation. Fractionation of the endonucleaseIN system into two macromolecular components; role of small molecule in nuclease activation. Biochem. Biophy Res. Commun. *81*, 947–954.

Rawlinson, W.D., Farrell, H.E., and Barrell, B.G. (1996 Analysis of the complete DNA sequence of murir cytomegalovirus. J. Virol. *70*, 8833–8849.

Rebsamen, M., Heinz, L.X., Meylan, E., Michallet, M.C Schroder, K., Hofmann, K., Vazquez, J., Benedict, C.A and Tschopp, J. (2009). DAI/ZBP1 recruits RIP1 an RIP3 through RIP homotypic interaction motifs activate NF-kappaB. EMBO Rep. *10*, 916–922.

Rhodes-Feuillette, A., Canivet, M., Champsaur, H Gluckman, E., Mazeron, M.C., and Peries, J. (1983 Circulating interferon in cytomegalovirus infecte bone-marrow- transplant recipients and in infants wit congenital cytomegalovirus disease. J. Interferon Res. 45–52.

Robek, M.D., Boyd, B.S., and Chisari, F.V. (2005). Lambo interferon inhibits hepatitis B and C virus replication. Virol. *79*, 3851–3854.

Sainz, B., Jr., LaMarca, H.L., Garry, R.F., and Morris, C.A (2005). Synergistic inhibition of human cytomegalovir replication by interferon-alpha/beta and interfero gamma. Virol. J. *2*, 14.

Scheu, S., Dresing, P., and Locksley, R.M. (2008 Visualization of IFNbeta production by plasmacytoi versus conventional dendritic cells under specif stimulation conditions in vivo. Proc. Natl. Acad. Sc U.S.A. *105*, 20416–20421.

Schneider, K., Loewendorf, A., De Trez, T.C., Fulton, Rhode, A., Shumway, H., Ha, S., Patterson, G., Pfeffer, K Nedospasov, S.A. et al. (2008). Lymphotoxin-mediate crosstalk between B cells and splenic stroma promotes th initial type I interferon response to cytomegalovirus. Cell Host Microbe 3, 67–76.

Schoggins, J.W., Wilson, S.J., Panis, M., Murphy, M.Y., Jones, C.T., Bieniasz, P., and Rice, C.M. (2011). A diverse range of gene products are effectors of the type I interferon antiviral response. Nature 472, 481–485.

Scott, I. (2009). Degradation of RIG-I following cytomegalovirus infection is independent of apoptosis. Microbes. Infect. 11, 973–979.

Sedmak, D.D., Chaiwiriyakul, S., Knight, D.A., and Waldmann, W.J. (1995). The role of interferon beta in human cytomegalovirus-mediated inhibition of HLA DR induction on endothelial cells. Arch. Virol. 140, 111–126.

Seo, J.Y., Yaneva, R., Hinson, E.R., and Cresswell, P. (2011). Human cytomegalovirus directly induces the antiviral protein viperin to enhance infectivity. Science 332, 1093–1097.

Sheppard, P., Kindsvogel, W., Xu, W., Henderson, K., Schlutsmeyer, S., Whitmore, T.E., Kuestner, R., Garrigues, U., Birks, C., Roraback, J., et al. (2003). IL-28, IL-29 and their class II cytokine receptor IL-28R. Nat. Immunol. 4, 63–68.

Shuai, K., Schindler, C., Prezioso, V.R., and Darnell, J.E., Jr. (1992). Activation of transcription by IFN-gamma: tyrosine phosphorylation of a 91-kD DNA-binding protein. Science 258, 1808–1812.

Shuai, K., Ziemiecki, A., Wilks, A.F., Harpur, A.G., Sadowski, H.B., Gilman, M.Z., and Darnell, J.E. (1993). Polypeptide signalling to the nucleus through tyrosine phosphorylation of Jak and Stat proteins. Nature 366, 580–583.

Silvennoinen, O., Ihle, J.N., Schlessinger, J., and Levy, D.E. (1993). Interferon-induced nuclear signalling by Jak protein tyrosine kinases. Nature 366, 583–585.

Simmen, K.A., Singh, J., Luukkonen, B.G., Lopper, M., Bittner, A., Miller, N.E., Jackson, M.R., Compton, T., and Fruh, K. (2001). Global modulation of cellular transcription by human cytomegalovirus is initiated by viral glycoprotein B. Proc. Natl. Acad. Sci. U.S.A. 98, 7140–7145.

Soderberg-Naucler, C., Fish, K.N., and Nelson, J.A. (1997). Interferon-gamma and tumor necrosis factor-alpha specifically induce formation of cytomegalovirus-permissive monocyte-derived macrophages that are refractory to the antiviral activity of these cytokines. J. Clin. Invest 100, 3154–3163.

Sommereyns, C., Paul, S., Staeheli, P., and Michiels, T. (2008). IFN-lambda (IFN-lambda) is expressed in a tissue-dependent fashion and primarily acts on epithelial cells in vivo. PLoS. Pathog. 4, e1000017.

Steinberg, C., Eisenacher, K., Gross, O., Reindl, W., Schmitz, F., Ruland, J., and Krug, A. (2009). The IFN regulatory factor 7-dependent type I IFN response is not essential for early resistance against murine cytomegalovirus infection. Eur. J. Immunol. 39, 1007–1018.

Strobl, B., Bubic, I., Bruns, U., Steinborn, R., Lajko, R., Kolbe, T., Karaghiosoff, M., Kalinke, U., Jonjic, S., and Muller, M. (2005). Novel functions of tyrosine kinase 2 in the antiviral defense against murine cytomegalovirus. J. Immunol. 175, 4000–4008.

Suh, H.S., Zhao, M.L., Rivieccio, M., Choi, S., Connolly, E., Zhao, Y., Takikawa, O., Brosnan, C.F., and Lee, S.C. (2007). Astrocyte indoleamine 2,3-dioxygenase is induced by the TLR3 ligand poly(I:C): mechanism of induction and role in antiviral response. J. Virol. 81, 9838–9850.

Swiecki, M., Gilfillan, S., Vermi, W., Wang, Y., and Colonna, M. (2010). Plasmacytoid dendritic cell ablation impacts early interferon responses and antiviral NK and CD8(+) T-cell accrual. Immunity 33, 955–966.

Szomolanyi-Tsuda, E., Liang, X., Welsh, R.M., Kurt-Jones, E.A., and Finberg, R.W. (2006). Role for TLR2 in NK cell-mediated control of murine cytomegalovirus in vivo. J. Virol. 80, 4286–4291.

Tabeta, K., Georgel, P., Janssen, E., Du, X., Hoebe, K., Crozat, K., Mudd, S., Shamel, L., Sovath, S., Goode, J. et al. (2004). Toll-like receptors 9 and 3 as essential components of innate immune defense against mouse cytomegalovirus infection. Proc. Natl. Acad. Sci. U.S.A. 101, 3516–3521.

Tailor, P., Tamura, T., Kong, H.J., Kubota, T., Kubota, M., Borghi, P., Gabriele, L., and Ozato, K. (2007). The feedback phase of type I interferon induction in dendritic cells requires interferon regulatory factor 8. Immunity 27, 228–239.

Tan, J.C., Avdic, S., Cao, J.Z., Mocarski, E.S., White, K.L., Abendroth, A., and Slobedman, B. (2011). Inhibition of 2′,5′-oligoadenylate synthetase expression and function by the human cytomegalovirus ORF94 gene product. J. Virol. 85, 5696–5700.

Taylor, G.A., Collazo, C.M., Yap, G.S., Nguyen, K., Gregorio, T.A., Taylor, L.S., Eagleson, B., Secrest, L., Southon, E.A., Reid, S.W., et al. (2000). Pathogen-specific loss of host resistance in mice lacking the IFN-gamma-inducible gene IGTP. Proc. Natl. Acad. Sci. U.S.A. 97, 751–755.

Taylor, R.T., and Bresnahan, W.A. (2005). Human cytomegalovirus immediate-early 2 gene expression blocks virus-induced beta interferon production. J. Virol. 79, 3873–3877.

Taylor, R.T., and Bresnahan, W.A. (2006a). Human cytomegalovirus IE86 attenuates virus- and tumor necrosis factor alpha-induced NFkappaB-dependent gene expression. J. Virol. 80, 10763–10771.

Taylor, R.T., and Bresnahan, W.A. (2006b). Human cytomegalovirus immediate-early 2 protein IE86 blocks virus-induced chemokine expression. J. Virol. 80, 920–928.

Trilling, M., Le, V.T., Fiedler, M., Zimmermann, A., Bleifuss, E., and Hengel, H. (2011). Identification of DNA-damage DNA-binding protein 1 as a conditional essential factor for cytomegalovirus replication in interferon-gamma-stimulated cells. PLoS. Pathog. 7, e1002069.

Unterholzner, L., Keating, S.E., Baran, M., Horan, K.A., Jensen, S.B., Sharma, S., Sirois, C.M., Jin, T., Latz, E., Xiao, T.S., et al. (2010). IFI16 is an innate immune sensor for intracellular DNA. Nat. Immunol. 11, 997–1004.

Upton, J.W., Kaiser, W.J., and Mocarski, E.S. (2008). Cytomegalovirus M45 cell death suppression requires receptor-interacting protein (RIP) homotypic interaction motif (RHIM)-dependent interaction with RIP1. J. Biol. Chem. 283, 16966–16970.

Vairo, D., Tassone, L., Tabellini, G., Tamassia, N., Gasperini, S., Bazzoni, F., Plebani, A., Porta, F., Notarangelo, L.D., Parolini, S., et al. (2011). Severe impairment of IFN-gamma and IFN-alpha responses in cells of a patient with a novel STAT1 splicing mutation. Blood 118, 1806–1817.

Valchanova, R.S., Picard-Maureau, M., Budt, M., and Brune, W. (2006). Murine cytomegalovirus m142 and m143 are both required to block protein kinase R-mediated shutdown of protein synthesis. J. Virol. 80, 10181–10190.

Veals, S.A., Schindler, C., Leonard, D., Fu, X.Y., Aebersold, R., Darnell, J.E., Jr., and Levy, D.E. (1992). Subunit of an alpha-interferon-responsive transcription factor is related to interferon regulatory factor and Myb families of DNA-binding proteins. Mol. Cell Biol. *12*, 3315–3324.

de Veer, M.J., Holko, M., Frevel, M., Walker, E., Der, S., Paranjape, J.M., Silverman, R.H., and Williams, B.R. (2001). Functional classification of interferon-stimulated genes identified using microarrays. J. Leukoc. Biol. *69*, 912–920.

Velazquez, L., Fellous, M., Stark, G.R., and Pellegrini, S. (1992). A protein tyrosine kinase in the interferon alpha/beta signaling pathway. Cell *70*, 313–322.

Vinkemeier, U. (2004). Getting the message across, STAT! Design principles of a molecular signaling circuit. J. Cell Biol. *167*, 197–201.

Watling, D., Guschin, D., Muller, M., Silvennoinen, O., Witthuhn, B.A., Quelle, F.W., Rogers, N.C., Schindler, C., Stark, G.R., Ihle, J.N., et al. (1993). Complementation by the protein tyrosine kinase JAK2 of a mutant cell line defective in the interferon-gamma signal transduction pathway. Nature *366*, 166–170.

Wenta, N., Strauss, H., Meyer, S., and Vinkemeier, U. (2008). Tyrosine phosphorylation regulates the partitioning of STAT1 between different dimer conformations. Proc. Natl. Acad. Sci. U.S.A. *105*, 9238–9243.

White, K.L., Slobedman, B., and Mocarski, E.S. (2000). Human cytomegalovirus latency- associated protein pORF94 is dispensable for productive and latent infection. J. Virol. *74*, 9333–9337.

Wu, S., and Kaufman, R.J. (1996). Double-stranded (ds) RNA binding and not dimerization correlates with the activation of the dsRNA-dependent protein kinase (PKR). J. Biol. Chem. *271*, 1756–1763.

Yamaguchi, T., Shinagawa, Y., and Pollard, R.B. (1988). Relationship between the production of murine cytomegalovirus and interferon in macrophages. J. Gen. Virol. *69*, 2961–2971.

Yamamoto, N., Shimokata, K., Maeno, K., and Nishiyama, Y. (1987). Effect of recombinant human interferon gamma against human cytomegalovirus. Arch. Virol. *94*, 323–329.

Yamamoto, M., Sato, S., Hemmi, H., Hoshino, K., Kaisho, T., Sanjo, H., Takeuchi, O., Sugiyama, M., Okabe, M., Takeda, K., et al. (2003). Role of adaptor TRIF in the MyD88-independent toll-like receptor signaling pathwa Science *301*, 640–643.

Yeow, W.S., Lai, C.M., and Beilharz, M.W. (1997). The in viv expression patterns of individual type I interferon gene in murine cytomegalovirus infections. Antiviral Res. *3* 17–26.

Yeow, W.S., Lawson, C.M., and Beilharz, M.W. (1998 Antiviral activities of individual murine IFN-alph subtypes in vivo: intramuscular injection of IFI expression constructs reduces cytomegaloviru replication. J. Immunol. *160*, 2932–2939.

Zhang, G., Raghavan, B., Kotur, M., Cheatham, J., Sedma D., Cook, C., Waldman, J., and Trgovcich, J. (2007 Antisense transcription in the human cytomegaloviru transcriptome. J. Virol. *81*, 11267–11281.

Zhang, S.Y., Boisson-Dupuis, S., Chapgier, A., Yang, K Bustamante, J., Puel, A., Picard, C., Abel, L., Jouangu E., and Casanova, J.L. (2008). Inborn errors of interfero (IFN)-mediated immunity in humans: insights into th respective roles of IFN-alpha/beta, IFN- gamma, an IFN-lambda in host defense. Immunol. Rev. *226*, 29–40.

Zhong, M., Henriksen, M.A., Takeuchi, K., Schaefer, O., Li B., ten Hoeve, J., Ren, Z., Mao, X., Chen, X., Shuai, K., et a (2005). Implications of an antiparallel dimeric structure c nonphosphorylated STAT1 for the activation-inactivatio cycle. Proc. Natl. Acad. Sci. U.S.A. *102*, 3966–3971.

Zhu, H., Cong, J.P., and Shenk, T. (1997). Use c differential display analysis to assess the effect of huma cytomegalovirus infection on the accumulation of cellul RNAs: induction of interferon-responsive RNAs. Pro Natl. Acad. Sci. U.S.A. *94*, 13985–13990.

Zilberstein, A., Kimchi, A., Schmidt, A., and Revel, M (1978). Isolation of two interferon- induced translation inhibitors: a protein kinase and an oligo-isoadenyla synthetase. Proc. Natl. Acad. Sci. U.S.A. *75*, 4734–4738.

Zimmermann, A., Trilling, M., Wagner, M., Wilborn, M Bubic, I., Jonjic, S., Koszinowski, U., and Hengel, H (2005). A cytomegaloviral protein reveals a dual role fc STAT2 in IFN-γ signaling and antiviral responses. J. Ex Med. *201*, 1543–1553.

Cytomegalovirus Interstrain Variance in Cell Type Tropism

Barbara Adler and Christian Sinzger

Abstract

Cytomegaloviruses (CMVs) are host species-specific pathogens that cause life-long persistent infections. Under conditions of reduced immune responses CMVs can cause acute systemic infections with replication in virtually any organ. The broad organ tropism is based on an equally broad range of target cell types. Epithelial cells, fibroblasts, endothelial cells and smooth muscle cells are the major target cells that support highly productive HCMV infection. Hepatocytes, trophoblasts, neurons, macrophages and dendritic cells are also susceptible to the full replication cycle of HCMV but are apparently less productive. Granulocytes and monocytes are non-productively infected by HCMV but are assumed to contribute to haematogenous dissemination as passive vehicles. Glycoprotein complexes containing gH-gL were identified as major determinants of the cell tropism of HCMV. They are assumed to recognize entry receptors and to trigger fusion of viral and cellular membranes during entry either directly at the plasma membrane or within endosomes. Virus strains that incorporate only gH–gL–gO in their envelope have a restricted target cell range excluding endothelial cells, epithelial cells and leucocytes whereas strains that also incorporate gH–gL–pUL128–pUL130–pUL131A have an extended target cell range including these cell types. HCMV progeny of the latter strains consists of distinct populations containing either high levels or low levels of the gH–gL–pUL128–pUL130–pUL131A complex thus allowing cells to navigate virus progeny by selectively releasing or retaining virion populations that differ in their tropism.

Introduction

Owing to their frequent isolation from the submaxillary or parotid gland of healthy appearing animals, CMVs were initially described as salivary gland viruses, and early attempts to find the characteristic inclusion body harbouring cells in other organs were without success (Cole and Kuttner, 1926; Kuttner and Wang, 1934). However, in humans it was soon noticed that 'in still-births and in the very youngest infants, the submaxillary and parotid glands are often negative, and the kidneys, lungs, and liver most frequently show the hypertrophied cells containing intranuclear inclusion bodies', which led to the conclusion that following transplacental infection 'the virus first attacks the viscera and then gradually localizes in the salivary glands' (Kuttner and Wang, 1934). Finally, McCordock and Smith (1936) succeeded to produce characteristic visceral lesions in liver, adrenals and spleen in mice by intraperitoneal inoculation with the respective salivary gland virus. Similarly, on the cellular level, cytomegalic inclusions were initially described mostly restricted to the epithelial layers of various organs (Farber and Wolbach, 1932) but have later been found in multiple cell types in various tissue layers (Vogel, 1958; Francis et al., 1989). It appears that it is the broad range of target tissues and target cell types that is characteristic for CMVs.

While the detection of cytomegalic cells with nuclear inclusions is indicative of productive infection, not all target cells necessarily proceed through the complete replicative cycle, yet, may contribute to pathogenesis of CMV infections. Hence, the term 'tropism' can describe quite different degrees of virus versus host cell interactions depending on the pathogenetic aspects of interest: (i) in the context of CMV-associated disease, virus-induced tissue damage is relevant regardless of whether it is due to necrosis in the final stage of permissive infection, apoptosis of an abortively infected cell or even damage of uninfected bystander cells; (ii) when viral in vivo replication sites are studied it is relevant whether the respective target cell is productive and capable of virus amplification; (iii) for aspects of viral dissemination, in contrast, low amounts of viral progeny or even passive transport of infectious particles in circulating cell types might be effective in initiating viral replication at a different site in the organism; (iv) in terms of immunopathology, the expression of cytokines, chemokines or adhesion molecules by infected cells is of interest, regardless of whether or not the infected cells are permissive for viral replication. Hence, for reasons of clarity, the underlying definition should always be mentioned when discussing aspects of CMV cell and tissue tropism.

From these considerations it is evident that unveiling the various aspects of cell and tissue tropism is fundamental to a profound pathogenetic understanding of CMV infections. The cell types which transmit the virus in tissue layers and in the circulation either as a result of passive transfer or active replication will greatly determine the route of the virus in the body during different states of infection. Cell type-specific entry pathways are of particular interest in developing strategies for blocking the earliest events of virus–cell interaction. Understanding the particular contribution of certain cell types to disease manifestations may open possibilities for therapy or vaccine development by specifically blocking the tropism for the respective cell types. It is noteworthy that beyond simple virus–cell interactions more complex virus–host interactions may also shape the apparent organ or cell tropism of CMV. For example, the restriction of replicating CMV to salivary glands in immunocompetent animals is obviously a result of the antiviral host defence, and it will be of interest to discuss which immune effectors can shape such a tissue restriction (Reddehase et al., 1985).

This review is meant to summarize what is currently known about the virus–host interactions underlying cell and tissue tropism of CMV but also to indicate which questions are still to be resolved by future research.

Tissue and cell tropism of CMV

The tissues and cell types that can be infected by CMV have been analysed in patient materials, animal models and cell culture systems. In synopsis of the numerous in vivo and in vitro analyses, CMVs appear as viruses that can disseminate haematogeneously throughout the host organism and are almost unrestricted with regard to their tissue and cell type tropism.

Broad range of tissues and cells targeted by CMV in vivo

Important sources of information are the clinical manifestation sites of HCMV-associated diseases. The CNS is among the most prominent targets of the virus. HCMV-associated encephalitis in the context of intrauterine infection (see also Chapter II.3) can result in cortical malformations, white matter atrophy, ventricular dilatation, cyst formation and calcification with corresponding functional deficits (Cheeran et al., 2009). In AIDS patients, HCMV-associated encephalitis is frequent and a major cause of death (Roullet 1999), whereas it is a rare complication in transplant recipients. Like the CNS, sensory organs are targeted by HCMV particularly in the context of congenital cytomegalic inclusion disease (CID) and the acquired immunodeficiency syndrome (AIDS). Affection of the inner ear in congenitally infected children can entail irreversible sensorineural hearing loss (Stagno et al., 1977; Barbi et al., 2003; Yamamoto et al., 2011). Retinitis in AIDS patients, if untreated, will inevitably lead to diminished visual acuity or complete destruction of the retina (Holbrook et al., 2003). Particularly in lung transplant recipients and BMT recipients (see also Chapter II.13 and Chapter II.16), CMV can cause an interstitial pneumonitis, which is a particularly serious manifestation with regard to its high lethality (Duncan et al., 1991; Enright et al., 1993). In the gastrointestinal tract HCMV infections can be associated with ulcerations in oesophagus, stomach, small intestine and colon (for a review, see Goodgame, 1993). Involvement of the liver is commonly indicated by elevation of hepatocyte-specific enzymes (Horwitz et al., 1980). Finally thrombocytopenia or neutropenia indicate that sites of haematopoiesis can be damaged by HCMV infection (Almeida-Porada and Ascensão, 1996).

There are three lines of evidence suggesting that tissues affected by HCMV are also sites of viral replication: (i) HCMV infected cells can regularly be detected in biopsy specimens or autopsy specimens from clinical manifestation sites by histological means, particularly by in situ detection of viral components. Cells showing the characteristic enlargement and nuclear inclusions so called 'owl eye cells', have been described in tissue sections from lung, kidney, liver, pancreas, adrenal gland, oesophagus, prostate, testes, thyroid gland, parathyroid gland, stomach, small intestine, large intestine and heart (Macasaet et al., 1975). HCMV antigens or genomes have been detected in cells of tissue sections from lung, liver, pancreas, spleen, kidney, adrenal gland, salivary gland, thyroid gland and small intestine of BMT patients who had died from interstitial pneumonia (Jiwa et al., 1989). In an analysis of 19 patients who died from various complications after BMT, lung, kidney, gastrointestinal tract, adrenals and lymph nodes were infected to some degree in about two thirds of the cases, whereas liver, spleen and thymus were infected in about one third of the cases (Rasing et al., 1990). Remarkably, even in patients who died due to non-CMV-related reasons viral antigens or genomes were detected in various organs, which highlights HCMV's potential to cause disseminated infection (Rasing et al., 1990); (ii) HCMV can be isolated from tissue samples of affected organs by inoculation of cell cultures, indicating the presence of infectious virus in these tissues (Macasaet et al., 1975; Bachman et al., 1982; Gertler et al., 1983; Morgello et al., 1987; Rasing et al., 1990; Strickler et al., 1990); (iii) Finally, resolution or prevention of disease manifestations by treatment with antiviral agents supports the assumption of a causal contribution of virus replication in the respective tissues (Jacobson and Mills, 1988; Forman and Zaia, 1994).

With regard to the cell types that are infected by HCMV, dual in situ staining of viral components and cell type-specific marker proteins in tissue sections from infected organs is the most powerful technique for an unequivocal identification of infected target cells, particularly when they are morphologically altered. Using this approach, epithelial cells, ECs, fibroblasts and smooth muscle cells were found to be the predominant targets of HCMV infection in lung, gastrointestinal and placental tissues (Roberts et al., 1989; Muhlemann et al., 1992; Sinzger et al., 1993, 1995; Ng-Bautista and Sedmak, 1995), whereas macrophages comprised a minor fraction of infected cells in these tissues (Sinzger et al., 1996). In cases of HCMV encephalitis, CMV infected cells were identified as capillary endothelia, astrocytes, and neurons (Morgello et al., 1987; Wiley and Nelson, 1988). In a fatal case of CID, the same cell types were infected in various organs including adrenal gland, bone marrow, diencephalon, heart, kidney, liver, lung, pancreas, placenta, small bowel and spleen, and in this study macrophages were among the major target cells in lung and pancreas tissues (Bissinger et al., 2002).

Data from small animal and non-human primate models indicate that the broad tissue and cell tropism found with HCMV is also a hallmark of other CMVs (see also Chapters II.5, II.15, II.17, and II.22 for GPCMV, RCMV, MCMV, and RhCMV, respectively). In MCMV-infected mice, virus could be detected by histological means in liver, salivary gland, spleen, lung, trachea, kidney, adrenals, skin, eye and ovaries (Henson et al., 1967; Mims and Gould, 1979; Staczek, 1990). A comprehensive investigation of various tissues from mice with simultaneous BMT and MCMV infection and subsequent depletion of T-cell subsets confirmed the broad organ tropism of MCMV and demonstrated that the apparent tropism was to some extent shaped by the immune response (Podlech et al., 1998). In CD8-depleted mice, a high density of MCMV infected cells was found in lung, liver, spleen, brain, lung, salivary gland, kidney,

adrenals and intestinal tract, whereas in the presence of a CD8 T-cell response, only salivary glands, adrenals and kidney contained high numbers of infected cells (Podlech et al., 1998). The predominant target cell types may vary between different organs. When the dynamics of distribution of MCMV during the first two days after peritoneal injection of virus was investigated with a GFP-expressing virus, a selective tropism for macrophages in mediastinal lymph nodes, for fibroblasts and DCs in the spleen and for hepatocytes in the liver was found (Henry et al., 2000; Hsu et al., 2009). In the lungs, interstitial fibroblasts, epithelial cells and ECs were predominant targets (Reddehase et al., 1985). However, despite preferences for certain cell types in certain tissues, the overall target cell range of MCMV is equally broad as of HCMV. For example, upon microinjection into neonatal or adult mouse brains, neurons and glial cells were major targets, but ECs, ependymal cells and cells from the meninges and choroid were also found infected (van Den Pol et al., 1999).

Upon RCMV-infection of immunosuppressed rats, immunohistochemical and hybridohistochemical analyses revealed infection of hepatocytes and macrophages in the liver, macrophages in the spleen, epithelial cells, ECs and monocytes in the lungs, mesangial cells, occasional vascular ECs and mesenchymal capsule cells in the kidneys and mononuclear cells in the bone marrow (Stals et al., 1990, 1996).

In guinea pigs, GPCMV was detected in the salivary glands (72%) and spleen (33%). Less frequently, virus was also detected in the brain, lung, pancreas and liver. Tissue lesions were most frequently observed in the brain and kidney, but also occurred in the salivary glands, liver, pancreas, thymus and spleen (Griffith et al., 1982).

In rhesus macaques infected with RhCMV via an oral or a parenteral route, viral DNA was detected in multiple tissues including bone marrow, liver, pancreas, ileum, kidney, lung, tonsils, thymus, submandibular gland, and lymph nodes, whereas infected cells were only found in the spleen by antigen detection and a minority of these cells were identified as ECs and macrophages (Lockridge et al., 1999). In SIV-infected immunodeficient rhesus monkeys with CMV reactivation, however, disseminated organ infection was proven by histological means in the brain, lung, lymph node, liver, spleen, small intestine, testicle, and nerves. Following intrauterine injection of RhCMV into the lateral ventricle of the fetal brain, numerous infected cells were detectable by immunohistochemistry during the acute phase of infection in various parts of the brain including cerebral cortex, cerebellum, striatum and choroid plexus but also in adrenals, kidney, liver, gonads, pancreas, lung, duodenum and colon (Chang et al., 2002).

Tropism of CMV towards sites of inflammation and proliferation?

In chronic ulcerative colitis there is some evidence suggesting that CMV is specifically localizing to areas of inflammation. While HCMV infection of the gastrointestinal tract is rare in patients with uncomplicated Crohn's disease or ulcerative colitis, HCMV antigens were detected at a frequency of 30% in actively inflamed colonic tissue of patients with steroid refractory ulcerative colitis (reviewed in Lawlor and Moss, 2010). The density of CMV-infected cells per tissue area was significantly higher in colon biopsies from patients with severe ulcerative colitis as compared to patients with moderate ulcerative colitis, with CMV-infected cells predominantly located around ulcer bases (Kuwabara et al., 2007). As inflammatory cytokines promote the differentiation of latently infected myeloid cells into macrophages that are permissive for viral replication and some of these cytokines are known to activate the major immediate-early promoter (MIEP)/enhancer of HCMV (see also Chapter I.10), an inflammatory environment could facilitate reactivation and replication of HCMV (Stein et al., 1993; Söderberg-Nauclér, 2006). On the other hand, the causal relationship could also be reverse with CMV increasing inflammation at initial sites of infection as suggested in the MCMV model (Saederup et al., 2001).

Similarly, it has been speculated whether atherosclerotic tissues may be particularly susceptible to replication of HCMV, based on the finding that HCMV strain Towne infected 8 out of 8 atherosclerotic plaque tissues but only 2 out of 14 non-atherosclerotic tissue explants (Nerheim et al., 2004). This notion is, however, in conflict with previous publications describing successful infection of non-atherosclerotic explants with HCMV strain TB40/E (Reinhardt et al., 2003) and lack of HCMV antigen and infectious virus in atherosclerotic and non-atherosclerotic vascular tissues from HCMV seropositive individuals (Hendrix et al., 1989). From the detection of viral DNA in the absence of viral replication (Hendrix et al., 1990) it was concluded that HCMV may be latently present in arterial walls and thus contributes to atherosclerosis, an issue that has been extensively studied in the RCMV model (see also Chapter II.15).

Repeatedly, the detection of HCMV DNA or antigens in glioblastoma tissues but not in adjacent non-tumour tissues has been reported (Cobbs et al., 2002; Mitchell et al., 2008; Scheurer et al., 2008; Lucas et al., 2011). Given the tropism of HCMV for various cell types present in the CNS this is an interesting issue, but lack of proper controls, unusually homogenous distribution of viral antigens in these tissues and unusual subcellular localization of the viral immediate-early (IE) antigen signals raises some concerns regarding the specificity of the signals obtained by highly amplified immunohistochemistry or by DNA-ISH. Immunohistochemical analysis of ependymomas, glioblastomas and oligodendrogliomas under more stringent conditions succeeded in detection of viral antigens with typical subcellular localization, yet only in scattered cells and in only a few cases of human gliomas, thus arguing against a particular tropism of HCMV for glioblastoma tissues (Lau et al., 2005; Sabatier et al., 2005).

CMV infections in cell culture: implications for the pathogenetic role of certain cell types

The remarkably broad cell tropism of CMV is well established and can easily explain why a variety of tissues and organs are usually involved during acute infection particularly in immunocompromised hosts. The particular contribution of certain cell types to defined steps in the pathogenesis of CMV-associated diseases, however, is largely a matter of speculation. Despite many efforts to define entry sites and sites of primary replication, neither cell types mediating the initial spread in lymphatic tissues and subsequent spread in the circulation, nor target cells stimulating specific anti-CMV responses, have been identified. It could also not be clarified in what respect infected cells contribute to structural and functional damage within infected tissues. Cell culture models are apt to answer basic questions of whether a given cell type supports the full replication cycle including production of viral progeny, whether virus is transmitted in a cell-free or cell-associated fashion, and whether viral replication results in lysis of the infected cell. Yet, more complex aspects like the exact routes of CMV through the organism or the contribution of direct and indirect cytopathic effects to tissue damage and the course of disease, can only be addressed in animal models.

Productive lytic replication in cell culture

The broad cell tropism of HCMV found in vivo is reflected by an equally broad range of cell types susceptible to HCMV in primary cell cultures (Sinzger et al., 2008). The typical course of infection in these culture systems is productive lytic replication, viral spread being mostly cell associated with recent clinical isolates and mostly cell free with cell culture-adapted strains (Yamane et al., 1983; Sinzger et al., 1999b).

Fibroblasts from lung or skin tissue are the standard cell types used for isolation and propagation of HCMV both for diagnostic and scientific reasons, due to their particular susceptibility to infection and their capability of producing high titres of viral progeny. Starting from single infected cells, recent isolates usually spread focally in fibroblast monolayers in a cell-associated manner (Yamane *et al.*, 1983; Sinzger *et al.*, 1999b). Foci of infected cells will finally become necrotic, resulting in plaque formation. Cell culture adapted strains can, in addition to focal growth, spread by release of virus particles that may infect distant cells, thus resulting in a comet shaped distribution of infected cells, which will also become necrotic after a number of days (Sinzger *et*

al., 1999b). If infected at a high infection multiplicity, cultured fibroblasts will proceed more or less synchronously through the replication cycle with sequential expression of regulatory viral IE proteins, of early proteins mediating DNA replication and of late proteins, which are typically contributing to encapsidation of viral genomes and further maturation, envelopment and release of progeny virions. Such productively infected cells can survive for some time and hence spread the infection for a period of several days but will finally become necrotic, resulting in a completely destroyed cell layer (Mocarski *et al.*, 2007). *In vivo*, fibroblasts appear to be similarly susceptible to productive infection, as late viral antigens can be detected in

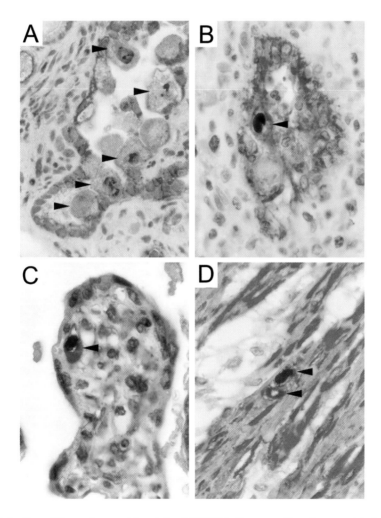

Figure I.17.1 Detection of the major capsid protein of HCMV in tissue sections from acutely infected patients. Formalin fixed, paraffin embedded tissue sections were processed by immunohistochemical techniques for detection of viral antigen (brown, nuclear, indicated by arrowheads) and cell type-specific marker proteins (red, cytoplasmic). Blue counter staining of nuclei was done with haematoxylin. (A): Stomach, epithelial cell marker cytokeratin; (B): Colon, endothelial cell marker F.VIII related antigen; (C): Placenta; fibroblast marker vimentin; (D): Colon, smooth muscle actin.

tissue fibroblasts by dual immunohistochemical stainings (Fig. I.17.1C), viral capsid formation can be found in nuclear inclusions by electron microscopy, and foci of infected fibroblasts can be found in various tissues, which is strongly indicative of virus production and dissemination (Fig. I.17.2B). Hence, fibroblasts can be assumed as ideal target cells for viral spread within the stroma of infected organs.

ECs of venous, microvascular and arterial origin have also extensively been used as cell culture models of HCMV infection. While differences in the course of infection were initially reported between ECs from different body sites (reviewed by Jarvis and Nelson, 2007), we and others have found that cultured ECs are in principle susceptible to productive lytic infection irrespective of their origin and also resemble each other with regard to functional alteration like, for example, modulation of adhesion molecules (Knight et al., 1997, 1999; Kahl et al., 2000). The finding of late viral antigen and infectious foci within vascular endothelial layers in tissue sections from infected organs demonstrates that this cell type is productively infected in vivo (Figs I.17.1B and I.17.2C). Consequently, ECs have been suggested as a source of infectious virus in the circulation and this concept has been confirmed by elegant cell culture approaches showing that leucocytes can acquire virus from an infected endothelial layer and then transmit this virus to another fully susceptible cell monolayer

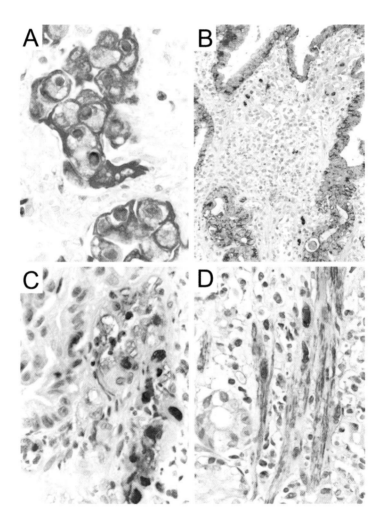

Figure I.17.2 Detection of foci of HCMV immediate-early antigen-positive cells in tissue sections from acutely infected patients. Formalin fixed, paraffin-embedded tissue sections were processed by immunohistochemical techniques for detection of viral antigen (brown, nuclear) and cell type-specific marker proteins (red, cytoplasmic) Blue counterstaining of nuclei was done with haematoxylin. (A) Lung, epithelial cell marker cytokeratin. (B) Stomach epithelial cell marker cytokeratin; stromal cells appear negative for keratin staining. (C) Stomach, endothelial cell marker F.VIII-related antigen. (D) Duodenum, smooth muscle actin.

(Waldman *et al.*, 1995; Revello *et al.*, 1998). The role of ECs for distribution of CMV within the organism has recently been extensively studied in the murine model. Sacher and colleagues infected Cre-transgenic mice with a recombinant MCMV BAC-derived virus carrying a loxP-flanked 'stop' cassette between promoter and coding sequence of the gene for EGFP. After cre recombination EGFP is activated and infected cells become EGFP-positive, which allows tracking the fate of virus progeny recombined in defined Cre-expressing cell types (Sacher *et al.*, 2008). Using systemic infection of Tie2-Cre mice in which cre-recombination specifically takes place in ECs, they revealed that most of the viral load in the blood and in organs represents progeny of virus that has before infected endothelial layers, an exception being organs with a discontinuous, fenestrated endothelium such as the liver where the majority of hepatocytes were infected by unrecombined virus prior to completion of one productive cycle (Sacher *et al.*, 2008). This strongly supports a role of ECs in virus production during virus amplification after primary viraemia. Yet, when the model was used to study virus spread from CMV-infected animals to organs transplanted from CMV-negative animals, EC-derived virus did not contribute preferentially to secondary viraemia (Sacher *et al.*, 2011).

Epithelial cell cultures from kidneys, gastrointestinal tract and retina have been successfully infected with HCMV (Smith, 1986; Heieren *et al.*, 1988; Tugizov *et al.*, 1996). In accordance with the concept of epithelial cells shedding infectious virus into various body secretions such as saliva or urine, epithelial cell cultures were described as high-level producers (Smith, 1986; Tugizov *et al.*, 1996; Wang and Shenk, 2005a). Not unexpectedly, foci of infected cells can be detected in epithelial layers in tissue sections from acutely infected patients, and infected epithelial cells contain late viral protein and intranuclear capsids, together highly indicative of productive infection (Figs. I.17.1A and I.17.2A).

In cell culture, RhCMV cell tropism closely resembles the cell tropism of HCMV, i.e. it efficiently replicates in fibroblasts, ECs and epithelial cells of rhesus and human origin (Lilja and Shenk, 2008). The contribution of infected epithelial cells is supported by a comparison of different RhCMV strains with regard to their potential for spread within the population. Whereas wild-type RhCMV was shed efficiently in saliva and urine, resulting in horizontal transmission, a strain lacking the UL/b' region, which contains genes relevant for epithelial and EC tropism (Lilja and Shenk, 2008), was not shed or transmitted to cagemates (Oxford *et al.*, 2011).

Hepatocytes, trophoblasts, neurons, macrophages and DCs have also been successfully infected in cell culture, and the course of infection was also productive and lytic. However, when compared to fibroblasts, these cell types were 100- to 1000-fold less productive despite similar infection rates (Poland *et al.*, 1994; Halwachs-Baumann *et al.*, 1998; Sinzger *et al.*, 1999a, 2006; Riegler *et al.*, 2000). Nevertheless, these cells may contribute to pathogenesis in a more localized fashion. DCs are mobile and can hence transfer the virus when moving from peripheral tissues towards lymphatic organs. Similarly, infected monocytes that invade inflamed tissues and subsequently differentiate to macrophages can initiate HCMV infection there. An organ-restricted infection has been documented for MCMV derived from hepatocytes using the strategy of traceable virus mentioned above (Sacher *et al.*, 2008; see also Chapter I.3).

Non-lytic infection in cell culture

A prominent example of a cell type in which HCMV replication is aborted on the level of entry or IE gene expression are polymorphonuclear leucocytes (PMNLs). In the circulating blood of patients with an acute HCMV infection, viral p65 antigen can be detected in the nuclei of up to several per cent of all PMNLs. In the vast majority of these cells, lack of viral IE antigen expression indicates that presence of the early-late pp65 protein is not due to *de novo* synthesis as the infection is obviously aborted after viral entry (Grefte *et al.*, 1994). The nuclear localization of pp65 indicates that at least some of the incoming virions, non-infectious enveloped particles or dense bodies (see also Chapter I.6) have fused their envelope with cellular membranes and released their content into the cytoplasm, from where pp65 is then translocated to the nucleus. Only a minority of pp65-positive PMNLs also contain IE antigens in the nucleus and the detection of the respective RNA proved that this actually represents some degree of viral gene expression in infected PMNLs (Grefte *et al.*, 1992, 1994). Early and late genes, however, are not expressed and no viral capsids are produced in the nucleus of these cells (Grefte *et al.*, 1994). Nevertheless, abortively infected PMNLs can transfer the infection to permissive cell types like fibroblasts, probably due to release of input virus harboured on the cell surface or in the cytoplasm (Revello *et al.*, 1998). Peripheral blood monocytes are non-productively infected with a block on the level of viral entry or on the level of IE gene expression (Grefte *et al.*, 1992; see also Chapter I.19) but like PMNLs they can transfer entry-derived virus to permissive cell types (Waldman *et al.*, 1995). Lymphocytes can also bind virus and contain viral transcripts to some extent as indicated by the detection of viral DNA and RNA in various lymphocyte populations in the peripheral blood (Schrier et al., 1985; von Laer et al., 1995; Hassan-Walker et al., 2001), but they do not express viral IE proteins to a detectable extent (Sinzger et al., 1996). It is unknown whether they can also serve as vehicles for virus transport. In the same line, DCs have been shown to bind and perhaps even take up non-DC tropic HCMV via DC-SIGN and transfer the virus to susceptible cells (Halary et al., 2002). Capture of virus particles on cell surfaces, which protects them from inactivation and thus enhances the chance of virus transmission, has also been shown in organotypic cultures of trophoblasts and has been discussed to play a role in intrauterine infection (Davey et al., 2011; see also Chapter II.4)

Non-productive infection of immortalized cells

In sharp contrast to primary cell cultures, HCMV replication in many immortalized cell lines is blocked after initiation of viral gene expression. Usually, these cell lines readily express IE proteins but are highly inefficient in progression towards the early and late phase of replication, resulting in release of very low amounts of infectious progeny, if at all (Smith, 1986; Wang and Shenk, 2005a). HeLa cells for example were 1000-fold less productive than ARPE-19 cells despite efficient entry and initiation of gene expression (Wang and Shenk, 2005a). In cultured glioblastoma cells HCMV also failed to establish productive lytic infection but remained in a latent state (Wolff et al., 1994). One possible explanation is that cell cycle dysregulation in these cell types may affect the efficiency of viral DNA replication (see also Chapter I.14). HCMV induces a cell cycle arrest with characteristics of G_1/S transition and a mitosis-like environment in the absence of cellular DNA replication to enable viral replication and maturation, and this feature is essentially conserved also in other CMVs (Jault et al., 1995; Lu and Shenk, 1996; Dittmer and Mocarski, 1997; Wiebusch and Hagemeier, 1999, 2001; Hertel and Mocarski, 2004; Petrik et al., 2006; Wiebusch et al., 2008). The finding that pUL117 of HCMV blocks cellular DNA replication by inhibiting proper assembly of the cellular pre-replication complex further corroborates the notion that HCMV competes with the cell for essential replication factors (Wiebusch et al., 2003; Qian et al., 2010). It is conceivable that failure to induce a similar cell cycle arrest in immortalized cell lines restrains viral replication (DeMarchi, 1983a,b). Whether such

an inhibition of viral growth can occur in vivo in natural target cells and may lead to 'low level replication' in tumour tissues is currently a matter of debate (Söderberg-Nauclér, 2006; Michaelis et al., 2011).

Mechanisms regulating the cell tropism of CMV

The cell tropism of viruses can be regulated at various stages of the viral replication cycle including attachment and entry, cell type-specific activation of viral promoters but also events during viral maturation and release (reviewed in Mims, 1989). These virus–cell interactions will determine important features of a viral infection like virus entry into or passage through cells, virus production, virus spread patterns and, in the case of herpesviruses, also the establishment of latency.

Viral envelope proteins mediating entry into target cells

Herpesvirus entry into a host cell is a complex multi-step process starting with binding and accumulation of particles on the cell surface, which is then followed by more specific interactions of viral glycoproteins with cellular receptors finally leading to conformational changes in glycoproteins that induce fusion of the viral envelope with cellular membranes (Connolly et al., 2011). Like other herpesviruses, CMVs depend on their envelope glycoproteins for successful entry into target cells (reviewed in Heldwein and Krummenacher, 2008). The essential glycoproteins of HCMV are organized within at least four glycoprotein complexes in the viral envelope: gB, gM–gN, gH–gL–gO and gH–gL–pUL128–pUL130–pUL131A (see also Chapters I.6, I.8, and II.10).

The contribution of gB to entry of HCMV is well documented by reports on the neutralizing activity of anti-gB antibodies (Pötzsch et al., 2011) and the attachment-blocking activity of soluble gB (Boyle and Compton, 1998). Neutralizing anti-gB antibodies did not inhibit virus attachment but prevented fusion of already bound virus, thus indicating an involvement of gB in the fusion process (Pötzsch et al., 2011). Accordingly, gB-negative virions have been shown to attach to the host cell, but not to be capable of entry and spread (Isaacson and Compton, 2009). In analogy to other herpesviruses, gB most probably forms a homotrimer and represents the fusion executor of CMVs (Heldwein et al., 2006; Hannah et al., 2007; Heldwein and Krummenacher, 2008; Backovic et al., 2009; Connolly et al., 2011). It has been speculated that sequence variations in gB may contribute to interstrain differences in tropism and virulence. Glycoprotein B is polymorphic and has been classified into four different genotypes (Chou and Dennison, 1991; Chou, 1992). Association of certain gB-genotypes with the clinical course of HCMV infection and with preferential sites of viral replication and shedding has been reported but the results of various studies are inconsistent. Initially, genotype gB3 was assumed to be associated with myelosuppression despite a lack of a particular bone marrow tropism over other genotypes (Fries et al., 1994; Torok-Storb et al., 1997; Randolph-Habecker et al., 2002). Others found genotype gB2 overrepresented in BMT recipients with CMV disease as compared to those without manifest disease (Woo et al., 1997). Repeated attempts to correlate gB genotypes with certain compartments in the body have also yielded contradicting results (Vogelberg et al., 1996; Rasmussen et al., 1997; Meyer-König et al., 1998; Peek et al., 1998; Aquino and Figueiredo, 2000; Fidouh-Houhou et al., 2001). More recently, it has been speculated that the presence of multiple gB genotypes, rather than the presence of a single gB genotype, could be a critical pathogenicity factor in immunocompromised patients (Coaquette et al., 2004). In conclusion, in the absence of convincing marker transfer studies, clear evidence for cell and tissue tropism differences associated with certain gB genotypes is still missing.

gM and gN form a heterodimer that is incorporated in the viral envelope, with gM being the most abundant glycoprotein in the HCMV virion (Mach et al., 2000; Varnum et al., 2004). gN is necessary for maturation and transport of gM during virion morphogenesis and *vice versa* (Mach et al., 2000, 2007). The presence of neutralizing antibodies directed against the gM–gN complex of HCMV indicates that this complex also plays a role during viral entry (Mach et al., 2000; Shimamura et al., 2006) although the nature of this contribution is unknown and cellular or viral interaction partners have not been defined. Clinicovirological studies have pointed towards a role of gN in tropism and pathogenesis of HCMV. The coding ORF is polymorphic and has been classified into four different genotypes (Pignatelli et al., 2001). gN subtypes can induce a strain-specific antibody response in the infected host, with genotype gN-4 being more sensitive to neutralization than the other genotypes (Burkhardt et al., 2009). On the other hand, genotype gN-4 has been found overrepresented in congenitally infected newborns with symptoms at birth, abnormal radiological imaging results, and sequelae, whereas in asymptomatically infected newborns with a favourable long-term outcome gN-1 and gN-3a genotypes were overrepresented (Pignatelli et al., 2010). Furthermore, overrepresentation of gN-1 in monocytes of healthy seropositive blood donors prompted the idea that gN may determine the cell tropism and influence the

establishment of latency (Pignatelli et al., 2006). Yet, as with gB, marker transfer experiments in defined cell culture models are required to corroborate this hypothesis.

Crystal structures of herpesvirus gH–gL showed a tight complex formation between these two glycoproteins (Chowdary et al., 2010; Matsuura et al., 2010). gH–gL of HCMV can form a three protein complex together with the accessory protein gO or a five protein complex together with pUL128, pUL130 and pUL131A (Li et al., 1997; Huber and Compton, 1998; Wang and Shenk, 2005b; Adler et al., 2006). Antibodies against gH of HCMV can neutralize the infectivity of particles for various cell types (Rasmussen et al., 1984; Pachl et al., 1989; Macagno et al., 2010; Jiang et al., 2011). Inhibition of gH was reported to act at a postattachment step, which pointed towards a participation of gH–gL complexes during the fusion step (Keay and Baldwin, 1991; Netterwald et al., 2004; Pötzsch et al., 2011). Expression of gH and gL from retroviral vectors was sufficient to mediate fusion of various immortalized cell lines but not of normal fibroblasts (Kinzler and Compton, 2005), indicating that an additional cellular factor was required for a fusion-mediating effect of gH–gL. In a similar experimental setting relying on adenoviral expression vectors, expression of neither gB nor gH–gL alone, but the combination of gB and gH–gL was sufficient to induce fusion in a number of physiologically relevant target cells of HCMV, including fibroblasts, epithelial cells and ECs (Vanarsdall et al., 2008). The fact that both gH–gL and gB are required for the induction of fusion suggests a cross talk between the two glycoprotein complexes within the viral envelope. Indeed, evidence for a transient physical interaction between gH and gB associated with a conformational change of gB upon entry into ECs has been reported (Patrone et al., 2007). The accessory proteins within the gH–gL complexes are not absolutely required for gH–gL mediated fusion activity. In the cell–cell fusion assays mentioned above, additional coexpression of the accessory proteins did not enhance fusion in any of the cell types tested (Kinzler and Compton, 2005; Vanarsdall et al., 2008).

Cellular receptors mediating entry into target cells

A number of cellular surface molecules have been shown to bind HCMV particles or specific viral envelope proteins (see also Chapter I.8).

Heparan sulfates are sugar moieties covalently bound to certain cell surface proteins on most cell types. Such heparan sulfate proteoglycans (HSPGs) bind to gB and gM–gN and can thus tether virions to the cell surface, which greatly enhances the chance of further interactions between virus and cell surface. If HSPGs are cleaved off, most of the initial attachment is abrogated and infectivity is greatly reduced, indicating that this interaction contributes significantly to HCMV infection on virtually all target cell types (Compton *et al.*, 1993, 2004; Boyle and Compton, 1998). Initial binding to HSPGs is followed by a higher affinity binding, which is independent of HPSGs (Compton *et al.*, 1993).

Owing to circumstantial evidence it has been suggested that HCMV particles coated with β2-microglobulin can bind to class I MHC molecules on the surface of target cells and that such binding occurs by displacement of the β2-microglobulin in the β2-microglobulin-heavy chain dimer (Grundy *et al.*, 1987a,b). Similar to heparan sulfates, MHC class I molecules are ubiquitously distributed, which could explain the broad cell tropism of HCMV. However, cell lines that do not express detectable levels of class I MHC molecules or β2-microglobulin supported HCMV entry and IE gene expression, and transfection of expression vectors for class I MHC molecules or β2-microglobulin did not enhance infection rates (Beersma *et al.*, 1991), thus arguing against class I MHC molecules as relevant HCMV entry receptors. The viral MHC class I homologue pUL18 was suggested as the viral protein that binds β2 microglobulin (Browne *et al.*, 1990). However, deletion of UL18, did not affect infectivity of the respective mutant viruses and pUL18 was not found in virions in a proteomic approach, thus arguing against a significant contribution of pUL18 to viral entry (Browne *et al.*, 1992; Varnum *et al.*, 2004).

Annexin II was identified as a 32-kDa HCMV-binding protein and has been suggested as a putative entry receptor, in particular as specific gB-binding to this molecule was demonstrated (Adlish *et al.*, 1990; Jowlin *et al.*, 1991; Wright *et al.*, 1995; Pietropaolo and Compton, 1997). Yet, neither anti-annexin II antibodies nor excess annexin II protein had an effect on infection efficiency, which argues against a direct role of annexin II as an entry receptor (Pietropaolo and Compton, 1999).

CD13 was reported as a common denominator of blood cells susceptible to HCMV infection, suggesting a functional role of this molecule in facilitating infection (Söderberg *et al.*, 1993b). In fact, transfection of human CD13 increased the amount of attachment of HCMV to mouse fibroblasts and this binding translated into an appropriate increase in viral IE protein expression as compared to non-transfected mouse fibroblasts (Söderberg *et al.*, 1993a), a finding which awaits independent confirmation. As removal of heparan sulfates abrogated most of the virus binding to fibroblasts it appears

unlikely that CD13-mediated attachment outnumbers HSPG-mediated attachment in HSPG-expressing cells. Nevertheless, it might account for virus binding to cell types low in HSPG expression.

In conclusion, each of these 'attachment receptors' may indirectly enhance infection rates by simply enhancing virus accumulation on the cell surface, yet from a biochemical point of view they are just HCMV-binding cell surface molecules rather than entry receptors. More recently, three additional HCMV binding proteins have been reported to facilitate HCMV infection at a postattachment level, thus qualifying them as candidate entry receptors.

Two research groups reported that the epidermal growth factor receptor (EGFR) was necessary for entry into fibroblasts and monocytes (Wang *et al.*, 2003; Chan *et al.*, 2009), whereas others found no contribution of the EGFR to successful HCMV infection of fibroblasts, ECs or epithelial cells (Cobbs *et al.*, 2007; Isaacson *et al.*, 2007). Neither interstrain differences nor cell type differences can account for the discrepant results as in both studies HCMV strain Towne was used to infect the same target cells (Wang *et al.*, 2003; Isaacson *et al.*, 2007).

Platelet-derived growth factor receptor (PDGFR) was recently reported to be activated by HCMV and the amount of IE gene expression correlated with the absence or presence of PDGFR both in human and murine cells. The structural protein pp65, representing input virus, was similarly reduced under conditions of PDGFR knockdown (Soroceanu *et al.*, 2008), which indicates decreased virus binding as the underlying mechanism. This is surprising, as HSPG-mediated attachment should be unaffected. A caveat comes from the fact that the high pp65-positivity rates in PDGFR-expressing cells given in the processed data are not convincingly supported by the scarce signals shown in the underlying primary immunofluorescence data. Further doubt regarding the relevance of PDGFR is raised by a retrospective clinical study that did not find an effect of the PDGFR inhibitor Imatinib (originally STI571) on the course of HCMV infections (Travi *et al.*, 2009).

Finally, a contribution of integrins during the entry process has been suggested. gB was reported to bind β1-integrins via a disintegrin-like domain, and this binding was found to mediate entry into fibroblasts, ECs and epithelial cells (Feire *et al.*, 2004, 2010). If this binding was blocked by a recombinant soluble version of the gB disintegrin-like domain, HCMV and MCMV entry into ECs and fibroblasts was inhibited (Feire *et al.*, 2010). Though this finding supports a role for a gB-integrin dependent entry pathway in both cell types, it argues against a cell type-specific effect of gB. With

HCMV, a minor contribution of αvβ3 integrin was also detectable, whereas MCMV primarily depended on β1 for viral entry (Feire et al., 2004). Others reported that αvβ3 integrins but not β1 integrins were necessary for successful HCMV infection of fibroblasts (Wang et al., 2005). Wang and colleagues also showed physical interactions between gB and EGFR and between gH and αvβ3 integrin.

The exact contribution of these putative entry receptors is still a matter of debate.

Evidence for the existence of cell type-specific HCMV receptors

The accessory proteins within gH–gL complexes of herpesviruses determine whether or not a target cell can be infected. For HCMV, the sole presence of gH–gL–gO allows infection of a restricted set of cell types including fibroblasts and smooth muscle cells but excluding other physiologically relevant target cells like ECs, epithelial cells and leucocytes. These can only be infected if gH–gL–pUL128–pUL130–pUL131A is also present in the envelope of progeny virions (Hahn et al., 2004; Gerna et al., 2005; Wang and Shenk, 2005a; Adler et al., 2006). Obviously, the two gH–gL complexes of HCMV are the prime candidates for recognition of cellular receptors determining the cell tropism of HCMV.

Like HCMV, also Epstein–Barr virus (EBV) and HHV-6 form two gH–gL complexes. EBV gH–gL–gp42 binds to HLA-class II molecules on the surface of B cells and promotes entry into B cells (Miller and Hutt-Fletcher, 1992; Spriggs et al., 1996; Hutt-Fletcher, 2007), whereas a gp42-negative gH–gL complex binds the integrins αvβ6 and αvβ8 on the surface of epithelial cells and promotes entry into this cell type (Chesnokova et al., 2009). It was suggested that gp42, the accessory protein of EBV gH–gL, restricts the cell tropism of the respective virus particles to B cells by blocking those receptor binding sites on gH–gL that are engaged upon contact with epithelial cells (Connolly et al., 2011). For HHV-6 it has been shown that the gH–gL–Q1–Q2 complex binds the HHV-6 cellular receptor CD46, whereas the HHV-6 gH–gL–gO complex binds to a yet unidentified cellular receptor (Mori et al., 2003a,b, 2004; Akkapaiboon et al., 2004).

Recent reports from the HCMV field also suggest the existence of at least two different cell type-specific receptors interacting with the five protein and the three protein gH–gL complex, respectively. Expression of gH–gL–pUL128–pUL130–pUL131A almost completely abrogated susceptibility of epithelial cells to HCMV strain TR, whereas expression of gH–gL–gO in epithelial cells had no such effect (Ryckman et al.,

2008; Vanarsdall et al., 2011). Conversely, expression of gH–gL–gO decreased susceptibility of fibroblasts to the same strain by 70% (Vanarsdall et al., 2011). In contrast, expression of gB or gH-gL alone did not significantly reduce infectivity (Vanarsdall et al., 2011). The expression of interfering gH–gL complexes did not block virus adsorption to the cell surface but subsequent stages of HCMV entry into cells (Ryckman et al., 2008), which fits well to the finding that anti-gH antibodies block infection on the postattachment level (Pötzsch et al., 2011). As an underlying mechanism of the interference the authors suppose physical interaction of the respective gH–gL complex with putative cell type-specific HCMV receptors either at the plasma membrane or at internal membranes, finally leading to sequestration of the receptor, which is hence no longer available for virus binding (Ryckman et al., 2008; Vanarsdall et al., 2011). With regard to interstrain variations in cell tropism and virulence, gO is interesting because it is by far the most polymorphic protein in gH–gL complexes of HCMV, with up to 45% variation on the protein level as compared to less than 2% variation for pUL128, pUL130 and pUL131A (Rasmussen et al., 2002; Baldanti et al., 2006; Sun et al., 2009). The gO sequences of clinical HCMV strains have been grouped into four genotypes, which, however, did not show differences in the clinical outcome of the respective patients (Paterson et al., 2002; Rasmussen et al., 2002; Roubalova et al., 2011). Data regarding the cell tropism of gO variants are not available to date. Deletion of gO resulted in strongly reduced virus production in fibroblasts (Jiang et al., 2008) and a drastic reduction of gH–gL complexes in virions and the authors concluded that gO acts only as a chaperone for gH–gL (Wille et al., 2010). For the HCMV strain TR, gO was found not to be incorporated into virions (Wille et al., 2010), whereas for the strains AD169 and TB40-BAC4 gO was found to be a virion constituent (Huber and Compton, 1998; Adler, B., unpublished observations). Thus, future research will have to show whether or not gO plays a direct role in receptor recognition.

Consistent with the assumption of cell type-specific receptors engaged by either gH–gL–gO or gH–gL–pUL128–pUL130–pUL131A, entry pathways appear to differ in fibroblasts and ECs (Fig. I.17.3). While entry of HCMV wild type strains into fibroblasts is considered to occur mainly by fusion of the viral envelope directly with the plasma membrane, which has been shown for the HCMV strain AD169 (Compton et al., 1992), entry of HCMV strain TR into endothelial and epithelial cells was found to be pH-dependent, which is suggestive of entry by fusion of the viral envelope with the membrane of an endolysosomal vesicle (Ryckman et al., 2006). Data obtained with HCMV strain VR1814 in ECs were not consistent with the assumption of an endocytic entry route (Patrone *et al.*, 2007). Using proteinase protection of viral envelope glycoproteins as a marker for endocytosis of virions, non-endotheliotropic strains were found protected upon entry into ECs, whereas the endotheliotropic strain VR1814 was not protected at any time during the entry process (Patrone *et al.*, 2007). The entry pathway of VR1814 was associated with a conformational change of gB preceding gB–gH interaction (Patrone *et al.*, 2007).

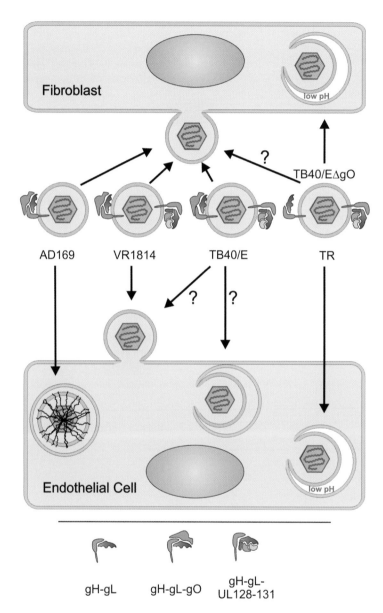

Figure I.17.3 Entry pathways as reported for various HCMV strains: Entry into fibroblasts is assumed to occur by direct fusion at the plasma membrane (Compton *et al.*, 1992) or, in the absence of gO, by an energy- and pH-dependent (probably endocytic) process (Scrivano *et al.*, 2010). Even more diverse pathways were reported for entry into endothelial cells by various HCMV strains, i.e. pH-dependent entry from endosomes with strain TR (Ryckman *et al.*, 2006), pH-independent entry after endosomal uptake with strain TB40/E (Sinzger, 2008), direct fusion with the plasma membrane with strain VR1814 (Patrone *et al.*, 2007), and inefficient entry despite endocytic uptake with strain AD169 (Bodaghi *et al.*, 1999). One explanation for the strain dependent usage of entry pathways may be the different equipment of virion envelopes with the various gH–gL complexes as indicated by the respective symbols. Arrows pointing towards an intracellular vesicle represent endocytic uptake.

A caveat comes here from the finding that fusion has been found to be executed within a few minutes after attachment to fibroblasts (Topilko and Michelson, 1994) and ECs (Sinzger, C., unpublished observations), whereas the conformational change of gB and the gH–gB interaction were observed at 30 minutes and 50 minutes, respectively, after shifting the culture to the penetration-permissive conditions (Patrone et al., 2007). VR1814 may be special in that it lacks a putative envelope component that is present in other strains and can trigger endocytosis. In any case, the data for VR1814 indicate that endocytosis is not an absolute requirement for successful entry into ECs. The endotheliotropic strain TB40/E differs from both TR and VR1814 as it is taken up into endocytic vesicles but does not depend on acidification for successful infection of ECs (Sinzger, 2008). Deletion of gO in TB40/E altered the entry pathway into fibroblasts from an energy- and pH-independent to an energy-and pH-dependent pathway (Scrivano et al., 2010). Interestingly, this gO-dependent switch of the entry pathway was shared by a gO-deletion mutant of MCMV (Scrivano et al., 2010). It is tempting to assume, that absence or presence of gO in the gH–gL complex determines whether virus–cell fusion is pH-dependent or independent and that a yet unidentified virion component determines if particles are taken up by endocytosis. These assumptions could reconcile the different findings obtained with different strains regarding entry pathways.

UL(128,130,131A) as mediators of HCMV's broad cell tropism

The fact that gH–gL–gO has been detected before gH–gL–pUL128–pUL130–pUL131A has biased our view on the two complexes in a way that the three protein complex has been regarded the primary complex suited for infection of fibroblasts, smooth muscle cells and others, and the 'new' five protein complex has been regarded as an alternative complex that additionally allows for infection of ECs, epithelial cells and leucocytes. However, there is some evidence suggesting a different point of view: apparently the five protein gH–gL complex is sufficient to mediate infection of a broad range of target cells (including both ECs and fibroblasts) and the three protein gH–gL complex is an additional variant allowing infection of a restricted set of target cells (excluding ECs, epithelial cells and leucocytes). In support of this view, HCMV mutants lacking gO are not restricted to EC cultures but can still grow focally in fibroblasts, whereas UL128–131 deletion mutants cannot form foci in ECs but only in fibroblast cultures (Jiang et al., 2008; Wille et al., 2010). For a TB40-based gO-deletion mutant

it was shown that spread in fibroblast culture could be blocked by anti-pUL131A, thus demonstrating that the residual focal spread in fibroblast culture is gH–gL–pUL128–pUL130–pUL131A-dependent (Scrivano et al., 2010).

At first glance this notion is challenged by recent data obtained with an infection interference assay testing for the potential of the different complexes to withdraw putative receptors needed for viral entry from the cell surface of fibroblasts or epithelial cells (Vanarsdall et al., 2011). Overexpression of gH–gL–pUL128–pUL130–pUL131A was found to inhibit infection by HCMV strain TR only in epithelial cells, and interference by overexpression of gH–gL–gO was similarly specific for fibroblasts. At second glance, however, the data are consistent with our hypothesis. While the five-protein complex completely abrogated infection of ARPE-19 cells, a residual infectivity of about 30% remained in fibroblasts after expression of gH–gL–gO. This could be due to the presence of a receptor for the five protein complex and, if so, only coexpression of both complexes would completely abrogate infection in fibroblasts. This assumption is supported by the finding that a TB40-derived gO-deletion mutant can still infect fibroblasts, which is apparently mediated by the five-protein gH–gL complex (Scrivano et al., 2011).

The question then arises why HCMV has evolved a gH–gL–gO complex allowing only for entry into a certain subset of cell types that is already targeted via the five protein complex? The available data suggest three potential growth advantages provided by the expression of gO: (1) a role of the gH–gL–gO complex for viral spread via cell-free infectivity, (2) a potential for navigation of HCMV from one cell type to another and (3) a role in the evasion of HCMV replication from neutralizing antibodies.

In the absence of gO, HCMV and MCMV can only grow efficiently in a cell-associated fashion with only very little cell-free infectivity, whereas expression of gO allows for efficient transmission via cell-free virions (Jiang et al., 2008; Scrivano et al., 2010; Wille et al., 2010). This might allow haematogenous dissemination over large distances at least in the absence of neutralizing antibodies. More importantly, this cell-free infectivity may greatly increase the probability of virus transmission to other hosts, e.g. by urine, saliva or breast milk (Tamura et al., 1980; Asanuma et al., 1996; Hamprecht et al., 2003, 2008).

HCMV progeny released from fibroblasts was reported to contain distinct populations of virions with high or low incorporation of the five protein gH–gL complex, which have a preference for infecting ECs or fibroblasts, respectively (Scrivano et al., 2011). Infected ECs preferentially release virions low in this complex while virions high in this complex are retained by these producer cells and consequently spread in a cell-associated manner (Scrivano *et al.*, 2011). It is currently not known which structure in or on ECs retains the latter type of virions. Translated into the *in vivo* situation this could mean that virus produced in the stroma of an organ tissue has the propensity to spread both to stromal cells and also to ECs of the vasculature. Within the vasculature local spread to neighbouring cells is favoured. Once virus has been disseminated haematogenously to distant body sites by circulating infected ECs (Grefte *et al.*, 1993), progeny released from these ECs has a clear preference for invasion of the adjacent organ stroma (Scrivano *et al.*, 2011). This may help HCMV navigate through the body (Griffiths, 2011) with infection routes being not only determined by the qualitative but also by the quantitative expression of postulated molecule or structure that retains the five protein gH–gL complex (Fig. I.17.4).

The expression and incorporation of gH–gL–gO in the viral envelope can rescue infection of fibroblasts

(and other cells that share the putative gH–gL–gO receptor) in the presence of antibodies against the five protein complex (Wang and Shenk, 2005b; Gerna *et al.*, 2008; Jiang *et al.*, 2008; Macagno *et al.*, 2010; Revello and Gerna, 2010; Saccoccio *et al.*, 2011). This could provide an advantage for the virus particularly during an early stage of primary infection when neutralizing antibodies against the proteins of the UL128 locus are already present while neutralizing antibodies against gH and gB have not yet been raised (Revello and Gerna, 2010). Infection of cells within the circulation (i.e. ECs and leucocytes) is already inhibited at this stage, but infection of stromal cells within tissues is still unrestricted. Later in the course of infection, the broader neutralizing activity of antibodies against gB and gH will also inhibit infection of these target cells. gO provides an additional advantage in this situation as it obviously protects the fusion apparatus of HCMV to some extent from the inhibitory action of gB- and gH-specific antibodies (Jiang *et al.*, 2011).

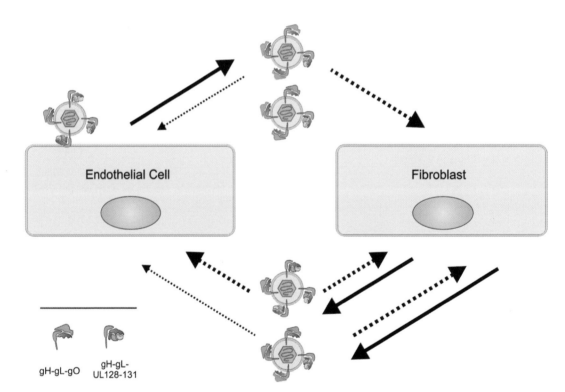

Figure I.17.4 Model for HCMV spread in cell culture: The model proposes the production of heterogeneous virus progeny from fibroblasts and endothelial cells. Fibroblasts release a virus progeny consisting of populations with high or low amounts of gH–gL–pUL128–pUL130–pUL131A complexes, whereas endothelial cells release only viruses containing low amounts of gH–gL–pUL128–pUL130–pUL131A complexes and retain viruses which contain high amounts of gH–gL–pUL128–pUL130–pUL131A. Thus, endothelial cell-derived virus progeny is either directed to a cell-associated spread, or, if released, to cells whose infection is not dependent on gH–gL–pUL128–pUL130–pUL131A. (solid arrow: release; dotted arrows indicate strength of infection efficiencies of virus released).

The role of the major IE gene for the broad cell tropism

Cell type-dependent activation of viral genes can modulate the cell tropism of viruses, which is well established for JC virus and other polyoma viruses (White et al., 2009; Marshall and Major, 2010), but may also play a role in infection with retroviruses (Agnarsdottir et al., 2000; Maury et al., 2005; Oskarsson et al., 2007) and some of the herpesviruses (West and Wood, 2003; Baiker et al., 2004). In contrast, CMVs appear widely unrestricted with regard to the initiation of viral lytic gene expression. The importance of both the major immediate-early promoter (MIEP) and the IE proteins for the establishment of viral replication has been unequivocally demonstrated for various CMVs (see Chapter I.10). A role of the IE 1 protein for determination of the target cell range was underscored by data showing that expression of HCMV IE1 renders human cells permissive to MCMV by increasing the amount of newly replicated viral DNA (Tang and Maul, 2006). Obviously, the MIEP-enhancer of HCMV can drive expression of IE proteins in multiple cell types and thus contributes to the broad cell tropism of HCMV. Hence, it was suggested that one role of the MIEP-enhancer is to optimize the efficiency of productive viral replication in the respective cell types (Stinski and Meier, 2007). This is not trivial, as in certain – undifferentiated – cell types the MIEP is silenced, resulting in abortive or latent rather than productive infection (Sinclair, 2010). On the other hand, MCMV in which the autologous MIEP-enhancer was replaced by the HCMV MIEP-enhancer replicated well even in cell types and tissues in which the isolated transgenic HCMV MIEP-enhancer was not active (Baskar et al., 1996; Grzimek et al., 1999), indicating that other viral factors can help overcome silencing of the MIEP-enhancer, with tegument proteins being good candidates.

Other viral genes with impact on cell tropism

In a comprehensive analysis of deletion mutants on the genetic background of the Towne strain of HCMV it has been found that the UL24-deletion mutant was significantly defective in growth in EC, whereas it grew as well as wild type virus in fibroblasts and epithelial cells (Dunn et al., 2003). pUL24 is a tegument protein of the US22 gene family, proteins which are assumed to participate in particle maturation (Adair et al., 2002). Intriguingly, such a finding has been predicted by analogy from the MCMV system where the US22 genes m140 and m141 function cooperatively and independently regulate MCMV replication in macrophages (Hanson et al., 2001). The authors assumed that the

HCMV homologues of m140 and m141 (US23 and US24, respectively) might confer similar functions in regulating cell or tissue tropism of HCMV (Hanson et al., 2001). However, although these genes are analogous in MCMV and HCMV with respect to a function in regulation of cell tropism, they are distinct in the cell types that are affected. In MCMV, deletion of m140 and m141 only affects growth in macrophages but has no effect on infection efficiency in an EC line (Menard et al., 2003). The m140 protein was recently shown to form a complex with the m139 and m141 proteins and to protect partners in the complex from proteasome-dependent degradation (Karabekian et al., 2005; Bolin et al., 2010). In the absence of m140, virion assembly in macrophages is defective, likely due to the reduced levels of the major capsid protein M86 and the tegument protein M25 (Hanson et al., 2009).

The MCMV protein M45 is a viral inhibitor of receptor-interacting-protein (RIP) activation (vIRA) that promotes viral replication by blocking various cell death pathways activated upon infection (see also Chapter I.15). M45 can block apoptosis through binding and modulating the cellular RIP1 via its RIP homotypic interaction motif (RHIM) (Brune et al., 2001; Mack et al., 2008; Upton et al., 2008). The cell death-preventing function of M45 is, however, not confined to caspase-dependent apoptotic pathways, as this protein also inhibits TNFα-induced necroptosis by blocking formation of a RHIM-dependent RIP3–RIP1 kinase complex and virus-induced necrosis by blocking a RIP3 RHIM-dependent, RIP1-independent step (Upton et al., 2010). A contribution of this gene to the cell tropism of MCMV was suggested by the finding that MCMV replication in an EC line and a macrophage cell line depended on M45 to avoid premature cell death, whereas the fibroblast cell line NIH-3T3 supported the complete replication cycle irrespective of M45 expression (Brune et al., 2001). However, NIH-3T3 cells are special in that they express RIP-3 at low levels (Upton et al., 2010). Actually, MCMV replication in RIP-3-expressing 3T3 cell lines and primary fibroblasts also depended on M45 (Upton et al., 2010), leaving the question whether reduced RIP-3 levels as observed in NIH 3T3 cells are a specific defect of this cell line or actually occur in natural cell types in mice. In HCMV, UL45 apparently does not regulate cell tropism. Deletion of the HCMV homologue UL45 in the context of HCMV Bacmid RVFIX did not affect the ability of the virus to replicate efficiently in EC, and an increase in levels of apoptotic death was not observed (Hahn et al., 2002). Still, when apoptosis was induced by exposing fibroblasts to proapoptotic stimuli, cell survival was reduced by 50% in a UL45 deletion mutant on the AD169 background, indicating a similar function of M45 and UL45 (Patrone *et al.*, 2003).

In the genome of the RhCMV a number of genes were recently described as critical determinants for replication in certain cell types (see Chapter II.15). Rh10 encodes an early viral protein with high homology to cellular COX-2 which is necessary for replication in ECs (Rue *et al.*, 2004). Rh01, Rh159, Rh160 and Rh203 (orthologues of HCMV TRL1, UL148, US22 and UL132, respectively) were found important for growth in epithelial cells but not in fibroblasts (Lilja *et al.*, 2008a). Finally, RhCMV encodes proteins homologous to pUL128, pUL130 and pUL131 of HCMV and these proteins contribute to cell tropism at least for epithelial cells (Lilja *et al.*, 2008b).

Host factors shaping the tropism of CMV

The clinical finding that preferential manifestation sites of HCMV disease depend on the underlying conditions indicates that the antiviral immune response, or host factors in general, can modify the apparent tissue tropism. For example, HCMV retinitis is common among AIDS patients but not among transplant recipients. While the factors modulating organ infection by HCMV are unknown, there is increasing information on how the tropism of MCMV is shaped by various effectors of the host defence system.

Podlech *et al.* (1998) demonstrated that restriction of the organ tropism in immunocompetent mice after intraplantar MCMV infection is due to an effective control by CD8 T-cells in most organs: haematoablated BALB/c mice that had received BM transplantation on the day of infection (see also Chapter II.17) later on showed a high density of infected cells only in salivary glands, adrenals and kidneys, whereas depletion of CD8 T-cells during the haematopoietic reconstitution revealed the complete pantropic potential of MCMV, with high density of infected cells in lung, liver, spleen, brain and small intestine in addition to the organs infected under conditions of intact immune reconstitution. In contrast, depletion of CD4 T-cells did not alter the restricted pattern of MCMV replication in such an experimental setting (Podlech *et al.*, 1998). Similarly, earlier work has shown that adoptive transfer of CD8 T-cells, but not of CD4 T-cells, restricts the apparent organ tropism of MCMV (Reddehase *et al.*, 1985). NK cells can also modify the distribution of MCMV in the organism. Ly49H+ NK cells were shown migrating to and protecting splenic white pulp stroma from MCMV infection, whereas in the absence of a protective NK cell response, MCMV causes destruction of splenic

white and red pulp areas in the first few days of infection (Bekiaris *et al.*, 2008).

Regarding the particular sensitivity of the brain under certain conditions, not only the immaturity of T-cell responses (Reuter *et al.*, 2004) but also an inherently higher susceptibility of the developing brain to MCMV infection has been suggested (van den Pol *et al.*, 2002).

The relevance of intrinsic response mechanisms is emphasized by the finding that the presence or absence of an antiapoptotic viral gene can influence the apparent organ tropism. Deletion of M36, a viral inhibitor of caspases, modulated the organ distribution of MCMV after intraperitoneal infection. Whereas viral growth in central organs like liver and spleen was only moderately attenuated, there was almost no dissemination towards peripheral organs like lung and salivary glands (Cicin-Sain *et al.*, 2008). Spread deficiency was even more pronounced after intraplantar infection when replication of delta-M36 MCMV was restricted to the local footpad tissue (Cicin-Sain *et al.*, 2005; Erlach *et al.*, 2006).

Concluding remarks

The broad range of susceptible target cells and tissues is apparently a hallmark of all CMVs and in principle provides an explanation for the very diverse manifestations of CMV infections. Detection of virus and viral components in infected tissues and the analyses of viral replication in various cell culture systems were a rich source of information, showing that epithelial cells, fibroblasts, smooth muscle cells, ECs, DCs, macrophages, granulocytes, but also specialized tissue cells like hepatocytes, neurons and glial cells are susceptible to CMV infection and may be pathogenetically relevant. Deletional approaches have identified a variety of viral genes in HCMV, MCMV, and RhCMV that are specifically required in different phases of viral replication for infection of certain cell types. Remarkably, HCMV and probably RhCMV share with the herpesviruses EBV and HHV-6 a particular role of gH–gL envelope complexes for the determination of cell tropism and entry pathways. We propose that the five protein complex of gH–gL–pUL128–pUL130–pUL131A mediates entry of HCMV in most cell types whereas the three protein complex gH–gL–gO targets a more restricted set of cell types, and there is recent evidence that both complexes engage different receptors.

Identification of these cellular receptors is a major task of HCMV research in the near future, and cell culture systems with deletions of either viral or cellular proteins will certainly be the tool of choice to identify receptors and dissect pathways of viral entry and exit in susceptible cell types. This may not only increase our understanding of virus–host interactions but have direct impact on the development of new antiviral drugs targeting the earliest steps in viral replication, i.e. attachment and entry. Consequently, different cell types have already been included in a recently published drug screen for entry inhibitors (Ibig-Rehm *et al.*, 2011). Similarly, antibody-based approaches to passive transfer of immunity or active immunization may benefit from knowledge of the different glycoprotein–receptor interactions engaged in the infection of different cell types as suggested recently (Macagno *et al.*, 2010; Revello and Gerna, 2010). Likewise, a particular contribution of certain target cells to T-cell stimulation may influence approaches for vaccine development (Mohr *et al.*, 2010).

However, while it appears obvious to consider aspects of cell tropism for the future development of antiviral strategies, there actually exist only few data concerning the relevance of certain cell types in the *in vivo* situation, and simple deduction from cell culture experiments may be misleading. For example, it has been shown that MHC class I molecules render cells susceptible for MCMV infection, which led to the claim that they serve as receptors for MCMV (Wykes *et al.*, 1993). Yet, when beta-2 microglobulin-deficient mice or fibroblasts derived from these mice were infected with MCMV neither organ virus titres nor virus production in fibroblasts was different from mice or cells expressing beta-2 microglobulin which strongly argues against a role of MHC class I in acting as a virus receptor (Polic *et al.*, 1996). For RhCMV it has been shown that the lack of genes UL128, UL130, and UL131 resulted in reduced replication efficiencies in epithelial and endothelial cell lines (Lilja and Shenk, 2008). Interestingly, infection of rhesus monkeys with this mutant showed a reduced virus secretion into the saliva and with it a loss of horizontal transmission, but spread within the infected animal was not affected (Oxford *et al.*, 2011). This means that the pathogenetic role of cell type-specific interactions of cellular receptors and envelope glycoproteins in the infected organism will have to be revealed by clinical studies of HCMV infections and by CMV infections in animal models. Genetically engineered, traceable MCMVs have already been developed (Sacher *et al.*, 2008) and represent powerful tools for such *in vivo* analyses while RhCMV-infected rhesus monkeys will be particularly promising because of the close relationship to HCMV-infection in humans (see also Chapter II.22).

Up to now the role of CMV cell and tissue tropism has mainly been evaluated by detecting infected cells or by determining infectious virus or viral DNA. We predict that the future of this field will be to identify cell types that route or channel infectious virus in the infected host for evaluating the particular role of such cells in the establishment of infection, the development of disease and, specifically, in intrauterine infection. We assume that studies of infections with CMV glycoprotein mutants in animal models and, where applicable, the combination with transgenic or knockout animals (see also Chapter II.12) could strongly push the progress in this field. In humans, the development and application of drugs and antibodies targeting only subsets of target cells will help clarify the role of these cells in the pathogenesis of HCMV infections.

References

Adair, R., Douglas, E.R., Maclean, J.B., Graham, S.Y., Aitken, J.D., Jamieson, F.E., and Dargan, D.J. (2002). The products of human cytomegalovirus genes UL23, UL24, UL43 and US22 are tegument components. J. Gen. Virol. *83*, 1315–1324.

Adler, B., Scrivano, L., Ruzcics, Z., Rupp, B., Sinzger, C., and Koszinowski, U. (2006). Role of human cytomegalovirus UL131A in cell type-specific virus entry and release. J. Gen. Virol. *87*, 2451–2460.

Adlish, J.D., Lahijani, R.S., and St Jeor, S.C. (1990). Identification of a putative cell receptor for human cytomegalovirus. Virology *176*, 337–345.

Agnarsdottir, G., Thorsteinsdottir, H., Oskarsson, T., Matthiasdottir, S., Haflidadottir, B.S., Andresson, O.S., and Andresdottir, V. (2000). The long terminal repeat is a determinant of cell tropism of maedi-visna virus. J. Gen. Virol. *81*, 1901–1905.

Akkapaiboon, P., Mori, Y., Sadaoka, T., Yonemoto, S., and Yamanishi, K. (2004). Intracellular processing of human herpesvirus 6 glycoproteins Q1 and Q2 into tetrameric complexes expressed on the viral envelope. J. Virol. *78*, 7969–7983.

Almeida-Porada, G.D., and Ascensão, J.L. (1996). Cytomegalovirus as a cause of pancytopenia. Leuk. Lymphoma *21*, 217–223.

Aquino, V.H., and Figueiredo, L.T. (2000). High prevalence of renal transplant recipients infected with more than one cytomegalovirus glycoprotein B genotype. J. Med. Virol. *61*, 138–142.

Asanuma, H., Numazaki, K., Nagata, N., Hotsubo, T., Horino, K., and Chiba, S. (1996). Role of milk whey in the transmission of human cytomegalovirus infection by breast milk. Microbiol. Immunol. *40*, 201–204.

Bachman, D.M., Rodrigues, M.M., Chu, F.C., Straus, S.E., Cogan, D.G., and Macher, A.M. (1982). Culture-proven cytomegalovirus retinitis in a homosexual man with the acquired immunodeficiency syndrome. Ophthalmology *89*, 797–804.

Backovic, M., Longnecker, R., and Jardetzky, T.S. (2009). Structure of a trimeric variant of the Epstein–Barr virus glycoprotein B. Proc. Natl. Acad. Sci. U.S.A. *106*, 2880–2885.

Baiker, A., Bagowski, C., Ito, H., Sommer, M., Zerboni, L., Fabel, K., Hay, J., Ruyechan, W., and Arvin, A.M. (2004). The immediate-early 63 protein of Varicella-Zoster virus: analysis of functional domains required for replication in vitro and for T-cell and skin tropism in the SCIDhu model in vivo. J. Virol. 78, 1181–1194.

Baldanti, F., Paolucci, S., Campanini, G., Sarasini, A., Percivalle, E., Revello, M.G., and Gerna, G. (2006). Human cytomegalovirus UL131A, UL130 and UL128 genes are highly conserved among field isolates. Arch. Virol. 151, 1225–1233.

Barbi, M., Binda, S., Caroppo, S., Ambrosetti, U., Corbetta, C., and Sergi, P. (2003). A wider role for congenital cytomegalovirus infection in sensorineural hearing loss. Pediatr. Infect. Dis. J. 22, 39–42.

Baskar, J.F., Smith, P.P., Nilaver, G., Jupp, R.A., Hoffmann, S., Peffer, N.J., Tenney, D.J., Colberg-Poley, A.M., Ghazal, P., and Nelson, J.A. (1996). The enhancer domain of the human cytomegalovirus major immediate-early promoter determines cell type-specific expression in transgenic mice. J. Virol. 70, 3207–3214.

Beersma, M.F., Wertheim-van Dillen, P.M., Geelen, J.L., and Feltkamp, T.E. (1991). Expression of HLA class I heavy chains and beta 2-microglobulin does not affect human cytomegalovirus infectivity. J. Gen. Virol. 72, 2757–2764.

Bekiaris, V., Timoshenko, O., Hou, T.Z., Toellner, K., Shakib, S., Gaspal, F., McConnell, F.M., Parnell, S.M., Withers, D., Buckley, C.D., et al. (2008). Ly49H+ NK cells migrate to and protect splenic white pulp stroma from murine cytomegalovirus infection. J. Immunol. 180, 6768–6776.

Bissinger, A.L., Sinzger, C., Kaiserling, E., and Jahn, G. (2002). Human cytomegalovirus as a direct pathogen: correlation of multiorgan involvement and cell distribution with clinical and pathological findings in a case of congenital inclusion disease. J. Med. Virol. 67, 200–206.

Bodaghi, B., Slobbe-van Drunen, M.E., Topilko, A., Perret, E., Vossen, R.C., van Dam-Mieras, M.C., Zipeto, D., Virelizier, J.L., LeHoang, P., Bruggeman, C.A., et al. (1999). Entry of human cytomegalovirus into retinal pigment epithelial and endothelial cells by endocytosis. Invest. Ophthalmol. Vis. Sci. 40, 2598–2607.

Bolin, L.L., Hanson, L.K., Slater, J.S., Kerry, J.A., and Campbell, A.E. (2010). Murine cytomegalovirus US22 protein pM140 protects its binding partner, pM141, from proteasome-dependent but ubiquitin-independent degradation. J. Virol. 84, 2164–2168.

Boyle, K.A., and Compton, T. (1998). Receptor-binding properties of a soluble form of human cytomegalovirus glycoprotein B. J. Virol. 72, 1826–1833.

Browne, H., Smith, G., Beck, S., and Minson, T. (1990). A complex between the MHC class I homologue encoded by human cytomegalovirus and beta 2 microglobulin. Nature 347, 770–772.

Browne, H., Churcher, M., and Minson, T. (1992). Construction and characterization of a human cytomegalovirus mutant with the UL18 (class I homolog) gene deleted. J. Virol. 66, 6784–6787.

Brune, W., Ménard, C., Heesemann, J., and Koszinowski, U.H. (2001). A ribonucleotide reductase homolog of cytomegalovirus and endothelial cell tropism. Science 291, 303–305.

Burkhardt, C., Himmelein, S., Britt, W., Winkler, T., and Mach, M. (2009). Glycoprotein N subtypes of human cytomegalovirus induce a strain-specific antibody response during natural infection. J. Gen. Virol. 90, 1951–1961.

Chan, G., Nogalski, M.T., and Yurochko, A.D. (2009). Activation of EGFR on monocytes is required for human cytomegalovirus entry and mediates cellular motility. Proc. Natl. Acad. Sci. U.S.A. 106, 22369–22374.

Chang, W.L.W., Tarantal, A.F., Zhou, S.S., Borowsky, A.D., and Barry, P.A. (2002). A recombinant rhesus cytomegalovirus expressing enhanced green fluorescent protein retains the wild-type phenotype and pathogenicity in fetal macaques. J. Virol. 76, 9493–9504.

Cheeran, M.C., Lokensgard, J.R., and Schleiss, M.R. (2009). Neuropathogenesis of congenital cytomegalovirus infection: disease mechanisms and prospects for intervention. Clin. Microbiol. Rev. 22, 99–126.

Chesnokova, L.S., Nishimura, S.L., and Hutt-Fletcher, L.M. (2009). Fusion of epithelial cells by Epstein–Barr virus proteins is triggered by binding of viral glycoproteins gHgL to integrins alphavbeta6 or alphavbeta8. Proc. Natl. Acad. Sci. U.S.A. 106, 20464–20469.

Chou, S. (1992). Comparative analysis of sequence variation in gp116 and gp55 components of glycoprotein B of human cytomegalovirus. Virology 188, 388–390.

Chou, S.W., and Dennison, K.M. (1991). Analysis of interstrain variation in cytomegalovirus glycoprotein B sequences encoding neutralization-related epitopes. J. Infect. Dis. 163, 1229–1234.

Chowdary, T.K., Cairns, T.M., Atanasiu, D., Cohen, G.H., Eisenberg, R.J., and Heldwein, E.E. (2010). Crystal structure of the conserved herpesvirus fusion regulator complex gH–gL. Nat. Struct. Mol. Biol. 17, 882–888.

Cicin-Sain, L., Podlech, J., Messerle, M., Reddehase, M.J., and Koszinowski, U.H. (2005). Frequent coinfection of cells explains functional in vivo complementation between cytomegalovirus variants in the multiply infected host. J. Virol. 79, 9492–9502.

Cicin-Sain, L., Ruzsics, Z., Podlech, J., Bubić, I., Menard, C., Jonjić, S., Reddehase, M.J., and Koszinowski, U.H. (2008). Dominant-negative FADD rescues the in vivo fitness of a cytomegalovirus lacking an antiapoptotic viral gene. J. Virol. 82, 2056–2064.

Coaquette, A., Bourgeois, A., Dirand, C., Varin, A., Chen, W., and Herbein, G. (2004). Mixed cytomegalovirus glycoprotein B genotypes in immunocompromised patients. Clin. Infect. Dis. 39, 155–161.

Cobbs, C.S., Harkins, L., Samanta, M., Gillespie, G.Y., Bharara, S., King, P.H., Nabors, L.B., Cobbs, C.G., and Britt, W.J. (2002). Human cytomegalovirus infection and expression in human malignant glioma. Cancer Res. 62, 3347–3350.

Cobbs, C.S., Soroceanu, L., Denham, S., Zhang, W., Britt, W.J., Pieper, R., and Kraus, M.H. (2007). Human cytomegalovirus induces cellular tyrosine kinase signaling and promotes glioma cell invasiveness. J. Neurooncol. 85, 271–280.

Cole, R., and Kuttner, A.G. (1926). A filterable virus present in the submaxillary glands of guinea pigs. J. Exp. Med. 44, 855–873.

Compton, T. (2004). Receptors and immune sensors: the complex entry path of human cytomegalovirus. Trends Cell Biol. 14, 5–8.

Compton, T., Nepomuceno, R.R., and Nowlin, D.M. (1992). Human cytomegalovirus penetrates host cells by pH-independent fusion at the cell surface. Virology 191, 387–395.

Compton, T., Nowlin, D.M., and Cooper, N.R. (1993). Initiation of human cytomegalovirus infection requires initial interaction with cell surface heparan sulfate. Virology 193, 834–841.

Connolly, S.A., Jackson, J.O., Jardetzky, T.S., and Longnecker, R. (2011). Fusing structure and function: a structural view of the herpesvirus entry machinery. Nat. Rev. Microbiol. 9, 369–381.

Davey, A., Eastman, L., Hansraj, P., and Hemmings, D.G. (2011). Human cytomegalovirus is protected from inactivation by reversible binding to villous trophoblasts. Biol. Reprod. 85, 198–207.

DeMarchi, J.M. (1983a). Correlation between stimulation of host cell DNA synthesis by human cytomegalovirus and lack of expression of a subset of early virus genes. Virology 129, 274–286.

DeMarchi, J.M. (1983b). Nature of the block in the expression of some early virus genes in cells abortively infected with human cytomegalovirus. Virology 129, 287–297.

van Den Pol, A.N., Mocarski, E., Saederup, N., Vieira, J., and Meier, T.J. (1999). Cytomegalovirus cell tropism, replication, and gene transfer in brain. J. Neurosci. 19, 10948–10965.

Dittmer, D., and Mocarski, E.S. (1997). Human cytomegalovirus infection inhibits G1/S transition. J. Virol. 71, 1629–1634.

Duncan, A.J., Dummer, J.S., Paradis, I.L., Dauber, J.H., Yousem, S.A., Zenati, M.A., Kormos, R.L., and Griffith, B.P. (1991). Cytomegalovirus infection and survival in lung transplant recipients. J. Heart Lung Transplant. 10, 638–644.

Dunn, W., Chou, C., Li, H., Hai, R., Patterson, D., Stolc, V., Zhu, H., and Liu, F. (2003). Functional profiling of a human cytomegalovirus genome. Proc. Natl. Acad. Sci. U.S.A. 100, 14223–14228.

Enright, H., Haake, R., Weisdorf, D., Ramsay, N., McGlave, P., Kersey, J., Thomas, W., McKenzie, D., and Miller, W. (1993). Cytomegalovirus pneumonia after bone marrow transplantation. Risk factors and response to therapy. Transplantation 55, 1339–1346.

Erlach, K.C., Bohm, V., Seckert, C.K., Reddehase, M.J., and Podlech, J. (2006). Lymphoma cell apoptosis in the liver induced by distant murine cytomegalovirus infection. J. Virol. 80, 4801–4819.

Farber, S., and Wolbach, S.B. (1932). Intranuclear and cytoplasmic inclusions ('protozoan-like bodies') in the salivary glands and other organs of infants. Am. J. Pathol. 8, 123–136 123.

Feire, A.L., Koss, H., and Compton, T. (2004). Cellular integrins function as entry receptors for human cytomegalovirus via a highly conserved disintegrin-like domain. Proc. Natl. Acad. Sci. U.S.A. 101, 15470–15475.

Feire, A.L., Roy, R.M., Manley, K., and Compton, T. (2010). The glycoprotein B disintegrin-like domain binds beta 1 integrin to mediate cytomegalovirus entry. J. Virol. 84, 10026–10037.

Fidouh-Houhou, N., Duval, X., Bissuel, F., Bourbonneux, V., Flandre, P., Ecobichon, J.L., Jordan, M.C., Vildé, J.L., Brun-Vézinet, F., and Leport, C. (2001). Salivary cytomegalovirus (CMV) shedding, glycoprotein B genotype distribution, and CMV disease in human immunodeficiency virus-seropositive patients. Clin. Infect. Dis. 33, 1406–1411.

Forman, S.J., and Zaia, J.A. (1994). Treatment and prevention of cytomegalovirus pneumonia after bone marrow transplantation: where do we stand? Blood 83, 2392–2398.

Francis, N.D., Boylston, A.W., Roberts, A.H., Parkin, J.M., and Pinching, A.J. (1989). Cytomegalovirus infection in gastrointestinal tracts of patients infected with HIV-1 or AIDS. J. Clin. Pathol. 42, 1055–1064.

Fries, B.C., Chou, S., Boeckh, M., and Torok-Storb, B. (1994). Frequency distribution of cytomegalovirus envelope glycoprotein genotypes in bone marrow transplant recipients. J. Infect. Dis. 169, 769–774.

Gerna, G., Percivalle, E., Lilleri, D., Lozza, L., Fornara, C., Hahn, G., Baldanti, F., and Revello, M.G. (2005). Dendritic-cell infection by human cytomegalovirus is restricted to strains carrying functional UL131–128 genes and mediates efficient viral antigen presentation to CD8+ T-cells. J. Gen. Virol. 86, 275–284.

Gerna, G., Sarasini, A., Patrone, M., Percivalle, E., Fiorina, L., Campanini, G., Gallina, A., Baldanti, F., and Revello, M.G. (2008). Human cytomegalovirus serum neutralizing antibodies block virus infection of endothelial/epithelial cells, but not fibroblasts, early during primary infection. J. Gen. Virol. 89, 853–865.

Gertler, S.L., Pressman, J., Price, P., Brozinsky, S., and Miyai, K. (1983). Gastrointestinal cytomegalovirus infection in a homosexual man with severe acquired immunodeficiency syndrome. Gastroenterology 85, 1403–1406.

Goodgame, R.W. (1993). Gastrointestinal cytomegalovirus disease. Ann. Intern. Med. 119, 924–935.

Grefte, A., Blom, N., van der Giessen, M., van Son, W., and The, T.H. (1993). Ultrastructural analysis of circulating cytomegalic cells in patients with active cytomegalovirus infection: evidence for virus production and endothelial origin. J. Infect. Dis. 168, 1110–1118.

Grefte, A., Harmsen, M.C., van der Giessen, M., Knollema, S., van Son, W.J., and The, T.H. (1994). Presence of human cytomegalovirus (HCMV) immediate-early mRNA but not ppUL83 (lower matrix protein pp65) mRNA in polymorphonuclear and mononuclear leukocytes during active HCMV infection. J. Gen. Virol. 75, 1989–1998.

Grefte, J.M., van der Gun, B.T., Schmolke, S., van der Giessen, M., van Son, W.J., Plachter, B., Jahn, G., and The, T.H. (1992). The lower matrix protein pp65 is the principal viral antigen present in peripheral blood leukocytes during an active cytomegalovirus infection. J. Gen. Virol. 73, 2923–2932.

Griffith, B.P., Lucia, H.L., and Hsiung, G.D. (1982). Brain and visceral involvement during congenital cytomegalovirus infection of guinea pigs. Pediatr. Res. 16, 455–459.

Griffiths, P.D. (2011). HCMV gets a Sat-Nav too. Rev. Med. Virol. 21, 137–138.

Grundy, J.E., McKeating, J.A., and Griffiths, P.D. (1987a). Cytomegalovirus strain AD169 binds beta 2 microglobulin in vitro after release from cells. J. Gen. Virol. 68, 777–784.

Grundy, J.E., McKeating, J.A., Ward, P.J., Sanderson, A.R., and Griffiths, P.D. (1987b). Beta 2 microglobulin enhances the infectivity of cytomegalovirus and when bound to the virus enables class I HLA molecules to be used as a virus receptor. J. Gen. Virol. 68, 793–803.

Grzimek, N.K., Podlech, J., Steffens, H.P., Holtappels, R., Schmalz, S., and Reddehase, M.J. (1999). In vivo replication of recombinant murine cytomegalovirus driven by the paralogous major immediate-early promoter-enhancer of human cytomegalovirus. J. Virol. 73, 5043–5055.

Hahn, G., Khan, H., Baldanti, F., Koszinowski, U.H., Revello, M.G., and Gerna, G. (2002). The human cytomegalovirus ribonucleotide reductase homolog UL45 is dispensable for growth in endothelial cells, as determined by a BAC-cloned clinical isolate of human cytomegalovirus with preserved wild-type characteristics. J. Virol. 76, 9551–9555.

Hahn, G., Revello, M.G., Patrone, M., Percivalle, E., Campanini, G., Sarasini, A., Wagner, M., Gallina, A., Milanesi, G., Koszinowski, U., et al. (2004). Human cytomegalovirus UL131–128 genes are indispensable for virus growth in endothelial cells and virus transfer to leukocytes. J. Virol. 78, 10023–10033.

Halary, F., Amara, A., Lortat-Jacob, H., Messerle, M., Delaunay, T., Houles, C., Fieschi, F., Arenzana-Seisdedos, F., Moreau, J.F., and Dechanet-Merville, J. (2002). Human cytomegalovirus binding to DC-SIGN is required for dendritic cell infection and target cell trans-infection. Immunity 17, 653–664.

Halwachs-Baumann, G., Wilders-Truschnig, M., Desoye, G., Hahn, T., Kiesel, L., Klingel, K., Rieger, P., Jahn, G., and Sinzger, C. (1998). Human trophoblast cells are permissive to the complete replicative cycle of human cytomegalovirus. J. Virol. 72, 7598–7602.

Hamprecht, K., Witzel, S., Maschmann, J., Dietz, K., Baumeister, A., Mikeler, E., Goelz, R., Speer, C.P., and Jahn, G. (2003). Rapid detection and quantification of cell free cytomegalovirus by a high-speed centrifugation-based microculture assay: comparison to longitudinally analyzed viral DNA load and pp67 late transcript during lactation. J. Clin. Virol. 28, 303–316.

Hamprecht, K., Maschmann, J., Jahn, G., Poets, C.F., and Goelz, R. (2008). Cytomegalovirus transmission to preterm infants during lactation. J. Clin. Virol. 41, 198–205.

Hannah, B.P., Heldwein, E.E., Bender, F.C., Cohen, G.H., and Eisenberg, R.J. (2007). Mutational evidence of internal fusion loops in herpes simplex virus glycoprotein B. J. Virol. 81, 4858–4865.

Hanson, L.K., Slater, J.S., Karabekian, Z., Ciocco-Schmitt, G., and Campbell, A.E. (2001). Products of US22 genes M140 and M141 confer efficient replication of murine cytomegalovirus in macrophages and spleen. J. Virol. 75, 6292–6302.

Hanson, L.K., Slater, J.S., Cavanaugh, V.J., Newcomb, W.W., Bolin, L.L., Nelson, C.N., Fetters, L.D., Tang, Q., Brown, J.C., Maul, G.G., et al. (2009). Murine cytomegalovirus capsid assembly is dependent on US22 family gene M140 in infected macrophages. J. Virol. 83, 7449–7456.

Hassan-Walker, A.F., Mattes, F.M., Griffiths, P.D., and Emery, V.C. (2001). Quantity of cytomegalovirus DNA in different leukocyte populations during active infection in vivo and the presence of gB and UL18 transcripts. J. Med. Virol. 64, 283–289.

Heieren, M.H., Kim, Y.K., and Balfour, H.H., Jr. (1988). Human cytomegalovirus infection of kidney glomerular visceral epithelial and tubular epithelial cells in culture. Transplantation 46, 426–432.

Heldwein, E.E., and Krummenacher, C. (2008). Entry of herpesviruses into mammalian cells. Cell. Mol. Life Sci. 65, 1653–1668.

Heldwein, E.E., Lou, H., Bender, F.C., Cohen, G.H., Eisenberg, R.J., and Harrison, S.C. (2006). Crystal structure of glycoprotein B from herpes simplex virus 1. Science 313, 217–220.

Hendrix, M.G., Dormans, P.H., Kitslaar, P., Bosman, F., and Bruggeman, C.A. (1989). The presence of cytomegalovirus nucleic acids in arterial walls of atherosclerotic and nonatherosclerotic patients. Am. J. Pathol. 134, 1151–1157.

Hendrix, M.G., Salimans, M.M., van Boven, C.P., and Bruggeman, C.A. (1990). High prevalence of latently present cytomegalovirus in arterial walls of patients suffering from grade III atherosclerosis. Am. J. Pathol. 136, 23–28.

Henry, S.C., Schmader, K., Brown, T.T., Miller, S.E., Howell, D.N., Daley, G.G., and Hamilton, J.D. (2000). Enhanced green fluorescent protein as a marker for localizing murine cytomegalovirus in acute and latent infection. J. Virol. Methods 89, 61–73.

Henson, D., Smith, R.D., Gehrke, J., and Neapolitan, C. (1967). Effect of cortisone on nonfatal mouse cytomegalovirus infection. Am. J. Pathol. 51, 1001–1011.

Hertel, L., and Mocarski, E.S. (2004). Global analysis of host cell gene expression late during cytomegalovirus infection reveals extensive dysregulation of cell cycle gene expression and induction of Pseudomitosis independent of US28 function. J. Virol. 78, 11988–12011.

Holbrook, J.T., Jabs, D.A., Weinberg, D.V., Lewis, R.A., Davis, M.D., and Friedberg, D. (2003). Visual loss in patients with cytomegalovirus retinitis and acquired immunodeficiency syndrome before widespread availability of highly active antiretroviral therapy. Arch. Ophthalmol. 121, 99–107.

Horwitz, C.A., Burke, M.D., Grimes, P., and Tombers, J. (1980). Hepatic function in mononucleosis induced by Epstein–Barr virus and cytomegalovirus. Clin. Chem. 26, 243–246.

Hsu, K.M., Pratt, J.R., Akers, W.J., Achilefu, S.I., and Yokoyama, W.M. (2009). Murine cytomegalovirus displays selective infection of cells within hours after systemic administration. J. Gen. Virol. 90, 33–43.

Huber, M.T., and Compton, T. (1998). The human cytomegalovirus UL74 gene encodes the third component of the glycoprotein H-glycoprotein L-containing envelope complex. J. Virol. 72, 8191–8197.

Hutt-Fletcher, L.M. (2007). Epstein–Barr virus entry. J. Virol. 81, 7825–7832.

Ibig-Rehm, Y., Gotte, M., Gabriel, D., Woodhall, D., Shea, A., Brown, N.E., Compton, T., and Feire, A.L. (2011). High-content screening to distinguish between attachment and post-attachment steps of human cytomegalovirus entry into fibroblasts and epithelial cells. Antiviral Res. 89, 246–256.

Isaacson, M.K., and Compton, T. (2009). Human cytomegalovirus glycoprotein B is required for virus entry and cell-to-cell spread but not for virion attachment, assembly, or egress. J. Virol. 83, 3891–3903.

Isaacson, M.K., Feire, A.L., and Compton, T. (2007). Epidermal growth factor receptor is not required for human cytomegalovirus entry or signaling. J. Virol. 81, 6241–6247.

Jacobson, M.A., and Mills, J. (1988). Serious cytomegalovirus disease in the acquired immunodeficiency syndrome (AIDS). Clinical findings, diagnosis, and treatment. Ann. Intern. Med. 108, 585–594.

Jarvis, M.A., and Nelson, J.A. (2007). Human cytomegalovirus tropism for endothelial cells: not all endothelial cells are created equal. J. Virol. 81, 2095–2101.

Jault, F.M., Jault, J.M., Ruchti, F., Fortunato, E.A., Clark, C., Corbeil, J., Richman, D.D., and Spector, D.H. (1995). Cytomegalovirus infection induces high levels of cyclins, phosphorylated Rb, and p53, leading to cell cycle arrest. J. Virol. 69, 6697–6704.

Jiang, X.J., Adler, B., Sampaio, K.L., Digel, M., Jahn, G., Ettischer, N., Stierhof, Y.-D., Scrivano, L., Koszinowski, U.,

Mach, M., *et al.* (2008). UL74 of human cytomegalovirus contributes to virus release by promoting secondary envelopment of virions. J. Virol. *82*, 2802–2812.

Jiang, X.J., Sampaio, K.L., Ettischer, N., Stierhof, Y.D., Jahn, G., Kropff, B., Mach, M., and Sinzger, C. (2011). UL74 of human cytomegalovirus reduces the inhibitory effect of gH-specific and gB-specific antibodies. Arch. Virol. *156*, 2145–2155.

Jiwa, N.M., Raap, A.K., van de Rijke, F.M., Mulder, A., Weening, J.J., Zwaan, F.E., The, T.H., and van der Ploeg, M. (1989). Detection of cytomegalovirus antigens and DNA in tissues fixed in formaldehyde. J. Clin. Pathol. *42*, 749–754.

Kahl, M., Siegel-Axel, D., Stenglein, S., Jahn, G., and Sinzger, C. (2000). Efficient lytic infection of human arterial endothelial cells by human cytomegalovirus strains. J. Virol. *74*, 7628–7635.

Karabekian, Z., Hanson, L.K., Slater, J.S., Krishna, N.K., Bolin, L.L., Kerry, J.A., and Campbell, A.E. (2005). Complex formation among murine cytomegalovirus US22 proteins encoded by genes M139, M140, and M141. J. Virol. *79*, 3525–3535.

Keay, S., and Baldwin, B. (1991). Anti-idiotype antibodies that mimic gp86 of human cytomegalovirus inhibit viral fusion but not attachment. J. Virol. *65*, 5124–5128.

Kinzler, E.R., and Compton, T. (2005). Characterization of human cytomegalovirus glycoprotein-induced cell–cell fusion. J. Virol. *79*, 7827–7837.

Knight, D.A., Waldman, W.J., and Sedmak, D.D. (1997). Human cytomegalovirus does not induce human leukocyte antigen class II expression on arterial endothelial cells. Transplantation *63*, 1366–1369.

Knight, D.A., Waldman, W.J., and Sedmak, D.D. (1999). Cytomegalovirus-mediated modulation of adhesion molecule expression by human arterial and microvascular endothelial cells. Transplantation *68*, 1814–1818.

Kuttner, A.G., and Wang, S.H. (1934). The problem of the significance of the inclusion bodies found in the salivary glands of infants, and the occurrence of inclusion bodies in the submaxillary glands of hamsters, white mice, and wild rats (peiping). J. Exp. Med. *60*, 773–791.

Kuwabara, A., Okamoto, H., Suda, T., Ajioka, Y., and Hatakeyama, K. (2007). Clinicopathologic characteristics of clinically relevant cytomegalovirus infection in inflammatory bowel disease. J. Gastroenterol. *42*, 823–829.

von Laer, D., Serr, A., Meyer-König, U., Kirste, G., Hufert, F.T., and Haller, O. (1995). Human cytomegalovirus immediate-early and late transcripts are expressed in all major leukocyte populations in vivo. J. Infect. Dis. *172*, 365–370.

Lau, S.K., Chen, Y.-Y., Chen, W.-G., Diamond, D.J., Mamelak, A.N., Zaia, J.A., and Weiss, L.M. (2005). Lack of association of cytomegalovirus with human brain tumors. Mod. Pathol. *18*, 838–843.

Lawlor, G., and Moss, A.C. (2010). Cytomegalovirus in inflammatory bowel disease: pathogen or innocent bystander? Inflamm. Bowel Dis. *16*, 1620–1627.

Li, L., Nelson, J.A., and Britt, W.J. (1997). Glycoprotein H-related complexes of human cytomegalovirus: identification of a third protein in the gCIII complex. J. Virol. *71*, 3090–3097.

Lilja, A.E., Chang, W.L.W., Barry, P.A., Becerra, S.P., and Shenk, T.E. (2008a). Functional genetic analysis of rhesus cytomegalovirus: Rh01 is an epithelial cell tropism factor. J. Virol. *82*, 2170–2181.

Lilja, A.E., and Shenk, T. (2008b). Efficient replication of rhesus cytomegalovirus variants in multiple rhesus and human cell types. Proc. Natl. Acad. Sci. U.S.A. *105*, 19950–19955.

Lockridge, K.M., Sequar, G., Zhou, S.S., Yue, Y., Mandell, C.P., and Barry, P.A. (1999). Pathogenesis of experimental rhesus cytomegalovirus infection. J. Virol. *73*, 9576–9583.

Lu, M., and Shenk, T. (1996). Human cytomegalovirus infection inhibits cell cycle progression at multiple points, including the transition from G1 to S. J. Virol. *70*, 8850–8857.

Lucas, K.G., Bao, L., Bruggeman, R., Dunham, K., and Specht, C. (2011). The detection of CMV pp65 and IE1 in glioblastoma multiforme. J. Neurooncol. *103*, 231–238.

Macagno, A., Bernasconi, N.L., Vanzetta, F., Dander, E., Sarasini, A., Revello, M.G., Gerna, G., Sallusto, F., and Lanzavecchia, A. (2010). Isolation of human monoclonal antibodies that potently neutralize human cytomegalovirus infection by targeting different epitopes on the gH/gL/UL128-131A complex. J. Virol. *84*, 1005–1013.

Macasaet, F.F., Holley, K.E., Smith, T.F., and Keys, T.F. (1975). Cytomegalovirus studies of autopsy tissue. II. Incidence of inclusion bodies and related pathologic data. Am. J. Clin. Pathol. *63*, 859–865.

McCordock, H.A., and Smith, M.G. (1936). The visceral lesions produced in mice by the salivary gland virus of mice. J. Exp. Med. *63*, 303–310.

Mach, M., Kropff, B., Dal Monte, P., and Britt, W. (2000). Complex formation by human cytomegalovirus glycoproteins M (gpUL100) and N (gpUL73). J. Virol. *74*, 11881–11892.

Mach, M., Osinski, K., Kropff, B., Schloetzer-Schrehardt, U., Krzyzaniak, M., and Britt, W. (2007). The carboxy-terminal domain of glycoprotein N of human cytomegalovirus is required for virion morphogenesis. J. Virol. *81*, 5212–5224.

Mack, C., Sickmann, A., Lembo, D., and Brune, W. (2008). Inhibition of proinflammatory and innate immune signaling pathways by a cytomegalovirus RIP1-interacting protein. Proc. Natl. Acad. Sci. U.S.A. *105*, 3094–3099.

Marshall, L.J., and Major, E.O. (2010). Molecular regulation of JC virus tropism: insights into potential therapeutic targets for progressive multifocal leukoencephalopathy. J. Neuroimmune Pharmacol. *5*, 404–417.

Matsuura, H., Kirschner, A.N., Longnecker, R., and Jardetzky, T.S. (2010). Crystal structure of the Epstein–Barr virus (EBV) glycoprotein H/glycoprotein L (gH/gL) complex. Proc. Natl. Acad. Sci. U.S.A. *107*, 22641–22646.

Maury, W., Thompson, R.J., Jones, Q., Bradley, S., Denke, T., Baccam, P., Smazik, M., and Oaks, J.L. (2005). Evolution of the equine infectious anemia virus long terminal repeat during the alteration of cell tropism. J. Virol. *79*, 5653–5664.

Menard, C., Wagner, M., Ruzsics, Z., Holak, K., Brune, W., Campbell, A.E., and Koszinowski, U.H. (2003). Role of murine cytomegalovirus US22 gene family members in replication in macrophages. J. Virol. *77*, 5557–5570.

Meyer-König, U., Vogelberg, C., Bongarts, A., Kampa, D., Delbrück, R., Wolff-Vorbeck, G., Kirste, G., Haberland, M., Hufert, F.T., and von Laer, D. (1998). Glycoprotein B genotype correlates with cell tropism in vivo of human cytomegalovirus infection. J. Med. Virol. *55*, 75–81.

Michaelis, M., Baumgarten, P., Mittelbronn, M., Driever, P.H., Doerr, H.W., and Cinatl, J., Jr. (2011). Oncomodulation by human cytomegalovirus: novel clinical findings open new roads. Med. Microbiol. Immunol. *200*, 1–5.

Miller, N., and Hutt-Fletcher, L.M. (1992). Epstein–Barr virus enters B cells and epithelial cells by different routes. J. Virol. *66*, 3409–3414.

Mims, C.A. (1989). The pathogenetic basis of viral tropism. Am. J. Pathol. *135*, 447–455.

Mims, C.A., and Gould, J. (1979). Infection of salivary glands, kidneys, adrenals, ovaries and epithelia by murine cytomegalovirus. J. Med. Microbiol. *12*, 113–122.

Mitchell, D.A., Xie, W., Schmittling, R., Learn, C., Friedman, A., McLendon, R.E., and Sampson, J.H. (2008). Sensitive detection of human cytomegalovirus in tumors and peripheral blood of patients diagnosed with glioblastoma. Neuro Oncol. *10*, 10–18.

Mocarski, E.S., Shenk, T., and Pass, R.F. (2007). Cytomegaloviruses. In Fields Virology, Knipe, D.M., ed. (Lippincott Williams and Wilkins, Philadelphia, PA), pp. 2701–2772.

Mohr, C.A., Arapovic, J., Muhlbach, H., Panzer, M., Weyn, A., Dolken, L., Krmpotic, A., Voehringer, D., Ruzsics, Z., Koszinowski, U., *et al.* (2010). A spread-deficient cytomegalovirus for assessment of first-target cells in vaccination. J. Virol. *84*, 7730–7742.

Morgello, S., Cho, E.S., Nielsen, S., Devinsky, O., and Petito, C.K. (1987). Cytomegalovirus encephalitis in patients with acquired immunodeficiency syndrome: an autopsy study of 30 cases and a review of the literature. Hum. Pathol. *18*, 289–297.

Mori, Y., Akkapaiboon, P., Yang, X., and Yamanishi, K. (2003a). The human herpesvirus 6 U100 gene product is the third component of the gH–gL glycoprotein complex on the viral envelope. J. Virol. *77*, 2452–2458.

Mori, Y., Yang, X., Akkapaiboon, P., Okuno, T., and Yamanishi, K. (2003b). Human herpesvirus 6 variant A glycoprotein H-glycoprotein L-glycoprotein Q complex associates with human CD46. J. Virol. *77*, 4992–4999.

Mori, Y., Akkapaiboon, P., Yonemoto, S., Koike, M., Takemoto, M., Sadaoka, T., Sasamoto, Y., Konishi, S., Uchiyama, Y., and Yamanishi, K. (2004). Discovery of a second form of tripartite complex containing gH-gL of human herpesvirus 6 and observations on CD46. J. Virol. *78*, 4609–4616.

Muhlemann, K., Miller, R.K., Metlay, L., and Menegus, M.A. (1992). Cytomegalovirus infection of the human placenta: an immunocytochemical study. Hum. Pathol. *23*, 1234–1237.

Nerheim, P.L., Meier, J.L., Vasef, M.A., Li, W.G., Hu, L., Rice, J.B., Gavrila, D., Richenbacher, W.E., and Weintraub, N.L. (2004). Enhanced cytomegalovirus infection in atherosclerotic human blood vessels. Am. J. Pathol. *164*, 589–600.

Netterwald, J.R., Jones, T.R., Britt, W.J., Yang, S.-J., McCrone, I.P., and Zhu, H. (2004). Postattachment events associated with viral entry are necessary for induction of interferon-stimulated genes by human cytomegalovirus. J. Virol. *78*, 6688–6691.

Ng-Bautista, C.L., and Sedmak, D.D. (1995). Cytomegalovirus infection is associated with absence of alveolar epithelial cell HLA class II antigen expression. J. Infect. Dis. *171*, 39–44.

Nowlin, D.M., Cooper, N.R., and Compton, T. (1991). Expression of a human cytomegalovirus receptor

correlates with infectibility of cells. J. Virol. *65*, 3114–3121.

Oskarsson, T., Hreggvidsdottir, H.S., Agnarsdottir, G., Matthiasdottir, S., Ogmundsdottir, M.H., Jonsson, S.R., Georgsson, G., Ingvarsson, S., Andresson, O.S., and Andresdottir, V. (2007). Duplicated sequence motif in the long terminal repeat of maedi-visna virus extends cell tropism and is associated with neurovirulence. J. Virol. *81*, 4052–4057.

Oxford, K.L., Strelow, L., Yue, Y., Chang, W.L., Schmidt, K.A., Diamond, D.J., and Barry, P.A. (2011). Open reading frames carried on UL/b′ are implicated in shedding and horizontal transmission of rhesus cytomegalovirus in rhesus monkeys. J. Virol. *85*, 5105–5114.

Pachl, C., Probert, W.S., Hermsen, K.M., Masiarz, F.R., Rasmussen, L., Merigan, T.C., and Spaete, R.R. (1989). The human cytomegalovirus strain Towne glycoprotein H gene encodes glycoprotein p86. Virology *169*, 418–426.

Paterson, D.A., Dyer, A.P., Milne, R.S., Sevilla-Reyes, E., and Gompels, U.A. (2002). A role for human cytomegalovirus glycoprotein O (gO) in cell fusion and a new hypervariable locus. Virology *293*, 281–294.

Patrone, M., Percivalle, E., Secchi, M., Fiorina, L., Pedrali-Noy, G., Zoppe, M., Baldanti, F., Hahn, G., Koszinowski, U.H., Milanesi, G., *et al.* (2003). The human cytomegalovirus UL45 gene product is a late, virion-associated protein and influences virus growth at low multiplicities of infection. J. Gen. Virol. *84*, 3359–3370.

Patrone, M., Secchi, M., Bonaparte, E., Milanesi, G., and Gallina, A. (2007). Cytomegalovirus UL131-128 products promote gB conformational transition and gB–gH interaction during entry into endothelial cells. J. Virol. *81*, 11479–11488.

Peek, R., Verbraak, F., Bruinenberg, M., Van der Lelij, A., Van den Horn, G., and Kijlstra, A. (1998). Cytomegalovirus glycoprotein B genotyping in ocular fluids and blood of AIDS patients with cytomegalovirus retinitis. Invest. Ophthalmol. Vis. Sci. *39*, 1183–1187.

Petrik, D.T., Schmitt, K.P., and Stinski, M.F. (2006). Inhibition of cellular DNA synthesis by the human cytomegalovirus IE86 protein is necessary for efficient virus replication. J. Virol. *80*, 3872–3883.

Pietropaolo, R., and Compton, T. (1999). Interference with annexin II has no effect on entry of human cytomegalovirus into fibroblast cells. J. Gen. Virol. *80*, 1807–1816.

Pietropaolo, R.L., and Compton, T. (1997). Direct interaction between human cytomegalovirus glycoprotein B and cellular annexin II. J. Virol. *71*, 9803–9807.

Pignatelli, S., Dal Monte, P., and Landini, M.P. (2001). gpUL73 (gN) genomic variants of human cytomegalovirus isolates are clustered into four distinct genotypes. J. Gen. Virol. *82*, 2777–2784.

Pignatelli, S., Dal Monte, P., Rossini, G., Camozzi, D., Toscano, V., Conte, R., and Landini, M.P. (2006). Latency-associated human cytomegalovirus glycoprotein N genotypes in monocytes from healthy blood donors. Transfusion *46*, 1754–1762.

Pignatelli, S., Lazzarotto, T., Gatto, M.R., Dal Monte, P., Landini, M.P., Faldella, G., and Lanari, M. (2010). Cytomegalovirus gN genotypes distribution among congenitally infected newborns and their relationship with symptoms at birth and sequelae. Clin. Infect. Dis. *51*, 33–41.

Podlech, J., Holtappels, R., Wirtz, N., Steffens, H.P., and Reddehase, M.J. (1998). Reconstitution of CD8 T-cells is essential for the prevention of multiple-organ cytomegalovirus histopathology after bone marrow transplantation. J. Gen. Virol. 79, 2099–2104.

Poland, S.D., Bambrick, L.L., Dekaban, G.A., and Rice, G.P. (1994). The extent of human cytomegalovirus replication in primary neurons is dependent on host cell differentiation. J. Infect. Dis. 170, 1267–1271.

Polic, B., Jonjic, S., Pavic, I., Crnkovic, I., Zorica, I., Hengel, H., Lucin, P., and Koszinowski, U.H. (1996). Lack of MHC class I complex expression has no effect on spread and control of cytomegalovirus infection in vivo. J. Gen. Virol. 77, 217–225.

van den Pol, A.N., Reuter, J.D., and Santarelli, J.G. (2002). Enhanced cytomegalovirus infection of developing brain independent of the adaptive immune system. J. Virol. 76, 8842–8854.

Pötzsch, S., Spindler, N., Wiegers, A.-K., Fisch, T., Rücker, P., Sticht, H., Grieb, N., Baroti, T., Weisel, F., Stamminger, T., et al. (2011). B cell repertoire analysis identifies new antigenic domains on glycoprotein B of human cytomegalovirus which are target of neutralizing antibodies. PLoS Pathog. 7, e1002172–e1002172.

Qian, Z., Leung-Pineda, V., Xuan, B., Piwnica-Worms, H., and Yu, D. (2010). Human cytomegalovirus protein pUL117 targets the mini-chromosome maintenance complex and suppresses cellular DNA synthesis. PLoS Pathog. 6, e1000814.

Randolph-Habecker, J., Iwata, M., and Torok-Storb, B. (2002). Cytomegalovirus mediated myelosuppression. J. Clin. Virol. 25(Suppl. 2), S51–S56.

Rasing, L.A., De Weger, R.A., Verdonck, L.F., van der Bij, W., Compier-Spies, P.I., De Gast, G.C., Van Basten, C.D., and Schuurman, H.J. (1990). The value of immunohistochemistry and in situ hybridization in detecting cytomegalovirus in bone marrow transplant recipients. APMIS 98, 479–488.

Rasmussen, L., Geissler, A., Cowan, C., Chase, A., and Winters, M. (2002). The genes encoding the gCIII complex of human cytomegalovirus exist in highly diverse combinations in clinical isolates. J. Virol. 76, 10841–10848.

Rasmussen, L., Hong, C., Zipeto, D., Morris, S., Sherman, D., Chou, S., Miner, R., Drew, W.L., Wolitz, R., Dowling, A., et al. (1997). Cytomegalovirus gB genotype distribution differs in human immunodeficiency virus-infected patients and immunocompromised allograft recipients. J. Infect. Dis. 175, 179–184.

Rasmussen, L.E., Nelson, R.M., Kelsall, D.C., and Merigan, T.C. (1984). Murine monoclonal antibody to a single protein neutralizes the infectivity of human cytomegalovirus. Proc. Natl. Acad. Sci. U.S.A. 81, 876–880.

Reddehase, M.J., Weiland, F., Münch, K., Jonjic, S., Lüske, A., and Koszinowski, U.H. (1985). Interstitial murine cytomegalovirus pneumonia after irradiation: characterization of cells that limit viral replication during established infection of the lungs. J. Virol. 55, 264–273.

Reinhardt, B., Vaida, B., Voisard, R., Keller, L., Breul, J., Metzger, H., Herter, T., Baur, R., Luske, A., and Mertens, T. (2003). Human cytomegalovirus infection in human renal arteries in vitro. J. Virol. Methods 109, 1–9.

Reuter, J.D., Gomez, D.L., Wilson, J.H., and Van Den Pol, A.N. (2004). Systemic immune deficiency necessary for cytomegalovirus invasion of the mature brain. J. Virol. 78, 1473–1487.

Revello, M.G., and Gerna, G. (2010). Human cytomegalovirus tropism for endothelial/epithelial cells: scientific background and clinical implications. Rev. Med. Virol. 20, 136–155.

Revello, M.G., Percivalle, E., Arbustini, E., Pardi, R., Sozzani, S., and Gerna, G. (1998). In vitro generation of human cytomegalovirus pp65 antigenemia, viremia, and leukoDNAemia. J. Clin. Invest. 101, 2686–2692.

Riegler, S., Hebart, H., Einsele, H., Brossart, P., Jahn, G., and Sinzger, C. (2000). Monocyte-derived dendritic cells are permissive to the complete replicative cycle of human cytomegalovirus. J. Gen. Virol. 81, 393–399.

Roberts, W.H., Sneddon, J.M., Waldman, J., and Stephens, R.E. (1989). Cytomegalovirus infection of gastrointestinal endothelium demonstrated by simultaneous nucleic acid hybridization and immunohistochemistry. Arch. Path. Lab. Med. 113, 461–464.

Roubalova, K., Strunecky, O., Vitek, A., Zufanova, S., and Prochazka, B. (2011). Genetic variability of cytomegalovirus glycoprotein O in hematopoietic stem cell transplant recipients. Transpl. Infect. Dis. 13, 237–243.

Roullet, E. (1999). Opportunistic infections of the central nervous system during HIV-1 infection (emphasis on cytomegalovirus disease). J. Neurol. 246, 237–243.

Rue, C.A., Jarvis, M.A., Knoche, A.J., Meyers, H.L., DeFilippis, V.R., Hansen, S.G., Wagner, M., Früh, K., Anders, D.G., Wong, S.W., et al. (2004). A cyclooxygenase-2 homologue encoded by rhesus cytomegalovirus is a determinant for endothelial cell tropism. J. Virol. 78, 12529–12536.

Ryckman, B.J., Chase, M.C., and Johnson, D.C. (2008). HCMV gH/gL/UL128-131 interferes with virus entry into epithelial cells: evidence for cell type-specific receptors. Proc. Natl. Acad. Sci. U.S.A. 105, 14118–14123.

Ryckman, B.J., Jarvis, M.A., Drummond, D.D., Nelson, J.A., and Johnson, D.C. (2006). Human cytomegalovirus entry into epithelial and endothelial cells depends on genes UL128 to UL150 and occurs by endocytosis and low-pH fusion. J. Virol. 80, 710–722.

Sabatier, J., Uro-Coste, E., Pommepuy, I., Labrousse, F., Allart, S., Trémoulet, M., Delisle, M.B., and Brousset, P. (2005). Detection of human cytomegalovirus genome and gene products in central nervous system tumours. Br. J. Cancer 92, 747–750.

Saccoccio, F.M., Sauer, A.L., Cui, X., Armstrong, A.E., Habib, E.-S.E., Johnson, D.C., Ryckman, B.J., Klingelhutz, A.J., Adler, S.P., and McVoy, M.A. (2011). Peptides from cytomegalovirus UL130 and UL131 proteins induce high titer antibodies that block viral entry into mucosal epithelial cells. Vaccine 29, 2705–2711.

Sacher, T., Podlech, J., Mohr, C.A., Jordan, S., Ruzsics, Z., Reddehase, M.J., and Koszinowski, U.H. (2008). The major virus-producing cell type during murine cytomegalovirus infection, the hepatocyte, is not the source of virus dissemination in the host. Cell Host Microbe 3, 263–272.

Saederup, N., Aguirre, S.A., Sparer, T.E., Bouley, D.M., and Mocarski, E.S. (2001). Murine cytomegalovirus CC chemokine homolog MCK-2 (m131-129) is a determinant of dissemination that increases inflammation at initial sites of infection. J. Virol. 75, 9966–9976.

Scheurer, M.E., Bondy, M.L., Aldape, K.D., Albrecht, T., and El-Zein, R. (2008). Detection of human cytomegalovirus in different histological types of gliomas. Acta Neuropathol. *116*, 79–86.

Schrier, R.D., Nelson, J.A., and Oldstone, M.B. (1985). Detection of human cytomegalovirus in peripheral blood lymphocytes in a natural infection. Science *230*, 1048–1051.

Scrivano, L., Esterlechner, J., Muhlbach, H., Ettischer, N., Hagen, C., Grunewald, K., Mohr, C.A., Ruzsics, Z., Koszinowski, U., and Adler, B. (2010). The m74 gene product of murine cytomegalovirus (MCMV) is a functional homolog of human CMV gO and determines the entry pathway of MCMV. J. Virol. *84*, 4469–4480.

Scrivano, L., Sinzger, C., Nitschko, H., Koszinowski, U.H., and Adler, B. (2011). HCMV spread and cell tropism are determined by distinct virus populations. PLoS Pathog. *7*, e1001256.

Shimamura, M., Mach, M., and Britt, W.J. (2006). Human cytomegalovirus infection elicits a glycoprotein M (gM)/gN-specific virus-neutralizing antibody response. J. Virol. *80*, 4591–4600.

Sinclair, J. (2010). Chromatin structure regulates human cytomegalovirus gene expression during latency, reactivation and lytic infection. Biochim. Biophys. Acta *1799*, 286–295.

Sinzger, C. (2008). Entry route of HCMV into endothelial cells. J. Clin. Virol. *41*, 174–179.

Sinzger, C., Muntefering, H., Loning, T., Stoss, H., Plachter, B., and Jahn, G. (1993). Cell types infected in human cytomegalovirus placentitis identified by immunohistochemical double staining. Virchows Arch. A Pathol. Anat. Histopathol. *423*, 249–256.

Sinzger, C., Grefte, A., Plachter, B., Gouw, A.S., The, T.H., and Jahn, G. (1995). Fibroblasts, epithelial cells, endothelial cells and smooth muscle cells are major targets of human cytomegalovirus infection in lung and gastrointestinal tissues. J. Gen. Virol. *76*, 741–750.

Sinzger, C., Plachter, B., Grefte, A., The, T.H., and Jahn, G. (1996). Tissue macrophages are infected by human cytomegalovirus in vivo. J. Infect. Dis. *173*, 240–245.

Sinzger, C., Bissinger, A.L., Viebahn, R., Oettle, H., Radke, C., Schmidt, C.A., and Jahn, G. (1999a). Hepatocytes are permissive for human cytomegalovirus infection in human liver cell culture and In vivo. J. Infect. Dis. *180*, 976–986.

Sinzger, C., Schmidt, K., Knapp, J., Kahl, M., Beck, R., Waldman, J., Hebart, H., Einsele, H., and Jahn, G. (1999b). Modification of human cytomegalovirus tropism through propagation in vitro is associated with changes in the viral genome. J. Gen. Virol. *80*, 2867–2877.

Sinzger, C., Eberhardt, K., Cavignac, Y., Weinstock, C., Kessler, T., Jahn, G., and Davignon, J.L. (2006). Macrophage cultures are susceptible to lytic productive infection by endothelial-cell-propagated human cytomegalovirus strains and present viral IE1 protein to CD4+ T-cells despite late down-regulation of MHC class II molecules. J. Gen. Virol. *87*, 1853–1862.

Sinzger, C., Digel, M., and Jahn, G. (2008). Cytomegalovirus cell tropism. Curr. Top. Microbiol. Immunol. *325*, 63–83.

Smith, J.D. (1986). Human cytomegalovirus: demonstration of permissive epithelial cells and nonpermissive fibroblastic cells in a survey of human cell lines. J. Virol. *60*, 583–588.

Söderberg-Nauclér, C. (2006). Does cytomegalovirus play a causative role in the development of various inflammatory diseases and cancer? J. Intern. Med. *259*, 219–246.

Söderberg, C., Giugni, T.D., Zaia, J.A., Larsson, S., Wahlberg, J.M., and Möller, E. (1993a). CD13 (human aminopeptidase N) mediates human cytomegalovirus infection. J. Virol. *67*, 6576–6585.

Söderberg, C., Larsson, S., Bergstedt-Lindqvist, S., and Möller, E. (1993b). Definition of a subset of human peripheral blood mononuclear cells that are permissive to human cytomegalovirus infection. J. Virol. *67*, 3166–3175.

Soroceanu, L., Akhavan, A., and Cobbs, C.S. (2008). Platelet-derived growth factor-alpha receptor activation is required for human cytomegalovirus infection. Nature *455*, 391–395.

Spriggs, M.K., Armitage, R.J., Comeau, M.R., Strockbine, L., Farrah, T., Macduff, B., Ulrich, D., Alderson, M.R., Mullberg, J., and Cohen, J.I. (1996). The extracellular domain of the Epstein–Barr virus BZLF2 protein binds the HLA-DR beta chain and inhibits antigen presentation. J. Virol. *70*, 5557–5563.

Staczek, J. (1990). Animal cytomegaloviruses. Microbiol. Rev. *54*, 247–265.

Stagno, S., Reynolds, D.W., Amos, C.S., Dahle, A.J., McCollister, F.P., Mohindra, I., Ermocilla, R., and Alford, C.A. (1977). Auditory and visual defects resulting from symptomatic and subclinical congenital cytomegaloviral and toxoplasma infections. Pediatrics *59*, 669–678.

Stals, F.S., Bosman, F., van Boven, C.P., and Bruggeman, C.A. (1990). An animal model for therapeutic intervention studies of CMV infection in the immunocompromised host. Arch. Virol. *114*, 91–107.

Stals, F.S., Steinhoff, G., Wagenaar, S.S., van Breda Vriesman, J.P., Haverich, A., Dormans, P., Moeller, F., and Bruggeman, C.A. (1996). Cytomegalovirus induces interstitial lung disease in allogeneic bone marrow transplant recipient rats independent of acute graft-versus-host response. Lab. Invest. *74*, 343–352.

Stein, J., Volk, H.D., Liebenthal, C., Krüger, D.H., and Prösch, S. (1993). Tumour necrosis factor alpha stimulates the activity of the human cytomegalovirus major immediate-early enhancer/promoter in immature monocytic cells. J. Gen. Virol. *74*, 2333–2338.

Stinski, M.F., and Meier, J.L. (2007). Immediate-early viral gene regulation and function. In Human Herpesviruses: Biology, Therapy, and Immunoprophylaxis (Cambridge University Press, Cambridge, UK).

Strickler, J.G., Manivel, J.C., Copenhaver, C.M., and Kubic, V.L. (1990). Comparison of in situ hybridization and immunohistochemistry for detection of cytomegalovirus and herpes simplex virus. Hum. Pathol. *21*, 443–448.

Sun, Z.R., Ji, Y.H., Ruan, Q., He, R., Ma, Y.P., Qi, Y., Mao, Z.Q., and Huang, Y.J. (2009). Structure characterization of human cytomegalovirus UL131A, UL130 and UL128 genes in clinical strains in China. Genet. Mol. Res. *8*, 1191–1201.

Tamura, T., Chiba, S., Chiba, Y., and Nakao, T. (1980). Virus excretion and neutralizing antibody response in saliva in human cytomegalovirus infection. Infect. Immun. *29*, 842–845.

Tang, Q., and Maul, G.G. (2006). Mouse cytomegalovirus crosses the species barrier with help from a few human cytomegalovirus proteins. J. Virol. *80*, 7510–7521.

Topilko, A., and Michelson, S. (1994). Hyperimmediate entry of human cytomegalovirus virions and dense bodies into human fibroblasts. Res. Virol. *145*, 75–82.

Torok-Storb, B., Boeckh, M., Hoy, C., Leisenring, W., Myerson, D., and Gooley, T. (1997). Association of specific cytomegalovirus genotypes with death from myelosuppression after marrow transplantation. Blood 90, 2097–2102.

Travi, G., Pergam, S.A., Xie, H., Carpenter, P., Kiem, H.P., Corey, L., and Boeckh, M.J. (2009). The effect of imatinib on cytomegalovirus reactivation in hematopoietic cell transplantation. Clin. Infect. Dis. 49, e120–123.

Tugizov, S., Maidji, E., and Pereira, L. (1996). Role of apical and basolateral membranes in replication of human cytomegalovirus in polarized retinal pigment epithelial cells. J. Gen. Virol. 77, 61–74.

Upton, J.W., Kaiser, W.J., and Mocarski, E.S. (2008). Cytomegalovirus M45 cell death suppression requires receptor-interacting protein (RIP) homotypic interaction motif (RHIM)-dependent interaction with RIP1. J. Biol. Chem. 283, 16966–16970.

Upton, J.W., Kaiser, W.J., and Mocarski, E.S. (2010). Virus inhibition of RIP3-dependent necrosis. Cell Host Microbe 7, 302–313.

Vanarsdall, A.L., Ryckman, B.J., Chase, M.C., and Johnson, D.C. (2008). Human cytomegalovirus glycoproteins gB and gH/gL mediate epithelial cell–cell fusion when expressed either in cis or in trans. J. Virol. 82, 11837–11850.

Vanarsdall, A.L., Chase, M.C., and Johnson, D.C. (2011). Human cytomegalovirus glycoprotein gO gomplexes with gH/gL, promoting interference with viral entry into human fibroblasts but not entry into epithelial cells. J. Virol. 85, 11638–11645.

Varnum, S.M., Streblow, D.N., Monroe, M.E., Smith, P., Auberry, K.J., Pasa-Tolic, L., Wang, D., Camp, D.G., 2nd, Rodland, K., Wiley, S., et al. (2004). Identification of proteins in human cytomegalovirus (HCMV) particles: the HCMV proteome. J. Virol. 78, 10960–10966.

Vogel, F.S. (1958). Enhanced susceptibility of proliferating endothelium to salivary gland virus under naturally occurring and experimental conditions. Am. J. Pathol. 34, 1069–1079.

Vogelberg, C., Meyer-König, U., Hufert, F.T., Kirste, G., and von Laer, D. (1996). Human cytomegalovirus glycoprotein B genotypes in renal transplant recipients. J. Med. Virol. 50, 31–34.

Waldman, W.J., Knight, D.A., Huang, E.H., and Sedmak, D.D. (1995). Bidirectional transmission of infectious cytomegalovirus between monocytes and vascular endothelial cells: an in vitro model. J. Infect. Dis. 171, 263–272.

Wang, D., and Shenk, T. (2005a). Human cytomegalovirus UL131 open reading frame is required for epithelial cell tropism. J. Virol. 79, 10330–10338.

Wang, D., and Shenk, T. (2005b). Human cytomegalovirus virion protein complex required for epithelial and endothelial cell tropism. Proc. Natl. Acad. Sci. U.S.A. 102, 18153–18158.

Wang, X., Huong, S.-M., Chiu, M.L., Raab-Traub, N., and Huang, E.-S. (2003). Epidermal growth factor receptor is a cellular receptor for human cytomegalovirus. Nature 424, 456–461.

Wang, X., Huang, D.Y., Huong, S.-M., and Huang, E.-S. (2005). Integrin alphavbeta3 is a coreceptor for human cytomegalovirus. Nat. Med. 11, 515–521.

West, J.T., and Wood, C. (2003). The role of Kaposi's sarcoma-associated herpesvirus/human herpesvirus-8 regulator of transcription activation (RTA) in control of gene expression. Oncogene 22, 5150–5163.

White, M.K., Safak, M., and Khalili, K. (2009). Regulation of gene expression in primate polyomaviruses. J. Virol. 83, 10846–10856.

Wiebusch, L., and Hagemeier, C. (1999). Human cytomegalovirus 86-kilodalton IE2 protein blocks cell cycle progression in G(1). J. Virol. 73, 9274–9283.

Wiebusch, L., and Hagemeier, C. (2001). The human cytomegalovirus immediate-early 2 protein dissociates cellular DNA synthesis from cyclin-dependent kinase activation. EMBO J. 20, 1086–1098.

Wiebusch, L., Uecker, R., and Hagemeier, C. (2003). Human cytomegalovirus prevents replication licensing by inhibiting MCM loading onto chromatin. EMBO Rep. 4, 42–46.

Wiebusch, L., Neuwirth, A., Grabenhenrich, L., Voigt, S., and Hagemeier, C. (2008). Cell cycle-independent expression of immediate-early gene 3 results in G1 and G2 arrest in murine cytomegalovirus-infected cells. J. Virol. 82, 10188–10198.

Wiley, C.A., and Nelson, J.A. (1988). Role of human immunodeficiency virus and cytomegalovirus in AIDS encephalitis. Am. J. Pathol. 133, 73–81.

Wille, P.T., Knoche, A.J., Nelson, J.A., Jarvis, M.A., and Johnson, D.C. (2010). A human cytomegalovirus gO-null mutant fails to incorporate gH/gL into the virion envelope and is unable to enter fibroblasts and epithelial and endothelial cells. J. Virol. 84, 2585–2596.

Wolff, D., Sinzger, C., Drescher, P., Jahn, G., and Plachter, B. (1994). Reduced levels of IE2 gene expression and shutdown of early and late viral genes during latent infection of the glioblastoma cell line U138-MG with selectable recombinants of human cytomegalovirus. Virology 204, 101–113.

Woo, P.C., Lo, C.Y., Lo, S.K., Siau, H., Peiris, J.S., Wong, S.S., Luk, W.K., Chan, T.M., Lim, W.W., and Yuen, K.Y. (1997). Distinct genotypic distributions of cytomegalovirus (CMV) envelope glycoprotein in bone marrow and renal transplant recipients with CMV disease. Clin. Diagn. Lab. Immunol. 4, 515–518.

Wright, J.F., Kurosky, A., Pryzdial, E.L., and Wasi, S. (1995). Host cellular annexin II is associated with cytomegalovirus particles isolated from cultured human fibroblasts. J. Virol. 69, 4784–4791.

Wykes, M.N., Shellam, G.R., McCluskey, J., Kast, W.M., Dallas, P.B., and Price, P. (1993). Murine cytomegalovirus interacts with major histocompatibility complex class I molecules to establish cellular infection. J. Virol. 67, 4182–4189.

Yamamoto, A.Y., Mussi-Pinhata, M.M., Isaac, M.D., Amaral, F.R., Carvalheiro, C.G., Aragon, D.C., Manfredi, A.K., Boppana, S.B., and Britt, W.J. (2011). Congenital cytomegalovirus infection as a cause of sensorineural hearing loss in a highly immune population. Pediatr. Infect. Dis. J. 30, 1043–1046.

Yamane, Y., Furukawa, T., and Plotkin, S.A. (1983). Supernatant virus release as a differentiating marker between low passage and vaccine strains of human cytomegalovirus. Vaccine 1, 23–25.

Molecular Basis of Cytomegalovirus Host Species Specificity

I.18

Wolfram Brune

Abstract

Cytomegaloviruses (CMVs) are highly species specific as they replicate almost exclusively in cells of their natural host species. However, the molecular basis of species specificity remains poorly understood. In cells of a foreign host a post-penetration block to viral gene expression and genome replication appears to restrict viral replication and spread. In some cases, infected cells of a foreign host undergo programmed cell death, indicating that apoptosis acts as a cellular antiviral defence mechanism to prevent viral replication. A few recent studies suggested that mediator and effector molecules of the interferon system and antiviral defences operating at PML nuclear bodies (PML-NBs) might also be involved in restricting the host range of CMVs. Moreover, a recently isolated spontaneous mutant of murine CMV, which is capable of replicating to high titres in human cells, provided a new opportunity to study the mechanisms of CMV host species specificity. In this spontaneously adapted virus, mutations in the region encoding the viral Early1 (E1) proteins were found to be responsible for the extended host range phenotype. Further investigations of the CMV host species specificity should lead to a better understanding of the viral replication machinery, interfering host cell factors, and viral countermeasures.

Introduction

Host specificity is an important concept that underlies the interactions of many pathogens with their hosts. Changes in the host specificity of animal pathogens can be dangerous due to their potential impact on human health (Kirzinger and Stavrinides, 2012). Sequence analyses of pathogen genomes have revealed that host switching or host jumps can often be traced to modifications of key microbial pathogenicity factors. In fact, the functional importance of some pathogenicity factors has only been fully recognized after the analysis of adaptive mutations that have occurred during host jumps.

CMVs are generally not thought to cause zoonotic infections, even though a few cases of human infection with CMVs of nonhuman primates have been reported (Huang et al., 1978; Martin et al., 1994; Michaels et al., 2001; see also Chapter II.22). Nevertheless, the question of why CMVs replicate almost exclusively in cells of their own host species has puzzled scientists for more than 50 years. In the 1950s, when CMVs – then called salivary gland virus (SGV) – were first isolated and propagated in cell culture, little was known about the biochemical and biophysical properties of these viruses. However, Margaret Smith, who described the discovery of murine SGV in 1954 and human SGV in 1956 (Smith, 1954, 1956), already realized that these two viruses have an obvious selectivity for cells of their own host species. As disclosed many years later in a review article by HCMV co-discoverer Thomas H. Weller (Weller, 1970), Margaret Smith had observed early on that MCMV could not be propagated in human cells, and HCMV could not be propagated in murine cells. In fact she was dismayed that her manuscript describing the first isolation of HCMV in cell culture was initially rejected because the reviewers thought she might have accidentally grown her previously described mouse SGV as a contaminant in her culture of human uterine tissue (see also Editor's Preface).

Over the years, Margaret Smith's observation was corroborated by other investigators and the host species specificity of the CMVs has become a truism that is regularly mentioned in textbooks, scientific articles, and presentations. It refers not only to observations that CMVs do not naturally infect and cause disease in a foreign host, but also to the inability of these viruses to replicate in cells of a foreign host in cell culture. However, this host cell specificity has exceptions. For instance, simian and rhesus CMVs can replicate in human fibroblasts (Black et al., 1963; Lafemina and Hayward, 1988; Alcendor et al., 1993; Rivailler et al.,

2006; Lilja and Shenk, 2008), and HCMV can replicate in chimpanzee skin fibroblasts (Perot et al., 1992; see also Chapter II.22). Similarly, murine cytomegalovirus (MCMV) productively infects rat cells (Reed et al., 1975; Bruggeman et al., 1982; Smith et al., 1986), but the rat cytomegalovirus (RCMV) Maastricht isolate does not replicate in murine fibroblasts (Bruggeman et al., 1982; see also Chapter II.15). In general, cells of other more distant species are non-permissive, even though a few exceptions have been reported: CMV isolates from a field mouse and a roof rat were reported to replicate and cause CPE in hamster kidney cells (Rabson et al., 1969; Raynaud et al., 1969), and a presumed CMV of horses was able to spread in rabbit cells (Hsiung et al., 1969). However, confirmations for these reports are missing. Moreover, it is unclear whether these viruses actually were CMVs as their classification was based merely on a typical CPE in infected cells and buoyant density of the viral particle.

There are also anecdotal reports of cross-species (zoonotic) infections of humans with simian or baboon CMVs (Huang et al., 1978; Martin et al., 1994; Michaels et al., 2001), but in two of the three cases the source of infection remained enigmatic. Moreover, phylogenetic analyses suggested interspecies transfer of CMVs between chimpanzees and gorillas (Leendertz et al., 2009; see also Chapter II.22). There is also a report of an experimental infection of laboratory mice (*Mus musculus*) with a field mouse (*Apodemus sylvaticus*) isolate (Raynaud and Barreau, 1965). Considering the fact that humans, chimpanzees, and gorillas represent, in terms of mammalian taxonomy, different genera within a subfamily (the same is true for house mice, field mice, and rats), it might seem more appropriate to call the CMVs 'subfamily specific' rather than species specific. However, since the term 'species specificity' has become a widely used and generally accepted term, I shall continue using it throughout this chapter.

In the 1970s and 1980s, a number of investigations on the mechanism of CMV species specificity were undertaken. These studies have shown that CMVs can enter cells of other species and express a subset of viral genes, predominantly of the immediate-early (IE) class (Kim and Carp, 1971, 1972; Fioretti et al., 1973; Kim et al., 1974; Lafemina and Hayward, 1988). These findings have led to the conclusion that the restriction to CMV replication in non-permissive cells is associated with a postpenetration block to viral gene expression and DNA replication, but is not due to a failure to enter the cell (Mocarski and Courcelle, 2001). Further in-depth studies of CMV species specificity were not published for many years, suggesting that investigations of this important question had been abandoned. The reason for abandoning this line of research was probably the difficulty to determine the underlying mechanism(s) with the tools available at that time.

More recently, this topic has again received increased attention as studies on the species specificity of other viruses, such as HIV or influenza virus, have provided intriguing new insights into cellular defence mechanisms and viral strategies to evade or counteract them (reviewed in Huthoff and Towers, 2008; Taubenberger and Kash, 2010). In recent years, a few studies have been published that addressed directly or indirectly the mechanisms of CMV host species specificity. Although significant new insights have been gained, we are still far away from completely understanding the molecular mechanisms that confine the CMVs to cells of their respective natural host. In the following paragraphs, the advances of the last few years will be summarized, and future directions will be discussed.

The role of apoptosis in host cell restriction

The first indication for a role of apoptosis in host cell restriction came from the observation that MCMV and RCMV induce apoptosis when they infect human fibroblasts or retinal epithelial cells (Jurak and Brune, 2006). Induction of apoptosis prevented a sustained replication of these viruses in human cells and reduced progeny production to insignificant levels. Apoptosis induction was strongly reduced when viral DNA replication was inhibited, indicating that viral DNA replication or events occurring after DNA replication were responsible for apoptosis induction. Surprisingly, much less apoptosis and a limited degree of MCMV replication was detected upon infection of HEK 293 (human embryonic kidney) cells or 911 human embryonic retinoblasts (Jurak and Brune, 2006). These two human cell lines have in common that they were produced by transformation with adenovirus type 5 DNA and express genes of the adenoviral E1 region (Graham et al., 1977; Fallaux et al., 1996). Further experiments demonstrated that expression of the adenoviral E1B-19k protein was sufficient to suppress apoptosis induced by MCMV infection and facilitate MCMV replication (Jurak and Brune, 2006). E1B-19k is an antiapoptotic protein with structural and functional similarity to Bcl-2. Indeed, upon overexpression of the antiapoptotic protein Bcl-2 or a functionally similar protein of cellular or viral origin, MCMV was able to replicate to significant titres in human cells (Fig. I.18.1). These results indicated that the induction of apoptosis is an important – albeit not the only – limitation to CMV cross-species infections (Jurak and Brune, 2006).

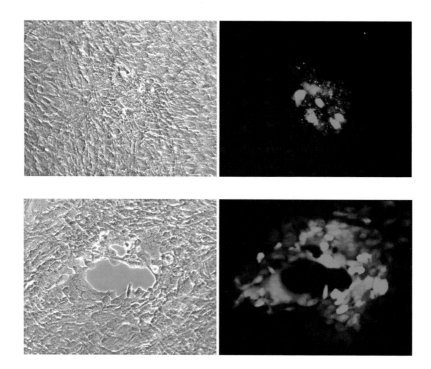

Figure I.18.1 Phase contrast and fluorescent images of human RPE-1 cells 8 days after a low-MOI infection with a GFP-expressing MCMV (upper panels) and a mutant virus expressing HCMV vMIA, a protein with a bcl-2-like function (lower panels). In the presence of HCMV vMIA, efficient spread of the infection and plaque formation was observed. In cells infected with the control virus (upper panels), spread was very limited, and cell fragmentation indicative of apoptosis was observed.

MCMV and RCMV express their own apoptosis inhibitors, among them proteins with functional similarity to Bcl-2 (Brune, 2011; see also Chapter I.15). So why do these viruses still induce apoptosis in human cells? Work from different laboratories has shown that the mitochondrial apoptosis inhibitors of MCMV (encoded by genes m38.5 and m41.1) have a weaker activity in human cells than in mouse cells (McCormick *et al.*, 2005; Arnoult *et al.*, 2008; Jurak *et al.*, 2008; Çam *et al.*, 2010; Brune, 2011). Conversely, the HCMV 'mitochondrion-localized inhibitor of apoptosis' (vMIA, pUL37x1) inhibits only Bax- but not Bak-mediated apoptosis in murine cells (Arnoult *et al.*, 2004; Poncet *et al.*, 2004; Çam *et al.*, 2010). However, increased apoptosis is not a common observation when, for instance, murine fibroblasts are infected with HCMV, suggesting that apoptosis induction might be of lesser importance during HCMV infection of murine cells. On the other hand, HCMV expresses only a limited set of IE and E genes and does not replicate its DNA in murine fibroblasts (Lafemina and Hayward, 1988). Hence the viral life cycle appears to be blocked before viral replication can induce apoptosis.

Adaptation to cells of a different host

In 1969, Raynaud and colleagues described a stepwise adaptation of a field mouse CMV isolate to human cells. The virus was initially grown on mouse embryonic fibroblasts. It was then propagated on embryonic hamster cells and baby hamster kidney cells of the BHK-21 line. After complete infection and lysis of BHK-21 cells, supernatant was used to inoculate primary monkey kidney cells. A typical CPE was observed after the third passage. The virus was subsequently titrated on the same cells and reached a titre of 10^7 $TCID_{50}/ml$ (Raynaud *et al.*, 1969). The virus that grew on monkey kidney cells was then used to infect human diploid cells. Characteristic nuclear inclusions and subsequent cell lysis was observed, and the virus was reported to replicate in these human cells (Raynaud *et al.*, 1969, 1972). A few years later, Kim and Carp compared the MCMV Raynaud strain (prior to its adaptation to non-murine cells) with the commonly used MCMV Smith strain, which was isolated from a laboratory mouse. The authors found that the two strains were indistinguishable antigenically (i.e. in neutralization assays), in terms of CPE morphology, replication in murine fibroblasts, and buoyant density of the viral particle (Kim *et al.*,

1974). As the authors had previously shown that the MCMV Smith strain infects human cells abortively (Kim and Carp, 1972), they raised the question whether the human cell-adapted Raynaud strain might have been a contaminant rather than a descendant of the original field mouse virus (Kim *et al.*, 1974). However, the authors did not report any attempt to reproduce the stepwise adaptation to human cells described by Raynaud and colleagues.

A recent study described the isolation and characterization of a mutant MCMV (Smith strain) that had spontaneously acquired the ability to replicate rapidly and to high titres in human retinal pigment epithelial (RPE-1) cells and, somewhat less efficiently, in human fibroblasts (Schumacher *et al.*, 2010). This virus was named MCMV/h. It induced less apoptosis in human cells and replicated its DNA faster than the parental wild-type MCMV.

Sequence analysis of the human cell-adapted MCMV strain, MCMV/h, revealed several alterations in comparison to the parental virus. Specifically, three mutations identified in the M112/113 coding region (Fig. I.18.2) were further analysed. When these mutations were introduced by site-directed mutagenesis into the wild-type MCMV genome, the resulting virus gained the capacity to replicate in human cells. This demonstrated that mutations in M112/113 are, in principle, sufficient to facilitate MCMV replication in human cells (Schumacher *et al.*, 2010). However, the recombinant virus replicated not as fast as did the spontaneously adapted virus, indicating that additional mutations must have contributed to the remarkably efficient replication of the adapted strain, MCMV/h.

The M112/113 region encodes the Early-1 (E1) proteins, which exist in at least four different isoforms (Bühler *et al.*, 1990; Ciocco-Schmitt *et al.*, 2002). Of the three mutations identified in the MCMV/h E1 region, one (a 3-amino acid deletion) affected all E1 proteins, whereas the two point mutations led to amino acid exchanges only in the large 87-kDa E1 protein (Fig. I.18.2). It is also possible that the point mutations have an impact on splicing efficiency and shift the balance between the various E1 isoforms. A higher abundance of the 87-kDa E1 protein at late times post infection (Schumacher *et al.*, 2010) might indeed reflect altered splicing, but might as well be based on differences in protein stability between the WT and the mutant 87-kDa proteins. To date it is not known which of the E1 mutations are responsible for the enhanced replication of MCMV/h.

The MCMV M112/113 and the HCMV UL112/113 region are highly similar in their location within the viral genome and share an almost identical splicing pattern (Ciocco-Schmitt *et al.*, 2002). Differentially spliced mRNAs encode at least four protein products. Although the E1 proteins have been identified many years ago, their mechanisms of action remains poorly understood. They become detectable shortly after the MIE proteins (Bühler *et al.*, 1990; Ahn *et al.*, 1999) and enhance the expression of other viral genes, particularly those involved in DNA replication (Iskenderian *et al.*, 1996; Kerry *et al.*, 1996). In addition, the E1 proteins might also have a more direct role in viral DNA replication as they can bind single- and double-stranded DNA and accumulate at nuclear replication sites (Iwayama *et al.*, 1994). Moreover, UL112/113 proteins are necessary for transient complementation of HCMV *oriLyt*-dependent DNA replication (Pari and Anders, 1993). Remarkably, the UL112/113 proteins can also activate the full lytic replication cycle of Kaposi's sarcoma-associated herpesvirus (Wells *et al.*, 2009), but the mechanism is yet to be elucidated.

It seems likely that the different E1 protein variants have non-identical functions, but how the different variants operate, and how they interact with other proteins has only begun to be understood. Recent studies have shown that the HCMV UL112/113 proteins interact with each other, and that the large 84-kDa protein binds and recruits UL44 to nuclear pre-replication foci (Park *et al.*, 2006; Kim and Ahn, 2010). In MCMV, the large 87-kDa E1 protein appears to be responsible for binding the major transactivator protein, IE3, and for relieving the repressive effect of IE3 on its own promoter (Tang *et al.*, 2005).

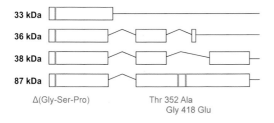

33 kDa

36 kDa

38 kDa

87 kDa

Δ(Gly-Ser-Pro) Thr 352 Ala
Gly 418 Glu

Figure I.18.2 Gene structure and splicing pattern of the four major E1 proteins encoded by the M112/113 region. The mutations identified in MCMV/h are indicated in grey by vertical bars. The apparent molecular mass of the respective protein is indicated.

A potential role of ND10 in host cell restriction

PML-NBs (also known as PODs or ND10) play a role in gene transcription, DNA replication, and regulation of apoptosis, and are also thought to function as intra-nuclear sites of intrinsic antiviral defence (reviewed in Tavalai and Stamminger, 2008, 2011; see also Chapter I.11). Interestingly, the human cell-adapted MCMV had an increased ability to disrupt PML-NBs (Schumacher et al., 2010). This observation was reminiscent of a previous study by Tang and Maul, who reported MCMV replication at very low levels in human cells in the presence of the HCMV IE1 protein and/or when cells were coinfected with an UV-inactivated HCMV, thereby providing HCMV tegument proteins in trans (Tang and Maul, 2006). Since the IE1 protein disrupts the structure of PML-NBs (Korioth et al., 1996) and the tegument protein pp71 induces degradation of Daxx, a major constituent of PML-NBs (Saffert and Kalejta, 2006; see also Chapter I.9), the authors speculated that an insufficient ability of MCMV to antagonize repressive effects of PML-NB constituents in human cells might restrict MCMV replication (Tang and Maul, 2006). Although a few pieces of evidence suggest that PML-NBs are involved in controlling CMV species specificity, further studies will be necessary to define their exact role.

Interferon response-mediated host cell restriction

Interferons (IFNs) play an important role in the control of viral infection as they can induce a so-called antiviral state (reviewed in Haller et al., 2006). Shortly after the discovery of IFNs it has been recognized that the induction of an antiviral state by IFNs is, in many cases, a species-specific phenomenon. IFN produced by cells of one species can induce an antiviral state in cells of the same species but not in cells of different species (Tyrrell, 1959).

Stimulation of the IFN-α/β receptor leads to an activation of Janus kinase and phosphorylation of STAT proteins. Phosphorylated STAT1 and STAT2 form a heterotrimeric complex with interferon regulatory factor (IRF) 9, which translocates to the nucleus, binds to interferon response elements, and activates interferon-stimulated gene (ISG) expression (see Chapter I.16). Many ISGs, such as the dsRNA-dependent protein kinase (PKR) and the oligoadenylate synthetase, have antiviral activity (Haller et al., 2006). Hence there is a strong selective pressure for viruses to evolve proteins that block IFN-dependent signalling pathways and the action of antiviral ISGs.

Even though the IFN-α/β antiviral system is evolutionarily conserved, small differences in the cellular signal transducer and effector proteins can result in a species-specific action of viral antagonists. For instance, the V proteins of the paramyxoviruses

human parainfluenza virus 2 and simian virus 5 induce the degradation of STAT proteins in human but not in mouse cells (Parisien et al., 2002). Similarly, the Marburg virus VP40 protein antagonizes IFN signalling in a species-specific manner (Valmas and Basler, 2011). It has also been demonstrated that the species specificity of myxoma virus, a poxvirus of rabbits, is a consequence of the virus' inability to inhibit the IFN response in cells of other species (Wang et al., 2004). Strikingly, myxoma virus can break the species barrier and replicate in murine cells provided that activation of the IFN response is blocked by inhibitory drugs or with IFN-neutralizing antibodies.

MCMV and HCMV both express a viral protein that interferes with STAT signalling (see Chapter I.16). However, the MCMV and HCMV proteins are not homologous and operate in a different fashion. While the MCMV M27 protein selectively binds and down-regulates STAT2 (Zimmermann et al., 2005), the HCMV IE1 protein forms a complex with STAT1 and STAT2 in infected cells and prevents activation of promoters of ISGs (Paulus et al., 2006). It is not known whether or not the actions of MCMV M27 or HCMV IE1 are species-specific, but it should be worthwhile investigating. In fact, indirect evidence suggests that MCMV is not capable of blocking the response to IFN-β: The human cell adapted MCMV/h replicated rapidly and to very high titres in human retinal pigment epithelial (RPE-1) cells, but spread less efficiently in human MRC-5 fibroblasts (Fig. I.18.3). After infection at very low MOI, the virus started to form plaques in human fibroblasts, but plaque expansion stalled after a few days, and plaques regressed (Schumacher et al., 2010). This observation suggested that cytokines secreted from infected cells might impair viral dissemination. Indeed, MCMV/h infection spread more efficiently through fibroblast monolayers when an IFN-β neutralizing antibody was added to the culture medium (U. Schumacher and W. Brune, unpublished data). Continued passage of MCMV/h in human MRC-5 fibroblasts increased its ability to replicate and spread in these cells (Fig. I.18.3B), suggesting that further adaptive mutations had occurred. These mutations might have increased the virus' ability to prevent the induction of the IFN response or to block IFN receptor dependent signalling.

PKR is an IFN-induced cellular protein with potent antiviral activity (Haller et al., 2006). It is activated by double-stranded RNA, which is frequently present in virus-infected cells (Weber et al., 2006). MCMV and HCMV infections also lead to the formation of dsRNA (Budt et al., 2009; Marshall et al., 2009), but both viruses express potent inhibitors of PKR. These viral inhibitors prevent the PKR-mediated protein synthesis

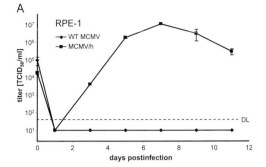

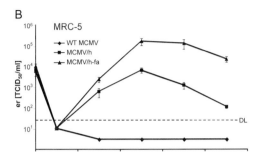

Figure I.18.3 Growth kinetics of WT MCMV and of human cell-adapted MCMV (MCMV/h) in human RPE-1 cell (A) and MRC-5 human embryonic lung fibroblasts (B). A fibroblast-adapted version of MCMV/h is labelled MCMV/h-fa. Titres were determined on murine fibroblasts using the TCID$_{50}$ method. DL, detection limit.

shutoff and are essential for viral replication. They are encoded by HCMV genes TRS1 and IRS1 and MCMV genes m142 and m143 (Child *et al.*, 2004, 2006; Valchanova *et al.*, 2006). The HCMV and MCMV genes share some sequence similarity suggesting that they have evolved from a common ancestor gene, but are not highly homologous (Valchanova *et al.*, 2006). Could PKR inhibition operate in a species-specific fashion? As mentioned before the human-cell adapted MCMV, MCMV/h, replicates to high titres in human cells (Schumacher *et al.*, 2010). This observation suggests that the PKR inhibitors of MCMV, m142 and m143, are fully functional in human cells. Conversely, MCMV mutants lacking m142 or m143 replicate only to moderate titres in murine fibroblasts if the HCMV TRS1 gene is expressed *in cis* or *in trans* (Valchanova *et al.*, 2006), suggesting that TRS1 inhibits murine PKR less efficiently than do the MCMV proteins. The notion that TRS1 might inhibit PKR in a species-specific manner was strongly supported by a recent study that compared HCMV TRS1 with the homologous gene of rhesus CMV (Child *et al.*, 2012). In addition to some differences in the exact mechanism of action, the study demonstrated that huTRS1 does not inhibit PKR in old

world monkey cells, and rhTRS1 does not function in human cells.

Conclusions

Our understanding of the molecular mechanisms underlying the CMV host species specificity is still fragmentary. It has, however, become evident that more than a single mechanism must be involved. In order to replicate and produce progeny, the virus needs to overcome many hurdles. It has to enter a target cell, deliver the viral genome to the nucleus, initiate and execute the viral gene expression cascade, replicate its DNA, and finally assemble and release new virions. During the entire process the virus also needs to counteract intrinsic and innate antiviral defences of the host cell. While entry, delivery of the viral genome to the nucleus, and initiation of IE gene expression do not seem to be strongly restricted in cells of a foreign host species, the timely execution of the viral gene expression cascade and replication of the viral genome are often massively impaired. Moreover, viral countermeasures against antiviral activities of the host cell sometimes function poorly in cells of a foreign host species, presumably because the CMVs have specifically adapted to their natural host species during host–virus coevolution resulting in cospeciation (see Chapters I.1 and I.2). Thus, it is to be expected that future investigations of the molecular mechanisms underlying the CMV host species specificity will provide new insights into the functionality of the innate immune system on the one hand and possibilities for viral interference on the other.

References

Ahn, J.H., Jang, W.J., and Hayward, G.S. (1999). The human cytomegalovirus IE2 and UL112–113 proteins accumulate in viral DNA replication compartments that initiate from the periphery of promyelocytic leukemia protein-associated nuclear bodies (PODs or ND10). J. Virol. *73*, 10458–10471.

Alcendor, D.J., Barry, P.A., Pratt-Lowe, E., and Luciw, P.A. (1993). Analysis of the rhesus cytomegalovirus immediate-early gene promoter. Virology *194*, 815–821.

Arnoult, D., Bartle, L.M., Skaletskaya, A., Poncet, D., Zamzami, N., Park, P.U., Sharpe, J., Youle, R.J., and Goldmacher, V.S. (2004). Cytomegalovirus cell death suppressor vMIA blocks Bax- but not Bak-mediated apoptosis by binding and sequestering Bax at mitochondria. Proc. Natl. Acad. Sci. U.S.A. *101*, 7988–7993.

Arnoult, D., Skaletskaya, A., Estaquier, J., Dufour, C., and Goldmacher, V.S. (2008). The murine cytomegalovirus cell death suppressor m38.5 binds Bax and blocks Bax-mediated mitochondrial outer membrane permeabilization. Apoptosis *13*, 1100–1110.

Black, P.H., Hartley, J.W., and Rowe, W.P. (1963). Isolation of a cytomegalovirus from African green monkey. Proc. Soc. Exp. Biol. Med. *112*, 601–605.

Bruggeman, C.A., Meijer, H., Dormans, P.H., Debie, W.M., Grauls, G.E., and van Boven, C.P. (1982). Isolation of a cytomegalovirus-like agent from wild rats. Arch. Virol. *73*, 231–241.

Brune, W. (2011). Inhibition of programmed cell death by cytomegaloviruses. Virus Res. *157*, 144–150.

Budt, M., Niederstadt, L., Valchanova, R.S., Jonjic, S., and Brune, W. (2009). Specific inhibition of the PKR-mediated antiviral response by the murine cytomegalovirus proteins m142 and m143. J. Virol. *83*, 1260–1270.

Bühler, B., Keil, G.M., Weiland, F., and Koszinowski, U.H. (1990). Characterization of the murine cytomegalovirus early transcription unit e1 that is induced by immediate-early proteins. J. Virol. *64*, 1907–1919.

Çam, M., Handke, W., Picard-Maureau, M., and Brune, W. (2010). Cytomegaloviruses inhibit Bak- and Bax-mediated apoptosis with two separate viral proteins. Cell Death Differ. *17*, 655–665.

Child, S.J., Hakki, M., De Niro, K.L., and Geballe, A.P. (2004). Evasion of cellular antiviral responses by human cytomegalovirus TRS1 and IRS1. J. Virol. *78*, 197–205.

Child, S.J., Hanson, L.K., Brown, C.E., Janzen, D.M., and Geballe, A.P. (2006). Double-stranded RNA binding by a heterodimeric complex of murine cytomegalovirus m142 and m143 proteins. J. Virol. *80*, 10173–10180.

Child, S.J., Brennan, G., Braggin, J.E., and Geballe, A.P. (2012). Species specificity of protein kinase R antagonism by cytomegalovirus TRS1 genes. J. Virol. *86*, 3880–3889.

Ciocco-Schmitt, G.M., Karabekian, Z., Godfrey, E.W., Stenberg, R.M., Campbell, A.E., and Kerry, J.A. (2002). Identification and characterization of novel murine cytomegalovirus M112–113 (e1) gene products. Virology *294*, 199–208.

Fallaux, F.J., Kranenburg, O., Cramer, S.J., Houweling, A., van Ormondt, H., Hoeben, R.C., and van der Eb, A.J. (1996). Characterization of 911: a new helper cell line for the titration and propagation of early region 1-deleted adenoviral vectors. Hum. Gene Ther. *7*, 215–222.

Fioretti, A., Furukawa, T., Santoli, D., and Plotkin, S.A. (1973). Nonproductive infection of guinea pig cells with human cytomegalovirus. J. Virol. *11*, 998–1003.

Graham, F.L., Smiley, J., Russell, W.C., and Nairn, R. (1977). Characteristics of a human cell line transformed by DNA from human adenovirus type 5. J. Gen. Virol. *36*, 59–74.

Haller, O., Kochs, G., and Weber, F. (2006). The interferon response circuit: induction and suppression by pathogenic viruses. Virology *344*, 119–130.

Hsiung, G.D., Fischman, H.R., Fong, C.K., and Green, R.H. (1969). Characterization of a cytomegalo-like virus isolated from spontaneously degenerated equine kidney cell culture. Proc. Soc. Exp. Biol. Med. *130*, 80–84.

Huang, E.S., Kilpatrick, B., Lakeman, A., and Alford, C.A. (1978). Genetic analysis of a cytomegalovirus-like agent isolated from human brain. J. Virol. *26*, 718–723.

Huthoff, H., and Towers, G.J. (2008). Restriction of retroviral replication by APOBEC3G/F and TRIM5α. Trends Microbiol. *16*, 612–619.

Iskenderian, A.C., Huang, L., Reilly, A., Stenberg, R.M., and Anders, D.G. (1996). Four of eleven loci required for transient complementation of human cytomegalovirus DNA replication cooperate to activate expression of replication genes. J. Virol. *70*, 383–392.

Iwayama, S., Yamamoto, T., Furuya, T., Kobayashi, R., Ikuta, K., and Hirai, K. (1994). Intracellular localization and DNA-binding activity of a class of viral early phosphoproteins in human fibroblasts infected with human cytomegalovirus (Towne strain). J. Gen. Virol. *75*, 3309–3318.

Jurak, I., and Brune, W. (2006). Induction of apoptosis limits cytomegalovirus cross-species infection. EMBO J. *25*, 2634–2642.

Jurak, I., Schumacher, U., Simic, H., Voigt, S., and Brune, W. (2008). Murine cytomegalovirus m38.5 protein inhibits Bax-mediated cell death. J. Virol. *82*, 4812–4822.

Kerry, J.A., Priddy, M.A., Jervey, T.Y., Kohler, C.P., Staley, T.L., Vanson, C.D., Jones, T.R., Iskenderian, A.C., Anders, D.G., and Stenberg, R.M. (1996). Multiple regulatory events influence human cytomegalovirus DNA polymerase (UL54) expression during viral infection. J. Virol. *70*, 373–382.

Kim, K.S., and Carp, R.I. (1971). Growth of murine cytomegalovirus in various cell lines. J. Virol. *7*, 720–725.

Kim, K.S., and Carp, R.I. (1972). Abortive infection of human diploid cells by murine cytomegalovirus. Infect. Immun. *6*, 793–797.

Kim, K.S., Sapienza, V., and Carp, R.I. (1974). Comparative studies of the Smith and Raynaud strains of murine cytomegalovirus. Infect. Immun. *10*, 672–674.

Kim, Y.E., and Ahn, J.H. (2010). Role of the specific interaction of UL112–113 p84 with UL44 DNA polymerase processivity factor in promoting DNA replication of human cytomegalovirus. J. Virol. *84*, 8409–8421.

Kirzinger, M.W., and Stavrinides, J. (2012). Host specificity determinants as a genetic continuum. Trends Microbiol. *20*, 88–93.

Korioth, F., Maul, G.G., Plachter, B., Stamminger, T., and Frey, J. (1996). The nuclear domain 10 (ND10) is disrupted by the human cytomegalovirus gene product IE1. Exp. Cell Res. *229*, 155–158.

Lafemina, R.L., and Hayward, G.S. (1988). Differences in cell-type-specific blocks to immediate-early gene expression and DNA replication of human, simian and murine cytomegalovirus. J. Gen. Virol. *69*, 355–374.

Leendertz, F.H., Deckers, M., Schempp, W., Lankester, F., Boesch, C., Mugisha, L., Dolan, A., Gatherer, D., McGeoch, D.J., and Ehlers, B. (2009). Novel cytomegaloviruses in free-ranging and captive great apes: phylogenetic evidence for bidirectional horizontal transmission. J. Gen. Virol. *90*, 2386–2394.

Lilja, A.E., and Shenk, T. (2008). Efficient replication of rhesus cytomegalovirus variants in multiple rhesus and human cell types. Proc. Natl. Acad. Sci. U.S.A. *105*, 19950–19955.

McCormick, A.L., Meiering, C.D., Smith, G.B., and Mocarski, E.S. (2005). Mitochondrial cell death suppressors carried by human and murine cytomegalovirus confer resistance to proteasome inhibitor-induced apoptosis. J. Virol. *79*, 12205–12217.

Marshall, E.E., Bierle, C.J., Brune, W., and Geballe, A.P. (2009). Essential role for either TRS1 or IRS1 in human cytomegalovirus replication. J. Virol. *83*, 4112–4120.

Martin, W.J., Zeng, L.C., Ahmed, K., and Roy, M. (1994). Cytomegalovirus-related sequence in an atypical cytopathic virus repeatedly isolated from a patient with chronic fatigue syndrome. Am. J. Pathol. *145*, 440–451.

Michaels, M.G., Jenkins, F.J., St George, K., Nalesnik, M.A., Starzl, T.E., and Rinaldo, C.R., Jr. (2001). Detection of infectious baboon cytomegalovirus after baboon-to-human liver xenotransplantation. J. Virol. 75, 2825–2828.

Mocarski, E.S., and Courcelle, C.T. (2001). Cytomegaloviruses and their replication. In Fields Virology, Knipe, D.M., and Howley, P.M., eds. (Lippincott-Williams & Wilkins, Philadelphia, PA), pp. 2629–2673.

Pari, G.S., and Anders, D.G. (1993). Eleven loci encoding trans-acting factors are required for transient complementation of human cytomegalovirus oriLyt-dependent DNA replication. J. Virol. 67, 6979–6988.

Parisien, J.P., Lau, J.F., and Horvath, C.M. (2002). STAT2 acts as a host range determinant for species-specific paramyxovirus interferon antagonism and simian virus 5 replication. J. Virol. 76, 6435–6441.

Park, M.Y., Kim, Y.E., Seo, M.R., Lee, J.R., Lee, C.H., and Ahn, J.H. (2006). Interactions among four proteins encoded by the human cytomegalovirus UL112–113 region regulate their intranuclear targeting and the recruitment of UL44 to prereplication foci. J. Virol. 80, 2718–2727.

Paulus, C., Krauss, S., and Nevels, M. (2006). A human cytomegalovirus antagonist of type I IFN-dependent signal transducer and activator of transcription signaling. Proc. Natl. Acad. Sci. U.S.A. 103, 3840–3845.

Perot, K., Walker, C.M., and Spaete, R.R. (1992). Primary chimpanzee skin fibroblast cells are fully permissive for human cytomegalovirus replication. J. Gen. Virol. 73, 3281–3284.

Poncet, D., Larochette, N., Pauleau, A.L., Boya, P., Jalil, A.A., Cartron, P.F., Vallette, F., Schnebelen, C., Bartle, L.M., Skaletskaya, A., et al. (2004). An antiapoptotic viral protein that recruits Bax to mitochondria. J. Biol. Chem. 279, 22605–22614.

Rabson, A.S., Edgcomb, J.H., Legallais, F.Y., and Tyrrell, S.A. (1969). Isolation and growth of rat cytomegalovirus in vitro. Proc. Soc. Exp. Biol. Med. 131, 923–927.

Raynaud, J., and Barreau, C. (1965). Transmission of salivary virus from the field mouse (Apodemus Sylvaticus) to the mouse. C. R. Hebd. Seances Acad. Sci. 260, 1034–1037.

Raynaud, J., Atanasiu, P., Barreau, C., and Jahkola, M. (1969). Adaptation d'un virus cytomégalique provenant do Mulot (Apodemus sylvaticus) sur différantes cellules hétérologiques, y compris les cellules humaines. C. R. Acad. Sci. Hebd. Seances Acad. Sci. D 269, 104–106.

Raynaud, J., Atanasiu, P., and Virat, J. (1972). Further observations on the adaptation to human cells, of a strain of cytomegalic virus from the field-mouse (Apodemus sylvaticus). C. R. Acad. Sci. Hebd. Seances Acad. Sci. D 274, 2920–2923.

Reed, J.M., Schiff, L.J., Shefner, A.M., and Poiley, S.M. (1975). Murine virus susceptibility of cell cultures of mouse, rat, hamster, monkey, and human origin. Lab. Anim. Sci. 25, 420–424.

Rivailler, P., Kaur, A., Johnson, R.P., and Wang, F. (2006). Genomic sequence of rhesus cytomegalovirus 180.92: insights into the coding potential of rhesus cytomegalovirus. J. Virol. 80, 4179–4182.

Saffert, R.T., and Kalejta, R.F. (2006). Inactivating a cellular intrinsic immune defense mediated by Daxx is the mechanism through which the human cytomegalovirus pp71 protein stimulates viral immediate-early gene expression. J. Virol. 80, 3863–3871.

Schumacher, U., Handke, W., Jurak, I., and Brune, W. (2010). Mutations in the M112/M113-coding region facilitate murine cytomegalovirus replication in human cells. J. Virol. 84, 7994–8006.

Smith, C.B., Wei, L.S., and Griffiths, M. (1986). Mouse cytomegalovirus is infectious for rats and alters lymphocyte subsets and spleen cell proliferation. Arch. Virol. 90, 313–323.

Smith, M.G. (1954). Propagation of salivary gland virus of the mouse in tissue culture. Proc. Soc. Exp. Biol. Med. 86, 434–440.

Smith, M.G. (1956). Propagation in tissue cultures of a cytopathogenic virus from human salivary gland virus (SGV) disease. Proc. Soc. Exp. Biol. Med. 92, 424–430.

Tang, Q., and Maul, G.G. (2006). Mouse cytomegalovirus crosses the species barrier with help from a few human cytomegalovirus proteins. J. Virol. 80, 7510–7521.

Tang, Q., Li, L., and Maul, G.G. (2005). Mouse cytomegalovirus early M112/113 proteins control the repressive effect of IE3 on the major immediate-early promoter. J. Virol. 79, 257–263.

Taubenberger, J.K., and Kash, J.C. (2010). Influenza virus evolution, host adaptation, and pandemic formation. Cell Host Microbe 7, 440–451.

Tavalai, N., and Stamminger, T. (2008). New insights into the role of the subnuclear structure ND10 for viral infection. Biochim. Biophys. Acta 1783, 2207–2221.

Tavalai, N., and Stamminger, T. (2011). Intrinsic cellular defense mechanisms targeting human cytomegalovirus. Virus Res. 157, 128–133.

Tyrrell, D.A. (1959). Interferon produced by cultures of calf kidney cells. Nature 184, 452–453.

Valchanova, R.S., Picard-Maureau, M., Budt, M., and Brune, W. (2006). Murine cytomegalovirus m142 and m143 are both required to block protein kinase R-mediated shutdown of protein synthesis. J. Virol. 80, 10181–10190.

Valmas, C., and Basler, C.F. (2011). Marburg virus VP40 antagonizes interferon signaling in a species-specific manner. J. Virol. 85, 4309–4317.

Wang, F., Ma, Y., Barrett, J.W., Gao, X., Loh, J., Barton, E., Virgin, H.W., and McFadden, G. (2004). Disruption of Erk-dependent type I interferon induction breaks the myxoma virus species barrier. Nat. Immunol. 5, 1266–1274.

Weber, F., Wagner, V., Rasmussen, S.B., Hartmann, R., and Paludan, S.R. (2006). Double-stranded RNA is produced by positive-strand RNA viruses and DNA viruses but not in detectable amounts by negative-strand RNA viruses. J. Virol. 80, 5059–5064.

Weller, T.H. (1970). Cytomegaloviruses: the difficult years. J. Infect. Dis. 122, 532–539.

Wells, R., Stensland, L., and Vieira, J. (2009). The human cytomegalovirus UL112–113 locus can activate the full Kaposi's sarcoma-associated herpesvirus lytic replication cycle. J. Virol. 83, 4695–4699.

Zimmermann, A., Trilling, M., Wagner, M., Wilborn, M., Bubic, I., Jonjic, S., Koszinowski, U., and Hengel, H. (2005). A cytomegaloviral protein reveals a dual role for STAT2 in IFN-γ signaling and antiviral responses. J. Exp. Med. 201, 1543–1553.

Epigenetic Regulation of Human Cytomegalovirus Gene Expression: Impact on Latency and Reactivation

Matthew Reeves and John Sinclair

Abstract

The myeloid lineage is now accepted to be an important site *in vivo* for the carriage of latent HCMV genomes, but the mechanisms underlying how the latent state is maintained and how latent virus reactivates are still far from clear. In this review, we discuss how analyses of promoter binding proteins and post-translational modifications of histones on viral promoters during virus infection have led to an understanding that the higher-order chromatin structure around the viral major immediate-early promoter region has profound effects on the control of viral latency and reactivation. We further discuss the role of chromatin during lytic infection and how this may also give insights into the cellular mechanisms important for the establishment and control of latent infection.

Introduction: chromatin – structure and function

It is now well established that the regulation of transcription is not solely dependent on regulating the formation of the transcriptional pre-initiation complex (PIC; comprising TFIIA, TFIIB, TFIID, TFIIE, TFIIF, and TFIIH) to position RNA polymerase over the transcription start site on naked DNA templates. Whilst the recruitment of basal transcription factors to promoters is clearly of crucial importance, it was known for some time that the regulation of gene expression also involved regulation of higher-order chromatin structure. However, the mechanism by which chromatin regulated transcription was largely left unaddressed until specific reagents, able to distinguish between modified chromatin components, became available. As our understanding of both chromatin structure and its role in the control of cellular gene expression increased, it became possible to directly assess the influence of chromatin modification both on the transcription of cellular genes and, as viral gene expression necessarily involved many of the same components as cellular

transcription, the effect of chromatin modification on viral gene expression as well.

The packaging of cellular DNA

The structure of the nucleosome, the basic unit of chromatin structure, was first elucidated nearly 40 years ago as a complex of chromosomal DNA and histone proteins (Kornberg, 1974; Oudet *et al.*, 1975). These nucleosomes, with their characteristic 'beads on a string' appearance under the electron microscope (Oudet *et al.*, 1975), typically protect a ~ 200 base pair fragment of DNA from the action of nucleases (Hewish and Burgoyne, 1973). Further data indicated that a very distinct arrangement of histone proteins (one copy of H1 and two copies each of H2A, H2B, H3 and H4) was associated with DNA (Fig. I.19.1; Chung *et al.*, 1978). This association between these basic cellular proteins and nucleic acid is clearly required for condensing the

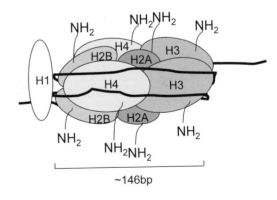

Figure I.19.1 Structure of the nucleosome. Shown is the characteristic octamer comprising two molecules each of histones H2A, H2B, H3 and H4. Each nucleosome has approximately 140 base pairs of DNA wrapped around it with histone H1 acting as a 'linker' holding the structure intact. NH_2 represents exposed N-terminal histone tails.

large amounts of DNA contained in the eukaryotic cell nucleus but this compaction does present problems in terms of accessibility of the DNA to, for example, the DNA replication and transcription apparatus. Indeed, it was initially proposed that DNA associated with histones was transcriptionally inactive and that only the free DNA observed between nucleosomes encoded transcribed genes (Oudet et al., 1975). However, it soon became clear that this was not the case and that gene transcription required some form of remodelling of chromatin structure. This led to the idea that the structure of chromatin at specific regions on DNA could play a role in the regulation of cellular gene transcription, as well as its more established role in the packaging of DNA (Weintraub and Groudine, 1976; Lilley and Pardon, 1979).

It has now become clear that, whilst transcriptionally active DNA is not necessarily free of histone association (Kuo and Allis, 1998), biochemical modifications of these histones on specific amino acid residues in their amino terminal domains are responsible for a more open, or a more closed, chromatin structure which leads to gene expression or gene silencing, respectively. These histone modifications include: methylation of lysine and arginine residues; (Zhang and Reinberg, 2001), acetylation of lysine residues; (Doenecke and Gallwitz, 1982), deacetylation of lysine residues; (Lopez-Rodas et al., 1993), sumoylation of lysine residues; (Shiio and Eisenman, 2003), ubiquitination of lysine and serine residues and phosphorylation of threonine residues (Lusser, 2002). This diverse array of modifications, which determines the overall structure of chromatin, is the basis for the 'histone code hypothesis' (Strahl and Allis, 2000) which stipulates that patterns, and not just levels, of enzyme-mediated modifications of histones result in the generation of a code that determines the binding of regulatory proteins. This, in turn, leads to down-stream effects which directly determine levels of gene expression based specifically on the histone modifications of the gene in question. The pattern generated by the different modifications exhibited by histones bound to a specific promoter is also referred to as the 'histone language' and, as such, the integration of these different marks is a key component of chromatin mediated regulation of gene expression (Bannister and Kouzarides, 2011).

Histone acetylation

One of the first histone modifications to be characterized was acetylation. To date, acetylation of lysine residues on histones H3 (at K9 and K14) and H4 (at K5, K8, K12 and K16) have been correlated with transcriptional activation (Kuo and Allis, 1998; Mizzen and Allis, 1998; Lusser, 2002). The acetylation of histone tails is catalysed by a family of enzymes known as histone acetyltransferases or 'HATs' (Fig. I.19.2; Eberharter and Becker, 2002) and these include p300, CREB-binding protein (CBP) and p300/CBP-associated factor (P/CAF). The removal of acetyl groups from histones is catalysed by histone deacetylases (HDACs; Fig. I.19.2), an evolutionarily conserved family of enzymes. The HDACs, structurally and functionally divided into distinct classes, are involved in transcriptional repression of gene expression (Khochbin et al., 2001).

It is probable that amino-terminal positively charged lysine residues within histones form multiple stable interactions with negatively charged DNA. This results in a close association between the histones and DNA, compaction of the chromatin, and a transcriptionally inactive state. Acetylation of these lysine residues,

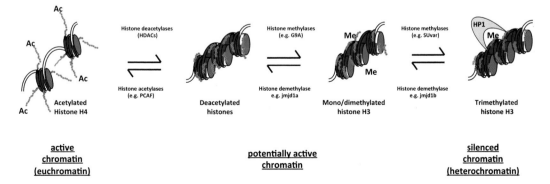

Figure I.19.2 The language of covalent histone modifications. Simplified model showing the interplay of histone acetyltransferases (HATs), histone deacetylases (HDACs), histone methyltransferases and histone demethylases in modulating chromatin structure. HP1 is the heterochromatin protein-1, Me are methyl groups, Ac are acetyl groups, and red structures represent histone tails.

particularly those of histones H3 and H4, is likely to disrupt the histone:DNA interaction leading to a looser association which facilitates access of transcription factors and the PIC to the DNA template (Fig. I.19.2; Cary *et al.*, 1982; Bode *et al.*, 1983; Morgan *et al.*, 1987). In addition, these acetyl-lysines may also provide binding or recognition sites for the so-called bromodomains of other proteins. These bromodomains are acetyl-lysine binding motifs (Zeng and Zhou, 2002) found in a diverse array of chromatin-associated proteins such as HATs (Dhalluin *et al.*, 1999), and TAFII250, a component of the TFIID transcription complex (Jacobson *et al.*, 2000). Conversely, the removal of acetyl groups from lysine residues, catalysed by HDACs, re-creates the basic charge on the lysine leading to a strong association between DNA and histones and hence the basis of a transcriptionally inactive state (Fig. I.19.2; Lusser, 2002).

However, it should be noted that acetylation of histones does not always correlate with transcriptional activation. Some promoters become active in the absence of significant acetylation changes (Dudley *et al.*, 1999), while other genes are activated despite no change in the hypersensitivity of the promoter (by definition, acetylation of histones results in a looser association with DNA and hence the DNA is more sensitive to nuclease activity) (Wong *et al.*, 1998; Urnov *et al.*, 2000). Furthermore, histone hyperacetylation of heterochromatin enhances the efficiency of origin firing during replication (Casas-Delucchi *et al.*, 2011), emphasizing a role for post-translational modifications of histones in regulation of other aspects of DNA function besides transcription (Burgess and Zhang, 2010; Bell *et al.*, 2011).

Histone methylation

Methylation of histones has historically been associated with transcriptionally silent regions of the genome (Razin, 1998) and is also linked with epigenetic inheritance where replication of DNA methylation patterns are used to propagate histone modifications associated with gene silencing (Lachner *et al.*, 2003). In particular, transcriptional repression and X chromosome inactivation is often associated with addition of methyl groups to lysine residues on histones H3 and H4 (Fig. I.19.2; Cao *et al.*, 2002; Mermoud *et al.*, 2002; Schotta *et al.*, 2002; Plath *et al.*, 2003). Lysine residues on these histones can be multiply methylated (up to three) by histone methyltransferases (HMTs). Tri-methylation of lysine residue 9 is performed by the SUvar family of histone methylases, whereas mono- and di-methylation is carried out by the G9a family of histone methyltransferases (Rice *et al.*, 2003). Interestingly, tri-methylation

appears to occur at heterochromatic regions (that is, densely staining chromatin regions believed generally to be transcriptionally inactive) of the genome, while mono- and di-methylation occurs at euchromatic regions of the genome (that is, regions believed generally to be transcriptionally active) (Rice *et al.*, 2003).

The marking of transcriptionally silent promoter regions of chromatin is also associated with the binding of specific transcriptional silencing proteins such as the heterochromatin protein-1 (HP1) (Kwon and Workman, 2011). As the name suggests, HP1 is often found at heterochromatic sites (James *et al.*, 1989) and is recruited to these regions by methylation events on lysine residues. It appears that there is a two-step process: (i) methylation of histone H3 on lysine residue 9 or 27 results in (ii) the creation of a target for the chromatin binding domain of HP1 or other polycomb proteins involved in long term gene silencing. In the case of HP1, increased binding is observed with simultaneous methylation of lysine 9 and 27 (Khorasanizadeh, 2004). Thus, heterochromatic regions can be identified by both methylated histone residues as well as binding by HP1 (Bannister *et al.*, 2001).

Furthermore, there is increasing evidence for a link between histone methylation and another mechanism of transcriptional silencing – DNA methylation (Cheng and Blumenthal, 2010). Fuks and co-workers first demonstrated a physical and functional interaction between components of the lysine methylation complex (such as HMTs) and methyl-CpG-binding protein (Fuks *et al.*, 2003a,b). Such cross-talk between repression mechanisms would serve to further enforce transcriptional silencing (Fuks, 2005) although, as always, there are instances where the relationship does not always hold (Wu *et al.*, 2007).

Essential to the concept of the 'histone code hypothesis' is the need for reversibility of histone post-translational modification: often repressed genes need to become active and active genes need to become repressed. As such, the antagonistic activities of HATs and HDACs had been characterized extensively. In contrast, whilst accumulating evidence clarified how histone methylation could silence gene expression, it was unclear how this modification was reversed to allow previously silenced genes to become activated. The hunt for histone demethylases eventually identified possible candidate demethylases. Interestingly, the first, a lysine-specific demethylase (LSD1), was reported to actually target the demethylation of lysine 4 on histone H3 (see below) and thus was not necessarily first thought to be responsible for de-repression (Shi *et al.*, 2005). However, further work identified that LSD1 was also active against H3-K9 and exhibited functional activity that correlated with the de-repression of silenced promoters (Metzger *et al.*, 2005). The belief that an array of histone demethylases would be identified has, so far, not occurred. Although, a second enzymatically distinct class of histone demethylases which can target multiple methylated lysine residues, has also now been described (Tsukada *et al.*, 2006; Whetstine *et al.*, 2006).

Whilst it is clear that methylation of lysine 9 on histones H3 is invariably a marker of transcriptional repression, methylation, per se, is not an absolute indicator of inactive chromatin. For example, methylation of arginine residue 3 on histone H4, catalysed by enzymes containing the protein R methyltransferase motif, supports histone H4 acetylation and transcriptional activation (Wang *et al.*, 2001). Similarly, di- and tri-methylation of lysine 4 on histone H3 is associated with active transcriptional elongation (Bernstein *et al.*, 2002). However, it is important to remember the limitations of such analyses: H3-K4 methylation can also identify a recently active promoter but is not, necessarily, indicative of ongoing transcription (Guenther *et al.*, 2007). Thus, when analysing the post-translational modifications of histones in relation to gene expression, it is important to appreciate that the analysis of specific histone marks is not totally definitive but needs to be part of an integrated approach to the analysis of transcription of genes of interest, be they cellular or viral.

Chromatin and the analysis of virus latency

It is clear that chromatin structure plays a profound role in the control of gene transcription and recent advances in molecular tools to analyse chromatin-mediated regulation of gene expression have, in turn, impacted on our understanding of the possible molecular mechanisms of control of virus gene expression which underpin viral latency and reactivation.

HCMV latency and the myeloid lineage

It is now accepted that that the myeloid lineage is one important site of carriage of latent viral genomes (reviewed in Sinclair and Sissons, 2006). Early experiments showed that HCMV could be transmitted in peripheral blood and that this transmission was reduced upon leucocyte depletion (Gilbert *et al.*, 1989). However, infectious HCMV cannot be directly isolated from the blood of healthy seropositive individuals which inferred that peripheral blood leucocytes contain reactivatable, but not necessarily infectious virus. Further experiments, utilizing highly sensitive DNA PCR identified viral genomes in the myeloid

lineage from primitive CD34+ progenitor cells through to monocytes (Taylor-Wiedeman *et al.*, 1991; Bevan *et al.*, 1993; Minton *et al.*, 1994; Mendelson *et al.*, 1996) but sensitive RT-PCR protocols failed to detect viral lytic gene expression in these cells (Taylor-Wiedeman *et al.*, 1994; Mendelson *et al.*, 1996). Intriguingly, terminal differentiation of monocytes to macrophages was shown to reactivate IE gene expression (Fig. I.19.3). However, under these conditions, infectious virus could not be recovered when macrophages were co-cultured with fully permissive fibroblasts (Taylor-Wiedeman *et al.*, 1994). However, subsequent reports have shown that *ex vivo* differentiation of monocytes, from healthy sero-positive carriers, resulting from treatment with supernatants from allogeneically stimulated T-cells, could result in virus reactivation and the release of infectious virions (Soderberg-Naucler *et al.*, 1997b). Attempts to identify the crucial components of the T-cell supernatants required for virus reactivation suggested that interferon-γ and TNF-α could be the important mediators (Soderberg-Naucler *et al.*, 1997a, 2001). However these cytokines, alone, were not able to recapitulate the effects seen with complete allogeneic media which suggested other undefined components were involved. One inference from these studies was that multiple events, perhaps both inflammation and differentiation, were important to promote reactivation. Consistent with this, the reactivating monocytes were observed to express differentiation markers specific for both macrophage and dendritic cell lineages (Soderberg-Naucler *et al.*, 1997b).

This strong link between reactivation of latent virus and the differentiation state of cells in the myeloid lineage appears to be a pervading theme in HCMV biology exemplified by a subsequent analysis of reactivation of HCMV in myeloid DCs (Reeves *et al.*, 2005b). In this study, the *ex vivo* differentiation of either CD34+ or CD14+ cells to specific populations of myeloid DCs (Sallusto and Lanzavecchia, 1994; Strobl *et al.*, 1997) resulted in the reactivation of HCMV gene expression and, in the case of CD34+-derived DCs, the recovery of infectious virus (Reeves *et al.*, 2005b) once again highlighting the importance of myeloid cell differentiation for HCMV reactivation.

Interestingly, this apparent dependence of myeloid differentiation for reactivation of viral lytic gene expression also correlates well with the known levels of permissiveness of myeloid cells for IE expression after exogenous infection with HCMV *in vitro* (Fig. I.19.2); many workers have shown that whilst monocytes are non-permissive, macrophages and, subsequently, dendritic cells, are fully permissive for infection (Lathey and Spector, 1991; Riegler *et al.*, 2000; Hertel *et al.*, 2003).

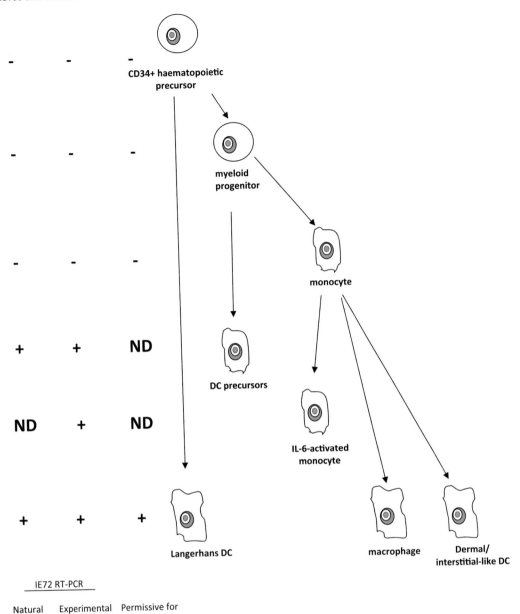

-     -   -

CD34+ haematopoietic precursor

-     -   -

myeloid progenitor

-     -   -

monocyte

+   +   **ND**

DC precursors

ND   +   **ND**

IL-6-activated monocyte

+   +   +

Langerhans DC   **macrophage**   **Dermal/ interstitial-like DC**

IE72 RT-PCR

Natural Latency	Experimental Latency	Permissive for lytic infection

Figure I.19.3 The correlation between experimental/natural latency and lytic infection with differentiation of the myeloid lineage. An overview of the data derived from a number of laboratories that have sought to address the role of myeloid cell differentiation in HCMV biology. The detection of IE72 RNA expression (+) in studies of natural and experimental latency is correlated with the reported permissiveness for lytic infection of the cells analysed. ND, no data.

The study of latency: model cell systems

Many studies, seeking to understand the molecular mechanisms regulating HCMV reactivation from latency, have used the differentiation-dependent permissiveness of myeloid cells for HCMV infection to carry out experiments that rely on infection of non-permissive cell types and the subsequent investigation of the effects of differentiation on viral IE gene expression and productive infection (Fig. I.19.2). These model systems include CD33$^+$ fetal liver stem cells (Kondo et al., 1994), myeloid lineage-committed progenitor cells which display both myeloid and DC markers (Hahn *et al.*, 1998), bone marrow progenitor cells. including CD34[+] cells (Maciejewski *et al.*, 1992; Minton *et al.*, 1994; Goodrum *et al.*, 2002, 2004), peripheral CD34[+] cells isolated from G-CSF mobilized patients (Reeves *et al.*, 2005a) as well as peripheral blood CD14[+] monocytes (Hargett and Shenk, 2010; Reeves and Compton, 2011).

Similarly, because of the difficulties of obtaining and working with primary material, such as monocytes and CD34[+] bone marrow progenitors, other studies have utilized cell lines that, *in vitro*, show differentiation state-dependent control of IE gene expression. The myelomonocytic cell line THP-1 and the NTera2 (T2) embryonal carcinoma cell line are non-permissive for HCMV infection due to a block in IE gene expression. However, following terminal differentiation of THP-1 cells to macrophage-like cells with phorbol esters or differentiation of T2 cells with retinoic acid, these cells become permissive for IE gene expression and productive infection (Gonczol *et al.*, 1984; Weinshenker *et al.*, 1988). Thus, these cell lines are good models for the differentiation state-dependence of IE gene expression and have given insight into the mechanisms regulating reactivation of viral IE gene expression upon differentiation of monocytes to macrophages.

The control of HCMV latency in the myeloid lineage

Model systems, based on exogenous infection of myeloid cells, which reproducibly confirm the differentiation-dependent regulation of viral IE gene expression after infection, have also been used to attempt to determine whether viral transcripts are associated with latent carriage of HCMV. Infected undifferentiated myeloid cells can maintain the HCMV genome in the absence of appreciable viral lytic gene expression. However, sensitive analyses of these experimentally latent cells have identified expression of a number of candidate gene products that may be considered latency-associated transcripts (Kondo *et al.*, 1994; Goodrum *et al.*, 2002; Jenkins *et al.*, 2004). Kondo and colleagues, using infection of granulocyte–macrophage progenitor cells (GMPs) as an experimental model of latent infection, identified both sense and anti-sense transcripts originating from the major IE region (Kondo *et al.*, 1994, 1996). These transcripts have also been found during natural latency in some healthy sero-positive carriers (Kondo *et al.*, 1996) and initially suggested that HCMV resembles HSV by encoding a transcript(s) that could, perhaps, provide anti-sense–mediated inhibition of critical lytic genes (Preston, 2000). Subsequent studies, however, have not identified a precise role for these latency-associated

transcripts. Only a small proportion of the infected GMPs express these transcripts (Slobedman and Mocarski, 1999) and viral deletion mutants unable to express these RNAs establish latency normally (White *et al.*, 2000). Similarly, these transcripts are also found during productive virus infection (Lunetta and Wiedeman, 2000) and thus their function remains unclear.

Interestingly, using the same model system, it has been shown that the viral IL-10 homologue (vIL-10) is expressed during both experimental and natural latency (Jenkins *et al.*, 2004). The vIL-10 shares some of the functions of cellular IL-10 conferring on the virus a capacity to modulate the immune response (Spencer *et al.*, 2002; Raftery *et al.*, 2004; Chang and Barry, 2010). Thus, latency-associated expression of vIL-10 may help the virus to avoid host immune system recognition by creating an immunosuppressed microenvironment around the latently infected cell (Avdic *et al.*, 2011).

Experimental infection of CD34[+] haematopoietic cells has also helped identify other putative latency-associated transcripts (Goodrum *et al.*, 2002). Although significant transcription was observed from a number of lytic loci immediately post infection, the prolonged expression of transcripts in the UL81 and UL138 regions was further investigated and led to detection of a transcript originally identified as arising from the UL81 region of the genome but transcribed on the anti-parallel strand of UL81–82 (Bego *et al.*, 2005). This transcript (UL81–82ast) can be detected in both monocytes and CD34[+] cells of some healthy sero-positive donors (Bego *et al.*, 2005; Reeves and Sinclair, 2010) but, to date, no function for it has been ascribed. Similarly, the polycistronic UL138 transcript has also been identified in healthy donors' cells (Goodrum *et al.*, 2007). However, most of the characterization of the gene product has been in the context of lytic infection (Petrucelli *et al.*, 2009; Montag *et al.*, 2011) and, thus, it remains to be seen what function UL138 plays during latency.

The precise role that HCMV gene products play during latent infection *in vivo* is still a crucial question that is under intense investigation by several laboratories (see Chapter I.20). What is evident, however, is that undifferentiated cells carrying viral genomes are generally silent with respect to viral IE gene expression. This suggests that the cell carrying the viral genome is not conducive to viral lytic gene expression. Specifically, the HCMV major IE promoter which drives expression of the viral major IE genes appears to be transcriptionally repressed in these undifferentiated cells. The molecular mechanism which mediates repression of the major immediate-early promoter/enhancer (MIEP) region in these undifferentiated cell types has been the subject of a number of studies and the emerging story is that repression of viral lytic gene expression during latency results from an orchestrated effect of transcriptional repressors and higher order chromatin structure of the MIEP in these cells.

Clearly, a thorough understanding of the molecular mechanisms involved in the establishment and regulation of latency by control of the viral MIEP will be important for a more comprehensive understanding of the molecular triggers of reactivation.

HCMV latency and chromatin-mediated regulation of transcription

The interaction of DNA with histones compacts the eukaryotic genome – essential for the architecture of the eukaryotic nucleus. It is, therefore, not surprising that viral genomes have been investigated for their level, if any, of chromatinisation; and early studies suggested that they were. Oudet et al. (1975) showed that lambda or adenovirus DNA, when incubated with histone proteins, reconstituted bona fide nucleosome structures in vitro. Later experiments, using the SV40 promoter-enhancer element (Cremisi et al., 1975; Cereghini and Yaniv, 1984), showed a similar association between viral DNA sequences and cellular histones. Although in many instances it is still unclear to what extent chromatinisation of viral genomes exactly mimics the structure of cellular chromatin, it has emerged that a number of herpesvirus genomes are associated with histones and that changes in chromatin structure appear to specifically regulate viral gene expression (for only a few examples, see Radkov et al., 1999; Krithivas et al., 2000; Kubat et al., 2004) and HCMV appears to be no exception (Murphy et al., 2002; Nevels et al., 2004; Reeves et al., 2005b, 2006; Cuevas-Bennett and Shenk, 2008; Nitzsche et al., 2008; Groves et al., 2009).

On this basis, an obvious mechanism for regulating HCMV IE gene expression and, therefore, reactivation from latency, was suggested to be changes in chromatin structure. Consistent with this, early experiments indicated that chromatin modifying enzymes had a critical role in regulating MIEP activity and virus infection. Treatment of ordinarily non-permissive T2 cells with the broad-spectrum HDAC inhibitor Trichostatin A (TSA) rendered these cells permissive for viral IE gene expression (Meier, 2001; Murphy et al., 2002). Moreover, overexpression of HDACs in normally permissive, differentiated, T2 cells (T2RA cells) inhibited viral IE gene expression following infection (Murphy et al., 2002). Similar results were observed in transient transfection assays of these cells with MIEP reporter constructs; TSA treatment of T2 cells rendered the

cells permissive for MIEP activity, whereas super-expression of HDACs in T2RA cells repressed MIEP activity (Murphy et al., 2002).

The apparent importance of HDACs in regulating the MIEP also strongly suggested that the MIEP might be subject to chromatinisation during latency. Chromatin immunoprecipitation (ChIP) assays are an extremely informative way to characterize the binding of proteins to DNA. Such an approach has now been used extensively to study the signature of different histone modifications around promoters of interest and how they are predicted to impact on viral gene expression. Using these assays it has been shown that, in experimentally infected undifferentiated non-permissive cells in which the MIEP is transcriptionally inactive, the MIEP is preferentially associated with deacetylated histones (Murphy et al., 2002; Reeves et al., 2005a). This hypoacetylation of histones associated with the MIEP was observed after infection of non-permissive T2 cells, peripheral blood monocytes and CD34$^+$ progenitor cells. Likewise, in primary monocytes and CD34$^+$ cells, similar analyses showed that not only was the MIEP associated with hypo-acetylated histones but it was also associated with the silencing protein HP1 (Murphy et al., 2002; Reeves et al., 2005a). These data are entirely consistent with the transcriptional silencing of the MIEP in these undifferentiated cells being due to a closed chromatin conformation. Conversely, infection of cells after they had been differentiated to a permissive phenotype showed an association of the MIEP with acetylated histones and a concomitant loss of association with HP1: all entirely consistent with transcriptional activation of the MIEP, expression of the IE proteins and a productive lytic infection (Murphy et al., 2002). Consequently, the differentiation–dependent state of chromatin around the MIEP in these experimentally infected cell lines was shown to play a pivotal role in regulating lytic gene expression and hence the outcome of infection.

Intriguingly, more recent data also suggest the presence of a pre-immediate-early phase of lytic viral gene expression in permissive cells that is intrinsically repressed (Groves et al., 2009) and that this repression is centred around the action of ND10 bodies (Maul, 2008). These nuclear bodies are sites of viral genome deposition and comprise a number of cellular proteins, including Daxx, PML and Sp100 as well as histone modifying enzymes, which are intrinsically able to repress viral gene expression (Tavalai and Stamminger, 2011). Consequently, it has now become clear that virus-induced inactivation of these repressive domains appears critical to allow lytic infection to proceed efficiently (Ishov et al., 1997; Hofmann *et al.*, 2002; Saffert and Kalejta, 2006; Tavalai *et al.*, 2006, 2008, 2011; Woodhall *et al.*, 2006; Lukashchuk *et al.*, 2008).

The question whether these repressive domains play any role in establishment or control of latency is a moot point. It has been suggested that the failure of the viral tegument protein pp71, which is able to disrupt ND10 (Kalejta, 2008), to translocate to the nucleus in CD34⁺ cells is a critical determinant of a latent outcome to infection (Saffert *et al.*, 2010). Although an appealing hypothesis, it does not lessen the importance of long standing observations that indicate the composition of the transcription factor milieu in non-permissive cells is intrinsically biased towards repression of the MIEP (Sinclair and Sissons, 2006) – a bias that could be augmented by a very specific signature of cellular gene expression triggered by HCMV during the initial stages of infection of haematopoietic cells (Slobedman *et al.*, 2004).

It is also noteworthy that a transient inhibition of lytic infection can also be observed in a cell cycle-dependent manner. Longstanding observations have illustrated that the infection of fibroblast cells in the S/G$_2$ phase of the cell cycle is temporarily non-permissive (Salvant *et al.*, 1998; Fortunato *et al.*, 2002). A partial rescue of lytic gene expression via the inhibition of the proteasome indicated that the regulation involved the activity of a dynamically regulated protein or family of proteins (Fortunato *et al.*, 2002). More recent data has suggested the repression of viral gene expression observed in S/G2 infected cells is due to cyclin-dependent kinase (cdk) activity (Zydek *et al.*, 2010) and not due to anti-viral components of ND10 bodies (Zydek *et al.*, 2011) and, as such, is overcome by the activity of the broad cdk inhibitor, Roscovitine. Interestingly, this cdk-mediated inhibition of IE gene expression is also argued to be important for promoting a quiescent infection of T2 cells and thus may have implications for the control of HCMV latency (Zydek *et al.*, 2010).

Clearly, many of these observations have been based on the experimental infection of cells *in vitro*. Consequently, of crucial importance is the question whether they can be re-capitulated in natural latency, and, tellingly, it has been shown they can. The direct analysis of either CD34⁺ cells or CD14⁺ monocytes isolated from healthy seropositive donors have also clearly shown that the MIEP of latent genomes is predominantly associated with methylated histones and the silencing protein, HP-1 (Reeves *et al.*, 2005b). Crucially, it was observed that the differentiation of either of these myeloid cells into terminally differentiated dendritic cells resulted in the reactivation of viral gene expression concomitant with the detection of the MIEP predominantly interacting with acetylated histone H4 (Reeves *et al.*, 2005b).

Studies on the state of chromatin surrounding IE, early and late genes during lytic infection have clearly shown a good correlation between histone marks associated with repression or activation on viral promoters and their temporal expression; at IE times of infection IE promoters are associated with histone marks of transcriptional activity whereas late promoters are associated with histone marks of transcriptional repression (Cuevas-Bennett and Shenk, 2008; Groves *et al.*, 2009). Similarly, as lytic infection progresses, late promoters become associated with acetylated histones (concomitant with induction of their expression) and the MIEP becomes re-associated with repressive histone marks consistent with MIEP repression at late times of infection (Reeves *et al.*, 2006; Cuevas-Bennett and Shenk, 2008; Groves *et al.*, 2009).

These observations would predict that during latent infection, the promoters of genes expressed during latency should be associated with acetylated histones; and this proved to be the case. Analysis of one latency-associated viral gene, expressing the UL81–82ast transcript, has shown that in naturally latently infected CD34⁺, cells where the MIEP is associated with histone marks of transcriptional repression as expected, the UL81–82ast promoter is indeed associated with acetylated histones (Reeves and Sinclair, 2010). Thus regulation of viral gene expression by chromatin structure during latency appears not to be restricted to just the transcriptional repression of the MIEP.

Finally, what of other mechanisms of epigenetic regulation such as DNA methylation? The regulation of eukaryotic promoters has been shown to involve cross-talk between histone modification and DNA methylation (Fuks, 2005). However, the case for the regulation of HCMV latency by DNA methylation is less clear. Early studies using transfection analyses suggested, to varying degrees, that the MIEP was subject to DNA methylation but with very different effects on functionality (Boom *et al.*, 1987; Honess *et al.*, 1989; Prosch *et al.*, 1996). More recent data using the murine CMV model has suggested that although MCMV latency recapitulates the observations seen with latent HCMV regarding extensive chromatin mediated regulation (Liu *et al.*, 2008), direct DNA methylation does not appear important for the regulation of MCMV latency (Hummel *et al.*, 2007).

Factors which direct chromatin-mediated regulation of the HCMV MIEP

Clearly the chromatin structure around the MIEP is likely to play an important role in regulating its activity during latency and reactivation, but this begs the question as to what regulates the acetylation/deacetylation state of MIEP-associated histones in cells carrying viral genomes?

Research on the regulation of the MIEP at the transcription factor level has identified a number of candidate proteins important for repression. A number of studies have shown that, using transient transfection assays with truncated versions of the MIEP as reporter constructs, distinct regions of the MIEP either positively or negatively regulate MIEP activity in non-permissive/permissive cell types. For example, the 21 base pair repeats and the dyad symmetry element (modulator) are negative regulatory elements for MIEP activity in non-permissive cell types (Nelson et al., 1987; Lubon et al., 1989; Kothari et al., 1991). Further analyses of such negative regulatory sequences, using in vitro and in vivo assays, has allowed the identification of binding sites for cellular transcription factors. Analyses of these factors confirm their binding to, and repression of, the MIEP. Factors so far identified include yin yang-1 (YY1; Liu et al., 1994), Ets2 repressor factor (ERF; Bain et al., 2003), growth factor independence-1 (Gfi-1; Zweidler-Mckay et al., 1996) and modulator recognition factor (MRF; Huang et al., 1996).

Current models suggest that some of these factors – YY1 for example – are preferentially expressed in non-permissive cell types where they repress MIEP activity and, following differentiation to the permissive phenotype, the expression levels of these factors decrease (Liu et al., 1994). In contrast, no such changes were observed in the absolute levels of expression of YY1 and ERF during dendritic cell differentiation (Reeves et al., 2005b), suggesting that the factors already identified may be part of a much more complex mechanism of regulation.

What is intriguing, though, is that several of the factors identified as binding to, and repressing, the MIEP are thought to mediate their effects through interactions with histone modifiers or their co-factors. For example, it is known that YY1 interacts with multiple HDAC family members (HDAC-1, -2 and -3; Yang et al., 1996). Similarly, ERF interacts with HDAC-1 and this interaction is important for ERF to mediate repression of the MIEP and may also serve to recruit histone methyltransferase SUvar to the MIEP (Wright et al., 2005). This suggests a multistep process of initial binding of repressor factors that sequentially recruit chromatin remodelling enzymes, eventually leading to transcriptional repression. Following differentiation to permissive phenotypes, however, the level of YY1, ERF and/or HDAC-1, -2 and -3 decreases; an observation which is common to many of the model systems discussed here including T2 cells, THP-1 cells, primary monocytes and CD34+ cells (Liu et al., 1994; Murphy et al., 2002; Reeves et al., 2005b). These observations, showing de-repression of the MIEP, support a model in which the MIEP in non-permissive or latent cell types is inhibited by the action of numerous cellular transcriptional repressors. These act through multiple sites within the MIEP and function by recruiting HDAC and methylase activities, leading to a more compact chromatin structure and transcriptional silencing (Fig. I.19.4). In contrast, in permissive cells or upon virus reactivation, levels of repressor complexes are reduced resulting in chromatin remodelling to a more open chromatin configuration and subsequent transcriptional activation of the MIEP.

HCMV reactivation and interactions with histone acetyltransferases

It is not unreasonable to predict that the chromatin-based mechanisms that control HCMV latency are also pivotal to the molecular switch needed to reactivate a latent genome. As outlined above, the current model for the differentiation state-dependent control of the MIEP relies on a delicate interplay between transcriptional repressors and transcriptional activators which, in part, dictate higher order chromatin structure. It is likely that reactivation of HCMV is concomitant with a number of different events that are acting cooperatively. At a cellular level, myeloid cell differentiation and inflammation are probably crucial events. At the molecular level, inflammation-associated signalling, an increase in the activation and/or levels of transcriptional activators (and probably their functional co-factors), which necessarily results in the modification of histone proteins bound to the MIEP are probably key players in reactivation. Perhaps the most studied activators are NF-κB and CREB which have numerous binding sites within the MIEP (see Chapter I.10). Interestingly, recent studies that have used the T2 quiescent infection model have suggested that CREB binding and function is an important trigger to break MIEP silencing in this model system (Keller et al., 2007; Yuan et al., 2009; Liu et al., 2010). Furthermore, a number of clinical studies have implicated inflammatory TNF-α and subsequent activation of NF-κB signalling as concomitant events associated with elevated HCMV reactivation in patients (Docke et al., 1994; Prösch et al., 2002). As research progresses, it will be intriguing to see how in vitro observations from a number of experimental systems can be integrated with the recent observations using the humanized mouse model (Smith et al., 2010) to help develop a broader understanding of how HCMV reactivates from latency so effectively in vivo.

It is of interest to note that at least some of these

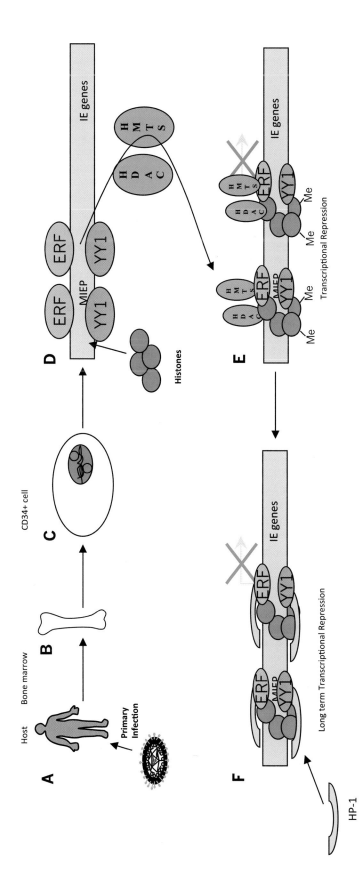

Figure I.19.4 Model for the regulation of the MIEP during the establishment of a latent state. Following primary infection of an immuno-competent host (A), the virus establishes a latent infection of cells resident in the bone marrow (B). Latent infection of the CD34+ haematopoietic progenitor cells (C) is defined by an absence of major IE gene expression and a lack of infectious virus production. At a molecular level, the repression of the MIEP is pivotal for the establishment of HCMV latency and is a concert of the function of cellular transcriptional repressors and higher order chromatin structure. Specifically, high levels of transcriptional repressors present in CD34+ cells bind to the MIEP (D) and promote the recruitment of histone modifying enzymes (E). The recruitment of histone deacetylases (HDACs) and histone methyltransferases (HMTs) promotes the methylation of histones bound to the MIEP. These tri-methylated histones are then targets for the recruitment of further silencing proteins such as heterochromatin protein-1 (HP-1) which augment the transcriptional repression of the MIEP (F). Reproduced with kind permission from Springer Science & Business Media: *Current Topics in Immunology and Microbiology*, Chapter: Aspects of Human Cytomegalovirus Latency and Reactivation 325 (2008), pp. 297–313 M. Reeves & J. Sinclair Fig. 2.

activating factors mediate their effects by the recruitment of HATs. For example, the p65 component of NF-kB interacts with CBP that has intrinsic HAT activity (Sheppard *et al.*, 1998; Wadgaonkar *et al.*, 1999). Thus, in an analogous manner to the recruitment of HDACs and HMTs to the MIEP by ERF and YY1, it is tempting to speculate that the recruitment of HAT activity could be equally as important in directing high-level MIEP activity as the action of HDACs are in causing MIEP repression. Such recruitment of activatory factors mediated by differentiation and/or inflammation may be of real importance in the context of HCMV reactivation from latency *in vivo*.

The importance of post-translational modification of histones in the establishment and maintenance of the latent HCMV genome, as well as in reactivation from latency and lytic infection is clear. However, it is also likely that changes in chromatin structure of cellular gene promoters are important during the course of a productive lytic infection. It is known that HCMV infection has profound effects on normal host gene expression. Precisely how all these cellular genes are affected is far from clear, but it is possible that at least some of them are activated/repressed in response to histone modification around their promoters. Such effects are likely to be mediated by viral factors in concert with cellular transcription factors. One such candidate viral factor is the potent and promiscuous transcriptional activator immediate-early factor IE2-p86 (see Chapter I.10). This protein is known to activate a wide range of viral genes, ensuring E and L viral gene expression, but is also known to promiscuously activate cellular gene expression (reviewed in Stenberg, 1996). The mechanisms by which IE2-p86 mediates such effects are not fully understood but it is probable that they also involve IE2-p86 targeted changes in histone modification. For instance, it has been shown that IE2-p86 interacts with the cellular protein P/CAF, a protein with HAT activity, and this interaction enhances IE2-p86-mediated activation of target cellular promoters (Bryant *et al.*, 2000).

Conversely, it has also been shown that IE2-p86 can also interact with the acetyltransferase domains of proteins such as p300 and CBP. These interactions block the intrinsic HAT activity of these proteins and, as a consequence, can affect the acetylation status, and therefore the function of, target proteins such as p53 (Hsu *et al.*, 2004). It is therefore likely that the acetylation status of promoters, and indeed transcription factors themselves, of many HCMV genes activated by IE2-p86 are also altered during the course of infection.

As discussed previously, there is significant regulation of all classes of HCMV genes by histones even during lytic infection (Cuevas-Bennett and Shenk,

2008; Groves *et al.*, 2009). Indeed, the growth defect associated with the IE72 deletion virus at low MOIs has been attributed to an absence of HDAC sequestration by IE72 (Nevels *et al.*, 2004). As well as IE72, IE86 also interacts with HDACs and HMTs (Reeves *et al.*, 2006) but these interactions are clearly not sufficient to rescue the IE72-deletion virus unlike, for example, TSA (Nevels *et al.*, 2004). Indeed, the IE86-mediated interaction is more likely to be important for the repression of the MIEP during the later phases of HCMV infection. Thus, two viral proteins target the same class of cellular proteins but with a very different functional outcome. In a microcosm, this exemplifies the complexity associated with understanding the intricacies of chromatin and the regulation of gene expression.

Concluding remarks

This review has detailed a number of studies that indicate a role for histone modification in the control of HCMV gene expression. It is clear that in many *in vitro* model systems, silencing of the MIEP in undifferentiated non-permissive cells appears to result from the association of this viral promoter with non-acetylated histones and silencing proteins such as HP1; heterochromatic proteins indicative of compacted transcriptionally inactive chromatin. Differentiation-induced activation of the viral MIEP, on the other hand, correlates with an open chromatin conformation. Similar changes in the chromatin structure of the viral MIEP of endogenous virus during natural latent infection are also seen in myeloid progenitors upon their differentiation to more mature myeloid DCs, resulting in reactivation of endogenous latent virus. Clearly, latency of HCMV in DCs during natural persistence, and this critical linkage of DCs with chromatin remodelling of the MIEP and subsequent reactivation, is likely to play an important role in the pathogenesis of HCMV infection and most likely is common theme during the terminal differentiation of myeloid progenitor cells to more differentiated cellular phenotypes (i.e. DCs and macrophages).

It is worth considering the role of other cell types and mechanisms beyond the regulation of the MIEP in the context of HCMV latency and reactivation. Studies on MCMV have highlighted that latency is also established in non-haematopoietic cell types and that the checkpoint controls during this and the reactivation phase are MIEP-independent (see Chapter I.22). Whether this is unique to MCMV biology or is actually illustrative of yet uncharacterised mechanisms governing HCMV latency remains to be investigated.

Consideration also needs to be given to the role of non-coding RNAs during epigenetic regulation. HCMV encodes a number of miRNAs that are expressed during lytic infection (Fannin Rider *et al.*, 2008; Grey and Nelson, 2008). Furthermore, HCMV also impacts on the expression profile of a targeted number of cellular encoded miRNAs during both lytic (Wang *et al.*, 2008) and latent infection (Poole *et al.*, 2011). The expression of miRNAs involves epigenetic regulation (Weber *et al.*, 2007) and, interestingly, a number of miRNA targets are histone modifying enzymes (Lewis *et al.*, 2005). Consequently, it has been hypothesized that miRNAs could also play a role in the regulation of gene expression through chromatin-mediated mechanisms. As such, a role for either cellular or viral miRNAs in the regulation of viral promoters via chromatin also becomes a possibility for further enquiry. Furthermore, the impact, if any, of recently identified small RNAs (Stark *et al.*, 2011) or the larger non-coding RNA transcripts (Gatherer *et al.*, 2011) on chromatin-mediated gene regulation needs to be assessed – particularly given that it has been reported the latency-associated transcript of herpes simplex virus facilitates heterochromatin formation on viral lytic promoters (Cliffe *et al.*, 2009).

It is also inevitable that the models described above will need to be revised or modified in the near future as our understanding of the histone code hypothesis develops. For example, no mention is made here of histone ubiquitination, phosphorylation or sumoylation and how they may influence HCMV biology. In addition, given the complexity of the HCMV MIEP and its multiplicity of binding sites for both cellular and viral factors, it may emerge that other, as yet uncharacterised, factors may use these processes for regulating HCMV gene expression. Nevertheless, the analysis of the molecular mechanisms regulating the viral MIEP with respect to chromatin structure has begun to shed new light on the mechanisms of viral latency and reactivation, which are a defining characteristic of the herpesvirus family, and will be crucial to our understanding of the biology of HCMV infection.

Acknowledgements

The authors thank past and present members of the laboratory whose work has contributed to the studies described here. The work from this laboratory was funded by the Medical Research Council and The Wellcome Trust.

References

Avdic, S., Cao, J.Z., Cheung, A.K., Abendroth, A., and Slobedman, B. (2011). Viral interleukin-10 expressed by human cytomegalovirus during the latent phase of infection modulates latently infected myeloid cell differentiation. J. Virol. 85, 7465–7471.

Bain, M., Mendelson, M., and Sinclair, J. (2003). Ets-2 Repressor Factor (ERF) mediates repression of the human cytomegalovirus major immediate-early promoter in undifferentiated non-permissive cells. J. Gen. Virol. 84, 41–49.

Bannister, A.J., and Kouzarides, T. (2011). Regulation of chromatin by histone modifications. Cell Res. 21, 381–395.

Bannister, A.J., Zegerman, P., Partridge, J.F., Miska, E.A., Thomas, J.O., Allshire, R.C., and Kouzarides, T. (2001). Selective recognition of methylated lysine 9 on histone H3 by the HP1 chromo domain. Nature 410, 120–124.

Bego, M., Maciejewski, J., Khaiboullina, S., Pari, G., and St Jeor, S. (2005). Characterization of an antisense transcript spanning the UL81–82 locus of human cytomegalovirus. J. Virol. 79, 11022–11034.

Bell, O., Tiwari, V.K., Thoma, N.H., and Schubeler, D. (2011). Determinants and dynamics of genome accessibility. Nat. Rev. Genet. 12, 554–564.

Bernstein, B.E., Humphrey, E.L., Erlich, R.L., Schneider, R., Bouman, P., Liu, J.S., Kouzarides, T., and Schreiber, S.L. (2002). Methylation of histone H3 Lys 4 in coding regions of active genes. Proc. Natl. Acad. Sci. U.S.A. 99, 8695–8700.

Bevan, I.S., Walker, M.R., and Daw, R.A. (1993). Detection of human cytomegalovirus DNA in peripheral blood leukocytes by the polymerase chain reaction. Transfusion 33, 783–784.

Bode, J., Gomez-Lira, M.M., and Schroter, H. (1983). Nucleosomal particles open as the histone core becomes hyperacetylated. Eur. J. Biochem. 130, 437–445.

Boom, R., Geelen, J.L., Sol, C.J., Minnaar, R.P., and van der Noordaa, J. (1987). Resistance to methylation de novo of the human cytomegalovirus immediate-early enhancer in a model for virus latency and reactivation in vitro. J. Gen. Virol. 68, 2839–2852.

Bryant, L.A., Mixon, P., Davidson, M., Bannister, A.J., Kouzarides, T., and Sinclair, J.H. (2000). The human cytomegalovirus 86-kilodalton major immediate-early protein interacts physically and functionally with histone acetyltransferase P/CAF. J. Virol. 74, 7230–7237.

Burgess, R.J., and Zhang, Z. (2010). Histones, histone chaperones and nucleosome assembly. Protein Cell 1, 607–612.

Cao, R., Wang, L., Wang, H., Xia, L., Erdjument-Bromage, H., Tempst, P., Jones, R.S., and Zhang, Y. (2002). Role of histone H3 lysine 27 methylation in Polycomb-group silencing. Science 298, 1039–1043.

Cary, P.D., Crane-Robinson, C., Bradbury, E.M., and Dixon, G.H. (1982). Effect of acetylation on the binding of N-terminal peptides of histone H4 to DNA. Eur. J. Biochem. 127, 137–143.

Casas-Delucchi, C.S., van Bemmel, J.G., Haase, S., Herce, H.D., Nowak, D., Meilinger, D., Stear, J.H., Leonhardt, H., and Cardoso, M.C. (2012). Histone hypoacetylation is required to maintain late replication timing of constitutive heterochromatin. Nucleic Acids Res. 40, 159–168.

Cereghini, S., and Yaniv, M. (1984). Assembly of transfected DNA into chromatin: structural changes in the origin-promoter-enhancer region upon replication. EMBO J. 3, 1243–1253.

Chang, W.L., and Barry, P.A. (2010). Attenuation of innate immunity by cytomegalovirus IL-10 establishes a long-term deficit of adaptive antiviral immunity. Proc. Natl. Acad. Sci. U.S.A. *107*, 22647–22652.

Cheng, X., and Blumenthal, R.M. (2010). Coordinated chromatin control: structural and functional linkage of DNA and histone methylation. Biochemistry *49*, 2999–3008.

Chung, S.Y., Hill, W.E., and Doty, P. (1978). Characterization of the histone core complex. Proc. Natl. Acad. Sci. U.S.A. *75*, 1680–1684.

Cliffe, A.R., Garber, D.A., and Knipe, D.M. (2009). Transcription of the herpes simplex virus latency-associated transcript promotes the formation of facultative heterochromatin on lytic promoters. J. Virol. *83*, 8182–8190.

Cremisi, C., Pignatti, P.F., Croissant, O., and Yaniv, M. (1975). Chromatin-like structures in polyoma virus and simian virus 10 lytic cycle. J. Virol. *17*, 204–211.

Cuevas-Bennett, C., and Shenk, T. (2008). Dynamic histone H3 acetylation and methylation at human cytomegalovirus promoters during replication in fibroblasts. J. Virol. *82*, 9525–9536.

Dhalluin, C., Carlson, J.E., Zeng, L., He, C., Aggarwal, A.K., and Zhou, M.M. (1999). Structure and ligand of a histone acetyltransferase bromodomain. Nature *399*, 491–496.

Docke, W.D., Prosch, S., Fietze, E., Kimel, V., Zuckermann, H., Klug, C., Syrbe, U., Kruger, D.H., von Baehr, R., and Volk, H.D. (1994). Cytomegalovirus reactivation and tumour necrosis factor. Lancet *343*, 268–269.

Doenecke, D., and Gallwitz, D. (1982). Acetylation of histones in nucleosomes. Mol. Cell Biochem. *44*, 113–128.

Dudley, A.M., Rougeulle, C., and Winston, F. (1999). The Spt components of SAGA facilitate TBP binding to a promoter at a post-activator-binding step in vivo. Genes Dev. *13*, 2940–2945.

Eberharter, A., and Becker, P.B. (2002). Histone acetylation: a switch between repressive and permissive chromatin. Second in review series on chromatin dynamics. EMBO Rep. *3*, 224–229.

Fannin Rider, P.J., Dunn, W., Yang, E., and Liu, F. (2008). Human cytomegalovirus microRNAs. Curr. Top. Microbiol. Immunol. *325*, 21–39.

Fortunato, E.A., Sanchez, V., Yen, J.Y., and Spector, D.H. (2002). Infection of cells with human cytomegalovirus during S phase results in a blockade to immediate-early gene expression that can be overcome by inhibition of the proteasome. J. Virol. *76*, 5369–5379.

Fuks, F. (2005). DNA methylation and histone modifications: teaming up to silence genes. Curr. Opin. Genet. Dev. *15*, 490–495.

Fuks, F., Hurd, P.J., Deplus, R., and Kouzarides, T. (2003a). The DNA methyltransferases associate with HP1 and the SUV39H1 histone methyltransferase. Nucleic Acids Res. *31*, 2305–2312.

Fuks, F., Hurd, P.J., Wolf, D., Nan, X., Bird, A.P., and Kouzarides, T. (2003b). The methyl-CpG-binding protein MeCP2 links DNA methylation to histone methylation. J. Biol. Chem. *278*, 4035–4040.

Gatherer, D., Seirafian, S., Cunningham, C., Holton, M., Dargan, D.J., Baluchova, K., Hector, R.D., Galbraith, J., Herzyk, P., Wilkinson, G.W., *et al.* (2011). High-resolution human cytomegalovirus transcriptome. Proc. Natl. Acad. Sci. U.S.A. *108*, 19755–19760.

Gilbert, G.L., Hayes, K., Hudson, I.L., and James, J. (1989). Prevention of transfusion-acquired cytomegalovirus infection in infants by blood filtration to remove leucocytes. Neonatal Cytomegalovirus Infection Study Group. Lancet *1*, 1228–1231.

Gonczol, E., Andrews, P.W., and Plotkin, S.A. (1984). Cytomegalovirus replicates in differentiated but not in undifferentiated human embryonal carcinoma cells. Science *224*, 159–161.

Goodrum, F., Jordan, C.T., Terhune, S.S., High, K., and Shenk, T. (2004). Differential outcomes of human cytomegalovirus infection in primitive hematopoietic cell subpopulations. Blood *104*, 687–695.

Goodrum, F., Reeves, M., Sinclair, J., High, K., and Shenk, T. (2007). Human cytomegalovirus sequences expressed in latently infected individuals promote a latent infection in vitro. Blood *110*, 937–945.

Goodrum, F.D., Jordan, C.T., High, K., and Shenk, T. (2002). Human cytomegalovirus gene expression during infection of primary hematopoietic progenitor cells: a model for latency. Proc. Natl. Acad. Sci. U.S.A. *99*, 16255–16260.

Grey, F., and Nelson, J. (2008). Identification and function of human cytomegalovirus microRNAs. J. Clin. Virol. *41*, 186–191.

Groves, I.J., Reeves, M.B., and Sinclair, J.H. (2009). Lytic infection of permissive cells with human cytomegalovirus is regulated by an intrinsic 'pre-immediate-early' repression of viral gene expression mediated by histone post-translational modification. J. Gen. Virol. *90*, 2364–2374.

Guenther, M.G., Levine, S.S., Boyer, L.A., Jaenisch, R., and Young, R.A. (2007). A chromatin landmark and transcription initiation at most promoters in human cells. Cell *130*, 77–88.

Hahn, G., Jores, R., and Mocarski, E.S. (1998). Cytomegalovirus remains latent in a common precursor of dendritic and myeloid cells. Proc. Natl. Acad. Sci. U.S.A. *95*, 3937–3942.

Hargett, D., and Shenk, T.E. (2010). Experimental human cytomegalovirus latency in CD14+ monocytes. Proc. Natl. Acad. Sci. U.S.A. *107*, 20039–20044.

Hertel, L., Lacaille, V.G., Strobl, H., Mellins, E.D., and Mocarski, E.S. (2003). Susceptibility of immature and mature Langerhans cell-type dendritic cells to infection and immunomodulation by human cytomegalovirus. J. Virol. *77*, 7563–7574.

Hewish, D.R., and Burgoyne, L.A. (1973). Chromatin sub-structure. The digestion of chromatin DNA at regularly spaced sites by a nuclear deoxyribonuclease. Biochem. Biophys. Res. Commun. *52*, 504–510.

Hofmann, H., Sindre, H., and Stamminger, T. (2002). Functional interaction between the pp71 protein of human cytomegalovirus and the PML-interacting protein human Daxx. J. Virol. *76*, 5769–5783.

Honess, R.W., Gompels, U.A., Barrell, B.G., Craxton, M., Cameron, K.R., Staden, R., Chang, Y.N., and Hayward, G.S. (1989). Deviations from expected frequencies of CpG dinucleotides in herpesvirus DNAs may be diagnostic of differences in the states of their latent genomes. J. Gen. Virol. *70*, 837–855.

Hsu, C.H., Chang, M.D., Tai, K.Y., Yang, Y.T., Wang, P.S., Chen, C.J., Wang, Y.H., Lee, S.C., Wu, C.W., and Juan, L.J. (2004). HCMV IE2-mediated inhibition of HAT activity down-regulates p53 function. EMBO J. *23*, 2269–2280.

Huang, T.H., Oka, T., Asai, T., Okada, T., Merrills, B.W., Gertson, P.N., Whitson, R.H., and Itakura, K. (1996). Repression by a differentiation-specific factor of the human cytomegalovirus enhancer. Nucleic Acids Res. 24, 1695–1701.

Hummel, M., Yan, S., Li, Z., Varghese, T.K., and Abecassis, M. (2007). Transcriptional reactivation of murine cytomegalovirus i.e. gene expression by 5-aza-2′-deoxycytidine and trichostatin A in latently infected cells despite lack of methylation of the major immediate-early promoter. J. Gen. Virol. 88, 1097–1102.

Ishov, A.M., Stenberg, R.M., and Maul, G.G. (1997). Human cytomegalovirus immediate-early interaction with host nuclear structures: definition of an immediate transcript environment. J. Cell Biol. 138, 5–16.

Jacobson, R.H., Ladurner, A.G., King, D.S., and Tjian, R. (2000). Structure and function of a human TAFII250 double bromodomain module. Science 288, 1422–1425.

James, T.C., Eissenberg, J.C., Craig, C., Dietrich, V., Hobson, A., and Elgin, S.C. (1989). Distribution patterns of HP1, a heterochromatin-associated nonhistone chromosomal protein of Drosophila. Eur. J. Cell Biol. 50, 170–180.

Jenkins, C., Abendroth, A., and Slobedman, B. (2004). A novel viral transcript with homology to human interleukin-10 is expressed during latent human cytomegalovirus infection. J. Virol. 78, 1440–1447.

Kalejta, R.F. (2008). Functions of human cytomegalovirus tegument proteins prior to immediate-early gene expression. Curr. Top. Microbiol. Immunol. 325, 101–115.

Keller, M.J., Wu, A.W., Andrews, J.I., McGonagill, P.W., Tibesar, E.E., and Meier, J.L. (2007). Reversal of human cytomegalovirus major immediate-early enhancer/promoter silencing in quiescently infected cells via the cyclic AMP signaling pathway. J. Virol. 81, 6669–6681.

Khochbin, S., Verdel, A., Lemercier, C., and Seigneurin-Berny, D. (2001). Functional significance of histone deacetylase diversity. Curr. Opin. Genet. Dev. 11, 162–166.

Khorasanizadeh, S. (2004). The nucleosome: from genomic organization to genomic regulation. Cell 116, 259–272.

Kondo, K., Kaneshima, H., and Mocarski, E.S. (1994). Human cytomegalovirus latent infection of granulocyte–macrophage progenitors. Proc. Natl. Acad. Sci. U.S.A. 91, 11879–11883.

Kondo, K., Xu, J., and Mocarski, E.S. (1996). Human cytomegalovirus latent gene expression in granulocyte–macrophage progenitors in culture and in seropositive individuals. Proc. Natl. Acad. Sci. U.S.A. 93, 11137–11142.

Kornberg, R.D. (1974). Chromatin structure: a repeating unit of histones and DNA. Science 184, 868–871.

Kothari, S., Baillie, J., Sissons, J.G., and Sinclair, J.H. (1991). The 21bp repeat element of the human cytomegalovirus major immediate-early enhancer is a negative regulator of gene expression in undifferentiated cells. Nucleic Acids Res. 19, 1767–1771.

Krithivas, A., Young, D.B., Liao, G., Greene, D., and Hayward, S.D. (2000). Human herpesvirus 8 LANA interacts with proteins of the mSin3 corepressor complex and negatively regulates Epstein–Barr virus gene expression in dually infected PEL cells. J. Virol. 74, 9637–9645.

Kubat, N.J., Tran, R.K., McAnany, P., and Bloom, D.C. (2004). Specific histone tail modification and not DNA methylation is a determinant of herpes simplex virus type 1 latent gene expression. J. Virol. 78, 1139–1149.

Kuo, M.H., and Allis, C.D. (1998). Roles of histone acetyltransferases and deacetylases in gene regulation. Bioessays 20, 615–626.

Kwon, S.H., and Workman, J.L. (2011). The changing faces of HP1: From heterochromatin formation and gene silencing to euchromatic gene expression: HP1 acts as a positive regulator of transcription. Bioessays 33, 280–289.

Lachner, M., O'Sullivan, R.J., and Jenuwein, T. (2003). An epigenetic road map for histone lysine methylation. J. Cell Sci. 116, 2117–2124.

Lathey, J.L., and Spector, S.A. (1991). Unrestricted replication of human cytomegalovirus in hydrocortisone-treated macrophages. J. Virol. 65, 6371–6375.

Lewis, B.P., Burge, C.B., and Bartel, D.P. (2005). Conserved seed pairing, often flanked by adenosines, indicates that thousands of human genes are microRNA targets. Cell 120, 15–20.

Lilley, D.M., and Pardon, J.F. (1979). Structure and function of chromatin. Annu. Rev. Genet. 13, 197–233.

Liu, R., Baillie, J., Sissons, J.G., and Sinclair, J.H. (1994). The transcription factor YY1 binds to negative regulatory elements in the human cytomegalovirus major immediate-early enhancer/promoter and mediates repression in non-permissive cells. Nucleic Acids Res. 22, 2453–2459.

Liu, X., Yuan, J., Wu, A.W., McGonagill, P.W., Galle, C.S., and Meier, J.L. (2010). Phorbol ester-induced human cytomegalovirus major immediate-early (MIE) enhancer activation through PKC-delta, CREB, and NF-kappaB desilences MIE gene expression in quiescently infected human pluripotent NTera2 cells. J. Virol. 84, 8495–8508.

Liu, X.F., Yan, S., Abecassis, M., and Hummel, M. (2008). Establishment of murine cytomegalovirus latency in vivo is associated with changes in histone modifications and recruitment of transcriptional repressors to the major immediate-early promoter. J. Virol. 82, 10922–10931.

Lopez-Rodas, G., Brosch, G., Georgieva, E.I., Sendra, R., Franco, L., and Loidl, P. (1993). Histone deacetylase. A key enzyme for the binding of regulatory proteins to chromatin. FEBS Lett. 317, 175–180.

Lubon, H., Ghazal, P., Hennighausen, L., Reynolds-Kohler, C., Lockshin, C., and Nelson, J. (1989). Cell-specific activity of the modulator region in the human cytomegalovirus major immediate-early gene. Mol. Cell Biol. 9, 1342–1345.

Lukashchuk, V., McFarlane, S., Everett, R.D., and Preston, C.M. (2008). Human cytomegalovirus protein pp71 displaces the chromatin-associated factor ATRX from nuclear domain 10 at early stages of infection. J. Virol. 82, 12543–12554.

Lunetta, J.M., and Wiedeman, J.A. (2000). Latency-associated sense transcripts are expressed uring in vitro human cytomegalovirus productive infection. Virology 278, 467–476.

Lusser, A. (2002). Acetylated, methylated, remodeled: chromatin states for gene regulation. Curr. Opin. Plant Biol. 5, 437–443.

Maciejewski, J.P., Bruening, E.E., Donahue, R.E., Mocarski, E.S., Young, N.S., and St Jeor, S.C. (1992). Infection of hematopoietic progenitor cells by human cytomegalovirus. Blood 80, 170–178.

Maul, G.G. (2008). Initiation of cytomegalovirus infection at ND10. Curr. Top. Microbiol. Immunol. 325, 117–132.

Meier, J.L. (2001). Reactivation of the human cytomegalovirus major immediate-early regulatory region and viral replication in embryonal NTera2 cells: role of trichostatin A, retinoic acid, and deletion of the 21-base-pair repeats and modulator. J. Virol. 75, 1581–1593.

Mendelson, M., Monard, S., Sissons, P., and Sinclair, J. (1996). Detection of endogenous human cytomegalovirus in

CD34+ bone marrow progenitors. J. Gen. Virol. *77*, 3099–3102.

Mermoud, J.E., Popova, B., Peters, A.H., Jenuwein, T., and Brockdorff, N. (2002). Histone H3 lysine 9 methylation occurs rapidly at the onset of random X chromosome inactivation. Curr. Biol. *12*, 247–251.

Metzger, E., Wissmann, M., Yin, N., Muller, J.M., Schneider, R., Peters, A.H., Gunther, T., Buettner, R., and Schule, R. (2005). LSD1 demethylates repressive histone marks to promote androgen-receptor-dependent transcription. Nature *437*, 436–439.

Minton, E.J., Tysoe, C., Sinclair, J.H., and Sissons, J.G. (1994). Human cytomegalovirus infection of the monocyte/macrophage lineage in bone marrow. J. Virol. *68*, 4017–4021.

Mizzen, C.A., and Allis, C.D. (1998). Linking histone acetylation to transcriptional regulation. Cell. Mol. Life Sci. *54*, 6–20.

Montag, C., Wagner, J.A., Gruska, I., Vetter, B., Wiebusch, L., and Hagemeier, C. (2011). The latency-associated UL138 gene product of human cytomegalovirus sensitizes cells to TNF{alpha} signaling by up-regulating TNF{alpha} receptor 1 cell surface expression. J. Virol. *85*, 11409–11421.

Morgan, J.E., Blankenship, J.W., and Matthews, H.R. (1987). Polyamines and acetylpolyamines increase the stability and alter the conformation of nucleosome core particles. Biochemistry *26*, 3643–3649.

Murphy, J.C., Fischle, W., Verdin, E., and Sinclair, J.H. (2002). Control of cytomegalovirus lytic gene expression by histone acetylation. EMBO J. *21*, 1112–1120.

Nelson, J.A., Reynolds-Kohler, C., and Smith, B.A. (1987). Negative and positive regulation by a short segment in the 5′-flanking region of the human cytomegalovirus major immediate-early gene. Mol. Cell. Biol. *7*, 4125–4129.

Nevels, M., Paulus, C., and Shenk, T. (2004). Human cytomegalovirus immediate-early 1 protein facilitates viral replication by antagonizing histone deacetylation. Proc. Natl. Acad. Sci. U.S.A. *101*, 17234–17239.

Nitzsche, A., Paulus, C., and Nevels, M. (2008). Temporal dynamics of cytomegalovirus chromatin assembly in productively infected human cells. J. Virol. *82*, 11167–11180.

Oudet, P., Gross-Bellard, M., and Chambon, P. (1975). Electron microscopic and biochemical evidence that chromatin structure is a repeating unit. Cell *4*, 281–300.

Petrucelli, A., Rak, M., Grainger, L., and Goodrum, F. (2009). Characterization of a novel Golgi apparatus-localized latency determinant encoded by human cytomegalovirus. J. Virol. *83*, 5615–5629.

Plath, K., Fang, J., Mlynarczyk-Evans, S.K., Cao, R., Worringer, K.A., Wang, H., de la Cruz, C.C., Otte, A.P., Panning, B., and Zhang, Y. (2003). Role of histone H3 lysine 27 methylation in X inactivation. Science *300*, 131–135.

Poole, E., McGregor Dallas, S.R., Colston, J., Joseph, R.S., and Sinclair, J. (2011). Virally induced changes in cellular microRNAs maintain latency of human cytomegalovirus in CD34 progenitors. J. Gen. Virol. *92*, 1539–1549.

Preston, C.M. (2000). Repression of viral transcription during herpes simplex virus latency. J. Gen. Virol. *81*, 1–19.

Prösch, S., Stein, J., Staak, K., Liebenthal, C., Volk, H.D., and Kruger, D.H. (1996). Inactivation of the very strong HCMV immediate-early promoter by DNA CpG methylation in vitro. Biol. Chem. Hoppe Seyler *377*, 195–201.

Prösch, S., Wuttke, R., Kruger, D.H., and Volk, H.D. (2002). NF-kappaB – a potential therapeutic target for inhibition of human cytomegalovirus (re)activation? Biol. Chem. *383*, 1601–1609.

Radkov, S.A., Touitou, R., Brehm, A., Rowe, M., West, M., Kouzarides, T., and Allday, M.J. (1999). Epstein–Barr virus nuclear antigen 3C interacts with histone deacetylase to repress transcription. J. Virol. *73*, 5688–5697.

Raftery, M.J., Wieland, D., Gronewald, S., Kraus, A.A., Giese, T., and Schonrich, G. (2004). Shaping phenotype, function, and survival of dendritic cells by cytomegalovirus-encoded IL-10. J. Immunol. *173*, 3383–3391.

Razin, A. (1998). CpG methylation, chromatin structure and gene silencing-a three-way connection. EMBO J. *17*, 4905–4908.

Reeves, M., Murphy, J., Greaves, R., Fairley, J., Brehm, A., and Sinclair, J. (2006). Autorepression of the human cytomegalovirus major immediate-early promoter/enhancer at late times of infection is mediated by the recruitment of chromatin remodeling enzymes by IE86. J. Virol. *80*, 9998–10009.

Reeves, M.B., and Compton, T. (2011). Inhibition of inflammatory interleukin-6 activity via ERK-MAPK signaling antagonizes human cytomegalovirus reactivation from dendritic cells from latency. J. Virol. *85*, 12750–12578.

Reeves, M.B., and Sinclair, J.H. (2010). Analysis of latent viral gene expression in natural and experimental latency models of human cytomegalovirus and its correlation with histone modifications at a latent promoter. J. Gen. Virol. *91*, 599–604.

Reeves, M.B., Lehner, P.J., Sissons, J.G., and Sinclair, J.H. (2005a). An in vitro model for the regulation of human cytomegalovirus latency and reactivation in dendritic cells by chromatin remodelling. J. Gen. Virol. *86*, 2949–2954.

Reeves, M.B., MacAry, P.A., Lehner, P.J., Sissons, J.G., and Sinclair, J.H. (2005b). Latency, chromatin remodeling, and reactivation of human cytomegalovirus in the dendritic cells of healthy carriers. Proc. Natl. Acad. Sci. U.S.A. *102*, 4140–4145.

Rice, J.C., Briggs, S.D., Ueberheide, B., Barber, C.M., Shabanowitz, J., Hunt, D.F., Shinkai, Y., and Allis, C.D. (2003). Histone methyltransferases direct different degrees of methylation to define distinct chromatin domains. Mol. Cell *12*, 1591–1598.

Riegler, S., Hebart, H., Einsele, H., Brossart, P., Jahn, G., and Sinzger, C. (2000). Monocyte-derived dendritic cells are permissive to the complete replicative cycle of human cytomegalovirus. J. Gen. Virol. *81*, 393–399.

Saffert, R.T., and Kalejta, R.F. (2006). Inactivating a cellular intrinsic immune defense mediated by Daxx is the mechanism through which the human cytomegalovirus pp71 protein stimulates viral immediate-early gene expression. J. Virol. *80*, 3863–3871.

Saffert, R.T., Penkert, R.R., and Kalejta, R.F. (2010). Cellular and viral control over the initial events of human cytomegalovirus experimental latency in CD34+ cells. J. Virol. *84*, 5594–5604.

Sallusto, F., and Lanzavecchia, A. (1994). Efficient presentation of soluble antigen by cultured human dendritic cells is maintained by granulocyte/macrophage colony-stimulating factor plus interleukin 4 and down-regulated by tumor necrosis factor alpha. J. Exp. Med. *179*, 1109–1118.

Salvant, B.S., Fortunato, E.A., and Spector, D.H. (1998). Cell cycle dysregulation by human cytomegalovirus: influence of the cell cycle phase at the time of infection and effects on cyclin transcription. J. Virol. *72*, 3729–3741.

Schotta, G., Ebert, A., Krauss, V., Fischer, A., Hoffmann, J., Rea, S., Jenuwein, T., Dorn, R., and Reuter, G. (2002). Central role of *Drosophila* SU(VAR)3–9 in histone H3-K9 methylation and heterochromatic gene silencing. EMBO J. *21*, 1121–1131.

Shi, Y.J., Matson, C., Lan, F., Iwase, S., Baba, T., and Shi, Y. (2005). Regulation of LSD1 histone demethylase activity by its associated factors. Mol. Cell *19*, 857–864.

Shiio, Y., and Eisenman, R.N. (2003). Histone sumoylation is associated with transcriptional repression. Proc. Natl. Acad. Sci. U.S.A. *100*, 13225–13230.

Sinclair, J., and Sissons, P. (2006). Latency and reactivation of human cytomegalovirus. J. Gen. Virol. *87*, 1763–1779.

Slobedman, B., and Mocarski, E.S. (1999). Quantitative analysis of latent human cytomegalovirus. J. Virol. *73*, 4806–4812.

Slobedman, B., Stern, J.L., Cunningham, A.L., Abendroth, A., Abate, D.A., and Mocarski, E.S. (2004). Impact of human cytomegalovirus latent infection on myeloid progenitor cell gene expression. J. Virol. *78*, 4054–4062.

Smith, M.S., Goldman, D.C., Bailey, A.S., Pfaffle, D.L., Kreklywich, C.N., Spencer, D.B., Othieno, F.A., Streblow, D.N., Garcia, J.V., Fleming, W.H., *et al.* (2010). Granulocyte-colony stimulating factor reactivates human cytomegalovirus in a latently infected humanized mouse model. Cell Host Microbe *8*, 284–291.

Soderberg-Naucler, C., Fish, K.N., and Nelson, J.A. (1997a). Interferon-gamma and tumor necrosis factor-alpha specifically induce formation of cytomegalovirus-permissive monocyte-derived macrophages that are refractory to the antiviral activity of these cytokines. J. Clin. Invest. *100*, 3154–3163.

Soderberg-Naucler, C., Fish, K.N., and Nelson, J.A. (1997b). Reactivation of latent human cytomegalovirus by allogeneic stimulation of blood cells from healthy donors. Cell *91*, 119–126.

Soderberg-Naucler, C., Streblow, D.N., Fish, K.N., Allan-Yorke, J., Smith, P.P., and Nelson, J.A. (2001). Reactivation of latent human cytomegalovirus in CD14(+) monocytes is differentiation dependent. J. Virol. *75*, 7543–7554.

Spencer, J.V., Lockridge, K.M., Barry, P.A., Lin, G., Tsang, M., Penfold, M.E., and Schall, T.J. (2002). Potent immunosuppressive activities of cytomegalovirus-encoded interleukin-10. J. Virol. *76*, 1285–1292.

Stark, T.J., Arnold, J.D., Spector, D.H., and Yeo, G.W. (2011). High-resolution profiling and analysis of viral and host small RNAs during human cytomegalovirus infection. J. Virol. *86*, 226–235.

Stenberg, R.M. (1996). The human cytomegalovirus major immediate-early gene. Intervirology *39*, 343–349.

Strahl, B.D., and Allis, C.D. (2000). The language of covalent histone modifications. Nature *403*, 41–45.

Strobl, H., Bello-Fernandez, C., Riedl, E., Pickl, W.F., Majdic, O., Lyman, S.D., and Knapp, W. (1997). flt3 ligand in cooperation with transforming growth factor-beta1 potentiates in vitro development of Langerhans-type dendritic cells and allows single-cell dendritic cell cluster formation under serum-free conditions. Blood *90*, 1425–1434.

Tavalai, N., and Stamminger, T. (2011). Intrinsic cellular defense mechanisms targeting human cytomegalovirus. Virus Res. *157*, 128–133.

Tavalai, N., Papior, P., Rechter, S., Leis, M., and Stamminger, T. (2006). Evidence for a role of the cellular ND10 protein PML in mediating intrinsic immunity against human cytomegalovirus infections. J. Virol. *80*, 8006–8018.

Tavalai, N., Papior, P., Rechter, S., and Stamminger, T. (2008). Nuclear domain 10 components promyelocytic leukemia protein and hDaxx independently contribute to an intrinsic antiviral defense against human cytomegalovirus infection. J. Virol. *82*, 126–137.

Tavalai, N., Adler, M., Scherer, M., Riedl, Y., and Stamminger, T. (2011). Evidence for a dual antiviral role of the major nuclear domain 10 component Sp100 during the immediate-early and late phases of the human cytomegalovirus replication cycle. J. Virol. *85*, 9447–9458.

Taylor-Wiedeman, J., Sissons, J.G., Borysiewicz, L.K., and Sinclair, J.H. (1991). Monocytes are a major site of persistence of human cytomegalovirus in peripheral blood mononuclear cells. J. Gen. Virol. *72*, 2059–2064.

Taylor-Wiedeman, J., Sissons, P., and Sinclair, J. (1994). Induction of endogenous human cytomegalovirus gene expression after differentiation of monocytes from healthy carriers. J. Virol. *68*, 1597–1604.

Tsukada, Y., Fang, J., Erdjument-Bromage, H., Warren, M.E., Borchers, C.H., Tempst, P., and Zhang, Y. (2006). Histone demethylation by a family of JmjC domain-containing proteins. Nature *439*, 811–816.

Urnov, F.D., Yee, J., Sachs, L., Collingwood, T.N., Bauer, A., Beug, H., Shi, Y.B., and Wolffe, A.P. (2000). Targeting of N-CoR and histone deacetylase 3 by the oncoprotein v-erbA yields a chromatin infrastructure-dependent transcriptional repression pathway. EMBO J. *19*, 4074–4090.

Wang, F.Z., Weber, F., Croce, C., Liu, C.G., Liao, X., and Pellett, P.E. (2008). Human cytomegalovirus infection alters the expression of cellular microRNA species that affect its replication. J. Virol. *82*, 9065–9074.

Weber, B., Stresemann, C., Brueckner, B., and Lyko, F. (2007). Methylation of human microRNA genes in normal and neoplastic cells. Cell Cycle *6*, 1001–1005.

Weinshenker, B.G., Wilton, S., and Rice, G.P. (1988). Phorbol ester-induced differentiation permits productive human cytomegalovirus infection in a monocytic cell line. J. Immunol. *140*, 1625–1631.

Weintraub, H., and Groudine, M. (1976). Chromosomal subunits in active genes have an altered conformation. Science *193*, 848–856.

Whetstine, J.R., Nottke, A., Lan, F., Huarte, M., Smolikov, S., Chen, Z., Spooner, E., Li, E., Zhang, G., Colaiacovo, M., *et al.* (2006). Reversal of histone lysine trimethylation by the JMJD2 family of histone demethylases. Cell *125*, 467–481.

White, K.L., Slobedman, B., and Mocarski, E.S. (2000). Human cytomegalovirus latency-associated protein pORF94 is dispensable for productive and latent infection. J. Virol. *74*, 9333–9337.

Wong, J., Patterton, D., Imhof, A., Guschin, D., Shi, Y.B., and Wolffe, A.P. (1998). Distinct requirements for chromatin assembly in transcriptional repression by thyroid hormone receptor and histone deacetylase. EMBO J. *17*, 520–534.

Woodhall, D.L., Groves, I.J., Reeves, M.B., Wilkinson, G., and Sinclair, J.H. (2006). Human Daxx-mediated repression of human cytomegalovirus gene expression correlates with a repressive chromatin structure around the major immediate-early promoter. J. Biol. Chem. *281*, 37652–37660.

Wright, E., Bain, M., Teague, L., Murphy, J., and Sinclair, J. (2005). Ets-2 repressor factor recruits histone deacetylase to silence human cytomegalovirus immediate-early gene expression in non-permissive cells. J. Gen. Virol. *86*, 535–544.

Wu, J., Wang, S.H., Potter, D., Liu, J.C., Smith, L.T., Wu, Y.Z., Huang, T.H., and Plass, C. (2007). Diverse histone modifications on histone 3 lysine 9 and their relation to DNA methylation in specifying gene silencing. BMC Genomics *8*, 131.

Yang, W.M., Inouye, C., Zeng, Y., Bearss, D., and Seto, E. (1996). Transcriptional repression by YY1 is mediated by interaction with a mammalian homolog of the yeast global regulator RPD3. Proc. Natl. Acad. Sci. U.S.A. *93*, 12845–12850.

Yuan, J., Liu, X., Wu, A.W., McGonagill, P.W., Keller, M.J., Galle, C.S., and Meier, J.L. (2009). Breaking human cytomegalovirus major immediate-early gene silence by vasoactive intestinal peptide stimulation of the protein kinase A-CREB-TORC2 signaling cascade in human pluripotent embryonal NTera2 cells. J. Virol. *83*, 6391–6403.

Zeng, L., and Zhou, M.M. (2002). Bromodomain: an acetyl-lysine binding domain. FEBS Lett. *513*, 124–128.

Zhang, Y., and Reinberg, D. (2001). Transcription regulation by histone methylation: interplay between different covalent modifications of the core histone tails. Genes Dev. *15*, 2343–2360.

Zweidler-Mckay, P.A., Grimes, H.L., Flubacher, M.M., and Tsichlis, P.N. (1996). Gfi-1 encodes a nuclear zinc finger protein that binds DNA and functions as a transcriptional repressor. Mol. Cell. Biol. *16*, 4024–4034.

Zydek, M., Hagemeier, C., and Wiebusch, L. (2010). Cyclin-dependent kinase activity controls the onset of the HCMV lytic cycle. PLoS Pathog. *6*, e1001096.

Zydek, M., Uecker, R., Tavalai, N., Stamminger, T., Hagemeier, C., and Wiebusch, L. (2011). General blockade of human cytomegalovirus immediate-early mRNA expression in the S/G2 phase by a nuclear, Daxx- and PML-independent mechanism. J. Gen. Virol. *92*, 2757–2769.

Transcription Associated with Human Cytomegalovirus Latency

Barry Slobedman, Selmir Avdic and Allison Abendroth

Abstract

Following primary infection, and despite the induction of a massive and sustained anti-viral immune response, human cytomegalovirus (HCMV) is never completely cleared from the human host, but rather establishes a life-long latent infection, during which time infectious virus becomes undetectable. Reactivation from latency results in re-initiation of productive virus replication, a process which often results in life-threatening disease in immunosuppressed individuals. The capacity of HCMV to establish, maintain and reactivate from a latent state contributes significantly to the success of this virus as a human pathogen, yet the molecular basis for latency remains relatively poorly understood. A major component of understanding how HCMV functions during latency has been identification and characterization of viral genes which are expressed during this phase of infection, with the underlying hypothesis being that viral genes expressed during latency are likely to play fundamental roles that enable HCMV to persist in a latent state in the healthy human host. Here we will focus on HCMV gene expression during latency, highlighting our current understanding of this challenging field, as well as areas which will require further focus. It is hoped that a better understanding of viral determinants of latency will provide a rational basis for the development of novel therapies to target HCMV during the latent phase, and so limit or prevent the devastating disease that often arises following reactivation from latency.

Introduction

In immunocompetent individuals, productively replicating human cytomegalovirus (HCMV) infection is eventually eliminated by the innate and adaptive arms of the host immune response, however the virus is never completely cleared. Rather, some HCMV persists in a life-long latent form, during which time the viral genome is maintained without the production of detectable infectious virus. Periodically, latent virus can reactivate, resulting in re-initiation of the productive replicative cycle with associated generation of new infectious virus. Whilst reactivation is usually asymptomatic in the immunocompetent host, in immunocompromised individuals such as allogeneic stem cell and solid organ transplant recipients, reactivation from latency often leads to life-threatening disease which is very difficult to treat (Mocarski et al., 2007). The capacity of HCMV to establish, maintain and reactivate from a latent state represents a critical factor in the life cycle of this virus which enables it to disseminate so widely in the community as well as to cause such serious disease in those who are immunocompromised. However, despite its importance to HCMV pathogenesis, latent infection remains relatively poorly understood, particularly in molecular terms. One of the key areas that has been the focus of increased research efforts in recent years has been definition of the extent of viral gene expression associated with latent infection, and the functions that latency associated viral genes play during this phase of infection. These have proven to be very challenging issues to address, due in part to the high species specificity of this virus, coupled with the complexities of different experimental models of latency, most of which necessitate the use of primary human myeloid lineage cells, as well as the logistical issues associated with studying a viral genome with such a large coding capacity. The presumption that viral genes expressed during latency are likely to represent those viral factors that play important functional roles during the establishment and maintenance of latency, or in facilitating the transition from latent to reactivated infections, have been important drivers of studies aimed at identifying and characterizing viral genes expressed during latency.

The need for experimental models to study HCMV latency

Studies aimed at the identification of viral transcripts expressed during HCMV latency have been complicated due to strict species specificity of HCMV, limiting these studies to use of human derived cells only. Owing to these limitations, despite extensive work undertaken to gain a greater understanding of the mechanisms regulating latent infection, the range of viral genes expressed during latency and their respective functions remain to be completely identified (Mocarski and Courcelle, 2001). The importance of the development of *in vitro* models to study HCMV latency was recognized well before identification of the cell types supporting latent HCMV infection. Early attempts included chemical treatment of permissive human cell types in order to prevent production of infectious virus (Gonczol and Vaczi, 1973), or extrapolation of data generated using animal (murine) CMV latency models (Olding *et al.*, 1976; Jordan *et al.*, 1977; Mayo *et al.*, 1978; Brautigam *et al.*, 1979; Hummel and Abecassis, 2002; Reddehase *et al.*, 2002, 2008) (see also Chapter I.22). However, the development of experimental models recognized as representing HCMV latency took much longer to be reported. A breakthrough came following the identification of cells of the myeloid lineage as carriers of latent HCMV (Taylor-Wiedeman *et al.*, 1991; Mendelson *et al.*, 1996; Sindre *et al.*, 1996), following which several human cell culture-based models of HCMV latency, involving the experimental infection of myeloid progenitor cells, have emerged (see also Chapter I.19). Indeed, it is now well established that myeloid lineage cells are a prominent site of HCMV latent infection, with myeloid progenitor cells supporting latency and terminal differentiation to either macrophages or myeloid dendritic cells (DCs) resulting in cell types from which virus can be induced to reactivate (Sinclair and Sissons, 2006; Slobedman *et al.*, 2010) (Fig. I.20.1).

Thus, models of latency and reactivation have relied heavily upon this link between myeloid cell differentiation state and the capacity to support either latency or reactivation.

The development of experimental models of latency was also necessitated by the very low frequency of latently infected cells during natural HCMV latency, with estimates of only 0.004% to 0.01% of mononuclear cells from mobilized peripheral blood and bone marrow donors harbouring viral genomes at a copy number of 2 to 13 per infected cell (Slobedman and Mocarski, 1999). Detection of viral DNA in peripheral blood mononuclear cells is even more challenging than detection in mobilized peripheral blood or bone marrow samples (which have a much higher proportion of myeloid progenitor cells), and the need to exclude any potential contribution of replicating virus (which is of particular relevance in individuals who may be immunocompromised) has added to the challenge of studying natural latency.

Cell culture-based latency models include the use of primary human myeloid progenitor cells derived from different sources such as human fetal liver (Kondo *et al.*, 1994, 1996; Hahn *et al.*, 1998; Slobedman and Mocarski, 1999; White *et al.*, 2000; Slobedman *et al.*, 2002, 2004; Jenkins *et al.*, 2004a; Cheung *et al.*, 2006, 2009; Stern and Slobedman, 2008; Avdic *et al.*, 2011), bone marrow, umbilical cord blood or mobilized peripheral blood (Minton *et al.*, 1994; Goodrum *et al.*, 2002; Khaiboullina *et al.*, 2004; Reeves *et al.*, 2005a; Cheung *et al.*, 2009; Reeves and Sinclair, 2010), myeloid progenitor cell lines (Beisser *et al.*, 2001) and primary human CD14+ monocytes (Hargett and Shenk, 2010; Reeves and Compton, 2011). These models not only

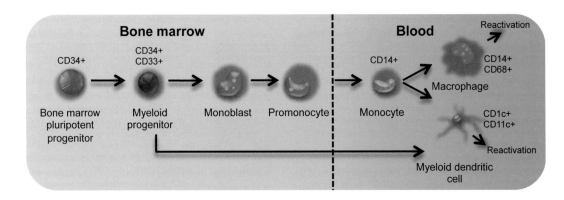

Figure I.20.1 Human cytomegalovirus (HCMV) latency and reactivation in the myeloid lineage. Haematopoiesis includes the differentiation of bone marrow derived pluripotent progenitor cells into myeloid-committed progenitors which subsequently undergo a well-coordinated series of stages of differentiation and enter the peripheral blood system. Latent HCMV infection is associated with these myeloid-committed progenitor cells. Reactivation from latency can occur when these myeloid-committed progenitor cells terminally differentiate to become macrophages or myeloid dendritic cells.

use myeloid cells from diverse sources but perhaps most importantly, they represent myeloid cells at different stages of cellular differentiation. The types of media, the amount of serum used, and incubation times also differ between models and this is likely to influence the differentiation state of latently infected cells in different experimental models. The importance of cellular differentiation to latency is underpinned by clear evidence that differentiation state plays a defining role in latency and reactivation. For example, myeloid progenitors are unable to support immediate-early (IE) gene expression leading to productive replication (Maciejewski et al., 1992; Minton et al., 1994; Taylor-Wiedeman et al., 1994; Zhuravskaya et al., 1997; Hahn et al., 1998), but terminally differentiated macrophages and mature DCs can support IE gene expression (Weinshenker et al., 1988; Ibanez et al., 1991; Lathey and Spector, 1991; Taylor-Wiedeman et al., 1994; Sinzger et al., 1996; Riegler et al., 2000; Hertel et al., 2003; Reeves et al., 2005a,b), leading to the initiation of productive infection.

Thus, differences between experimental models of latency in terms of the source of cells and their culture conditions add to the complexity when it comes to interpreting results from different models. Some models also employ a 'feeder' layer of murine stromal cells to support myeloid cell culture, whilst others do not utilize any supporting cell type. The topic is further complicated by identification of low levels of productively replicating virus in a small proportion of cells within latently infected cultures that has been indicated in at least one model (Goodrum et al., 2002, 2004). Other experimental models of HCMV latency have been reported to be free of infectious virus in cell cultures prior to reactivation (Reeves et al., 2005a; Cheung et al., 2006). Despite differences of various in vitro models of latency used, all have been reported to include some capacity to reactivate virus from latency, accomplished by co-culturing latently infected cells with fibroblasts or by addition of various stimuli to induce differentiation of latently infected cells. Other fundamental features of latency, like the ability of in vitro infected cells to sustain at least a proportion of viral genomes in the absence of infectious virus production, are also maintained in all models.

In addition to cell culture-based latency models using cultured myeloid cells, an HCMV latency model in vivo has been described using severe combined immunodeficient (SCID) mice implanted with human CD34+ haematopoietic cells (Smith et al., 2010) (see also Chapter I.23). Specifically, for this humanized mouse model system, non-obese diabetic (NOD)/SCID IL2R$\gamma_c$null mice were stably engrafted with human umbilical cord-derived CD34+ cells before intraperitoneal injection with HCMV infected fibroblasts. This led to a carriage of quiescent HCMV DNA which could be induced to reactivate as indicated by expansion of viral genome copy number and expression of productive infection-associated viral gene products following haematopoietic cell mobilization with granulocyte colony stimulating factor (G-CSF). The authors also demonstrated that this reactivated infection was restricted to the reconstituted human monocytes/macrophages. This represents the first animal-based model which displays key features of systemic HCMV latency and reactivation, and is one which has been used to provide an additional means to examine HCMV gene expression and functions during latency and reactivation (Umashankar et al., 2011).

At this stage, it is not clear how well each model reflects all features of natural latency and the collective list of features required to unify results from different models, or at least to characterize required features of newly developing models, remains to be agreed upon. It is clear, however, that some of the questions related to the viral gene expression during HCMV latency are starting to be answered, in particular in instances where in vitro findings have been confirmed in naturally infected cells. These models will therefore continue to play an important role in defining molecular mechanisms of HCMV latency and reactivation.

Is viral gene expression required for HCMV latency?

The HCMV genome contains a large number of predicted open reading frames (ORFs) (Murphy et al., 2003b; Dolan et al., 2004; Murphy and Shenk, 2008) (see also Chapter I.1), of which many are transcriptionally active during productive infection (Gatherer et al., 2011), and a considerable proportion are required for efficient productive virus replication (Dunn et al., 2003). However, the extent to which viral gene expression is required for latent infection has been subjected to much less study. In particular, whether viral gene expression is even needed to facilitate the establishment of latency has only been relatively recently explored. In considering how HCMV enters into a latent state from the perspective of viral gene expression, there are three hypotheses encompassing three different pathways that may lead to the establishment of latency (Cheung et al., 2006). In the first pathway, it is hypothesized that after binding and entry, the virus establishes a latent infection without any requirement for de novo viral gene expression. In this scenario, either the type of cell encountered by the virus and/or virion component(s) would be sufficient for successful latency. In the second pathway, following binding and entry, the virus initiates a productive infection (typified by a temporally regulated cascade of viral gene expression) that is prematurely halted, prior to the generation of new infectious virus, thus leading to the establishment of latency. The third pathway involves expression of a specific subset of viral genes that is not associated with the productive replicative cycle, but is necessary for the successful establishment of latency.

In a study of the establishment of latency, primary human myeloid progenitor cells were exposed to viable or UV-inactivated virus, and the efficiency of latent viral genome maintenance over time measured by PCR (Cheung *et al.*, 2006). Whilst both viable virus and UV-inactivated virus were able to initially infect myeloid progenitor cells with the same efficiency, by day 3 post infection only ~ 1% of cells exposed to UV-inactivated virus remained viral genome positive whereas the viral genome was maintained in almost all cells latently infected with viable virus. Thus, the capacity of HCMV to efficiently express viral gene products is associated with the capacity to efficiently establish and/or maintain latency. Furthermore, broad analyses of viral gene transcription in various models of latent infection indicate that viral gene expression during latency does not follow the pattern of coordinated expression of viral genes assigned to different kinetic classes which is characteristic of productive replication (Goodrum *et al.*, 2002, 2004; Cheung *et al.*, 2006). Thus, the molecular pathway that leads to latency is inconsistent with the interruption of a productive infection gene expression cascade, suggesting that the establishment of latency is facilitated by the expression of a unique subset of viral RNAs. The role of the host cell type and cellular factors, such as post-translational modifications of cellular histones associated with the viral genome, also appear to play critical roles in latency and reactivation (see Chapter I.19). Ultimately, further studies focusing on the dynamic interplay between viral and cellular factors will be required to fully define the mechanistic basis of the establishment, maintenance and reactivation phases of latency. The following sections will focus on those viral gene regions that have been associated with transcriptional activity during latency (Fig. I.20.2).

Latency associated gene expression from the major immediate-early (MIE) region

Although it was known early on that productive gene expression was repressed during latency (Taylor-Wiedeman *et al.*, 1993; Minton *et al.*, 1994; Mendelson *et al.*, 1996), it was unclear whether the latent viral genome was transcriptionally inactive or whether a low level of transcription of a subset of genes occurred

(Mocarski and Courcelle, 2001). It became clear that the latter was the case when several studies using an experimental latency model detected a novel class of latent transcripts, termed CMV latency-associated transcripts (CLTs) (Kondo *et al.*, 1994, 1996; Kondo and Mocarski, 1995).

The CLTs are encoded from both sense and antisense strands from the major immediate-early (MIE) region of the HCMV genome (Fig. I.20.2) and found in only 2–5% of latently infected cells (Kondo *et al.*, 1994, 1996; Kondo and Mocarski, 1995; Slobedman and Mocarski, 1999). Sense CLTs are encoded in the same direction as the IE products of productive infection but utilize two novel transcriptional start sites upstream of the productive infection start site, resulting in three putative ORFs: ORF45, ORF55 and ORF94. On the other hand, the antisense CLT is an unspliced polycistronic transcript encoded from the opposite strand to productive phase transcripts, and predicted to code for three putative ORFs: ORF59, ORF154 and ORF152/UL124 (Kondo *et al.*, 1996). Sense CLTs have also been detected during *in vitro* productive infection of human fibroblasts (HFs) (Lunetta and Wiedeman, 2000), although whether the antisense CLTs are expressed in this setting remains to be examined. Whilst MIE region CLTs have not been universally detected during natural latency (Bego *et al.*, 2005), expression of these transcripts has been reported in bone marrow-derived haematopoietic cells from healthy seropositive carriers (Kondo *et al.*, 1996; Hahn *et al.*, 1998) and antibodies to one or more of putative ORFs encoded by the MIE region CLTs have been detected in more than 85% of blood samples from healthy seropositive individuals, implying the products of these ORFs are expressed (Landini *et al.*, 2000). The predominant protein appears to be the sense CLT-derived pORF94, with anti-pORF94 serum antibody detected in 44–47% of healthy seropositive carriers (Kondo *et al.*, 1996; Landini *et al.*, 2000). When expressed in mammalian cells, pORF94 localizes to the nucleus in transiently transfected cells, and it is dispensable for both latent and productive infection *in vitro*, as demonstrated by analysis of an ORF94 deletion virus, which was able to replicate normally in HFs and was able to establish and reactivate from experimental latency in a manner comparable to parental virus (White *et al.*, 2000). Recently, the first identification of a function for pORF94 was reported, where pORF94 expressed in transiently transfected cells was able to inhibit constitutive expression and functional activity of 2′,5′-oligoadenylate synthetase 1 (OAS1) (Tan *et al.*, 2011), an enzyme involved in degradation of viral and cellular RNA and inhibition of virus replication (Silverman, 2007). OAS1 expression and function was also modulated

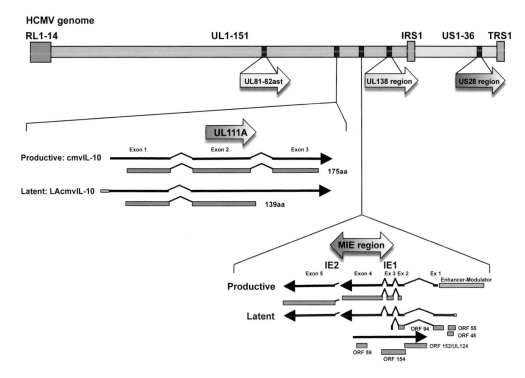

Figure I.20.2 Schematic representation of latency associated HCMV transcripts. Top line represents the full-length HCMV genome encoding genes RL1-TRS1. The positions of the UL81–82ast, UL111A, major immediate-early (MIE), UL138, and US28 regions shown to be expressed during latency are indicated by black boxes, expanded to indicate the direction of latency-associated transcription. The UL111A and MIE regions are further expanded to show splicing patterns of transcripts (solid black arrows) and their associated open reading frames (ORFs; coloured boxes) from both latent and productive infection settings. The region encompassing the start site of the UL111A region transcript expressed during latency, and the MIE enhancer-modulator region are shown as grey boxes. The sense latency-associated transcripts from the MIE region and the latency-associated transcripts from the UL111A region have also been shown to be expressed during productive infection (Lunetta and Wiedeman, 2000; Jenkins et al., 2008a).

by pORF94 during productive infection of HFs, demonstrated by significantly higher levels of OAS1 in cells infected with ORF94 mutant virus compared to the infection with parental virus, accompanied by increased sensitivity of the ORF94 mutant virus to the anti-viral effects of IFN-γ compared to infection with the parental virus (Tan *et al.*, 2011). Whether pORF94 inhibits OAS1 in latently infected cells or possibly enhances virus replication capacity following reactivation from latency remains to be investigated. Apart from ORF94, the functions of other CLTs arising from the MIE region and their putative protein products remain to be determined.

Expression of viral IL-10 during latent HCMV infection

In addition to a functional viral homologue of the immunomodulatory cytokine interleukin 10 (IL-10)

identified during productive infection, termed cmvIL-10 (Kotenko *et al.*, 2000; Lockridge *et al.*, 2000), transcriptional activity from the UL111A gene region has been detected by RT-PCR in human myeloid progenitor cells latently infected with HCMV *in vitro* and in mononuclear cells from healthy stem cell donors (Jenkins *et al.*, 2004) (Fig. I.20.2). However, in contrast to a doubly spliced cmvIL-10 transcript that encodes a 175-aa protein, the latency associated cmvIL-10 (LAcmvIL-10) transcript has a different splicing pattern where only the first of the two introns found in cmvIL-10 is spliced whilst the second intron remains in the LAcmvIL-10 mRNA, resulting in an in-frame stop codon. As a consequence, the LAcmvIL-10 is predicted to be a truncated protein of 139-aa, sharing the first 127 amino acids with cmvIL-10 but then diverging for its final 12 amino acids at the C terminus (Jenkins *et al.*, 2004). It remains to be determined whether cmvIL-10 transcripts are also expressed during latency. However, antibodies against cmvIL-10 were detected in 28% of healthy human seropositive blood donors. These antibodies efficiently neutralized cmvIL-10 function *in vitro* and were specific to cmvIL-10 only, with no cross-reactivity with either ebvIL-10 or human cellular interleukin 10 (hIL-10) (de Lemos Rieper *et al.*, 2011). Intriguingly, the authors of this study could not rule out the possibility that these anti-cmvIL-10 antibodies recognize LAcmvIL-10 as well, so the expression of cmvIL-10 during latency remains to be confirmed. Nevertheless, it is known now that LAcmvIL-10 transcripts are expressed during productive infection, although their expression kinetics appear to differ to that of cmvIL-10 transcripts during this phase of infection (Jenkins *et al.*, 2008a).

Treatment of myeloid cells with recombinant LAcmvIL-10 protein reduces cell surface major histocompatibility complex (MHC) class II expression and transcription of components of the MHC class II biosynthesis pathway, including HLA-DRα chain, HLA-DRβ chain and invariant chain (Jenkins *et al.*, 2008b). Transcription of the master regulator of expression of MHC class II components, the class II transactivator (CIITA), is also suppressed by LAcmvIL-10 pointing to a mechanism of MHC class II down-regulation at the level of mRNA transcription. However, the observation that MHC class II molecules accumulate in cytoplasmic vesicles in a small percentage (3–5%) of monocytes treated with LAcmvIL-10 protein implies that LAcmvIL-10 may also exert a repressive effect on MHC class II at the post-translational level during assembly and/or transport (Jenkins *et al.*, 2008b).

Whilst LAcmvIL-10 mimics the function of cmvIL-10 in downmodulation of MHC class II by myeloid cells (Jenkins *et al.*, 2008b), LAcmvIL-10 does not appear to share the full range of immunomodulatory functions of cmvIL-10. For example, the immunosuppressive effects of cmvIL-10 on LPS induced DC maturation (Chang *et al.*, 2004; Raftery *et al.*, 2004) cannot be replicated by LAcmvIL-10 (Jenkins *et al.*, 2008b), and unlike cmvIL-10 and hIL-10, LAcmvIL-10 does not stimulate proliferation of B lymphocytes (Spencer *et al.*, 2008) nor does it increase Fcγ receptor mediated phagocytosis by human monocytes (Jaworowski *et al.*, 2009). This limited biological activity of LAcmvIL-10 could be due to apparent incapacity of LAcmvIL-10 to bind and signal through the hIL-10 receptor (hIL-10R), whereas cmvIL-10 has been shown to exert its immunomodulatory function via signalling through hIL-10R. In this respect, cmvIL-10 rapidly induces Stat3 phosphorylation, but this and the immunosuppressive functions of cmvIL-10 are blocked when monocytes are pre-treated with hIL-10R neutralizing

antibodies (Spencer *et al.*, 2002; Jenkins *et al.*, 2008b), while LAcmvIL-10 does not induce phosphorylation of Stat3 and still down-regulates MHC class II even in the presence of neutralizing antibodies to hIL-10R (Jenkins *et al.*, 2008b). The inability of LAcmvIL-10 to induce phosphorylation of Stat3 associated with the hIL-10R signalling pathway may be due to the absence of amino acid residues coded by exon 3 of the UL111A transcript, resulting in the loss of the α-helices which are present in the C-terminal portion of cmvIL-10 and have been reported to be important for binding to the hIL-10R (Jones *et al.*, 2002). Thus, it has been suggested that the truncation in LAcmvIL-10 protein results in either a loss of capacity to bind to the IL-10R in the same manner to that of cmvIL-10 protein, or in a complete failure to bind to hIL-10R, accounting for the differences in functions of LAcmvIL-10 and cmvIL-10 (Jenkins *et al.*, 2008b).

In the context of latent infection it has been reported that the UL111A gene is dispensable for HCMV latency *in vitro*, as demonstrated by the ability of a UL111A deletion virus (unable to express either cmvIL-10 or LAcmvIL-10 and collectively referred to as viral IL-10) to establish, maintain and reactivate from experimental latency in CD34+ myeloid progenitor cells in a manner comparable to parental virus (Cheung *et al.*, 2009). The same study identified that the UL111A gene functions to render latently infected cells refractory to CD4+ T-cell recognition, where myeloid progenitors latently infected with a UL111A deletion virus stimulated proliferation of both allogeneic and autologous CD4+ T-cells, unlike myeloid progenitor cells latently infected with parental virus. As surface MHC class II levels on myeloid progenitors latently infected with the UL111A deletion virus were significantly higher compared to cells infected with the parental virus, the authors concluded that the capacity of parental virus but not UL111A deletion virus to evade CD4+ T-cell recognition was likely a consequence of viral IL-10 mediated suppression of the capacity of latently infected cells to present antigenic peptides via MHC class II (Fig. I.20.3).

A subsequent study utilizing the same UL111A deletion virus reported that viral IL-10 inhibits proinflammatory cytokine mRNA and protein expression by latently infected myeloid progenitor cells, accompanied by viral IL-10 controlled inhibition of differentiation of latently infected cells towards a DC phenotype (Avdic *et al.*, 2011). As DCs are the most potent antigen presenting cells, suppression of differentiation of latently infected myeloid progenitors towards a DC phenotype by viral IL-10 is likely to enhance the ability of latent virus to limit presentation of viral peptides to HCMV specific T-cells. These two studies provide the first

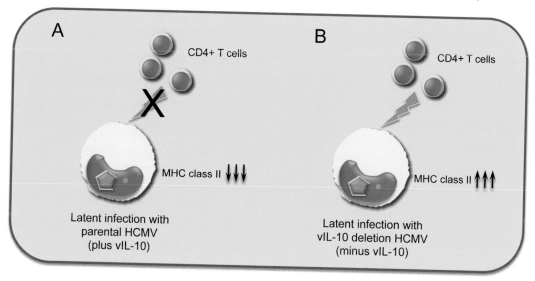

Figure I.20.3 Role of HCMV UL111A in evading CD4+ T-cell recognition of latently infected cells. The HCMV UL111A region expresses viral IL-10 during latent infection of primary human myeloid lineage cells. (A) This latency associated viral IL-10 suppresses the up-regulation of cell-surface MHC class II by latently infected cells and also acts to prevent recognition of these latently infected cells by CD4+ T-cells. (B) In contrast, latent infection of myeloid cells with an HCMV unable to express viral IL-10 results in an up-regulation of MHC class II and recognition of these cells by CD4+ T-cells. Thus, viral IL-10 expressed during latency functions to render latently infected cells refractory to CD4+ T-cell recognition.

evidence of a viral gene that has been shown to function in an immunomodulatory capacity during the latent phase of infection.

UL138 expression during HCMV latency

DNA sequence analysis of the prototypic laboratory HCMV strain AD169 and sequence comparison with clinical HCMV isolates revealed significant differences in their genomes, including at least 19 genes within the ULb' region present in low passage strains such as Toledo or clinical isolates, but absent in AD169 (Cha *et al.*, 1996). Transcription from UL138, which is found within the ULb' region, has been reported during experimental latency (Goodrum *et al.*, 2002, 2004, 2007; Reeves and Sinclair, 2010) and in natural latency in CD14+ monocytes or CD34+ myeloid progenitor cells from HCMV seropositive donors, but not from HCMV seronegative donors (Goodrum *et al.*, 2007).

UL138 has been reported to be an important determinant of HCMV latency. A UL138 deletion mutant productively replicated in an experimental model of HCMV latent infection using primary CD34+ cells, whereas the parental virus with intact UL138 was efficient in its capacity to establish latency (Goodrum *et al.*, 2007). Further investigation of UL138 function in latency was performed by quantification of infectious

centres formed in HFs following incubation with CD34+ cells infected with (i) virus with a deletion of a 5-kb segment of the ULb' region encompassing UL136-UL142, (ii) a UL138 null virus and (iii) virus able to express UL138 transcripts but not UL138 protein (Petrucelli *et al.*, 2009). Results of this study described only a partial role for the UL138 protein in the establishment of latency, suggesting that UL138-region transcripts or other ORFs in the deleted segment of the ULb' region may also contribute to latent infection. In this respect, it has also been reported that similar to UL138, UL133 (encoded by the UL133–138 locus of the ULb' region) suppresses virus replication in CD34+ myeloid progenitor cells (Umashankar *et al.*, 2011). The mechanism by which UL138 contributes to the establishment of latency remains to be determined, but the UL138 protein does not suppress expression from the MIE promoter in transient transfection or productive infection settings (Petrucelli *et al.*, 2009). In transient transfection experiments and during productive infection, the UL138 protein localizes to the Golgi apparatus (Petrucelli *et al.*, 2009), however it is not known whether the pUL138 localizes to the Golgi apparatus during latency.

Additional investigation of transcripts coding for UL138 demonstrated that HCMV uses multiple mechanisms to ensure the expression of pUL138. In addition to two previously reported polycistronic transcripts of 3.6 kb and 2.7 kb coding for pUL138 (Petrucelli *et al.*, 2009), a 1.4-kb transcript was also identified during both productive infection of fibroblasts and during latent infection of CD34[+] haematopoietic progenitor cells (Grainger *et al.*, 2010). Interestingly, initiation of UL138 translation from the 3.6-kb and 2.7-kb transcripts is inducible under serum deprived stress conditions, while the 1.4-kb transcript translation is suppressed under the same conditions, suggesting that different mechanisms are employed to ensure UL138 expression under a variety of conditions encountered by HCMV. An internal ribosome entry site (IRES)-like sequence located in the 5' end of the UL136 ORF was identified to promote translation of a downstream UL138 (Grainger *et al.*, 2010). UL138 is highly conserved at both the nucleotide and the amino acid level across multiple clinical isolates, with 97.41% to 99.41% nucleotide identity and 98.24% to 99.42% amino acid identity compared to the Toledo strain (Qi *et al.*, 2009). In addition, *ex vivo* stimulation of PBMCs from HCMV seropositive blood samples with overlapping peptide pools or recombinant adenovirus encoding pUL138 revealed the presence of pUL138-specific CD8[+] T-cells in seropositive individuals (Tey *et al.*, 2010). Further experiments showed that endogenous pUL138-derived antigenic peptide can be efficiently presented to T-cells during productive HCMV infection, but it appears that this is the case only for newly synthesized pUL138 and not for stable long-lived pUL138, which induced very little T-cell response in this study (Tey *et al.*, 2010).

Two recent reports suggest indirectly that UL138 may prime latently infected cells towards reactivation of latent HCMV, mediated by TNF-α (Le *et al.*, 2011; Montag *et al.*, 2011). Indeed, HCMV strains which lack the ULb' region can effectively down-regulate TNF-α mediated NFκB activation while strains with an intact ULb' region maintain TNF-α induced NFκB activation (Montag *et al.*, 2006, 2011). During productive HCMV infection, surface expression of TNF receptor 1 (TNFR1) is maintained in the presence of UL138 while surface TNFR1 levels are significantly suppressed in HCMV infection in absence of UL138 (Montag *et al.*, 2011). Transfection of HeLa cells with a lentiviral expression vector containing the UL138 ORF resulted in up-regulation of TNFR1 surface expression in HeLa cells. None of the other genes encoded by ULb' region had any effect on the TNFR1 expression, identifying UL138 as the key mediator of increased surface expression of TNFR1 (Montag *et al.*, 2011). These findings, together with results indicating that UL138 stimulates TNF-α induced expression of IE transcripts may indicate that UL138 could be a factor in reactivation

of latent HCMV (Montag *et al.*, 2011), although this remains to be demonstrated in the context of latency. Whilst it has been shown that pUL138 increased expression of surface TNFR1, other positive modulators of TNF-α signal transduction encoded by the ULb' region have also been identified (Le *et al.*, 2011). Immunoprecipitation experiments demonstrated that pUL138 interacts directly with TNFR1, resulting in increased surface expression levels, responsiveness to TNF-α and increased TNFR1 half-life (Le *et al.*, 2011). These findings remain to be recapitulated during latent infection, and so it is still uncertain whether UL138 mediated control of TNFR1 represents a mechanism that operates during, or impacts upon, latency and/or reactivation.

There are a number of studies using different models of experimental latency with laboratory strains which are missing the ULb' region (and therefore lack UL138) which demonstrate that the ULb' region is not essential for the establishment of HCMV latency (Kondo *et al.*, 1994, 1996; Hahn *et al.*, 1998; Slobedman and Mocarski, 1999; White *et al.*, 2000; Slobedman *et al.*, 2002; Jenkins *et al.*, 2004; Cheung *et al.*, 2006, 2009; Avdic *et al.*, 2011). It is not clear whether this discrepancy in reports on the importance of UL138 in the establishment of latency reflects differences in the experimental models of latency used to characterize infection, and/or provides evidence that other HCMV gene products can play a similar role in these settings in the absence of UL138. It is clear, however, that additional studies are required to fully delineate viral determinants of the establishment of latency.

Expression of antisense UL81–82 during HCMV latency

Apart from three antisense MIE CLTs that encode ORF59, ORF154 and ORF152, a polyadenylated transcript that overlaps part of the UL81–82 region of the HCMV genome is the only other transcript identified so far to be transcribed from the antisense DNA strand during latency. This transcript was identified by RACE (rapid amplification of cDNA ends) from a cDNA library prepared from monocytes from a healthy HCMV seropositive donor, and termed UL81–82ast (Bego *et al.*, 2005). Expression of UL81–82ast was confirmed in bone marrow samples from five HCMV seropositive donors, but not from HCMV seronegative donors. Interestingly, UL81–82ast transcripts were also detected at early stages of HCMV productive infection of HFs, but expression was transient, disappearing after the initiation of late gene transcription (Bego *et al.*, 2005). UL81–82ast transcript is predicted to encode a 133-aa serine-rich protein termed the UL82as protein or either 'latency unique natural antigen' or 'latency unidentified nuclear antigen' (LUNA), which appears to be highly conserved across laboratory and clinical isolates of HCMV as well as chimpanzee CMV. Expression of LUNA was detected during productive infection of HFs from 8 hours to 6 days post infection (Bego et al., 2005). Although no detection of LUNA was identified during latent HCMV infection in this study, a recent publication from the same group proposed that LUNA is present during natural HCMV latency, as evidenced by detection of specific antibodies against LUNA protein by co-immunoprecipitation and ELISA assays using sera of HCMV-seropositive donors (Bego et al., 2011). However, it is not yet certain whether this represents LUNA expression during latency as it remains possible that LUNA specific antibodies could be derived from initial productive infection or, alternatively, from episodes of HCMV reactivation.

The function of UL81–82ast or LUNA during productive or latent phases of infection has not been published but it has been suggested that because the UL81–82ast is antisense to the gene UL82, UL81–82ast and UL82 transcripts may be mutually exclusive or UL81–82ast may function to regulate UL82 expression by affecting its mRNA stability or translation. The tegument protein pp71, encoded by the UL82 gene, functions as an IE gene transactivator that takes part in initiation of productive HCMV infection (Bresnahan and Shenk, 2000) (see also Chapter I.9 and I.10). Thus, modulation of pp71 expression by UL81–82ast or LUNA may assist the virus in maintaining latency or controlling reactivation (Bego et al., 2005; Reeves and Sinclair, 2008), although more recent work suggests that pp71 does not act as a trigger for reactivation, as no pp71 expression was detected preceding reactivation of latent HCMV (Reeves and Sinclair, 2010).

Investigation of LUNA expression kinetics has been performed during experimental latent infection of CD34$^+$ myeloid progenitor cells (Reeves and Sinclair, 2010). Transcripts coding for LUNA were first detected 2 days post infection, but transcript levels decreased as latently infected CD34$^+$ myeloid progenitors differentiated towards an immature DC phenotype. LPS stimulated maturation of these immature DCs then again increased transcript levels. The authors presumed that this second phase of expression was due to reactivation of the lytic replication cycle in mature DCs. Investigation of post-translational modifications of histones in the same study found the LUNA promoter during both experimental and natural latency to be associated predominantly with acetylated histones, consistent with activation of this promoter during latency and in contrast to the major immediate-early promoter (MIEP), which is associated with markers of transcriptional repression during latent infection (Reeves et al., 2005a,b; Reeves and Sinclair, 2010) (see also Chapter I.19).

Expression of US28 during HCMV latency

US28 is one of four HCMV encoded homologues of G-protein-coupled chemokine receptors (GPCR) (Chee et al., 1990). US28 is expressed during productive HCMV infection (Welch et al., 1991; Zipeto et al., 1999) but is dispensable for productive infection in vitro (Vieira et al., 1998; Dunn et al., 2003; Yu et al., 2003). It has a broad spectrum of specificity, binding with high affinity to the small inducible CC chemokines CCL3 (MIP-1α), CCL4 (MIP-1β), CCL5 (RANTES) and CCL2 (MCP-1) (Gao and Murphy, 1994; Kuhn et al., 1995; Randolph-Habecker et al., 2002) and even with higher affinity to the membrane bound chemokine CX3CL1 (fractalkine) (Kledal et al., 1998). US28 also alters the production of the CXC chemokine IL-8 (Randolph-Habecker et al., 2002). These findings suggested a possible function for US28 as a 'chemokine sink' by sequestering endogenously and exogenously produced chemokines (Randolph-Habecker et al., 2002).

There are a limited number of publications reporting US28 expression during HCMV latent infection. Using the THP-1 monocytic cell line infected with HCMV strain Toledo as a model of latency, transcription from the US28 region was detected by RT-PCR, starting from day 8 post infection (Beisser et al., 2001). The authors suggested that US28 may play a role in the dissemination of latent HCMV via interaction with the membrane bound fractalkine expressed on endothelial tissue and a number of DC subtypes that support productive HCMV infection. This proposition, however, was not further investigated in this study. Transcription from the US28 region has also been indicated in viral microarray-based studies during experimental HCMV latent infection of myeloid progenitor cells (Goodrum et al., 2002, 2004; Cheung et al., 2006) and in naturally infected donors (Patterson et al., 1998). In the study of expression during natural infection, US28 transcripts were detected by quantitative RT-PCR in peripheral blood leucocytes from 5 out of 12 women, but two of the US28-positive samples were also HIV-positive, and it is not clear whether these donors may have been productively infected with HCMV. Nonetheless, these existing reports do provide reasonable evidence to suggest that US28 region transcripts are expressed during latent infection, and so further investigation of US28 and the role it may play in latency is warranted.

Viral transcriptome profiling during experimental HCMV latent infection

Owing to the large number of genes encoded by HCMV genome and limitations of experimental models of HCMV latency, determining the full repertoire of viral genes expressed during latency has been a substantial challenge. In an attempt to provide a broader analysis of the viral transcriptome during latency, HCMV gene-specific microarrays were utilized to screen myeloid progenitor cells for the expression of viral transcripts at multiple time points during experimental latent infection (Goodrum *et al.*, 2002, 2004; Cheung *et al.*, 2006). A combined list of viral ORFs showing evidence of transcriptional activity at any time point in any of these studies would seem to indicate that the viral genome is far more transcriptionally active during latency than originally thought. However, these studies were carried out using different models of latent infection and different types of viral gene microarrays with different criteria for data analysis, so combining data from these studies may represent an over-estimation of the number of transcriptionally active viral genes.

At least one study, however, indicated that viral gene expression was repressed over time following HCMV latent infection of myeloid progenitor cells. From a total of 37 viral RNAs detected at some point during an 11-day time course, 25 of these were expressed at day 1 post infection, followed by a significant decrease in the number of HCMV genes expressed at later time points, with transcripts from only 12 and 7 genes being detected at days 5 and 11 post infection, respectively (Cheung *et al.*, 2006). In this study, the viral genome was maintained throughout the time course and virus could be reactivated by co-culture of latently infected cells with HFs. The authors concluded that the number of viral genes expressed during latency maybe small relative to the total HCMV coding capacity, although additional viral genes are expressed initially after infection, possibly to assist in the establishment of latency, before being silenced.

Future perspectives and challenges

Due to the high species specificity of HCMV and the very low proportion of latently infected cells in naturally infected individuals, experimental models of latent HCMV infection based upon cultured human myeloid cells, and humanized mice engrafted with human haematopoietic cells which can support latency, will continue to play critical roles in defining the key molecular determinants of latent infection. The multiple stages of myeloid progenitor cell differentiation, and the link between myeloid cell differentiation state and permissiveness to HCMV replication and reactivation from latency re-enforces the need to assign latency associated viral gene expression patterns to different stages in myeloid cell differentiation during latent infection and also during the transition to reactivation. It is clear that whilst cell culture-based models of latency have proven to be a very useful means to address some of these fundamental issues, with often good correlation of findings between different models, there also remain significant differences between different models, which highlights the complexity of the field and some of the challenges that are still to be overcome. Despite evidence showing a number of viral genes are expressed during latency, these transcripts have also been reported to be expressed during productive infection. Thus, these are more accurately defined as 'latency associated', rather than 'latency specific' transcripts. In this respect, latency-associated transcripts reported to be expressed by other herpesviruses such as herpes simplex virus type 1, varicella zoster virus, Epstein–Barr virus and pseudorabies virus are also expressed during productive infection (Spivack and Fraser, 1987; Lear *et al.*, 1992; Debrus *et al.*, 1995; Schaefer *et al.*, 1995; Jin and Scherba, 1999; Xia and Straus, 1999; Cohrs *et al.*, 2002; Xia *et al.*, 2003). In addition to efforts to better define HCMV transcriptional activity during latency using experimental models, there are a number of related considerations which will likely shape future research efforts into this important area of HCMV biology. These are discussed in the following sections.

Repertoire of viral genes expressed during natural latent infection

To complement the analysis of viral gene expression using experimental models of latency, a much better analysis of latency in healthy, naturally infected individuals will be required before the full repertoire of latency associated HCMV gene expression can be accurately defined. As well as the inherent difficulty of analysing a virus with such a large coding capacity, analysis of latency from naturally infected individuals poses additional hurdles, including the very low level of latently infected cells and the corresponding issue of sensitivity of detection of viral gene expression, as well as the increased likelihood of variation of gene expression profiles between different latently infected individuals. Thus, a systematic approach, utilizing high-throughput, specific and sensitive screening techniques to detect viral RNAs will be needed. The development of real time quantitative RT-PCR assays, with robotic-based sample handling to increase throughput, may prove to be very useful in addressing this issue. Similarly, next generation nucleic acid sequencing technologies (also known as deep sequencing), which have revolutionized the capacity to sequence large amounts of nucleic acid, may also represent an approach to define the HCMV transcriptome during latency. Next generation sequencing has recently been applied to the study of the HCMV transcriptome during productive infection of fibroblasts (Gatherer *et al.*, 2011). This study demonstrated the strength of this technology in further unravelling the complex transcription patterns of HCMV during productive replication. The lower levels of viral transcripts expressed during latency may significantly complicate analysis of the latent HCMV transcriptome using next generation sequencing, but this technology still holds much promise to better define HCMV transcriptional activity during experimental and perhaps also natural latent infection settings (particularly if latently infected cells or transcripts could be enriched prior to sequencing).

Functional analyses of latency associated viral gene products

To date, most viral gene regions identified as being transcriptionally active during latency have not yet undergone any functional analysis in the context of latent infection. Thus, future research will also require additional focus on the construction and testing (in experimental models of latent infection) of recombinant HCMV with targeted mutations of the region of interest to elucidate any function in the context of establishment, maintenance and reactivation phases of latency. The capacity to rapidly generate recombinant viruses with targeted gene deletions or other modifications is already upon us, largely due to the advent of bacterial artificial chromosome (BAC)-mediated recombinant virus construction techniques, with BAC clones of a range of HCMV strains and clinical isolates now constructed (see also Chapter I.3). However, this will still be far from a trivial exercise, given the relatively large number of viruses to be tested, the nature of the analyses required, and the intrinsic limitations of those latency models that require primary human myeloid progenitor cells. Further development of experimental models of latent infection to encompass a measure of immune function (e.g. T-cell recognition) has been reported in the study of latency associated viral IL-10 function (Cheung *et al.*, 2009), and this may also be required in evaluating any role of other latency-associated viral gene products in immune evasion.

Structural analyses of latency associated viral transcripts

Although the configuration of a few viral transcripts expressed during latency has been determined, the sites of transcriptional initiation or termination or possible splicing events of the majority of viral transcripts detected during latency have not yet been analysed. Whilst these viral RNAs identified during latency may be transcribed in an identical manner and therefore code for the same protein as corresponding transcripts expressed during productive infection, it remains possible that some of these transcripts encode latency-specific ORFs. Indeed, HCMV latent transcripts from both MIE region CLTs and the UL111A region (Fig. I.20.2) differ from those originally identified during productive infection due to different splicing patterns and/or transcription start sites (Kondo *et al.*, 1996; Jenkins *et al.*, 2004). In addition, as the viral gene microarrays used to identify virus transcription during latency were derived from double stranded PCR products (Goodrum *et al.*, 2002, 2004; Cheung *et al.*, 2006), expression of viral transcripts antisense to known ORFs cannot be excluded. So far, HCMV latency-associated transcripts that are antisense to ORFs expressed during productive infection include the HCMV MIE region antisense CLTs (Kondo *et al.*, 1996) and the HCMV UL81–82ast that is antisense to ORFs UL81 and UL82 (Bego *et al.*, 2005).

Does the presence of latent viral DNA predict latency associated viral gene expression?

Examination of the HCMV genome distribution during latency suggests that there is a dissociation between the presence of viral DNA and latency-associated transcription, emphasized by detection of MIE region CLTs in only 2–5% of latently infected myeloid progenitor cells (Kondo *et al.*, 1994; Slobedman and Mocarski, 1999), and the LAcmvIL-10 transcripts in approximately 12% of latently infected cells (Jenkins *et al.*, 2004). Although this may be a consequence of differences in the relative sensitivities of the assays used to detect viral DNA and RNA, these findings are also consistent with the notion that different subpopulations of latently infected myeloid cells support differential expression of viral transcripts, as cells differentiate down the myeloid lineage towards macrophages and DCs. Evidence of complex dynamics of latency associated viral gene transcription was demonstrated by viral microarray analysis of experimental latent infection of different CD34[+] cell subpopulations. These studies demonstrated that the outcome of infection was influenced by the nature of the CD34$^+$ cell populations infected (Goodrum *et al.*, 2004). In addition, time course experiments demonstrated noticeable differences in the subsets of latency-associated transcripts expressed at different time points during experimental latent infection (Goodrum *et al.*, 2002; Cheung *et al.*, 2006).

Latency associated viral protein detection

Despite the detection of a number of latency associated viral transcripts, viral protein detection from most of these transcripts has not been reported during latent infection. Lack of protein detection may be due to low level of translation of these proteins during HCMV latency that is below detection limits of current protein detection methodologies, or a deficiency of suitable antibodies. However, it remains possible that some transcripts may not be translated into proteins, but function in a similar manner to host cell non-coding RNA. In this respect, a number of regions within HCMV genome that appear to be transcriptionally active during latency are not predicted to yield a translated gene product (Davison *et al.*, 2003; Murphy *et al.*, 2003a,b; Dolan *et al.*, 2004). Therefore, HCMV may express non-coding latency-associated transcripts that themselves may function in latent infection, and this has been postulated previously (Cheung *et al.*, 2006) and supported to some degree by a study of UL138 expression (Petrucelli *et al.*, 2009). There is a precedent for functions of viral transcripts that are not translated as demonstrated for the latency-associated transcripts (LATs) encoded by HSV, which have been reported to play roles in the establishment and maintenance of latency (Sawtell and Thompson, 1992; Thompson and Sawtell, 1997; Kang *et al.*, 2003), spontaneous reactivation (Perng *et al.*, 1994, 1999), as well as the inhibition of apoptosis (Perng *et al.*, 2000; Ahmed *et al.*, 2002; Jin *et al.*, 2003) and the enhancement of encephalitis in mice (Jones *et al.*, 2005).

HCMV encoded microRNAs in latency

In addition to identifying and characterizing latency associated viral transcripts that encode latency associated viral proteins, the role of viral small non-coding microRNA molecules (miRNAs) in latency and reactivation represents an area of emerging interest which may provide additional insights into virus-mediated control of these phases of infection. As discussed in detail in Chapter I.5, during productive replication, HCMV expresses a number of miRNAs. Recent analyses of fibroblasts productively infected with HCMV

have identified no fewer than 12 miRNA precursors, giving rise to 22 mature miRNAs (Stark *et al.*, 2012). Whether virally encoded miRNAs are expressed and function during HCMV latency remains to be examined directly, although studies of HCMV-encoded miRNA functions have demonstrated important roles in regulating productive infection, including the suppression of immediate-early gene expression (Grey *et al.*, 2007; Murphy *et al.*, 2008). Thus, there is clear merit in addressing the hypothesis that HCMV miRNAs that suppress the replicative cycle may promote latency, as well as determining whether HCMV expresses functional miRNAs during the latent phase of infection.

Concluding remarks

The medical significance of HCMV latency and reactivation is high, and preventative or therapeutic treatment options remain limited, so there is a clear need to better understand these phases of infection. Virus-mediated control of latency and reactivation is a key component of HCMV pathogenesis, yet latency associated viral gene expression and functions are still relatively poorly understood. The pioneering work identifying the main cell types associated with latency has spawned the development of experimental models to study latency in greater detail. As newer experimental models are developed, and existing models are further characterized, our capacity to more fully define viral gene expression and functions during latency and reactivation will increase. Concomitant with the application of experimental models of latency, and a greater range of molecular techniques that can be applied to characterize viral gene function, advances in technologies to detect latency associated viral gene expression during natural infection will serve as critical adjuncts to build a more comprehensive picture of the latent phase of infection.

Acknowledgements

The authors thank both current and past colleagues in the field who have contributed to the work included here. Whilst we have sought to encompass a broad spectrum of published work on HCMV gene expression during the latent phase of infection we apologize to any of our colleagues whose work may not have been separately cited due to space constraints or the specific focus of this chapter. We also wish to thank John Cao for his assistance with generation of figures. Our work has been supported by Australian National Health and Medical Research Council (NHMRC) Project Grants awarded to B.S. and A.A.

References

Ahmed, M., Lock, M., Miller, C.G., and Fraser, N.W. (2002). Regions of the herpes simplex virus type 1 latency-associated transcript that protect cells from apoptosis in vitro and protect neuronal cells in vivo. J. Virol. *76*, 717–729.

Avdic, S., Cao, J.Z., Cheung, A.K., Abendroth, A., and Slobedman, B. (2011). Viral interleukin-10 expressed by human cytomegalovirus during the latent phase of infection modulates latently infected myeloid cell differentiation. J. Virol. *85*, 7465–7471.

Bego, M., Maciejewski, J., Khaiboullina, S., Pari, G., and St Jeor, S. (2005). Characterization of an antisense transcript spanning the UL81–82 locus of human cytomegalovirus. J. Virol. *79*, 11022–11034.

Bego, M.G., Keyes, L.R., Maciejewski, J., and St Jeor, S.C. (2011). Human cytomegalovirus latency-associated protein LUNA is expressed during HCMV infections in vivo. Arch. Virol. *156*, 1847–1851.

Beisser, P.S., Laurent, L., Virelizier, J.L., and Michelson, S. (2001). Human cytomegalovirus chemokine receptor gene US28 is transcribed in latently infected THP-1 monocytes. J. Virol. *75*, 5949–5957.

Brautigam, A.R., Dutko, F.J., Olding, L.B., and Oldstone, M.B. (1979). Pathogenesis of murine cytomegalovirus infection: the macrophage as a permissive cell for cytomegalovirus infection, replication and latency. J. Gen. Virol. *44*, 349–359.

Bresnahan, W.A., and Shenk, T.E. (2000). UL82 virion protein activates expression of immediate-early viral genes in human cytomegalovirus-infected cells. Proc. Natl. Acad. Sci. U.S.A. *97*, 14506–14511.

Cha, T.A., Tom, E., Kemble, G.W., Duke, G.M., Mocarski, E.S., and Spaete, R.R. (1996). Human cytomegalovirus clinical isolates carry at least 19 genes not found in laboratory strains. J. Virol. *70*, 78–83.

Chang, W.L., Baumgarth, N., Yu, D., and Barry, P.A. (2004). Human cytomegalovirus-encoded interleukin-10 homolog inhibits maturation of dendritic cells and alters their functionality. J. Virol. *78*, 8720–8731.

Chee, M.S., Bankier, A.T., Beck, S., Bohni, R., Brown, C.M., Cerny, R., Horsnell, T., Hutchison, C.A., 3rd, Kouzarides, T., Martignetti, J.A., et al. (1990). Analysis of the protein-coding content of the sequence of human cytomegalovirus strain AD169. Curr. Top. Microbiol. Immunol. *154*, 125–169.

Cheung, A.K., Abendroth, A., Cunningham, A.L., and Slobedman, B. (2006). Viral gene expression during the establishment of human cytomegalovirus latent infection in myeloid progenitor cells. Blood *108*, 3691–3699.

Cheung, A.K., Gottlieb, D.J., Plachter, B., Pepperl-Klindworth, S., Avdic, S., Cunningham, A.L., Abendroth, A., and Slobedman, B. (2009). The role of the human cytomegalovirus UL111A gene in down-regulating CD4+ T-cell recognition of latently infected cells: implications for virus elimination during latency. Blood *114*, 4128–4137.

Cohrs, R.J., Wischer, J., Essman, C., and Gilden, D.H. (2002). Characterization of varicella-zoster virus gene 21 and 29 proteins in infected cells. J. Virol. *76*, 7228–7238.

Davison, A.J., Dolan, A., Akter, P., Addison, C., Dargan, D.J., Alcendor, D.J., McGeoch, D.J., and Hayward, G.S. (2003). The human cytomegalovirus genome revisited: comparison with the chimpanzee cytomegalovirus genome. J. Gen. Virol. *84*, 17–28.

Debrus, S., Sadzot-Delvaux, C., Nikkels, A.F., Piette, J., and Rentier, B. (1995). Varicella-zoster virus gene 63 encodes an immediate-early protein that is abundantly expressed during latency. J. Virol. *69*, 3240–3245.

Dolan, A., Cunningham, C., Hector, R.D., Hassan-Walker, A.F., Lee, L., Addison, C., Dargan, D.J., McGeoch, D.J., Gatherer, D., Emery, V.C., et al. (2004). Genetic content of wild-type human cytomegalovirus. J. Gen. Virol. *85*, 1301–1312.

Dunn, W., Chou, C., Li, H., Hai, R., Patterson, D., Stolc, V., Zhu, H., and Liu, F. (2003). Functional profiling of a human cytomegalovirus genome. Proc. Natl. Acad. Sci. U.S.A. *100*, 14223–14228.

Gao, J.L., and Murphy, P.M. (1994). Human cytomegalovirus open reading frame US28 encodes a functional beta chemokine receptor. J. Biol. Chem. *269*, 28539–28542.

Gatherer, D., Seirafian, S., Cunningham, C., Holton, M., Dargan, D.J., Baluchova, K., Hector, R.D., Galbraith, J., Herzyk, P., Wilkinson, G.W., et al. (2011). High-resolution human cytomegalovirus transcriptome. Proc. Natl. Acad. Sci. U.S.A. *108*, 19755–19760.

Gonczol, E., and Vaczi, L. (1973). Cytomegalovirus latency in cultured human cells. J. Gen. Virol. *18*, 143–151.

Goodrum, F., Jordan, C.T., High, K., and Shenk, T. (2002). Human cytomegalovirus gene expression during infection of primary hematopoietic progenitor cells: a model for latency. Proc. Natl. Acad. Sci. U.S.A. *99*, 16255–16260.

Goodrum, F., Jordan, C.T., Terhune, S.S., High, K., and Shenk, T. (2004). Differential outcomes of human cytomegalovirus infection in primitive hematopoietic cell subpopulations. Blood *104*, 687–695.

Goodrum, F., Reeves, M., Sinclair, J., High, K., and Shenk, T. (2007). Human cytomegalovirus sequences expressed in latently infected individuals promote a latent infection in vitro. Blood *110*, 937–945.

Grainger, L., Cicchini, L., Rak, M., Petrucelli, A., Fitzgerald, K.D., Semler, B.L., and Goodrum, F. (2010). Stress-inducible alternative translation initiation of human cytomegalovirus latency protein pUL138. J. Virol. *84*, 9472–9486.

Grey, F., Meyers, H., White, E.A., Spector, D.H., and Nelson, J. (2007). A human cytomegalovirus-encoded microRNA regulates expression of multiple viral genes involved in replication. PLoS Pathog. *3*, e163.

Hahn, G., Jores, R., and Mocarski, E.S. (1998). Cytomegalovirus remains latent in a common precursor of dendritic and myeloid cells. Proc. Natl. Acad. Sci. U.S.A. *95*, 3937–3942.

Hargett, D., and Shenk, T.E. (2010). Experimental human cytomegalovirus latency in CD14+ monocytes. Proc. Natl. Acad. Sci. U.S.A. *107*, 20039–20044.

Hertel, L., Lacaille, V.G., Strobl, H., Mellins, E.D., and Mocarski, E.S. (2003). Susceptibility of immature and mature Langerhans cell-type dendritic cells to infection and immunomodulation by human cytomegalovirus. J. Virol. *77*, 7563–7574.

Hummel, M., and Abecassis, M.M. (2002). A model for reactivation of CMV from latency. J. Clin. Virol. *25*, 123–136.

Ibanez, C.E., Schrier, R., Ghazal, P., Wiley, C., and Nelson, J.A. (1991). Human cytomegalovirus productively infects primary differentiated macrophages. J. Virol. *65*, 6581–6588.

Jaworowski, A., Cheng, W.J., Westhorpe, C.L., Abendroth, A., Crowe, S.M., and Slobedman, B. (2009). Enhanced monocyte Fc phagocytosis by a homologue of interleukin-10 encoded by human cytomegalovirus. Virology *391*, 20–24.

Jenkins, C., Abendroth, A., and Slobedman, B. (2004). A novel viral transcript with homology to human interleukin-10 is expressed during latent human cytomegalovirus infection. J. Virol. *78*, 1440–1447.

Jenkins, C., Garcia, W., Abendroth, A., and Slobedman, B. (2008a). Expression of a human cytomegalovirus latency-associated homolog of interleukin-10 during the productive phase of infection. Virology *370*, 285–294.

Jenkins, C., Garcia, W., Godwin, M.J., Spencer, J.V., Stern, J.L., Abendroth, A., and Slobedman, B. (2008b). Immunomodulatory properties of a viral homolog of human interleukin-10 expressed by human cytomegalovirus during the latent phase of infection. J. Virol. *82*, 3736–3750.

Jin, L., and Scherba, G. (1999). Expression of the pseudorabies virus latency-associated transcript gene during productive infection of cultured cells. J. Virol. *73*, 9781–9788.

Jin, L., Peng, W., Perng, G.C., Brick, D.J., Nesburn, A.B., Jones, C., and Wechsler, S.L. (2003). Identification of herpes simplex virus type 1 latency-associated transcript sequences that both inhibit apoptosis and enhance the spontaneous reactivation phenotype. J. Virol. *77*, 6556–6561.

Jones, B.C., Logsdon, N.J., Josephson, K., Cook, J., Barry, P.A., and Walter, M.R. (2002). Crystal structure of human cytomegalovirus IL-10 bound to soluble human IL-10R1. Proc. Natl. Acad. Sci. U.S.A. *99*, 9404–9409.

Jones, C., Inman, M., Peng, W., Henderson, G., Doster, A., Perng, G.C., and Angeletti, A.K. (2005). The herpes simplex virus type 1 locus that encodes the latency-associated transcript enhances the frequency of encephalitis in male BALB/c mice. J. Virol. *79*, 14465–14469.

Jordan, M.C., Shanley, J.D., and Stevens, J.G. (1977). Immunosuppression reactivates and disseminates latent murine cytomegalovirus. J. Gen. Virol. *37*, 419–423.

Kang, W., Mukerjee, R., and Fraser, N.W. (2003). Establishment and maintenance of HSV latent infection is mediated through correct splicing of the LAT primary transcript. Virology *312*, 233–244.

Khaiboullina, S.F., Maciejewski, J.P., Crapnell, K., Spallone, P.A., Dean Stock, A., Pari, G.S., Zanjani, E.D., and Jeor, S.S. (2004). Human cytomegalovirus persists in myeloid progenitors and is passed to the myeloid progeny in a latent form. Br. J. Haematol. *126*, 410–417.

Kledal, T.N., Rosenkilde, M.M., and Schwartz, T.W. (1998). Selective recognition of the membrane-bound CX3C chemokine, fractalkine, by the human cytomegalovirus-encoded broad-spectrum receptor US28. FEBS Lett. *441*, 209–214.

Kondo, K., and Mocarski, E.S. (1995). Cytomegalovirus latency and latency-specific transcription in hematopoietic progenitors. Scand. J. Infect. Dis. Suppl. *99*, 63–67.

Kondo, K., Kaneshima, H., and Mocarski, E.S. (1994). Human cytomegalovirus latent infection of granulocyte–macrophage progenitors. Proc. Natl. Acad. Sci. U.S.A. *91*, 11879–11883.

Kondo, K., Xu, J., and Mocarski, E.S. (1996). Human cytomegalovirus latent gene expression in granulocyte–macrophage progenitors in culture and in seropositive individuals. Proc. Natl. Acad. Sci. U.S.A. *93*, 11137–11142.

Kotenko, S.V., Saccani, S., Izotova, L.S., Mirochnitchenko, O.V., and Pestka, S. (2000). Human cytomegalovirus harbors its own unique IL-10 homolog (cmvIL-10). Proc. Natl. Acad. Sci. U.S.A. *97*, 1695–1700.

Kuhn, D.E., Beall, C.J., and Kolattukudy, P.E. (1995). The cytomegalovirus US28 protein binds multiple CC chemokines with high affinity. Biochem. Biophys. Res. Commun. *211*, 325–330.

Landini, M.P., Lazzarotto, T., Xu, J., Geballe, A.P., and Mocarski, E.S. (2000). Humoral immune response to proteins of human cytomegalovirus latency-associated transcripts. Biol. Blood Marrow Transplant. *6*, 100–108.

Lathey, J.L., and Spector, S.A. (1991). Unrestricted replication of human cytomegalovirus in hydrocortisone-treated macrophages. J. Virol. *65*, 6371–6375.

Le, V.T., Trilling, M., and Hengel, H. (2011). The cytomegaloviral protein pUL138 acts as potentiator of tumor necrosis factor (TNF) receptor 1 surface density to enhance ULb'-encoded modulation of TNF-alpha signaling. J. Virol. *85*, 13260–13270.

Lear, A.L., Rowe, M., Kurilla, M.G., Lee, S., Henderson, S., Kieff, E., and Rickinson, A.B. (1992). The Epstein–Barr virus (EBV) nuclear antigen 1 BamHI F promoter is activated on entry of EBV-transformed B cells into the lytic cycle. J. Virol. *66*, 7461–7468.

de Lemos Rieper, C., Galle, P., Pedersen, B.K., and Hansen, M.B. (2011). Characterization of specific antibodies against cytomegalovirus (CMV)-encoded interleukin 10 produced by 28% of CMV-seropositive blood donors. J. Gen. Virol. *92*, 1508–1518.

Lockridge, K.M., Zhou, S.S., Kravitz, R.H., Johnson, J.L., Sawai, E.T., Blewett, E.L., and Barry, P.A. (2000). Primate cytomegaloviruses encode and express an IL-10-like protein. Virology *268*, 272–280.

Lunetta, J.M., and Wiedeman, J.A. (2000). Latency-associated sense transcripts are expressed during in vitro human cytomegalovirus productive infection. Virology *278*, 467–476.

Maciejewski, J.P., Bruening, E.E., Donahue, R.E., Mocarski, E.S., Young, N.S., and St Jeor, S.C. (1992). Infection of hematopoietic progenitor cells by human cytomegalovirus. Blood *80*, 170–178.

Mayo, D., Armstrong, J.A., and Ho, M. (1978). Activation of latent murine cytomegalovirus infection: cocultivation, cell transfer, and the effect of immunosuppression. J. Infect. Dis. *138*, 890–896.

Mendelson, M., Monard, S., Sissons, P., and Sinclair, J. (1996). Detection of endogenous human cytomegalovirus in CD34+ bone marrow progenitors. J. Gen. Virol. *77*, 3099–3102.

Minton, E.J., Tysoe, C., Sinclair, J.H., and Sissons, J.G. (1994). Human cytomegalovirus infection of the monocyte/macrophage lineage in bone marrow. J. Virol. *68*, 4017–4021.

Mocarski, E.S., and Courcelle, C.T. (2001). Cytomegaloviruses and their replication, p. 2629–73. In Fields Virology (4th ed.), vol. 2, Knipe, D.M., Howley, P.M., Griffin, D.E., Lamb, R.A., Martin, M.A., Roizman, B., and Straus, S. eds. (Lippincott Williams & Wilkins, Philadelphia, PA).

Mocarski, E.S., Shenk, T., and Pass, R.F. (2007). Cytomegaloviruses. In Fields Virology, Knipe, D.M., and Howley, P.M., eds. (Lippincott Williams & Wilkins, Philadelphia, PA), pp. 2701–2772.

Montag, C., Wagner, J., Gruska, I., and Hagemeier, C. (2006). Human cytomegalovirus blocks tumor necrosis factor alpha- and interleukin-1beta-mediated NF-kappaB signaling. J. Virol. *80*, 11686–11698.

Montag, C., Wagner, J.A., Gruska, I., Vetter, B., Wiebusch, L., and Hagemeier, C. (2011). The latency-associated UL138 gene product of human cytomegalovirus sensitizes cells to tumor necrosis factor alpha (TNF-alpha) signaling by up-regulating TNF-alpha receptor 1 cell surface expression. J. Virol. *85*, 11409–11421.

Murphy, E., and Shenk, T. (2008). Human cytomegalovirus genome. Curr. Top. Microbiol. Immunol. *325*, 1–19.

Murphy, E., Rigoutsos, I., Shibuya, T., and Shenk, T.E. (2003a). Reevaluation of human cytomegalovirus coding potential. Proc. Natl. Acad. Sci. U.S.A. *100*, 13585–13590.

Murphy, E., Yu, D., Grimwood, J., Schmutz, J., Dickson, M., Jarvis, M.A., Hahn, G., Nelson, J.A., Myers, R.M., and Shenk, T.E. (2003b). Coding potential of laboratory and clinical strains of human cytomegalovirus. Proc. Natl. Acad. Sci. U.S.A. *100*, 14976–14981.

Murphy, E., Vanicek, J., Robins, H., Shenk, T., and Levine, A.J. (2008). Suppression of immediate-early viral gene expression by herpesvirus-coded microRNAs: implications for latency. Proc. Natl. Acad. Sci. U.S.A. *105*, 5453–5458.

Olding, L.B., Kingsbury, D.T., and Oldstone, M.B. (1976). Pathogenesis of cytomegalovirus infection. Distribution of viral products, immune complexes and autoimmunity during latent murine infection. J. Gen. Virol. *33*, 267–280.

Patterson, B.K., Landay, A., Andersson, J., Brown, C., Behbahani, H., Jiyamapa, D., Burki, Z., Stanislawski, D., Czerniewski, M.A., and Garcia, P. (1998). Repertoire of chemokine receptor expression in the female genital tract: implications for human immunodeficiency virus transmission. Am. J. Pathol. *153*, 481–490.

Perng, G.C., Dunkel, E.C., Geary, P.A., Slanina, S.M., Ghiasi, H., Kaiwar, R., Nesburn, A.B., and Wechsler, S.L. (1994). The latency-associated transcript gene of herpes simplex virus type 1 (HSV-1) is required for efficient in vivo spontaneous reactivation of HSV-1 from latency. J. Virol. *68*, 8045–8055.

Perng, G.C., Slanina, S.M., Yukht, A., Drolet, B.S., Keleher, W., Jr., Ghiasi, H., Nesburn, A.B., and Wechsler, S.L. (1999). A herpes simplex virus type 1 latency-associated transcript mutant with increased virulence and reduced spontaneous reactivation. J. Virol. *73*, 920–929.

Perng, G.C., Jones, C., Ciacci-Zanella, J., Stone, M., Henderson, G., Yukht, A., Slanina, S.M., Hofman, F.M., Ghiasi, H., Nesburn, A.B., et al. (2000). Virus-induced neuronal apoptosis blocked by the herpes simplex virus latency-associated transcript. Science *287*, 1500–1503.

Petrucelli, A., Rak, M., Grainger, L., and Goodrum, F. (2009). Characterization of a novel Golgi apparatus-localized latency determinant encoded by human cytomegalovirus. J. Virol. *83*, 5615–5629.

Qi, Y., He, R., Ma, Y.P., Sun, Z.R., Ji, Y.H., and Ruan, Q. (2009). Human cytomegalovirus UL138 open reading frame is highly conserved in clinical strains. Chin. Med. Sci. J. *24*, 107–111.

Raftery, M.J., Wieland, D., Gronewald, S., Kraus, A.A., Giese, T., and Schonrich, G. (2004). Shaping phenotype, function, and survival of dendritic cells by cytomegalovirus-encoded IL-10. J. Immunol. *173*, 3383–3391.

Randolph-Habecker, J.R., Rahill, B., Torok-Storb, B., Vieira, J., Kolattukudy, P.E., Rovin, B.H., and Sedmak, D.D. (2002). The expression of the cytomegalovirus chemokine receptor homolog US28 sequesters biologically active CC chemokines and alters IL-8 production. Cytokine *19*, 37–46.

Reddehase, M.J., Podlech, J., and Grzimek, N.K. (2002). Mouse models of cytomegalovirus latency: overview. J. Clin. Virol. *25(Suppl. 2)*, S23–36.

Reddehase, M.J., Simon, C.O., Seckert, C.K., Lemmermann, N., and Grzimek, N.K. (2008). Murine model of cytomegalovirus latency and reactivation. Curr. Top. Microbiol. Immunol. *325*, 315–331.

Reeves, M., Lehner, P.J., Sissons, J.G., and Sinclair, J.H. (2005a). An in vitro model for the regulation of human cytomegalovirus latency and reactivation in dendritic cells by chromatin remodelling. J. Gen. Virol. *86*, 2949–2954.

Reeves, M., MacAry, P.A., Lehner, P.J., Sissons, J.G., and Sinclair, J.H. (2005b). Latency, chromatin remodeling, and reactivation of human cytomegalovirus in the dendritic cells of healthy carriers. Proc. Natl. Acad. Sci. U.S.A. *102*, 4140–4145.

Reeves, M., and Sinclair, J. (2008). Aspects of human cytomegalovirus latency and reactivation. Curr. Top. Microbiol. Immunol. *325*, 297–313.

Reeves, M., and Sinclair, J. (2010). Analysis of latent viral gene expression in natural and experimental latency models of human cytomegalovirus and its correlation with histone modifications at a latent promoter. J. Gen. Virol. *91*, 599–604.

Reeves, M.B., and Compton, T. (2011). Inhibition of inflammatory interleukin-6 activity via extracellular signal-regulated kinase-mitogen-activated protein kinase signaling antagonizes human cytomegalovirus reactivation from dendritic cells. J. Virol. *85*, 12750–12758.

Riegler, S., Hebart, H., Einsele, H., Brossart, P., Jahn, G., and Sinzger, C. (2000). Monocyte-derived dendritic cells are permissive to the complete replicative cycle of human cytomegalovirus. J. Gen. Virol. *81*, 393–399.

Sawtell, N.M., and Thompson, R.L. (1992). Herpes simplex virus type 1 latency-associated transcription unit promotes anatomical site-dependent establishment and reactivation from latency. J. Virol. *66*, 2157–2169.

Schaefer, B.C., Strominger, J.L., and Speck, S.H. (1995). Redefining the Epstein–Barr virus-encoded nuclear antigen EBNA-1 gene promoter and transcription initiation site in group I Burkitt lymphoma cell lines. Proc. Natl. Acad. Sci. U.S.A. *92*, 10565–10569.

Silverman, R.H. (2007). Viral encounters with 2',5'-oligoadenylate synthetase and RNase L during the interferon antiviral response. J. Virol. *81*, 12720–12729.

Sinclair, J., and Sissons, P. (2006). Latency and reactivation of human cytomegalovirus. J. Gen. Virol. *87*, 1763–1779.

Sindre, H., Tjoonnfjord, G.E., Rollag, H., Ranneberg-Nilsen, T., Veiby, O.P., Beck, S., Degre, M., and Hestdal, K. (1996). Human cytomegalovirus suppression of and latency in early hematopoietic progenitor cells. Blood *88*, 4526–4533.

Sinzger, C., Plachter, B., Grefte, A., The, T.H., and Jahn, G. (1996). Tissue macrophages are infected by human cytomegalovirus in vivo. J. Infect. Dis. *173*, 240–245.

Slobedman, B., and Mocarski, E.S. (1999). Quantitative analysis of latent human cytomegalovirus. J. Virol. *73*, 4806–4812.

Slobedman, B., Mocarski, E.S., Arvin, A.M., Mellins, E.D., and Abendroth, A. (2002). Latent cytomegalovirus down-regulates major histocompatibility complex class II expression on myeloid progenitors. Blood *100*, 2867–2873.

Slobedman, B., Stern, J.L., Cunningham, A.L., Abendroth, A., Abate, D.A., and Mocarski, E.S. (2004). Impact of human cytomegalovirus latent infection on myeloid progenitor cell gene expression. J. Virol. *78*, 4054–4062.

Slobedman, B., Cao, J.Z., Avdic, S., Webster, B., McAllery, S., Cheung, A.K., Tan, J.C., and Abendroth, A. (2010). Human cytomegalovirus latent infection and associated viral gene expression. Future Microbiol. *5*, 883–900.

Smith, M.S., Goldman, D.C., Bailey, A.S., Pfaffle, D.L., Kreklywich, C.N., Spencer, D.B., Othieno, F.A., Streblow, D.N., Garcia, J.V., Fleming, W.H., *et al.* (2010). Granulocyte-colony stimulating factor reactivates human cytomegalovirus in a latently infected humanized mouse model. Cell Host Microbe 8, 284–291.

Spencer, J.V., Lockridge, K.M., Barry, P.A., Lin, G., Tsang, M., Penfold, M.E., and Schall, T.J. (2002). Potent immunosuppressive activities of cytomegalovirus-encoded interleukin-10. J. Virol. *76*, 1285–1292.

Spencer, J.V., Cadaoas, J., Castillo, P.R., Saini, V., and Slobedman, B. (2008). Stimulation of B lymphocytes by cmvIL-10 but not LAcmvIL-10. Virology *374*, 164–169.

Spivack, J.G., and Fraser, N.W. (1987). Detection of herpes simplex virus type 1 transcripts during latent infection in mice. J. Virol. *61*, 3841–3847.

Stark, T.J., Arnold, J.D., Spector, D.H., and Yeo, G.W. (2012). High-resolution profiling and analysis of viral and host small RNAs during human cytomegalovirus infection. J. Virol. *86*, 226–235.

Stern, J.L., and Slobedman, B. (2008). Human cytomegalovirus latent infection of myeloid cells directs monocyte migration by up-regulating monocyte chemotactic protein-1. J. Immunol. *180*, 6577–6585.

Tan, J.C., Avdic, S., Cao, J.Z., Mocarski, E.S., White, K.L., Abendroth, A., and Slobedman, B. (2011). Inhibition of 2′,5′-oligoadenylate synthetase expression and function by the human cytomegalovirus ORF94 gene product. J. Virol. *85*, 5696–5700.

Taylor-Wiedeman, J., Sissons, J.G., Borysiewicz, L.K., and Sinclair, J.H. (1991). Monocytes are a major site of persistence of human cytomegalovirus in peripheral blood mononuclear cells. J. Gen. Virol. *72*, 2059–2064.

Taylor-Wiedeman, J., Hayhurst, G.P., Sissons, J.G., and Sinclair, J.H. (1993). Polymorphonuclear cells are not sites of persistence of human cytomegalovirus in healthy individuals. J. Gen. Virol. *74*, 265–268.

Taylor-Wiedeman, J., Sissons, P., and Sinclair, J. (1994). Induction of endogenous human cytomegalovirus gene

expression after differentiation of monocytes from healthy carriers. J. Virol. *68*, 1597–1604.

Tey, S.K., Goodrum, F., and Khanna, R. (2010). CD8+ T-cell recognition of human cytomegalovirus latency-associated determinant pUL138. J. Gen. Virol. *91*, 2040–2048.

Thompson, R.L., and Sawtell, N.M. (1997). The herpes simplex virus type 1 latency-associated transcript gene regulates the establishment of latency. J. Virol. *71*, 5432–5440.

Umashankar, M., Petrucelli, A., Cicchini, L., Caposio, P., Kreklywich, C., Rak, M., Bughio, F., Goldman, D., Hamlin, K., Nelson, J., *et al.* (2011). A novel human cytomegalovirus locus modulates cell type-specific outcomes of infection. PLoS Pathog. 7, e1002444.

Vieira, J., Schall, T.J., Corey, L., and Geballe, A.P. (1998). Functional analysis of the human cytomegalovirus US28 gene by insertion mutagenesis with the green fluorescent protein gene. J. Virol. *72*, 8158–8165.

Weinshenker, B.G., Wilton, S., and Rice, G.P. (1988). Phorbol ester-induced differentiation permits productive human cytomegalovirus infection in a monocytic cell line. J. Immunol. *140*, 1625–1631.

Welch, A.R., McGregor, L.M., and Gibson, W. (1991). Cytomegalovirus homologs of cellular G protein-coupled receptor genes are transcribed. J. Virol. *65*, 3915–3918.

White, K.L., Slobedman, B., and Mocarski, E.S. (2000). Human cytomegalovirus latency-associated protein pORF94 is dispensable for productive and latent infection. J. Virol. *74*, 9333–9337.

Xia, D., and Straus, S.E. (1999). Transcript mapping and transregulatory behavior of varicella-zoster virus gene 21, a latency-associated gene. Virology *258*, 304–313.

Xia, D., Srinivas, S., Sato, H., Pesnicak, L., Straus, S.E., and Cohen, J.I. (2003). Varicella-zoster virus open reading frame 21, which is expressed during latency, is essential for virus replication but dispensable for establishment of latency. J. Virol. *77*, 1211–1218.

Yu, D., Silva, M.C., and Shenk, T. (2003). Functional map of human cytomegalovirus AD169 defined by global mutational analysis. Proc. Natl. Acad. Sci. U.S.A. *100*, 12396–12401.

Zhuravskaya, T., Maciejewski, J.P., Netski, D.M., Bruening, E., Mackintosh, F.R., and St Jeor, S. (1997). Spread of human cytomegalovirus (HCMV) after infection of human hematopoietic progenitor cells: model of HCMV latency. Blood *90*, 2482–2491.

Zipeto, D., Bodaghi, B., Laurent, L., Virelizier, J.L., and Michelson, S. (1999). Kinetics of transcription of human cytomegalovirus chemokine receptor US28 in different cell types. J. Gen. Virol. *80*, 543–547.

Myeloid Cell Recruitment and Function in Cytomegalovirus Dissemination and Immunity

I.21

Lisa P. Daley-Bauer and Edward S. Mocarski

Abstract

Cytomegalovirus pathogenesis, dissemination and immunity are tied to the behaviour of myelomonocytic cells. Investigations of the roles of monocyte subsets in the dissemination of cytomegalovirus revealed that the MCMV-encoded chemokine, MCK2, controls recruitment patterns of the two major monocyte subsets (inflammatory and patrolling) to sites of infection. Monocytes give rise to both macrophages and dendritic cells that populate tissues. Mice deficient in the chemokine axis (CCR2 and CCL2/MCP-1) do not support inflammatory monocyte emigration from bone marrow. Inflammatory monocyte-derived lineages, which are non-permissive for MCMV, are dispensable for pathogenesis and dissemination as well as for the establishment of latency set-points. Nevertheless, recruitment of inflammatory monocytes is enhanced by elaboration of MCK2 and impairs the CTL response to delay viral clearance. In contrast, patrolling monocytes support MCMV replication and contribute to dissemination patterns in the host. Studies in CX$_3$CR1-deficient infected mice show reduced viral dissemination to salivary glands, consistent with the reduced survival of patrolling monocytes in these animals. HCMV studies suggest myelomonocytic progenitor cells, including inflammatory and patrolling monocyte subsets, are associated with acute as well as latent infections. MCMV latency in mice occurs in myeloid as well as epithelial and endothelial cell lineages. In this chapter, we review the current understanding of myelomonocytic lineage cells in the establishment of cytomegalovirus infections based on human and murine studies.

Introduction

Cytomegalovirus (CMV) engages myeloid cells in the host for dissemination as well as for efficient establishment of latency. Human CMV (HCMV) and the related betaherpesviruses infecting rats, rhesus macaques, guinea pigs, and, in particular, mice, have proved useful in dissecting pathogen–host interaction. Murine CMV (MCMV) and HCMV share many characteristics, including the expression of potent immune modulators to deflect as well as orchestrate the host response to infection. Gene products encoded by these viruses modulate three broad levels of host defence: (1) cell-intrinsic response attributes such as pathogen sensing, epigenetic gene regulation and cell death, (2) innate immune attributes such as cytokine and chemokine elaboration, natural killer cell activity and antiviral mediators, and (3) adaptive immune attributes involved in antibody and cell-mediated immune surveillance. Together, these virus-encoded modulators increase the efficiency of primary infection, dissemination, and persistent infection of the host, key elements that are only readily dissected in small laboratory animals such as mice. In addition, both HCMV and MCMV exploit the host inflammatory response to infection, inducing the migration and activation of leucocytes through elaboration of virus-encoded chemokines. Opportunistic HCMV infection and disease during tissue and organ transplantation unveiled the benefit of a strong inflammatory response to the virus. Viral chemokines such as HCMV vCXCL-1 (encoded by UL146) (Penfold et al., 1999), and MCMV MCK2 (encoded by m131-m129) (Fleming et al., 1999) attract myeloid cells (Saederup et al., 1999, 2001; Noda et al., 2006). Studies on MCMV, in particular, reveal the role of the viral chemokine in enhancing systemic leucocyte mobilization from the bone marrow (BM) and recruitment via the peripheral blood (PB) to sites of infection, where patrolling monocytes (PMs) acquire virus and promote dissemination through PB to salivary glands (SGs) and inflammatory monocytes (IMs) modulate the quality of the antiviral CD8 T-cell response to delay viral clearance from tissues (Daley-Bauer et al., 2012). Together, these contribute to establishing a reservoir of latent virus. The ability of proinflammatory viral functions expressed during productive infection to successfully manipulate the host response to infection is one of many diverse and surprising characteristics that ensure the success of this pathogen during evolution in the remarkably wide variety of mammalian species that various cytomegaloviruses infect.

Monocytes are part of the host immune response to HCMV or MCMV. In mice, two populations of PB CD11b$^+$CD115$^+$ F4/80$^+$ monocytes give rise to tissue macrophages (MACs) and dendritic cells (DCs). These may be divided into BM-resident CX$_3$CR1^{int}CCR2$^+$Ly6C^{hi} IMs and long-lived CX$_3$CR1^{hi}CCR2$^-$Ly6C^{lo} PMs (Geissmann et al., 2003; Sunderkotter et al., 2004; Serbina et al., 2006). CCR2-deficient mice fail to mobilize IMs from BM into the bloodstream and tissues in response to infection (Serbina et al., 2006), but continue to exhibit normal levels of PMs. PMs are resident within blood vessels where long-term survival is dependent on constitutive CX$_3$CR1-mediated signals provided by endothelial contact (Geissmann et al., 2003; Sunderkotter et al., 2004; Landsman et al., 2009). Thus, CX$_3$CR1-deficient mice have reduced circulating PMs but retain normal levels of IMs and IM mobilization from BM (Geissmann et al., 2003). In humans, CD14 and CD16 mark paralogous human PB monocyte subsets (Serbina et al., 2006; Robbins et al., 2010). IMs give rise to diverse cells that have been referred to as MACs, tumour necrosis factor (TNF) and inducible nitric oxide (NO)-producing DCs (TipDCs) (Serbina et al., 2006) and mononuclear myeloid-derived suppressor cells (MDSCs) (Movahedi et al., 2008), all of which carry out diverse immune roles ranging from immunostimulation of innate immunity and suppression of adaptive immunity to immunopathogenesis (Serbina et al., 2008; van Ginderachter et al., 2010; Chioda et al., 2011). IMs are correlated with protection during infection with *Listeria monocytogenes* (Serbina et al., 2006), *Toxoplasma gondii* (Robben et al., 2005; Dunay et al., 2008), *Cryptosporidium neoformans* (Traynor et al., 2002) or *Aspergillus fumigatus* (Blease et al., 2001). IM-like MDSCs downmodulate the activity and numbers of CD8 T-cells (Kusmartsev et al., 2005; Sinha et al., 2007; Movahedi et al., 2008). The behaviour of IMs in virus infection is also varied. IMs protect from West Nile virus infection (Lim et al., 2011), but promote influenza disease (Lin et al., 2008) whereas they play an MDSC-like role in Theiler's murine encephalomyelitis virus (Bowen et al., 2009). During MCMV infection MCK2 synergizes with host CCR2-mediated signalling to enhance the natural IM mobilization (Noda et al., 2006). Mobilization of IMs during MCMV infection in C57BL/6 mice stimulates a robust antiviral natural killer (NK) cell response (Hokeness et al., 2005; Crane et al., 2009); whereas, mobilization of IMs in BALB/c mice leads to inhibition of CD8 T-cell activation, contributing to slower viral clearance by producing an iNOS-dependent suppression of CTL activity (Daley-Bauer et al., 2012).

Myeloid cell populations and their function in HCMV and MCMV

Myeloid cell types in CMV infection

CMV cell tropism starts with attachment and entry, using envelope glycoprotein complexes that mediate infection through fusion at the plasma membrane (Compton et al., 1992 and Chapter I.8) or following endocytosis (Ryckman et al., 2008; Revello et al., 2010), that is largely determined by the cell type. In the case of HCMV, the list of susceptible cells supporting viral replication ranges from parenchymal to connective tissue to haematopoietic cells (Sinzger et al., 2008 and Chapter I.17). Of note, although supported by *in vitro* studies, *in vivo* evaluations of acute infection have been conducted primarily in immunocompromised patients, leaving the nature of HCMV cell tropism in immunocompetent individuals yet to be evaluated. Haematopoietic cells, in particular leucocytes, are recognized in HCMV studies due to their contribution to transmission (Lang et al., 1977; Adler 1983), pathogenesis (Gerna et al., 2004) and latency (Reeves et al., 2008; Sinclair 2010). Myeloid lineage leucocytes have garnered attention due to accumulating evidence supporting a role in viral dissemination (Reeves et al., 2008; Sinclair 2010). On the one hand, polymorphonuclear leucocytes (PMNL) in PB are capable of taking up and disseminating HCMV even though the replicative cycle is not supported by these cells (Turtinen et al., 1987; Grefte et al., 1994; Gerna et al., 2000). On the other hand, non-permissive PB monocytes may also harbour virus (Taylor-Wiedeman et al., 1991; Sinclair et al., 1996) and are likewise capable of disseminating virus, undergo a MAC/DC differentiation programme that renders these cells permissive to productive viral replication (Hertel et al., 2003; Reeves et al., 2005; Sinclair 2008). Additionally, monocytes are a major cell type harbouring viral DNA long term in blood and bone marrow (Taylor-Wiedeman et al., 1991, 1994; Manez et al., 1996; Gerna et al., 2000; Saez-Lopez et al., 2005), and in tissue transplants monocyte-derived MACs may be latently infected with HCMV (Gnann et al., 1988). The aggregate evidence is consistent with monocytes having a major role in HCMV infection. The body of data inculpating myelomononuclear cells in HCMV infection has been corroborated by *in vivo* studies of MCMV infection where these cells are the predominant cell-type that disseminates virus in PB during acute infection (Bale *et al.*, 1989; Collins *et al.*, 1993; Collins *et al.*, 1994; Stoddart *et al.*, 1994), contributes significantly in transmission via blood transfusion (Cheung *et al.*, 1977) and has been implicated as a reservoir during latency (Pollock *et al.*, 1997; Koffron *et al.*, 1998).

CMV encoded functions modulate migratory behaviour of myelomonocytic cells

It is evident that a robust inflammatory response is favoured and likely advantageous in an infection setting given that the immunomodulatory function to regulate leucocyte migration patterns is maintained in both HCMV and MCMV. To this end, both viruses possess genes encoding homologues of chemokine receptors as well as ligands (Table I.21.1). Four G-protein coupled chemokine receptor homologues are encoded by HCMV; UL33, UL78, US27 and US28. The US28 protein is a potent chemokine-activated and constitutive receptor expressed on the surface of infected cells that serves primarily to sequester extracellular host chemokines such as RANTES, MCP-1, MIP-1α and MIP-1β, but may also function to promote smooth muscle cell migration (Bodaghi *et al.*, 1998; Streblow *et al.*, 2003); even though, US28-regulated host gene expression was not detected when assessed during infection (Hertel *et al.*, 2004). Although the functions of UL33, UL78 and US27 remain largely undefined, these proteins co-localize and may heterodimerize with US28 (Tschische *et al.*, 2011). Monocytes are responsive to RANTES and MCP1 and are a major source for MIP-1α that recruits NK cells, so US28 may play a specific modulatory role in these cells. HCMV encodes one *bona fide* CXC chemokine (UL146-encoded vCXCL1) (Penfold *et al.*, 1999) and two homologues; UL147 (Penfold *et al.*, 1999) and UL128 (Akter *et al.*, 2003),

Table I.21.1 CMV-encoded chemokine and chemokine receptor homologues

	HCMV	MCMV
Receptor	UL33	M33
	UL78	M78
	US27	
	US28	
Chemokine	UL128	m131–m129
	UL146	
	UL147	

the latter of which has been recognized as a component of a gH-gL envelope glycoprotein complex that facilitates entry into certain cell types (Ryckman *et al.*, 2008; Revello *et al.*, 2010). Although the UL146 gene product, vCXCL1, exhibits very limited ELRCXC sequence homology to the neutrophil chemoattractant, IL-8, it is nevertheless a potent agonist for human CXCR2 (Penfold *et al.*, 1999). UL146 shows the greatest strain-to-strain variability of any HCMV gene, although even diverse sequence variants in chimpanzee CMV retain function (Miller-Kittrell *et al.*, 2007). MCMV encodes two chemokine receptor homologues, M33 and M78. M33 has been compared to US28 because it binds RANTES to promote vascular smooth muscle cell migration (Saederup *et al.*, 2002; Melnychuk *et al.*, 2005). Although efforts have been undertaken to elucidate the biochemical properties of M78 (Sharp *et al.*, 2009), a UL78 homologue (Rawlinson *et al.*, 1996), little concrete information is available. Expression of the MCMV-encoded CC chemokine homologue, MCK2, has been implicated in enhancing recruitment of leucocytes to the portal of entry, facilitating recruitment of monocytes that act as vehicles for viraemia and dissemination to the SGs during acute infection (Saederup *et al.*, 1999, 2001; Noda *et al.*, 2006). MCK2 expression also increases viral titres and persistence of productive infection (Fleming *et al.*, 1999). Myeloid cells implicated in pathogenesis and latency contribute significantly to the initial establishment of infection within the host.

Monocyte subsets in CMV infection

CMV virulence and pathogenesis depend on the behaviour of myelomonocytic lineage cells. Although the processes and cellular factors involved in CMV dissemination have not been fully elucidated, a preponderance of evidence implicates monocyte-derived cells, including MACs and DCs, in many aspects of viral pathogenesis, from dissemination to latency. MCMV recapitulates HCMV behaviour in the implication of monocytes in CMV dissemination (Bale *et al.*, 1989; Stoddart *et al.*, 1994). The most prevalent cell type with markers of virus infection in PB exhibits MAC- or DC-like characteristics (Stoddart *et al.*, 1994). In addition to a role in dissemination, monocyte-derived MACs have also been implicated in the control of virus infection (Presti *et al.*, 2001; Noda *et al.*, 2006) and the susceptibility to MCMV disease (Hokeness *et al.*, 2005). Interestingly, MCK2 was shown to promote recruitment of a monocytic cell population that drives local inflammation at the inoculation site and to contribute to cell-associated viraemia and dissemination (Stoddart *et al.*, 1994;

Saederup et al., 1999, 2001; Noda et al., 2006). Initial characterization of these MCK2-responsive cells by FACS analyses employed Gr-1 as a critical marker instead of Ly6C, identifying a CD11b+Gr-1intPECAM1+Sca1+ late myeloid progenitor as a cell type dramatically mobilized by viral MCK2 (Noda et al., 2006). We have recently clarified the situation and identified these cells as IMs, showing that these monocytes do not control dissemination but rather suppress the CD8 T-cell response (Daley-Bauer et al., 2012).

Human monocyte subsets are defined as CD14+CD16− IMs or CD14−CD16+ PMs (Passlick et al., 1989; Fingerle-Rowson et al., 1998; Grage-Griebenow et al., 2001a,b; Ancuta et al., 2003; Skrzeczynska-Moncznik et al., 2008). The cognate subsets in mice co-express CD11b, CD115 and F4/80 and are subdivided into Ly6ChiGr-1hiCR2+ IMs and Ly6Clo/−Gr-1−CX3CR1hi PMs (Geissmann et al., 2003; Sunderkotter et al., 2004). In MCMV-infected BALB/c mice, Gr-1 is poorly expressed on both subsets, so the division is into CX3CR1intCCR2+Ly6Chi IMs and CX3CR1hiCCR2−Ly6Clo PM, in agreement with others (Nagendra et al., 2004; Dolcetti et al., 2010). IM and PM subsets differ in their cytokine profiles induced by toll-like receptor stimulation as well as by their migratory propensities (Hume et al., 2002; Geissmann et al., 2003; Sunderkotter et al., 2004; Cros et al., 2010). Monocyte-derived MACs and DCs have multifaceted roles in shaping both innate and adaptive immune responses to all infectious pathogens.

IMs are mobilized from the BM in a CCR2-dependent manner to enter the bloodstream in response to inflammatory signals and traffic to inflamed tissue. IM-derived cells carry out immune functions ranging from protective to pathogenic to suppressive, conferring protection against bacterial, fungal, protozoal and viral pathogens (Blease et al., 2001; Traynor et al., 2002; Robben et al., 2005; Serbina et al., 2006; Getts et al., 2008), disease pathology in influenza pulmonary infection (Lin et al., 2008) and MDSC activity to impede development of an antigen-specific CD8 T-cell response to tumours. IMs are a hallmark of acute CMV infection and are important in recruiting early responding NK cells that protect in C57BL6 mice (Hokeness et al., 2005) via MCMV m157-driven activation of Ly49H receptors (Arase et al., 2002; Smith et al., 2002). This robust NK cell response may dampen the T-cell response by killing off cells that express viral antigens (Andrews et al., 2010; Mitrovic et al., 2012). Recently, IM-derived cells have been implicated in suppressing the

anti-viral CD8 T-cell immune response to MCMV (Daley-Bauer et al., 2012).

A lot less is understood about the role of PM function in pathogenic infections; however, their inherent migratory nature places these cells in an opportune position to constantly survey endothelium for inflammatory signals and contribute significantly to the early host inflammatory response (Geissmann et al., 2003; Auffray et al., 2007). Mouse studies show that PMs crawl in a multidirectional pattern along the luminal side of blood vessels in a CX3CR1- and LFA-1-dependent fashion remaining in contact with the vasculature endothelium (Auffray et al., 2007). For reasons not yet understood, although present on many other leucocyte populations, CX3CR1 interacts with the CX3CL1 (fractalkine) constitutively expressed on endothelium to confer survival signals only to PMs. Comparative functional assessments of human monocytes reveal that, unlike IMs, PMs do not respond to bacterial stimuli but instead respond primarily to nucleic acid-mediated TLR7/8 stimuli to produce inflammatory cytokines.

Modulation of IMs by MCMV

MCK2 regulates recruitment of CCR2-dependent IMs

Mobilization of myelomonocytic cells is a hallmark of the inflammatory response to productive infection with cytomegalovirus (Saederup et al., 2001; Noda et al., 2006; Crane et al., 2009; Daley-Bauer et al., 2012). In vivo studies showed that the intensity of this myelomonocytic cell response is determined by the expression of the viral chemokine, MCK2, which also controls patterns of viraemia that establish the viral titre set point in salivary glands (Stoddart et al., 1994; Saederup et al., 1999, 2001). Although Noda et al. (2006) described an MCK2 impact on a population termed late myeloid progenitors that seemed to be connected to viral dissemination, the reliance on Gr-1 instead of Ly6C as a marker failed to adequately distinguish cells in FACS analyses. Based on the understanding of CCR2 at the time, the data was interpreted as preventing IM/macrophage migration into tissues rather than the currently understood role of CCR2 in mobilization from BM (Serbina et al., 2006). This misled the conclusion that recruitment of MCK2-responsive, late myeloid progenitors into inoculated footpads was normal in Ccr2−/−Ccl2−/− mice. It is now clear that IM mobilization from BM during MCMV infection is CCR2-dependent as well as MCK2-dependent, although these systems are independent.

A secondary role of CCR2 in the transendothelial migration of IM-derived lineages into tissues remains a possibility.

MCK2-enhanced, CCR2-dependent inflammatory monocyte recruitment impairs CD8 T-cell immunity

Although MCK2 expression enhances IM recruitment in cooperation with CCR2 to dramatically enhance inflammation at initial sites of infection (Saederup et al., 2001; Noda et al., 2006; Daley-Bauer et al., 2012), this MCK2-enhanced, CCR2-dependent IM response is dispensable for efficient viral dissemination and latency. Viral dissemination to the SGs and latency set points remain unperturbed in *Ccr2⁻/⁻Ccl2⁻/⁻* mice (Daley-Bauer et al., 2012). Instead, data from our studies (Daley-Bauer et al., 2012) and others (Hokeness et al., 2005; Crane et al., 2009) reveal discrete yet overlapping immunomodulatory effects of IMs on the host antiviral response. From *in vitro* and adoptive transfer studies in BALB/c mice, we observed that the MCK2-enhanced, CCR2-dependent IM recruitment is associated with impaired activation, differentiation and cytolytic function of virus-specific CD8 T-cells

(Fig. I.21.1). CCR2/CCL2-deficiency improves the quality and proliferative potential of virus-specific IFNγ-producing CTLs and contributes to improved control and more rapid clearance of MCMV from tissues (Fig. I.21.2). Inducible nitric-oxide synthase (iNOS) production of NO by the MCMV mobilized IMs mediates this impaired CD8 T-cell state (Fig. I.21.1). Thus, in a setting where viral clearance is primarily mediated by CD8 T-cells, virus modulation of IM recruitment serves to dampen immunity and fosters viral persistence. In C57BL/6 mice, IM mobilization is essential for recruiting the early NK responders that are central to the resistance shown to MCMV in these mice (Hokeness et al., 2005; Crane et al., 2009). Depletion and transgenic mouse studies show that the robust NK cell response during MCMV infection serves to limit the amount of viral antigens available, ultimately resulting in an impaired adaptive immune response and greater viral persistence in SGs (Su et al., 2001; Andrews et al., 2010). NK responses are much less dominant in BALB/c mice as well as in wild-derived mice due to the distribution of the Ly49H allele that responds to MCMV m157 to drive the aggressive NK cell response (see also Chapter I.2). In addition, inhibitory receptors, such as Ly49I that also responds to m157 (Arase et

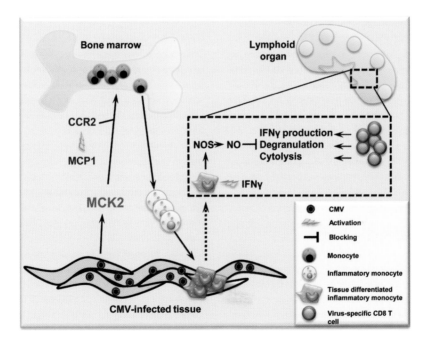

Figure I.21.1 MCK2 cooperates with host CCR2-mediated signals to harness the immunosuppressive effects of IMs. MCMV infection induces a natural CCR2-dependent monocyte mobilization from BM that is enhanced by MCK2. These IMs produce NO via the NO synthase (NOS) pathway to impair the Ag-specific CD8 T-cell response thus creating a dampened immune environment that promotes viral persistence.

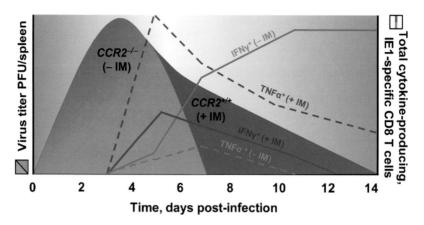

Figure I.21.2 Virus titres and CD8 T-cell response in spleens in the presence or absence of IMs. The absence of IM mobilization (–IM) coincides with faster viral clearance (red histogram), greater numbers of IFNγ-producing, IE1-specific CD8 T-cells (red, solid line) and fewer numbers of TNF-producing, IE1-specific CD8 T-cells (red, dashed lines). In contrast, a robust IM response (+IM) correlates with delayed viral clearance (blue histogram), greater numbers of TNF-producing, IE1-specific CD8 T-cells (blue, dashed lines) and fewer numbers of IFNγ-producing, IE1-specific CD8 T-cells (blue, solid line).

al., 2002), may dominate and prevent NK cell activation. Thus, virus modulation of IM recruitment in the C57BL/6 setting employs a different pathway that nevertheless generates an impaired immune state to facilitate viral persistence. In HCMV, any role of UL146-encoded vCXCL1, a CXCR2 agonist, has not been fully elucidated. CXCR2 is distributed on a both mononuclear and PMNL, although it is rapidly lost in culture, leaving open the possibility that IMs (or PMNL, another myeloid subset possessing immunosuppressive traits) are recruited via this viral chemokine to impair adaptive immunity in HCMV infection. This again reflects the type of suppressive activity that has been shown for MDSC populations in tumour immunology (Dolcetti *et al.*, 2010). Given that NK, CD4 and CD8 T-cells all contribute to HCMV control, it is likely that a combination of the immunomodulatory strategies described in mice may be relevant to human infection settings.

Modulation of PMs by MCMV

Current findings on IMs (Daley-Bauer *et al.*, 2012) add to previous investigations showing that dissemination occurs independent of CCR2-signalling or recruitment (Noda *et al.*, 2006), leaving the open question regarding the cell type that controls dissemination. The body of work demonstrating dramatic increases in leucocyte recruitment, myelomonocytic cell-associated viraemia, and dissemination when the viral-encoded chemokine is expressed, unequivocally support a role for MCK2 in the enhancement of viral

dissemination independent of host adaptive immunity (Fleming *et al.*, 1999; Saederup *et al.*, 2001). We set out to specifically address whether PMs, a monocyte that distributes independent of CCR2, may function as vehicles for virus dissemination through the PB to SGs (Stoddart *et al.*, 1994).

PMs and viral dissemination

PMs may be distinguished by a high-level expression of the chemokine receptor CX_3CR1, the unique partner of the agonist, CX_3CL1 (fractalkine). CX_3CL1 is synthesized as a transmembrane protein bearing an extended mucin-like stalk and chemokine head region. This protein confers a dual role as an adhesion molecule as well as chemokine on the surface of endothelial cells as well as other cell types. Cleavage of the stalk by metalloproteases present in blood and inflamed tissues results in the release of soluble CX_3CL1 and the resultant generation of a chemokine gradient. In addition to these activities, recent studies (Jakubzick *et al.*, 2008; Landsman *et al.*, 2009) suggest that the CX_3CR1–CX_3CL1 interaction also induces prosurvival signals. $Cx_3cr1^{gfp/gfp}$ reporter mice (Jung *et al.*, 2000) afford a tool to investigate PMs in MCMV infection. In these mice, rapid recruitment of PMs to MCMV inoculated footpads proceeds in an MCK2- and CX_3CR1-dependent manner, peaking at 12 hours post inoculation (Fig. I.21.3A). The absence of CX_3CR1-mediated signalling leads to significantly fewer PMs in the blood during the first 5 days of infection as well as a 50% reduction in the total numbers of PMs recruited to infected footpads

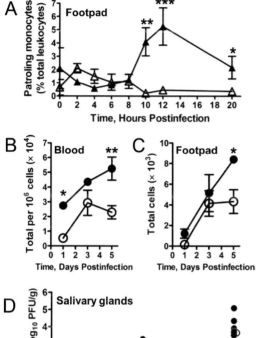

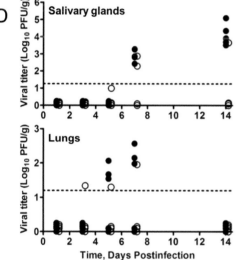

Figure I.21.3 Recruitment of MCK2-responsive PMs is a determinant of the establishment of MCMV infection. (A) MCK2 recruits PMs to infected sites independent of CX₃CR1-mediated chemotaxis. $CX_3CR1^{-/-}$ mice infected via footpad route with 10^6 PFU MCMV. Open triangle, MCK2 mutant; closed triangle, MCK2 rescue. (B, C) CX₃CR1 deficiency reduces PM levels in blood (B) and at the inoculation site (C). Open circle, $CX_3CR1^{-/-}$; closed circle, $CX_3CR1^{+/-}$. (D) PM recruitment influences MCMV dissemination. Viral titres in SG (top panel) and lungs (bottom panel) where open circle represent $CX_3CR1^{-/-}$ and closed circle, $CX_3CR1^{+/-}$. *$P<0.05$; **$P<0.01$; ***$P<0.001$.

by 5 dpi (Fig. I.21.3B and C). In contrast, levels and function of other CX₃CR1-expressing immune cells were unaffected compared to immunocompetent mice (unpublished data). The defect in PMs correlated with an absence of virus in SGs of mutant mice at 14 dpi (Fig. I.21.3D). Virus detected at 7 dpi is typically located in the adjacent fat tissues. Although the recruitment is not as dramatic as IM mobilization from BM, we hypothesize that the virus-encoded chemokine influences PM recruitment to the initial inoculation site to carry out dissemination (Fig. I.21.4). The observations thus far warrant further investigation.

Implications for future directions

Monocytes are the most versatile of leucocytes. Improving markers of monocyte subsets have already provided key insights into how IMs and PMs contribute to MCMV infection. Future investigation must draw comparisons to HCMV and seek to better define the myeloid reservoir in both MCMV and HCMV involved in latency and reactivation.

Conclusion

Despite the fact that immune control by CD8 T-cells dominates MCMV infection in BALB/c mice, the pathogen–host détente plays out with the help of monocyte-derived lineages whose recruitment is facilitated by virus-encoded chemokine (Fig. I.21.4). While the most responsive population of monocytes, IMs, is not susceptible to MCMV infection and does not directly facilitate dissemination, this population disables CD8 T-cells just enough to prolong infection in host tissues (Fig. I.21.4). As would be predicted, suppression of adaptive effector T-cell immunity allows greater viral persistence without impacting the latency set point. The immunosuppressive effect of IMs on CD8 T-cells was demonstrated directly in a natural mouse infection without the manipulations necessary in tumour models that enable evaluation of T-cell immunity. In contrast to the dramatic mobilization of IMs, blood-resident PMs mount a more modest response to MCK2 but nevertheless are crucial to the dissemination of virus from initial infected sites to the SGs where increased shedding would enhance viral transmission (Fig. I.21.4).

Acknowledgement

This work was supported by Public Health Service Grants R01 AI030363 and AI020211.

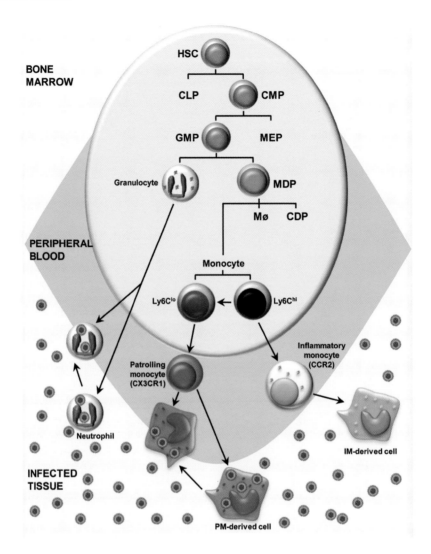

Figure I.21.4 Differentiation of myeloid cells and their contributions to MCMV infection. Haematopoietic stem cells (HSC) give rise to the common lymphoid (CLP) and myeloid (CMP) progenitors. CMPs give rise to megakaryocyte–erythrocyte progenitors (MEP) and granulocyte-mononuclear cell progenitors (GMP), the latter of which give rise to granulocytes. Of these, PMNL from PB and infected tissues have been identified in infectious centre assays to harbour infectious viral particles. Reverse transendothelial migration possibly enables these cells to disseminate virus to distal organs. GMP also give rise to monocyte-MAC-DC progenitors (MDP), the precursors of monocytes, subsets of MAC (Mø) and common DC progenitors (CDP). Ly6C^{hi} monocytes in BM are thought to give rise to Ly6C^{lo} monocytes. Ly6C^{hi} and Ly6C^{lo} monocytes enter peripheral blood as IMs and PMs, respectively. In response to MCMV infection, IMs are modulated by MCK2 to dampen antiviral adaptive immunity while being dispensable for dissemination. On the other hand, there is a positive correlation between functional MCK2-responsive PMs and viral spread suggesting a role for these cells in virus dissemination.

References

Adler, S.P. (1983). Transfusion-associated cytomegalovirus infections. Rev. Infect. Dis. 5, 977–993.

Akter, P., Cunningham, C., McSharry, B.P., Dolan, A., Addison, C., Dargan, D.J., Hassan-Walker, A.F., Emery, V.C., Griffiths, P.D., Wilkinson, G.W., et al. (2003). Two novel spliced genes in human cytomegalovirus. J. Gen. Virol. 84, 1117–1122.

Ancuta, P., Rao, R., Moses, A., Mehle, A., Shaw, S.K., Luscinskas, F.W., and Gabuzda, D. (2003). Fractalkine preferentially mediates arrest and migration of CD16+ monocytes. J. Exp. Med. 197, 1701–1707.

Andrews, D.M., Estcourt, M.J., Andoniou, C.E., Wikstrom, M.E., Khong, A., Voigt, V., Fleming, P., Tabarias, H., Hill, G.R., van der Most, R.G., et al. (2010). Innate immunity defines the capacity of antiviral T-cells to limit persistent infection. J. Exp. Med. 207, 1333–1343.

Arase, H., Mocarski, E.S., Campbell, A.E., Hill, A.B., and Lanier, L.L. (2002). Direct recognition of cytomegalovirus by activating and inhibitory NK cell receptors. Science 296, 1323–1326.

Auffray, C., Fogg, D., Garfa, M., Elain, G., Join-Lambert, O., Kayal, S., Sarnacki, S., Cumano, A., Lauvau, G., and Geissmann, F. (2007). Monitoring of blood vessels and tissues by a population of monocytes with patrolling behavior. Science 317, 666–670.

Bale, J.F., Jr., and O'Neil, M.E. (1989). Detection of murine cytomegalovirus DNA in circulating leukocytes harvested during acute infection of mice. J. Virol. 63, 2667–2673.

Blease, K., Mehrad, B., Lukacs, N.W., Kunkel, S.L., Standiford, T.J., and Hogaboam, C.M. (2001). Antifungal and airway remodeling roles for murine monocyte chemoattractant protein-1/CCL2 during pulmonary exposure to Asperigillus fumigatus conidia. J. Immunol. 166, 1832–1842.

Bodaghi, B., Jones, T.R., Zipeto, D., Vita, C., Sun, L., Laurent, L., Arenzana-Seisdedos, F., Virelizier, J.L., and Michelson, S. (1998). Chemokine sequestration by viral chemoreceptors as a novel viral escape strategy, withdrawal of chemokines from the environment of cytomegalovirus-infected cells. J. Exp. Med. 188, 855–866.

Bowen, J.L., and Olson, J.K. (2009). Innate immune CD11b+Gr-1+ cells, suppressor cells, affect the immune response during Theiler's virus-induced demyelinating disease. J. Immunol. 183, 6971–6980.

Cheung, K.S., and Lang, D.J. (1977). Transmission and activation of cytomegalovirus with blood transfusion: a mouse model. J. Infect. Dis. 135, 841–845.

Chioda, M., Peranzoni, E., Desantis, G., Papalini, F., Falisi, E., Samantha, S., Mandruzzato, S., and Bronte, V. (2011). Myeloid cell diversification and complexity: an old concept with new turns in oncology. Cancer Metastasis Rev. 30, 27–43.

Collins, T., Pomeroy, C., and Jordan, M.C. (1993). Detection of latent cytomegalovirus DNA in diverse organs of mice. J. Infect. Dis. 168, 725–729.

Collins, T.M., Quirk, M.R., and Jordan, M.C. (1994). Biphasic viremia and viral gene expression in leukocytes during acute cytomegalovirus infection of mice. J. Virol. 68, 6305–6311.

Compton, T., Nepomuceno, R.R., and Nowlin, D.M. (1992). Human cytomegalovirus penetrates host cells by pH-independent fusion at the cell surface. Virology 191, 387–395.

Crane, M.J., Hokeness-Antonelli, K.L., and Salazar-Mather, T.P. (2009). Regulation of inflammatory monocyte/macrophage recruitment from the bone marrow during murine cytomegalovirus infection: role for type I Interferons in localized induction of CCR2 ligands. J. Immunol. 183, 2810–2817.

Cros, J., Cagnard, N., Woollard, K., Patey, N., Zhang, S.Y., Senechal, B., Puel, A., Biswas, S.K., Moshous, D., Picard, C., J., et al. (2010). Human CD14dim monocytes patrol and sense nucleic acids and viruses via TLR7 and TLR8 receptors. Immunity 33, 375–386.

Daley-Bauer, L.P., Wynn, G.M., and Mocarski, E.S. (2012). Cytomegalovirus impairs antiviral CD8+ T-cell immunity by recruiting inflammatory monocytes. Immunity 37, 122–133.

Dolcetti, L., Peranzoni, E., Ugel, S., Marigo, I., Fernandez Gomez, A., Mesa, C., Geilich, M., Winkels, G., Traggiai, E., Casati, A., et al. (2010). Hierarchy of immunosuppressive strength among myeloid-derived suppressor cell subsets is determined by GM-CSF. Eur. J. Immunol. 40, 22–35.

Dunay, I.R., Damatta, R.A., Fux, B., Presti, R., Greco, S., Colonna, M., and Sibley, L.D. (2008). Gr1(+) inflammatory monocytes are required for mucosal resistance to the pathogen Toxoplasma gondii. Immunity 29, 306–317.

Fingerle-Rowson, G., Auers, J., Kreuzer, E., Fraunberger, P., Blumenstein, M., and Ziegler-Heitbrock, L.H. (1998). Expansion of CD14+CD16+ monocytes in critically ill cardiac surgery patients. Inflammation 22, 367–379.

Fleming, P., Davis-Poynter, N., Degli-Esposti, M., Densley, E., Papadimitriou, J., Shellam, G., and Farrell, H. (1999). The murine cytomegalovirus chemokine homolog, m131/129, is a determinant of viral pathogenicity. J. Virol. 73, 6800–6809.

Geissmann, F., Jung, S., and Littman, D.R. (2003). Blood monocytes consist of two principal subsets with distinct migratory properties. Immunity 19, 71–82.

Gerna, G., Percivalle, E., Baldanti, F., Sozzani, S., Lanzarini, P., Genini, E., Lilleri, D., and Revello, M.G. (2000). Human cytomegalovirus replicates abortively in polymorphonuclear leukocytes after transfer from infected endothelial cells via transient microfusion events. J. Virol. 74, 5629–5638.

Gerna, G., Baldanti, F., and Revello, M.G. (2004). Pathogenesis of human cytomegalovirus infection and cellular targets. Hum. Immunol. 65, 381–386.

Getts, D.R., Terry, R.L., Getts, M.T., Muller, M., Rana, S., Shrestha, B., Radford, J., Van Rooijen, N., Campbell, I.L., and King, N.J. (2008). Ly6c+ 'inflammatory monocytes' are microglial precursors recruited in a pathogenic manner in West Nile virus encephalitis. J. Exp. Med. 205, 2319–2337.

Gnann, J.W., Jr., Ahlmen, J., Svalander, C., Olding, L., Oldstone, M.B., and Nelson, J.A. (1988). Inflammatory cells in transplanted kidneys are infected by human cytomegalovirus. Am. J. Pathol. 132, 239–248.

Grage-Griebenow, E., Flad, H.D., and Ernst, M. (2001a). Heterogeneity of human peripheral blood monocyte subsets. J. Leukoc. Biol. 69, 11–20.

Grage-Griebenow, E., Zawatzky, R., Kahlert, H., Brade, L., Flad, H., and Ernst, M. (2001b). Identification of a novel dendritic cell-like subset of CD64(+)/CD16(+) blood monocytes. Eur. J. Immunol. 31, 48–56.

Grefte, A., Harmsen, M.C., van der Giessen, M., Knollema, S., van Son, W.J., and The, T.H. (1994). Presence of human cytomegalovirus (HCMV) immediate-early mRNA but not ppUL83 (lower matrix protein pp65) mRNA in polymorphonuclear and mononuclear leukocytes during active HCMV infection. J. Gen. Virol. 75, 1989–1998.

Hertel, L., and Mocarski, E.S. (2004). Global analysis of host cell gene expression late during cytomegalovirus infection reveals extensive dysregulation of cell cycle gene expression and induction of Pseudomitosis independent of US28 function. J. Virol. 78, 11988–12011.

Hertel, L., Lacaille, V.G., Strobl, H., Mellins, E.D., and Mocarski, E.S. (2003). Susceptibility of immature and mature Langerhans cell-type dendritic cells to infection and immunomodulation by human cytomegalovirus. J. Virol. 77, 7563–7574.

Hokeness, K.L., Kuziel, W.A., Biron, C.A., and Salazar-Mather, T.P. (2005). Monocyte chemoattractant protein-1 and CCR2 interactions are required for

IFN-alpha/beta-induced inflammatory responses and antiviral defense in liver. J. Immunol. *174*, 1549–1556.

Hume, D.A., Ross, I.L., Himes, S.R., Sasmono, R.T., Wells, C.A., and Ravasi, T. (2002). The mononuclear phagocyte system revisited. J. Leukoc. Biol. *72*, 621–627.

Jakubzick, C., Tacke, F., Ginhoux, F., Wagers, A.J., van Rooijen, N., Mack, M., Merad, M., and Randolph, G.J. (2008). Blood monocyte subsets differentially give rise to CD103+ and CD103- pulmonary dendritic cell populations. J. Immunol. *180*, 3019–3027.

Jung, S., Aliberti, J., Graemmel, P., Sunshine, M.J., Kreutzberg, G.W., Sher, A., and Littman, D.R. (2000). Analysis of fractalkine receptor CX(3)CR1 function by targeted deletion and green fluorescent protein reporter gene insertion. Mol. Cell. Biol. *20*, 4106–4114.

Koffron, A.J., Hummel, M., Patterson, B.K., Yan, S., Kaufman, D.B., Fryer, J.P., Stuart, F.P., and Abecassis, M.I. (1998). Cellular localization of latent murine cytomegalovirus. J. Virol. *72*, 95–103.

Kusmartsev, S., Nagaraj, S., and Gabrilovich, D.I. (2005). Tumor-associated CD8+ T-cell tolerance induced by bone marrow-derived immature myeloid cells. J. Immunol. *175*, 4583–4592.

Landsman, L., Bar-On, L., Zernecke, A., Kim, K.W., Krauthgamer, R., Shagdarsuren, E., Lira, S.A., Weissman, I.L., Weber, C., and Jung, S. (2009). CX3CR1 is required for monocyte homeostasis and atherogenesis by promoting cell survival. Blood *113*, 963–972.

Lang, D.J., Ebert, P.A., Rodgers, B.M., Boggess, H.P., and Rixse, R.S. (1977). Reduction of postperfusion cytomegalovirus-infections following the use of leukocyte depleted blood. Transfusion *17*, 391–395.

Lim, J.K., Obara, C.J., Rivollier, A., Pletnev, A.G., Kelsall, B.L., and Murphy, P.M. (2011). Chemokine receptor Ccr2 is critical for monocyte accumulation and survival in West Nile virus encephalitis. J. Immunol. *186*, 471–478.

Lin, K.L., Suzuki, Y., Nakano, H., Ramsburg, E., and Gunn, M.D. (2008). CCR2+ monocyte-derived dendritic cells and exudate macrophages produce influenza-induced pulmonary immune pathology and mortality. J. Immunol. *180*, 2562–2572.

Manez, R., Kusne, S., Rinaldo, C., Aguado, J.M., St George, K., Grossi, P., Frye, B., Fung, J.J., and Ehrlich, G.D. (1996). Time to detection of cytomegalovirus (CMV) DNA in blood leukocytes is a predictor for the development of CMV disease in CMV-seronegative recipients of allografts from CMV-seropositive donors following liver transplantation. J. Infect. Dis. *173*, 1072–1076.

Melnychuk, R.M., Smith, P., Kreklywich, C.N., Ruchti, F., Vomaske, J., Hall, L., Loh, L., Nelson, J.A., Orloff, S.L., and Streblow, D.N. (2005). Mouse cytomegalovirus M33 is necessary and sufficient in virus-induced vascular smooth muscle cell migration. J. Virol. *79*, 10788–10795.

Miller-Kittrell, M., Sai, J., Penfold, M., Richmond, A., and Sparer, T.E. (2007). Functional characterization of chimpanzee cytomegalovirus chemokine, vCXCL-1(CCMV). Virology *364*, 454–465.

Mitrovic, M., Arapovic, J., Jordan, S., Fodil-Cornu, N., Ebert, S., Vidal, S.M., Krmpotic, A., Reddehase, M.J., and Jonjic, S. (2012). The NK cell response to mouse cytomegalovirus infection affects the level and kinetics of the early CD8(+) T-cell response. J. Virol. *86*, 2165–2175.

Movahedi, K., Guilliams, M., Van den Bossche, J., Van den Bergh, R., Gysemans, C., Beschin, A., De Baetselier, P., and Van Ginderachter, J.A. (2008). Identification of discrete tumor-induced myeloid-derived suppressor cell subpopulations with distinct T-cell-suppressive activity. Blood *111*, 4233–4244.

Nagendra, S., and Schlueter, A.J. (2004). Absence of cross-reactivity between murine Ly-6C and Ly-6G. Cytometry A *58*, 195–200.

Noda, S., Aguirre, S.A., Bitmansour, A., Brown, J.M., Sparer, T.E., Huang, J., and Mocarski, E.S. (2006). Cytomegalovirus MCK-2 controls mobilization and recruitment of myeloid progenitor cells to facilitate dissemination. Blood *107*, 30–38.

Passlick, B., Flieger, D., and Ziegler-Heitbrock, H.W. (1989). Identification and characterization of a novel monocyte subpopulation in human peripheral blood. Blood *74*, 2527–2534.

Penfold, M.E., Dairaghi, D.J., Duke, G.M., Saederup, N., Mocarski, E.S., Kemble, G.W., and Schall, T.J. (1999). Cytomegalovirus encodes a potent alpha chemokine. Proc. Natl. Acad. Sci. U.S.A. *96*, 9839–9844.

Pollock, J.L., Presti, R.M., Paetzold, S., and Virgin, H.W. (1997). Latent murine cytomegalovirus infection in macrophages. Virology *227*, 168–179.

Presti, R.M., Popkin, D.L., Connick, M., Paetzold, S., and Virgin, H.W. (2001). Novel cell type-specific antiviral mechanism of interferon gamma action in macrophages. J. Exp. Med. *193*, 483–496.

Rawlinson, W.D., Farrell, H.E., and Barrell, B.G. (1996). Analysis of the complete DNA sequence of murine cytomegalovirus. J. Virol. *70*, 8833–8849.

Reeves, M., and Sinclair, J. (2008). Aspects of human cytomegalovirus latency and reactivation. Curr. Top. Microbiol. Immunol. *325*, 297–313.

Reeves, M., Sissons, P., and Sinclair, J. (2005). Reactivation of human cytomegalovirus in dendritic cells. Discov. Med. *5*, 170–174.

Revello, M.G., and Gerna, G. (2010). Human cytomegalovirus tropism for endothelial/epithelial cells, scientific background and clinical implications. Rev. Med. Virol. *20*, 136–155.

Robben, P.M., LaRegina, M., Kuziel, W.A., and Sibley, L.D. (2005). Recruitment of Gr-1+ monocytes is essential for control of acute toxoplasmosis. J. Exp. Med. *201*, 1761–1769.

Robbins, C.S., and Swirski, F.K. (2010). The multiple roles of monocyte subsets in steady state and inflammation. Cell. Mol. Life Sci. *67*, 2685–2693.

Ryckman, B.J., Chase, M.C., and Johnson, D.C. (2008). HCMV gH/gL/UL128-131 interferes with virus entry into epithelial cells: evidence for cell type-specific receptors. Proc. Natl. Acad. Sci. U.S.A. *105*, 14118–14123.

Saederup, N., and Mocarski, E.S., Jr. (2002). Fatal attraction: cytomegalovirus-encoded chemokine homologs. Curr. Top. Microbiol. Immunol. *269*, 235–256.

Saederup, N., Lin, Y.C., Dairaghi, D.J., Schall, T.J., and Mocarski, E.S. (1999). Cytomegalovirus-encoded beta chemokine promotes monocyte-associated viremia in the host. Proc. Natl. Acad. Sci. U.S.A. *96*, 10881–10886.

Saederup, N., Aguirre, S.A., Sparer, T.E., Bouley, D.M., and Mocarski, E.S. (2001). Murine cytomegalovirus CC chemokine homolog MCK-2 (m131–129) is a determinant of dissemination that increases inflammation at initial sites of infection. J. Virol. *75*, 9966–9976.

Saez-Lopez, C., Ngambe-Tourere, E., Rosenzwajg, M., Petit, J.C., Nicolas, J.C., and Gozlan, J. (2005). Immediate-early antigen expression and modulation of apoptosis after in vitro infection of polymorphonuclear leukocytes by human cytomegalovirus. Microbes Infect. 7, 1139–1149.

Serbina, N.V., and Pamer, E.G. (2006). Monocyte emigration from bone marrow during bacterial infection requires signals mediated by chemokine receptor CCR2. Nat. Immunol. 7, 311–317.

Serbina, N.V., Jia, T., Hohl, T.M., and Pamer, E.G. (2008). Monocyte-mediated defense against microbial pathogens. Annu. Rev. Immunol. 26, 421–452.

Sharp, E.L., Davis-Poynter, N.J., and Farrell, H.E. (2009). Analysis of the subcellular trafficking properties of murine cytomegalovirus M78, a 7 transmembrane receptor homologue. J. Gen. Virol. 90, 59–68.

Sinclair, J. (2008). Manipulation of dendritic cell functions by human cytomegalovirus. Expert Rev. Mol. Med. 10, e35.

Sinclair, J. (2010). Chromatin structure regulates human cytomegalovirus gene expression during latency, reactivation and lytic infection. Biochim. Biophys. Acta. 1799, 286–295.

Sinclair, J., and Sissons, P. (1996). Latent and persistent infections of monocytes and macrophages. Intervirology 39, 293–301.

Sinha, P., Clements, V.K., Bunt, S.K., Albelda, S.M., and Ostrand-Rosenberg, S. (2007). Cross-talk between myeloid-derived suppressor cells and macrophages subverts tumor immunity toward a type 2 response. J. Immunol. 179, 977–983.

Sinzger, C., Digel, M., and Jahn, G. (2008). Cytomegalovirus cell tropism. Curr. Top. Microbiol. Immunol. 325, 63–83.

Skrzeczynska-Moncznik, J., Bzowska, M., Loseke, S., Grage-Griebenow, E., Zembala, M., and Pryjma, J. (2008). Peripheral blood CD14high CD16+ monocytes are main producers of IL-10. Scand. J. Immunol. 67, 152–159.

Smith, H.R., Heusel, J.W., Mehta, I.K., Kim, S., Dorner, B.G., Naidenko, O.V., Iizuka, K., Furukawa, H., Beckman, D.L., Pingel, J.T., et al. (2002). Recognition of a virus-encoded ligand by a natural killer cell activation receptor. Proc. Natl. Acad. Sci. U.S.A. 99, 8826–8831.

Stoddart, C.A., Cardin, R.D., Boname, J.M., Manning, W.C., Abenes, G.B., and Mocarski, E.S. (1994). Peripheral blood mononuclear phagocytes mediate dissemination of murine cytomegalovirus. J. Virol. 68, 6243–6253.

Streblow, D.N., Vomaske, J., Smith, P., Melnychuk, R., Hall, L., Pancheva, D., Smit, M., Casarosa, P., Schlaepfer, D.D., and Nelson, J.A. (2003). Human cytomegalovirus chemokine receptor US28-induced smooth muscle cell migration is mediated by focal adhesion kinase and Src. J. Biol. Chem. 278, 50456–50465.

Su, H.C., Nguyen, K.B., Salazar-Mather, T.P., Ruzek, M.C., Dalod, M.Y., and Biron, C.A. (2001). NK cell functions restrain T-cell responses during viral infections. Eur. J. Immunol. 31, 3048–3055.

Sunderkotter, C., Nikolic, T., Dillon, M.J., Van Rooijen, N., Stehling, M., Drevets, D.A., and Leenen, P.J. (2004). Subpopulations of mouse blood monocytes differ in maturation stage and inflammatory response. J. Immunol. 172, 4410–4417.

Taylor-Wiedeman, J., Sissons, J.G., Borysiewicz, L.K., and Sinclair, J.H. (1991). Monocytes are a major site of persistence of human cytomegalovirus in peripheral blood mononuclear cells. J. Gen. Virol. 72, 2059–2064.

Taylor-Wiedeman, J., Sissons, P., and Sinclair, J. (1994). Induction of endogenous human cytomegalovirus gene expression after differentiation of monocytes from healthy carriers. J. Virol. 68, 1597–1604.

Traynor, T.R., Herring, A.C., Dorf, M.E., Kuziel, W.A., Toews, G.B., and Huffnagle, G.B. (2002). Differential roles of CC chemokine ligand 2/monocyte chemotactic protein-1 and CCR2 in the development of T1 immunity. J. Immunol. 168, 4659–4666.

Tschische, P., Tadagaki, K., Kamal, M., Jockers, R., and Waldhoer, M. (2011). Heteromerization of human cytomegalovirus encoded chemokine receptors. Biochem. Pharmacol. 82, 610–619.

Turtinen, L.W., Saltzman, R., Jordan, M.C., and Haase, A.T. (1987). Interactions of human cytomegalovirus with leukocytes in vivo: analysis by in situ hybridization. Microb. Pathog. 3, 287–297.

Van Ginderachter, J.A., Beschin, A., De Baetselier, P., and Raes, G. (2010). Myeloid-derived suppressor cells in parasitic infections. Eur. J. Immunol. 40, 2976–2985.

Immune Surveillance of Cytomegalovirus Latency and Reactivation in Murine Models: Link to 'Memory Inflation'

I.22

Christof K. Seckert, Marion Grießl, Julia K. Büttner, Kirsten Freitag and Niels A.W. Lemmermann

&

Mary A. Hummel, Xue-Feng Liu and Michael I. Abecassis

&

Ana Angulo and Martin Messerle

&

Charles H. Cook

&

Matthias J. Reddehase

Abstract

Cytomegalovirus (CMV) disease with cytopathogenic viral replication and multiple organ involvement is typically confined to the immunocompromised or immunologically immature host. In the immunocompetent host, productive primary CMV infection is efficiently controlled, and is eventually resolved at all tissue sites, by well-orchestrated mechanisms of the innate and adaptive branches of the immune system in due time to prevent overt disease manifestations. At the earliest stages of an acute infection, NK cells rapidly followed by virus epitope-specific CD8+ T-cells play major antiviral roles, and recent findings indicate that these two effector systems are cross-talking for keeping the virus in check despite the fact that during co-speciation with their specific host species all CMVs have evolved strategies to reduce the infected cells' susceptibility to both NK cell-mediated and CD8+ T-cell-mediated antiviral immune functions. The outcome of this virus–host struggle for survival is a ceasefire in which the viral genome is not cleared but is maintained for the lifetime of the individual host in the presence of a fully developed, protective antiviral 'immune memory' without producing infectious viral progeny but retaining the functional capacity to complete the productive replication cycle under conditions of waning immune surveillance and transcription factor-mediated viral gene desilencing as a result of inflammatory cytokine signalling.

These phenomena, known as 'latency' and 'reactivation' are biological hallmarks that CMVs share with all other members of the herpesvirus family. Notably, while immune surveillance appears to play a central role in maintaining latency, that is in preventing the virus from completing the productive replication cycle and, if it nevertheless should happen locally, preventing recurrent virus from further rounds of infection and spreading, increasing evidence suggests that the establishment of latency on the molecular level may not be immune-driven. Rather, molecular latency results from the cells' intrinsic antiviral defence by epigenetic silencing of viral gene expression associated with rapid circularization and chromatinization of incoming linear viral genomes within repressive nuclear domains. In this view, 'latency' is the default state, whereas productive infection, from the hosts' perspective, is the accident when viral genomes evade epigenetic silencing, with the chance for this being dependent on cell type, cell differentiation stage, cell cycle stage, and an overall nuclear environment that favours open chromatin structures, collectively defining what we describe as 'permissivity' for productive infection.

It is proposed that during latency stochastic episodes of promiscuous desilencing of single or combinatorial sets of viral genes can lead to the expression of transcripts (<u>t</u>ranscript <u>e</u>xpressed in <u>l</u>atency, TEL), which, when translated into proteins, can result in the presentation of antigenic peptides sensed by tissue-patrolling effector-memory T-cells. Importantly, promiscuous gene desilencing, unlike reactivation, does not usually initiate the productive viral replication cycle and can affect any viral gene regardless of its temporal expression in the kinetic classes immediate-early (IE), early (E) and late (L), and regardless of the function it takes during lytic infection. It is our current understanding that these limited desilencing episodes are the molecular motor that drives the CMV-typic expansion of T-cells, of CD8[+] T-cells in particular, a phenomenon commonly known under the catchphrase 'memory inflation'.

The 'classical era' of research in diverse murine models of CMV latency and reactivation has been reviewed by Jordan (1983) and authors of this book chapter have provided updates (Hummel and Abecassis, 2002; Reddehase et al., 2002, 2008). Here, independent research groups have joined to review their more recent results and current views on CMV latency and reactivation based on murine CMV models with focus on neonatal infection, haematopoietic (stem) cell transplantation (HCT), solid organ transplantation (SOT) and sepsis.

Introduction: the definition of latency in a historical perspective

The discussion on CMV latency has been, and in part still is, complicated by a semantic problem in that some authors understand 'latency' and 'persistence' as synonyms describing lifelong viral genome carriage. Here, and in line with most herpesvirologists, we use the definition of latency that is since long generally held for alpha-herpesviruses (reviewed by Roizman and Sears, 1987). Accordingly, 'latency' is a state in the viral biology during which functionally intact, reactivation-competent viral genome is maintained without completion of the productive replication cycle, so that infectious viral progeny is not formed or released. In contrast, 'persistence' is equivalent to ongoing productive infection, though this might occur on a quantitatively very low level and at restricted tissue sites in the host's body.

For decades of research, experimental distinction between these two states was hampered by the limited sensitivity of assays for detecting infectious virions, so that protagonists of the 'latency hypothesis' could not formally prove the absence of minute amounts of infectious virions below the level of detection, whereas advocates of the 'low-level persistence hypothesis' could not formally prove the presence of infectious virions but were in the comfortable position to use the limited assay sensitivity as an excuse. As a further layer of complication, persistent infection, reasoned in the murine model from extended periods with detectable infectivity at certain immune evasion-privileged tissue sites and in particular cell types, such as in the secretory glandular epithelial cells of salivary glands, specifically after depletion of CD4[+] T-cells (Jonjic et al., 1989; reviewed in Campbell et al., 2008; Walton et al., 2011) or under certain host-genetic conditions of local immune depression (see Chapter II.12), could theoretically be mimicked by frequent episodes of 'intermittent reactivation' from latency. Viral genes involved directly or indirectly in the persistent virus replication in the salivary glands have been discussed previously (reviewed in Reddehase et al., 2008). However, even at this privileged site, productive infection eventually ceases (Henson and Strano, 1972) without the viral genome being cleared (Reddehase et al., 1994).

For murine CMV (mCMV), detection of functionally competent, reactivatable viral genome by PCR combined with improved assay sensitivity for excluding an ongoing low-level productive infection identified the lungs (Balthesen et al., 1993) as well as spleen and kidney (Pollock and Virgin, 1995) as organ sites of latency in absence of persistence. Specifically, Pollock and Virgin argued against low-level persistent infection in viral DNA-harbouring spleen and kidney based on the finding that transfer of sonicated tissue failed to infect highly susceptible, severe combined immunodeficient (SCID) mice, an in vivo assay that was otherwise capable of detecting ~ 3 PFU. A quantum leap forward in sensitivity was reached by a cell culture assay employing centrifugal enhancement of infectivity combined with prolonged cultivation to allow for several rounds of virus amplification (then called 'focus expansion') followed by RT-PCR specific for viral transcripts generated in these indicator cultures (Kurz *et al.*, 1997). This assay was found to detect infectivity in a dilution of a purified virion preparation containing roughly five viral genomic molecules, which, based on an experimentally determined genome-to-infectivity ratio of ~ 500:1 (for the Smith strain), equals the viral genome equivalent of 0.01 cell culture infectious units (0.01 PFU *in toto*). Absence of infectivity in homogenates of perfused lungs led to the conclusion that latently infected lungs that harbour significant amounts of viral DNA, specifically in the order of 10^3–10^4 genome copies per 10^6 lung cells (Balthesen *et al.*, 1993; and subsequent publications reviewed by Reddehase *et al.*, 2002, 2008) do not contain infectious virus, whereas replication competence of the latent viral genomes (at least of a fraction of those) was verified by *in vivo* reactivation to productive infection of the lungs upon immunoablative treatment of the latently infected mice (Balthesen *et al.*, 1993; Kurz *et al.*, 1997; Kurz and Reddehase, 1999) as well as *in vitro* in lung tissue explant cultures (Böhm *et al.*, 2009; Marquardt *et al.*, 2011).

A hint to suggest independence of local latency from a preceding local productive infection was given by the observation that CD8+ T-cell immunotherapy, preventing productive infection of the spleen, did not prevent the establishment of latency in the spleen (Balthesen *et al.*, 1994). Establishment of latency and long-term maintenance of the viral genome independent of acute and persistent viral replication, respectively, have also been indicated by a temperature-sensitive (ts) mutant of mCMV, tsm5 (at that time with the mutation not mapped). Upon acute 'infection' of mice, tsm5 did not detectably replicate but nonetheless entered the latent stage, out of which reactivation to expression of a set of genes was inducible until a stage in the viral transcriptional programme was reached at which the ts mutation prevented the completion of the productive cycle, so that infectious virus was not recovered (Bevan *et al.*, 1996). These early findings were corroborated more recently with a spread-deficient BAC-cloned mCMV (see Chapter I.3) containing a targeted deletion of the gene that codes for the essential virion structural protein pM94 (Mohr *et al.*, 2010), the genome of which was maintained long-term in 'first-hit' target cells of the lungs. Similarly, long-term maintenance of a spread-deficient 'single-cycle' ΔgL mutant of mCMV was suggested from immunological data, although the copy number of latent viral genome in this study was close to or below the detection level of qPCR (Snyder *et al.*, 2011). Inherent to the approach, which excludes completion of the productive cycle under the

non-permissive conditions in host tissues, the criterion of reactivation competence of the latent viral tsm5, ΔM94, and ΔgL genomes was unaccomplishable. Interestingly, in much earlier studies originally aimed at preventing the establishment of mCMV latency by adoptive transfer of antiviral CD8+ T-cells, Steffens and colleagues found that control of viral spread and replication also reduced the subsequent latent viral DNA load in host tissues in a cell-dose dependent fashion down to a basal level that appeared to be refractory (Steffens *et al.*, 1998). In the light of today's knowledge, the refractory fraction of the load likely reflected latency established in 'first-hit' target cells that resist cytolysis by CD8+ T-cells, whereas the susceptible fraction of the load resulted from secondary viral spread.

Combined, all these data provide more than *bona fide* evidence for maintenance of viral genome in absence of low-level, persistently productive infection, and are thus strongly supportive of the 'latency hypothesis'. In addition, the data with spread-deficient viruses also imply that latent viral genome is not lost by cell death, including immune-mediated cell death, either because cells maintaining the genome are very long-lived and not attacked or not lysed by immune effector cells, or because the latent viral genome is replicated by the host cell during cell division.

Advocates of the 'low-level persistence hypothesis' (Yuhasz *et al.*, 1994) questioned the earlier findings by Balthesen and colleagues (1993) and thought to refute the interpretation that lungs represent a major organ site of mCMV latency. Like other investigators had seen before in the spleen (Henry and Hamilton, 1993), by using RT-PCR, Yuhasz and colleagues (1994) detected the 'productive cycle' major immediate-early (MIE) transcript *IE1* in the lungs of otherwise latently infected mice in absence of detectable virus, and took this as an evidence for low-level persistent infection of the lungs, arguing that detection of this single transcript species by RT-PCR is a more sensitive indicator of productive infection than any direct assay for infectious virus can be. Hence, these authors proposed that infectious virus detected by Balthesen and colleagues (1993) after immunoablative treatment represented an amplification of persistent infection in the lungs rather than true reactivation of latent virus.

As we will discuss in greater detail later in this chapter, whilst the definition of 'latency' demands absence of infectious virus, it does not formally exclude limited transcriptional activity from certain viral promoters, generating so-called 'latency-associated transcripts', known in herpesvirus virology as LATs, in CMV research also named 'CMV latency-associated transcripts', CLTs (see Chapter I.20). Many authors still adhere to the opinion that transcripts from 'lytic cycle genes' cannot be LATs/CLTs and thus compulsorily indicate productive infection, but actually, true 'latency-specific' transcripts have, to our knowledge, not been identified for CMVs. Instead, all known CLTs of human CMV (HCMV) are also expressed in acute infection, though their role in either acute or latent infection is, in most cases, not well understood (see Chapter I.20). Although regulatory MIE proteins are of key relevance in starting the programme of productive cycle gene expression (see also Chapter I.10), many other essential viral genes need to be de-repressed and released from the host cell's intrinsic defence mechanisms for completion of the productive cycle even under conditions of acute infection (see also Chapter I.11) and thus likely constitute also molecular checkpoints in the transition from latency to full reactivation. These reflections make it clear that simply equalling the detection of MIE gene expression, and even more so of the non-essential *IE1* transcript alone, with productive infection, is a misconception.

As a more academic footnote, the restriction of virus replication in cells of a heterologous host, one that has not co-speciated with the virus in evolution (see Chapters I.1 for primate CMVs and Chapter I.2 for mCMV), mostly occurs at a stage beyond virus entry into the cell, viral genome delivery into the cell nucleus, and MIE gene expression, as discussed by W. Brune in the chapter on the host-species specificity of CMVs (Chapter I.18), Thus, in a way, these cells are 'latently' infected despite expressing MIE genes. However, whereas during reactivation in a latently infected autologous host cell checkpoints downstream of MIE gene expression in the productive programme can be passed through, these checkpoints remain closed in the heterologous host cell, unless specific mutations in the viral genome suspend the blockade (see Chapter I.18 for details).

In the light of the aforesaid one might assume that lytic cycle transcripts found during latency must indicate an interrupted attempt of reactivation in the regular, forward-leading IE-E-L path of productive viral gene expression. However, as we will discuss later in this chapter, viral transcripts expressed in latency, TELs, might reflect a probabilistic phenomenon known in genetics as transcriptional 'noise in eukaryotic gene expression' (Blake *et al.*, 2003; Kaern *et al.*, 2005) that is proposed to apply also to viral genes evolved in adaptation to a eukaryotic host.

Another problem in the definition of 'latency' concerns the question if latency can be established at one site in the host while productive infection continues at a different site. From a medical and epidemiological point of view, which sees the organism in its entirety, 'clinical viral latency' is established when virus is no longer shed in bodily fluids, so that the carrier person is no longer infectious and is free of risk to transmit the virus. Understandably enough, absence of infectious virus in organs is not usually verified clinically, as this would require biopsies and is prone to false negative results due to sampling error, but animal models allow a more strict analysis. In general, CMVs show a broad cell-type and tissue tropism as it becomes evident from multiple organ involvement during CMV disease and from histopathological studies, including the identification of infected cells by detecting viral proteins or nucleic acids (see Chapter I.17 and Chapter II.17). Whereas clinical strains of HCMV can differ significantly in their cell-type tropism and also adapt genetically to cell types during propagation in cell culture (see Chapter I.1 and Chapter I.17), mCMV is genetically more stable (Chapter I.2). All cell culture and wild-derived strains of mCMV analysed so far share poly-cell tropism – being capable of infecting, for instance, fibroblasts, endothelial cells (ECs), smooth muscle cells (SMCs), cardiac myocytes, macrophages, dendritic cells (DCs) and various epithelial cell types, including hepatocytes, but not T or B lymphocytes – and retain this feature also after many generations in cell culture. Hence, mCMV replicates in a wide array of organs. Although major organ sites of viral replication (such as spleen, liver, and lungs) are usually tested for clearance of productive infection, it is difficult to formally exclude ongoing viral replication somewhere else, but as salivary glands are a known site of prolonged productive infection (see above), cessation of virus replication at that site is usually taken for an acceptable evidence of virus latency established in the entire organism.

This strict, 'organismal' definition of viral latency, however, is too dogmatic, as it deflects from a more conceptual, molecular understanding of latency. In a wide array of tissues mCMV not only replicates, but also becomes latent (reviewed in Jordan, 1983; Hummel and Abecassis, 2002; Reddehase *et al.*, 2002), which either indicates latency established in different tissue-typic cell types or in one or more widely distributed cell type(s) (see below). Productive primary infection is cleared with different kinetics in different organs, with rapid clearance in spleen and liver, delayed clearance in the lungs, and salivary glands being the last. The cell type responsible for long-lasting virus productivity in the salivary glands is the glandular epithelial cell that is not a cellular site of latency but undergoes cell death, then called 'necrosis', as viral replication ceases (Henson and Strano, 1972). So, why should – in the definition – viral latency not be established in spleen and lungs and at other sites at a time when these highly specialized glandular epithelial cells still release infectious virus for local shedding in the saliva and host-to-host transmission? The conclusion that latency and productive infection are not mutually exclusive on an organismal level also becomes evident from reactivation, a situation that is reciprocal to the establishment of latency. As shown in the model of viral latency after neonatal infection, mCMV recurrence after immunoablative treatment *in vivo* is a stochastic event that occurs independently in different organs (Reddehase *et al.*, 1994). Specifically, out of 30 latently infected mice tested in this report, seven reactivated virus in the salivary glands but not in the lungs, six reactivated virus in the lungs but not in the salivary glands, and only four reactivated virus in both organs simultaneously, while 13 did not reactivate virus in either salivary glands or lungs. This result is in accordance with the null hypothesis of independent distribution (Fig. I.22.1) and demonstrates (1) that reactivation occurs locally and independently in any organ, and (2) that reactivated productive infection in any organ does not exclude maintenance of latency in another organ. The original work also included the spleen, showing all random combinations of reactivation in the three organs, with only one mouse out of 30 having reactivated virus in all three organs and 11 mice out of 30 in none (Reddehase *et al.*, 1994).

It is thus obvious that latency can be established as well as maintained at particular organ sites of latency while, at the same time, productive infection continues or recurs elsewhere. This important principle not only applies to different organs but also to different cell types within the same organ. For instance, in the liver,

hepatocytes account for most of the virus production (Sacher *et al.*, 2008), whereas latency is established not in the hepatocytes but in the liver sinusoidal endothelial cells (LSECs) (Seckert *et al.*, 2009), which are very inefficient and transient virus producers (Sacher *et al.*, 2008). Accordingly, latency could be established in LSECs while, at the same time, acute infection of the liver continues in hepatocytes. We can scale down this principle even to the level of a single cell type. Individual cells could possibly repress all incoming viral genomes by chromatinization and become latently infected while, at the same time, viral genomes might evade the cell's intrinsic defence in other individual cells of the same cell type in the same tissue. There is indeed evidence to propose that productive infection may originate from single viral genome molecules that evade epigenetic repression (see the Editor's Preface).

In conclusion, we propose a molecular definition of latency that is independent of productive infection and describes the situation of long-term carriage of non-replicating – but replication competent and thus reactivatable – viral genomes in individual cells.

Cellular sites of CMV latency

As CMV entry into cells does not appear to be the main restriction for infection, there is no *a priori* rationale to assume that there exists just a single cellular site of latency. For HCMV, haematopoietic progenitor cells of the myeloid differentiation lineage are a canonical site(s) of latency, and reactivation appears to be

2x2 Contingency table observed (O)

Reactivation in SG

		+	-	
Reactivation in lungs	+	4	6	10
	-	7	13	20
		11	19	30

2x2 Contingency table expected (E)

Reactivation in SG

		+	-	
Reactivation in lungs	+	3.7	6.3	10
	-	7.3	12.7	20
		11	19	30

Figure I.22.1 Stochastic nature of virus reactivation in latently infected host organs. *In vivo* virus reactivation from latently infected salivary glands (SG) and lungs of 30 BALB/c mice at 12 months after neonatal primary infection was determined by PFU assay of organ homogenates 14 days after haematoablative total-body γ-irradiation. Incidence data for the four possible combinations of positive and negative results in the two organs are arranged in a 2×2 contingency table (O, observed data) and compared with the 2×2 contingency table expected under the assumption of independent distribution (E, expected). The one sided probability value, p(O≥E) was calculated by Fisher's exact probability test and was found to be 0.548, which indicates independent distribution.

linked to global viral gene desilencing by remodelling/opening of the higher-order chromatin-like structure of the latent viral episome upon progenitor cell differentiation into end-stage descendants, such as subpopulations of macrophages and DCs, as well as cytokine-triggered, transcription-factor mediated signalling to the viral MIE enhancer in these mature cells (see Chapters I.10 and I.19). Other human cell types, specifically microvascular ECs and SMCs were in discussion but considered unlikely to be important sites of HCMV latency *in vivo* because of viral genome carriage being far below that in myeloid lineage cells from the same subjects (Reeves *et al.*, 2004). As far as ECs are concerned, there remains some uncertainty if different subpopulations of the highly diverse ECs might differ in this respect, but, more regularly, ECs were implicated in persistent productive replication (for a review, see Jarvis and Nelson, 2007). An unexplained phenomenon suggesting an additional cellular site of HCMV latency is the clinical observation that virus variants harboured by transplant donors (D) are frequently transmitted to recipients (R) with donor organs in SOT (highest risk/viral load for serological status D⁺R⁻), but less frequently with the haematopoietic stem and progenitor cells in HCT (lowest risk/viral load for status D⁺R⁻), although latently infected myeloid lineage-committed progenitor cells should be enriched in HCT and although immunosuppression, favouring virus reactivation, is more intense in HCT compared to SOT recipients (Emery, 1998). This might, however, be alternatively explained by transfer of donor immunity in HCT but not or less in SOT, although a beneficial effect of transferred donor immunity is not consistently reported and co-transfer of immune cells can differ largely with the cell depletion regimen and the source and purity of haematopoietic cells (HCs) used in HCT (see Chapter II.16). As an alternative explanation for efficient HCMV transmission in SOT one might discuss the possibility of a superior cytokine-triggered reactivation propensity of latent virus in already differentiated, tissue-resident histiocytes (macrophages or DCs) compared to progenitor cells in bone marrow (BM). Nonetheless, the question if HCs are the dominant source of reactivating HCMV is not definitely settled. In 'humanized mouse models' (see Chapter I.23), HCMV latency could be established in human HCs and their progeny following experimental primary infection of HCT xenograft chimeras (Smith *et al.*, 2010), but intrinsic reactivation in latently infected HCs derived from D⁺ xenograft donors awaits to be demonstrated.

For mCMV, latency in HCs of the myeloid lineage has been investigated much less intensely, which may prompt the misinterpretation that human and murine CMV differ essentially in their cellular sites of latency. However, in the murine models of neonatal infection (Balthesen *et al.*, 1993) and infection after experimental HCT (Kurz *et al.*, 1997), viral DNA was found to be harboured in blood leucocytes for prolonged periods. Specifically, in the HCT model (Kurz *et al.*, 1997), this period lasted for ~4 months after clearance of productive infection in all tested organs, including the salivary glands, which corresponded to ~10 months after HCT and infection; actually longer than half-lives of subsets of circulating mature leucocytes could explain. These findings suggested renewal from a latently infected haematopoietic stem or progenitor cell or presence in a cell type that undergoes cell division by homeostatic or induced proliferation in lymphoid organs and replicates the episomal latent viral genome independent of viral replication. In line with latency in HC progeny, Pollock and colleagues demonstrated the presence of reactivatable latent viral genomes in Ly71(F4/80)⁺ peritoneal exudate macrophages recruited from the BM, with the frequency of latently infected cells estimated to be ~1/50,000 among all peritoneal exudate cells (PECs) and a latent viral genome copy number of 1–10 per latently infected PEC (Pollock *et al.*, 1997). A histological approach performed to localize latent viral genome in host tissues by PCR *in situ* hybridization (PISH) indicated latent viral genome carriage in pulmonary alveolar macrophages (Koffron *et al.*, 1998).

This 'haematopoietic latency' of mCMV, however, though it likely exists, is apparently only transient. Specifically, in the HCT model (Kurz *et al.*, 1999), longitudinal analysis of latent viral genome load by qPCR revealed a chronology of viral genome clearance after resolution of acute infection first from the BM followed by clearance from circulating blood leucocytes, whereas a high viral genome load was maintained for the lifespan in diverse organs, particularly in the lungs (Balthesen *et al.*, 1993; Reddehase *et al.*, 1994; Kurz *et al.*, 1999). These findings excluded the BM compartment, and thus the BM resident haematopoietic myeloid lineage stem and progenitor cells, as a perpetual source of latent mCMV genome. In accordance with an only transient nature of latency in HCs, latent mCMV was not transferable to mCMV-naïve female (*sry⁻*) recipients with male (*sry⁺*) donor HCs in a model of sex-mismatched HCT (Seckert *et al.*, 2008), and reactivatable latent mCMV genome in the lungs did not localize to CD11b⁺ and CX3CR1⁺ subsets of myeloid lineage cells (Marquardt *et al.*, 2011).

It is, of course, always worth a discussion if animal models are of predictive value for HCMV. However, before one argues that murine and human CMVs differ fundamentally in the cellular site(s) of latency, one should concede that systematic, longitudinal analyses of viral genome clearance in the BM in the progression from primary infection to latency were, for understandable logistic reasons, never undertaken in humans. This does not say that the mouse is right, but it also does not justify to claim that the mouse is wrong. As advocates of mice, we would like to draw attention to the diagnostic clinical routine in pre-emptive antiviral therapy of HCMV infection in HCT recipients (see Chapter II.16) in which, in a longitudinal monitoring by qPCR, occurrence of viral DNA in blood serves to diagnose productive virus reactivation and its clearance from blood serves as an end-point in therapy control. So, leucocytes in the blood circulation, including monocytes principally known to harbour latent HCMV (see Chapters I.19 through I.21), do not seem to be a quantitatively dominant site of lifelong latent HCMV carriage. Taken together, cellular sites of latent CMV genome carriage, in both murine and human CMVs, must be correlated with the time that has passed after clearance of productive infection. The problem is that this important information on the individual's infection history is not usually available for human healthy carriers.

So, what finally are the cellular sites of non-transient, durable mCMV latency? Pioneering classical work in that direction came from the groups of D.H. Spector (Mercer *et al.*, 1988) and M.C. Jordan (Pomeroy *et al.*, 1991) providing evidence for mCMV latency in MHC class-II (MHC-II)⁻ stromal cells of the spleen, whereas viral DNA was absent from macrophage-enriched non-stromal cell preparations. As productive infection in the spleen was localized to MHC-II⁻Ly71(F4/80)⁻ sinusoidal lining cells staining positive for VIIIR:Ag (factor VIII-related antigen) expressed also by ECs, these endothelial sinusoidal lining cells qualified as candidates for a cellular site of latency as well (Mercer *et al.*, 1988). The possibility of viral latency in ECs was subsequently corroborated by combining cell-type identification by *in situ* immunofluorescence with PISH to reveal ECs in diverse organs, including spleen and liver, as cells co-localizing with cells carrying mCMV DNA in adjacent tissue sections in latently infected mice (Koffron *et al.*, 1998). Notably, in an earlier study with human tissues, PISH led to the detection also of HCMV DNA in cells lining the hepatic sinusoids, supposedly in the LSECs, in the liver of latently infected SOT donors (Koffron *et al.*, 1997), thus indicating that murine and human CMVs may be less different with respect to cellular sites of latency than it was assumed.

More recently, murine LSECs were definitely identified as a cellular site of mCMV latency by qPCR-based detection of the viral genome in purified LSECs and by virus reactivation from LSEC-enriched but not LSEC-depleted non-parenchymal liver cells (NPLCs) upon cell transfer into mCMV-naïve recipients (Seckert *et al.*, 2009) (Fig. I.22.2). In that study, latently infected, Y-chromosome chimeric mice were generated by sex-mismatched HCT with male (*sry*⁺) donors and infected, female (*sry*⁻) recipients for distinguishing between donor-derived HC progeny and recipient-derived resident tissue cells. Latent viral genome was found to localize to NPLCs, thus excluding hepatocytes, and within the NPLCs to a recipient-derived (genotype *sry*⁻) cell type characterized as LSEC functionally by endocytotic uptake of acetylated low-density lipoprotein (acLDL) and phenotypically by the cell surface marker combination CD31⁺CD146⁺L-SIGN⁺. Missing expression of CD11b, CD11c, and MHC-II excluded macrophages and DCs. The frequency of latently infected LSECs among all LSECs was estimated to be ∼1/15,000 with a copy number of ∼10 genomes per latently infected LSEC (Seckert *et al.*, 2009).

It is proposed in the literature that LSECs are actually also of haematopoietic origin, specifically, that they are derived from BM-resident hemangioblastic endothelial progenitor cells (Choi *et al.*, 1998; Gao *et al.*, 2001). In this context it is important to note that latent mCMV genome was not detected in a minor NPLC subpopulation that was donor-derived (genotype *sry*⁺), and thus of haematopoietic lineage, and expressed the LSEC antigen CD146 as well as MHC-II (Seckert *et al.*, 2009). As LSECs are, by definition, confined to liver sinusoids, they cannot account for viral latency in other organs, but the *in situ* localization data (Koffron *et al.*, 1998), maintenance of spread-deficient ΔM94 genome in the lungs where ECs are the main 'first-hit' target cells (Mohr *et al.*, 2010), as well as localization of latent viral DNA to ECs in sorted lung cell types (authors' unpublished data) collectively indicate viral latency in additional EC subsets.

Viral latency in ECs, at least in subsets thereof, makes sense in that ECs have an extremely low turnover, being among cell types exhibiting the lowest proliferation level in the body, with only 0.01% of cells engaged in cell division at any time (Ortega *et al.*, 1999). In addition, the wide tissue distribution of ECs can explain the correspondingly wide distribution of latent CMV genomes and the ease of transmission by SOT in both, humans and animal models. Whilst we are now convinced that subsets of ECs are cellular sites of CMV latency, we will not exclude identification of additional, latently infected cell types in investigations to come.

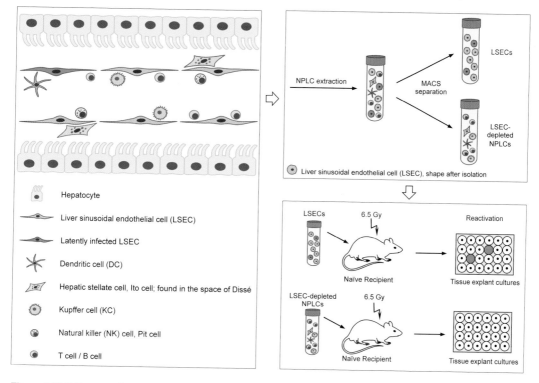

Figure I.22.2 Process flow diagram for identifying liver sinusoidal endothelial cells (LSECs) as a cellular site of mCMV latency. NPLC, non-parenchymal liver cells including all liver cells except hepatocytes.

Epigenetic mechanisms involved in virus latency

A vain search for true 'latency genes'?

Based on an understanding that the establishment of latency is a specific virtue and evolutionary accomplishment of CMVs, the identification of 'latency genes' – genes that are causally involved in the establishment and/or maintenance of latency – is a reasonable aim in research (see Chapter I.20). For mCMV, to our knowledge, no such gene has been defined.

Trivially, viral genes involved in efficient virus replication and/or dissemination from a portal of virus entry into the organism to distant organ sites of latency will indirectly make an impact on latency in that they have an effect on reaching the cells in which latency can be established, on the latent viral genome load, and thus also on the incidence of reactivation. This is, however, not what we understand by a 'latency gene' that is mechanistically involved in the property of lifelong viral genome maintenance within the cells once those have been entered. To illustrate the problem with a few selected examples from mCMV, deletion of the MIE ORF m123 (IE1) attenuates the virus *in vivo* (Ghazal *et al.*, 2005; Wilhelmi *et al.*, 2008), resulting in low latent viral DNA load and low

incidence of reactivation/recurrence, but compensating the growth deficit of the deletion mutant by increasing the initial dose of infection made it clear that IE1 is needed neither for the establishment of latency nor for reactivation (Busche *et al.*, 2009). Similarly, deletion of the M33 chemokine receptor homologue (Cardin *et al.*, 2009) and of mCMV 'immune evasion' genes that interfere with innate and adaptive immune control (see Chapter II.17) (Böhm *et al.*, 2009) interfere indirectly also with an efficient establishment of latency by attenuating the virus, but are not mechanistically involved in the viral genome attaining or maintaining a latent state. Approaching this insight from another direction, controlling the acute infection and spread of wild-type (WT) virus by adoptive transfer of CD8[+] T-cells also reduced the latent viral DNA burden in organs and, linked to this, reduced the incidence/risk of recurrence (Steffens *et al.*, 1998), but, obviously, this does not mean that CD8[+] T-cells play a role in the molecular mechanism that leads to latency. Confusing indirect with direct influences on the establishment of latency is a frequent cause of misconception.

A search for viral 'latency genes' might become obsolete if an emerging new view on the establishment of latency proves correct. It appears possible that viral latency is not primarily a property of the virus for which particular viral genes are responsible; instead, viral latency may result mainly from the cell's nuclear intrinsic defence function to silence incoming foreign DNA by packaging it with histones to a closed chromatin-like structure, similar to epigenetic silencing of cellular genes. From the viewpoint of the virus, latency is thus rather its own incapability to evade the host's repressive mechanisms, at least temporarily and in only conditionally permissive cell types, likely dependent upon cell differentiation stage, cell cycle stage, cell metabolism supporting or not supporting the virus' metabolic needs in DNA replication and protein synthesis, and microenvironmental conditions of cytokine-induced, transcription factor-mediated signalling to viral cis-regulatory genetic elements, such as the enhancer element within the MIE promoter-enhancer (MIEP) (see also Chapter I.10). Events of gene silencing and desilencing associated with chromatin remodelling are thus key to understanding viral latency and reactivation, respectively. A viral 'latency gene' thus should be a gene coding for a regulatory RNA or protein that assists the cell in wrapping the incoming viral genome and/or maintaining it silenced by a closed chromatin structure.

The implication of chromatin remodelling in HCMV latency and reactivation is reviewed in Chapter I.19; here we review insights obtained from murine *in vivo* models.

Epigenetic control of gene expression

Expression of some cellular genes is repressed through methylation of cytosines in CpG dinucleotides (Zaidi *et al.*, 2011). Studies of latent mCMV, as well as other herpesviruses, have shown that the viral DNA is not methylated in latently infected mice (Kubat *et al.*, 2004; Hummel *et al.*, 2007). Eukaryotic DNA is wrapped around histone proteins H3, H4, H2A, and H2B in a repeating nucleosome pattern, which provides a means of compacting the DNA. Post-translational modifications of the histones and the positioning of nucleosomes on the DNA play critical roles in determining gene structure and accessibility to the transcriptional apparatus. Histone proteins are not detectable in the virions of either human or murine CMV (Streblow *et al.*, 2006), but are rapidly acquired by the viral DNA during HCMV infection of fibroblasts in cell culture (Nitzsche *et al.*, 2008). The mCMV genomes were shown to acquire histones very early during the course of acute infection *in vivo*, and viral genomes are highly enriched in histones in latent infection (Liu *et al.*, 2008, 2010).

Histones bound to the major immediate-early promoter in latency have modifications characteristic of repressed chromatin

Histone proteins are subject to a wide variety of post-translational modifications, which serve as docking sites for non-histone proteins. Many of these proteins are enzymes, which further modify chromatin, either through histone modification or nucleosome remodelling (Kouzarides, 2007). Some types of histone modifications are associated with active transcription, while others are associated with repression. Of the many known histone modifications, acetylation of histone H3 lysines 9 and 14, acetylation of H4 lysines 5, 8, 12, and 16, methylation of H3 lysine 4 (H3K4), which are all associated with active transcription, and methylation of H3 lysine 9 (H3K9), which is associated with repression, have been studied in chromatin isolated from kidneys of acutely and latently infected mice (Liu *et al.*, 2008, 2010). These studies have focused primarily on histones bound to the mCMV MIEP, since changes in MIE gene expression are likely to be key events in governing latency and reactivation. These studies show that, during acute infection, RNA polymerase II is recruited to the MIEP, and histones bound to this region have modifications associated with active transcription. In contrast, chromatin isolated from latently infected mice shows features of transcriptional repression: binding of RNA polymerase II to the MIEP is lost, and histones bound to the MIEP are hypo-acetylated and have methylation patterns consistent with repression. In addition, histone de-acetylases (HDACs), HDAC2 and HDAC3, and the repressors Daxx, CIR (CBF-1-interacting corepressor), and HP-1γ (heterochromatin protein-1 gamma) are bound to the MIEP in latently infected mice.

Mechanisms of recruitment of repressors in latency

HDACs do not bind directly to DNA, but rather, are recruited through interaction with co-repressor complexes assembled through interaction with transcription factors that recognize specific DNA sequences (Jepsen and Rosenfeld, 2002). Thus, identification of DNA-sequence specific factors bound to the MIEP in latency is crucial to understanding the mechanism for establishment of latency. The enhancer element of the mCMV MIEP has multiple potential binding sites for two proteins with dual roles as repressors and activators of transcription, Ying-Yang 1 (YY1) and CBF-1 (RBPJK) (Liu *et al.*, 2010) (Fig. I.22.3). YY1 binds to a core CCAT sequence flanked by variable regions (Hyde-DeRuyscher *et al.*, 1995; Kim and Kim, 2009).

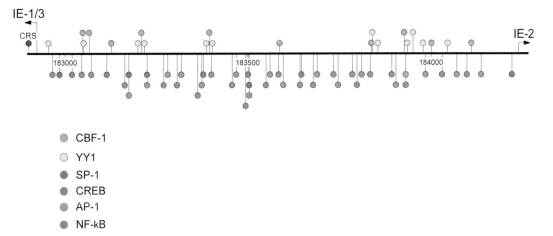

Figure I.22.3 Potential transcription factor binding sites in the mCMV MIE enhancer. Sites were identified by *in silico* analysis using MatInspector (Genomatix, release 8.01) (Cartharius *et al.*, 2005) software. Arrows indicate transcription initiation sites for the *IE1/3* and *IE2* transcripts. Numbers indicate sequence coordinates of the mCMV genome (Rawlinson *et al.*, 1996). CRS, cis-repression sequence.

It acts as a repressor in multiple ways (Gordon *et al.*, 2006), including interaction with polycomb repressor complexes 2 and 3 (Ko *et al.*, 2008), which mediate methylation of H3K27, de-methylation of H3K4, and chromatin compaction (Simon and Kingston, 2009), and the NCoR/SMRT complexes, which mediate histone de-acetylation (Yang *et al.*, 1996; Jepsen and Rosenfeld, 2002; Huang *et al.*, 2003; Le May *et al.*, 2008). CBF-1 is the major downstream effector of the Notch signalling pathway (Lai, 2002). In the absence of Notch ligands, CBF-1 binds to its target sequence and represses transcription through interaction with CIR, which recruits NCoR/SMRT co-repressor complexes (Hsieh *et al.*, 1999).

The mCMV MIE enhancer element has 12 potential YY1 sites and 10 CBF-1 sites identified by *in silico* analysis (Liu *et al.*, 2010). Chromatin immunoprecipitation (ChIP) analyses showed that both YY1 and CBF-1 are bound to the MIEP in latently infected cells. In addition, YY1 was bound to the M100 promoter region, which also has YY1 sites identified *in silico*, but not to the M112 promoter, which lacks these sites. The M100 and M112 promoters lack CBF-1 sites, and no binding of CBF-1 to these promoters was observed. Thus, these repressors bind specifically to promoters with their cognate sites, and recruitment of transcriptional co-repressor complexes to the MIEP is likely mediated through interaction with YY1 and CBF-1. The HCMV enhancer also has YY1 binding sites, and YY1 acts as a repressor of the HCMV enhancer (Liu *et al.*, 1994).

Relative to cellular DNA, the mCMV genome is highly enriched in histones. Binding of histones and HDACs was not restricted to the MIEP, but was also observed in two other regions of the genome analysed, the M100 and M112 promoters (Liu *et al.*, 2010). These observations suggest that the entire genome is in a highly condensed chromatinized configuration, which is inaccessible to the transcription apparatus. HDACs and other repressors may be recruited to other regions of the mCMV genome through interactions with additional DNA-binding transcription factors.

Repressors are bound to the MIEP at very early times post infection

Analysis of the kinetics of binding of several different repressors consistently showed that binding to the MIEP was detectable as early as day 2 post infection in chromatin from kidney cells of infected mice, prior to activation of MIE gene expression, and then fell to a nadir at day 7, which correlated with peak activation of MIE gene expression (Liu *et al.*, 2010). Cell culture studies with both human and murine CMVs have shown that viral genomes become rapidly associated with ND10 bodies in the nucleus. These multiprotein complexes are thought to act as a form of host intrinsic defence against infection through repression of viral gene expression (Kalejta, 2008; Maul, 2008), and are dissociated by IE1 protein to facilitate viral replication (Korioth *et al.*, 1996; Ahn and Hayward, 1997, 2000; Ishov *et al.*, 1997; Ahn *et al.*, 1998; Wilkinson *et al.*, 1998; Ghazal *et al.*, 2005; Wilhelmi *et al.*, 2008) (see also Chapter I.9 and Chapter I.11). More recently, an analysis of many individual cells employing an enhancer deletion mutant of mCMV, mCMV-ΔEnh.Luc, revealed the importance of the enhancer for both IE1 expression and ND10 dissociation, as well as a correlation between ND10 dissociation and the variable level of IE1 expression resulting from basal promoter activity (Podlech *et al.*, 2010). The rapid recruitment of repressors to the MIEP prior to activation of MIE gene expression, and the subsequent loss of these repressors, is consistent with the hypothesis that host intrinsic defence mechanisms operate *in vivo* as well as *in vitro*, and that the virus has means to counteract this defence (Fig. I.22.4). Loss of repressor binding correlated with activation of MIE gene expression and changes in modifications of histones bound to the MIEP (Liu *et al.*, 2010). However, even at the peak of MIE gene expression, some binding of repressors was still detectable. These observations suggest that in some cells, activation of viral gene expression may fail to occur, and these cells may become the reservoir

for latency. Thus, latency is likely established in some cells at the earliest phase of infection.

Biphasic pattern of repressor binding to the MIEP during the course of infection

Following activation of viral gene expression in infected kidneys at day 7, the percentage of genomes bound to repressors was found to increase starting at day 10, which coincided with loss of RNA polymerase II and acetylated histones, and changes in methylation of histones bound to the MIEP. Productively infected cells, that is cells in which viral genomes successfully evaded the cells' intrinsic repression mechanisms, either undergo cell death due to the cytopathogenic viral replication or are cleared by the host immune response. In response to danger signals emitted by infected cells, tissue-resident immature dendritic cells

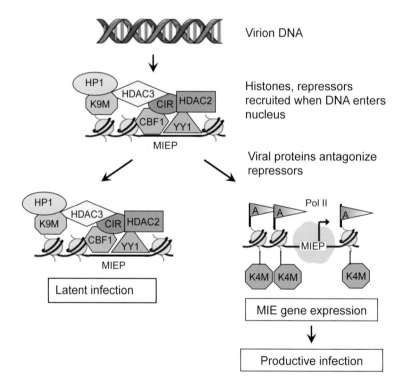

Figure I.22.4 Model for establishment of mCMV latency at the MIE locus. Viral DNA is not complexed to histones in the virion, but acquires histones early in infection. Cellular repressors are also bound to viral DNA at very early times, but binding of these factors decreases as activation of MIE gene expression occurs, leading to productive infection. Activation of MIE gene expression is associated with binding of RNA polymerase II and changes in modifications of histones bound to the MIEP, including acetylation of H3K9/14 and methylation of H3K4. Studies with HCMV suggest a similar pattern of repression prior to activation of MIE gene expression (Groves *et al.*, 2009). Co-repressor complexes may be recruited to mCMV DNA through interaction with transcription factors that bind directly to sequences in the viral MIE enhancer, including YY1 and CBF-1. We speculate that loss of repressors is due to inactivation of host cell defences by virion proteins analogous to the HCMV pp71 tegument protein (Kalejta, 2008). Activation of viral gene expression may fail to occur in some cells (see the main text), and these cells are likely reservoirs for latency.

differentiate and migrate to draining lymph nodes, where they present antigen to naïve T-cells, which differentiate into effector cells and memory cells that are recruited back to the site of infection for exerting antiviral effector functions such as killing of infected cells. This process requires several days. By day 10, many of the productively infected cells will have been cleared, and thus, the apparent increase in binding of repressors starting at day 10 likely reflects a shift in the population of cells, as productively infected cells are eliminated, while latently infected cells are spared.

Viral gene expression in latency

As discussed earlier in this chapter, the definition of viral latency does not demand transcriptional silencing of all latent viral genome molecules at all genetic loci and all the time. Although, as described above, latency is established by repression of viral gene expression in the closed higher-order chromatin-like structure of the circularized viral episome (Chapter I.19), local and only transient opening of the 'viral chromatin' at certain loci may lead to episodes of limited viral gene expression without completing or even without ever entering the viral productive cycle. Under such conditions latency is maintained despite transcriptional activity. Only if desilencing involves all essential loci, reactivation can proceed to the formation of infectious virus. In fact, studies on HCMV latency in myeloid lineage cell culture models have revealed what is called 'latency-associated' transcription, resulting in CLTs from a number of viral gene loci, although these are not evenly distributed over the viral genome but indicate 'hot spots' of transcriptional activity based on a transiently opened chromatin structure, possibly related to promoter strength. As discussed in more detail in Chapter I.20, viral transcriptome profiling in myeloid progenitor cell models of HCMV latency indicated that the viral genome is far more transcriptionally active during latency than originally thought. If all these transcripts are 'latency-associated' CLTs with the implicit understanding that they take an active part in the establishment and/or maintenance of latency is an open question.

Notably, the MIE locus, though essential for starting the viral productive cycle, is also a particularly active site during HCMV latency, leading to a number of alternatively spliced sense and antisense CLTs (reviewed in Chapter I.20, also providing a map). It can be speculated that this high activity at the MIE locus relates to the presence of a very strong enhancer element that favours opening of the closed chromatin structure around that locus (Chapter I.10). So far, however, a function selectively in latency could not be attributed to any of these MIE CLTs, and none appears to be latency-specific, since all, as far as looked at, are expressed also during productive infection. It is important to note that specific CLTs analysed are not present in all individual cells that carry latent viral genome(s) but only in a low proportion (2–5%) of cells at any time (Chapter I.20), which suggests that most latent viral genomes remain silenced at particular loci or, at least, that not all CLTs are present simultaneously in a latently infected individual cell. That latency-associated transcription is sporadically occurring in only few cells argues against a causal role for CLTs in the maintenance of latency and may, instead, rather reflect episodes of local gene desilencing by transient chromatin opening. This might be interpreted as a primordial stage in reactivation that remains below a threshold required for triggering the replicative cycle. Alternatively, this might reflect stochastic transcriptional noise with no specific role in establishing or maintaining latency. Indeed, absence of the ORF94 (UL126a)-encoded MIE locus sense CLT after infection with a ΔORF94 virus did not show a phenotype in establishing or reactivating from experimental latency (White et al., 2000). Thus, based on its presence also in acute infection and apparent lack of a role in latency, the MIE ORF94 sense CLT, though transcribed also during latency, does not show a functional association with latency and might thus better be named a 'transcript expressed in latency', TEL (see above). We use here the neutral designation TEL with no other meaning than that the transcript can be found also in cells in which the viral replicative cycle fails to proceed to virus production.

In mCMV, much less is known about genome-wide transcriptional activity during latency; however, it would be preliminary to propose a fundamental difference to HCMV as lack of cell culture models of mCMV latency has, for a long time, discouraged investigations on mCMV transcripts expressed during latency.

MIE locus transcripts of murine CMV

Bidirectional gene pair architecture of the MIE locus

Like its HCMV counterpart, the MIE locus of mCMV is governed by a strong enhancer element and appears to be a 'hot spot' of transcription in latency. As sketched in Fig. I.22.5A, the MIE locus of mCMV shows the architecture of a bidirectional gene pair with transcription unit *ie1/3* (ORFs m123-M122, corresponding to HCMV *ie1*/UL123-*ie2*/UL122) and gene *ie2* (ORF m128, no correspondence in HCMV) located on opposite strands of the viral dsDNA and flanking a dual tail-to-tail oriented

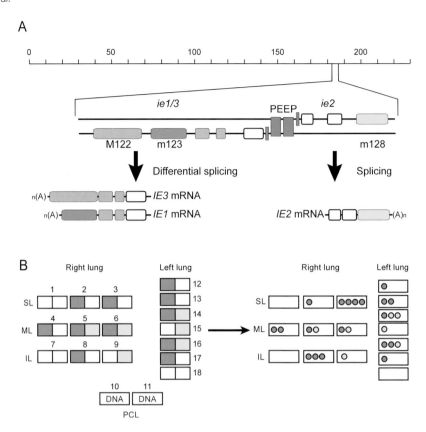

Figure I.22.5 Stochastic transcription from the MIE locus. (A) Bidirectional gene pair architecture, exon–intron structure, and spliced MIE transcripts. PEEP, Tail-to-tail oriented promoter-enhancer-enhancer-promoter element. (B) Left panel: Stochastic patterns of *IE1* and *IE2* transcripts detected by splice-specific RT-qPCR in 18 pieces of a statistically averaged latently infected lung, based on data compiled from several lungs. Right panel: Number of actual transcription events per piece of the statistically averaged lung estimated from the observed incidence of transcript-negative pieces for each type of transcript (zero fraction) by using the Poisson distribution function with the understanding that a transcript-positive piece may comprise more than one transcription event. SL, superior lobe; ML, middle lobe; IL, inferior lobe; PCL, postcaval lobe, in schematic anatomical view.

promoter-enhancer-enhancer-promoter region (MIE-PEEP; equivalent to MIEP) to the left and to the right, respectively, with transcription occurring in opposite directions (Dorsch-Häsler *et al.*, 1985; Keil *et al.*, 1987a,b; Chatellard *et al.*, 2007; Simon *et al.*, 2007; Kropp *et al.*, 2009). Gene *ie2* consists of three exons, with exon 3 encoding the 43-kDa IE2 regulatory protein (Messerle *et al.*, 1991) whose role in mCMV biology still awaits to be unveiled (Cardin *et al.*, 1995). Transcription unit *ie1/3* consists of five exons of which exon 1 is non-coding. Alternative splicing of an *IE1/3* precursor transcript generates *IE1* mRNA (exons 1–2–3–4) and *IE3* mRNA (exons 1–2–3–5) coding for the pleiotropic adjunct trans-activator of viral early (E)-phase transcription, 89/76-kDa IE1 (Keil *et al.*, 1985, 1987a,b; Ghazal *et al.*, 2005) and the essential transactivator of E promoters, 88–90-kDa IE3 (Messerle *et al.*, 1992;

Angulo *et al.*, 2000), respectively. For mCMV MIE protein functions, see a more recent review (Busche *et al.*, 2008).

Quantitative assessment of MIE locus transcription in latently infected lungs: discovery of stochasticity of MIE gene expression during latency

In an *in vivo* analysis of latency in the lungs in the HCT model of mCMV infection, splice-specific RT-PCR identified *IE1* and *IE2* transcripts during latency confirmed by absence of infectious virus and absence of transcripts coding for the essential viral transactivator IE3 and the essential virion envelope protein gpM55 (gB). This showed that correctly spliced productive cycle MIE sense transcripts can be expressed also in latency and that progression of the viral gene expression programme can be prevented at a molecular checkpoint(s) downstream of MIE promoter activity (Kurz et al., 1999; Grzimek et al., 2001). As the main function of MIE transcripts/proteins is in the lytic cycle, and as we do not wish to imply a function for latency, we use the designation TEL for MIE transcripts found in the absence of essential transcripts downstream in the IE-E-L kinetic path of acute or reactivated viral gene expression. In the 'twilight zone' of subthreshold IE1/3 splice precursor transcription that generates IE1 transcripts but undetectable levels of essential IE3 transcripts, it is a matter of faith and semantics if we view the IE1 transcripts as TELs or as the result of a failed attempt of the virus to reactivate.

In these studies, highly sensitive, splice-specific RT-PCR was performed on poly(A)+ RNA isolated from tissue of latently infected lungs, and combined with a statistical approach allowing an in vivo clonal analysis and prevalence estimation for TEL-generating desilencing events at the MIE locus (for the principle, see Fig. I.22.5B). Specifically, a geometrical grid was applied to the lungs subdividing the organ into 18 fields (physically, 3D tissue pieces representing $3–4 \times 10^6$ lung cells each) tested individually for viral DNA load or presence of TELs. Although latent viral genomes were present in all individual fields in relatively high and roughly comparable numbers, ranging between 6000 and 9000 copies per 10^6 lung cells (Kurz et al., 1999), there existed fields positive (containing) or negative (not-containing) for IE1 TELs, a binary on–off phenomenon that can be mathematically treated with the Poisson distribution function based on the understanding that absence of TELs indicates the absence of a transcription event in that field, whereas presence of TELs indicates the presence of at least one transcription event, but can also result from two, or three, or more events. With experimental knowledge of the proportion of negative fields, the Poisson distribution function gives an estimate of the total number of transcription events in the whole organ. To give an idea of the order of magnitude (precise numbers can vary with experimental conditions), a statistically averaged latently infected lung harboured ~ 10–20 'foci' of IE1 TELs at the time of tissue shock-freezing for analysis (Kurz et al., 1999; Grzimek et al., 2001). This value represents the 'point prevalence' of IE1 TEL events. The dynamics of this transcription is currently unknown, but it is predictable that the number of transcription events occurring over a longer period of time for a particular TEL, its 'period prevalence', will be higher. Based on the number of latent viral genomes in the whole lungs (~ 500,000) and the copy number of viral genomes of ~ 10 per latently infected cell, a 'point prevalence' of ~ 10 IE1 TEL events in the whole lungs

(as in the example of Kurz et al., 1999) corresponds to one cell expressing IE1 TELs among ~ 5000 latently infected lung cells, that is only 0.02% of cells that carry latent viral DNA express IE1 TELs at any moment. This low proportion also explains why chromatin isolated from latently infected mice predominantly revealed features of transcriptional repression at the MIE locus (Liu et al., 2008, 2010; see above).

Notably, correlative analysis of IE1 and IE2 TELs showed that expression flanking the dual enhancer is not linked but occurs independent of each other, which implies that the ie2 core promoter can remain silenced while the ie1/3 core promoter is active and vice versa (Fig. I.22.5B) (Grzimek et al., 2001; Simon et al., 2007). Other splice variants or antisense transcripts from the MIE locus, as well as TELs from other loci, were so far not described for mCMV, not because they were not found but because they were not systematically searched for.

What drives MIE gene expression in latency? The 'transcriptional noise hypothesis'

From the aforesaid it became evident that at any time the vast majority of latent viral genomes are silenced at the MIE locus, a situation that we from now on will refer to as 'MIE locus latency' with the understanding that at the same time other viral loci might be open but cannot lead to entering the lytic cycle because MIE transcripts, specifically IE3, are essential. The observation that the rare MIE locus activity is 'stochastic' in that it can be described by the Poisson distribution function should not be misinterpreted as an event that happens 'spontaneously'. Genes respond stochastically to inducers and stimuli can by themselves occur stochastically. This might be a 'cytokine noise' as it always present in tissues under non-germless conditions by responding to environmental antigenic challenge, particularly at exposed surfaces like in the lungs or in the gastrointestinal tract.

One should also consider that within a complex organism there is constant management of very localized injury and repair in diverse organs. This might lead to local and finite 'spikes' in inflammatory cytokines as part of this injury/repair process that would appear from a broader perspective to have a stochastic pattern. This putative pattern might match the stochastic transcriptional activity seen in latently infected hosts. Such triggering events would imply that TEL activity is a response to extrinsic signals. Indeed, tumour necrosis factor (TNF)-α, as one example, signals to the MIE enhancer and increases MIE gene transcription in a dose-dependent manner, but without inducing complete reactivation of the productive cycle (Hummel *et al.*, 2001; Simon *et al.*, 2005, 2007), thus suggesting that besides the expression of transactivating MIE proteins additional conditions must be fulfilled for full reactivation (discussed below; see Fig. I.22.11).

The phenomenon is related to the observation in cellular developmental genetics that cellular genes epigenetically silenced in the course of cell differentiation can be reversibly desilenced at a very low frequency. Transient 'promiscuous gene expression' has been discussed as a possible mechanism by which medullary thymic epithelial cells (mTECs) present antigenic peptides derived from tissue-restricted autoantigens that are not usually expressed in the thymus for inducing central tolerance in the thymic maturation of T-cells that imprints the individual's T-cell repertoire. This process involves acquisition of active histone marks and local chromatin decontraction occurring transiently in single mTECs only (Derbinski *et al.*, 2001; Klein *et al.*, 2009; Tykocinski *et al.*, 2010), a situation that is reminiscent of the desilencing of the mCMV MIE locus at either the *ie1/3* or the *ie2* promoter in single latently infected cells (Grzimek *et al.*, 2001; Simon *et al.*, 2007). In a broader perspective, stochastic expression of genes of eukaryote-adapted viruses mirrors the inherently stochastic nature of eukaryotic gene expression known to be a major factor in the heterogeneous response of individual cells within a clonal population to an inducing stimulus (Blake *et al.*, 2003).

We thus speculate that the incoming viral, *id est* foreign, DNA is silenced in latently infected cells like cellular genes that need to be shut-off in the course of cell differentiation and specialization, and that episodes of transient promiscuous gene expression from any one viral gene or a theoretically almost infinite number (2^{180}) of possible on–off combinations for the at least 180 ORFs of mCMV (Chapter I.2) occur in latent mCMV genomes, with the extremes being all-open and all-closed. So, it is assumed that any viral gene may be expressed at some time during latency. However, the probability for desilencing may not be the same for all genes but likely is higher for genes that possess a strong promoter or even an enhancer that facilitates local opening of the viral chromatin. Unevenly distributed desilencing probabilities will reduce the complexity and number of the combinatorial expression patterns, but since a latently infected lung was found to harbour viral genomes in an order of magnitude of only 500,000 (see above), there is no capacity for all theoretical combinations, and of existing combinations each will be rare. Theoretically, it even might be that in no single viral genome molecule all loci remain closed (for the concept, see Fig. I.22.6).

Impact of promiscuous viral gene expression during latency on the immune response: the phenomenon of T-cell 'memory inflation'

It is an intriguing question if viral proteins that are encoded by TELs alter the respective latently infected cells functionally and/or immunologically. A question central to this is if TELs are translated and thus lead to proteins at all (see Chapter I.20 for HCMV CLTs). In the case of the mCMV *IE1* TEL-expression in latently infected lungs, attempts to detect the IE1 protein biochemically or *in situ* by immunohistology were not successful; however, this may be explained by the overall low number of 10–20 expressing cells in a total of 60 million lung cells. Firm evidence for presence of a MIE locus TEL-derived protein was reported only recently for a transgenic diphtheria-toxin receptor (DTR) expressed in place of the mCMV IE2 protein from a BAC-cloned recombinant mCMV. As a single complex formed between DTR and its ligand DT at the cell surface triggers cell death, specific loss of latently infected cells in the presence of DT, measured as reduction of viral DNA load in the lungs, provided a most sensitive means to confirm expression of a TEL-derived protein (S. Scheller, Abstract 6.02, 35th Annual International Herpesvirus Workshop, 2010, Salt Lake City, Utah, USA).

The IE1 protein of mCMV was the first protein of a CMV for which an immunodominant antigenic peptide was identified in its amino acid sequence: the IE1 peptide *YPHFMPTNL* that is presented at the cell surface of infected cells of haplotype *H-2^d* by the MHC-I molecule L^d (Reddehase and Koszinowski, 1984; Reddehase *et al.*, 1989; reviewed by Reddehase, 2002) (see also Chapter II.17). It was therefore at hand to ask if the expression of *IE1* TELs in only few cells at any time, but continuously over a months-long period of viral latency, might leave an imprint in the pool size and activation state of IE1 epitope-specific CD8+ T-cells.

Evidence for TEL-encoded antigenic peptides driving the expansion of CD8+ T-cell memory came from work in the syngeneic HCT model with BALB/c mice as HC donors and immunocompromised and infected BALB/c mice as HCT recipients (for the HCT model, see Chapter II.17) (reviewed in Simon *et al.*, 2006b; Reddehase *et al.*, 2008). While focal CD8+ T-cell infiltrates confined and eventually terminated acute infection of the lungs, elevated numbers of pulmonary interstitial CD8+ T-cells persisted after resolution of the productive infection (Fig. I.22.7A). These long-term tissue-resident CD8+ T-cells proved to be not exhausted but fully functional, capable of controlling productive virus infection upon adoptive

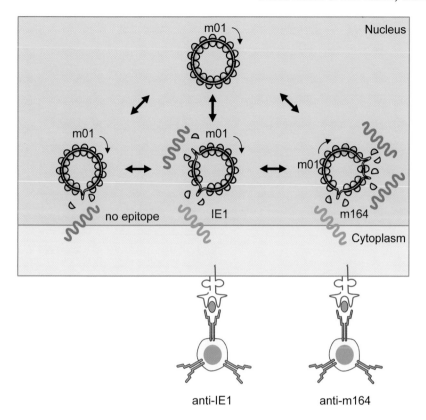

anti-IE1   anti-m164

Figure I.22.6 Model illustrating the principle of promiscuous viral epitope-encoding gene expression during viral latency. It is proposed that latent viral genomes exist in fluctuating stages, a stage in which all viral gene loci are silenced in closed viral chromatin and in numerous stages with local and only transient desilencing of single genes or combinations of genes, which reflect a transcriptional noise that generates 'transcripts expressed stochastically in latency', TELs, but does not lead into the coordinated, progressive IE-E-L gene expression path that characterizes acute or reactivated productive infection. The desilencing probability is likely enhanced by transcription factor-mediated signalling events and may not be the same for all viral genes, rather showing 'hot spots' formed by genes with strong promoters. The MIE locus might be such a 'hot spot' because of its strong enhancer element. Local desilencing of epitope-encoding genes may lead to the presentation of antigenic peptides, which are recognized by cognate T-cells.

transfer into immunocompromised and infected indicator recipients (Podlech et al., 2000). Notably, in the work by Podlech and colleagues all pulmonary CD8+ T-cells were antigen-experienced, as they all expressed CD44, a marker that distinguishes CD44high 'primed' antigen-experienced cells from mature but antigen-naïve CD44low T-cells (reviewed in Boyman et al., 2009). Based on these early findings it was already proposed that the CD8+ T-cells stay at this extralymphoid site (or continuously return there) to stand guard for surveillance of viral latency (Podlech et al., 2000). Subsequent analysis of the peptide-specificity of pulmonary CD8+ T-cells in acute and latent infection revealed a relative and even absolute increase in the numbers of responsive (secreting IFN-γ upon TCR signalling) IE1 epitope-specific cells, while the total numbers of responsive CD8+ T-cells as well as the

numbers of those specific for another mCMV epitope, M83 *YPSKEPFNF*, declined at the same time, thus showing selectivity of accumulation of IE1-specific CD8+ T-cells (Fig. I.22.7B,a and b). Importantly, sorting of CD44highCD8+ T-cells from latently infected lungs into CD62Lhigh central memory cells (T_{CM}) and a mixed population of CD62Llow effector-memory cells (T_{EM}), early effector cells (EEC), and short-lived effector cells (SLEC) (for reviews on memory CD8+ T-cell differentiation, see Obar and Lefrancois, 2010a,b) revealed an enrichment of responsive, IE1 epitope-specific cells in the CD62Llow populations (Holtappels et al., 2000) (Fig. I.22.7B,c), which is confirmed today also by cytofluorometric analysis using IE1-pMHC-I multimers for the detection of CD8+ T-cells expressing IE1-TCRs (Fig. I.22.7C). From all this it was proposed that an epitope-selective expansion, here of IE1 but not

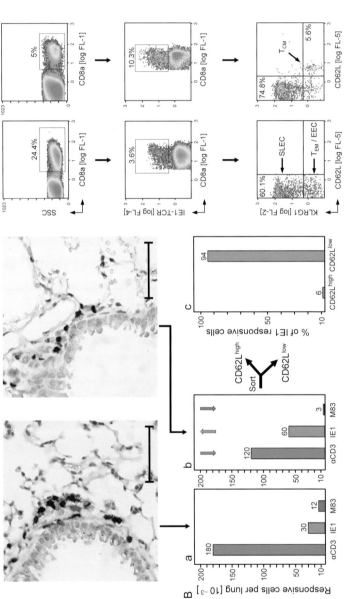

Figure I.22.7 Epitope-specific CD8+ T-cell 'memory inflation' in latently infected lungs in the model of HCT and mCMV infection. (A) Pulmonary T-cell infiltrates and infected cells in the lungs. Left panel: Inflammatory peribronchiolar focus during acute infection at 4 weeks after HCT. Right panel: Persistence of peribronchiolar T-cell infiltrate in absence of IE1+ cells in latently infected lungs at 3 months after HCT. 2-colour IHC staining of CD3ε+ T-cells (black) and IE1+ productively infected cells (red). Bar markers represent 50µm. (B, a and b) Absolute numbers of isolated pulmonary CD8+ T-cells capable of secreting IFN-γ in an ELISpot assay after epitope-independent polyclonal stimulation of TCR-CD3 signalling by anti-CD3ε antibody (positive system control) or after epitope-specific TCR-CD3 signalling induced by stimulator cells presenting IE1 or M83 peptides. (B, c) Relative proportions of IE1-specific responsive cells in sorted CD62L^high and CD62L^low fractions of pulmonary CD8+ T-cells isolated from latently infected lungs. (C) Cytofluorometric phenotyping of CD8+ T-cells from lungs comparing acute and latent infection. Top panels: CD8a expression of cells in the lymphocyte gate. Centre panels: Cells expressing an IE1-specific TCR (IE1-TCR) in the CD8+IE1-TCR+ lymphocyte gate. Bottom panels: expression of cell surface markers CD62L and KLRG1 by cells in the CD8+IE1-TCR+ gate. SSC, side scatter within a lymphocyte gate. FL, fluorescence channel and log intensity. Dots representing cells are displayed as colour-coded density plots (with red and blue representing highest and lowest cell numbers, respectively). Percentages of main interest are indicated. SLEC, short-lived effector cells; EEC, early effector cells (distinguished from T_EM by being CD127^low and from SLEC by being KLRG1^low; not shown); T_EM, effector-memory T-cells; T_CM, central memory T-cells.

of M83 epitope-specific CD44[high]CD8[+]CD62L[low] cells, is driven by IE1 peptide presented on latently infected cells sporadically expressing *IE1* TELs (Holtappels *et al.*, 2000). Shortly after that, a newly defined immunodominant D[d]-presented mCMV peptide *AGPPRYSRI* derived from the m164 protein, meanwhile identified as an E phase protein gp36.5 (Däubner *et al.*, 2010), was shown to also drive the expansion of memory CD8[+] T-cells (Holtappels *et al.*, 2002).

These original findings were later confirmed by Karrer and colleagues in the model of high-dose systemic infection of immunocompetent BALB/c mice by a longitudinal analysis showing that after a peak of acute, primary immune response followed by a contraction in cell numbers, CD8[+] T-cells specific for the epitopes IE1 and m164, but not M83, more or less steadily increased in numbers over time – not only in the lungs but also at other non-lymphoid organ sites as well as in the spleen as a main peripheral lymphatic organ – and they renamed this phenomenon with the catchphrase 'memory inflation' (Karrer *et al.*, 2003). It is worth calling to mind (see above) that all these organs were previously identified also as sites of mCMV latency (Reddehase *et al.*, 1994). In the years since its discovery, 'memory inflation' has become a 'hot topic' on which many groups have worked to decipher the mechanisms also in other mouse strains and other host species and their respective CMVs, and have contributed many refining details, mainly regarding the immunological aspects including an advanced phenotyping of the cells involved, the type of antigen-presenting cells involved, and the source of cells continuously replenishing the SLEC pool (for more recent reviews, see Klenerman and Dunbar, 2008; Snyder, 2011; O'Hara *et al.*, 2012). A whole chapter of this book (Chapter II.21) is dedicated to the exploitation of the phenomenon for 'memory cell vaccines' against other pathogens or even malignancies with CMVs as vectors, and the dynamics of HCMV memory as well as a possible link between 'memory inflation' and a proposed HCMV-associated immunosenescence are discussed in Chapter II.7; so, we refrain from re-reviewing all these aspects here.

Model linking viral gene expression in latency to 'memory inflation'

Our current view of how local gene desilencing events during latency are sensed at an extralymphoid site of latency, such as the lungs, by patrolling memory T-cells to induce 'memory inflation' and stand guard for controlling latency and preventing full reactivation is sketched in Fig. I.22.8 for the example of the IE1 epitope, the prototype of a memory inflation-inducing antigenic peptide of mCMV. As 'late and enduring'

mCMV latency is established predominantly in MHC-II[−] tissue cells (see above), we focus here on CD8[+] T-cell memory inflation.

It is proposed that in latently infected host tissue cells, which, as discussed above, are mostly not haematopoietic lineage-derived professional antigen-presenting cells (profAPCs) but are MHC-II[−] host tissue cells, for instance subtypes of ECs, *IE1* TELs are indeed translated into IE1 protein. Like in acutely infected cells, the IE1 protein is subjected to proteasomal processing, which is known to occur principally by the constitutive proteasome present in all cell types but more efficiently by the immunoproteasome that is constitutively present in profAPCs or is assembled in non-profAPCs in response to IFNs, to IFN-γ in particular, but not exclusively (Kloetzel, 2001; Knuehl *et al.*, 2001; Strehl *et al.*, 2005; Shin *et al.*, 2006; Sijts and Kloetzel, 2011). An N-terminally elongated precursor peptide, *DM-YPHFMPTNL*, is transported through the 'transporter associated with antigen processing' (TAP) into the ER where it is N-terminally trimmed to the final antigenic peptide *YPHFMPTNL* (Knuehl *et al.*, 2001) that is loaded onto nascent MHC-I molecules to form a pMHC-I complex (IE1-L[d] in the specific case of the example). This complex traffics with the constitutive vesicular flow via the Golgi apparatus to the cell surface where it is presented to CD8[+] T-cells bearing a cognate T-cell receptor (IE1-TCR) (reviewed in Reddehase, 2002).

The first experimental evidence for a link between IE1 epitope-specific immune surveillance in latently infected lungs and *IE1* TELs was provided with an IE1 antigenicity loss mutant L176A of mCMV in which the amino acid leucine at the MHC-I anchor position no. 9 of the IE1 peptide is replaced with alanine (Simon *et al.*, 2006a,b; for the mutation principle, see also Lemmermann *et al.*, 2011). Lack of IE1-specific T-cells after infection with this virus mutant led to a significantly increased number of detectable *IE1* TEL events and progression to alternatively spliced *IE3* transcripts, suggesting that the lower number observed with WT and revertant viruses expressing the antigenic A176L IE1 peptide resulted from CD8[+] T-cell effector function either preventing transcription or eliminating cells that express *IE1* TELs (Simon *et al.*, 2006a; reviewed by Reddehase *et al.*, 2008). This finding also implied that the frequency of desilencing events at the MIE locus is underestimated in the presence of surveillance by CD8[+] T-cells.

CD8[+] memory T-cells, T[CM] and T[EM], are generated from thymus-educated but naïve T-cells during the primary immune response to acute infection and are capable of homeostatic, cognate antigen-independent cell division, T[CM] more than T[EM], for balancing cell

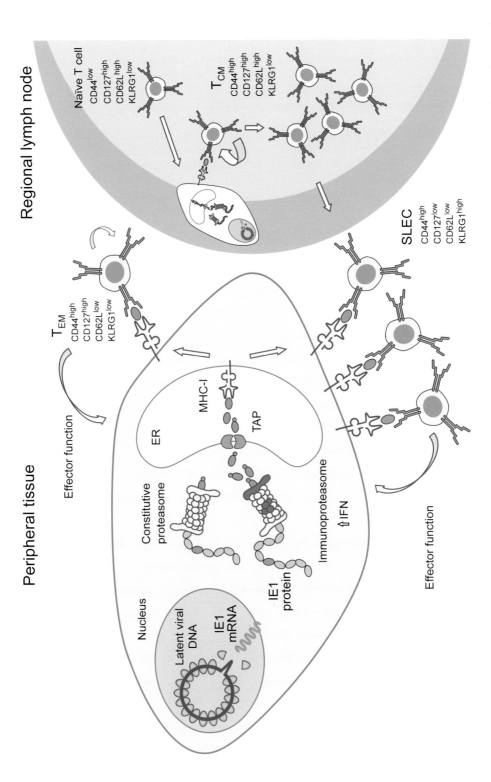

Figure I.22.8 Model linking stochastic viral gene expression during latency to 'memory inflation'. MIE locus desilencing resulting in *IE1* TEL expression, and processing and presentation of the antigenic IE1 peptide are used as an example for the generation of 'memory inflation'-inducing viral epitopes. See the main text for detailed explanation. SLEC, short-lived effector cells; T_{EM}, effector-memory T-cells; T_{CM}, central memory T-cells. TAP, transporter associated with antigen processing. Circuit arrows indicate cell proliferation and their different sizes symbolize proliferation potentials, which is high for T_{CM} and lower for T_{EM}.

death to maintain pool size (reviewed in Boyman et al., 2009). Only T_{EM} can exert immediate antigen-specific cytolytic activity upon encountering pMHC-I. Whilst CD8$^+$ T_{EM} are patrolling non-lymphoid tissues seeking their cognate pMHC-I on tissue cells, T_{CM} usually circulate between blood and secondary lymphoid organs (SLOs), such as lymph nodes (LNs), but can migrate also to peripheral sites under conditions of inflammation (for a review, see Weninger et al., 2002). As latently infected organs are at best sites of 'micro-inflammation' we assume in our model (Fig. I.22.8) that cells sensing TEL-associated antigenic peptide presentation for immune surveillance are primarily the T_{EM} characterized by the phenotype CD44^{high}CD127^{high}CD62L^{low}KLRG1^{low}. Of note, it is difficult to retrieve from the literature definite information on the fate and migratory pattern of T_{EM} after peripheral antigen encounter, though lack of the 'homing' receptor CD62L (L-selectin) would not predict that they migrate to regional lymph nodes for antigen-triggered cell proliferation. However, as T_{EM} do have proliferative potential, though lower than T_{CM}, one can assume limited on-site expansion. The special situation with mCMV is the establishment of latency also in non-haematopoietic MHC-II$^-$ cells in SLOs (see above), which may engage also CD44^{high}CD127^{high}CD62L^{high}KLRG1^{low} T_{CM} in TEL-associated antigen recognition. As T_{CM} have a high proliferative potential, this may drive 'memory inflation'. Recent work, performed in parallel by two groups in independent models of BM chimeric mice, has indeed provided evidence for 'memory inflation' being dependent upon direct antigen presentation by recipient-genotype, non-haematopoietic cells (Seckert et al., 2011; Torti et al., 2011a) and expansion of T_{CM} in lymph nodes (Torti et al., 2011a). These cells are supposed to generate SLECs (Snyder et al., 2008), which are characterized by the phenotype CD44^{high}CD127^{low}CD62L^{low}KLRG1^{high} (Thimme et al., 2005). SLECs are recruited to peripheral sites of TEL-associated antigen presentation and account for the majority of CD8$^+$ T-cells found in the lungs. The immunohistological analysis of pulmonary T-cell infiltrates in latently infected lungs expressing IE1 TELs provided direct in situ evidence for their extra-lymphoid localization in lung interstitium to exclude a misinterpreted analysis of inadvertently co-dissected pulmonary regional LNs (Podlech et al., 2000) (Fig. I.22.7A, showing peribronchiolar localization of T-cells). Indeed, by far most of the pMHC-I multimer-stained IE1 epitope-specific pulmonary interstitial CD8$^+$ T-cells show the typical, highly activated SLEC phenotype being CD62L^{low}KLRG1^{high} (Seckert et al., 2011) (Fig. I.22.7C). So, to be more precise, the 'inflated' number of antigen-specific CD8$^+$ T-cells detected at an extralymphoid site of viral

latency might reflect massive recruitment of SLECs to that site, rather than 'memory inflation' as such.

Why do recent thymic emigrants contribute so little to 'memory inflation'?

In addition to T_{CM}, recent thymic emigrants, that is thymus-educated naïve CD44^{low}CD127^{high}CD62L^{high}KLRG1^{low} T-cells, were discussed as a cellular source for 'memory inflation' generating new memory cells by permanent priming; so to speak repetitive episodes of primary immune responses superimposed on the memory response and thereby escalating the overall response over time (Snyder et al., 2008). Data from thymectomized mice, however, indicated that priming of recent thymic emigrants is not essential for 'memory inflation' to occur and also contributes little to the magnitude of the response (Loewendorf et al., 2011). This, at first glance, unexpected result is explainable by the requirement of antigen (cross-) presentation by profAPCs, in particular CD8$^+$ DCs, for the priming of naïve CD8$^+$ T-cells, a condition not likely fulfilled in mCMV latency where TELs encode nominal amounts of antigenic protein in non-haematopoietic tissue cells that do not qualify as profAPCs for T-cell priming (Seckert et al., 2011; Torti et al., 2011a).

Which factors determine if an epitope drives 'memory inflation'?

In the original work discovering the expansion of CD8$^+$ T-cells in the memory phase of the immune response to mCMV (Holtappels et al., 2000) it was already proposed that this expansion, meanwhile renamed 'memory inflation', must be antigen-driven, involving repetitive restimulation of memory cells. This 'restimulation hypothesis' was based on (i) the accumulation at a non-lymphoid tissue site selectively of IE1 epitope-specific CD8$^+$ T-cells for which expression of the epitope-encoding gene during viral latency was verified by RT-qPCR demonstrating the presence of IE1 TELs (Kurz et al., 1999) as well as (ii) from the CD62L^{low} phenotype of the accumulating CD8$^+$ T-cells, a marker excluding T_{CM}. With progress in available tools and methods, and based on an advanced general immunological understanding, work in the past 12 years has refined our view in details, but, in essence, the original hypothesis has survived the test of time. Whilst enhanced intermittent homeostatic, γ_C-cytokine (IL-7 and, in particular, IL-15)-driven proliferation of memory cells (Boyman et al., 2009) specific for certain epitopes, putatively facilitated by properties imprinted by the naïve CD8$^+$ T-cell repertoire and/or the avidity of the TCR/pMHC-I/

costimulatory molecule interaction at the immunological synapse in the 'priming phase', has been considered an alternative explanation, current opinion has returned to the 'restimulation hypothesis' (Snyder *et al.*, 2008; Hutchinson *et al.*, 2011; Seckert *et al.*, 2011; Torti *et al.*, 2011a,b). Two significant recent findings proved to be difficult to explain by homeostatic proliferation of memory subsets: (*i*) the predominance of CD62L-^{low}KLRG1^{high} SLECs in peripheral tissue infiltrates, and (*ii*) the requirement for direct antigen presentation by tissue cells that are not profAPCs usually involved in priming, thus indicating differential rules for CD8$^+$ T-cell priming and 'memory inflation' (references as above).

Yet, we still owe an explanation for the epitope-selectivity. It appears that several conditions need to be fulfilled for an epitope to be able to drive 'memory inflation'. In the following, we will discuss and speculate about some of these conditions in their expected temporal hierarchy.

Does the magnitude of the acute immune response to a particular epitope imprint its 'inflation phenotype'?

As memory subsets are generated during the acute immune response (for reviews, see Weninger *et al.*, 2002; Boyman *et al.*, 2009; Obar and Lefrancois, 2010a,b), the 'fate' of the memory cells, becoming 'inflationary' or 'non-inflationary', might already be determined at the stage of the acute immune response. A very recent review article (O'Hara *et al.*, 2012) gives the unfortunate impression that inflationary CD8$^+$ T-cells were characterized by having low frequencies during the acute immune response with CD8$^+$ T-cells continuously increasing in numbers over time, whereas non-inflationary CD8$^+$ T-cells would be present in high frequencies during the acute immune response and contract in numbers over time resulting in a small but stable T$_{CM}$ pool. This suggested an epitope-dependent difference in the balance between effector and memory subsets generated during the acute immune response predetermines later memory. This view, however, is not compatible with the data on mCMV. Specifically, for the prototypes of 'inflationary' and 'non-inflationary' CD8$^+$ T-cells in the immune response to mCMV in mice of MHC haplotype *H-2^d* (e.g. BALB/c), specific for epitopes IE1 and M83 both presented by the same MHC-I molecule L^d, frequencies in the acute response were high and low, respectively (Holtappels *et al.*, 2000) (see Fig. I.22.7B). Like with IE1, CD8$^+$ T-cells specific for the D^d-presented m164 epitope, the second known example for 'memory inflation', was characterized by high frequencies in both acute response and memory

(Holtappels *et al.*, 2002). These early examples also already indicated that the type of epitope-presenting MHC-I molecule does not predict the 'inflation phenotype' (see also Chapter II.17). In the MHC haplotype *H-2^b* (e.g. C57BL/6 mice), five kinetic patterns of acute versus memory frequencies were observed: high–high (m139, showing transient contraction; M38, remaining high), high–low (M45), low–high (IE3/M122) and low–low (most other epitopes) (Munks *et al.*, 2006).

In conclusion, results from two groups obtained independently in two different mouse MHC (*H-2*) haplotypes show that the magnitude of the acute CD8$^+$ T-cell response to mCMV epitopes does not predict the 'inflation phenotype'. This also argues against a major impact of the naïve TCR repertoire on the 'inflation phenotype' of an epitope.

Not all existing TELs can give rise to antigenic peptides driving 'memory inflation' in an individual

For an immunologist it may be a trivial note that a peptide can only be antigenic when it is generated by protein processing and when it binds to an MHC-I molecule for presentation at the cell surface. Peptides that do not fulfil the latter condition are possibly generated by the proteasome, yet, are then degraded. Thus, for instance, the L^d-presented IE1 peptide cannot be isolated from cells that do not express L^d in the *L^d* gene deletion mutant BALB/c-H-2^{dm2} (Del Val *et al.*, 1991) and, consequently, IE1-specific CD8$^+$ T-cells primed during the acute response by BALB/c donor-derived profAPCs in BALB/c X BALB/c-H-2^{dm2} BM chimeras failed to undergo 'memory inflation' that depends on antigen presentation by recipient-derived tissue cells (Seckert *et al.*, 2011). In a broader perspective, based on MHC polymorphism on the host population level, each individual (or inbred mouse strain) selects its unique repertoire of antigenic peptides even if the transcriptional activity from the viral genome were identical. Whether host genetics, including all non-MHC loci, has an influence directly on the viral TEL activity is not known.

It has been proposed that 'memory inflation'-inducing epitopes predict viral loci with TEL activity. This is probably correct, but, when based on polymorphic MHC-I coverage, the overall TEL activity of the latent viral mCMV genome will be much higher than the currently predicted 5 mCMV TEL loci (IE1/3, M38, M102, m164, and m139) based on studies in BALB/c and C57BL/6 mice (Holtappels *et al.*, 2002; Munks *et al.*, 2006). Whilst a positive result for 'memory inflation' likely indicates TEL activity, a negative result does not argue against TEL activity. For instance, m164 protein-derived K^b- and D^b-presented peptides *GTTDFLWM* and *WAVNNQAIV*, respectively, both failed to promote 'memory inflation' in C57BL/6 mice (Munks et al., 2006), which could have indicated lack of TEL activity from the m164 locus. However, as was known before, the D^d-presented m164 peptide *AGP-PRYSRI* induced 'memory inflation' in BALB/c mice (Holtappels et al., 2002), indicating TEL activity from the m164 locus. This may prompt the idea that TEL activity differs genetically in these two haplotypes. Notably, however, a non-CMV epitope, the ovalbumin-derived K^b-presented peptide *SIINFEKL* integrated by site-directed BAC mutagenesis into the m164 protein in place of the endogenous inflation-inducing D^d-presented antigenic peptide *AGPPRYSRI* (Lemmermann et al., 2010, 2011) assumed an inflationary phenotype in C57BL/6 mice in accordance with TEL activity from the m164 locus in both BALB/c and C57BL/6 mice (Lemmermann, N.A.W., unpublished data).

Requirement for recent epitope-encoding gene expression

Due to the constitutive turnover of pMHC-I complexes at the cell surface (Lemmermann et al., 2010; and references therein) it appears irrational to assume that pMHC-I presented by latently infected cells (or by cells that survived acute infection without retaining latent viral genome) would have been acquired already during the acute phase of infection and stayed at the cell surface for many months to drive memory inflation; – in a formal sense, however, this idea is not formally excluded by evidence. One might speculate that long-lived, quiescent cells or cell types with constitutively low metabolic activity might have a very low pMHC-I turnover and also resist lysis by $CD8^+$ T-cells. Even more speculative, long-lived cells might store acute infection-derived antigenic protein in aggresome-like induced structures (ALIS), protein storage compartments for substrates of the proteasome (Herter et al., 2005; Pierre, 2005; Szeto et al., 2006), for processing it later upon triggering stimuli. However, both these possibilities would depend on a fixed amount of antigenic protein synthesized during acute infection, and such a source should get exhausted over time. We consider these possibilities unlikely.

Thus, provided that the 'restimulation hypothesis' holds true, new viral gene expression must take place either during latency or during episodes of incomplete, interrupted reactivation. The conclusion that 'memory inflation' does not depend on episodes of productive reactivation is meanwhile current opinion, based on absence of transcripts encoding

essential proteins, e.g. IE3 and M55/gB, in the lungs of latently infected mice in which IE1-SLECs accumulate (reviewed by Reddehase et al., 2008), as well as the more recent observation of 'memory inflation' after infection with a spread-deficient recombinant mCMV (see above) (Snyder et al., 2011).

Although the 'restimulation hypothesis' is meanwhile current opinion, more than a decade after this hypothesis has been raised (Holtappels et al., 2000), the *IE1* TEL remained the only mCMV epitope-encoding transcript molecularly identified in latently infected tissue. However, though without molecular detection of transcripts, indirect evidence for 'memory inflation' being driven by *IE2* TELs (Grzimek et al., 2001; Simon et al., 2007), for which no natural epitopes have been described in haplotypes $H-2^d$ and $H-2^b$, was provided by transgenic expression of foreign epitopes in *ie2* gene recombinant mCMVs (Karrer et al., 2004) (see also Chapter II.21). The lack of evidence for the expression of TELs corresponding to other epitopes already known to drive memory inflation (e.g. m164 in $H-2^d$ and five epitopes in $H-2^b$) (Holtappels et al., 2002; Munks et al., 2006) has been raised as an argument against a dependence on TELs. One must consider, however, two aspects: (i) the predicted TELs have not been intensely searched for (so, not having been studied is no argument against or for their existence) and (ii) transcription events may be so rare that they escape detection. As discussed earlier in this chapter, measuring transcripts by RT-qPCR performed on shock-frozen tissue describes the situation at one moment in latency, which is known as 'point prevalence', and this was estimated to be just 10–20 transcription events per whole lungs for *IE1* TELs (see above) despite the MIE locus transcription being facilitated by a very strong enhancer element. If the 'point prevalence' of transcription from other loci is much lower, which is reasonable to assume from the absence of an enhancer, we will not be able to detect it. Nonetheless, even extremely rare events accumulating over months-long periods of time, the 'period prevalence', may suffice for presentation of antigenic peptides that drive 'memory inflation'. Recent work on transgenic recombinant *IE2* TELs coding for the DTR (see above) indeed indicates more frequent expression when viewed over longer periods (Scheller, 2010). We propose that cytokine-induced promiscuous gene desilencing during latency (discussed above) is the molecular motor for 'memory inflation' (Fig. I.22.6). Importantly, this hypothesis of stochastic and fluctuating gene expression from variable loci of the latent genome, which is independent of the temporally directional IE, E, and L phases of the productive viral gene expression cascade, can explain 'memory inflation' for E and L epitopes in latently infected lungs not expressing IE3, the essential transactivator of E phase (and thus in the temporal consequence also of L phase) transcripts.

Importance of latent viral DNA load for TEL activity and 'memory inflation'

It is almost needless to say that the probability for TEL activity increases with the number of latent viral genomes from which TEL activity can arise; so, establishing a high latent viral genome load during acute infection is also a prerequisite for efficient 'memory inflation' during latent infection. In accordance with this, an enhanced antiviral control after infection with an mCMV gene deletion mutant devoid of MHC-I pathway 'immune evasion genes' (see Chapter II.17) led to reduced viral DNA load and consequently a reduced MIE locus TEL activity in the lungs concomitant with curtailed accumulation of IE1 epitope-specific CD8+ SLECs in the lungs (Böhm *et al.*, 2009). This may explain why 'memory inflation' is quite variable depending on the infection history, such as dose and route of primary infection. The observation of 'memory inflation' after infection with spread-deficient virus must be cautiously interpreted, as very high virus doses and systemic infection route needed to establish latency in a reasonable number of 'first-hit' target cells (Mohr *et al.*, 2010; Snyder *et al.*, 2011) are unlikely to have a correlate in infection by natural or iatrogenic virus transmission. This needs to be considered for vaccine approaches (see Chapter II.21).

Importance of antigen processing, presentation, and recognition efficacy

In view of the low TEL transcription and the failure to detect the resultant proteins during latency by biochemical methods, the generation and presentation of antigenic peptides is likely to critically depend on efficient processing of minute amounts of antigenic protein. profAPCs, such as CD8+ DCs capable of cross-presenting antigen (Heath and Carbone, 2001; Allan *et al.*, 2003; Belz *et al.*, 2004; Schnorrer *et al.*, 2006) constitutively express the immunoproteasome, whereas the non-haematopoietic cell types known to be sites of latent mCMV infection (discussed earlier in this chapter) express the constitutive proteasome but can conditionally express the immunoproteasome upon induction by IFNs, by IFN-γ in particular. Principally, both types of proteasomes can generate different, though overlapping, sets of antigenic peptides, but as shown for the IE1 peptide of mCMV, a

particular peptide can be generated in higher amounts by the immunoproteasome (Knuehl *et al.*, 2001). So, on the one hand, generation of antigenic peptides by the immunoproteasome ought to favour 'memory inflation'. On the other hand, however, dependence on the immunoproteasome limits the generation of antigenic peptides in non-profAPC tissue cells to inflammatory conditions needed to induce the immunoproteasome, and this prerequisite may reduce the overall chance for memory cell restimulation. Indeed, a recent study using immunoproteasome-deficient LMP7−/− mutants of C57BL/6 mice suggested antigenic peptides driving 'memory inflation' are less dependent upon the immunoproteasome (Hutchinson *et al.*, 2011). Specifically, in this study, the largely immunoproteasome-dependent and *in vivo* unprotective M45-D^b peptide *HGIRNASFI* (Gold *et al.*, 2002; Holtappels *et al.*, 2004, 2009; Fink *et al.*, 2012) (see also Chapter II.17) failed in inducing 'memory inflation' in both WT C57BL/6 mice and LMP7−/− mutants, consistent with the interpretation that immunoproteasome-dependence may disfavour memory cell restimulation and thus disfavour also 'memory inflation'. However, this conclusion is only valid if the first-rank condition in the hierarchy of conditions for 'memory inflation', namely the expression of M45 protein, is fulfilled. Trivially, without any antigenic protein being available for processing, the type of protein processing does not matter. In fact, the immunoproteasome-independent and *in vivo* protective M45-D^d epitope *VGPALGRGL* (Holtappels *et al.*, 2009) also fails to promote 'memory inflation' (Sierro *et al.*, 2005), indicating that the M45 protein may not be expressed during latency in the first place. Nonetheless, an advantage of low prerequisites for efficient processing is reasonable, and in accordance with this view, peptides inducing 'memory inflation' in latently infected BALB/c mice, IE1 and m164 (Holtappels *et al.*, 2002), are produced in productively infected cells in high numbers not dependent upon IFN-γ pretreatment (see Chapter II.17). Interestingly, generation of IE1 peptide by proteasomal processing was shown to occur also independent of preceding ubiquitination of the protein substrate (Voigt *et al.*, 2007); so, one might speculate that this particular feature of reduced prerequisites for processing may contribute to efficient processing of minute amounts of IE1 protein produced in latently infected cells. If protein ubiquitination-independence is a feature of all inflation-promoting epitopes awaits analysis.

For efficient presentation at the cell surface, post-proteasomal processes, such as TAP transport, N-terminal trimming, peptide-loading on MHC-I, and pMHC-I complex trafficking to the cell surface (Fig. I.22.8), must as well be efficient. So, probably, the peptide-binding affinity to MHC-I may matter too. Finally, memory CD8[+] T-cells with high functional avidity of TCR binding to presented cell surface pMHC-I, as well as low costimulatory signal requirements (Arens et al., 2011) should have a restimulation advantage over those cells with low-avidity binding. Considering all these variables, it comes no longer as a surprise that the same viral protein may contain inflation-promoting and non-promoting antigenic peptides, as has been shown for closely neighbouring D^d- and D^b-presented m164 protein-derived peptides discussed in greater detail above (Holtappels et al., 2002; Munks et al., 2006).

Why is there no apparent role for viral proteins that interfere with antigen presentation?

Like other CMV species, mCMV codes for 'immune evasion' proteins that inhibit direct antigen presentation to CD8[+] T-cells (viral regulators of antigen presentation; vRAPs) by interfering with the cell surface transport of recent pMHC-I complexes (see Chapter II.17). So, in particular when considering the low level of pMHC-I to be expected in latency, why do these proteins not prevent presentation of antigenic peptides in mice latent of WT virus? In fact, deletion of these proteins did not even enhance 'memory inflation' in the spleen in both the BALB/c model (Holtappels et al., 2006) and the C57BL/6 model (Munks et al., 2007) (see also a preceding meeting review, Yewdell and Del Val, 2004).

Again, the hypothesis of stochastic and fluctuating gene expression during viral latency (Fig. I.22.6) can provide an answer. Though it is not excluded that viral genes coding for vRAPs, m152/gp40 and m06/gp48 (see Chapter II.17), are also stochastically desilenced during latency in a very low frequency, this will statistically only rarely coincide in the same cell with desilencing of epitope-encoding genes involved in 'memory inflation'. Logically, physical separation of the antigenic protein and the presentation inhibitor in different cells allows peptide presentation and 'memory inflation'.

Requirements for 'memory inflation' beyond antigen sensing on latently infected cells

Long-term maintenance of high numbers of memory cells that continuously replenish the SLEC pool may not only depend on repeated stimulation by presented antigenic peptides but also by factors that promote memory cell longevity and proliferation capacity. Two

publications on that issue gave partially conflicting results for the requirement of CD4[+] T-cell help. Whilst Walton and colleagues described abolishment of CD8[+] 'memory inflation' in latently infected CD4[+] T-cell-deficient MHC-II[-/-] mice (Walton et al., 2011), Snyder and colleagues, more in accordance with already historical data on a largely CD4[+] helper cell independent generation and long-term maintenance of protective CD8[+] T-cells controlling mCMV infection in all host organs except in the salivary glands (Jonjic et al., 1989), did not observe such a substantial impact of CD4[+] T-cells on 'memory inflation', with the notable exception of an IE3 epitope, which is known to differ from other inflation-promoting epitopes by a late onset of 'memory inflation' (Snyder et al., 2009). Notably, IE3 is also special in that its associated 'memory inflation' appears to depend on cross-presentation by MHC-II[+] profAPCs (Torti et al., 2011b), which are co-deleted in MHC-II[-/-] mice, and this might explain lack of 'memory inflation' in these mice regardless of the additional lack of CD4[+] T-cell help.

In our own experience, IE3 'memory inflation' is not consistently observable (unpublished data) and may thus depend on particular conditions that are not always fulfilled in latently infected C57BL/6 mice. From a molecular point of view, IE3 transcripts differ from TELs since their generation does not only depend on a gene desilencing event but also depends on alternative splicing from an unspliced IE1/3 precursor, a process that is subject to splicing regulation (see Fig. I.22.5A), unless we assume independent expression of a still functional short version of IE3 from an 'exon 5 only' transcript by alternative promoter usage, a possibility that has not been formally excluded for mCMV. In any case, whilst IE3 is essential for initiating the viral productive cycle (Messerle et al., 1992; Angulo et al., 2000), during latency the IE1 TEL is generated preferentially, if not exclusively (Kurz et al., 1999; Grzimek et al., 2001), so that one would not predict the existence of an IE3 peptide able to drive 'memory inflation'. Yet, TNF-α signalling promotes splicing to generate IE3 transcripts (Simon et al., 2005) and so do other triggers of virus reactivation (Kurz and Reddehase, 1999) (see the subsequent sections). One may thus speculate that only occasional and usually late appearance of IE3-specific CD8[+] T-cells reflects induction-dependent reactivation of productive cycle viral gene expression rather than TEL activity; in other words, late expression of IE3 may be 'reactivation-associated'. This might explain the deviant antigen-presentation requirements for IE3 'memory inflation' (Torti et al., 2011b) as well as dependence on cytokine signalling events that are not always present in all individual mice and under all experimental conditions, and even may reflect the microbial status of the mouse colony (Tanaka *et al.*, 2007).

Further, memory levels are not constant but are shaped by episodes of competing antigen exposure, a phenomenon long known in immunology (for reviews, see Welsh and Selin, 2002; Welsh *et al.*, 2004). This may explain why 'continuous memory inflation' is *de facto* often less continuous than usually shown in publications. Finally, absence of IL-10 counter-regulation in IL-10$^{-/-}$ mice was found to expedite 'memory inflation' (Jones *et al.*, 2010), thus indicating, by *argumentum e contrario*, that the 'memory inflation' in WT mice already represents a curtailed situation.

Mechanisms and triggers of virus reactivation

The 'immune compromise hypothesis' of reactivation: evidence and caveats

The capability of the silenced, latent virus genome to reactivate to the recurrence of infectious virions is an integral part of the definition of latency, distinguishing functional latency from carriage of replication-incompetent, defective viral DNA (see above). Thus, detection of viral DNA in host tissues *per se* is not unequivocal evidence for latent infection. Recurrence of mCMV from tissue explants, historically from the spleen (Wise *et al.*, 1979; Jordan and Mar, 1982), salivary glands and prostate (Cheung and Lang, 1977), kidneys (Porter *et al.*, 1985), and ocular tissue (Kercher and Mitchell, 2002), or *in vivo* after experimental organ transplantation and various modes of immunosuppressive or haematoablative treatment, served to prove replication competence of latent viral genomes (reviewed by Jordan, 1983; Hummel and Abecassis, 2002; Reddehase *et al.*, 2002, 2008).

Productive reactivation observed after immunosuppression was the basis for the 'immune compromise hypothesis' of virus reactivation or, in *argumentum e contrario*, the 'immune surveillance' hypothesis, postulating that immune cells interrupt the reactivation process whenever an antigenic protein comes to its expression in the IE-E-L time course of the productive viral cycle. The most direct evidence in support of this idea was provided by *in vivo* mCMV reactivation in latently infected B-cell-deficient C57BL/6 μ$^{-/-}$ mice following selective depletion of innate and adaptive immune cell subsets (Polic *et al.*, 1998), a system in which absence of antibodies allowed secondary virus spread after reactivation for an enhanced sensitivity of detection (Jonjic *et al.*, 1994). Despite a strong NK cell response in C57BL/6 mice based on ligation of

the activatory NK cell receptor Ly49H by the mCMV protein m157 (for more explanation, see Chapter II.9), selective depletion of NK cells largely failed in inducing reactivation, whereas combined depletion of NK cells and CD8$^+$ T-cells compared to combined depletion of NK cells and CD4$^+$ T-cells revealed a dominant role for CD8$^+$ T-cells in controlling latency, further improved by CD4$^+$ T-cell help (Polic *et al.*, 1998). Though these findings were intriguing to conclude on a cooperative function of lymphocyte subsets in controlling latency, there is a caveat. Conspicuously, while in presence of NK cells (likely producing much IFN-γ) neither selective depletion of CD8$^+$ or CD4$^+$ T-cells alone, nor blockade of IFN-γ alone, induced notable reactivation, blockade of IFN-γ in concert with depletion of either T-cell subset induced reactivation regardless of which subset was depleted. These data are also consistent with the alternative interpretation of virus reactivation being triggered on the transcription factor signalling level by a 'cytokine storm' – possibly mainly involving TNF-α – that is elicited by cell death inherent to depletion of whatever lymphocyte subset, but that blockade of IFN-γ is a second condition for reactivation to proceed. So, combined conditions of TNF-α^{high}IFN-γ^{low} may favour virus reactivation. This might explain the failure to achieve full reactivation by TNF-α alone (Hummel *et al.*, 2001; Simon *et al.*, 2005, 2007).

Generally, *in vivo* reactivation studies based on measuring infectious virus have difficulty distinguishing between immune control of the virus reactivation process within the cell in which reactivation is triggered and immune control of subsequent virus spread and consequent amplification. As will be shown below, most transcriptional reactivation events actually stop before completion of the productive cycle even in the absence of cellular immune control in a model of global cellular immune ablation. As far as CD8$^+$ T-cells are concerned, an immune control of reactivation should wane after the point is reached when MHC-I pathway 'immune evasion genes', *m152* and *m06* (see Chapter II.17), become expressed.

It has been postulated that virus recurrence in explant cultures may also be the consequence of release from immune control, supporting the 'immune compromise hypothesis', but metabolic and cell-cycle alterations due to hypoxic stress might play a role as well, which is supported by the empirical evidence of virus recurrence induced in tissue cell suspensions much less efficiently than in explants with preserved tissue architecture. If cessation of immune control were the key mechanism, one rather would have expected less reactivation in the explants where cell–cell interaction between tissue-resident T$_{EM}$, which are capable of exerting direct effector function, and latently infected tissue cells, is less disturbed.

In what follows, we review some more recent approaches aimed at understanding the mechanisms of mCMV reactivation.

Virus reactivation in lung tissue explant cultures

The advances in CMV mutagenesis strategies (see Chapter I.3) also made it possible to colour-tag mCMV for single cell imaging of the infection. So, recently, we have described reporter viruses all coding simultaneously for *Gaussia* luciferase and the red fluorescence protein mCherry, some equipped with additional features of interest (Marquardt *et al.*, 2011). Tissue culture of precision-cut lung slices (PCLS) from lungs of mice latently infected with reporter virus allowed rapid detection of virus reactivation based on *Gaussia* luminescence in the tissue culture supernatants of few PCLS on day 4 post-explantation and microscopical detection of mCherry fluorescence in single cells (Fig. I.22.9A). The conclusion that this reactivation did not just indicate reporter gene expression but yielded infectious virus was supported by focal spread of mCherry signal and an ~20-fold increased *Gaussia* luciferase activity by day 7 post-explantation (Fig. I.22.9B), which is with comparable kinetics of reactivation events but more sensitive detection compared to virus reactivation from explanted solid lung tissue pieces measured by direct infectivity/PFU assay (Böhm *et al.*, 2009). Importantly, early detection of reactivation at the stage of single cells by mCherry fluorescence gives a chance to identify the cell type, presumably identical with the latently infected cell type, by colocalization with cell-type markers. Using fractalkine receptor-GFP (green) reporter mice latent with mCherry (red) reporter virus did not reveal colocalization of fluorescence signals in PCLS (Fig. I.22.9C), thus providing no evidence for virus reactivation in CX3CR1+ subsets of leucocytes, including pulmonary DCs and patrolling monocytes (PM). This does not exclude virus reactivation in inflammatory monocytes (IMs) known to be independent of CX3CR1 signalling. Absence of reactivation from PMs is of interest as these cells are supposed to disseminate the virus during acute infection (see the directly preceding Chapter I.21).

Although positive identification of colocalization of mCherry with cell-type markers, such as CD31 and CD146 for ECs, is waiting, this is a most promising approach, as it is based on functional reactivation instead of mere viral genome carriage.

Virus reactivation in lung tissue *in vivo*

Virus reactivation in the lungs of BALB/c mice that were latent of WT mCMV proved to be particularly instrumental in identifying reactivation as a stepwise, sequential and stochastic process (Kurz and Reddehase, 1999) (Fig. I.22.10). Specifically, latency was established in the lungs after syngeneic HCT and clearance of productive infection, and reactivation was triggered by genotoxic stress and haematoablation caused by total-body γ-irradiation. Like in earlier work in the neonatal latency model discussed above (Reddehase *et al.*, 1994), virus recurrence was observed in a proportion of the mice when the whole lungs were tested for infectivity (PFU). New insights came from the 'statistical grid approach' subdividing each lung into 18 statistical fields (physically: tissue pieces), a principle explained in more detail earlier in this chapter (recall Fig. I.22.5B). This again revealed the stochastic nature of productive reactivation (i.e. virus recurrence) in that a high proportion of pieces (15 out of 18 pieces of a statistically averaged lung or 67 out of 81 pieces tested in reality) were still found to be negative for infectious virus; that is infection had remained latent in these particular pieces, whereas directly neighbouring pieces (only 3 out of 18 or 14 out of 81) contained infectious virus. In view of the fact that in the reported set of experiments each lung piece contained ~5000 latent viral genomes prior to reactivation, full reactivation is apparently an exceedingly rare event. When tissue pieces were tested for transcripts *IE1*, *IE3*, and *M55/gB*, 5 out of 18 pieces (or 23 out of 81) did not contain any of these three transcripts, indicating still silenced MIE and M55/gB loci despite induction of productive reactivation on the level of the whole lung. Out of the 18 (or 81) pieces, 2 (or 12) expressed only *IE1*, 4 (or 15) went on to *IE3* splicing expressing *IE1* and *IE3*, and in the remaining 4 (or 17) pieces transcription proceeded to *M55/gB*, expressing *IE1* and *IE3* and *M55/gB* but still not producing infectious virus (Fig. I.22.10A). Control lungs of mice of the same group but not induced for virus reactivation contained IE1 transcripts in 12 out of 27 pieces tested, whereas no piece contained *IE3* or *M55/gB* transcripts (Kurz and Reddehase, 1999).

Importantly, this gene expression pattern observed during reactivation was strictly sequential in that *M55/gB* transcripts were never found in absence of *IE1* and *IE3* transcripts, and *IE3* transcripts were found only concomitant with *IE1* transcripts (Fig. I.22.10A). In this context it is worth noting that these data on reactivation of latent mCMV in the lungs gave no indication for 'lateral entry' into the lytic programme bypassing MIE gene expression, a mechanism that appears to exist, however, in the reactivation of latent herpes simplex virus from neurons. Specifically, reactivated viral

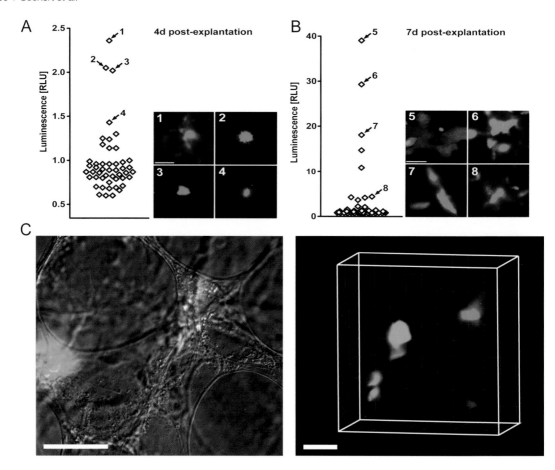

Figure I.22.9 Single-cell detection of virus reactivation in explanted precision-cut lung slices (PCLS) latently infected with recombinant mCMV dual-reporter virus encoding *Gaussia* luciferase and mCherry. *Gaussia* luciferase activity in PCLS explant culture supernatants (*n*=48; diamond symbols) was quantitated in relative luminescence units (RLU), and mCherry expressing cells in explants are shown as red fluorescence images from the arrow-marked PCLS (numbered 1–8) that displayed significant *Gaussia* luciferase activity. (A) Reporter gene activity on day 4 post-explantation. (B) Reporter gene activity on day 7 post-explantation, showing focal virus spread as evidence of productive virus reactivation. Bar markers in images 1 and 5 represent 50 µm, which applies to all eight images. (C) Lack of co-localization between GFP (green) fluorescent CX3CR1-expressing myeloid lineage cells and mCherry (red) fluorescent lung cells in a PCLS explant derived from a fractalkine receptor-GFP reporter mouse latently infected with the dual-reporter virus. Left panel: Fluorescence microscopic image. Right panel: Computed 3D-reconstruction. Bar markers represent 40 µm. Reproduced in a rearranged form with permission of Journal of General Virology from 'Single cell detection of latent cytomegalovirus reactivation in host tissue'. Marquardt *et al.*, *92*, 1279–1291; permission conveyed through Copyright Clearance Centre, Inc.

gene expression in explanted ganglia was found to be disordered rather than sequentially ordered (Du *et al.*, 2011), and exit from latency *in vivo* in trigeminal ganglion neurons following stress is initiated by stochastic derepression of the VP16 promoter and consequent expression of VP16, a virion protein with transactivating function normally expressed late in the viral lytic programme (Thompson *et al.*, 2009). Though mCMV currently appears not to encode a transactivator of lytic gene expression for bypassing MIE genes in the exit from latency, the example of herpes simplex virus VP16

shows that viral gene expression during latency can be induced by signals in absence of MIE gene expression, thus not complying with the IE-E-L cascade of lytic gene expression. This principle may apply also to CMV latency in support of the model proposed in Fig. I.22.6.

Altogether, the reactivation experiment led to three major conclusions: (i) even under conditions favouring reactivation, only very few viral genomes actually enter and complete the productive cycle; (ii) reactivation proceeds along the temporally directional, sequential programme of productive cycle gene expression, and

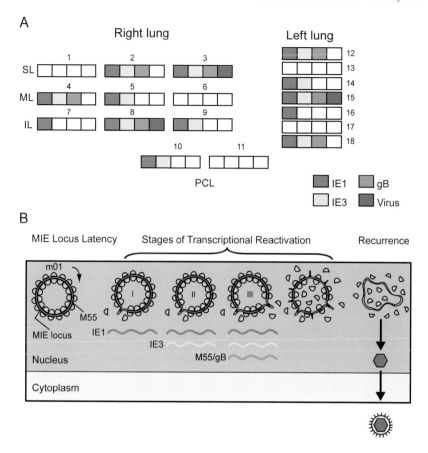

Figure I.22.10 Stochastic nature of directional reactivation of viral gene expression along the kinetic IE-E-L path of the productive viral cycle induced in latently infected lungs by genotoxic and immune suppressive haematoablative treatment. (A) Colour-coded results for presence of the indicated transcripts (RT-qPCR data) and of infectious virus (PFU data) in 18 individual pieces of a statistically averaged lung, based on data compiled from several lungs. SL, superior lobe; ML, middle lobe; IL, inferior lobe; PCL, postcaval lobe, in schematic anatomical view. (B) Model illustrating and interpreting the results.

(iii) there must exist molecular checkpoints that can halt the reactivation independent of innate or adaptive cellular immune control ('checkpoint model'; Fig. I.22.10B).

Transplant models for reactivation

Reactivation of latent HCMV is a significant infectious complication in SOT and HCT (Razonable and Paya, 2002; Pescovitz, 2006; Freeman, 2009), and is frequently associated with episodes of acute graft rejection or infection (Meyers et al., 1986; Fietze et al., 1994; Lao et al., 1997; Prösch et al., 1998; Razonable et al., 2001; Dmitrienko et al., 2009; Freeman, 2009) (see also clinical Chapters II.13, II.14, and II.16). Owing to the species-specificity of CMVs (see Chapter I.18), murine models have been developed to study reactivation of mCMV in the context of transplantation (Rubin

et al., 1984; Hummel et al., 2001; Gosselin et al., 2005; Forster et al., 2009). Forster and colleagues showed that reactivation of virus can be induced by transplanting skin from allogeneic, naïve mice onto latently infected recipients. Syngeneic transplantation was ineffective in inducing reactivation, and reactivation in allogeneic recipients could be prevented by treating recipients with cortisol, an inhibitor of NF-κB activation (Forster et al., 2009).

The most serious complications of HCMV infection arise in transplant patients when a seronegative recipient receives an organ from a seropositive donor (D^+R^-). An mCMV D^+R^- kidney transplant model for reactivation, in which kidneys from latently infected BALB/c mice are transplanted into immunocompetent, naïve mice, has been developed (Hummel et al., 2001). These studies show that reactivation of MIE gene expression occurs within 2 days when latently infected BALB/c (*H-2^d*) kidneys are transplanted into allogeneic C57BL/6 (*H-2^b*) recipients, but not when kidneys are transplanted into genetically identical, syngeneic BALB/c mice.

Mechanisms of MIE transcriptional reactivation induced by transplantation

Allogeneic transplantation induces an inflammatory immune response due to T-cell recognition of a non-tolerated set of cellular peptides presented by foreign MHC molecules (Le Moine *et al.*, 2002; Walsh *et al.*, 2004). Hundreds of genes are up-regulated in kidney allografts as early as at day 2 post transplantation, including genes encoding the inflammatory cytokines TNF-α and IFN-γ (Hummel *et al.*, 2009). These cytokines are important in inducing expression of additional cytokines, chemokines, MHC molecules, adhesion molecules, and transcription factors that are involved in innate and adaptive immunity in response to allogeneic transplantation ultimately leading to rejection (Famulski *et al.*, 2006; Hummel *et al.*, 2009; Ishii *et al.*, 2010). Although syngeneic transplantation also induces an inflammatory response due to ischaemia and reperfusion injury (de Groot and Rauen, 2007), the response is transient and largely undetectable 2 days after transplant (Hummel *et al.*, 2001; El-Sawy *et al.*, 2004; Schenk *et al.*, 2008).

Reactivation of mCMV MIE gene expression in allogeneic transplants correlated with increased expression of inflammatory cytokines, including TNF-α and IFN-γ, and activation of transcription factors NF-κB and AP-1 (Hummel *et al.*, 2001). Recent studies show that transplantation of latently infected kidneys into allogeneic recipients also results in changes in viral chromatin within 2 days after transplant that are consistent with gene desilencing (Liu, X.F., unpublished observations). These include recruitment of phosphorylated RNA polymerase II to the MIE promoter and coding regions, changes in histone modifications, and recruitment of NF-κB subunits p65 and p50 to the MIE enhancer. However, TNF-α is not indispensable for reactivation of MIE gene expression in latent kidneys transplanted into allogeneic recipients (Zhang *et al.*, 2009), indicating redundance in signalling to the MIE enhancer (see also Chapter I.10).

Enhanced MIE gene expression was observed within two days in kidneys transplanted into allogeneic recipients. The observation that allogeneic, but not syngeneic transplantation induced this expression suggests that recognition of foreign antigens by recipient T-cells is required. The observations that expression of inflammatory cytokines is detected in kidneys transplanted into allogeneic, but not syngeneic recipients,

and that TNF-α expression is largely derived from the recipient in this model (Hummel *et al.*, 2001; Zhang *et al.*, 2009) supports the hypothesis that the inflammatory response is primarily due to allorecognition, rather than injury to the kidney itself. Since the *de novo* activation/priming of antigen-inexperienced, naïve T-cells takes a few days, it is unlikely that reactivation observed on day 2 results from pre-existing alloreactive effector T-cells (EECs and SLECs; see above) generated from naïve precursors. Rather, a rapid alloreactive response is based on TCR recognition promiscuity of 'cross-reactive' memory T-cells derived during 'immune history' from previous encounters with unrelated environmental or pathogen-derived antigens, a phenomenon known as 'heterologous immunity' (Selin *et al.*, 2006). Indeed, recent studies have shown that alloreactive memory T-cells are present in naïve mice, infiltrate allografts within one day after transplant, and secrete inflammatory cytokines (Schenk *et al.*, 2008). We thus hypothesize that allo-cross-reactive memory T-cells, in particular tissue patrolling T$_{EM}$ that are capable of exerting effector functions immediately and on-site, trigger reactivation of MIE gene expression in latently infected kidneys transplanted into allogeneic recipients. However, in latently infected kidneys transplanted into immunocompetent, allogeneic recipients, MIE gene expression disappeared by day 5, and expression of genes associated with later stages of viral replication was not observed at all (Hummel *et al.*, 2001). This finding could relate to the 'checkpoint model' discussed above for latently infected lungs in which virus reactivation was triggered by global cellular immunoablation but nevertheless rarely proceeded to recurrence of infectious virus (Fig. I.22.10). So, the allostimulus alone might not be sufficient to reactivate the full productive cycle. Alternatively, or maybe in addition, failure to achieve complete reactivation may result from rapid alloimmune rejection of foreign cells combined with the slow viral replication cycle. An argument against this explanation, however, is the very low number of latently infected cells in an outnumbering amount of uninfected cells likely to competitively inhibit the allorecognition of the latently infected cells, unless we propose that latently infected allogeneic cells selectively attract the alloreactive cells.

Reactivation of virus in immunocompromised transplant recipients

An SOT model for full reactivation/recurrence of infectious virus has been developed using NOD. Cg-*Prkdc^{scid}IL2rg^{tm1Wjl}*/Szj (NSG) mice (Shultz *et al.*, 2005) as recipients of latently infected kidneys (Li *et al.*, 2012). These mice are deficient in adaptive immunity due to the presence of the *scid* mutation in the *Prkdc* (protein kinase, DNA activated, catalytic polypeptide) gene and in NK cell function due to targeted mutation of the IL-2 receptor gamma chain, the common gamma chain γ_C that is shared by several interleukin receptors, including those for IL-2, 4, 7, 9, 15, and 21. The alloimmune response is therefore blunted in these mice due to deficiency in T, B, and NK cells, but is not entirely absent, due to leakiness of the *scid* mutation (Bosma *et al.*, 1988; Carroll *et al.*, 1989; Nonoyama *et al.*, 1993). NSG mice have an *H-2^{g7}* haplotype, and therefore have a partial mismatch with BALB/c (*H-2^d*) mice in the MHC region. Transplantation of latently infected kidneys into NSG recipients results in a slow and sporadic recurrence of infectious virus, which spreads from the donor kidney to organs of the recipient (Li *et al.*, 2012). Reactivation was observed in most mice by 6 weeks after transplant. Thus, reactivation of latent virus can occur in the absence of a robust alloimmune response, pretty much in line with the sporadic recurrence observed in the HCT lung latency model after immunoablative treatment (discussed above). The mechanisms of reactivation in the SOT NSG model, and the connection between this model and the previously described model of transcriptional reactivation of MIE gene expression induced by transplantation of latently infected kidneys into allogeneic, immunocompetent mice, are currently under investigation.

The sepsis model for reactivation

There is a decades old association between bacterial sepsis and CMV reactivation that was first observed in burn patients (Kagan *et al.*, 1985). Subsequent work by Prösch and colleagues confirmed this observation in both non-immunosuppressed and transplant patients with sepsis, both of whom had elevated TNF-α levels and a high incidence of reactivation (Docke *et al.*, 1994; Fietze *et al.*, 1994). This connection has since been corroborated by a number of subsequent investigations in humans (Mutimer *et al.*, 1997; Cook *et al.*, 1998; Heininger *et al.*, 2001, 2011; von Müller and Mertens, 2008; Kalil and Florescu, 2009; Kalil *et al.*, 2010) as well as mice (Cook *et al.*, 2002). In the following section we will outline how sepsis influences epigenetic regulation, MIEP stimulation, and host immunity to allow replicative viral reactivation.

Sepsis induced epigenetic regulation of MIE gene expression

How sepsis might influence epigenetic regulation of latent CMVs is currently unknown. It is becoming increasingly clear that sepsis exerts widespread influence on epigenetic regulation of host DNA (reviewed in Cornell *et al.*, 2010; Li and Alam, 2011). One could infer from its association with CMV reactivation that sepsis might also favourably influence acetylation or deacetylation at the MIE promoter. This hypothesis is supported by a single report from experiments using the HCMV-MIE promoter-enhancer (see Chapter I.10) in a gene delivery system. In this system, MIE transcription becomes repressed *in vivo* by cellular factors causing loss of target gene expression (Loser *et al.*, 1998). Interestingly, systemic LPS treatment temporarily reverses this MIEP suppression and restores transcription of the delivered gene. These authors confirmed that LPS does not influence MIE locus methylation, but did not study acetylation. It is also possible that LPS induces more transcription activators to bind the stochastically accessible MIEP. Further study will be required therefore to determine if sepsis can directly influence epigenetic regulation by promoting viral chromatin acetylation/deacetylation at the MIE locus.

Sepsis induced inflammatory stimulation of the MIEP

The clinical association between bacterial sepsis and CMV reactivation was confirmed *in vivo* using the mCMV model (Cook *et al.*, 2002), but the triggering mechanism remained unclear. Early work from the Prösch lab was foundational in defining this mechanism. Bacterial sepsis is a strong inducer of TNF-α, and Prösch's group first showed *in vitro* that TNF-α is stimulatory to the HCMV-MIEP (Stein *et al.*, 1993). Subsequent *in vitro* studies confirmed that such stimulation was secondary to NF-κB activation (Prösch *et al.*, 1995). Using the murine CMV model, it has since been confirmed *in vivo* that transcriptional reactivation of mCMV MIE gene expression can be induced in lungs of latent mice in a dose-dependent manner by treatment with TNF-α (Hummel *et al.*, 2001; Simon *et al.*, 2005, 2007). Using a slightly different model, it has also been shown that replicative reactivation of latent virus can be induced by sepsis-associated cytokines TNF-α and IL-1β (Cook *et al.*, 2006).

Within the complex host response to sepsis, one major mechanism of pathogen recognition is activation of pattern recognition receptor (PRR) systems such as toll like receptors (TLRs). TLR signalling is known to be a strong activator of NF-κB, and therefore was a likely candidate for MIEP activation. There are somewhat limited data that support this TLR signalling/reactivation hypothesis. The clearest evidence comes from studies of HCMV-MIEP showing that *in vitro* stimulation of TLR-4 and TLR-9 activate the MIEP through an NF-κB/c-Jun mechanism (Lee *et al.*, 2004). *In vivo* data on TLR-signalling influencing MIEP activity are lacking, but it is clear that LPS binding to TLR-4 can reactivate mCMV from latency (Cook *et al.*, 2006). Thus, similar to transplant models, recognition of bacterial antigens by PRRs might trigger transcription regulated by the MIEP.

Bacteria may also stimulate CMVs outside of the septic condition. As described in detail earlier in this chapter, mCMV infection causes CD8[+] T-cell 'memory inflation', a phenomenon thought to be consequent to transcriptional leak. Interestingly, germ-free mice do not develop such expansion of the memory pool during mCMV latency (Tanaka *et al.*, 2007). This observation suggests that host microbiota are somehow 'tickling' the CMV-specific CD8[+] T-cell niche, but where and how exactly this occurs is unknown. Tanaka and colleagues (2007) speculated that CMV-specific memory T-cells cross-recognize bacterial antigens in the commensal bacterial flora due to TCR promiscuity, which repeatedly pushes 'heterologous immunity' (Selin *et al.*, 2006) expanding the memory pool, similar to what we have discussed above for allo-crossreactive memory cells. However, this hypothesis does not conclusively explain why the phenomenon of 'memory inflation' does not apply to all pathogens or even all antigens; why should only CMV-specific memory cells be able to cross-recognize antigens of the indigenous microbiota? Alternatively, we would rather propose that the presence of commensal bacteria may provide a constant low-level inflammatory stimulus that causes sporadic TEL activity and associated antigenic peptide presentation. Differences in inflammatory mediators or TLR expression between specific pathogen-free and germ-free mice were not detectable by mRNA array in the lungs (Tanaka *et al.*, 2007), but viral MIE transcription and TNF-α levels were conspicuously not reported. It will be extremely interesting to evaluate *IE1* TEL expression to determine if stochastic MIE locus desilencing is reduced or absent in latently infected germ-free mice.

Sepsis induced immune suppression

Unlike in murine models of sepsis-induced CMV reactivation, not all immune competent septic patients reactivate virus during their illness (reviewed in Kalil and Florescu, 2009). This is no doubt a consequence of their underlying individually different viral load based on personal infection history (Reddehase *et al.*, 1994) and virus variants involved, magnitude of their pre-existing CMV-specific immunity, and the strength of the septic stimulus that they receive. Because sepsis induces a variable period of compensatory immune suppression, replicative reactivation may ultimately be determined by the degree to which that septic challenge impacts host immunity (Hotchkiss *et al.*, 2009). It stands to reason that – even if all other parameters were equal – two challenges that differ in the magnitude of induced immune compromise might have substantially different virus reactivation outcomes. This is actually supported by clinical data that suggest that some septic patients have their virus reactivation events intercepted by the immune system, and therefore never go on to full replicative reactivation (von Müller *et al.*, 2007).

Given the importance of CD8[+] T-cells to control of infection and likely reactivation, we were interested to study the influence of sepsis on CMV-specific immunity. To this end, we have monitored CMV-specific T-cells in latently infected mice after an LPS stimulus. We found that LPS triggered a contraction of all CD8[+] CMV-specific immunity, both inflationary and non-inflationary cells (Campbell *et al.*, 2012). This contraction was followed by transcriptional reactivation. Eventually, there was recovery of CD8[+] T-cells recognizing the immunodominant 'memory inflation'-inducing epitopes IE1/m123 and m164, followed by resumption of latency. Contraction of CD8[+] T-cell memory from TLR-3 signalling has been previously shown to be mediated by type-I interferons (McNally *et al.*, 2001), which are also released in response to TLR-4 activation (Karaghiosoff *et al.*, 2003). Thus, in addition to whatever epigenetic and MIEP stimulating properties LPS has, it (and very likely bacterial sepsis) undoubtedly influences host immunity to allow reactivation to proceed.

Concluding thoughts and perspectives

In this chapter, different groups working on mCMV latency have reviewed findings in their specific fields of expertise and have tried to bring down their views to a common denominator. While writing, what became more and more clear as a consensus is the key importance of epigenetic events of viral gene silencing and desilencing in virus latency and reactivation, respectively. What remained in lively discussion is to what extent CMV is an active partaker in these processes, one that has evolved specific functions during co-speciation with its host to survive by establishing and maintaining latency and reactivate for host-to-host transmission in response to inflammation, or if the host cell and its signalling pathways in response to microenvironmental cytokines decide it all.

Virus-host relation in latency

In one view, CMV latency is considered primarily a result of the host cell's intrinsic defence, silencing the incoming viral genomes by transcriptional repressors in nuclear domains and packaging the 'foreign' DNA in a condensed, higher-order chromatin-like structure to prevent cytopathogenic productive infection. The virus, in counteraction, has adopted a strong enhancer element, the MIE enhancer, for recruiting transcription factors to enhance the transcription of the transactivatory MIE genes, and has adopted mechanisms to dissolve the cell's repressive domains for initiating lytic infection resulting in viral progeny (Tang and Maul, 2006; Maul, 2008) (see also Chapters I.9–I.11). Accordingly, viral latency might be interpreted as the virus's incompetence in evading the host cell's intrinsic defence, at least in certain cell types, dependent on parameters such as differentiation stage, metabolic stage, and cell cycle stage or combinations of those. If so, we should not expect that CMVs have evolved genes particularly dedicated to establish latency, i.e. genes that actively assist the host cell in silencing the viral genome and/or in keeping it silenced 'on purpose'.

Besides activating MIE genes, CMVs can also repress them. Autoregulatory transcriptional repression exerted by the essential viral IE2 (murine IE3) transactivator is the prominent example (see Chapter I.10). Nonetheless, no one would seriously propose that it were IE2's primary evolutionary dedication to help establish latency; – rather, repressive autoregulation may be important for a coordinated moving forward along the IE-E-L path. One can speculate that under circumstances of transcriptional blockade downstream of IE2, the autoregulatory MIE gene repression by IE2 inadvertently contributes to an 'accidental latency'. Such a scenario might contribute to host-species specificity of CMVs, where MIE genes are expressed in a heterologous host cell non-permissive for completing the productive cycle (see Chapter I.18).

The strongest arguments in favour of an active role of CMV gene expression in the establishment of latency, as well as for the existence of true 'latency genes', come from the CD34+ HC culture model of HCMV latency (discussed in detail in Chapter I.20). Rapid loss of UV-inactivated viral genomes, in contrast to maintenance of viable viral genomes, served as an argument to suggest a requirement for viral gene expression in the establishment of HCMV latency (Cheung et al., 2006). However, the possibility that defective viral genomes are efficiently degraded remains a caveat. Closest to fulfil the definition of a gene causally involved in the establishment of latency is HCMV UL138, as a UL138 deletion mutant productively replicated in CD34+ HCs, whereas WT virus established latency

by a mechanism operating downstream of MIE locus transcription and possibly involving UL138-region transcripts rather than the UL138 protein (Goodrum et al., 2007; Petrucelli et al., 2009). Similarly, HCMV UL133 appears to suppress virus replication in CD34+ HCs (Umashankar et al., 2011). The strength of these models is that a 'latency gene' is defined positively as one in whose presence latency is established in a certain cell type, whereas its absence permits productive infection of the very same cell type.

Is this really ultimate proof for viral 'latency genes' that have specifically evolved under a selection advantage gained by establishing latency? As long as we do not know all functions of UL138 and UL133 in the viral biology in vivo, it is difficult to definitely exclude the possibility that they play an essential role for the virus being able to replicate efficiently in specific tissue cell types, to evade the host's immune defence, or to perceive signals for sensing the microenvironment; – and that this all occurs at the expense of having lost the capacity to replicate in HCs as a sort of evolutionarily tolerated side effect. Indeed, UL138's proposed role in potentiating TNF receptor 1 (TNFR1) cell surface density for TNF-α signalling (Le et al., 2011) in response to inflammation (see below) already indicates an important function of UL138 in clinical strains of HCMV, which have maintained the UL138-containing ULb' region that is lost in highly attenuated cell culture strains (see Chapter I.1).

A further argument to propose an active role for CMV in establishing latency comes from the finding that latent viral genomes have apparently recruited repressive transcription factors YY1 and CBF-1 (the latter being repressive in absence of Notch ligands) to corresponding binding motifs present in several copies in MIE enhancers (R. Liu et al., 1994; X.F. Liu et al., 2008, 2010). So, why has the enhancer adopted these motifs if not for silencing? However, YY1 and CBF-1 (the latter in presence of Notch ligands) can also be activatory. The question thus remains whether the Yin or the Yang was the driving force in CMVs' evolution. Deleting the respective motifs in MIE enhancers may give the answer. If we take Yin and Yang by their meaning in Chinese philosophy, we should expect to find their role in balancing MIE enhancer activity.

Virus–host relation in reactivation

There is consensus that desilencing of viral genes, kick-starting from the MIE locus, in concert with immune suppression is key to virus reactivation, and that inflammatory conditions are decisively involved. An idea linking promiscuous gene expression in latency to reactivation may be that the stochastic opening of the MIE locus generates only subthreshold levels of MIE transcripts insufficient to generate the essential *IE3* transcript by alternative splicing of the *IE1/3* precursor RNA, whereas intensified signalling to the MIE enhancer by inflammatory cytokines increases the amount of *IE1/3* transcript up to a 'flash point' at which *IE3* is spliced to initiate the lytic cycle. This hypothesis is supported by the absence of *IE3* transcripts in latently infected lungs expressing low amounts of *IE1* transcript (Kurz *et al.*, 1999; Grzimek *et al.*, 2001) and by presence of *IE3* transcripts and elevated amounts of *IE1* transcript after TNF-α signalling (Simon *et al.*, 2005). *IE3* splicing, rather than transcription of the MIE genes, is therefore regarded as the decisive checkpoint that distinguishes uncoordinated transcription in latency from the coordinated gene expression in the IE-E-L path of reactivation (reviewed in Reddehase *et al.*, 2008). If alternative promoter usage might lead to a functional isoform of IE3 protein, for instance from an 'exon 5 only' transcript independent of *IE1/3* splicing, is speculative at the moment.

The question remains if we expect an existence of viral genes that have been adopted during virus–host co-speciation to facilitate reactivation specifically. Though genes and regulatory DNA elements that determine viral fitness in acute productive infection will almost certainly also support reactivation, we would not define those as 'virus reactivation'-specific. Rather, latent virus may take the opportunity to enter the productive cycle whenever the stranglehold of the host's intrinsic and immune defences wanes – so, CMVs act as opportunists.

Desilencing of the MIE locus by opening of the viral chromatin in the course of cell differentiation and/or cytokine-induced transcription factor-mediated signalling to the MIE enhancer is the initial spark for reactivation by providing essential transactivator proteins (Stinski and Isomura, 2008); though this is a necessary condition, it is not a sufficient condition. Proceeding further along the entire viral replication cycle to the eventual release of infectious virus requires (*i*) derepression of all essential viral gene loci, for instance under inflammatory signalling conditions – alloreactive immune response and sepsis discussed herein as examples – and (*ii*) compromise of antiviral immune control that otherwise would interrupt reactivation as soon as viral peptides become presented. Predictably, from the viewpoint of the virus, the reactivation process is most vulnerable at stages preceding the expression of viral 'immune evasion' genes (see Chapter II.17), and expression of early and late viral cell death inhibitors (Chapter I.15)

also probably are critical steps in moving forward to virus production. This is particularly plausible if we consider that TNF-α, the main driver of MIE locus desilencing for initiating reactivation simultaneously is a major inducer of cell death terminating virus production (Chapter I.15). Thus, both the viral 'immune evasion' genes, though operating with some leakiness (see Chapter II.17), and the cell death inhibitory genes may have evolved not only for supporting productive acute infection but also for facilitating productive reactivation (for a summary of the concept, see Fig. I.22.11).

An argument to propose CMVs' evolutionary adaptation to inflammatory conditions by responding with reactivation is the presence of multiple NF-κB and AP-1 binding sites in the enhancer, both involved in TNFR–mediated signalling (Barnes and Karin, 1997). Notably, as discussed above, and in more detail in Chapter I.20, HCMV UL138 may contribute to HCMV's responsiveness to TNF-α by preventing the down-regulation of its receptor TNFR1 on the surface of infected cells (Le *et al.*, 2011; Montag *et al.*, 2011). However, besides being a key cytokine in inflammation inducing virus reactivation, TNF-α is also an integral part of a cytokine network induced early in acute infection (see Chapters II.11 and II.12). So, it remains open if the driver in the evolution of viral determinants involved in TNFR signalling pathways has been acute infection or reactivation; unfortunately, this will be difficult to distinguish. What stays is the notion that CMVs are particularly well equipped to respond to inflammatory signals with MIE gene activation, the kick-starter of the productive viral cycle.

Linking latency to 'memory inflation'

Whilst reactivation is characterized by a coordinated, progressive transcription along the kinetic IE-E-L path (sketched in Fig. I.22.10 and Fig. I.22.11), promiscuous gene desilencing does not follow the IE-E-L path but occurs in the latent viral genome locally and stochastically (sketched in Fig. I.22.6), giving rise to TELs. Though a specific TEL may be expressed from a very low proportion of latent viral genomes and only in a few latently infected cells at any point in time, fluctuating expression can occur for long periods. It is proposed that genes in the latent viral genome and silent cellular genes obey the same rules. Cellular heterogeneity arising from stochastic expression of genes, proteins, and metabolites is increasingly recognized as a fundamental principle of cell biology calling for single cell analysis of biological processes (Wang and Bodovitz, 2010).

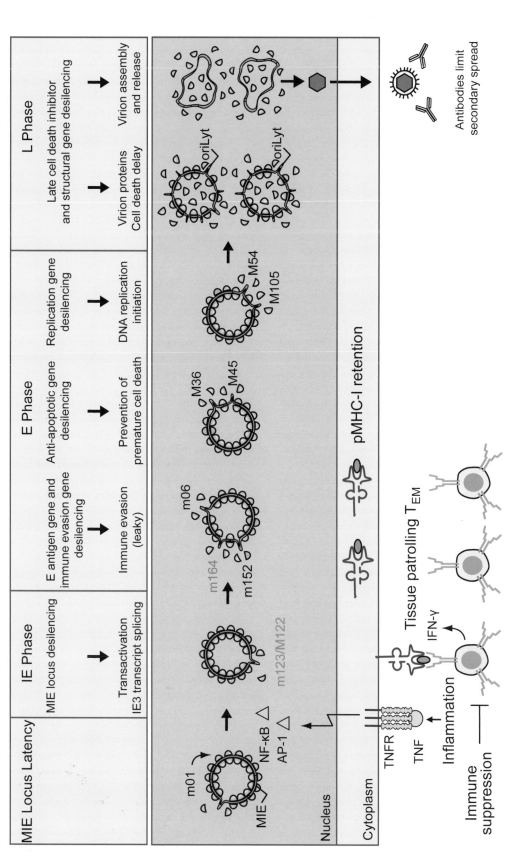

Figure I.22.11 Concluding model for triggers of reactivation and proposed critical steps in the transition from virus latency to recurrence of infectious virus.

Accounted for by the low number of TEL events at any time point, little is known about TEL-derived viral proteins and functional consequences of their uncoordinated expression in latently infected cells. Experimentally, transgenic DTR expressed in place of the mCMV *IE2* TEL renders latently infected cells susceptible to cell-death induced by diphtheria toxin; so, viral TEL-derived proteins expressed out of their productive cycle context could also exert functions of currently undefined roles.

Particularly intriguing is the expression of viral TELs that code for antigens, thus linking latency to immunity. As a related phenomenon, a link between stochastic cellular gene expression and immunity is discussed as a mechanism of self antigen presentation in thymic precursor T-cell selection for central tolerance (Derbinski *et al.*, 2001; Klein *et al.*, 2009; Tykocinski *et al.*, 2010). The prototype of a molecularly documented antigenic viral TEL coding for a known antigenic peptide is mCMV *IE1*. This finding, in concert with the accumulation of IE1 epitope-specific short-lived CD8+CD62L^{low}KLRG1^{high} effector cells (SLECs) in latently infected lungs, implies presentation of the IE1 peptide and links latency to 'memory inflation', which is a focus of discussion in this chapter. It is inferred from this documented example that other 'memory inflation'-associated antigenic peptides indicate TEL activity from the corresponding viral genes, though formal proof awaits molecular detection of the respective TELs. As discussed in great detail in this chapter, MHC polymorphism selects TELs that can lead to antigenic peptides, so that each individual is expected to show a unique pattern of 'memory inflation'-inducing antigenic peptides during latency, even if TEL activity were the same in different individuals. Obviously, genetic differences between viral strains that translate into different viral proteomes may contribute as well to patterns of antigenic peptides in virus–host pairs. In addition, besides being expressed at all during latency as the *conditio sine qua non*, a number of conditions must be fulfilled for the presentation of peptides derived from vanishingly low amounts of protein, with efficient processing and optimal MHC-I binding ranking foremost.

It should be emphasized that 'memory inflation' induced by promiscuous and stochastic TEL activity independent of the IE-E-L path does of course not exclude an additive contribution by antigenic peptides expressed in the course of a complete or interrupted reactivation along the IE-E-L path. The conceptual difference is that promiscuous gene expression, though it affects only few viral genes at any time, occurs at a very low level constantly and should lead to a more or less continuous 'memory inflation', whereas incomplete

transcriptional reactivation along the productive cycle should give less regular 'pulses' of peptide presentation with a probability decreasing from IE to L gene-derived epitopes. As we have discussed earlier in this chapter, 'memory inflation' specific for the IE3 epitope in C57BL/6 mice differs from all other known 'memory inflation'-inducing epitopes in that it depends on presentation by profAPCs instead of on direct antigen presentation by host tissue cells. IE3 thus might be the prototype of a 'reactivation-associated' driver of 'memory inflation'.

It is an open question why SLECs do not appear to reduce latent viral DNA load. Non-cytolytic effector function of SLECs, metabolic resistance of latently infected cells to cell death by cytolysis, and mechanisms of replication of latent viral episomes by the host cell concomitant with cell division are discussed possibilities. Expression of antigenic transgenes for inducing 'memory inflation' specific for any kind of unrelated pathogen and also tumours has a tremendous potential for use of CMVs as new generation vaccine vectors to which book Chapter II.21 is dedicated specifically.

Perspectives

We have begun to understand the precise mechanisms of reactivation on the level of MIE locus desilencing, but much less is known of how molecular checkpoints downstream of MIE gene transactivator expression in the viral productive cycle are hurdled during reactivation. Deciphering the precise conditions might help to develop a therapeutic intervention aimed at preventing CMV reactivation in transplant patients. The now available reporter viruses allowing the early visualization of virus reactivation in viable single cells of explanted tissue slices (Marquardt *et al.*, 2011) open the opportunity to study reactivation-promoting conditions in the mouse model by single cell 'omics'.

Acknowledgements

We would like to thank members of our laboratories past and present whose work over the years has contributed to this chapter. We would also like to apologize to colleagues in the field whose work may have escaped our attention or did not fit to the chapter's focus and is hopefully appreciated in cross-referenced other chapters of this book.

The work of the Mainz group (C.K.S., M.G., J.K.B., K.F., N.A.W.L. and M.J.R.) was funded by the Deutsche Forschungsgemeinschaft, Collaborative Research Grant (Sonderforschungsbereich) 490, individual project E2 'Immunological control of latent cytomegalovirus infection', the young investigators program MAIFOR (C.K.S. and N.A.W.L.) and the Gender Equality Program of the Research Focus Immunology (J.K.B.) at the University Medical Center of the Johannes Gutenberg-University Mainz, as well as by the Clinical Research Group KFO 183 of the Deutsche Forschungsgemeinschaft (M.J.R. and N.A.W.L.). Work by C.H. Cook was funded by NIH R01 GM066115 'Bacterial sepsis and reactivation of latent cytomegalovirus'. A. Angulo was supported by the Spanish Ministry of Science and Innovation (SAF2011–25155). Work by M. Messerle was funded by the Deutsche Forschungsgemeinschaft, Collaborative Research Grant (Sonderforschungsbereich) 587, individual project A13 'Mechanisms of latency and reactivation of cytomegalovirus in the lungs'. M.A. Hummel and M.I. Abecassis were supported by the NIH, grant NIH R21 AI097867–01.

References

Ahn, J.H., and Hayward, G.S. (1997). The major immediate-early proteins IE1 and IE2 of human cytomegalovirus colocalize with and disrupt PML-associated nuclear bodies at very early times in infected permissive cells. J. Virol. *71*, 4599–4613.

Ahn, J.H., and Hayward, G.S. (2000). Disruption of PML-associated nuclear bodies by IE1 correlates with efficient early stages of viral gene expression and DNA replication in human cytomegalovirus infection. Virology *274*, 39–55.

Ahn, J.H., Brignole, E.J., 3rd, and Hayward, G.S. (1998). Disruption of PML subnuclear domains by the acidic IE1 protein of human cytomegalovirus is mediated through interaction with PML and may modulate a RING finger-dependent cryptic transactivator function of PML. Mol. Cell. Biol. *18*, 4899–4913.

Allan, R.S., Smith, C.M., Belz, G.T., van Lint, A.L., Wakim, L.M., Heath, W.R., and Carbone, F.R. (2003). Epidermal viral immunity induced by CD8alpha+ dendritic cells but not by Langerhans cells. Science *301*, 1925–1928.

Angulo, A., Ghazal, P., and Messerle, M. (2000). The major immediate-early gene ie3 of mouse cytomegalovirus is essential for viral growth. J. Virol. *74*, 11129–11136.

Arens, R., Loewendorf, A., Redeker, A., Sierro, S., Boon, L., Klenerman, P., Benedict, C.A., and Schoenberger, S.P. (2011). Differential B7-CD28 costimulatory requirements for stable and inflationary mouse cytomegalovirus-specific memory CD8 T-cell populations. J. Immunol. *186*, 3874–3881.

Balthesen, M., Messerle, M., and Reddehase., M.J. (1993). Lungs are a major organ site of cytomegalovirus latency and recurrence. J. Virol. *67*, 5360–5366.

Balthesen, M., Dreher, L., Lucin, P., and Reddehase, M.J. (1994). The establishment of cytomegalovirus latency in organs is not linked to local virus production during primary infection. J. Gen. Virol. *75*, 2329–2336.

Barnes, P.J., and Karin, M. (1997). Nuclear factor-kappaB: a pivotal transcription factor in chronic inflammatory diseases. N. Engl. J. Med. *336*, 1066–1071.

Belz, G.T., Smith, C.M., Eichner, D., Shortman, K., Karupiah, G., Carbone, F.R., and Heath, W.R. (2004). Cutting edge: conventional CD8 alpha+ dendritic cells are generally involved in priming CTL immunity to viruses. J. Immunol. *172*, 1996–2000.

Bevan, I.S., Sammons, C.C., and Sweet, C. (1996). Investigation of murine cytomegalovirus latency and reactivation in mice using viral mutants and the polymerase chain reaction. J. Med. Virol. *48*, 308–320.

Blake, W.J., Kaern, M., Cantor, C.R., and Collins, J.J. (2003). Noise in eukaryotic gene expression. Nature *422*, 633–637.

Böhm, V., Seckert, C.K., Simon, C.O., Thomas, D., Renzaho, A., Gending, D., Holtappels, R., and Reddehase, M.J. (2009). Immune evasion proteins enhance cytomegalovirus latency in the lungs. J. Virol. *83*, 10293–10298.

Bosma, G.C., Fried, M., Custer, R.P., Carroll, A., Gibson, D.M., and Bosma, M.J. (1988). Evidence of functional lymphocytes in some (leaky) scid mice. J. Exp. Med. *167*, 1016–1033.

Boyman, O., Letourneau, S., Krieg, C., and Sprent, J. (2009). Homeostatic proliferation and survival of naïve and memory T-cells. Eur. J. Immunol. *39*, 2088–2094.

Busche, A., Angulo, A., Kay-Jackson, P., Ghazal, P., and Messerle, M. (2008). Phenotypes of major immediate-early gene mutants of mouse cytomegalovirus. Med. Microbiol. Immunol. *197*, 233–240.

Busche, A., Marquardt, A., Bleich, A., Ghazal, P., Angulo, A., and Messerle, M. (2009). The mouse cytomegalovirus immediate-early 1 gene is not required for establishment of latency or for reactivation in the lungs. J. Virol. *83*, 4030–4038.

Campbell, A.E., Cavanaugh, V.J., and Slater, J.S. (2008). The salivary glands as a privileged site of cytomegalovirus immune evasion and persistence. Med. Microbiol. Immunol. *197*, 205–213.

Campbell, J.E., Trgovcich, J., Kincaid, M., Zimmermann, P.D., Klenerman, P., Sims, S., and Cook, C.H. (2012). Transient CD8-memory contraction: A potential contributor to latent cytomegalovirus reactivation. J. Leukoc. Biol. *92*, 933–937.

Cardin, R.D., Abenes, G.B., Stoddart, C.A., and Mocarski, E.S. (1995). Murine cytomegalovirus IE2, an activator of gene expression, is dispensable for growth and latency in mice. Virology *209*, 236–241.

Cardin, R.D., Schaefer, G.C., Allen, J.R., Davis-Poynter, N.J., and Farrell, H.E. (2009). The M33 chemokine receptor homolog of murine cytomegalovirus exhibits a differential tissue-specific role during in vivo replication and latency. J. Virol. *83*, 7590–7601.

Carroll, A.M., Hardy, R.R., and Bosma, M.J. (1989). Occurrence of mature B (IgM+, B220+) and T (CD3+) lymphocytes in scid mice. J. Immunol. *143*, 1087–1093.

Cartharius, K., Frech, K., Grote, K., Klocke, B., Haltmeier, M., Klingenhoff, A., Frisch, M., Bayerlein, M., and Werner, T. (2005). MatInspector and beyond: promoter analysis based on transcription factor binding sites. Bioinformatics *21*, 2933–2942.

Chatellard, P., Pankiewicz, R., Meier, E., Durrer, L., Sauvage, C., and Imhof, M.O. (2007). The IE2 promoter/enhancer region from mouse CMV provides high levels of therapeutic protein expression in mammalian cells. Biotechnol. Bioeng. *96*, 106–117.

Cheung, A.K.L., Abendroth, A., Cunningham, A.L., and Slobedman, B. (2006). Viral gene expression during the establishment of human cytomegalovirus latent infection in myeloid progenitor cells. Blood *108*, 3691–3699.

Cheung, K.S., and Lang, D.J. (1977). Detection of latent cytomegalovirus in murine salivary and prostate explant cultures and cells. Infect. Immun. *15*, 568–574.

Choi, K., Kennedy, M., Kazarov, A., Papadimitriou, J.C., and Keller, G. (1998). A common precursor for hematopoietic and endothelial cells. Development *125*, 725–732.

Cook, C.H., Yenchar, J.K., Kraner, T.O., Davies, E.A., and Ferguson, R.M. (1998). Occult herpes family viruses may increase mortality in critically ill surgical patients. Am. J. Surg. *176*, 357–360.

Cook, C.H., Zhang, X., McGuinness, B., Lahm, M., Sedmak, D., and Ferguson, R. (2002). Intra-abdominal bacterial infection reactivates latent pulmonary cytomegalovirus in immunocompetent mice. J. Infect. Dis. *185*, 1395–1400.

Cook, C.H., Trgovcich, J., Zimmerman, P.D., Zhang, Y., and Sedmak, D.D. (2006). Lipopolysaccharide, tumor necrosis factor alpha, or interleukin-1(beta) triggers reactivation of latent cytomegalovirus in immunocompetent mice. J. Virol. *80*, 9151–9158.

Cornell, T.T., Wynn, J., Shanley, T.P., Wheeler, D.S., and Wong, H.R. (2010). Mechanisms and regulation of the gene expression response to sepsis. Pediatrics *125*, 1248–1258.

Däubner, T., Fink, A., Seitz, A., Tenzer, S., Müller, J., Strand, D., Seckert, C.K., Janssen, C., Renzaho, A., Grzimek, N.K.A., *et al.* (2010). Identification of a novel transmembrane domain mediating retention of a highly motile herpesviral glycoprotein in the endoplasmic reticulum. J. Gen. Virol. *91*, 1524–1534.

Del Val, M., Hengel, H., Hacker, H., Hartlaub, U., Ruppert, T., Lucin, P., and Koszinowski, U.H. (1991). Cytomegalovirus prevents antigen presentation by blocking the transport of peptide-loaded major histocompatibility complex class I molecules into the medial-Golgi compartment. J. Exp. Med. *176*, 729–738.

Derbinski, J., Schulte, A., Kyewski, B., and Klein, J. (2001). Promiscuous gene expression in medullary thymic epithelial cells mirrows the peripheral self. Nat. Immunol. *2*, 1032–1039.

Dmitrienko, S., Balshaw, R., Machnicki, G., Shapiro, R.J., and Keown, P.A. (2009). Probabilistic modeling of cytomegalovirus infection under consensus clinical management guidelines. Transplantation *87*, 570–577.

Docke, W.D., Prosch, S., Fietze, E., Kimel, V., Zuckermann, H., Klug, C., Syrbe, U., Kruger, D.H., von Baehr, R., and Volk, H.D. (1994). Cytomegalovirus reactivation and tumour necrosis factor. Lancet *343*, 268–269.

Dorsch-Häsler, K., Keil, G.M., Weber, F., Jasin, M., Schaffner, W., and Koszinowski, U.H. (1985). A long and complex enhancer activates transcription of the gene coding for the highly abundant immediate-early mRNA in murine cytomegalovirus. Proc. Natl. Acad. Sci. U.S.A. *82*, 8325–8329.

Du, T., Zhou, G., and Roizman, B. (2011). HSV-1 gene expression from reactivated ganglia is disordered and concurrent with suppression of latency-associated transcript and miRNAs. Proc. Natl. Acad. Sci. U.S.A. *108*, 18820–18824.

El-Sawy, T., Miura, M., and Fairchild, R. (2004). Early T-cell response to allografts occurring prior to alloantigen priming up-regulates innate-mediated inflammation and graft necrosis. Am. J. Pathol. *165*, 147–157.

Emery, V.C. (1998). Relative importance of cytomegalovirus load as a risk factor for cytomegalovirus disease in the immunocompromised host, p. 288–301. In Monographs in Virology 21: CMV-related Immunopathology, Scholz, M., Rabenau, H.F., Doerr, H.W., and Cinatl, J., Jr., eds. (Karger, Basel).

Famulski, K.S., Einecke, G., Reeve, J., Ramassar, V., Allanach, K., Mueller, T., Hidalgo, L.G., Zhu, L.F., and Halloran, P.F. (2006). Changes in the transcriptome in allograft rejection: IFN-gamma-induced transcripts in mouse kidney allografts. Am. J. Transplant. *6*, 1342–1354.

Fietze, E., Prösch, S., Reinke, P., Stein, J., Docke, W.D., Staffa, G., Loning, S., Devaux, S., Emmrich, F., and von Baehr, R. (1994). Cytomegalovirus infection in transplant recipients. The role of tumor necrosis factor. Transplantation *58*, 675–680.

Fink, A., Lemmermann, N.A.W., Gillert-Marien, D., Thomas, D., Freitag, K., Böhm, V., Wilhelmi, V., Reifenberg, K., Reddehase, M.J., and Holtappels, R. (2012). Antigen presentation under the influence of 'immune evasion' proteins and its modulation by interferon-gamma: implications for immunotherapy of cytomegalovirus infection with antiviral CD8 T cells. Med. Microbiol. Immunol. *201*, 513–525.

Forster, M.R., Bickerstaff, A.A., Wang, J.J., Zimmerman, P.D., and Cook, C.H. (2009). Allogeneic stimulation causes transcriptional reactivation of latent murine cytomegalovirus. Transplant. Proc. *41*, 1927–1931.

Freeman, R.B., Jr. (2009). The 'indirect' effects of cytomegalovirus infection. Am. J. Transplant. *9*, 2453–2458.

Gao, Z., McAlister, V.C., and Williams, G.M. (2001). Repopulation of liver endothelium by bone-marrow-derived cells. Frequency of protein Z deficiency in patients with ischaemic stroke. Lancet *357*, 932–933.

Ghazal, P., Visser, A.E., Gustems, M., Garcia, R., Borst, E.M., Sullivan, K., Messerle, M., and Angulo, A. (2005). Elimination of ie1 significantly attenuates murine cytomegalovirus virulence but does not alter replicative capacity in cell culture. J. Virol. *79*, 7182–7194.

Gold, M.C., Munks, M.W., Wagner, M., Koszinowski, U.H., Hill, A.B., and Fling, S.P. (2002). The murine cytomegalovirus immunomodulatory gene m152 prevents recognition of infected cells by M45-specific CTL but does not alter the immunodominance of the M45-specific CD8 T-cell response in vivo. J. Immunol. *169*, 359–365.

Goodrum, F., Reeves, M., Sinclair, J., High, K., and Shenk, T. (2007). Human cytomegalovirus sequences expressed in latently infected individuals promote a latent infection in vitro. Blood *110*, 937–945.

Gordon, S., Akopyan, G., Garban, H., and Bonavida, B. (2006). Transcription factor YY1: structure, function, and therapeutic implications in cancer biology. Oncogene *25*, 1125–1142.

Gosselin, J., Borgeat, P., and Flamand, L. (2005). Leukotriene B4 protects latently infected mice against murine cytomegalovirus reactivation following allogeneic transplantation. J. Immunol. *174*, 1587–1593.

de Groot, H., and Rauen, U. (2007). Ischemia-reperfusion injury: processes in pathogenetic networks: a review. Transplant. Proc. *39*, 481–484.

Groves, I.J., Reeves, M.B., and Sinclair, J.H. (2009). Lytic infection of permissive cells with human cytomegalovirus is regulated by an intrinsic 'pre-immediate-early' repression of viral gene expression mediated by histone post-translational modification. J. Gen. Virol. *90*, 2364–2374.

Grzimek, N.K., Dreis, D., Schmalz, S., and Reddehase, M.J. (2001). Random, asynchronous, and asymmetric transcriptional activity of enhancer-flanking major immediate-early genes ie1/3 and ie2 during murine cytomegalovirus latency in the lungs. J. Virol. 75, 2692–2705.

Heath, W.R., and Carbone, F.R. (2001). Cross-presentation, dendritic cells, tolerance and immunity. Annu. Rev. Immunol. 19, 47–64.

Heininger, A., Jahn, G., Engel, C., Notheisen, T., Unertl, K., and Hamprecht, K. (2001). Human cytomegalovirus infections in nonimmunosuppressed critically ill patients. Crit. Care. Med. 29, 541–547.

Heininger, A., Haeberle, H., Fischer, I., Beck, R., Riessen, R., Rohde, F., Meisner, C., Jahn, G., Koenigsrainer, A., Unertl, K., et al. (2011). Cytomegalovirus reactivation and associated outcome of critically ill patients with severe sepsis. Crit. Care 15, R77.

Henry, S.C., and Hamilton, J.D. (1993). Detection of murine cytomegalovirus immediate-early 1 transcripts in the spleens of latently infected mice. J. Infect. Dis. 167, 950–954.

Henson, D., and Strano, A.J. (1972). Mouse cytomegalovirus. Necrosis of infected and morphologically normal submaxillary gland acinar cells during termination of chronic infection. Am. J. Pathol. 68, 183–202.

Herter, S., Osterloh, P., Hilf, N., Rechtsteiner, G., Hohfeld, J., Rammensee, H.G., and Schild, H. (2005). Dendritic cell aggresome-like-induced structure formation and delayed antigen presentation coincide in influenza virus-infected dendritic cells. J. Immunol. 175, 891–898.

Holtappels, R., Pahl-Seibert, M.F., Thomas, D., and Reddehase, M.J. (2000). Enrichment of immediate-early 1 (m123/pp89) peptide-specific CD8 T-cells in a pulmonary CD62L(lo) memory-effector cell pool during latent murine cytomegalovirus infection of the lungs. J. Virol. 74, 11495–11503.

Holtappels, R., Thomas, D., and Reddehase, M.J. (2002). Two antigenic peptides from genes m123 and m164 of murine cytomegalovirus quantitatively dominate CD8 T-cell memory in the H-2 d haplotype. J. Virol. 76, 151–164.

Holtappels, R., Podlech, J., Pahl-Seibert, M.F., Jülch, M., Thomas, D., Simon, C.O., Wagner, M., and Reddehase, M.J. (2004). Cytomegalovirus misleads its host by priming of CD8 T-cells specific for an epitope not presented in infected tissues. J. Exp. Med. 199, 131–136.

Holtappels, R., Munks, M.W., Podlech, J., and Reddehase, M.J. (2006). CD8 T-cell-based immunotherapy of cytomegalovirus disease in the mouse model of the immunocompromised bone marrow transplantation recipient. In Cytomegaloviruses: Molecular Biology and Immunology, Reddehase, M.J., ed. (Caister Academic Press, Norfolk, UK), pp. 383–418.

Holtappels, R., Thomas, D., and Reddehase, M.J. (2009). The efficacy of antigen processing is critical for protection against cytomegalovirus disease in the presence of viral immune evasion proteins. J. Virol. 83, 9611–9615.

Hotchkiss, R.S., Coopersmith, C.M., McDunn, J.E., and Ferguson, T.A. (2009). The sepsis seesaw: tilting toward immunosuppression. Nat. Med. 15, 496–497.

Hsieh, J.J., Zhou, S., Chen, L., Young, D.B., and Hayward, S.D. (1999). CIR, a corepressor linking the DNA binding factor CBF1 to the histone deacetylase complex. Proc. Natl. Acad. Sci. U.S.A. 96, 23–28.

Huang, N.E., Lin, C.H., Lin, Y.S., and Yu, W.C. (2003). Modulation of YY1 activity by SAP30. Biochem. Biophys. Res. Commun. 306, 267–275.

Hummel, M., and Abecassis, M.M. (2002). A model for reactivation of CMV from latency. J. Clin. Virol. 25, S123-S136.

Hummel, M., Zhang, Z., Yan, S., DePlaen, I., Golia, P., Varghese, T., Thomas, G., and Abecassis, M.I. (2001). Allogeneic transplantation induces expression of cytomegalovirus immediate-early genes in vivo: a model for reactivation from latency. J. Virol. 75, 4814–4822.

Hummel, M., Yan, S., Li, Z., Varghese, T.K., and Abecassis, M. (2007). Transcriptional reactivation of murine cytomegalovirus ie gene expression by 5-aza-2′-deoxycytidine and trichostatin A in latently infected cells despite lack of methylation of the major immediate-early promoter. J. Gen. Virol. 88, 1097–1102.

Hummel, M., Kurian, S.M., Lin, S., Borodyanskiy, A., Zhang, Z., Li, Z., Kim, S.J., Salomon, D.R., and Abecassis, M. (2009). Intragraft TNF receptor signaling contributes to activation of innate and adaptive immunity in a renal allograft model. Transplantation 87, 178–188.

Hutchinson, S., Sims, S., O'Hara, G., Silk, J., Gileadi, U., Cerundolo, V., and Klenerman, P. (2011). A dominant role for the immunoproteasome in CD8+ T-cell responses to murine cytomegalovirus. PLoS One 6, e14646.

Hyde-DeRuyscher, R.P., Jennings, E., and Shenk, T. (1995). DNA binding sites for the transcriptional activator/repressor YY1. Nucl. Acids Res. 23, 4457–4465.

Ishii, D., Schenk, A.D., Baba, S., and Fairchild, R.L. (2010). Role of TNFalpha in early chemokine production and leukocyte infiltration into heart allografts. Am. J. Transplant. 10, 59–68.

Ishov, A.M., Stenberg, R.M., and Maul, G.G. (1997). Human cytomegalovirus immediate-early interaction with host nuclear structures: definition of an immediate transcript environment. J. Cell. Biol. 138, 5–16.

Jarvis, M.A., and Nelson, J.A. (2007). Human cytomegalovirus tropism for endothelial cells: not all endothelial cells are created equal. J. Virol. 81, 2095–2101.

Jepsen, K., and Rosenfeld, M.G. (2002). Biological roles and mechanistic actions of co-repressor complexes. J. Cell. Sci. 115, 689–698.

Jones, M., Ladell, K., Wynn, K.K., Stacey, M.A., Quigley, M.F., Gostick, E., Price, D.A., and Humphreys, I.R. (2010). IL-10 restricts memory T-cell inflation during cytomegalovirus infection. J. Immunol. 185, 3583–3592.

Jonjic, S., Mutter, W., Weiland, F., Reddehase, M.J., and Koszinowski, U.H. (1989). Site-restricted persistent cytomegalovirus infection after selective long-term depletion of CD4+ T lymphocytes. J. Exp. Med. 169, 1199–1212.

Jonjic, S., Pavic, I., Polic, B., Crnkovic, I., Lucin, P., and Koszinowski, U.H. (1994). Antibodies are not essential for the resolution of primary cytomegalovirus infection but limit dissemination of recurrent virus. J. Exp. Med. 179, 1713–1717.

Jordan, M.C. (1983). Latent infection and the elusive cytomegalovirus. Rev. Infect. Dis. 5, 205–215.

Jordan, M.C., and Mar, V.L. (1982). Spontaneous activation of latent cytomegalovirus from murine spleen explants. Role of lymphocytes and macrophages in release and replication of virus. J. Clin. Invest. 70, 762–768.

Kaern, M., Elston, T.C., Blake, W.J., and Collins, J.J. (2005). Stochasticity in gene expression: from theories to phenotypes. Nat. Rev. Genet. *6*, 451–464.

Kagan, R.J., Naraqi, S., Matsuda, T., and Jonasson, O.M. (1985). Herpes simplex virus and cytomegalovirus infections in burned patients. J. Trauma *25*, 40–45.

Kalejta, R.F. (2008). Functions of human cytomegalovirus tegument proteins prior to immediate-early gene expression. Curr. Top. Microbiol. Immunol. *325*, 101–115.

Kalil, A.C., and Florescu, D.F. (2009). Prevalence and mortality associated with cytomegalovirus infections in non-immunosuppressed ICU patients. Crit. Care Med. *37*, 2350–2358.

Kalil, A.C., Sun, J., and Florescu, D.F. (2010). The importance of detecting cytomegalovirus infections in studies evaluating new therapies for severe sepsis. Crit. Care Med. *38*, 663–667.

Karaghiosoff, M., Steinborn, R., Kovarik, P., Kriegshauser, G., Baccarini, M., Donabauer, B., Reichart, U., Kolbe, T., Bogdan, C., Leanderson, T., *et al.* (2003). Central role for type I interferons and Tyk2 in lipopolysaccharide-induced endotoxin shock. Nat. Immunol. *4*, 471–477.

Karrer, U., Sierro, S., Wagner, M., Oxenius, A., Hengel, H., Koszinowski, U.H., Phillips, R.E., and Klenerman, P. (2003). Memory inflation: continuous accumulation of antiviral CD8+ T-cells over time. J. Immunol. *170*, 2022–2029 [Correction appeared in J. Immunol. *171*, 3895 (2003)].

Karrer, U., Wagner, M., Sierro, S., Oxenius, A., Hengel, H., Dumrese, T., Freigang, S., Koszinowski, U.H., Phillips, R.E., and Klenerman, P. (2004). Expansion of protective CD8+ T-cell responses driven by recombinant cytomegaloviruses. J. Virol. *78*, 2255–2264.

Keil, G.M., Fibi, M.R., and Koszinowski, U.H. (1985). Characterization of the major immediate-early polypeptides encoded by murine cytomegalovirus. J. Virol. *54*, 422–428.

Keil, G.M., Ebeling-Keil, A., and Koszinowski, U.H. (1987a). Sequence and structural organization of murine cytomegalovirus immediate-early gene 1. J. Virol. *61*, 1901–1908.

Keil, G.M., Ebeling-Keil, A., and Koszinowski, U.H. (1987b). Immediate-early genes of murine cytomegalovirus: location, transcripts, and translation products. J. Virol. *61*, 526–533.

Kercher, L., and Mitchell, B.M. (2002). Persisting murine cytomegalovirus can reactivate and has unique transcriptional activity in ocular tissue. J. Virol. *76*, 9165–9175.

Kim, J., and Kim, J. (2009). YY1's longer DNA-binding motifs. Genomics *93*, 152–158.

Klein, L., Hinterberger, M., Wirnsberger, G., and Kyewski, B. (2009). Antigen presentation in the thymus for positive selection and central tolerance induction. Nat. Rev. Immunol. *9*, 833–844.

Klenerman, P., and Dunbar, P.R. (2008). CMV and the art of memory inflation. Immunity *29*, 520–522.

Kloetzel, P.M. (2001). Antigen processing by the proteasome. Nat. Rev. Mol. Cell Biol. *2*, 179–187.

Knuehl, C., Spee, P., Ruppert, T., Kuckelkorn, U., Henklein, P., Neefjes, J., and Kloetzel, P.M. (2001). The murine cytomegalovirus pp89 immunodominant H-2Ld epitope is generated and translocated into the endoplasmic reticulum as an 11-mer precursor peptide. J. Immunol. *167*, 1515–1521.

Ko, C.Y., Hsu, H.C., Shen, M.R., Chang, W.C., and Wang, J.M. (2008). Epigenetic silencing of CCAAT/enhancer-binding protein delta activity by YY1/polycomb group/DNA methyltransferase complex. J. Biol. Chem. *283*, 30919–30932.

Koffron, A.J., Patterson, B.K., Yan, S., Kaufman, D.B., Fryer, J.P., Stuart, F.P., and Abecassis, M.I. (1997). Latent human cytomegalovirus: a functional study. Transplant. Proc. *1997*, 793–795.

Koffron, A.J., Hummel, M., Patterson, B.K., Yan, S., Kaufman, D.B., Fryer, J.P., Stuart, F.P., and Abecassis, M.I. (1998). Cellular localization of latent murine cytomegalovirus. J. Virol. *72*, 95–103.

Korioth, F., Maul, G.G., Plachter, B., Stamminger, T., and Frey, J. (1996). The nuclear domain 10 (ND10) is disrupted by the human cytomegalovirus gene product IE1. Exp. Cell Res. *229*, 155–158.

Kouzarides, T. (2007). Chromatin modifications and their function. Cell *128*, 693–705.

Kropp, K.A., Simon, C.O., Fink, A., Renzaho, A., Kühnapfel, B., Podlech, J., Reddehase, M.J., and Grzimek, N.K.A. (2009). Synergism between the components of the bipartite major immediate-early transcriptional enhancer of murine cytomegalovirus does not accelerate virus replication in cell culture and host tissues. J. Gen. Virol. *90*, 2395–2401.

Kubat, N.J., Tran, R.K., McAnany, P., and Bloom, D.C. (2004). Specific histone tail modification and not DNA methylation is a determinant of herpes simplex virus type 1 latent gene expression. J. Virol. *78*, 1139–1149.

Kurz, S.K., and Reddehase, M.J. (1999). Patchwork pattern of transcriptional reactivation in the lungs indicates sequential checkpoints in the transition from murine cytomegalovirus latency to recurrence. J. Virol. *73*, 8612–8622.

Kurz, S.K., Steffens, H.-P., Mayer, A., Harris, J.R., and Reddehase, M.J. (1997). Latency versus persistence or intermittent recurrences: evidence for a latent state of murine cytomegalovirus in the lungs. J. Virol. *71*, 2980–2987.

Kurz, S.K., Rapp, M., Steffens, H.P., Grzimek, N.K., Schmalz, S., and Reddehase, M.J. (1999). Focal transcriptional activity of murine cytomegalovirus during latency in the lungs. J. Virol. *73*, 482–494.

Lai, E.C. (2002). Keeping a good pathway down: transcriptional repression of Notch pathway target genes by CSL proteins. EMBO Rep. *3*, 840–845.

Lao, W.C., Lee, D., Burroughs, A.K., Lanzani, G., Rolles, K., Emery, V.C., and Griffiths, P.D. (1997). Use of polymerase chain reaction to provide prognostic information on human cytomegalovirus disease after liver transplantation. J. Med. Virol. *51*, 152–158.

Le, V.T., Trilling, M., and Hengel, H. (2011). The cytomegaloviral protein pUL138 acts as potentiator of tumor necrosis factor (TNF) receptor 1 surface density to enhance Ulb'-encoded modulation of TNF-alpha signaling. J. Virol. *85*, 13260–13270.

Le May, N., Mansuroglu, Z., Leger, P., Josse, T., Blot, G., Billecocq, A., Flick, R., Jacob, Y., Bonnefoy, E., and Bouloy, M. (2008). A SAP30 complex inhibits IFN-beta expression in Rift Valley fever virus infected cells. PLoS Pathog. *4*, e13.

Le Moine, A., Goldman, M., and Abramowicz, D. (2002). Multiple pathways to allograft rejection. Transplantation 73, 1373–1381.

Lee, Y., Sohn, W.-J., Kim, D.-S., and Kwon, H.-J. (2004). NF-κB- and c-Jun-dependent regulation of human cytomegalovirus immediate-early gene enhancer/promoter in response to lipopolysaccharide and bacterial CpG-oligodeoxynucleotides in macrophage cell line RAW 264.7. Eur. J. Biochem. 271, 1094–1105.

Lemmermann, N.A.W., Gergely, K., Böhm, V., Deegen, P., Däubner, T., and Reddehase, M.J. (2010). Immune evasion proteins of murine cytomegalovirus preferentially affect cell surface display of recently generated peptide presentation complexes. J. Virol. 84, 1221–1236.

Lemmermann, N.A.W., Kropp, K.A., Seckert, C.K., Grzimek, N.K.A., and Reddehase, M.J. (2011). Reverse genetics modification of cytomegalovirus antigenicity and immunogenicity by CD8 T-cell epitope deletion and insertion. J. Biomed. Biotechnol. 2011:812742.

Li, Y., and Alam, H.B. (2011). Modulation of acetylation: creating a pro-survival and anti-inflammatory phenotype in lethal hemorrhagic and septic shock. J. Biomed. Biotechnol. 2011:523481.

Li, Z., Wang, X., Yan, S., Zhang, Z., Jie, C., Sustento-Reodica, N., Hummel, M., and Abecassis, M. (2012). A mouse model of CMV transmission following kidney transplantation. Am. J. Transplant. 12, 1024–1028.

Liu, R., Baillie, J., Sissons, J.G., and Sinclair, J.H. (1994). The transcription factor YY1 binds to negative regulatory elements in the human cytomegalovirus major immediate-early enhancer/promoter and mediates repression in non-permissive cells. Nucl. Acids Res. 22, 2453–2459.

Liu, X.F., Yan, S., Abecassis, M., and Hummel, M. (2008). Establishment of murine cytomegalovirus latency in vivo is associated with changes in histone modifications and recruitment of transcriptional repressors to the major immediate-early promoter. J. Virol. 82, 10922–10931.

Liu, X.F., Yan, S., Abecassis, M., and Hummel, M. (2010). Biphasic recruitment of transcriptional repressors to the murine cytomegalovirus major immediate-early promoter during the course of infection in vivo. J. Virol. 84, 3631–3643.

Loewendorf, A.I., Arens, R., Purton, J.F., Surh, C.D., and Benedict, C.A. (2011). Dissecting the requirements for maintenance of the CMV-specific memory T-cell pool. Viral Immunol. 24, 351–355.

Loser, P., Jennings, G.S., Strauss, M., and Sandig, V. (1998). Reactivation of the previously silenced cytomegalovirus major immediate-early promoter in the mouse liver: involvement of NFkappaB. J. Virol. 72, 180–190.

McNally, J.M., Zarozinski, C.C., Lin, M.-Y., Brehm, M.A., Chen, H.D., and Welsh, R.M. (2001). Attrition of bystander CD8 T-cells during virus-induced T-cell and interferon responses. J. Virol. 75, 5965–5976.

Marquardt, A., Halle, S., Seckert, C.K., Lemmermann, N.A., Veres, T.Z., Braun, A., Maus, U.A., Förster, R., Reddehase, M.J., Messerle, M., et al. (2011). Single cell detection of latent cytomegalovirus reactivation in host tissue. J. Gen. Virol. 92, 1279–1291.

Maul, G.G. (2008). Initiation of cytomegalovirus infection at ND10. Curr. Top. Microbiol. Immunol. 325, 117–132.

Mercer, J.A., Wiley, C.A., and Spector, D.H. (1988). Pathogenesis of murine cytomegalovirus infection: identification of infected cells in the spleen during acute and latent infections. J. Virol. 62, 987–997.

Messerle, M., Keil, G.M., and Koszinowski, U.H. (1991). Structure and expression of murine cytomegalovirus immediate-early gene 2. J. Virol. 65, 1638–1643.

Messerle, M., Bühler, B., Keil, G.M., and Koszinowski, U.H. (1992). Structural organization, expression, and functional characterization of the murine cytomegalovirus immediate-early gene 3. J. Virol. 66, 27–36.

Meyers, J.D., Flournoy, N., and Thomas, E.D. (1986). Risk factors for cytomegalovirus infection after human marrow transplantation. J. Infect. Dis. 153, 478–488.

Mohr C.A., Arapovic, J., Mühlbach, H., Panzer, M., Weyn, A., Dölken, L., Krmpotic, A., Voehringer, D., Ruzsics, Z., Koszinowski, U., et al. (2010). A spread-deficient cytomegalovirus for assessment of first-target cells in vaccination. J. Virol. 84, 7730–7742.

Montag, C., Wagner, J.A., Gruska, I., Vetter, B., Wiebusch, L., and Hagemeier, C. (2011). The latency-associated UL138 gene product of human cytomegalovirus sensitizes cells to TNF(alpha) signaling by up-regulating TNF(alpha) receptor 1 cell surface expression. J. Virol. 85, 11409–11421.

von Müller, L., and Mertens, T. (2008). Human cytomegalovirus infection and antiviral immunity in septic patients without canonical immunosuppression. Med. Microbiol. Immunol. 197, 75–82.

von Müller, L., Klemm, A., Durmus, N., Weiss, M., Suger-Wiedeck, H., Schneider, M., Hampl, W., and Mertens, T. (2007). Cellular immunity and active human cytomegalovirus infection in patients with septic shock. J. Infect. Dis. 196, 1288–1295.

Munks, M.W., Cho, K.S., Pinto, A.K., Sierro, S., Klenerman, P., and Hill, A.B. (2006). Four distinct patterns of memory CD8 T-cell responses to chronic murine cytomegalovirus infection. J. Immunol. 177, 450–458.

Munks, M.W., Pinto, A.K., Doom, C.M., and Hill, A.B. (2007). Viral interference with antigen presentation does not alter acute or chronic CD8 T-cell immunodominance in murine cytomegalovirus infection. J. Immunol. 178, 7235–7241.

Mutimer, D., Mirza, D., Shaw, J., O'Donnell, K., and Elias, E. (1997). Enhanced (cytomegalovirus) viral replication associated with septic bacterial complications in liver transplant recipients. Transplantation 63, 1411–1415.

Nitzsche, A., Paulus, C., and Nevels, M. (2008). Temporal dynamics of cytomegalovirus chromatin assembly in productively infected human cells. J. Virol. 82, 11167–11180.

Nonoyama, S., Smith, F.O., Bernstein, I.D., and Ochs, H.D. (1993). Strain-dependent leakiness of mice with severe combined immune deficiency. J. Immunol. 150, 3817–3824.

Obar, J.J., and Lefrancois, L. (2010a). Memory CD8+ T-cell differentiation. Ann. N.Y. Acad. Sci. 1183, 251–266.

Obar, J.J., and Lefrancois, L. (2010b). Early events governing memory CD8+ T-cell differentiation. Int. Immunol. 22, 619–625.

O'Hara, G.A., Welten, S.P., Klenerman, P., and Arens, R. (2012). Memory T-cell inflation: understanding cause and effect. Trends Immunol. 33, 84–90.

Ortega, N., Hutchings, H., and Plouët, J. (1999). Signal relays in the VEGF system. Front. Biosci. 4, D141-D152.

Pescovitz, M.D. (2006). Benefits of cytomegalovirus prophylaxis in solid organ transplantation. Transplantation 82, S4–8.

Petrucelli, A., Rak, M., Grainger, L., and Goodrum, F. (2009). Characterization of a novel Golgi apparatus-localized latency determinant encoded by human cytomegalovirus. J. Virol. *83*, 5615–5629.

Pierre, P. (2005). Dendritic cells, DRiPs, and DALIS in the control of antigen processing. Immunol. Rev. *207*, 184–190.

Podlech, J., Holtappels, R., Pahl-Seibert, M.F., Steffens, H.P., and Reddehase, M.J. (2000). Murine model of interstitial cytomegalovirus pneumonia in syngeneic bone marrow transplantation: persistence of protective pulmonary CD8-T-cell infiltrates after clearance of acute infection. J. Virol. *74*, 7496–7507.

Podlech, J., Pintea, R., Kropp, K.A., Fink, A., Lemmermann, N.A.W., Erlach, K.C., Isern, E., Angulo, A., Ghazal, P., and Reddehase, M.J. (2010). Enhancerless cytomegalovirus is capable of establishing a low-level maintenance infection in severe immunodeficient host tissues but fails in exponential growth. J. Virol. *84*, 6254–6261.

Polic, B., Hengel, H., Krmpotic, A., Trgovcich, J., Pavic, I., Luccaronin, P., Jonjic, S., and Koszinowski, U.H. (1998). Hierarchical and redundant lymphocyte subset control precludes cytomegalovirus replication during latent infection. J. Exp. Med. *188*, 1047–1054.

Pollock, J.L., and Virgin, H.W. (1995). Latency, without persistence, of murine cytomegalovirus in the spleen and kidney. J. Virol. *69*, 1762–1768.

Pollock, J.L., Presti, R.M., Paetzold, S., and Virgin, H.W. (1997). Latent murine cytomegalovirus infection in macrophages. Virology *227*, 168–179.

Pomeroy, C., Hilleren, P.J., and Jordan, M.C. (1991). Latent murine cytomegalovirus DNA in splenic stromal cells of mice. Microbiology *65*, 3330–3334.

Porter, K.R., Starnes, D.M., and Hamilton, J.D. (1985). Reactivation of latent murine cytomegalovirus from kidney. Kidney Int. *28*, 922–925.

Prösch, S., Staak, K., Stein, J., Liebenthal, C., Stamminger, T., Volk, H.D., and Krüger, D.H. (1995). Stimulation of the human cytomegalovirus IE enhancer/promoter in HL-60 cells by TNFalpha is mediated via induction of NF-kappaB. Virology *208*, 197–206.

Prösch, S., Volk, H.-D., Reinke, P., Pioch, K., Docke, W.-D., and Krüger, D.H. (1998). Human cytomegalovirus infection in transplant recipients: role of TNF-alpha for reactivation and replication of human cytomegalovirus. In CMV-Related Immunopathology, Scholz, M., Rabenau, H.F., Doerr, H.W., and Cinatl, J., eds. (Karger, Basel), pp. 29–41.

Rawlinson, W.D., Farrell, H.E., and Barrell, B.G. (1996). Analysis of the complete DNA sequence of murine cytomegalovirus. J. Virol. *70*, 8833–8849.

Razonable, R.R., and Paya, C.V. (2002). Beta-Herpesviruses in transplantation. Rev. Med. Microbiol. *13*, 163–176.

Razonable, R.R., Rivero, A., Rodriguez, A., Wilson, J., Daniels, J., Jenkins, G., Larson, T., Hellinger, W.C., Spivey, J.R., and Paya, C.V. (2001). Allograft rejection predicts the occurrence of late-onset cytomegalovirus (CMV) disease among CMV-mismatched solid organ transplant patients receiving prophylaxis with oral ganciclovir. J. Infect. Dis. *184*, 1461–1464.

Reddehase, M.J. (2002). Antigens and immunoevasins: opponents in cytomegalovirus immune surveillance. Nat. Rev. Immunol. *2*, 831–844.

Reddehase, M.J., and Koszinowski, U.H. (1984). Significance of herpesvirus immediate-early gene expression in cellular immunity to cytomegalovirus infection. Nature *312*, 369–371.

Reddehase, M.J., Rothbard, J.B., and Koszinowski, U.H. (1989). A pentapeptide as minimal antigenic determinant for MHC class I-restricted T lymphocytes. Nature *337*, 651–653.

Reddehase, M.J., Balthesen, M., Rapp, M., Jonjic, S., Pavic, I., and Koszinowski, U.H. (1994). The conditions of primary infection define the load of latent viral genome in organs and the risk of recurrent cytomegalovirus disease. J. Exp. Med. *179*, 185–193.

Reddehase, M.J., Podlech, J., and Grzimek, N.K.A. (2002). Mouse models of cytomegalovirus latency: overview. J. Clin. Virol. *25*, 23–36.

Reddehase, M.J., Simon, C.O., Seckert, C.K., Lemmermann, N., and Grzimek, N.K. (2008). Murine model of cytomegalovirus latency and reactivation. Curr. Top. Microbiol. Immunol. *325*, 315–331.

Reeves, M.B., Coleman, H., Chadderton, J., Goddard, M., Sissons, J.G.P., and Sinclair, J.H. (2004). Vascular endothelial and smooth muscle cells are unlikely to be major sites of latency of human cytomegalovirus in vivo. J. Gen. Virol. *85*, 3337–3341.

Roizman, B., and Sears, A.E. (1987). An inquiry into the mechanisms of herpes simplex virus latency. Ann. Rev. Microbiol. *41*, 543–571.

Rubin, R.H., Wilson, E.J., Barrett, L.V., and Medearis, D.N. (1984). Primary cytomegalovirus infection following cardiac transplantation in a murine model. Transplantation *37*, 306–310.

Sacher, T., Podlech, J., Mohr, C.A., Jordan, S., Ruzsics, Z., Reddehase, M.J., and Koszinowski, U.H. (2008). The major virus-producing cell type during murine cytomegalovirus infection, the hepatocyte, is not the source of virus dissemination in the host. Cell Host Microbe *3*, 263–272.

Scheller, S. (2010). Abstract 6.02, 35th Annual International Herpesvirus Workshop, 2010, Salt Lake City, Utah, USA.

Schenk, A.D., Nozaki, T., Rabant, M., Valujskikh, A., and Fairchild, R.L. (2008). Donor-reactive CD8 memory T-cells infiltrate cardiac allografts within 24-h posttransplant in naive recipients. Am. J. Transplant. *8*, 1652–1661.

Schnorrer, P., Behrens, G.M., Wilson, N.S., Pooley, J.L., Smith, C.M., El-Sukkari, D., Davey, G., Kupresanin, F., Li, M., Maraskovsky, E., *et al.* (2006). The dominant role of CD8+ dendritic cells in cross-presentation is not dictated by antigen capture. Proc. Natl. Acad. Sci. U.S.A. *103*, 10729–10734.

Seckert, C.K., Renzaho, A., Reddehase, M.J., and Grzimek, N.K.A. (2008). Hematopoietic stem cell transplantation with latently infected donors does not transmit virus to immunocompromised recipients in the murine model of cytomegalovirus infection. Med. Microbiol. Immunol. *197*, 251–259.

Seckert, C.K., Renzaho, A., Tervo, H.-M., Krause, C., Deegen, P., Kühnapfel, B., Reddehase, M.J., and Grzimek, N.K.A. (2009). Liver sinusoidal endothelial cells are a site of murine cytomegalovirus latency and reactivation. J. Virol. *83*, 8869–8884.

Seckert, C.K., Schader, S.I., Ebert, S., Thomas, D., Freitag, K., Renzaho, A., Podlech, J., Reddehase, M.J., and Holtappels, R. (2011). Antigen-presenting cells of haematopoietic origin prime cytomegalovirus-specific CD8 T-cells but are not sufficient for driving memory inflation during viral latency. J. Gen. Virol. *92*, 1994–2005.

Selin, L.K., Brehm, M.A., Naumov, Y.N., Cornberg, M., Kim, S.K., Clute, S.C., and Welsh, R.M. (2006). Memory of mice and men: CD8+ T-cell cross-reactivity and heterologous immunity. Immunol. Rev. *211*, 164–181.

Shin, E.C., Seifert, U., Kato, T., Rice, C.M., Feinstone, S.M., Kloetzel, P.M., and Rehermann, B. (2006). Virus induced type I IFN stimulates generation of immunoproteasomes at the site of infection. J. Clin. Invest. *116*, 3006–3014.

Shultz, L.D., Lyons, B.L., Burzenski, L.M., Gott, B., Chen, X., Chaleff, S., Kotb, M., Gillies, S.D., King, M., Mangada, J., et al. (2005). Human lymphoid and myeloid cell development in NOD/LtSz-scid IL2R gamma null mice engrafted with mobilized human hemopoietic stem cells. J. Immunol. *174*, 6477–6489.

Sierro, S., Rothkopf, R., and Klenerman, P. (2005). Evolution of diverse antiviral CD8+ T-cell populations after murine cytomegalovirus infection. Eur. J. Immunol. *35*, 1113–1123.

Sijts, E.J., and Kloetzel, P.M. (2011). The role of the proteasome in the generation of MHC class I ligands and immune responses. Cell Mol. Life Sci. *68*, 1491–1502.

Simon, C.O., Seckert, C.K., Dreis, D., Reddehase, M.J., and Grzimek, N.K. (2005). Role for tumor necrosis factor alpha in murine cytomegalovirus transcriptional reactivation in latently infected lungs. J. Virol. *79*, 326–340.

Simon, C.O., Holtappels, R., Tervo, H.-M., Böhm, V., Däubner, T., Oehrlein-Karpi, S.A., Kühnapfel, B., Renzaho, A., Strand, D., Podlech, J., et al. (2006a). CD8 T-cells control cytomegalovirus latency by epitope-specific sensing of transcriptional reactivation. J. Virol. *80*, 10436–10456.

Simon, C.O., Seckert, C.K., Grzimek, N.K.A., and Reddehase, M. (2006b). Murine model of cytomegalovirus latency and reactivation: the silencing/desilencing and immune sensing hypothesis. In Cytomegaloviruses: Molecular Biology and Immunology, Reddehase, M.J., ed. (Caister Academic Press, Norfolk, UK), pp. 483–500.

Simon, C.O., Kühnapfel, B., Reddehase, M.J., and Grzimek, N.K.A. (2007). Murine cytomegalovirus major immediate-early enhancer region operating as a genetic switch in bidirectional gene pair transcription. J. Virol. *81*, 7805–7810.

Simon, J.A., and Kingston, R.E. (2009). Mechanisms of polycomb gene silencing: knowns and unknowns. Nat. Rev. Mol. Cell. Biol. *10*, 697–708.

Smith, M.S., Goldman, D.C., Bailey, A.S., Pfaffle, D.L., Kreklywich, C.N., Spencer, D.B., Othieno, F.A., Streblow, D.N., Garcia, J.V., Fleming, W.H., et al. (2010). Granulocyte-colony stimulating factor reactivates human cytomegalovirus in a latently infected humanized mouse model. Cell Host Microbe *8*, 284–291.

Snyder, C.M. (2011). Buffered memory: a hypothesis for the maintenance of functional, virus-specific CD8(+) T-cells during cytomegalovirus infection. Immunol. Res. *51*, 195–204.

Snyder, C.M., Cho, K.S., Bonnett, E.L., van Dommelen, S., Shellam, G.R., and Hill, A.B. (2008). Memory inflation during chronic viral infection is maintained by continuous production of short-lived, functional T-cells. Immunity *29*, 650–659.

Snyder, C.M., Loewendorf, A., Bonnett, E.L., Croft, M., Benedict, C.A., and Hill, A.B. (2009). CD4+ T-cell help has an epitope-dependent impact on CD8+ T-cell

memory inflation during murine cytomegalovirus infection. J. Immunol. *183*, 3932–3941.

Snyder, C.M., Cho, K.S., Allan, J.E., and Hill, A.B. (2011). Sustained CD8+ T-cell memory inflation after infection with a single-cycle cytomegalovirus. PLoS Pathog. *7*, e1002295.

Steffens, H.-P., Kurz, S., Holtappels, R., and Reddehase, M.J. (1998). Preemptive CD8 T-cell immunotherapy of acute cytomegalovirus infection prevents lethal disease, limits the burden of latent viral genomes, and reduces the risk of virus recurrence. J. Virol. *72*, 1797–1804.

Stein, J., Volk, H.D., Liebenthal, C., Krüger, D.H., and Prösch, S. (1993). Tumour necrosis factor alpha stimulates the activity of the human cytomegalovirus major immediate-early enhancer/promoter in immature monocytic cells. J. Gen. Virol. *74*, 2333–2338.

Stinski, M.F., and Isomura, H. (2008). Role of the cytomegalovirus major immediate-early enhancer in acute infection and reactivation from latency. Med. Microbiol. Immunol. *197*, 223–231.

Streblow, D.N., Varnum, S.M., Smith, R.D., and Nelson, J.A. (2006). A proteomics analysis of human cytomegalovirus particles. In Cytomegaloviruses: Molecular Biology and Immunology, Reddehase, M.J., ed. (Caister Academic Press, Norfolk, UK), pp. 91–110.

Strehl, B., Seifert, U., Kruger, E., Heink, S., Kuckelkorn, U., and Kloetzel, P.M. (2005). Interferon-gamma, the functional plasticity of the ubiquitin-proteasome system, and MHC class I antigen processing. Immunol. Rev. *207*, 19–30.

Szeto, J., Kaniuk, N.A., Canadien, V., Nisman, R., Mizushima, N., Yoshimori, T., Bazett-Jones, D.P., and Brumell, J.H. (2006). ALIS are stress-induced protein storage compartments for substrates of the proteasome and autophagy. Autophagy *2*, 189–199.

Tanaka, K., Sawamura, S., Satoh, T., Kobayashi, K., and Noda, S. (2007). Role of the indigenous microbiota in maintaining the virus-specific CD8 memory T-cells in the lung of mice infected with murine cytomegalovirus. J. Immunol. *178*, 5209–5216.

Tang, Q., and Maul, G. (2006). Immediate-early interactions and epigenetic defense mechanisms. In Cytomegaloviruses: Molecular Biology and Immunology, Reddehase, M.J., ed. (Caister Academic Press, Norfolk, UK), pp. 131–150.

Thimme, R., Appay, V., Koschella, M., Panther, E., Roth, E., Hislop, A.D., Rickinson, A.B., Rowland-Jones, S.L., Blum, H.E., and Pircher, H. (2005). Increased expression of the NK cell receptor KLRG1 by virus-specific CD8 T-cells during persistent antigen stimulation. J. Virol. *79*, 12112–12116.

Thompson, R.L., Preston, C.M., and Sawtell, N.M. (2009). De novo synthesis of VP16 coordinates the exit from HSV latency in vivo. PloS Pathog. *5*, e1000352.

Torti, N., Walton, S.M., Brocker, T., Rulicke, T., and Oxenius, A. (2011a). Non-hematopoietic cells in lymph nodes drive memory CD8 T-cell inflation during murine cytomegalovirus infection. PLoS Pathog. *7*, e1002313.

Torti, N., Walton, S.M., Murphy, K.M., and Oxenius, A. (2011b). Batf3 transcription factor-dependent DC subsets in murine CMV infection: differential impact on T-cell priming and memory inflation. Eur. J. Immunol. *41*, 2612–2618.

Tykocinski, L.-O., Sinemus, A., Rezavandy, E., Weiland, Y., Baddeley, D., Cremer, C., Sonntag, S., Willecke, K., Derbinski, J., and Kyewski, B. (2010). Epigenetic regulation of promiscuous gene expression in thymic medullary epithelial cells. Proc. Nat. Acad. Sci. U.S.A. *107*, 19426–19431.

Umashankar, M., Petrucelli, A., Cicchini, L., Caposio, P., Kreklywich, C., Rak, M., Bughio, F., Goldman, D., Hamlin, K., Nelson, J., *et al.* (2011). A novel human cytomegalovirus locus modulates cell type-specific outcomes of infection. PLoS Pathog. *7*, e1002444.

Voigt, A., Salzmann, U., Seifert, U., Dathe, M., Soza, A., Kloetzel, P.M., and Kuckelkorn, U. (2007). 20S proteasome-dependent generation of an IEpp89 murine cytomegalovirus-derived H-2L(d) epitope from a recombinant protein. Biochem. Biophys. Res. Commun. *355*, 549–554.

Walsh, P.T., Strom, T.B., and Turka, L.A. (2004). Routes to transplant tolerance versus rejection; the role of cytokines. Immunity *20*, 121–131.

Walton, S.M., Mandaric, S., Torti, N., Zimmermann, A., Hengel, H., and Oxenius, A. (2011). Absence of cross-presenting cells in the salivary gland and viral immune evasion confine cytomegalovirus immune control to effector CD4 T-cells. PLoS Pathog. *7*, e1002214.

Wang, D., and Bodovitz, S. (2010). Single cell analysis: the new frontier in 'omics'. Trends Biotechnol. *28*, 281–290.

Welsh, R.M., and Selin, L.K. (2002). No one is naive: the significance of heterologous T-cell immunity. Nat. Rev. Immunol. *2*, 417–426.

Welsh, R.M., Selin, L.K., and Szomolanyi-Tsuda, E. (2004). Immunological memory to viral infections. Annu. Rev. Immunol. *22*, 711–743.

Weninger, W., Manjunath, N., and von Andrian, U.H. (2002). Migration and differentiation of CD8+ T-cells. Immunol. Rev. *186*, 221–233.

White, K.L., Slobedman, B., and Mocarski, E.S. (2000). Human cytomegalovirus latency-associated protein pORF94 is dispensable for productive and latent infection. J. Virol. *74*, 9333–9337.

Wilhelmi, V., Simon, C.O., Podlech, J., Böhm, V., Däubner, T., Emde, S., Strand, D., Renzaho, A., Lemmermann, N.A.W., Seckert, C.K., *et al.* (2008). Transactivation of cellular genes involved in nucleotide metabolism by the regulatory IE1 protein of murine cytomegalovirus is not critical for viral replicative fitness in quiescent cells and host tissues. J. Virol. *82*, 9900–9916.

Wilkinson, G.W., Kelly, C., Sinclair, J.H., and Rickards, C. (1998). Disruption of PML-associated nuclear bodies mediated by the human cytomegalovirus major immediate-early gene product. J. Gen. Virol. *79*, 1233–1245.

Wise, T.G., Manischewitz, J.E., Quinnan, G.V., Aulakh, G.S., and Ennis, F.A. (1979). Latent cytomegalovirus infection of BALB/c mouse spleens detected by an explant culture technique. J. Gen. Virol. *44*, 551–556.

Yang, W.M., Inouye, C., Zeng, Y., Bearss, D., and Seto, E. (1996). Transcriptional repression by YY1 is mediated by interaction with a mammalian homolog of the yeast global regulator RPD3. Proc. Natl. Acad. Sci. U.S.A. *93*, 12845–12850.

Yewdell, J.W., and Del Val, M. (2004). Immunodominance in TCD8+ responses to viruses: cell biology, cellular immunology, and mathematical models. Immunity *21*, 149–153.

Yuhasz, S.A., Dissette, V.B., Cook, M.L., and Stevens, J.G. (1994). Murine cytomegalovirus is present in both chronic active and latent states in persistently infected mice. Virology *202*, 272–280.

Zaidi, S.K., Young, D.W., Montecino, M., van Wijnen, A.J., Stein, J.L., Lian, J.B., and Stein, G.S. (2011). Bookmarking the genome: maintenance of epigenetic information. J. Biol. Chem. *286*, 18355–18361.

Zhang, Z., Li, Z., Yan, S., Wang, X., and Abecassis, M. (2009). TNF-alpha signaling is not required for in vivo transcriptional reactivation of latent murine cytomegalovirus. Transplantation *88*, 640–645.

Humanized Mouse Models of Cytomegalovirus Pathogenesis and Latency

M. Shane Smith, Daniel N. Streblow, Patrizia Caposio and Jay A. Nelson

Abstract

The generation of mice engrafted with human haematopoietic stem cells (HSC) has allowed, for the first time, the study of human specific viruses in an *in vivo* setting. These humanized mouse models have been developed and improved over the past 30 years. It is now possible to achieve high levels of human cell engraftment producing human myeloid and lymphoid lineage cells. Humanized mouse models have been increasingly utilized in the study of human cytomegalovirus, a human-specific beta-herpesvirus that infects myeloprogenitor cells and establishes a life-long latency in the infected host. Upon mobilization and differentiation of infected bone marrow progenitor cells the latent virus reactivates and disseminates to other tissues. In this chapter, we review the current status of the HSC-engrafted mouse models used to study HCMV latency and reactivation. We will first highlight the role myeloid lineage cells plays in HCMV biology and then describe the types of humanized mouse models that have been used in HIV, EBV, KSHV and HCMV anti-viral therapy studies. We will then describe recent studies utilizing the latest generation of humanized mice for the study of HCMV latency and reactivation and outline the future role that these models may play in the study of human-specific viruses.

Introduction

Lack of small animal models for human viruses with restricted cellular and tissue tropism, including human immunodeficiency virus (HIV), hepatitis C virus (HCV), human cytomegalovirus (HCMV) and many other human herpesviruses, has impeded the understanding of viral pathogenesis and the development of antiviral therapies. Over the last two decades, the development of humanized mouse model systems in which immune deficient mice are engrafted with human tissues has opened the door for the direct *in vivo* investigation of such human-restricted viruses. Advancements relating to xenograft tolerance and xenograft tissue function have allowed high levels of human chimerism, especially with respect to immune cells and liver tissue. Stable human hepatocyte engraftment in mouse liver tissue has, for example, allowed the development of the first murine model system suitable for studying HCV (Mercer *et al.*, 2001). Owing to the critical role immune cells play in the latency, persistence, and/or in the pathobiology of many human herpesviruses, the field of herpesvirus research has benefited tremendously over the last decade from the continued improvements in human immune system (HIS) mouse technology. HIS mice as generally defined are immunodeficient mice in which the murine immune cell compartments, most notably the BM, are depleted, typically by irradiation, and reconstituted with human haematopoietic progenitor cells (HPCs). The human haematopoietic progenitor cells home primarily to the BM and subsequently reconstitute nearly all human haematopoietic cell lineages found in circulation. Although humanized murine model systems for HCMV are the primary focus of this chapter, an overview of humanized murine model systems is provided herein for HIV, Dengue virus (DV), KSHV, and EBV, given that the development of these model systems has in many instances paved the way for the development and improvement of HCMV humanized mouse model systems.

Role of myeloid lineage cells in HCMV latency and persistence

To fully appreciate the advances and discoveries relating to HCMV humanized mouse model systems, one must first appreciate the critical role that haematopoietic progenitor cells and mature haematopoietic lineage cells play in the HCMV life cycle within the host. Of the haematopoietic lineage cells, which comprise all haematopoietic stem cell-derived myeloid and lymphoid lineages, it is commonly accepted that the myeloid cell lineage is the most important lineage with respect to HCMV latency, reactivation, and persistence (Table I.23.1). The term myeloid cell lineage implies any leucocyte that is not a lymphocyte and includes monocytes, macrophages, dendritic cells, neutrophils, eosinophils, and basophils. Cells of the myeloid lineage are not only critical targets in the HCMV life cycle but are also involved in the pathogenesis of HCMV-associated inflammatory and transplant diseases.

A characteristic of HCMV infection is the ability of the virus to spread to and persist within multiple host organs (Gnann *et al.*, 1988; Sinzger and Jahn, 1996; Sinzger *et al.*, 1996). Monocytes are the primary targets for infection in the blood (Taylor-Wiedeman *et al.*, 1991), and a large body of evidence suggests that monocytes are the cell type responsible for viral spread to organ tissues, and establishing persistence (Booss *et al.*, 1989; Gerna *et al.*, 1992; Grefte *et al.*, 1992, 1994; van der Strate *et al.*, 2003). Monocytes acutely infected *in vivo* and infected *in vitro* are non-permissive for viral gene expression (Taylor-Wiedeman *et al.*, 1991, 1994; Brytting *et al.*, 1995; Sinclair and Sissons, 1996). Macrophages, however, are productively infected in patients with HCMV disease (Sinclair and Sissons, 1996; Jahn *et al.*, 1999), and *in vitro* studies have confirmed that macrophages are permissive for HCMV replication (Söderberg-Naucler *et al.*, 1997; Sissons *et al.*, 2002). HCMV dissemination is proposed to occur, therefore, after infected monocytes migrate into tissue and differentiate into permissive macrophages (Söderberg-Naucler *et al.*, 1998). HCMV replication in organ tissues is associated with significant morbidity in immunocompromised hosts, while HCMV persists at low levels in immunocompetent hosts with periodic viral shedding through the mucosal epithelium (Sinzger and Jahn, 1996; Smith *et al.*, 2004a). Furthermore, it has been shown that HCMV utilizes an inflammatory response to drive monocyte migration into host tissue as a mechanism of viral dissemination (Yurochko and Huang, 1999; Smith *et al.*, 2004a,b, 2007). It is proposed that this mode of viral spread could promote inflammatory diseases associated with HCMV

infection through aberrant monocyte migration and immune mediator release (Yurochko and Huang, 1999; Smith *et al.*, 2004b; Chan *et al.*, 2008).

Cells of the myeloid lineage are not only critical for viral spread and persistence but are also critical cells involved in viral latency. The monocyte was the first site of HCMV latency identified (Taylor-Wiedeman *et al.*, 1994; Söderberg-Naucler *et al.*, 1997, 2001). A small percentage of circulating monocytes in the peripheral blood of healthy seropositive hosts carry latent HCMV DNA, and HCMV can be reactivated *ex vivo* in latently infected monocytes isolated from seropositive hosts through allogeneic stimulation (Taylor-Wiedeman *et al.*, 1994; Söderberg-Naucler *et al.*, 1997, 2001). These findings not only demonstrated that monocytes are sites of viral latency, but also suggest that reactivation due to allogeneic stimulation during organ transplantation is a contributor to HCMV-associated disease in transplant recipients.

Monocytes are short-lived cells in circulation and thus cannot be the latency reservoir in the host (Hume *et al.*, 2002). Significant evidence indicates that latently infected peripheral blood monocytes are generated from latently infected HPCs of the BM (Sinclair and Sissons, 1996). Latent infection was first detected in lineage-committed CD33$^+$ macrophage progenitors in the BM of healthy seropositive hosts (Kondo *et al.*, 1994; Hahn *et al.*, 1998;). Later evidence revealed that latent HCMV DNA resides in a more primitive CD34$^+$ HPC (Mendelson *et al.*, 1996). HCMV replication can be reactivated by coculture of both CD14$^+$ CD15$^+$ CD33$^+$ progenitor cells with human fibroblasts (Hahn *et al.*, 1998). Although these primitive HPCs have the capacity to mature into a number of cell lineages, latent HCMV DNA is strictly associated with myelomonocytic lineage cells in healthy hosts (Minton *et al.*, 1994). This suggests that either latent infection of myeloid stem cells promotes maturation into the myeolomonocytic lineage or that only cells of the myelomonocytic lineage are capable of maintaining the latent viral genome. During GM-CSF-stimulated proliferation of CD34$^+$ stem cells in culture, monocytic cells are derived that harbour latent HCMV DNA (Minton *et*

Table I.23.1 Role of myeloid lineage cells in HCMV pathogenesis

Myeloid cell type	Viral genome (Hudnall *et al.*, 2008)	Viral transcription (mRNA)	Permissive for HCMV replication
CD34$^+$ progenitor cells	+	–	–
Monocyte	+	–	–
Macrophage	+++	+	+

al., 1994). This finding indicates that the latent viral genome is passed to daughter cells during progenitor cell proliferation.

Surrogate animal models of CMV latency and pathogenesis

The use of surrogate animal models to directly examine HCMV pathobiology and anti-HCMV therapies *in vivo* has been hindered due to the strict species specificity of HCMV (see also Chapter I.18). HCMV infection of murine, rat, and guinea pig cells is abortive, and HCMV replication appears to be strictly limited to cells of human origin. Consequently, non-xenograft animal models are not available to directly examine mechanisms of HCMV pathogenesis and reactivation from latency, especially in the context of BM HSCs that are considered the primary site of HCMV latency. Although models using rat CMV (RCMV), murine CMV (MCMV), and guinea pig CMV (GPCMV) infections in their respective hosts have not provided significant insight into the mechanisms of HCMV latency and reactivation within the BM compartment, such surrogate animal model systems have been instrumental in the development and testing of drugs targeting CMV and in the understanding of other facets of CMV pathogenesis, including viral persistence, atherosclerosis and solid organ graft rejection (see Chapter II.15) as well as congenital infection (See Chapter II.5).

MCMV is the most widely used surrogate CMV for evaluating anti-CMV drug candidates and the immunology of CMV infection. A non-lethal MCMV infection in Swiss–Webster mice results in a high-titre disseminated MCMV infection of the lungs, liver, spleen, kidney, and blood within 24 hours post infection. Within 24–72 hours post infection, a high viral load is detected in the salivary glands. The virus persists in the lung, liver, kidney, and spleen for 45–60 days post infection and persists in the salivary glands for months. In contrast to HCMV infection of humans, MCMV is cleared from murine haematopoietic cells and does not establish latency in the mice (Reddehase *et al.*, 2002; Seckert *et al.*, 2008). Interestingly, MCMV models have demonstrated the life-long maintenance of latency in stromal and/or parenchymal cellular sites of diverse organ tissues, including the spleen, lungs, liver, and salivary glands (Balthesen *et al.*, 1993; Reddehase *et al.*, 1994; Seckert *et al.*, 2008, 2009). Such findings suggest that either a non-HSC latency component has not been realized for HCMV or that a fundamental difference in cellular latency sites exists between MCMV and HCMV infected hosts. Such discrepancies underscore the need for animal models capable of supporting HCMV latency and reactivation.

Despite the aforementioned differences between the HCMV and MCMV viral life cycle in their respective hosts, MCMV mouse model systems have been indispensable for the development of the antiviral drugs ganciclovir (GCV), cidofovir (CDV), and foscarnet (FOS). However, MCMV mouse model systems are inadequate for the development and evaluation of drug candidates that are uniquely specific to HCMV or more generally to human herpesviruses (Weber *et al.*, 2000; Williams *et al.*, 2003). MCMV and other nonhuman CMVs do not have consistently high nucleotide sequence homology with HCMV across the viral genome, and CMVs of various species and even intraspecies CMV strains encode unique viral gene products (Rawlinson *et al.*, 1996) (see also Chapters I.1 and I.2). Thus, surrogate CMVs are inadequate for assessing drugs targeting unique HCMV gene products or operating at the genetic level, such as drugs based upon small interfering (si)RNAs. In addition, MCMV mouse models may be hampered by differences with HCMV in terms of drug susceptibility. For example, HCMV is resistant to acyclovir, while MCMV is highly susceptible to acyclovir (Glasgow *et al.*, 1982).

Other surrogate CMV models have been instrumental to our understanding of distinct niches of CMV pathogenesis. The RCMV rat model stands out for its ability to model HCMV-associated vascular disease and solid organ rejection (see Chapter II.15), while the GPCMV model (see Chapter II.5) and potentially also primate CMV models are models of CMV congenital infections. However, the RCMV and GPCMV models have similar limitations as the MCMV models, in that the genetic dissimilarities of these viruses to HCMV and apparent differences in sites of viral latency in relation to HCMV do not make these models ideal for evaluation of HCMV specific drug candidates and for studies of HCMV latency and reactivation. Primate CMV models have the disadvantages of high costs and limited or restricted availability of primates. As such, these models potentially face the issue of limited reproducibility due to small experimental cohorts and genetic divergence within cohorts of primates.

SCID-hu models for HIV

The strict cellular tropism of HCMV and the aforementioned limitations of surrogate CMV animal models have driven the development of humanized murine models in which mice are engrafted with human cells or tissues capable of supporting local HCMV infection. One of the earliest and most widely used humanized mouse models of viral pathogenesis involved the co-implantation of human fetal thymus and fetal liver fragments (Thy/Liv) under the kidney capsule of severe combined immunodeficiency (SCID) mice. Owing to a defect in double-stranded DNA break repair activity in SCID mice, the V(D)J coding regions of T and B-cell antigen receptors do not properly join, resulting in a lack of functionally mature T and B cells and resistance to xenogeneic tissue rejection (Bosma and Carroll, 1991; Greiner *et al.*, 1998) (see Table I.23.2). These Thy/Liv SCID-human (hu) mice exhibit stable human tissue engraftment and support long-term reconstitution of human HPCs for up to 11 months post transplantation (Mocarski *et al.*, 1993; Brown *et al.*, 1995; Greiner *et al.*, 1998). T-cell lymphopoiesis occurs within the graft tissues of Thy/Liv SCID-hu mice, and the Thy/Liv grafts provide microenvironments capable of sustaining the HPCs and HPC-derived T lymphoid, B lymphoid, myelomonocytic, erythroid, and megakaryocytic lineages (Namikawa *et al.*, 1990). Outside of the Thy/Liv graft, Thy/Liv SCID-hu mice exhibit human circulating CD4$^+$ and CD8$^+$ T-cells (McCune *et al.*, 1988). Other early SCID-hu mouse models alternatively involved the implantation of human fetal bone fragments, retinal tissue, or lymph nodes, which could similarly support long-term human multilineage haematopoiesis within the engrafted tissue (Kaneshima *et al.*, 1991; Kyoizumi *et al.*, 1992).

Infection of SCID-hu mice with HIV-1 closely resembles what has been seen in human patients. Only human tissues containing CD4$^+$ T-cells are infected in SCID-hu mice (McCune *et al.*, 1991). SCID-hu mice engrafted with human lymph nodes and infected i.v. with HIV became viremic, and cells of both lymphoid and myelomonocytic lineages are specifically targeted (Kaneshima *et al.*, 1991). Given the exquisite human cellular topism of HIV, the early SCID-hu mouse models have been integral in evaluating anti-HIV drug treatments and have bridged the gap between laboratory studies and clinical studies of anti-HIV compounds. For example, 3'-azido-3'-deoxythymidine (AZT) or 2'3'-dideoxyinosine (ddIno) protected SCID-hu mice at dose ranges comparable to those used in human patients. SCID-hu mice engrafted with haematolymphoid tissues were also used to show that AZT was effective in suppressing HIV-1 infection when given between 2 to 36 hours post exposure, suggesting the use of AZT to treat occupational exposure to HIV (Shih *et al.*, 1991).

Thy/Liv SCID-hu mouse models for HCMV

The success of the early SCID-hu mouse studies of HIV infection provided an impetus to develop SCID-hu model systems for HCMV. Mocarski and colleagues utilized a Thy/Liv SCID-hu mouse model to assess the ability of the Toledo strain of HCMV to replicate within fetal human tissue implants (Mocarski *et al.*, 1993). Virus replication was consistently detected from 5 to 35 days post inoculation and continued for up to 9 months in some mice. HCMV was localized to the Thy/Liv implanted tissue and was not found in surrounding mouse tissue or in peripheral blood leucocytes. The Thy/Liv implants of infected mice exhibited no signs of viral damage to tissue and had no distinguishing morphological features compared to control mice.

Within thymic implants, HCMV-infected cells were identified in the thymic medulla rather than cortical region and were rarely clustered (Mocarski *et al.*, 1993). Immunofluorescence co-localization of HCMV antigens with human keratin, which is an epithelial cell marker, but not with markers for thymocytes, thymic

Table I.23.2 Evolution of NOD/SCID mouse models

Mouse model	Genetic defect	Effect	Human cell engraftment potential	Human cell types produced following engraftment
C.B-17 SCID	prkdc gene	No T-cells or B cells	15% in BM	Myeloid cells, B cells, monocytes
NOD/SCID	prkdc gene; impaired NKG2D; C5 deficiency	No T or B cells, defective NK cell and complement activities, defective macrophage activities	5% in PBL; 40% in BM	Myeloid cell, B cells, monocytes, T-cells
*NOD/SCID/IL2Rγc^null^*	prkdc gene; impaired NKG2D; C5 deficiency; deficient IL2Rγc	No T or B cells, less NK cells and defective complement activities	70% in PBL; 40% in BM	Myeloid cell, IgG$^+$ B cells, monocytes, functional T-cells, NK cells, Plasmacytoid Dendritic cells

dendritic cells, or myelomonocytic cells, suggested that epithelial cells were the primary site of HCMV infection within the implanted tissue (Mocarski *et al.*, 1993). Although the lack of detectably infected myeloid lineage cells within the implant is surprising given the critical role of myeloid lineage cells in HCMV pathobiology, epithelial cells in ducts of the salivary glands and renal tubules are permissive for HCMV replication and play critical roles in HCMV persistence and transmission (Cowdry and Scott, 1935; McGavran and Smith, 1965; Ho, 1991). Mocarski and colleagues further evaluated the ability of human lung, colon, and skin implants to support HCMV infection in SCID-hu mice. Human lung and colon exhibited greater variability and a shorter duration in HCMV replication in comparison to Thy/Liv tissue implants, while human skin implants exhibited low levels of infection (Mocarski *et al.*, 1993). A notable feature of this Thy/Liv SCID-hu mouse model was the ability to reduce HCMV replication within Thy/Liv implants by both oral and i.p. routes of GCV treatment.

In a separate study, Brown and colleagues utilized a Thy/Liv SCID-hu mouse model to evaluate and compare the replicative capacity of a low-passage Toledo strain of HCMV and high-passage, laboratory-adapted HCMV strains AD169 and Towne (Brown *et al.*, 1995). Low-passage Toledo grew 2–3 orders of magnitude better than any laboratory strain. Laboratory strains of HCMV either failed to grow or replicated at low levels. Based on these results, these authors predicted the existence of viral genetic determinants for growth in tissues that are lost during propagation in culture. The differences in tropism of these strains seen in this study was likely due to the fact that the Toledo viral genome, like many clinical isolates, contains a 15-kbp DNA segment composed of 18 open reading frames (UL133–UL150) that are absent in AD169 and contains a 13-kbp fragment absent in the Towne genome (Cha *et al.*, 1996). It is believed that such gene segments are tissue tropism determinants lost during cell culture passage by large-scale deletion and rearrangement of the UL/b' region (see also Chapter I.1).

Using a Thy/Liv SCID-hu mouse model, Wang and colleagues tested the hypothesis that the 15-kbp DNA segment present in Toledo and all virulent HCMV strains but deleted in AD169 and other attenuated strains is required for in vivo replication (Wang *et al.*, 2005). Thy/Liv SCID-hu mice infected with AD169 were compared to wild-type Toledo, bacterial artificial chromosome (BAC)-derived Toledo (Toledo$_{BAC}$), or a BAC-derived Toledo deleted for the 15-kbp DNA segment (Toledo$_{\Delta15kbp}$). While Toledo$_{BAC}$ grew within the tissue implant to high titres comparable to wild-type Toledo, both AD169 and Toledo$_{\Delta15kbp}$ failed to replicate in engrafted tissue. Given that Toledo$_{\Delta15kbp}$ exhibited only a minor growth defect *in vitro*, this study suggested that this 15-kbp DNA segment contains tissue-tropism determinants. A number of putative glycoproteins are encoded in this region in addition to the CXC chemokines UL146 and UL147 and the tumour necrosis factor receptor homologue UL144 (Prichard *et al.*, 2001).

HCMV studies in retinal transplant SCID-hu mice

One of the most common AIDS-related CMV diseases in HIV patients is HCMV retinitis, which can result in visual impairment or blindness. Human neural retina transplant models whereby fetal human neural retina is transplanted into the anterior chamber of the eye of immunosuppressed rats or SCID mice were developed as *in vivo* models of HIV-associated neural retina pathogenesis (Epstein *et al.*, 1994). These neural retina xenografts were shown to survive for many months and to produce a blood–eye barrier following vascularization. While these xenografts proved to be refractory to HIV-1 infection, it was reported that co-engraftment with HIV-1-infected human monocytes resulted in pathological changes of these tissues consistent with HIV-1-associated pathology seen in human patients.

DiLoreto and colleagues adapted such model systems to examine HCMV-associated retinitis in SCID mice engrafted with fetal human retinal tissue (DiLoreto *et al.*, 1994). These investigators transplanted fetal human retinal tissue into the anterior chambers of SCID mice and inoculated with HCMV at 1 week post transplantation. HCMV-infected retinal grafts exhibited pathology characteristic of HCMV-infected neural tissue such as intranuclear and intracytoplasmic inclusions. Infected cells expressed HCMV immediate-early, early, and late gene products, and the majority of infected cells co-expressed protein gene product 9.5, which is a neuronal marker. They were also able to isolate and culture infectious HCMV from graft tissue harvested at 45 days post infection (DiLoreto *et al.*, 1994).

Given that the blood–eye barrier impedes the bioavailability of systemically administered drugs in retinal tissue, the Thy/Liv SCID-hu mouse model, which contains highly vascularized human tissue, is not ideal for predicting the efficacy of anti-HCMV drug therapies for the treatment of ocular infections. Therefore, Bidanset and colleagues utilized a human retinal implant SCID-hu model similar to that developed by DiLoreto and colleagues to predict the efficacy and potential side effects of using antiviral drugs to treat HCMV ocular infections (Bidanset *et al.*, 2001). In this study, HCMV was injected into the anterior chamber containing the fetal retinal implant 2–18 weeks after implantation. Virus replication was localized to glial cells in the xenograft and was first detected 7 days post infection. It peaked at 21–28 days post infection and was undetectable at 8 weeks post infection. Similar to the study of Mocarski and colleagues, the Toledo clinical HCMV isolate replicated to higher titres in comparison to AD169 and Towne strains, suggesting the loss of tissue tropism determinants in laboratory-adapted strains of HCMV (Mocarski *et al.*, 1993; Bidanset *et al.*, 2001). Immunofluorescent staining with antibodies against IE1 and glial fibrillary acidic protein (GFAP), which is a glial cell marker, showed colocalization, while no colocalization was seen with IE1 and neurofilament 200 antigen, which is a marker for cells of neuronal origin (Bidanset *et al.*, 2001). The cause for the discrepancy between the predominant infection of glial cells in the Bidanset study and the predominant infection of neuronal cells in the DiLoreto study is unknown.

In a follow-up study, Bidanset and colleagues tested the effect of GCV on HCMV replication in retinal implant tissues (Bidanset *et al.*, 2004b). Four weeks post implantation, mice were infected with HCMV and administered 15 or 45 mg GCV/kg twice daily for the first 14 days and once daily for an additional 14 days. GCV resulted in a significant decrease in HCMV titres and infection rates in a dose-dependent manner. HCMV replication resumed when GCV treatment was terminated after 28 days of treatment. Furthermore, viral titres also increased within 14 days of decreasing dosing from twice daily to once daily. This study thus revealed that reducing the dose, frequency of administration, or termination of treatment results in increased HCMV replication to levels comparable to vehicle-treated controls (Bidanset *et al.*, 2004b). Therefore GCV therapy was only effective for suppressing viral replication in ocular implants but ineffective for clearing HCMV infection from the tissue. A similar dose-dependent reduction in viral titres and rates of infection were also seen with CDV treatments.

The use of SCID-hu model systems to test the efficacy, toxicity, and bioavailability of novel anti-HCMV drug candidates

Current drugs that have been licensed in the United States for treatment of systemic cytomegalovirus infections include FOS, CDV, GCV, and the GCV oral prodrug valganciclovir (VGCV). The desire for drugs with improved efficacy, oral availability, and reduced toxicity and the emergence of drug resistant HCMV strains drive the development of both analogues of

these drugs and altogether new classes of anti-viral drugs with activity against HCMV (see also Chapter II.19).

It has been suggested that the Thy/Liv implant SCID-hu model and the retinal implant SCID-hu model should be used concurrently to evaluate novel antiviral drugs against HCMV, because both models would represent different bioavailability characteristics. That is, the Thy/Liv model would represent blood delivery to visceral tissue, while the retinal implant model would represent a tissue requiring that systemic therapy can (or will) cross the blood-eye or blood-brain barrier (Kern *et al.*, 2004a,b). Studies where GCV and CDV treatments were compared between retinal implant and Thy/Liv implant SCID-hu mice indeed revealed that while GCV and CDV reduced viral titres in SCID-hu retinal tissue between days 14 and 28 post infection, GCV and CDV treatments of Thy/Liv SCID-hu mice resulted in either a complete inhibition of viral replication or a significantly higher viral titre reduction between days 14 and 28 post infection as compared to GCV and CDV-treated retinal implant SCID-hu mice (Kern *et al.*, 2001). These findings support the notion that the blood–eye barrier impedes the bioavailability of these drugs in retinal tissue and that intravitreal administration may be a practical clinical therapeutic route of application.

Despite the promising results obtained by Bidanset and colleagues with CDV in regard to treatment of retinal tissue in SCID-hu mice, a drawback to CDV is that its lack of oral activity requires administration by periodic i.v. infusion in humans or by i.p. injection in mice (Bidanset *et al.*, 2004b). Orally active ether lipid ester CDV analogues hexadecyloxypropyl-CDV (HDP-CDV) and octadecyloxyethyl-CDV (ODE-CDV) have been developed, and *in vitro* studies suggest substantially greater anti-viral activity against HCMV in comparison to CDV and GCV (Beadle *et al.*, 2002). Compared to i.p. administered CDV, both ether lipid CDV analogues were 4- to 8-fold more efficacious on a molar basis when administered orally to both Thy/Liv and retinal implant SCID-hu mice (Bidanset *et al.*, 2004a).

Kern and colleagues evaluated the efficacy of benzimidazole D- and L-ribonucleosides using both the SCID-hu ocular implant and a SCID-hu thy/liv implant model system (Kern *et al.*, 2004b). 2,5,6-trichloro-(1-β-D-ribofuranosyl) benzimidazole and its homologue 2-bromo-5,6-dichloro (1-beta-D-ribofuranosyl) benzimidazole (BDCRB) are potent and selective inhibitors of HCMV replication. Owing to the short plasma half-life of these compounds, more stable analogues of these compounds have been synthesized and include the ribopyranosyl analogue of BDCRB (GW 275175X or 175X) and 2-isopropylamino-5,6-dichloro-(1-β-L-ribofuranosyl) benzimidazole (1263W94 or maribavir).

BDCRB blocks the processing and maturation of viral DNA. BDCRB inhibits HCMV DNA maturation and processing by interaction with the UL89 and UL56 gene products (Chulay et al., 1999), and 175X is believed to operate by this same mechanism (Underwood et al., 2004). Maribavir (MBV) inhibits the viral enzyme pUL97 and blocks DNA synthesis and has also been shown to block nuclear egress. Both MBV and 175X have increased stability compared to BDCRB, exhibit antiviral activity against HCMV at a comparable or greater level than GCV, and are active against EBV. However, MBV and 175X have no effect on murine, rat, and guinea pig CMV strains, which necessitated their evaluation of efficacy in SCID-hu mouse model systems.

Kern and colleagues found that doses up to 75 mg of BDCRB/kg administered i.p. did not significantly reduce viral titres in SCID-hu retinal tissue (Kern et al., 2004a). MBV administered orally at 25 or 75 mg/kg significantly reduced virus titres by approximately 3-fold in SCID-hu retinal tissue compared to vehicle controls. 175X administered orally at 25 or 75 mg/kg was effective at 14 days post infection but had no significant effect on viral titres at either dose on days 21 and 28 post infection. These data indicate that MBV administered orally can effectively cross the blood–eye barrier and suggest that MBV would be efficacious for treating patients with HCMV complications in the eye or brain. In contrast to those results obtained with the SCID-hu retinal transplant mice, i.p. administration of 33 or 100 mg of BDCRB/kg inhibited viral replication in Thy/Liv SCID-hu mice by 2–3 $\log_{10}$ PFU/g compared to vehicle control, suggesting more effective penetration into the heavily vascularized Thy/Liv tissue. Twice daily administrations of 75 mg/kg of both MBV and 175X significantly reduced HCMV replication, while lower doses of 33 mg/kg of these drugs did not significantly inhibit HCMV replication. In conclusion, this study suggests that similar compounds such as BDCRB, 175X, and MBV exhibit differing bioavailability characteristics and should be used differentially depending on the specific tissue infected and whether or not the tissue to be treated is highly supplied with blood vessels.

A side-by-side comparison of (Z)-9-{[2,2-bis-(hydroxymethyl)cyclopropylidene]methyl} guanine (ZSM-I-62, cyclopropavir), which is a second-generation purine 2-(hydroxymethyl)methylenecyclopropane analogue, was also conducted by Kern and colleagues in both Thy/Liv implant and retinal tissue implant SCID-hu mice (Kern et al., 2004a).

The mechanism of HCMV-inhibition by cyclopropavir (CPV) is believed to be similar to GCV, which inhibits DNA synthesis following phosphorylation by UL97-encoded phosphotransferase (Kern et al., 2005). This study revealed that CPV administered orally at 45 or 15 mg/kg initiated 24 hours post infection was more effective than GCV in retinal implant SCID-hu mice and reduced viral titres to undetectable levels (Kern et al., 2004a). A similar superior, dose-dependent reduction in viral titre following CPV treatment was observed in Thy/Liv SCID-hu mice, with a minimum effective dose of 10 mg of CPV/kg. These results indicate that CPV administered orally has superior bioavailability in both highly vascularized tissue and tissue with limited vascular permeability compared with GCV administered intraperitoneally.

The use of cellular implant SCID-hu models to test the efficacy and toxicity of novel anti-HCMV drug candidates

SCID-hu model systems relying on the implantation of in vitro-derived human cells rather than human tissue can be broadly classified as cellular implant SCID-hu model systems. Generally, such cellular implant model systems rely on the implantation of human cells that have been cultured to grow on or within a physical support construct or matrix. Cellular implant SCID-hu models have distinct advantages over tissue xenograft SCID-hu models. First, the materials for generating the cellular implant and the immunocompetent mice are generally inexpensive and readily available, unlike human fetal tissue. Second, surgery in many cases is not required since support construct or matrix can often be injected subcutaneously or physically implanted under a small incision to the skin. Third, the cellular implants are more amenable to manipulation, because the site of implantation is often accessible for infection, for treatment with various compounds, and for cellular implant recovery and ex vivo culture. In the case of implants consisting of cells embedded in cellular matrix components, ex vivo treatments, such as collagenase digestion, can be utilized to yield a relatively pure population of living human cells for further ex vivo analysis. While these cellular implant model systems have the disadvantage of lacking a tissue-specific context to evaluate viral infection, these cellular implant models are useful for initial evaluations of drug efficacy and toxicity.

The most basic cell implant model was developed by Pari and colleagues and does not require a solid cellular support (Pari et al., 1998). Based upon the fact that HCMV can replicate in a limited number of transformed cell lines capable of forming solid tumours in nude mice, these authors utilized the U373MG transformed cell line, which is derived from human cells of astrocytoma origin, to generate subcutaneous tumours infected with HCMV. Initially, they found that U373MG cells formed tumours in nude mice only when cells embedded in matrigel were subcutaneously injected. Following multiple passage of matrigel-embedded U373MG cells as tumours in mice, the U373MGTU clone was isolated which retained permissiveness to HCMV replication in tissue culture and could efficiently form subcutaneous tumours in the absence of matrigel. They determined that these tumours supported replication of both AD169 and two HCMV clinical isolates obtained from HIV-infected patients. Immunohistochemical staining of tumours revealed IE1 and IE2 nuclear staining and gB cytosolic staining in discrete foci, suggesting cell to cell spread of HCMV in these tumour cells. To evaluate the effects of GCV on viral replication in this model system, GCV was injected at dosages of 8 mg/kg or 40 mg/kg per day from four hours post infection to seven days post infection. HCMV DNA levels were significantly reduced at 8 days post infection in both the low-dose and high-dose GCV-treated groups compared to control mice.

Bravo and colleagues developed a cellular implant SCID-hu model system in which SCID mice are subcutaneously implanted with Gelfoam implants containing HCMV-infected human foreskin fibroblasts (HFFs) (Bravo *et al.*, 2007). Gelfoam is a porous three-dimensional matrix composed of gelatin, which is used as a haemostasis medical device in surgical procedures. Gelfoam is capable of supporting cell growth *in vitro* and *in vivo* following implantation into animals (Centra *et al.*, 1992). Twice daily i.p. injections of 50 mg/kg GCV or CDV began on day 0 or day 7 post implantation and continued until day 5 or 14 post implantation (Bravo *et al.*, 2007). These authors determined that while CDV and GCV were ineffective at reducing HCMV titres within the implant at 5 days post implantation, this was due to lack of implant vascularization. At 14 days post implantation, the Gelfoam implant became encapsulated by a tissue membrane and vascularized, resulting in the bioavailability of CDV and GCV within the implant and a significant reduction in HCMV titre in mice receiving GCV and CDV treatment initiated at either day 0 or day 7 post implantation (Bravo *et al.*, 2007).

Lischka and colleagues utilized a similar approach to test the *in vivo* efficacy of the anti-CMV compound AIC246 (Lischka *et al.*, 2010). By screening a compound library in a high-throughput manner, these authors identified 3,4-dihydro-quinazoline-4-yl-acetic acid derivatives as a class of compounds with activity against HCMV but not MCMV or GPCMV. In

unpublished studies relating to structure–activity relationship studies and pharmacological analyses of this class of compounds, Lischka and colleagues elected the novel anti-CMV drug AIC246 as a development candidate. To evaluate AIC246 *in vivo*, mice, were implanted with Gelfoam containing AD169-infected normal human dermal fibroblasts (NHDF). At 4 hours post implantation, mice were treated orally once daily with either AIC246 or VGCV for 9 days at doses ranging from 1 mg/kg/day to 100 mg/kg/day. AIC246 exhibited antiviral activity comparable to VGCV at the ED_{50} level and surpassed the *in vivo* activity of VGCV at the ED_{90} level.

Allen and colleagues developed an encapsulation system in which MRC-5 human embryonic lung fibroblasts were infected 4 to 6 hours pre-implantation, subsequently mixed with agarose solution containing media, and allowed to solidify (Allen *et al.*, 1992). CD-1 mice (immunocompetent) were implanted with the agarose implants by s.c. and i.p. routes. The i.p. implants were reported to produce higher titres of virus and were more easily recovered than s.c. implants. These authors evaluated the efficacy of GCV administered s.c. at 100, 50 and 25 mg per kg body weight per day. GCV treatment began 4 hours before implantation and was maintained until day 4 post implantation. Virus titres were reduced in a dose-dependent manner, with 100 mg/kg/day dose reducing virus titres by 92.5%. No toxicity was seen in the mice at all doses tested. A drawback of this model system was the high level of GCV needed to reduce the viral titre (Allen *et al.*, 1992). Freitas and colleagues for instance reported that 9 mg/kg/day was sufficient to significantly increase the survival of MCMV-infected mice (Freitas *et al.*, 1985). The large amount of drug needed in the model system used by Allen and colleagues probably reflects poor drug penetration into the implant owing to lack of implant vascularization and interference of drug influx resulting from encapsulation by murine peritoneal cells. Weber and colleagues used a similar agarose plug implant model system to demonstrate that the novel peptide aldehyde PA8, which exhibited no activity against MCMV, was effective against HCMV *in vivo* with antiviral activity similar to GCV (Weber *et al.*, 2000).

Weber and colleagues utilized an alternative cellular implant model system originally developed by Casciari and colleagues as a solid tumour model to test the anti-HCMV activity of a novel compound (Casciari *et al.*, 1994; Weber *et al.*, 2001). This model system involves the culture of human cells within polyvinylidine fluoride hollow fibres, which can be implanted IP or SC in SCID mice (Weber *et al.*, 2001). It was found that a novel non-nucleosidic compound

3-hydroxy-2,2-dimethyl-N-[4((([5-(dimethylamino)–1-naphthyl]sulfonyl)amino)-phenyl]pronanamide (BAY38–4766) exhibited antiviral activity against HCMV comparable to GCV.

Human immune system (HIS) mouse models of viral infection

A key drawback to both human tissue implant and cellular implant SCID-hu model systems is that viral infection is localized to the human cells contained within the engrafted tissue or implant. Therefore, these SCID-hu model systems are inadequate for examining inter-organ viral spread, systemic viral pathogenesis, and potentially other elements of the virus life cycle, such as viral latency and reactivation in the case of human herpesviruses. Owing to these drawbacks, human tissue implant and cellular implant SCID-hu model systems are not useful for the discovery and development of antiviral drugs that operate by inhibiting establishment of viral latency, reactivation from viral latency, or inter-organ viral spread.

Recently, non-obese diabetic (NOD)/SCID mice engrafted with huCD34+ HSCs have been useful tools to study human viruses that involve haematopoietic cells in their life cycle. The NOD strain of mice, which has been used as a model of spontaneous autoimmune T-cell-mediated insulin-dependent diabetes mellitus, is characterized by defects in innate immunity, including NK cell functional deficiencies, defective complement-dependent haemolytic activity, and defects in the differentiation and function of antigen presenting cells (Legrand et al., 2006). NOD/SCID mice can be stably engrafted with human CD34+ HSCs following sub-lethal irradiation due to a lack of an adaptive immune system and innate immunity defects. Engrafted huCD34+ HSCs in HIS mice self-renew in the BM and provide long-term repopulation of mice with multilineage human myeloid and lymphoid cell populations. As described below, HIS mouse models have been used to recapitulate aspects of pathogenesis and viral life cycle for DV, HIV, EBV, KSHV, and most recently HCMV.

HIS model systems of HIV pathogenesis, immunology and therapy

As discussed above, SCID-hu mice engrafted with human fetal tissue support HIV infection of lymphoid and myelomonocytic lineages and have been integral in bridging the gap between laboratory studies and clinical studies of anti-HIV compounds (Kaneshima et al., 1991). Similar to the fetal tissue engrafted SCID-hu mouse models for human herpesviruses,

HIV-1 infection of fetal tissue xenograft SCID mice is restricted to the transplanted organ (Berkowitz et al., 1998). SCID mice engrafted with HIV-infected human peripheral blood lymphocytes from HIV patients (hu-PBL-SCID) were developed which support systemic HIV infection and are amenable to evaluation of anti-retroviral therapies (Boyle et al., 1995). However, both fetal tissue engrafted SCID-hu mice and hu-PBL-SCID mice do not support human haematopoietic cell multilineage differentiation and de novo human haematopoiesis thereby limiting the duration of HIV infection in the mice and preventing the evaluation of progenitor and broad multilineage haematopoietic cell involvement in HIV pathogenesis (Legrand et al., 2009).

The need for long-term HIV infection of mice having a broad spectrum of multilineage human haematopoietic cells has driven the development of numerous HIS mouse model systems of HIV pathogenesis. HSC-engrafted NOD/SCID mice have low to undetectable levels of human progenitor engraftment within the murine thymus, resulting in a lack of T-cell maturation (Legrand et al., 2006). NOD-*scid* IL2rγ*null* (NOG) mice have additional defects in innate immunity resulting from IL2R deficiency, and this correlates with higher HSC engraftment levels, engraftment of human progenitors within the murine thymus, T lymphocyte maturation, and improved human haematolymphoid development in comparison to conventional NOD/SCID mice (Ito et al., 2002; Legrand et al., 2006). Given that human HSC-engrafted NOG mice support differentiation of human T lymphocytes, albeit at low levels, NOG mice have been the preferred HSC transplant recipient for a number of HIV studies having shown that HSC-engrafted NOG mice support both long-term HIV infection and high levels of HIV replication (Watanabe et al., 2007a,b).

Although HSC-engrafted NOG mice exhibit some level of human progenitor engraftment of the murine thymus, the levels of mature T-cell populations in these mice are reported to be low (Legrand et al., 2006, 2009). Therefore, the conventional approach of using HSC-engrafted NOG mice for HIV studies is suboptimal for the evaluation of HIV-specific CD8+ lymphocyte responses and infection of mature CD4+ lymphocyte populations. A recently developed technique to improve both the level of human HSC engraftment and the degree of human T-cell maturation in NOG mice is to engraft newborn mice with human HSCs (Legrand et al., 2006, 2009). The enhanced ability of HSC-engrafted newborn mice to develop mature T-cells is proposed to occur by two potential mechanisms. First, thymic involution may be too advanced in adult NOG mice to adequately support human T-cell development (Legrand *et al.*, 2006). Second, myeloid lineage cells of newborn mice could be more tolerant to xenogeneic engraftment. Sato and colleagues report that NOG mice engrafted with CD34+ HSCs at 0–2 days of age and infected IP with HIV at 12 to 13 weeks post engraftment supported a persistently high viral load in the peripheral blood, disseminated infection to the spleen, the *de novo* generation and proliferation of effector/memory CD8+ T lymphocytes in response to infection, and the presence of HIV-specific human IgG in two of three mice (Sato *et al.*, 2012).

Melkus and colleagues utilized a different approach to overcome the T-cell maturation deficiency of NOG mice (Melkus *et al.*, 2006). These authors developed a HIS variant model system to evaluate specific adaptive and innate immune responses to EBV and the superantigen toxic shock syndrome toxin 1 (TSST-1). The goal in designing this HIS variant model system was to take advantage of the systemic repopulation of human haematopoietic cells of the human HSC-engrafted HIS model systems and the ability of the thymic organoid of Thy/Liv SCID-hu mice to support positive and negative autologous human T-cell selection. These authors found that NOD/SCID mice dually engrafted with a fetal human Thy/Liv organoid and autologous fetal human HSCs, which have been termed bone marrow/liver/thymus (BLT) mice, exhibited both a marked increase in human haematopoietic engraftment and the development of human T-cells capable of mounting potent MHC-I- and MHC-II-restricted immune responses to EBV and TSST-1. While NOG mice engrafted with human HSCs as newborns support T-cell selection within the mouse thymus, Melkus and colleagues report that BLT mice support long-term thymopoiesis and T-cell maturation within the human thymic organoid. Using enhanced green fluorescence protein (EGFP)-expressing HSCs as the source for HSC reconstitution of the Thy/Liv implanted mice, these authors further demonstrated that the reconstituted HSCs seed the thymic organoid resulting in *de novo* thymopoiesis and T-cell homeostasis.

Owing to the distinct advantage of the BLT model in terms of human T-cell maturation, Sun and colleagues as well as Denton and colleagues utilized BLT mice to develop intrarectal and vaginal HIV transmission model systems, respectively (Sun *et al.*, 2007; Denton *et al.*, 2008). Prior to these studies, small animal model systems were not available to investigate the processes of intrarectal and vaginal HIV transmission. These studies revealed that both the female reproductive tract and the gastrointestinal tract of BLT mice exhibited reconstitution with HSC-derived human CD4+ T-cells. HIV-infected BLT mice exhibited plasma viraemia, disseminated HIV infection to multiple organs

accompanied by tissue damage, the presence of human IgG anti-HIV-specific antibodies in the plasma of two out of three mice, and systemic CD4+ T-cell depletion. BLT mice have great potential for evaluating mucosal HIV transmission and prevention, including the preclinical evaluation of anti-HIV therapeutic agents and pre-exposure prophylaxis. These recent improvements in HIS mouse technology are leading investigators closer to small animal models suitable for HIV vaccine testing.

HIS model systems of herpesvirus pathogenesis, immunology and therapy

Epstein–Barr virus (HHV4)

The restricted tropism of EBV has hindered the development of an adequate animal system to study EBV-induced lymphoproliferative disease. Marmosets have previously been shown to be susceptible to EBV infection, and the cottontop marmoset develops multifocal large-cell lymphomas following i.p. or i.m. EBV infection (Shope and Miller, 1973). However, marmosets are endangered species and are costly, therefore studies in these New World primate species are generally limited to small numbers of primates.

To overcome the limitations of primate model systems, a number of SCID-hu mouse model systems have been developed to study EBV pathogenesis. A number of groups have shown that PBMC isolated from EBV-seropositive donors can be injected into SCID mice to form EBV+ B-cell lymphomas (Mosier, 1990, 1992a,b). Such EBV SCID-hu models, however, have several drawbacks, including variability resulting from human donor-derived PBMC and the inability to determine the effects of *in vivo* infection of mice in the development of EBV-induced lymphoproliferative disease (see Table I.23.3). To overcome these limitations, Islas-Ohlmayer and colleagues developed a HIS model system in which NOD/SCID mice engrafted with human CD34+ HSCs were infected with EGFP-tagged EBV by direct injection into the spleen (Islas-Ohlmayer *et al.*, 2004). These authors reported that EBV infected HIS mice exhibited high levels of viral DNA in the peripheral blood and developed large EBV+ immunoblastic lymphomas. Interestingly, tumour cells exhibited a type II latency programme, which is associated with nasopharyngeal carcinoma, Hodgkin's disease, and peripheral T-cell lymphoma.

Yajima and colleagues developed an alternative HIS mouse model using NOG mice (Yajima *et al.*, 2008, 2009). This group demonstrated that HSC-engrafted NOG mice exhibited B cell lymphoproliferative

Table I.23.3 Humanized NOD/SCID mouse models of virus infection

Virus	Infection	Disease outcome	Proven utility
Epstein–Barr virus (EBV)	Viral DNA detected in spleen and B cells of peripheral blood	Develop lymphomas with type II latency programme	EBV-induced lymphoproliferative disease model
Dengue virus	Viral infection detected in spleen	Develop viraemia and clinical signs of infection (rash, fever, thrombocytopenia)	Development of Dengue disease animal model
Kaposi's sarcoma virus (HHV8)	Virus detected in B cells and macrophages of the spleen	No T- or B-cells, fewer NK cells and defective complement activities	Development of KS animal model
Human Immunodeficiency virus (HIV)	Systemic infections are possible	Global reduction in CD4+ T-cells	Antiviral testing

disorder when infected with a high dose of EBV, and an asymptomatic persistent infection when infected with a low dose of EBV (Yajima *et al.*, 2008). Furthermore, they determined that NOG mice exhibited an EBV-specific T-cell response and produced IgM specific to EBV BFRF3. By depleting CD3+ and CD8+ T-cells with anti-CD3 and anti-CD8 antibodies, respectively, they found that the life span of infected HIS mice was reduced (Yajima *et al.*, 2009). CD8+ T-cells isolated from infected HIS mice were also able to suppress the outgrowth of *in vitro* EBV-infected autologous B cells isolated from uninfected mice. Together, these studies suggest that human CD8+ T-cells have the capacity to control EBV infection and suppress EBV-induced lymphocyte transformation in HIS mice infected with low doses of EBV.

As discussed above, Melkus and colleagues developed and utilized BLT mice to evaluate MHC-I and MHC-II-restricted immune responses to EBV (Melkus *et al.*, 2006). The BLT model has the advantages of superior human HSC engraftment levels, positive and negative selection of thymocytes within the autologous Thy/Liv organoid, and the *de novo* generation of both adaptive and innate immune responses. The BLT hybrid model system is a promising model for evaluating new therapeutic modalities for EBV and could potentially serve as an EBV vaccine challenge model system.

Kaposi's sarcoma virus (HHV8)

Wu and colleagues as well as Parsons and colleagues independently developed two different HIS mouse model systems for KSHV with the hope that the presence of human haematopoietic lineage cells would support whole-virus infection, latent and lytic viral gene expression, and infection of a wide array of tissue types (Parsons *et al.*, 2006; Wu *et al.*, 2006). Previous

studies have suggested that monocytes/macrophages and B-lymphocytes may be key in the process of viral dissemination. KSHV has also been detected in circulating haematopoietic progenitor cells in KS patients, further suggesting a potential role for haematopoietic lineage cells in viral latency or persistence.

The strategy of Wu and colleagues involved the inoculation of *in vitro* KSHV.219-infected CD34+ HSCs into NOD/SCID mice at 2 days post sublethal irradiation (Wu *et al.*, 2006). KSHV.219 is a recombinant KSHV expressing GFP from the latent elongation factor 1-α promoter and Ds red fluorescence protein (RFP) from the polyadenylated nuclear (PAN) promoter (active during lytic replication). The advantage of this approach is that a high percentage of human haematopoietic lineage cells would be infected if the engraftment is successful. However, this approach does not account for alterations in HSC phenotype following *in vitro* infection and does not reproduce a natural infection, whereby HSCs are infected following viral dissemination from the initial site of infection. In a comparison of mock-infected, UV-inactivated KSHV.219-treated, and KSHV.219-infected HSCs by *in vitro* colony forming assays, Wu and colleagues found that KSHV.219 infection *in vitro* suppressed haematopoiesis with a significant reduction in clonogenic colony formation activity (CFA) compared to UV-inactivated KSHV.219-treated and mock-infected HSC controls. Notably, UV-inactivated KSHV.219-treated HSCs exhibited elevated CFA compared to mock-infected HSC controls, suggesting that cellular or viral cytokines or growth factors in the inoculums stimulate CFA and/or that viral proteins of inactivated virions directly stimulate cellular signal transduction associated with CFA. 21% of granulocyte/monocyte precursors (CFU-GM) and 40% of high proliferative potential (HPP) progenitor cells tested positive for

KSHV.219 DNA with approximately 5.3 and 1.7 KSHV genomic copies per human cell, respectively. Erythroid precursors (BFU-E) did not test positive for KSHV.219 DNA. KSHV.219-infected clonogenic colonies further expressed both GFP and RFP at 28 days post infection, suggesting that colonies derived from KSHV.219-infected HSCs become fully permissive for KSHV gene expression. These authors did not report the phenotype of these permissive clonogenic populations. However, it would have been interesting to determine if cellular permissiveness for lytic gene expression was linked to differentiation of the HSCs into specific mature lineages similar to what is seen with HCMV.

Wu and colleagues further found that KSHV.219 persistently infected NOD/SCID mice following engraftment with KSHV.219-infected CD34+ HSCs. KSHV.219 DNA was detected in the spleen and BM of 14/16 mice and the peripheral blood of 10/12 mice. Ten mice engrafted with KSHV.219-infected HSCs expressed detectable levels of GFP in cells isolated from the spleen and BM, and GFP expression colocalized with either CD14 or CD19 expression, indicating KSHV.219 gene expression in monocytes/macrophages and B cells, respectively. These authors confirmed viral gene expression by showing detectable levels of viral transcripts for LANA-1 (ORF73) and Rta (ORF50). Overall, this study suggests that B-lymphocytes and monocytes/macrophages may participate in viral persistence and dissemination. They propose that infected monocytes transport KSHV to tissues, whereupon neighbouring cells are infected and subsequently differentiate into the latently infected spindle-like endothelial macrophages found in KS lesions. Also of note, Wu and colleagues found that three mice engrafted with KSHV.219-infected HSCs developed pleural effusions at 14 to 23 weeks post-engraftment. While these pleural effusions consisted primarily of murine B or pro-B lymphocytes, GFP+ monocytes/macrophages were present. Inoculation of cells from the pleural effusions into naïve NOD/SCID mice resulted in the development of neoplasms. These authors further propose to determine whether or not the infected monocytes/macrophages of these pleural effusions play a causative role for lymphomagenesis.

In contrast to that study, Parsons and colleagues directly injected KSHV virions i.v. into HIS mice at 3-weekly intervals (Parsons *et al.*, 2006). This approach would presumably recapitulate viral dissemination to and infection of HSCs, but would likely have a lower incidence of HSC infection. These investigators found that KSHV-infected NOD/SCID-hu mice exhibited increasing amounts of KSHV genomic DNA, latent viral transcripts (ORF73), and lytic viral transcripts (ORF50 and ORF65) in the spleen between 1–4 months post infection. Immunofluorescence-based assays revealed that 0.5% to 1% of host spleen cells in infected HIS mice exhibited punctate intranuclear staining for latency-associated nuclear antigen (LANA; ORF73), while only 0.001% to 0.1% of host spleen cells exhibited membrane or cytoplasmic staining for the lytic viral gene product K8.1. Electron microscopy (TEM) confirmed lytic replication with the detection of mature virions within splenocytes at a low frequency. Intranuclear LANA staining was seen in B220+, Ly49+, CD11b+, and CD11c+ splenocytes, suggesting KSHV infection of B cells, NK cells, monocytes/macrophages, and dendritic cells, respectively. Co-staining of LANA and CD3 was negligible indicating that T-cells were not infected in the spleens of HIS mice. Parsons and colleagues reported that pre-emptive GCV treatment of HIS mice prior to infection significantly reduced KSHV DNA levels, delayed LANA (ORF73) mRNA expression, and suppressed expression of lytic transcripts (ORF50 and ORF65) compared with non-GCV-treated KSHV-infected HIS mice. However, the findings that KSHV DNA levels increased at later time points and that latent gene expression was merely delayed in HIS mice pre-treated with GCV prior to infection indicate that GCV inhibits lytic replication but is unable to block the establishment and maintenance of latency. These findings suggest that the initial establishment of latency in the model system either occurs independent of productive lytic replication or requires only low levels of lytic replication.

Human cytomegalovirus (HHV5)

As discussed in detail above, myeloid lineage cells play an integral role in viral latency, persistence, dissemination to organ tissues, and pathogenesis. HCMV SCID-hu mouse models containing human fetal tissue xenografts are not adequate for addressing questions relating to latency, persistence, and dissemination because they do not support *de novo* generation of human BM-derived myeloid precursor cells and mature myeloid lineage cells and do not support spread beyond the xenograft tissue. Although SCID-hu mice engrafted with human fetal tissue do contain myeloid lineage cells within the engrafted tissue, myeloid cells within the engrafted tissue were not detectably infected in either the Thy/Liv model or the retinal transplant model. While Bidanset and colleagues reported that HCMV IE1 colocalized with the glial cell marker GFAP, GFAP+ populations in the brain do not include the myeloid lineage microglial cells (Bidanset *et al.*, 2001).

To overcome such limitations in HCMV SCID-hu model systems, Smith and colleagues developed an HSC-engrafted NOG mouse model for HCMV latency and reactivation (Smith *et al.*, 2010). In this model, 7 to 10 week old NOG mice were sublethally irradiated and engrafted with CD34$^+$ cord blood cells (see Fig. I.23.1). At 8 weeks post engraftment, approximately 5% of PBMC were human monocytes (huCD45$^+$ huCD33$^+$ huCD14$^+$), which is approximately half of the percentage of monocytes comprising the PBMC of healthy humans (Smith *et al.*, 2010). In testing multiple routes of HCMV infection, these authors report that i.p. injection of HCMV-infected fibroblasts at 4 weeks post engraftment was the only route of infection leading to detectable viral DNA in organ tissues. HCMV DNA was detected in the BM of all three HCMV-infected, engrafted NOG mice tested and in the spleen and kidney of two-thirds and one-third of mice, respectively. HCMV DNA was not detected in HCMV-infected, non-engrafted NOG mice or in mock-infected, engrafted NOG mice, indicating that HCMV virions or DNA was not persisting in murine cells and that the donor stem cells were not contaminated with HCMV. While an average of approximately 300 HCMV DNA copies per μg of BM tissue was detected in HCMV-infected, engrafted NOD mice at 6 weeks

post infection, HCMV mRNA expression associated with lytic infections was absent in these mice, which was suggestive of a latent HCMV infection of the BM (Smith *et al.*, 2010).

HCMV DNA was not detected in the peripheral blood of HCMV-infected, engrafted NOG mice, which is consistent with the low to undetectable levels of HCMV in the blood of healthy HCMV-infected humans (Hudnall *et al.*, 2008; Smith *et al.*, 2010). Smith and colleagues hypothesized that granulocyte colony-stimulating factor (G-CSF) would promote viral spread to the peripheral blood and organ tissues and that G-CSF would trigger reactivation of HCMV from latency based on two lines of evidence. First, G-CSF is known to robustly mobilize monocytes in human HSC-engrafted NOG mice and in humans (Gyger *et al.*, 2000; Hess *et al.*, 2007). As discussed above, it has been proposed that a population of latently infected peripheral blood monocytes derived from a latently infected myeloid progenitor cells of the BM reactivate HCMV from latency during their migration into organ tissues and their subsequent differentiation into tissue macrophages (Smith *et al.*, 2004a). Second,

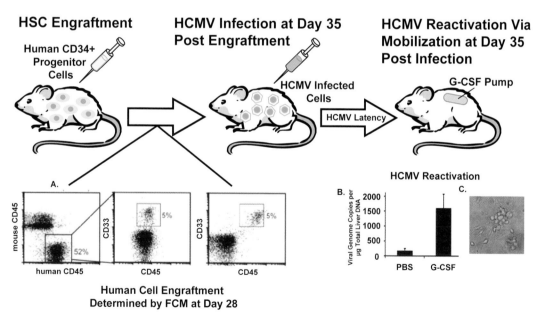

Figure I.23.1 HSC-engrafted NOG mouse model for HCMV latency and reactivation. NOG mice are engrafted with human CD34$^+$ stem cells isolated from cord blood. At 28 days the mice are checked for the level of human cell engraftment by flow cytometry (FCM) using antibodies specific for human CD45 and cell type specific markers such as CD14 (monocyte marker; as shown in panel A). At 35 days, the engrafted mice are infected with HCMV-infected human fibroblasts (1 × 10^7) by intraperitoneal injection. The infection establishes a latent infection in the bone marrow. To induce viral reactivation the mice are treated with G-CSF via an osmotic pump at 35 days post infection. After 7–14 days post G-CSF treatment viral reactivation can be detected in bone marrow, liver and spleen. Animals receiving osmotic pumps containing PBS failed to reactivate virus (shown in panel B). Mice infected with HCMV containing a GFP expression cassette can be detected in bone marrow cells post-mobilization (shown in panel C).

studies have suggested that the use of G-CSF mobilized peripheral blood stem cells isolated from HCMV⁺ donors doubles the risk of late-onset HCMV disease in stem cell transplant recipients in comparison to patients receiving BM transplants from HCMV⁺ donors and that recipients of CD34⁺ progenitor cells from HCMV⁺ donors have a marked increase in HCMV disease and HCMV-associated deaths but the precise mechanisms for these effect are still unknown (Holmberg *et al.*, 1999; Anderson *et al.*, 2003) (see Table I.23.4).

Smith and colleagues report that G-CSF mobilization of human HSC-engrafted NOG mice robustly increased the percentage of monocytes comprising the PBMC to approximately 24% (Smith *et al.*, 2010). G-CSF mobilization in HCMV-infected, engrafted NOG mice correlated with HCMV spread to the peripheral blood, spleen, liver and kidney in all mice tested and with detectable HCMV spread to the lung, submandibular salivary gland, and bladder in some mice (Smith *et al.*, 2010). Moreover, early and late HCMV transcripts were present in the liver tissue of all HCMV-infected, G-CSF mobilized NOG mice and absent in HCMV-infected, non-mobilized NOG mice. Immunofluorescence staining of liver tissue revealed that HCMV early and late proteins colocalized exclusively with human monocyte and macrophage markers (Smith *et al.*, 2010).

The same study also examined the effects of co-treatment of HCMV-infected NOG mice with AMD3100 (Plerixafor) and G-CSF. AMD3100 increases haematopoietic cell mobilization by blocking stromal cell-derived factor-1 (SDF-1) binding to CXCR4 (Broxmeyer *et al.*, 2005; Pitchford *et al.*, 2009). SDF-1 is constitutively expressed within the bone marrow, and its interaction with CXCR4 promotes retention of haematopoietic stem cells within the bone marrow compartment (Aiuti *et al.*, 1997). Clinically, AMD3100 can be used in conjunction with G-CSF treatments to boost HSC mobilization in stem cell donors who respond poorly to G-CSF alone. Smith and colleagues found that G-CSF and AMD3100 co-treatment significantly increased the number of human monocytes per µl of blood at day 3 of treatment over G-CSF treatment alone (Smith *et al.*, 2010). This increase in mobilized human monocytes correlated with a significant increase in viral load over G-CSF treatment alone, thus further supporting the link between increased mobilization of monocytes from the BM and increased viral dissemination to organ tissues.

In summary, Smith and colleagues demonstrated that HSC-engrafted NOG mice support a latent HCMV infection that reactivates in disseminated tissue macrophages following G-CSF-induced mobilization of haematopoietic cells of the BM. They propose that G-CSF mobilized blood products contain myeloid cells primed for HCMV replication and that their use could pose an elevated risk for HCMV transmission and disease in transplant recipients (Smith *et al.*, 2010). These findings are significant because they provide the first HCMV animal model capable of supporting a systemic and latent infection. Furthermore, this study could have a direct impact on current protocols for stem cell transplant procedures. Smith and colleagues further propose to explore the elevated risk for HCMV transmission to G-CSF-mobilized stem cell recipients in further detail by comparing HCMV transmission to NOG mice receiving a stem cell transplant with either BM or G-CSF-mobilized cells collected from HCMV-infected, engrafted NOG mice (Smith *et al.*, 2010).

Perspectives: future directions for humanized mouse models of HCMV

Substantial advances in SCID-hu mouse technology over the last two decades have brought researchers closer to mouse models of human-tropic viruses that faithfully recapitulate viral pathogenesis and immunology in humans. Virologists with a research focus on small animal models of human-tropic viruses are

Table I.23.4 G-CSF administration mediates efficient mobilization

Cellular differentiation marker	Per cent in blood pre-G-CSF	Per cent in blood post-G-CSF
CD33	2.26±0.7	13.9±2.87
CD14	1.11±0.21	12.89±2.09
CD20	37.18±11.0	27.07±5.96
CD19	52.93±6.69	42.45±12.39
CD45	65.36±5.77	47.74±12.42
CD34	ND	0.49±0.19
Ly5.2	40.62±8.32	43.7±14.65

transitioning from a phase consisting primarily of model development and optimization into a phase in which key issues of viral pathobiology and immunity can be addressed *in vivo*. It will be interesting to see how HCMV virologists implement current and future findings relating to HIS mouse technology into the continued development of an HCMV small animal model system and how this will enhance or transform our understanding of HCMV latency, persistence, reactivation, immunology, and therapy (see Table I.23.5).

With respect to the current state of HIS mouse technology, one can envisage that optimization of the HCMV HIS mouse model developed by Smith and colleagues could address a number of key questions regarding HCMV latency in the near future. While it is known that $CD34^+$ stem cells support HCMV latency, $CD34^+$ stem cells are heterogeneous and comprise pluripotent stem cell, lymphoid lineage committed stem cell, and myeloid committed stem cell populations. Although an effort has been made to better define the precise stem cell populations harbouring latent HCMV, reproducing the conditions in which stem cell infection, latency, and reactivation would occur *in vivo* is technically challenging and subject to experimental bias. Therefore, the $CD34^+$ progenitor cell phenotype harbouring latent HCMV is still uncertain.

HIS mice capable of supporting latent HCMV infection such as in the model reported by Smith and colleagues will undoubtedly be valuable tools in delineating the latent stem cell phenotype from the heterogeneous $CD34^+$ stem cell population. It is possible that this model could be modified specifically to address this question. For instance, it can be anticipated that the quantity of latently infected stem cells in the BM would be a limiting factor in the isolation and characterization of latently infected cell populations. This potential obstacle could be resolved with the utilization of BLT mice or NOG mice engrafted with HSCs as newborns, both of which have higher engraftment levels, as discussed above. Alternatively or in conjunction with the use of BLT mice or newborn engrafted NOG mice, the expression of human genes in HIS mice could augment the homeostasis, multilineage development, or tissue distribution and levels of human cells. For example, the expression of human growth factors such as human stem cell factor (SCF), which promotes

the establishment and maintenance of HSCs, could improve human cell homeostasis, engraftment levels, and tissue distribution.

One can also envisage that the use of recombinant HCMV tagged with or expressing a fluorescent protein such as EGFP could enable the isolation of infected haematopoietic progenitor cells from the BM (see Fig. I.23.1C). This could not only enable the determination of the earliest stem cell lineage capable of supporting latent infection, but could also allow the *ex vivo* determination of triggers for viral reactivation.

Of equal importance to our understanding of the earliest cell lineage reservoir of latency is identifying the roles of specific HCMV gene products in the establishment and maintenance of latency and in reactivation (see also the latency-related Chapters I.19-I.22). HIS mice will be a valuable tool in evaluating putative genes associated with latency and reactivation identified in both *in vitro* and *ex vivo* studies, because viral genes of interest can be evaluated in HIS mice in the context of deletion viruses or recombinant viruses conditionally expressing the gene of interest. Umashankar and colleagues recently utilized the HCMV HIS model developed by Smith and colleagues to extend *in vitro* findings regarding the role of the UL133–UL138 locus in modulating cell type-specific outcomes of infection to an *in vivo* model of viral reactivation and spread (Umashankar *et al.*, 2012). The UL133-UL138 locus is located in the ULb' region, which is common to all clinical isolates and dispensable for growth *in vitro* (Cha *et al.*, 1996; Murphy *et al.*, 2003; Dolan *et al.*, 2004). Petrucelli and colleagues determined that disruption of the UL133–UL138 locus results in a loss of latency phenotype in HPCs infected *in vitro* (Petrucelli *et al.*, 2009). In this context, Umashankar and colleagues report that human HSC-engrafted NOG mice infected with UL133–UL138$_{null}$ HCMV exhibited increased viral replication and dissemination following G-CSF-induced stem cell mobilization in comparison to those mice infected with the wild type HCMV clinical isolate TB40E (Umashankar *et al.*, 2012). These results suggest an important restricting role for the UL133-UL138 locus in HCMV reactivation and dissemination *in vivo*.

Given that HIS mouse models overcome the limited access to organ tissues, BM, and peripheral blood of HCMV$^+$ human patients and allow the pooling of

---

Table I.23.5 Possible uses of humanized NOD/SCID HCMV mouse model

Determine sites of HCMV latency and identify gene products expressed during latency *in vivo*
Determine the mechanisms of HCMV reactivation and dissemination
Generate model for HCMV-associated transplant disease
Evaluate antiviral modalities (drugs, vaccines, gene therapy)

multiple, genetically identical mouse tissue samples, HIS mouse models are ideally suited for bio-informatic analyses of both HCMV gene expression and cellular gene expression of HCMV infected cells and tissues. Viral genes associated with latency could be identified, for instance, by determining the latency-associated viral gene expression profiles of FACS-sorted EGFP⁺-infected BM cells from NOG mice infected with EGFP-expressing HCMV. Candidate latency-associated viral genes could then be evaluated *in vivo* by infecting HSC-engrafted NOG mice with a BAC-derived HCMV deleted for the gene of interest or by infecting mice with BAC-derived HCMV expressing the gene of interest under an inducible promoter.

Klenovsek and colleagues demonstrated that mice infected with the MCMV expressing luciferase can be subjected to live *in vivo* imaging to evaluate viral spread to organ tissue in the presence or absence of an antiviral therapy (Klenovsek *et al.*, 2007) (see Chapter II.10). Likewise, a recombinant luciferase-expressing HCMV could also be utilized for live *in vivo* imaging of infected NOG mice to evaluate the kinetics of viral spread to specific tissues and to evaluate how HCMV gene products or anti-HCMV therapies affect tissue-specific spread and persistence.

As noted with respect to the BLT mouse studies for EBV and HIV and the newborn HSC-engrafted NOG mouse studies for HIV and DV, the potential to evaluate HCMV-specific immune responses exists, thus HCMV HIS mouse models could be utilized to evaluate the efficacy of candidate HCMV vaccines. Given that CMVs are highly immunogenic viruses and that CMVs can re-infect and disseminate in hosts with a prior history of CMV infection, CMV is now considered to be an attractive vaccine vector (Farroway *et al.*, 2005; Sylwester *et al.*, 2005; Hansen *et al.*, 2009; Ross *et al.*, 2010; Tsuda *et al.*, 2012; see Chapter II.21). Tsuda and colleagues recently developed a replicating MCMV-based vaccine encoding a CD8⁺ T-cell epitope from the nucleoprotein of *Zaire ebolavirus* that protected mice from lethal Ebola virus challenge (Tsuda *et al.*, 2012). Hansen and colleagues report that rhesus macaques infected with a rhesus cytomegalovirus (RhCMV)-based simian immunodeficiency virus (SIV) vaccine vector encoding SIV Gag, Rev/Nef/Tat, and Env were resistant to acquisition of progressive SIV infection following repeated intrarectal challenge (Hansen *et al.*, 2009) (for non-human primate models of CMV infections, see also Chapter II.22). Lack of animal model systems permissive to HCMV infection has impeded the preclinical testing of HCMV-based vaccine vectors. With the here reviewed development of an HCMV HIS mouse model and with continued advancements in HIS mouse technology, it can be expected that HCMV-based vaccine vectors for human pathogens and cancers will be explored in the near future.

Acknowledgement

We would like to thank Andrew Townsend for graphics assistance.

References

Aiuti, A., Webb, I.J., Bleul, C., Springer, T., and Gutierrez-Ramos, J.C. (1997). The chemokine SDF-1 is a chemoattractant for human CD34+ hematopoietic progenitor cells and provides a new mechanism to explain the mobilization of CD34+ progenitors to peripheral blood. J. Exp. Med. *185*, 111–120.

Allen, L.B., Li, S.X., Arnett, G., Toyer, B., Shannon, W.M., and Hollingshead, M.G. (1992). Novel method for evaluating antiviral drugs against human cytomegalovirus in mice. Antimicrob. Agents. Chemother. *36*, 206–208.

Anderson, D., DeFor, T., Burns, L., McGlave, P., Miller, J., Wagner, J., and Weisdorf, D. (2003). A comparison of related donor peripheral blood and bone marrow transplants: importance of late-onset chronic graft-versus-host disease and infections. Biol. Blood Marrow Transplant. *9*, 52–59.

Balthesen, M., Messerle, M., and Reddehase, M.J. (1993). Lungs are a major organ site of cytomegalovirus latency and recurrence. J. Virol. *67*, 5360–5366.

Beadle, J.R., Hartline, C., Aldern, K.A., Rodriguez, N., Harden, E., Kern, E.R., and Hostetler, K.Y. (2002). Alkoxyalkyl esters of cidofovir and cyclic cidofovir exhibit multiple-log enhancement of antiviral activity against cytomegalovirus and herpesvirus replication in vitro. Antimicrob. Agents Chemother. *46*, 2381–2386.

Berkowitz, R.D., Alexander, S., Bare, C., Linquist-Stepps, V., Bogan, M., Moreno, M.E., Gibson, L., Wieder, E.D., Kosek, J., Stoddart, C.A., et al. (1998). CCR5- and CXCR4-utilizing strains of human immunodeficiency virus type 1 exhibit differential tropism and pathogenesis in vivo. J. Virol. *72*, 10108–10117.

Bidanset, D.J., Rybak, R.J., Hartline, C.B., and Kern, E.R. (2001). Replication of human cytomegalovirus in severe combined immunodeficient mice implanted with human retinal tissue. J. Infect. Dis. *184*, 192–195.

Bidanset, D.J., Beadle, J.R., Wan, W.B., Hostetler, K.Y., and Kern, E.R. (2004a). Oral activity of ether lipid ester prodrugs of cidofovir against experimental human cytomegalovirus infection. J. Infect. Dis. *190*, 499–503.

Bidanset, D.J., Rybak, R.J., Hartline, C.B., and Kern, E.R. (2004b). Efficacy of ganciclovir and cidofovir against human cytomegalovirus replication in SCID mice implanted with human retinal tissue. Antiviral Res. *63*, 61–64.

Booss, J., Dann, P.R., Griffith, B.P., and Kim, J.H. (1989). Host defense response to cytomegalovirus in the central nervous system. Predominance of the monocyte. Am. J. Pathol. *134*, 71–78.

Bosma, M.J., and Carroll, A.M. (1991). The SCID mouse mutant: definition, characterization, and potential uses. Annu. Rev. Immunol. *9*, 323–350.

Boyle, M.J., Connors, M., Flanigan, M.E., Geiger, S.P., Ford, H., Jr., Baseler, M., Adelsberger, J., Davey, R.T., Jr., and

Lane, H.C. (1995). The human HIV/peripheral blood lymphocyte (PBL)-SCID mouse. A modified human PBL-SCID model for the study of HIV pathogenesis and therapy. J. Immunol. *154*, 6612–6623.

Bravo, F.J., Cardin, R.D., and Bernstein, D.I. (2007). A model of human cytomegalovirus infection in severe combined immunodeficient mice. Antiviral. Res. *76*, 104–110.

Brown, J.M., Kaneshima, H., and Mocarski, E.S. (1995). Dramatic interstrain differences in the replication of human cytomegalovirus in SCID-hu mice. J. Infect. Dis. *171*, 1599–1603.

Broxmeyer, H.E., Orschell, C.M., Clapp, D.W., Hangoc, G., Cooper, S., Plett, P.A., Liles, W.C., Li, X., Graham-Evans, B., Campbell, T.B., *et al.* (2005). Rapid mobilization of murine and human hematopoietic stem and progenitor cells with AMD3100, a CXCR4 antagonist. J. Exp. Med. *201*, 1307–1318.

Brytting, M., Mousavi-Jazi, M., Bostrom, L., Larsson, M., Lunderberg, J., Ljungman, P., Ringden, O., and Sundqvist, V.A. (1995). Cytomegalovirus DNA in peripheral blood leukocytes and plasma from bone marrow transplant recipients. Transplantation *60*, 961–965.

Casciari, J.J., Hollingshead, M.G., Alley, M.C., Mayo, J.G., Malspeis, L., Miyauchi, S., Grever, M.R., and Weinstein, J.N. (1994). Growth and chemotherapeutic response of cells in a hollow-fiber in vitro solid tumor model. J. Natl. Cancer. Inst. *86*, 1846–1852.

Centra, M., Ratych, R.E., Cao, G.L., Li, J., Williams, E., Taylor, R.M., and Rosen, G.M. (1992). Culture of bovine pulmonary artery endothelial cells on Gelfoam blocks. FASEB J. *6*, 3117–3121.

Cha, T.A., Tom, E., Kemble, G.W., Duke, G.M., Mocarski, E.S., and Spaete, R.R. (1996). Human cytomegalovirus clinical isolates carry at least 19 genes not found in laboratory strains. J. Virol. *70*, 78–83.

Chan, G., Bivins-Smith, E.R., Smith, M.S., and Yurochko, A.D. (2008). Transcriptome analysis of NF-kappaB- and phosphatidylinositol 3-kinase-regulated genes in human cytomegalovirus-infected monocytes. J. Virol. *82*, 1040–1046.

Chulay, J., Biron, K., Wang, L., Underwood, M., Chamberlain, S., Frick, L., Good, S., Davis, M., Harvey, R., Townsend, L., *et al.* (1999). Development of novel benzimidazole riboside compounds for treatment of cytomegalovirus disease. Adv. Exp. Med. Biol. *458*, 129–134.

Cowdry, E., and Scott, G. (1935). Nuclear inclusions suggestive of virus action in salivary glands of the monkey, *Cebus fatuellus* L. Proc. Soc. Exp. Biol. Med *32*, 709–711.

Denton, P.W., Estes, J.D., Sun, Z., Othieno, F.A., Wei, B.L., Wege, A.K., Powell, D.A., Payne, D., Haase, A.T., and Garcia, J.V. (2008). Antiretroviral pre-exposure prophylaxis prevents vaginal transmission of HIV-1 in humanized BLT mice. PLoS Med. *5*, e16.

DiLoreto, D., Jr., Epstein, L.G., Lazar, E.S., Britt, W.J., and del Cerro, M. (1994). Cytomegalovirus infection of human retinal tissue: an in vivo model. Lab. Invest. *71*, 141–148.

Dolan, A., Cunningham, C., Hector, R.D., Hassan-Walker, A.F., Lee, L., Addison, C., Dargan, D.J., McGeoch, D.J., Gatherer, D., Emery, V.C., *et al.* (2004). Genetic content of wild-type human cytomegalovirus. J. Gen. Virol. *85*, 1301–1312.

Epstein, L.G., Cvetkovich, T.A., Lazar, E.S., DiLoreto, D., Saito, Y., James, H., del Cerro, C., Kaneshima, H., McCune, J.M., Britt, W.J., *et al.* (1994). Human neural xenografts: progress in developing an in-vivo model to study human immunodeficiency virus (HIV) and human cytomegalovirus (HCMV) infection. Adv. Neuroimmunol. *4*, 257–260.

Farroway, L.N., Gorman, S., Lawson, M.A., Harvey, N.L., Jones, D.A., Shellam, G.R., and Singleton, G.R. (2005). Transmission of two Australian strains of murine cytomegalovirus (MCMV) in enclosure populations of house mice (Mus domesticus). Epidemiol. Infect. *133*, 701–710.

Freitas, V.R., Smee, D.F., Chernow, M., Boehme, R., and Matthews, T.R. (1985). Activity of 9-(1,3-dihydroxy-2-propoxymethyl)guanine compared with that of acyclovir against human, monkey, and rodent cytomegaloviruses. Antimicrob. Agents. Chemother. *28*, 240–245.

Gerna, G., Zipeto, D., Percivalle, E., Parea, M., Revello, M.G., Maccario, R., Peri, G., and Milanesi, G. (1992). Human cytomegalovirus infection of the major leukocyte subpopulations and evidence for initial viral replication in polymorphonuclear leukocytes from viremic patients. J. Infect. Dis. *166*, 1236–1244.

Glasgow, L.A., Richards, J.T., and Kern, E.R. (1982). Effect of acyclovir treatment on acute and chronic murine cytomegalovirus infection. Am. J. Med. *73*, 132–137.

Gnann, J.W., Jr., Ahlmen, J., Svalander, C., Olding, L., Oldstone, M.B., and Nelson, J.A. (1988). Inflammatory cells in transplanted kidneys are infected by human cytomegalovirus. Am. J. Pathol. *132*, 239–248.

Grefte, J.M., van der Gun, B.T., Schmolke, S., van der Giessen, M., van Son, W.J., Plachter, B., Jahn, G., and The, T.H. (1992). The lower matrix protein pp65 is the principal viral antigen present in peripheral blood leukocytes during an active cytomegalovirus infection. J. Gen. Virol. *73*, 2923–2932.

Grefte, A., Harmsen, M.C., van der Giessen, M., Knollema, S., van Son, W.J., and The, T.H. (1994). Presence of human cytomegalovirus (HCMV) immediate-early mRNA but not ppUL83 (lower matrix protein pp65) mRNA in polymorphonuclear and mononuclear leukocytes during active HCMV infection. J. Gen. Virol. *75*, 1989–1998.

Greiner, D.L., Hesselton, R.A., and Shultz, L.D. (1998). SCID mouse models of human stem cell engraftment. Stem Cells *16*, 166–177.

Gyger, M., Stuart, R.K., and Perreault, C. (2000). Immunobiology of allogeneic peripheral blood mononuclear cells mobilized with granulocyte-colony stimulating factor. Bone Marrow Transplant. *26*, 1–16.

Hahn, G., Jores, R., and Mocarski, E.S. (1998). Cytomegalovirus remains latent in a common precursor of dendritic and myeloid cells. Proc. Natl. Acad. Sci. U.S.A. *95*, 3937–3942.

Hansen, S.G., Vieville, C., Whizin, N., Coyne-Johnson, L., Siess, D.C., Drummond, D.D., Legasse, A.W., Axthelm, M.K., Oswald, K., Trubey, C.M., *et al.* (2009). Effector memory T-cell responses are associated with protection of rhesus monkeys from mucosal simian immunodeficiency virus challenge. Nat. Med. *15*, 293–299.

Hess, D.A., Bonde, J., Craft, T.P., Wirthlin, L., Hohm, S., Lahey, R., Todt, L.M., Dipersio, J.F., Devine, S.M., and Nolta, J.A. (2007). Human progenitor cells rapidly mobilized by AMD3100 repopulate NOD/SCID mice with increased frequency in comparison to cells from the same donor mobilized by granulocyte colony stimulating factor. Biol. Blood Marrow Transplant. *13*, 398–411.

Ho, M. (1991). Cytomegalovirus: biology and infection (Plenum, New York).

Holmberg, L.A., Boeckh, M., Hooper, H., Leisenring, W., Rowley, S., Heimfeld, S., Press, O., Maloney, D.G., McSweeney, P., Corey, L., *et al.* (1999). Increased incidence of cytomegalovirus disease after autologous CD34-selected peripheral blood stem cell transplantation. Blood 94, 4029–4035.

Hudnall, S.D., Chen, T., Allison, P., Tyring, S.K., and Heath, A. (2008). Herpesvirus prevalence and viral load in healthy blood donors by quantitative real-time polymerase chain reaction. Transfusion 48, 1180–1187.

Hume, D.A., Ross, I.L., Himes, S.R., Sasmono, R.T., Wells, C.A., and Ravasi, T. (2002). The mononuclear phagocyte system revisited. J. Leukoc. Biol. 72, 621–627.

Islas-Ohlmayer, M., Padgett-Thomas, A., Domiati-Saad, R., Melkus, M.W., Cravens, P.D., Martin Mdel, P., Netto, G., and Garcia, J.V. (2004). Experimental infection of NOD/SCID mice reconstituted with human CD34+ cells with Epstein–Barr virus. J. Virol. 78, 13891–13900.

Ito, M., Hiramatsu, H., Kobayashi, K., Suzue, K., Kawahata, M., Hioki, K., Ueyama, Y., Koyanagi, Y., Sugamura, K., Tsuji, K., *et al.* (2002). NOD/SCID/gamma(c) (null) mouse: an excellent recipient mouse model for engraftment of human cells. Blood 100, 3175–3182.

Jahn, G., Stenglein, S., Riegler, S., Einsele, H., and Sinzger, C. (1999). Human cytomegalovirus infection of immature dendritic cells and macrophages. Intervirology 42, 365–372.

Kaneshima, H., Shih, C.C., Namikawa, R., Rabin, L., Outzen, H., Machado, S.G., and McCune, J.M. (1991). Human immunodeficiency virus infection of human lymph nodes in the SCID-hu mouse. Proc. Natl. Acad. Sci. U.S.A. 88, 4523–4527.

Kern, E.R., Rybak, R.J., Hartline, C.B., and Bidanset, D.J. (2001). Predictive efficacy of SCID-hu mouse models for treatment of human cytomegalovirus infections. Antivir. Chem. Chemother. 12, 149–156.

Kern, E.R., Bidanset, D.J., Hartline, C.B., Yan, Z., Zemlicka, J., and Quenelle, D.C. (2004a). Oral activity of a methylenecyclopropane analog, cyclopropavir, in animal models for cytomegalovirus infections. Antimicrob. Agents Chemother. 48, 4745–4753.

Kern, E.R., Hartline, C.B., Rybak, R.J., Drach, J.C., Townsend, L.B., Biron, K.K., and Bidanset, D.J. (2004b). Activities of benzimidazole D- and L-ribonucleosides in animal models of cytomegalovirus infections. Antimicrob. Agents Chemother. 48, 1749–1755.

Kern, E.R., Kushner, N.L., Hartline, C.B., Williams-Aziz, S.L., Harden, E.A., Zhou, S., Zemlicka, J., and Prichard, M.N. (2005). In vitro activity and mechanism of action of methylenecyclopropane analogs of nucleosides against herpesvirus replication. Antimicrob. Agents Chemother. 49, 1039–1045.

Klenovsek, K., Weisel, F., Schneider, A., Appelt, U., Jonjic, S., Messerle, M., Bradel-Tretheway, B., Winkler, T.H., and Mach, M. (2007). Protection from CMV infection in immunodeficient hosts by adoptive transfer of memory B cells. Blood 110, 3472–3479.

Kondo, K., Kaneshima, H., and Mocarski, E.S. (1994). Human cytomegalovirus latent infection of granulocyte–macrophage progenitors. Proc. Natl. Acad. Sci. U.S.A. 91, 11879–11883.

Kyoizumi, S., Baum, C.M., Kaneshima, H., McCune, J.M., Yee, E.J., and Namikawa, R. (1992). Implantation and maintenance of functional human bone marrow in SCID-hu mice. Blood 79, 1704–1711.

Legrand, N., Weijer, K., and Spits, H. (2006). Experimental models to study development and function of the human immune system in vivo. J. Immunol. 176, 2053–2058.

Legrand, N., Ploss, A., Balling, R., Becker, P.D., Borsotti, C., Brezillon, N., Debarry, J., de Jong, Y., Deng, H., Di Santo, J.P., *et al.* (2009). Humanized mice for modeling human infectious disease: challenges, progress, and outlook. Cell Host Microbe 6, 5–9.

Lischka, P., Hewlett, G., Wunberg, T., Baumeister, J., Paulsen, D., Goldner, T., Ruebsamen-Schaeff, H., and Zimmermann, H. (2010). In vitro and in vivo activities of the novel anticytomegalovirus compound AIC246. Antimicrob. Agents. Chemother. 54, 1290–1297.

McCune, J., Kaneshima, H., Krowka, J., Namikawa, R., Outzen, H., Peault, B., Rabin, L., Shih, C.C., Yee, E., Lieberman, M., *et al.* (1991). The SCID-hu mouse: a small animal model for HIV infection and pathogenesis. Annu. Rev. Immunol. 9, 399–429.

McCune, J.M., Namikawa, R., Kaneshima, H., Shultz, L.D., Lieberman, M., and Weissman, I.L. (1988). The SCID-hu mouse: murine model for the analysis of human hematolymphoid differentiation and function. Science 241, 1632–1639.

McGavran, M.H., and Smith, M.G. (1965). Ultrastructural, cytochemical, and microchemical observations on cytomegalovirus (salivary gland virus) infection of human cells in tissue culture. Exp. Mol. Pathol. 76, 1–10.

Melkus, M.W., Estes, J.D., Padgett-Thomas, A., Gatlin, J., Denton, P.W., Othieno, F.A., Wege, A.K., Haase, A.T., and Garcia, J.V. (2006). Humanized mice mount specific adaptive and innate immune responses to EBV and TSST-1. Nat. Med. 12, 1316–1322.

Mendelson, M., Monard, S., Sissons, P., and Sinclair, J. (1996). Detection of endogenous human cytomegalovirus in CD34+ bone marrow progenitors. J. Gen. Virol. 77, 3099–3102.

Mercer, D.F., Schiller, D.E., Elliott, J.F., Douglas, D.N., Hao, C., Rinfret, A., Addison, W.R., Fischer, K.P., Churchill, T.A., Lakey, J.R., *et al.* (2001). Hepatitis C virus replication in mice with chimeric human livers. Nat. Med. 7, 927–933.

Minton, E.J., Tysoe, C., Sinclair, J.H., and Sissons, J.G. (1994). Human cytomegalovirus infection of the monocyte/macrophage lineage in bone marrow. J. Virol. 68, 4017–4021.

Mocarski, E.S., Bonyhadi, M., Salimi, S., McCune, J.M., and Kaneshima, H. (1993). Human cytomegalovirus in a SCID-hu mouse: thymic epithelial cells are prominent targets of viral replication. Proc. Natl. Acad. Sci. U.S.A. 90, 104–108.

Mosier, D.E. (1990). Immunodeficient mice xenografted with human lymphoid cells: new models for in vivo studies of human immunobiology and infectious diseases. J. Clin. Immunol. 10, 185–191.

Mosier, D.E., Picchio, G.R., Baird, S.M., Kobayashi, R., and Kipps, T.J. (1992a). Epstein–Barr virus-induced human B-cell lymphomas in SCID mice reconstituted with human peripheral blood leukocytes. Cancer Res. 52, 5552–5553.

Mosier, D.E., Picchio, G.R., Kirven, M.B., Garnier, J.L., Torbett, B.E., Baird, S.M., Kobayashi, R., and Kipps, T.J. (1992b). EBV-induced human B cell lymphomas in hu-PBL-SCID mice. AIDS Res. Hum. Retroviruses 8, 735–740.

Murphy, E., Rigoutsos, I., Shibuya, T., and Shenk, T.E. (2003). Reevaluation of human cytomegalovirus coding potential. Proc. Natl. Acad. Sci. U.S.A. *100*, 13585–13590.

Namikawa, R., Weilbaecher, K.N., Kaneshima, H., Yee, E.J., and McCune, J.M. (1990). Long-term human hematopoiesis in the SCID-hu mouse. J. Exp. Med. *172*, 1055–1063.

Parsons, C.H., Adang, L.A., Overdevest, J., O'Connor, C.M., Taylor, J.R., Jr., Camerini, D., and Kedes, D.H. (2006). KSHV targets multiple leukocyte lineages during long-term productive infection in NOD/SCID mice. J. Clin. Invest. *116*, 1963–1973.

Petrucelli, A., Rak, M., Grainger, L., and Goodrum, F. (2009). Characterization of a novel Golgi apparatus-localized latency determinant encoded by human cytomegalovirus. J. Virol. *83*, 5615–5629.

Pinkoski, M.J., Hobman, M., Heibein, J.A., Tomaselli, K., Li, F., Seth, P., Froelich, C.J., and Bleackley, R.C. (1998). Entry and trafficking of granzyme B in target cells during granzyme B-perforin-mediated apoptosis. Blood *92*, 1044–1054.

Pitchford, S.C., Furze, R.C., Jones, C.P., Wengner, A.M., and Rankin, S.M. (2009). Differential mobilization of subsets of progenitor cells from the bone marrow. Cell Stem Cell *4*, 62–72.

Prichard, M.N., Penfold, M.E., Duke, G.M., Spaete, R.R., and Kemble, G.W. (2001). A review of genetic differences between limited and extensively passaged human cytomegalovirus strains. Rev. Med. Virol. *11*, 191–200.

Rawlinson, W.D., Farrell, H.E., and Barrell, B.G. (1996). Analysis of the complete DNA sequence of murine cytomegalovirus. J. Virol. *70*, 8833–8849.

Reddehase, M.J., Balthesen, M., Rapp, M., Jonjic, S., Pavic, I., and Koszinowski, U.H. (1994). The conditions of primary infection define the load of latent viral genome in organs and the risk of recurrent cytomegalovirus disease. J. Exp. Med. *179*, 185–193.

Reddehase, M.J., Podlech, J., and Grzimek, N.K. (2002). Mouse models of cytomegalovirus latency: overview. J. Clin. Virol. *25*, 23–36.

Ross, S.A., Arora, N., Novak, Z., Fowler, K.B., Britt, W.J., and Boppana, S.B. (2010). Cytomegalovirus reinfections in healthy seroimmune women. J. Infect. Dis. *201*, 386–389.

Sato, K., Nie, C., Misawa, N., Tanaka, Y., Ito, M., and Koyanagi, Y. (2012). Dynamics of memory and naive CD8+ T lymphocytes in humanized NOD/SCID/IL-2Rgammanull mice infected with CCR5-tropic HIV-1. Vaccine *28*, 32–37.

Seckert, C.K., Renzaho, A., Reddehase, M.J., and Grzimek, N.K. (2008). Hematopoietic stem cell transplantation with latently infected donors does not transmit virus to immunocompromised recipients in the murine model of cytomegalovirus infection. Med. Microbiol. Immunol. *197*, 251–259.

Seckert, C.K., Renzaho, A., Tervo, H.M., Krause, C., Deegen, P., Kuhnapfel, B., Reddehase, M.J., and Grzimek, N.K. (2009). Liver sinusoidal endothelial cells are a site of murine cytomegalovirus latency and reactivation. J. Virol. *83*, 8869–8884.

Shih, C.C., Kaneshima, H., Rabin, L., Namikawa, R., Sager, P., McGowan, J., and McCune, J.M. (1991). Postexposure prophylaxis with zidovudine suppresses human immunodeficiency virus type 1 infection in SCID-hu mice in a time-dependent manner. J. Infect. Dis. *163*, 625–627.

Shope, T., and Miller, G. (1973). Epstein–Barr virus. Heterophile responses in squirrel monkeys inoculated with virus-transformed autologous leukocytes. J. Exp. Med. *137*, 140–147.

Sinclair, J., and Sissons, P. (1996). Latent and persistent infections of monocytes and macrophages. Intervirology *39*, 293–301.

Sinzger, C., and Jahn, G. (1996). Human cytomegalovirus cell tropism and pathogenesis [Review]. Intervirology 39, 302–319.

Sinzger, C., Plachter, B., Grefte, A., The, T.H., and Jahn, G. (1996). Tissue macrophages are infected by human cytomegalovirus in vivo. J. Infect. Dis. *173*, 240–245.

Sissons, J.G., Bain, M., and Wills, M.R. (2002). Latency and reactivation of human cytomegalovirus. J. Infect. *44*, 73–77.

Smith, M.S., Bentz, G.L., Alexander, J.S., and Yurochko, A.D. (2004a). Human cytomegalovirus induces monocyte differentiation and migration as a strategy for dissemination and persistence. J. Virol. *78*, 4444–4453.

Smith, M.S., Bentz, G.L., Smith, P.M., Bivins, E.R., and Yurochko, A.D. (2004b). HCMV activates PI(3)K in monocytes and promotes monocyte motility and transendothelial migration in a PI(3)K-dependent manner. J. Leukoc. Biol. *76*, 65–76.

Smith, M.S., Bivins-Smith, E.R., Tilley, A.M., Bentz, G.L., Chan, G., Minard, J., and Yurochko, A.D. (2007). Roles of phosphatidylinositol 3-kinase and NF-kappaB in human cytomegalovirus-mediated monocyte diapedesis and adhesion: strategy for viral persistence. J. Virol. *81*, 7683–7694.

Smith, M.S., Goldman, D.C., Bailey, A.S., Pfaffle, D.L., Kreklywich, C.N., Spencer, D.B., Othieno, F.A., Streblow, D.N., Garcia, J.V., Fleming, W.H., et al. (2010). Granulocyte-colony stimulating factor reactivates human cytomegalovirus in a latently infected humanized mouse model. Cell Host Microbe 8, 284–291.

Söderberg-Naucler, C., Fish, K.N., and Nelson, J.A. (1997). Reactivation of latent human cytomegalovirus by allogeneic stimulation of blood cells from healthy donors. Cell *91*, 119–126.

Söderberg-Naucler, C., Fish, K.N., and Nelson, J.A. (1998). Growth of human cytomegalovirus in primary macrophages. Methods *16*, 126–138.

Söderberg-Naucler, C., Streblow, D.N., Fish, K.N., Allan-Yorke, J., Smith, P.P., and Nelson, J.A. (2001). Reactivation of latent human cytomegalovirus in CD14(+) monocytes is differentiation dependent. J. Virol. *75*, 7543–7554.

van der Strate, B.W., Hillebrands, J.L., Lycklama a Nijeholt, S.S., Beljaars, L., Bruggeman, C.A., Van Luyn, M.J., Rozing, J., The, T.H., Meijer, D.K., Molema, G., et al. (2003). Dissemination of rat cytomegalovirus through infected granulocytes and monocytes in vitro and in vivo. J. Virol. *77*, 11274–11278.

Sun, Z., Denton, P.W., Estes, J.D., Othieno, F.A., Wei, B.L., Wege, A.K., Melkus, M.W., Padgett-Thomas, A., Zupancic, M., Haase, A.T., et al. (2007). Intrarectal transmission, systemic infection, and CD4+ T-cell depletion in humanized mice infected with HIV-1. J. Exp. Med. *204*, 705–714.

Sylwester, A.W., Mitchell, B.L., Edgar, J.B., Taormina, C., Pelte, C., Ruchti, F., Sleath, P.R., Grabstein, K.H., Hosken, N.A., Kern, F., et al. (2005). Broadly targeted human cytomegalovirus-specific CD4+ and CD8+ T-cells dominate the memory compartments of exposed subjects. J. Exp. Med. *202*, 673–685.

Taylor-Wiedeman, J., Sissons, J.G., Borysiewicz, L.K., and Sinclair, J.H. (1991). Monocytes are a major site of persistence of human cytomegalovirus in peripheral blood mononuclear cells. J. Gen. Virol. 72, 2059–2064.

Taylor-Wiedeman, J., Sissons, P., and Sinclair, J. (1994). Induction of endogenous human cytomegalovirus gene expression after differentiation of monocytes from healthy carriers. J. Virol. *68*, 1597–1604.

Tsuda, Y., Caposio, P., Parkins, C.J., Botto, S., Messaoudi, I., Cicin-Sain, L., Feldmann, H., and Jarvis, M.A. (2012). A replicating cytomegalovirus-based vaccine encoding a single Ebola virus nucleoprotein CTL epitope confers protection against Ebola virus. PLoS Negl. Trop. Dis. *5*, e1275.

Umashankar, M., Petrucelli, A., Cicchini, L., Caposio, P., Kreklywich, C.N., Rak, M., Bughio, F., Goldman, D.C., Hamlin, K.L., Nelson, J.A., *et al.* (2012). A novel human cytomegalovirus locus modulates cell type-specific outcomes of infection. PLoS Pathog. 7, e1002444.

Underwood, M.R., Ferris, R.G., Selleseth, D.W., Davis, M.G., Drach, J.C., Townsend, L.B., Biron, K.K., and Boyd, F.L. (2004). Mechanism of action of the ribopyranoside benzimidazole GW275175X against human cytomegalovirus. Antimicrob. Agents Chemother. *48*, 1647–1651.

Wang, W., Taylor, S.L., Leisenfelder, S.A., Morton, R., Moffat, J.F., Smirnov, S., and Zhu, H. (2005). Human cytomegalovirus genes in the 15-kilobase region are required for viral replication in implanted human tissues in SCID mice. J. Virol. 79, 2115–2123.

Watanabe, S., Ohta, S., Yajima, M., Terashima, K., Ito, M., Mugishima, H., Fujiwara, S., Shimizu, K., Honda, M., Shimizu, N., *et al.* (2007a). Humanized NOD/SCID/IL2Rgamma(null) mice transplanted with hematopoietic stem cells under nonmyeloablative conditions show prolonged life spans and allow detailed analysis of human immunodeficiency virus type 1 pathogenesis. J. Virol. *81*, 13259–13264.

Watanabe, S., Terashima, K., Ohta, S., Horibata, S., Yajima, M., Shiozawa, Y., Dewan, M.Z., Yu, Z., Ito, M., Morio, T.,

et al. (2007b). Hematopoietic stem cell-engrafted NOD/SCID/IL2Rgamma null mice develop human lymphoid systems and induce long-lasting HIV-1 infection with specific humoral immune responses. Blood *109*, 212–218.

Weber, O., Reefschlager, J., Rubsamen-Waigmann, H., Raddatz, S., Hesseling, M., and Habich, D. (2000). A novel peptide aldehyde with activity against human cytomegalovirus in two different in vivo models. Antivir. Chem. Chemother. *11*, 51–59.

Weber, O., Bender, W., Eckenberg, P., Goldmann, S., Haerter, M., Hallenberger, S., Henninger, K., Reefschlager, J., Trappe, J., Witt-Laido, A., *et al.* (2001). Inhibition of murine cytomegalovirus and human cytomegalovirus by a novel non-nucleosidic compound in vivo. Antiviral Res. *49*, 179–189.

Williams, S.L., Hartline, C.B., Kushner, N.L., Harden, E.A., Bidanset, D.J., Drach, J.C., Townsend, L.B., Underwood, M.R., Biron, K.K., and Kern, E.R. (2003). In vitro activities of benzimidazole D- and L-ribonucleosides against herpesviruses. Antimicrob. Agents. Chemother. *47*, 2186–2192.

Wu, W., Vieira, J., Fiore, N., Banerjee, P., Sieburg, M., Rochford, R., Harrington, W., Jr., and Feuer, G. (2006). KSHV/HHV-8 infection of human hematopoietic progenitor (CD34+) cells: persistence of infection during hematopoiesis in vitro and in vivo. Blood *108*, 141–151.

Yajima, M., Imadome, K., Nakagawa, A., Watanabe, S., Terashima, K., Nakamura, H., Ito, M., Shimizu, N., Honda, M., Yamamoto, N., *et al.* (2008). A new humanized mouse model of Epstein–Barr virus infection that reproduces persistent infection, lymphoproliferative disorder, and cell-mediated and humoral immune responses. J. Infect. Dis. *198*, 673–682.

Yajima, M., Imadome, K., Nakagawa, A., Watanabe, S., Terashima, K., Nakamura, H., Ito, M., Shimizu, N., Yamamoto, N., and Fujiwara, S. (2009). T-cell-mediated control of Epstein–Barr virus infection in humanized mice. J. Infect. Dis. *200*, 1611–1615.

Yurochko, A.D., and Huang, E.S. (1999). Human cytomegalovirus binding to human monocytes induces immunoregulatory gene expression. J. Immunol. *162*, 4806–4816.

Subject index

Open reading frame index

See specific chapters for CMVs other than HCMV and MCMV.

Cluster of differentiation index

CMV strains index

See specific chapters for CMVs other than HCMV and MCMV.

Antiviral drugs index

Frequently discussed molecules

Cell-death related proteins

Bcl-2	I	19, 266–267, –, 323–324
Bak	I	217, 266–268, 272, 324
Bax	I	212, 217, 266–268, 272, 324
Caspase	I	47, 264–273, 311
	II	245, 248, 271
Caspase-1	II	197, 248
Caspase-3	I	104, 266
	II	129
Caspase-7	I	104
Caspase-8	I	181, 266, 268–269
	II	130, 278, 471
Caspase-9	I	266
FADD	I	265, 269
Fas(L)	I	269
vICA	I	106, 181, 264–266, 268–271
	II	471
vIRA	I	264–265, 269–270, 311
vMIA	I	106, 181, 212–218, 264–272, 285, 324
	II	130, 471

Chemokines and their receptors

CCL2	I	104, 355, 363, 366–367
	II	237, 244, 291,
CCL3/MIP1α	I	101, 355, 365
	II	94, 196, 204, 237, 244, 270, 291
CCL4/MIP1β	I	101, 355, 365
	II	148, 196
CCL5/RANTES	I	101, 104, 355, 365
	II	131, 133, 242–243, 291, 326
CCL7/MCP-3	I	101
	II	194, 237
CCL8/MCP-2	I	101
CCL12	II	194, 237
CCL20/MIP3α	I	101
CCL21	II	131
CCL22	I	211
CCL23	I	101
CCR1	I	104
CCR2	I	363–368
CCR5	II	148–150, 336
CCR7	II	452
CX3CR1	I	369, 379, 399
	II	356
CXCL1	I	101, 106, 363, 365, 368
CXCL3	II	131

CXCL5	I	101
CXCL6	I	101
CXCL8	I	101–102, 104
CXCL9	II	131, 237, 243–244
CXCL10	II	131, 244
CXCL11	I	101
	II	243–244
CXCL16	I	79, 11
CXCR2	II	473
G-protein coupled chemokine receptor (GPCR)	I	15, 96, 232, 355, 365
	II	94, 98, 318, 322, 326, 330, 472
MCK2	I	27, 105, 363–369
	II	257–259, 278

Cytokines and their receptors

Epidermal growth factor receptor (EGFR)	I	127–128, 132, 134, 306–307
	II	81, 415–416, 498
Granulocyte colony stimulating factor (G-CSF)	I	335, 349, 429–430
	II	360
Granulocyte–macrophage colony stimulating factor (GM-CSF)	I	101–102, 104, 418
	II	160, 198, 237, 243, 248, 251
Stem cell factor (SCF)	I	252, 431
	II	354, 360–361
TNF receptor (TNFR)	I	101, 160, 211, 265, 269, 282, 354, 405–406
	II	271, 274
Tumor necrosis factor α (TNF-α)	I	79, 101, 160, 269–270, 282–284, 311, 333, 354, 364, 387, 397–398, 402–406
	II	127–128, 131, 133, 145, 148, 155, 174, 193, 195–196, 198, 204, 207, 233–236, 243–244, 247, 250, 261, 269, 271, 276, 291, 328, 453, 479, 501, 506
TGF-β	I	98, 101
	II	15, 77, 152, 155, 302
Vascular endothelial cell growth factor (VEGF)	I	104–105
	II	76, 78, 83, 500

Interferons and their receptors

IFN	I	129, 278–290, 326–327
	II	131, 176–177, 183, 232–253, 312, 426, 471
IFN-α/ β (Type I)	I	80, 112, 118, 128–130, 133–134, 145–146, 163, 278–290, 326–327
	II	175, 193, 195–197, 249–250, 260–264, 270, 277–279
IFN-γ (Type II)	I	112, 278–290, 333, 351, 389, 391, 396, 398, 402
	II	13, 128, 131, 133–134, 143–162, 174, 183, 195, 198, 202–205, 207, 233, 264–265, 269, 287, 292, 304, 364–366, 369, 372, 396, 477, 479
IFN (Type III)	I	278–282
	II	196

Interleukins and their receptors

IL-1	I	160, 403
	II	126, 194–198, 234–237, 243–245, 247–248, 253
IL-2	I	403
	II	155–156, 175, 183, 194, 275, 287, 291, 370, 479
IL-6	I	88, 101–106, 134, 160
	II	4, 83, 196, 233–237, 243–244, 360–361, 500–501, 505–506
IL-7	I	393
	II	148, 348
IL-8	I	101–102, 104, 134, 283, 355, 365
IL-10/cmvIL-10	I	80, 111, 182, 281, 285, 335, 351–352, 357, 396
	II	76–77, 145, 155, 157, 195, 234, 237, 242–243, 247, 250–251, 271, 275, 278, 322, 338, 357, 473, 476–479, 500
IL-12	II	183, 193–196, 233–237, 243–244
IL-13	II	152
IL-15	I	393
	II	175, 183, 194, 196, 237, 244–246, 264
IL-17	II	152, 155
IL-18	II	183, 194, 196–198, 233–235, 237, 244–248, 252–253, 264, 269
IL-22	I	281